Foundations of
Electrodynamics

# Foundations of Electrodynamics

S. R. de Groot and L. G. Suttorp

*Institute of Theoretical Physics*
*University of Amsterdam*

1972

NORTH-HOLLAND PUBLISHING COMPANY – AMSTERDAM
AMERICAN ELSEVIER PUBLISHING CO. INC. – NEW YORK

Library of Congress Catalog Card Number: 75-166303
North-Holland ISBN: 0 7204 0248 4
American Elsevier ISBN: 0 444 10370 8

Publishers:
NORTH-HOLLAND PUBLISHING COMPANY – AMSTERDAM

Sole distributors for the U.S.A. and Canada:
AMERICAN ELSEVIER PUBLISHING COMPANY, INC.
52 VANDERBILT AVENUE, NEW YORK, N.Y. 10017



PRINTED IN THE NETHERLANDS

# Preface

Electrodynamics may be said to consist of two parts, at different levels: microscopic and macroscopic theory. The first contains the laws that govern the interaction of fields and point particles – often grouped into stable sets such as atoms and molecules – and the second those that describe the interaction of fields and continuous media. The two theories are linked together, since the phenomena at the macroscopic level may be looked upon as being the result of the interplay of many particles. Therefore one should be able to obtain the electromagnetic laws for continuous media from those for point particles. Such a derivation, together with a discussion of the microscopic starting points, forms the subject of this monograph.

The programme will be carried out in the framework of both classical and quantum theory. The classical theory is given in the non-relativistic approximation and then in covariant formulation. In the latter various topics will receive special attention: among these figure the covariant description of composite particles, the obtention of statistical averages in a relativistically invariant way and a discussion of the energy–momentum tensor for continuous media. The quantum-mechanical theory will be formulated in such a fashion that the analogy with classical theory can be exploited as far as possible. This is achieved by representing the physical quantities by ordinary functions rather than by operators. Again the non-relativistic approximation will be studied first. Subsequently magnetic effects are discussed in a 'semi-relativistic' theory, which goes one step beyond the non-relativistic treatment. The completely covariant extension of quantum theory will be confined to the discussion of the motion of single particles with and without spin in slowly varying external fields. The covariant generalization to statistical assemblies of particles moving in each other's fields would require quantization of the electromagnetic field together with its sources: this forms the subject of quantum electrodynamics not dealt with here.

The subject matter of the various chapters is, roughly spoken, of two kinds.

Part is meant especially to serve as textbook material for graduate students who take courses in electromagnetic theory. By reading the first two chapters they will get acquainted with the way in which the macroscopic laws of electrodynamics are obtained from a microscopic basis, albeit in the framework of classical, non-relativistic theory. In the relativistic part the third chapter may be useful as an exposé of the covariant equations for fields and particles with the inclusion of the effects of radiation damping, while the final results of the fourth and fifth chapters give an idea of the way in which the non-relativistic laws may be generalized. Similarly the results of chapters VI and VII show the consequences of the use of quantum mechanics. The special formulation of quantum mechanics in terms of Weyl transforms and Wigner functions can be studied independently from the appendix of chapter VI.

More advanced students will be interested in the covariant formulation of the equations of motion for composite particles in chapter IV, relativistic statistics as discussed in chapter V, the covariant quantum-mechanical equations of motion for particles with spin 0 and $\frac{1}{2}$ in chapter VIII, and in the semi-relativistic treatment of magnetic effects given in chapters IX and X.

We are greatly indebted to Miss A. Kitselar, and Messrs. A.J. Kox and M.A.J. Michels for their help in preparing the manuscript.

S.R. de G.

L.G.S.

Amsterdam 1971

# Table of contents

# PART A

# *Non-relativistic classical electrodynamics*

CHAPTER I

# Particles: their fields and motion

## 1 Introduction

The aim of this chapter is to obtain a description of the electromagnetic behaviour of composite particles in the framework of classical, non-relativistic theory. Such composite particles, like atoms, molecules or ions are supposed to consist of charged point particles: the electrons and nuclei. The equations which govern their motion and describe their fields will be derived from the corresponding basic equations valid for charged point particles without structure. The latter microscopic equations are the Maxwell–Lorentz field equations and the Newton equation with the Lorentz force inserted. A series expansion in terms of multipoles leads then to the field equations and the momentum, energy and angular momentum equations for the composite particles.

## 2 The microscopic field equations

The electric and magnetic fields $\boldsymbol{e}(\boldsymbol{R}, t)$ and $\boldsymbol{b}(\boldsymbol{R}, t)$ at the point with coordinates $\boldsymbol{R}$ and at time $t$, generated by a collection of point particles $i = 1, 2, \ldots$ with charges $e_i$, positions $\boldsymbol{R}_i(t)$ and velocities $\dot{\boldsymbol{R}}_i(t)$, satisfy the Maxwell–Lorentz field equations (in the rationalized Gauss system[1])

$$
\begin{aligned}
\nabla\cdot\boldsymbol{e} &= \sum_i e_i\,\delta(\boldsymbol{R}_i-\boldsymbol{R}),\\
-\partial_0\,\boldsymbol{e}+\nabla\wedge\boldsymbol{b} &= c^{-1}\sum_i e_i\,\dot{\boldsymbol{R}}_i\,\delta(\boldsymbol{R}_i-\boldsymbol{R}),\\
\nabla\cdot\boldsymbol{b} &= 0,\\
\partial_0\,\boldsymbol{b}+\nabla\wedge\boldsymbol{e} &= 0,
\end{aligned}
\tag{1}
$$

[1] In the Giorgi system different numerical coefficients appear: the factors $c^{-1}$ (both explicitly and in $\partial_0$) are absent, while in the first two equations $\boldsymbol{e}$ and $\boldsymbol{b}$ are replaced by $\varepsilon_0\boldsymbol{e}$ and $\mu_0^{-1}\boldsymbol{b}$ respectively.

where $\nabla$ and $\partial_0$ are differentiations with respect to $\boldsymbol{R}$ and $ct$ (with $c$ the speed of light) and the dot and the symbol $\wedge$ scalar and vector products of vectors. The sources contain the three-dimensional delta functions of $\boldsymbol{R}_i-\boldsymbol{R}$.

In non-relativistic theory one is interested in solutions of these equations up to order $c^{-1}$. To find them it is convenient to introduce potentials. From the third equation it follows that

$$\boldsymbol{b} = \nabla \wedge \boldsymbol{a} \tag{2}$$

with the vector potential $\boldsymbol{a}(\boldsymbol{R}, t)$. Then with the fourth equation one has

$$\boldsymbol{e} = -\nabla\varphi - \partial_0 \boldsymbol{a}, \tag{3}$$

where $\varphi(\boldsymbol{R}, t)$ is the scalar potential. Insertion of these expressions into the first two equations of (1) gives, if one omits terms in $c^{-2}$,

$$\begin{aligned} \Delta\varphi + \partial_0 \nabla\cdot\boldsymbol{a} &= -\sum_i e_i \delta(\boldsymbol{R}_i - \boldsymbol{R}), \\ \Delta\boldsymbol{a} - \nabla(\nabla\cdot\boldsymbol{a} + \partial_0\varphi) &= -c^{-1}\sum_i e_i \dot{\boldsymbol{R}}_i \delta(\boldsymbol{R}_i - \boldsymbol{R}), \end{aligned} \tag{4}$$

where $\Delta = \nabla\cdot\nabla$ is the Laplace operator. The potentials are not fixed in a unique way by the relations (2) and (3). The same electromagnetic fields are described by potentials $\boldsymbol{a}'$ and $\varphi'$ which are related to the original potentials $\boldsymbol{a}$ and $\varphi$ by a gauge transformation

$$\begin{aligned} \varphi' &= \varphi - \partial_0 \psi, \\ \boldsymbol{a}' &= \boldsymbol{a} + \nabla\psi \end{aligned} \tag{5}$$

with an arbitrary function $\psi$. This property is utilized to choose the potentials in such a way that they satisfy

$$\partial_0\varphi + \nabla\cdot\boldsymbol{a} = 0, \tag{6}$$

the Lorentz condition. The reason for imposing this condition is that then the equations (4) become two uncoupled Poisson equations for $\varphi$ and $\boldsymbol{a}$:

$$\begin{aligned} \Delta\varphi &= -\sum_i e_i \delta(\boldsymbol{R}_i - \boldsymbol{R}), \\ \Delta\boldsymbol{a} &= -c^{-1}\sum_i e_i \dot{\boldsymbol{R}}_i \delta(\boldsymbol{R}_i - \boldsymbol{R}), \end{aligned} \tag{7}$$

where again a term of order $c^{-2}$ has been dismissed. The solutions follow from the property for the delta function:

$$\Delta \frac{1}{|\boldsymbol{r}|} = -4\pi\delta(\boldsymbol{r}). \tag{8}$$

Using this property, one finds from (7) the non-relativistic potentials in the Lorentz gauge:

$$\varphi = \sum_i \frac{e_i}{4\pi|\boldsymbol{R}_i - \boldsymbol{R}|}, \qquad \boldsymbol{a} = c^{-1} \sum_i \frac{e_i \dot{\boldsymbol{R}}_i}{4\pi|\boldsymbol{R}_i - \boldsymbol{R}|}, \tag{9}$$

so that the non-relativistic fields (2) and (3) are

$$\boldsymbol{e} = \sum_i \boldsymbol{e}_i, \qquad \boldsymbol{e}_i = -\nabla \frac{e_i}{4\pi|\boldsymbol{R}_i - \boldsymbol{R}|}, \qquad \boldsymbol{b} = \sum_i \boldsymbol{b}_i, \qquad \boldsymbol{b}_i = c^{-1}\nabla \wedge \frac{e_i \dot{\boldsymbol{R}}_i}{4\pi|\boldsymbol{R}_i - \boldsymbol{R}|}. \tag{10}$$

These formulae show that the non-relativistic electric field is of order $c^0$ (a term in $c^{-1}$ does not appear), while the non-relativistic magnetic field is of order $c^{-1}$ (no term in $c^0$ arises). From the first line of (10) it follows that $\boldsymbol{e}$ is irrotational. This is in agreement with the fourth field equation in (1), since $\partial_0 \boldsymbol{b}$ is of order $c^{-2}$ and hence has to be neglected in non-relativistic theory. So strictly spoken one should write in a non-relativistic theory the truncated equation $\nabla \wedge \boldsymbol{e} = 0$ instead of the fourth field equation.

## 3 *The equation of motion for a point particle*

The equation of motion for a particle with charge $e$, mass $m$, position $\boldsymbol{R}_1(t)$, velocity $\dot{\boldsymbol{R}}_1(t)$ and acceleration $\ddot{\boldsymbol{R}}_1(t)$ in an external electromagnetic field $(\boldsymbol{E}_e, \boldsymbol{B}_e)$ is:

$$m\ddot{\boldsymbol{R}}_1 = e\{\boldsymbol{E}_e(\boldsymbol{R}_1, t) + c^{-1}\dot{\boldsymbol{R}}_1 \wedge \boldsymbol{B}_e(\boldsymbol{R}_1, t)\}, \tag{11}$$

where at the right-hand side the Lorentz force appears. The equation of motion of one particle of a set labelled by the index $i = 1, 2, \ldots, N$ reads

$$m_i \ddot{\boldsymbol{R}}_i = e_i\{\boldsymbol{e}_t(\boldsymbol{R}_i, t) + c^{-1}\dot{\boldsymbol{R}}_i \wedge \boldsymbol{b}_t(\boldsymbol{R}_i, t)\}, \tag{12}$$

where the total electric and magnetic fields are the sums of the external fields and the fields (10) generated by the other particles:

$$\boldsymbol{e}_t(\boldsymbol{R}_i, t) = \sum_{j(\neq i)} \boldsymbol{e}_j(\boldsymbol{R}_i, t) + \boldsymbol{E}_e(\boldsymbol{R}_i, t), \qquad \boldsymbol{b}_t(\boldsymbol{R}_i, t) = \sum_{j(\neq i)} \boldsymbol{b}_j(\boldsymbol{R}_i, t) + \boldsymbol{B}_e(\boldsymbol{R}_i, t). \tag{13}$$

Since in the equation of motion (12) the magnetic field is accompanied by a factor $c^{-1}$, one needs there as fields

$$\begin{aligned} \boldsymbol{e}_t(\boldsymbol{R}_i, t) &= -\sum_{j(\neq i)} \nabla_i \frac{e_j}{4\pi|\boldsymbol{R}_i - \boldsymbol{R}_j|} + \boldsymbol{E}_e(\boldsymbol{R}_i, t), \\ \boldsymbol{b}_t(\boldsymbol{R}_i, t) &= \boldsymbol{B}_e(\boldsymbol{R}_i, t), \end{aligned} \tag{14}$$

instead of the complete expressions (13) with (10).

The equations of motion (12) with (14) may be written in Hamiltonian form

$$\frac{\partial H}{\partial \boldsymbol{P}_i} = \dot{\boldsymbol{R}}_i, \qquad \frac{\partial H}{\partial \boldsymbol{R}_i} = -\dot{\boldsymbol{P}}_i \tag{15}$$

with the Hamiltonian

$$H = \sum_i \frac{\boldsymbol{P}_i^2}{2m_i} + \sum_{i, j(i \neq j)} \frac{e_i e_j}{8\pi|\boldsymbol{R}_i - \boldsymbol{R}_j|} + \sum_i e_i \left\{\varphi_e(\boldsymbol{R}_i, t) - c^{-1} \frac{\boldsymbol{P}_i}{m_i} \cdot \boldsymbol{A}_e(\boldsymbol{R}_i, t)\right\}, \tag{16}$$

with $\varphi_e$ and $\boldsymbol{A}_e$ potentials for the external fields. Indeed insertion of (16) into (15) leads to (12) with (14).

## 4 *The equations for the fields due to composite particles*

### a. *The atomic series expansion*

Charged point particles (electrons and nuclei) are often grouped into stable sets, like atoms, molecules or ions. (For convenience we shall sometimes refer to such composite particles simply as 'atoms'.) The starting point for the derivation of the equations for the fields due to such atoms is the set of microscopic field equations (1). It will be convenient in the present case to replace the numbering $i$ of the point particles by a numbering $k$ of the stable groups and $i$ of their constituent particles. The position vector $\boldsymbol{R}_i$, written as $\boldsymbol{R}_{ki}$ now, can be split into two parts:

$$\boldsymbol{R}_{ki} = \boldsymbol{R}_k + \boldsymbol{r}_{ki}. \tag{17}$$

Here $\boldsymbol{R}_k$ is the position of some privileged point of the stable group $k$ (e.g. the nucleus of an atom or the centre of mass, etc.), while the $\boldsymbol{r}_{ki}$ $(i = 1, 2, \ldots)$ are the internal coordinates, which specify the positions of the constituent particles $ki$ with respect to that of the privileged point of the stable group $k$.

The case will now be studied in which the solutions $\boldsymbol{e}$ and $\boldsymbol{b}$ of the field equations can be considered as converging series expansions in $|\boldsymbol{r}_{ki}|/|\boldsymbol{R}_k - \boldsymbol{R}|$.

(This situation is realized if the expansion parameter is smaller than unity. Physically this means that the atomic dimension $\boldsymbol{r}_{ki}$ has to be smaller than the distance $|\boldsymbol{R}_k-\boldsymbol{R}|$ from the observation point $\boldsymbol{R}$ of the fields to the central point $\boldsymbol{R}_k$ of the atom. In other words one is now interested in observing the fields outside the stable atomic structure. On the other hand this is also the physical situation in which it is useful to speak of atoms, characterized by a small number of physical quantities which are still to be specified.) Accordingly the delta functions may now be developed in powers of $\boldsymbol{r}_{ki}$ around $\boldsymbol{R}_k-\boldsymbol{R}$. Using (17) one gets then for the field equations (1):

$$\begin{aligned}
\nabla\cdot\boldsymbol{e} &= \sum_{k,i} e_{ki}\sum_{n=0}^{\infty}\frac{(-1)^n}{n!}(\boldsymbol{r}_{ki}\cdot\nabla)^n\delta(\boldsymbol{R}_k-\boldsymbol{R}),\\
-\partial_0\,\boldsymbol{e}+\nabla\wedge\boldsymbol{b} &= \frac{1}{c}\sum_{k,i} e_{ki}(\dot{\boldsymbol{R}}_k+\dot{\boldsymbol{r}}_{ki})\sum_{n=0}^{\infty}\frac{(-1)^n}{n!}(\boldsymbol{r}_{ki}\cdot\nabla)^n\delta(\boldsymbol{R}_k-\boldsymbol{R}),\\
\nabla\cdot\boldsymbol{b} &= 0,\\
\partial_0\,\boldsymbol{b}+\nabla\wedge\boldsymbol{e} &= 0,
\end{aligned}\tag{18}$$

since $\partial/\partial\boldsymbol{R}_k$ acting on the delta function is the same as $-\nabla$ acting on it. The first of these equations can alternatively be written as

$$\nabla\cdot\boldsymbol{e} = \rho^{\mathrm{e}}-\nabla\cdot\boldsymbol{p},\tag{19}$$

if the following abbreviations are used

$$\rho^{\mathrm{e}} = \sum_k \rho_k^{\mathrm{e}}\,;\qquad \rho_k^{\mathrm{e}} = \sum_i e_{ki}\,\delta(\boldsymbol{R}_k-\boldsymbol{R}),\tag{20}$$

$$\boldsymbol{p} = \sum_k \boldsymbol{p}_k\,;\qquad \boldsymbol{p}_k = \sum_i e_{ki}\sum_{n=1}^{\infty}\frac{(-1)^{n-1}}{n!}\boldsymbol{r}_{ki}(\boldsymbol{r}_{ki}\cdot\nabla)^{n-1}\delta(\boldsymbol{R}_k-\boldsymbol{R}).\tag{21}$$

The second field equation can be written as

$$-\partial_0\,\boldsymbol{e}+\nabla\wedge\boldsymbol{b} = \frac{1}{c}\left\{\boldsymbol{j}-\sum_k \dot{\boldsymbol{R}}_k\nabla\cdot\boldsymbol{p}_k+\sum_{k,i} e_{ki}\sum_{n=0}^{\infty}\frac{(-1)^n}{n!}\dot{\boldsymbol{r}}_{ki}(\boldsymbol{r}_{ki}\cdot\nabla)^n\delta(\boldsymbol{R}_k-\boldsymbol{R})\right\}\tag{22}$$

with the abbreviation

$$\boldsymbol{j} = \sum_k \boldsymbol{j}_k;\qquad \boldsymbol{j}_k = \sum_i e_{ki}\,\dot{\boldsymbol{R}}_k\,\delta(\boldsymbol{R}_k-\boldsymbol{R}).\tag{23}$$

Taking the derivative $\partial_0 = \partial/\partial ct$ of (21), one finds (replacing $n$ by $n+1$)

$$\partial_0\,\boldsymbol{p}_k+\frac{1}{c}\,\dot{\boldsymbol{R}}_k\cdot\nabla\boldsymbol{p}_k-\frac{1}{c}\sum_i e_{ki}\sum_{n=0}^{\infty}\frac{(-1)^n}{(n+1)!}(\dot{\boldsymbol{r}}_{ki}\boldsymbol{r}_{ki}\cdot\nabla+n\boldsymbol{r}_{ki}\dot{\boldsymbol{r}}_{ki}\cdot\nabla)(\boldsymbol{r}_{ki}\cdot\nabla)^{n-1}\,\delta(\boldsymbol{R}_k-\boldsymbol{R}) = 0,\tag{24}$$

because the time derivative of the delta function is equal to $-\dot{\boldsymbol{R}}_k\cdot\nabla$ acting on it. (In the term $n = 0$ the operator product $\nabla\nabla^{-1}$ should be understood as unity.) Upon adding this (vanishing) expression, summed over $k$, to the right-hand side of (22), and using the vector identity

$$\nabla\cdot(\boldsymbol{v}\boldsymbol{w}-\boldsymbol{w}\boldsymbol{v}) = \nabla\wedge(\boldsymbol{w}\wedge\boldsymbol{v}), \tag{25}$$

one obtains for the field equation the form

$$-\partial_0\boldsymbol{e}+\nabla\wedge\boldsymbol{b} = c^{-1}\boldsymbol{j}+\partial_0\boldsymbol{p}+\nabla\wedge\boldsymbol{m}, \tag{26}$$

where the following abbreviation has been used:

$$\boldsymbol{m} = \sum_k \boldsymbol{m}_k, \tag{27}$$

$$\boldsymbol{m}_k = \frac{1}{c}\,\boldsymbol{p}_k\wedge\dot{\boldsymbol{R}}_k+\frac{1}{c}\sum_i e_{ki}\sum_{n=1}^{\infty}\frac{(-1)^{n-1}n}{(n+1)!}\,\boldsymbol{r}_{ki}\wedge\dot{\boldsymbol{r}}_{ki}(\boldsymbol{r}_{ki}\cdot\nabla)^{n-1}\delta(\boldsymbol{R}_k-\boldsymbol{R}).$$

The forms of the field equations may be further simplified and the abbreviations interpreted, if first physical quantities are introduced, which characterize the internal atomic structure.

## b. *Multipole moments*

The stable groups (which were called atoms here) may carry total charges

$$e_k = \sum_i e_{ki}. \tag{28}$$

This is the case for ions and also for single electrons. Other properties which characterize the atoms are the electromagnetic multipole moments. These atomic multipole moments are useful combinations of the internal atomic parameters. The electric dipole and quadrupole moments, for instance, are defined as

$$\overline{\boldsymbol{\mu}}_k^{(1)} = \sum_i e_{ki}\boldsymbol{r}_{ki}, \qquad \overline{\boldsymbol{\mu}}_k^{(2)} = \tfrac{1}{2}\sum_i e_{ki}\boldsymbol{r}_{ki}\boldsymbol{r}_{ki}. \tag{29}$$

These are a vector and a symmetric tensor respectively[1]. The latter is written as a 'dyadic product' $\boldsymbol{r}_{ki}\boldsymbol{r}_{ki}$, which means that it has the Cartesian tensor components $r_{ki}^{\alpha}r_{ki}^{\beta}$ with $\alpha, \beta = 1, 2, 3$. The magnetic dipole and quadrupole moments have the forms

$$\overline{\boldsymbol{\nu}}_k^{(1)} = \tfrac{1}{2}c^{-1}\sum_i e_{ki}\boldsymbol{r}_{ki}\wedge\dot{\boldsymbol{r}}_{ki}, \qquad \overline{\boldsymbol{\nu}}_k^{(2)} = \tfrac{1}{3}c^{-1}\sum_i e_{ki}\boldsymbol{r}_{ki}\boldsymbol{r}_{ki}\wedge\dot{\boldsymbol{r}}_{ki}. \tag{30}$$

[1] Vectors will be denoted by boldface italics; tensors by boldface roman (upright) symbols.

Again the dipole moment is a vector, the quadrupole moment a tensor written here as a dyadic product of the vectors $\boldsymbol{r}_{ki}$ and $\boldsymbol{r}_{ki} \wedge \dot{\boldsymbol{r}}_{ki}$. It should be noticed that $\overline{\boldsymbol{\mu}}_k^{(1)}$ is of the first order in the internal coordinates, but $\overline{\boldsymbol{\mu}}_k^{(2)}$ and $\overline{\boldsymbol{\nu}}_k^{(1)}$ of second order.

The definitions of multipoles of arbitrary order are the following: the electric $2^n$-pole moment is

$$\overline{\boldsymbol{\mu}}_k^{(n)} = \frac{1}{n!} \sum_i e_{ki} \boldsymbol{r}_{ki}^n, \qquad (n = 1, 2, \ldots), \tag{31}$$

where the power $n$ stands for a polyad of $n$ vectors $\boldsymbol{r}_{ki}$; the magnetic $2^n$-pole moment is

$$\overline{\boldsymbol{\nu}}_k^{(n)} = \frac{n}{(n+1)!} \sum_i e_{ki} \boldsymbol{r}_{ki}^n \wedge \frac{\dot{\boldsymbol{r}}_{ki}}{c}, \qquad (n = 1, 2, \ldots), \tag{32}$$

where again the power is a polyad. The electric $2^n$-pole moment is of order $n$ in the internal coordinates, whereas the magnetic $2^n$-pole is of order $n+1$ in the internal coordinates.

The values of the multipole moments may depend on the choice of the privileged point $\boldsymbol{R}_k$, but their forms will always be the same combinations of the internal coordinates $\boldsymbol{r}_{ki}$. The electric dipole moment, for instance, depends in general on the choice of $\boldsymbol{R}_k$, but not if the total charge $e_k$ vanishes.

If external electromagnetic fields are acting on the system it may be polarized and thus get induced electromagnetic multipole densities. This is the situation which one studies usually in the theories of the dielectric constant and the magnetic permeability. These theories are however completely independent of the derivation of the Maxwell equations. In the derivation the multipole moments are in fact arbitrary, that is they may be induced by fields exerted on the atoms, they may be permanent moments or they may be sums of both.

If the expansion parameter $|\boldsymbol{r}_{ki}|/|\boldsymbol{R}_k - \boldsymbol{R}|$ is small compared to unity, then only a few terms of the multipole expansion have to be taken into account. Such a situation will occur in diluted systems, where the observation point $\boldsymbol{R}$ of the fields can easily be chosen at a distance $|\boldsymbol{R}_k - \boldsymbol{R}|$ from the atom which is large compared to the atomic dimension $|\boldsymbol{r}_{ki}|$.

### c. *The field equations*

With the atomic charges (28) one can now write the expressions (20) and (23) as

$$\rho^{e} = \sum_{k} e_k \, \delta(\boldsymbol{R}_k - \boldsymbol{R}),$$

$$\boldsymbol{j} = \sum_{k} e_k \, \boldsymbol{v}_k \, \delta(\boldsymbol{R}_k - \boldsymbol{R}), \tag{33}$$

where we have written $\boldsymbol{v}_k$ for the velocity $\dot{\boldsymbol{R}}_k$. The quantities (33) are the *atomic charge* and *current* densities. Similarly with (31) and (32) one obtains from (21) and (27) the expressions

$$\boldsymbol{p} = \sum_{n=1}^{\infty} (-1)^{n-1} \nabla^{n-1} \vdots \sum_{k} \overline{\boldsymbol{\mu}}_k^{(n)} \delta(\boldsymbol{R}_k - \boldsymbol{R}),$$

$$\boldsymbol{m} = \sum_{n=1}^{\infty} (-1)^{n-1} \nabla^{n-1} \vdots \sum_{k} \overline{\boldsymbol{\nu}}_k^{(n)} \delta(\boldsymbol{R}_k - \boldsymbol{R}) + c^{-1} \sum_{k} \boldsymbol{p}_k \wedge \boldsymbol{v}_k \tag{34}$$

$$= \sum_{n=1}^{\infty} (-1)^{n-1} \nabla^{n-1} \vdots \sum_{k} (\overline{\boldsymbol{\nu}}_k^{(n)} + c^{-1} \overline{\boldsymbol{\mu}}_k^{(n)} \wedge \boldsymbol{v}_k) \delta(\boldsymbol{R}_k - \boldsymbol{R}),$$

where the dots stand for $(n-1)$-fold contractions. They may be called the *atomic electric* and *magnetic polarization* densities.

The quantities given here are thus expressed in the internal properties $e_k$, $\overline{\boldsymbol{\mu}}_k^{(n)}$ and $\overline{\boldsymbol{\nu}}_k^{(n)}$ and the external properties $\boldsymbol{R}_k$ and $\boldsymbol{v}_k \equiv \dot{\boldsymbol{R}}_k$ of the atoms. Since these properties depend themselves on time the atomic densities (33) and (34) are functions of the space and time coordinates $\boldsymbol{R}$ and $t$. The magnetization vector $\boldsymbol{m}$ contains contributions from the electric multipoles in motion. A similar term, showing the influence of moving magnetic moments on the electric polarization vector $\boldsymbol{p}$ is absent in the present – non-relativistic – treatment.

The atomic charge, current and polarization densities occur in the field equations (19) and (26) so that the total set now reads:

$$\nabla \cdot \boldsymbol{e} = \rho^{e} - \nabla \cdot \boldsymbol{p},$$

$$-\partial_0 \boldsymbol{e} + \nabla \wedge \boldsymbol{b} = c^{-1} \boldsymbol{j} + \partial_0 \boldsymbol{p} + \nabla \wedge \boldsymbol{m},$$

$$\nabla \cdot \boldsymbol{b} = 0, \tag{35}$$

$$\partial_0 \boldsymbol{b} + \nabla \wedge \boldsymbol{e} = 0.$$

Although they have the same form as the macroscopic Maxwell equations, they are still equations for the microscopic fields $\boldsymbol{e}$ and $\boldsymbol{b}$ in which the existence of atoms (stable groups of point particles) has been taken into account: they will be referred to as *atomic field equations*. Instead of $\boldsymbol{p}$ and $\boldsymbol{m}$ one could alternatively use '*atomic displacement vectors*', defined as

$$\boldsymbol{d} = \boldsymbol{e} + \boldsymbol{p}, \qquad \boldsymbol{h} = \boldsymbol{b} - \boldsymbol{m}. \tag{36}$$

Then the atomic field equations (35) can be written as

$$\begin{aligned} \nabla \cdot \boldsymbol{d} &= \rho^{\mathrm{e}}, \\ -\partial_0 \boldsymbol{d} + \nabla \wedge \boldsymbol{h} &= c^{-1}\boldsymbol{j}, \\ \nabla \cdot \boldsymbol{b} &= 0, \\ \partial_0 \boldsymbol{b} + \nabla \wedge \boldsymbol{e} &= 0. \end{aligned} \tag{37}$$

The atomic field equations for $\boldsymbol{e}$ and $\boldsymbol{b}$ were derived from the microscopic field equations for these same fields without any averaging or smoothing of the quantities. But while in the microscopic field equations properties of the separate electrons and nuclei occur, in the atomic field equations matter is characterized by parameters pertaining to the atoms (stable groups of electrons and nuclei). The atomic field equations can therefore be said to be valid on the '*kinetic level*' of the theory for electromagnetic fields in the presence of matter.

Finally it may be noted that from (33) follows the law of conservation of atomic charge

$$\partial \rho^{\mathrm{e}}/\partial t = -\nabla \cdot \boldsymbol{j}. \tag{38}$$

It may be derived by taking the time derivative of $\rho^{\mathrm{e}}$. The fact that $\partial/\partial \boldsymbol{R}_k$ acting on the delta function is the same as minus the nabla operator acting on it, together with the expression for $\boldsymbol{j}$, then leads directly to the right-hand side of (38).

## 5 *The momentum and energy equations for composite particles*

### a. *The equation of motion*

The equation of motion for a constituent particle $ki$ (of the $k$th atom) with charge $e_{ki}$, mass $m_{ki}$, position $\boldsymbol{R}_{ki}(t)$, velocity $\dot{\boldsymbol{R}}_{ki}(t)$ and acceleration $\ddot{\boldsymbol{R}}_{ki}(t)$ in an electromagnetic field $(\boldsymbol{e}_{\mathrm{t}}, \boldsymbol{b}_{\mathrm{t}})$ has been given in (12):

$$m_{ki}\ddot{\boldsymbol{R}}_{ki} = e_{ki}\{\boldsymbol{e}_{\mathrm{t}}(\boldsymbol{R}_{ki}, t) + c^{-1}\dot{\boldsymbol{R}}_{ki} \wedge \boldsymbol{b}_{\mathrm{t}}(\boldsymbol{R}_{ki}, t)\}. \tag{39}$$

The expressions for the total fields $\boldsymbol{e}_{\mathrm{t}}$ and $\boldsymbol{b}_{\mathrm{t}}$ have been written in (14); they read in the present case:

$$\begin{aligned} \boldsymbol{e}_{\mathrm{t}}(\boldsymbol{R}_{ki}, t) &= -\sum_{j(\neq i)} \nabla_{ki} \frac{e_{kj}}{4\pi|\boldsymbol{R}_{ki}-\boldsymbol{R}_{kj}|} - \sum_{l(\neq k)j} \nabla_{ki} \frac{e_{lj}}{4\pi|\boldsymbol{R}_{ki}-\boldsymbol{R}_{lj}|} + \boldsymbol{E}_{\mathrm{e}}(\boldsymbol{R}_{ki}, t), \\ \boldsymbol{b}_{\mathrm{t}}(\boldsymbol{R}_{ki}, t) &= \boldsymbol{B}_{\mathrm{e}}(\boldsymbol{R}_{ki}, t). \end{aligned} \tag{40}$$

Here $\nabla_{ki}$ stands for the space derivative with respect to $\boldsymbol{R}_{ki}$. The fields $(\boldsymbol{e}_t, \boldsymbol{b}_t)$ consist of three parts: the intra-atomic fields due to the other constituent particles $kj$ $(j \neq i)$ of the same atom $k$, the interatomic fields due to the other atoms and the external fields $(\boldsymbol{E}_e, \boldsymbol{B}_e)$ due to sources outside the system.

The motion of atom $k$ as a whole may be described by introducing as a central point the centre of mass

$$\boldsymbol{R}_k = \sum_i m_{ki}\boldsymbol{R}_{ki}/m_k, \tag{41}$$

where $m_k = \sum_i m_{ki}$. Summation of (39) over the index $i$, which labels the constituent particles, leads to

$$m_k\ddot{\boldsymbol{R}}_k = \sum_i e_{ki}\{\boldsymbol{e}_t(\boldsymbol{R}_{ki}, t)+c^{-1}\dot{\boldsymbol{R}}_{ki}\wedge \boldsymbol{b}_t(\boldsymbol{R}_{ki}, t)\}. \tag{42}$$

If (40) is inserted into (42) it appears that the contribution of the *intra-atomic* fields drops out, as should be expected for central forces. In this way (42) gets the form

$$m_k\ddot{\boldsymbol{R}}_k = -\sum_{l(\neq k)i,j}\nabla_{ki}\frac{e_{ki}e_{lj}}{4\pi|\boldsymbol{R}_{ki}-\boldsymbol{R}_{lj}|}+\sum_i e_{ki}\left\{\boldsymbol{E}_e(\boldsymbol{R}_{ki}, t)+\frac{\dot{\boldsymbol{R}}_{ki}}{c}\wedge \boldsymbol{B}_e(\boldsymbol{R}_{ki}, t)\right\}. \tag{43}$$

If the *external* fields are developed in Taylor series in terms of the internal coordinates

$$\boldsymbol{r}_{ki} \equiv \boldsymbol{R}_{ki}-\boldsymbol{R}_k, \tag{44}$$

the last term of (43) gets the form

$$\sum_i e_{ki}\sum_{n=0}^{\infty}\frac{1}{n!}(\boldsymbol{r}_{ki}\cdot\nabla_k)^n\{\boldsymbol{E}_e(\boldsymbol{R}_k, t)+c^{-1}\dot{\boldsymbol{R}}_k\wedge \boldsymbol{B}_e(\boldsymbol{R}_k, t)+c^{-1}\dot{\boldsymbol{r}}_{ki}\wedge \boldsymbol{B}_e(\boldsymbol{R}_k, t)\}, \tag{45}$$

which might be expressed in terms of the atomic charge (28) and the atomic multipole moments (31), (32). Since the external fields vary slowly over the atoms we shall limit ourselves in (45) to the contributions of the charges and the electric and magnetic dipole moments. (Cf. problem 2 for the general case.) Using the identity

$$\boldsymbol{r}_{ki}\cdot\nabla_k\dot{\boldsymbol{r}}_{ki} = \frac{1}{2}\frac{\mathrm{d}}{\mathrm{d}t}(\boldsymbol{r}_{ki}\boldsymbol{r}_{ki})\cdot\nabla_k+\tfrac{1}{2}(\boldsymbol{r}_{ki}\wedge\dot{\boldsymbol{r}}_{ki})\wedge\nabla_k, \tag{46}$$

one gets for (45):

$$(e_k+\overline{\boldsymbol{\mu}}_k^{(1)}\cdot\nabla_k)\{\boldsymbol{E}_e(\boldsymbol{R}_k, t)+c^{-1}\boldsymbol{v}_k\wedge \boldsymbol{B}_e(\boldsymbol{R}_k, t)\}+(c^{-1}\dot{\overline{\boldsymbol{\mu}}}_k^{(1)}+\overline{\boldsymbol{v}}_k^{(1)}\wedge\nabla_k)\wedge \boldsymbol{B}_e(\boldsymbol{R}_k, t), \tag{47}$$

where $\boldsymbol{v}_k$ is the atomic velocity $\dot{\boldsymbol{R}}_k$.

The *interatomic* contribution to the force may likewise be expanded in terms of $\boldsymbol{r}_{ki}$ and $\boldsymbol{r}_{lj}$ if the atoms are outside each other, i.e. if the interatomic distance $|\boldsymbol{R}_k - \boldsymbol{R}_l|$ is greater than the sum of the largest $|\boldsymbol{r}_{ki}|$ and $|\boldsymbol{r}_{lj}|$. In that case the first term at the right-hand side of (43) reads

$$-\sum_{l(\neq k)i,\,j}\ \sum_{n,\,m=0}^{\infty} \frac{1}{n!m!} (\boldsymbol{r}_{ki}\cdot\nabla_k)^n (\boldsymbol{r}_{lj}\cdot\nabla_l)^m \nabla_k \frac{e_{ki}\,e_{lj}}{4\pi|\boldsymbol{R}_k - \boldsymbol{R}_l|}, \tag{48}$$

or, if the electric multipole moments (31) are used,

$$-\sum_{l(\neq k)}\ \sum_{n,\,m=0}^{\infty} \overline{\boldsymbol{\mu}}_k^{(n)} : \nabla_k^n \overline{\boldsymbol{\mu}}_l^{(m)} : \nabla_l^m \nabla_k \frac{1}{4\pi|\boldsymbol{R}_k - \boldsymbol{R}_l|}. \tag{49}$$

In the general case of arbitrary interatomic distances we may write the right-hand side of (43) as the sum of a long range force $\boldsymbol{f}_k^{\mathrm{L}}$, which consists of (47) and (49) together, and a remaining short range force $\boldsymbol{f}_k^{\mathrm{S}}$. In this way we have

$$m_k \dot{\boldsymbol{v}}_k = \boldsymbol{f}_k^{\mathrm{L}} + \boldsymbol{f}_k^{\mathrm{S}}, \tag{50}$$

$$\begin{aligned}\boldsymbol{f}_k^{\mathrm{L}} = &-\sum_{l(\neq k)}\ \sum_{n,\,m=0}^{\infty} \overline{\boldsymbol{\mu}}_k^{(n)} : \nabla_k^n \overline{\boldsymbol{\mu}}_l^{(m)} : \nabla_l^m \nabla_k \frac{1}{4\pi|\boldsymbol{R}_k - \boldsymbol{R}_l|} \\ &+ (e_k + \overline{\boldsymbol{\mu}}_k^{(1)}\cdot\nabla_k)\{\boldsymbol{E}_{\mathrm{e}}(\boldsymbol{R}_k, t) + c^{-1}\boldsymbol{v}_k \wedge \boldsymbol{B}_{\mathrm{e}}(\boldsymbol{R}_k, t)\} \\ &\qquad + (c^{-1}\dot{\overline{\boldsymbol{\mu}}}_k^{(1)} + \overline{\boldsymbol{v}}_k^{(1)} \wedge \nabla_k) \wedge \boldsymbol{B}_{\mathrm{e}}(\boldsymbol{R}_k, t),\end{aligned} \tag{51}$$

$$\boldsymbol{f}_k^{\mathrm{S}} = -\sum_{l(\neq k)i,\,j} \nabla_{ki} \frac{e_{ki}\,e_{lj}}{4\pi|\boldsymbol{R}_{ki} - \boldsymbol{R}_{lj}|} + \sum_{l(\neq k)}\ \sum_{n,\,m=0}^{\infty} \overline{\boldsymbol{\mu}}_k^{(n)} : \nabla_k^n \overline{\boldsymbol{\mu}}_l^{(m)} : \nabla_l^m \nabla_k \frac{1}{4\pi|\boldsymbol{R}_k - \boldsymbol{R}_l|}. \tag{52}$$

Indeed the contribution $\boldsymbol{f}_k^{\mathrm{S}}$ has short range character, since it vanishes if the atoms are outside each other.

The external fields obey the homogeneous field equations

$$\nabla\cdot\boldsymbol{B}_{\mathrm{e}} = 0, \qquad c^{-1}\partial\boldsymbol{B}_{\mathrm{e}}/\partial t + \nabla \wedge \boldsymbol{E}_{\mathrm{e}} = 0. \tag{53}$$

They permit us to write the long range force (51) as

$$\begin{aligned}\boldsymbol{f}_k^{\mathrm{L}} = &-\sum_{l(\neq k)}\ \sum_{n,\,m=0}^{\infty} \overline{\boldsymbol{\mu}}_k^{(n)} : \nabla_k^n \overline{\boldsymbol{\mu}}_l^{(m)} : \nabla_l^m \nabla_k \frac{1}{4\pi|\boldsymbol{R}_k - \boldsymbol{R}_l|} \\ &+ e_k\{\boldsymbol{E}_{\mathrm{e}}(\boldsymbol{R}_k, t) + c^{-1}\boldsymbol{v}_k \wedge \boldsymbol{B}_{\mathrm{e}}(\boldsymbol{R}_k, t)\} + \{\nabla_k \boldsymbol{E}_{\mathrm{e}}(\boldsymbol{R}_k, t)\}\cdot\overline{\boldsymbol{\mu}}_k^{(1)} \\ &+ \{\nabla_k \boldsymbol{B}_{\mathrm{e}}(\boldsymbol{R}_k, t)\}\cdot(\overline{\boldsymbol{v}}_k^{(1)} + c^{-1}\overline{\boldsymbol{\mu}}_k^{(1)} \wedge \boldsymbol{v}_k) + c^{-1}\frac{\mathrm{d}}{\mathrm{d}t}\{\overline{\boldsymbol{\mu}}_k^{(1)} \wedge \boldsymbol{B}_{\mathrm{e}}(\boldsymbol{R}_k, t)\},\end{aligned} \tag{54}$$

where d/d$t$ is the total time derivative. In this way the *equation of motion* (50) with (54) and (52) for an atom in an electromagnetic field of interatomic and external origin is found. If only one atom in an external field is present it gets the simple form:

$$m_k \dot{\boldsymbol{v}}_k = e_k(\boldsymbol{E}_e + c^{-1}\boldsymbol{v}_k \wedge \boldsymbol{B}_e) + (\nabla_k \boldsymbol{E}_e)\cdot\bar{\boldsymbol{\mu}}_k^{(1)} + (\nabla_k \boldsymbol{B}_e)\cdot(\bar{\boldsymbol{v}}_k^{(1)} + c^{-1}\bar{\boldsymbol{\mu}}_k^{(1)} \wedge \boldsymbol{v}_k) + c^{-1}\frac{\mathrm{d}}{\mathrm{d}t}(\bar{\boldsymbol{\mu}}_k^{(1)} \wedge \boldsymbol{B}_e). \quad (55)$$

The first term is the Lorentz force exerted on a charged composite particle, the second is the Kelvin force on an electric dipole moment, the third is the analogous magnetic term which contains the magnetic dipole moment and the electric dipole moment in motion, while the last term describes an electrodynamic effect with a time derivative.

### b. *The energy equation*

The energy equation for the constituent particle $ki$ follows if (39) is multiplied by the velocity $\dot{\boldsymbol{R}}_{ki}$:

$$m_{ki}\dot{\boldsymbol{R}}_{ki}\cdot\ddot{\boldsymbol{R}}_{ki} = e_{ki}\dot{\boldsymbol{R}}_{ki}\cdot\boldsymbol{e}_t(\boldsymbol{R}_{ki}, t). \quad (56)$$

Taking the sum over $i$, introducing the centre of mass (41) and the internal coordinates (44) and inserting the expression (40) for the field $\boldsymbol{e}_t$ one gets

$$\frac{\mathrm{d}}{\mathrm{d}t}(\tfrac{1}{2}m_k \boldsymbol{v}_k^2 + \tfrac{1}{2}\sum_i m_{ki}\dot{\boldsymbol{r}}_{ki}^2) = -\sum_{i,j(i\neq j)} \dot{\boldsymbol{R}}_{ki}\cdot\nabla_{ki}\frac{e_{ki}e_{kj}}{4\pi|\boldsymbol{R}_{ki}-\boldsymbol{R}_{kj}|} - \sum_{l(\neq k)i,j} \dot{\boldsymbol{R}}_{ki}\cdot\nabla_{ki}\frac{e_{ki}e_{lj}}{4\pi|\boldsymbol{R}_{ki}-\boldsymbol{R}_{lj}|} + \sum_i e_{ki}\dot{\boldsymbol{R}}_{ki}\cdot\boldsymbol{E}_e(\boldsymbol{R}_{ki}, t). \quad (57)$$

The first term at the right-hand side is an *intra-atomic* contribution. It may be transformed into the total time derivative of the intra-atomic Coulomb energy of the atom:

$$-\sum_{i,j(i\neq j)}(\dot{\boldsymbol{R}}_{ki}\cdot\nabla_{ki} + \dot{\boldsymbol{R}}_{kj}\cdot\nabla_{kj})\frac{e_{ki}e_{kj}}{8\pi|\boldsymbol{R}_{ki}-\boldsymbol{R}_{kj}|} = -\frac{\mathrm{d}}{\mathrm{d}t}\sum_{i,j(i\neq j)}\frac{e_{ki}e_{kj}}{8\pi|\boldsymbol{r}_{ki}-\boldsymbol{r}_{kj}|}, \quad (58)$$

where (44) has been used. This term will be shifted to the left-hand side.

The *external* field term of (57) may be expanded in terms of $\boldsymbol{r}_{ki}$. One obtains then

$$\sum_i e_{ki}\sum_{n=0}^{\infty}\frac{1}{n!}(\boldsymbol{r}_{ki}\cdot\nabla_k)^n(\dot{\boldsymbol{R}}_k + \dot{\boldsymbol{r}}_{ki})\cdot\boldsymbol{E}_e(\boldsymbol{R}_k, t). \quad (59)$$

Introducing the electric and magnetic multipole moments and confining ourselves to terms with the external fields and their first derivatives we are left with an expression containing charges and dipole moments only:

$$(e_k+\overline{\boldsymbol{\mu}}_k^{(1)}\cdot\nabla_k)\boldsymbol{v}_k\cdot\boldsymbol{E}_e(\boldsymbol{R}_k,t)+(\dot{\overline{\boldsymbol{\mu}}}_k^{(1)}+c\overline{\boldsymbol{v}}_k^{(1)}\wedge\nabla_k)\cdot\boldsymbol{E}_e(\boldsymbol{R}_k,t), \tag{60}$$

where (46) has been used. (Cf. problem 3 for the general expression with all multipole moments.)

The *interatomic* contribution in (57) may also be expanded in terms of $\boldsymbol{r}_{ki}$ and $\boldsymbol{r}_{lj}$ in case the atoms are outside each other:

$$-\sum_{l(\neq k)i,j}\sum_{n,m=0}^{\infty}\frac{1}{n!m!}(\boldsymbol{r}_{ki}\cdot\nabla_k)^n(\boldsymbol{r}_{lj}\cdot\nabla_l)^m(\dot{\boldsymbol{R}}_k+\dot{\boldsymbol{r}}_{ki})\cdot\nabla_k\frac{e_{ki}e_{lj}}{4\pi|\boldsymbol{R}_k-\boldsymbol{R}_l|}, \tag{61}$$

or, in terms of the multipoles (31),

$$-\sum_{l(\neq k)}\sum_{m=0}^{\infty}\Big(\sum_{n=0}^{\infty}\overline{\boldsymbol{\mu}}_k^{(n)}:\nabla_k^n\boldsymbol{v}_k\cdot\nabla_k+\sum_{n=1}^{\infty}\dot{\overline{\boldsymbol{\mu}}}_k^{(n)}:\nabla_k^n\Big)\overline{\boldsymbol{\mu}}_l^{(m)}:\nabla_l^m\frac{1}{4\pi|\boldsymbol{R}_k-\boldsymbol{R}_l|}. \tag{62}$$

In the general case the second and third term at the right-hand side of (57) may be written as the sum of a long range contribution $\psi_k^{\mathrm{L}}$, which consists of (60) and (62), and a remaining short range contribution $\psi_k^{\mathrm{S}}$:

$$\frac{\mathrm{d}}{\mathrm{d}t}\left(\tfrac{1}{2}m_k v_k^2+\tfrac{1}{2}\sum_i m_{ki}\dot{\boldsymbol{r}}_{ki}^2+\sum_{i,j(i\neq j)}\frac{e_{ki}e_{kj}}{8\pi|\boldsymbol{r}_{ki}-\boldsymbol{r}_{kj}|}\right)=\psi_k^{\mathrm{L}}+\psi_k^{\mathrm{S}}, \tag{63}$$

$$\psi_k^{\mathrm{L}}=-\sum_{l(\neq k)}\sum_{m=0}^{\infty}\Big(\sum_{n=0}^{\infty}\overline{\boldsymbol{\mu}}_k^{(n)}:\nabla_k^n\boldsymbol{v}_k\cdot\nabla_k+\sum_{n=1}^{\infty}\dot{\overline{\boldsymbol{\mu}}}_k^{(n)}:\nabla_k^n\Big)\overline{\boldsymbol{\mu}}_l^{(m)}:\nabla_l^m\frac{1}{4\pi|\boldsymbol{R}_k-\boldsymbol{R}_l|}$$
$$+(e_k+\overline{\boldsymbol{\mu}}_k^{(1)}\cdot\nabla_k)\boldsymbol{v}_k\cdot\boldsymbol{E}_e(\boldsymbol{R}_k,t)+(\dot{\overline{\boldsymbol{\mu}}}_k^{(1)}+c\overline{\boldsymbol{v}}_k^{(1)}\wedge\nabla_k)\cdot\boldsymbol{E}_e(\boldsymbol{R}_k,t), \tag{64}$$

$$\psi_k^{\mathrm{S}}=-\sum_{l(\neq k)i,j}\dot{\boldsymbol{R}}_{ki}\cdot\nabla_{ki}\frac{e_{ki}e_{lj}}{4\pi|\boldsymbol{R}_{ki}-\boldsymbol{R}_{lj}|}$$
$$+\sum_{l(\neq k)}\sum_{m=0}^{\infty}\Big(\sum_{n=0}^{\infty}\overline{\boldsymbol{\mu}}_k^{(n)}:\nabla_k^n\boldsymbol{v}_k\cdot\nabla_k+\sum_{n=1}^{\infty}\dot{\overline{\boldsymbol{\mu}}}_k^{(n)}:\nabla_k^n\Big)\overline{\boldsymbol{\mu}}_l^{(m)}:\nabla_l^m\frac{1}{4\pi|\boldsymbol{R}_k-\boldsymbol{R}_l|}. \tag{65}$$

With the second equation of (53) one gets for (64)

$$\psi_k^{\mathrm{L}}=-\sum_{l(\neq k)}\sum_{m=0}^{\infty}\Big(\sum_{n=0}^{\infty}\overline{\boldsymbol{\mu}}_k^{(n)}:\nabla_k^n\boldsymbol{v}_k\cdot\nabla_k+\sum_{n=1}^{\infty}\dot{\overline{\boldsymbol{\mu}}}_k^{(n)}:\nabla_k^n\Big)\overline{\boldsymbol{\mu}}_l^{(m)}:\nabla_l^m\frac{1}{4\pi|\boldsymbol{R}_k-\boldsymbol{R}_l|}$$
$$+e_k\boldsymbol{v}_k\cdot\boldsymbol{E}_e(\boldsymbol{R}_k,t)+\boldsymbol{v}_k\cdot\{\nabla_k\boldsymbol{E}_e(\boldsymbol{R}_k,t)\}\cdot\overline{\boldsymbol{\mu}}_k^{(1)}+\dot{\overline{\boldsymbol{\mu}}}_k^{(1)}\cdot\boldsymbol{E}_e(\boldsymbol{R}_k,t)$$
$$-(\overline{\boldsymbol{v}}_k^{(1)}+c^{-1}\overline{\boldsymbol{\mu}}_k^{(1)}\wedge\boldsymbol{v}_k)\cdot\frac{\partial\boldsymbol{B}_e(\boldsymbol{R}_k,t)}{\partial t}. \tag{66}$$

The *atomic energy equation* (63) is thus completely specified. It contains at the left-hand side kinetic and potential energy terms of which the last two are of intra-atomic character. The right-hand side contains contributions, specified by (66) and (65), in which the fields of interatomic and external origin occur.

For a single atom in an external field the atomic energy equation reduces to

$$\frac{\mathrm{d}}{\mathrm{d}t}\left(\tfrac{1}{2}m_k \boldsymbol{v}_k^2+\tfrac{1}{2}\sum_i m_{ki}\dot{\boldsymbol{r}}_{ki}^2+\sum_{i,\,j(i\neq j)}\frac{e_{ki}e_{kj}}{8\pi|\boldsymbol{r}_{ki}-\boldsymbol{r}_{kj}|}\right)$$
$$= e_k\boldsymbol{v}_k\cdot\boldsymbol{E}_{\mathrm{e}}+\boldsymbol{v}_k\cdot(\nabla_k\boldsymbol{E}_{\mathrm{e}})\cdot\overline{\boldsymbol{\mu}}_k^{(1)}+\boldsymbol{v}_k\cdot(\nabla_k\boldsymbol{B}_{\mathrm{e}})\cdot(\overline{\boldsymbol{v}}_k^{(1)}+c^{-1}\overline{\boldsymbol{\mu}}_k^{(1)}\wedge\boldsymbol{v}_k)+\dot{\overline{\boldsymbol{\mu}}}_k^{(1)}\cdot\boldsymbol{E}_{\mathrm{e}}$$
$$-(\overline{\boldsymbol{v}}_k^{(1)}+c^{-1}\overline{\boldsymbol{\mu}}_k^{(1)}\wedge\boldsymbol{v}_k)\cdot\frac{\mathrm{d}\boldsymbol{B}_{\mathrm{e}}}{\mathrm{d}t},\qquad(67)$$

where the total time derivative $\mathrm{d}/\mathrm{d}t \equiv \partial/\partial t+\boldsymbol{v}_k\cdot\nabla_k$ has been introduced in the last term. In this way at the right-hand side the 'power' terms due to the Lorentz and Kelvin forces appear, as well as two terms with total time derivatives. It may be noted that the latter show a remarkable asymmetry: the first contains the derivative of a dipole moment multiplied by a field, while the second contains the derivative of a field multiplied by a dipole moment expression.

## 6 *The inner angular momentum equation for composite particles*

The inner angular momentum of an atom $k$ may be defined as

$$\overline{\boldsymbol{s}}_k \equiv \sum_i m_{ki}\boldsymbol{r}_{ki}\wedge\dot{\boldsymbol{r}}_{ki}\qquad(68)$$

in terms of the masses $m_{ki}$ of the constituent particles and the internal coordinates $\boldsymbol{r}_{ki}$ (44). Its time derivative is

$$\dot{\overline{\boldsymbol{s}}}_k = \sum_i m_{ki}\boldsymbol{r}_{ki}\wedge\ddot{\boldsymbol{r}}_{ki} = \sum_i m_{ki}\boldsymbol{r}_{ki}\wedge\ddot{\boldsymbol{R}}_{ki},\qquad(69)$$

where (41) has been used.

Introducing the equation of motion (39) with the fields (40) we get

$$\dot{\overline{\boldsymbol{s}}}_k = -\sum_{l(\neq k)i,\,j}\boldsymbol{r}_{ki}\wedge\nabla_{ki}\frac{e_{ki}e_{lj}}{4\pi|\boldsymbol{R}_{ki}-\boldsymbol{R}_{lj}|}$$
$$+\sum_i e_{ki}\boldsymbol{r}_{ki}\wedge\{\boldsymbol{E}_{\mathrm{e}}(\boldsymbol{R}_{ki},t)+c^{-1}\dot{\boldsymbol{R}}_{ki}\wedge\boldsymbol{B}_{\mathrm{e}}(\boldsymbol{R}_{ki},t)\}.\qquad(70)$$

The *intra-atomic* contributions to the fields have dropped out in this expression, as should be expected for central forces.

If the *external* fields change slowly, they can be expanded in terms of $\boldsymbol{r}_{ki}$, so that the external field contribution in (70) may then be written as:

$$\sum_i e_{ki}\boldsymbol{r}_{ki}\wedge\sum_{n=0}^{\infty}\frac{1}{n!}(\boldsymbol{r}_{ki}\cdot\boldsymbol{\nabla}_k)^n\{\boldsymbol{E}_e(\boldsymbol{R}_k,t)+c^{-1}(\dot{\boldsymbol{R}}_k+\dot{\boldsymbol{r}}_{ki})\wedge\boldsymbol{B}_e(\boldsymbol{R}_k,t)\}. \qquad (71)$$

Assuming that the external fields are sufficiently smooth we limit ourselves to electric and magnetic dipole moments in this expression. We obtain then, with the help of the vector identity

$$\boldsymbol{r}_{ki}\wedge(\dot{\boldsymbol{r}}_{ki}\wedge\boldsymbol{B}_e)=\tfrac{1}{2}(\boldsymbol{r}_{ki}\wedge\dot{\boldsymbol{r}}_{ki})\wedge\boldsymbol{B}_e+\frac{1}{2}\frac{\mathrm{d}}{\mathrm{d}t}(\boldsymbol{r}_{ki}\boldsymbol{r}_{ki})\cdot\boldsymbol{B}_e-\frac{1}{2}\frac{\mathrm{d}}{\mathrm{d}t}(r_{ki}^2)\boldsymbol{B}_e, \qquad (72)$$

instead of (71):

$$\overline{\boldsymbol{\mu}}_k^{(1)}\wedge\{\boldsymbol{E}_e(\boldsymbol{R}_k,t)+c^{-1}\boldsymbol{v}_k\wedge\boldsymbol{B}_e(\boldsymbol{R}_k,t)\}+\overline{\boldsymbol{v}}_k^{(1)}\wedge\boldsymbol{B}_e(\boldsymbol{R}_k,t), \qquad (73)$$

where we have written $\boldsymbol{v}_k$ for the atomic velocity $\dot{\boldsymbol{R}}_k$.

In case the atoms are outside each other the *interatomic* contribution to the right-hand side of (70) may likewise be expanded, with the result

$$-\sum_{l(\neq k)i,j}\sum_{n,m=0}^{\infty}\frac{1}{n!m!}\boldsymbol{r}_{ki}\wedge\boldsymbol{\nabla}_k(\boldsymbol{r}_{ki}\cdot\boldsymbol{\nabla}_k)^n(\boldsymbol{r}_{lj}\cdot\boldsymbol{\nabla}_l)^m\frac{e_{ki}e_{lj}}{4\pi|\boldsymbol{R}_k-\boldsymbol{R}_l|}, \qquad (74)$$

or, in terms of the electric multipole moments (31),

$$\sum_{l(\neq k)}\sum_{n,m=0}^{\infty}n\boldsymbol{\nabla}_k\wedge\overline{\boldsymbol{\mu}}_k^{(n)}:\boldsymbol{\nabla}_k^{n-1}\overline{\boldsymbol{\mu}}_l^{(m)}:\boldsymbol{\nabla}_l^m\frac{1}{4\pi|\boldsymbol{R}_k-\boldsymbol{R}_l|}. \qquad (75)$$

In the general case the right-hand side of (70) may be written as the sum of a long range moment $\boldsymbol{d}_k^{\mathrm{L}}$ and a short range moment $\boldsymbol{d}_k^{\mathrm{S}}$:

$$\dot{\boldsymbol{s}}_k=\boldsymbol{d}_k^{\mathrm{L}}+\boldsymbol{d}_k^{\mathrm{S}}, \qquad (76)$$

$$\boldsymbol{d}_k^{\mathrm{L}}=\sum_{l(\neq k)}\sum_{n,m=0}^{\infty}n\boldsymbol{\nabla}_k\wedge\overline{\boldsymbol{\mu}}_k^{(n)}:\boldsymbol{\nabla}_k^{n-1}\overline{\boldsymbol{\mu}}_l^{(m)}:\boldsymbol{\nabla}_l^m\frac{1}{4\pi|\boldsymbol{R}_k-\boldsymbol{R}_l|}$$
$$+\overline{\boldsymbol{\mu}}_k^{(1)}\wedge\{\boldsymbol{E}_e(\boldsymbol{R}_k,t)+c^{-1}\boldsymbol{v}_k\wedge\boldsymbol{B}_e(\boldsymbol{R}_k,t)\}+\overline{\boldsymbol{v}}_k^{(1)}\wedge\boldsymbol{B}_e(\boldsymbol{R}_k,t), \qquad (77)$$

$$\boldsymbol{d}_k^{\mathrm{S}}=-\sum_{l(\neq k)i,j}\boldsymbol{r}_{ki}\wedge\boldsymbol{\nabla}_{ki}\frac{e_{ki}e_{lj}}{4\pi|\boldsymbol{R}_{ki}-\boldsymbol{R}_{lj}|}$$
$$-\sum_{l(\neq k)}\sum_{n,m=0}^{\infty}n\boldsymbol{\nabla}_k\wedge\overline{\boldsymbol{\mu}}_k^{(n)}:\boldsymbol{\nabla}_k^{n-1}\overline{\boldsymbol{\mu}}_l^{(m)}:\boldsymbol{\nabla}_l^m\frac{1}{4\pi|\boldsymbol{R}_k-\boldsymbol{R}_l|}. \qquad (78)$$

For a single atom in an external field the equation (76) with (77–78) reduces to

$$\dot{\boldsymbol{s}}_k = \overline{\boldsymbol{\mu}}_k^{(1)} \wedge (\boldsymbol{E}_e + c^{-1} \boldsymbol{v}_k \wedge \boldsymbol{B}_e) + \overline{\boldsymbol{v}}_k^{(1)} \wedge \boldsymbol{B}_e, \tag{79}$$

which shows that in non-relativistic theory the electric dipole moment is coupled to both the electric and the magnetic field, while the magnetic dipole moment is coupled to the magnetic field only.

# PROBLEMS

**1.** Prove by expansion of the non-relativistic field expressions (10) in powers of atomic parameters $\boldsymbol{r}_{ki}$ and introduction of the multipole moments (31–32), that the fields generated by a set of atoms with positions $\boldsymbol{R}_k$, velocities $\boldsymbol{v}_k$, charges $e_k$ and multipole moments $\overline{\boldsymbol{\mu}}_k^{(n)}$, $\overline{\boldsymbol{\nu}}_k^{(n)}$ are:

$$\boldsymbol{e}(\boldsymbol{R}, t) = -\sum_k \nabla \frac{e_k}{4\pi|\boldsymbol{R}_k - \boldsymbol{R}|} - \sum_k \sum_{n=1}^{\infty} (-1)^n \overline{\boldsymbol{\mu}}_k^{(n)} \vdots \nabla^n \nabla \frac{1}{4\pi|\boldsymbol{R}_k - \boldsymbol{R}|},$$

$$\boldsymbol{b}(\boldsymbol{R}, t) = c^{-1} \sum_k \nabla \wedge \frac{e_k \boldsymbol{v}_k}{4\pi|\boldsymbol{R}_k - \boldsymbol{R}|} + \sum_k \sum_{n=1}^{\infty} (-1)^n \nabla \wedge (c^{-1} \overline{\boldsymbol{\mu}}_k^{(n)} \vdots \nabla^n \boldsymbol{v}_k$$
$$- c^{-1} \dot{\overline{\boldsymbol{\mu}}}_k^{(n)} \vdots \nabla^{n-1} + \nabla^{n-1} \vdots \overline{\boldsymbol{\nu}}_k^{(n)} \wedge \nabla) \frac{1}{4\pi|\boldsymbol{R}_k - \boldsymbol{R}|}.$$

Show, by adding a delta function $\delta(\boldsymbol{R}' - \boldsymbol{R}_k)$ and an integration over $\boldsymbol{R}'$, that these formulae may be written as:

$$\boldsymbol{e}(\boldsymbol{R}, t) = -\nabla \int \{\rho^{\mathrm{e}}(\boldsymbol{R}', t) - \nabla' \cdot \boldsymbol{p}(\boldsymbol{R}', t)\} \frac{1}{4\pi|\boldsymbol{R} - \boldsymbol{R}'|} \, \mathrm{d}\boldsymbol{R}',$$

$$\boldsymbol{b}(\boldsymbol{R}, t) = \nabla \wedge \int \{c^{-1} \boldsymbol{j}(\boldsymbol{R}', t) + \partial_0 \boldsymbol{p}(\boldsymbol{R}', t) + \nabla' \wedge \boldsymbol{m}(\boldsymbol{R}', t)\} \frac{1}{4\pi|\boldsymbol{R} - \boldsymbol{R}'|} \, \mathrm{d}\boldsymbol{R}',$$

with $\rho^{\mathrm{e}}$, $\boldsymbol{j}$, $\boldsymbol{p}$ and $\boldsymbol{m}$ given by (33) and (34). Verify, by using (8), that these expressions for the electric and magnetic fields are the solutions of the atomic field equations (35).

**2.** Prove from (45) that the general expression for the force on a composite particle with position $\boldsymbol{R}_k$, velocity $\boldsymbol{v}_k$, charge $e_k$ and multiple moments $\overline{\boldsymbol{\mu}}_k^{(n)}$, $\overline{\boldsymbol{\nu}}_k^{(n)}$ in an external field $\boldsymbol{E}_{\mathrm{e}}$, $\boldsymbol{B}_{\mathrm{e}}$ is:

$$\boldsymbol{f}_k = (e_k + \sum_{n=1}^{\infty} \overline{\boldsymbol{\mu}}_k^{(n)} \vdots \nabla_k^n)(\boldsymbol{E}_{\mathrm{e}} + c^{-1} \boldsymbol{v}_k \wedge \boldsymbol{B}_{\mathrm{e}})$$
$$+ \sum_{n=1}^{\infty} (c^{-1} \dot{\overline{\boldsymbol{\mu}}}_k^{(n)} \wedge \nabla_k^{n-1} + \nabla_k^{n-1} \vdots \overline{\boldsymbol{\nu}}_k^{(n)} \wedge \nabla_k) \wedge \boldsymbol{B}_{\mathrm{e}}.$$

This expression is a generalization to all multipoles of (47). By using the field equations (53) for the external field it follows that the equation of motion for a composite particle in an external field may be cast into the form

$$m_k \dot{\boldsymbol{v}}_k = e_k(\boldsymbol{E}_e + c^{-1}\boldsymbol{v}_k \wedge \boldsymbol{B}_e) + \sum_{n=1}^{\infty} \Big[ (\boldsymbol{\nabla}_k \boldsymbol{E}_e)\cdot({}^{\leftarrow}\boldsymbol{\nabla}_k^{n-1} : \overline{\boldsymbol{\mu}}_k^{(n)})$$
$$+ (\boldsymbol{\nabla}_k \boldsymbol{B}_e)\cdot\{{}^{\leftarrow}\boldsymbol{\nabla}_k^{n-1} : (\overline{\boldsymbol{\nu}}_k^{(n)} + c^{-1}\overline{\boldsymbol{\mu}}_k^{(n)} \wedge \boldsymbol{v}_k)\} + c^{-1}\frac{\mathrm{d}}{\mathrm{d}t}(\boldsymbol{\nabla}_k^{n-1} : \overline{\boldsymbol{\mu}}_k^{(n)} \wedge \boldsymbol{B}_e)\Big],$$

where ${}^{\leftarrow}\boldsymbol{\nabla}_k$ is the differentiation operator $\partial/\partial \boldsymbol{R}_k$ acting to the left i.e. acting on the external fields. Furthermore the total time derivative $\mathrm{d}/\mathrm{d}t$ is $\partial/\partial t + \boldsymbol{v}_k\cdot\boldsymbol{\nabla}_k$. This second equation is a generalization to all multipoles of (55).

**3.** Prove from (59) that the work per unit of time exerted on a composite particle in an external field is:

$$\psi_k = (e_k + \sum_{n=1}^{\infty} \overline{\boldsymbol{\mu}}_k^{(n)} : \boldsymbol{\nabla}_k^n)\boldsymbol{v}_k\cdot\boldsymbol{E}_e + \sum_{n=1}^{\infty} (\dot{\overline{\boldsymbol{\mu}}}_k^{(n)} : \boldsymbol{\nabla}_k^{n-1} + c\boldsymbol{\nabla}_k^{n-1} : \overline{\boldsymbol{\nu}}_k^{(n)} \wedge \boldsymbol{\nabla}_k)\cdot\boldsymbol{E}_e,$$

with the same notation as in problem 2. This expression generalizes (60) to all multipole orders. By means of the second field equation (53) it follows that the energy equation for a composite particle in an external field is:

$$\frac{\mathrm{d}}{\mathrm{d}t}\Big(\tfrac{1}{2}m_k \boldsymbol{v}_k^2 + \tfrac{1}{2}\sum_i m_{ki}\dot{\boldsymbol{r}}_{ki}^2 + \sum_{i,j(i\neq j)} \frac{e_{ki}e_{kj}}{8\pi|\boldsymbol{r}_{ki} - \boldsymbol{r}_{kj}|}\Big)$$
$$= e_k \boldsymbol{v}_k\cdot\boldsymbol{E}_e + \sum_{n=1}^{\infty} \Big[ \boldsymbol{v}_k\cdot(\boldsymbol{\nabla}_k \boldsymbol{E}_e)\cdot({}^{\leftarrow}\boldsymbol{\nabla}_k^{n-1} : \overline{\boldsymbol{\mu}}_k^{(n)})$$
$$+ \boldsymbol{v}_k\cdot(\boldsymbol{\nabla}_k \boldsymbol{B}_e)\cdot\{{}^{\leftarrow}\boldsymbol{\nabla}_k^{n-1} : (\overline{\boldsymbol{\nu}}_k^{(n)} + c^{-1}\overline{\boldsymbol{\mu}}_k^{(n)} \wedge \boldsymbol{v}_k)\}$$
$$+ \dot{\overline{\boldsymbol{\mu}}}_k^{(n)} : \boldsymbol{\nabla}_k^{n-1}\boldsymbol{E}_e - \boldsymbol{\nabla}_k^{n-1} : (\overline{\boldsymbol{\nu}}_k^{(n)} + c^{-1}\overline{\boldsymbol{\mu}}_k^{(n)} \wedge \boldsymbol{v}_k)\cdot\frac{\mathrm{d}\boldsymbol{B}_e}{\mathrm{d}t}\Big]$$

as a generalization of equation (67) to all multipole orders.

**4.** Prove from (71) that the angular momentum equation for a composite particle in an external field is:

$$\dot{\boldsymbol{s}}_k = \sum_{n=1}^{\infty} [n(\boldsymbol{\nabla}_k^{n-1} : \overline{\boldsymbol{\mu}}_k^{(n)}) \wedge (\boldsymbol{E}_e + c^{-1}\boldsymbol{v}_k \wedge \boldsymbol{B}_e) + \boldsymbol{\nabla}_k^{n-1} : \overline{\boldsymbol{\nu}}_k^{(n)} \wedge \boldsymbol{B}_e$$
$$+ (n-1)(\boldsymbol{\nabla}_k^{n-2} : \overline{\boldsymbol{\nu}}_k^{(n)}\cdot\boldsymbol{B}_e) \wedge {}^{\leftarrow}\boldsymbol{\nabla}_k$$
$$+ c^{-1}(n-1)\{\boldsymbol{\nabla}_k^{n-2} : \dot{\overline{\boldsymbol{\mu}}}_k^{(n)}\cdot\boldsymbol{B}_e - \boldsymbol{\nabla}_k^{n-2} : \dot{\overline{\boldsymbol{\mu}}}_k^{(n)} : \mathbf{U}\boldsymbol{B}_e\}],$$

where $\mathbf{U}$ is the unit tensor of the second rank and where the field equations (53) have been used. Furthermore ${}^{\leftarrow}\boldsymbol{\nabla}_k$ is the nabla operator acting to the left. This equation is the generalization of (79) to all multipole orders.

CHAPTER II

# Statistical description of field and matter

## 1 *Macroscopic laws*

The electromagnetic and material quantities that occur in the atomic equations show rapid changes in space and time. To describe macroscopic situations one has to find laws which contain physical quantities that change much slower in space and time. The reason is that such quantities are measured by means of macroscopic devices. These instruments do not yield information on individual particles, but averages over large numbers of particles contained in domains which are small compared to the total system.

So, if one wants to derive the macroscopic laws from the atomic equations some averaging procedure must be used. To that purpose one must define macroscopic quantities as statistical averages over a number of atoms contained in a mass element which is large enough so that the principles of statistical mechanics may be applied, but which is still small from a macroscopic point of view. As is implied by the definition of the macroscopic quantities the spatial dimensions of the mass elements should be on the one hand large compared to the distance between neighbouring atoms and on the other hand small compared to macroscopic distances. The possibility to realize such a situation requires the system to fulfil suitable physical conditions. Gases, liquids and solids will satisfy these conditions under wide ranges of physical circumstances. In gases the density should not be so small that the dimension of the mass cell would have to be excessive in order to fulfil the condition that the cell must contain many atoms.

The macroscopic quantities are thus rid of the extremely rapid changes in space and time which the corresponding microscopic quantities show. In fact, just as in the rest of macroscopic physics, the physical quantities can then be considered as continuous functions of the space-time coordinates, except at boundaries.

After having indicated how such macroscopic quantities may be described with the help of distribution functions, we shall derive in this chapter the macroscopic equations that govern the behaviour of fields and matter in bulk: the Maxwell equations, the momentum, energy and angular momentum balances and the laws of thermodynamics, all in the framework of classical non-relativistic theory.

## 2 *Average quantities*

Formally one may represent statistical averages with the help of distribution functions. In the non-relativistic approximation all dynamical quantities depend only on the positions $\boldsymbol{R}_{ki}$ and velocities $\dot{\boldsymbol{R}}_{ki}$ of the particles *ki*. Thus the averages of a microscopic quantity $a(\boldsymbol{R}_{ki}, \dot{\boldsymbol{R}}_{ki}, t)$ can be considered as an average in a 'fluxion' space spanned by the positions and velocities of the particles:

$$A(\boldsymbol{R}, t) = \langle a \rangle = \int af\,\mathrm{d}\varphi, \tag{1}$$

where $f = f(\boldsymbol{R}_{ki}, \dot{\boldsymbol{R}}_{ki}; t)$ depends on the particle positions $\boldsymbol{R}_{ki}$ and velocities $\dot{\boldsymbol{R}}_{ki}$ and where $\mathrm{d}\varphi = \prod_{ki} \mathrm{d}\boldsymbol{R}_{ki}\mathrm{d}\dot{\boldsymbol{R}}_{ki}$ is the fluxion space element. The probability to find the system in the volume element $\mathrm{d}\varphi$ is $f\mathrm{d}\varphi$.

Now from the conservation of probability one may conclude that time differentiation and averaging of a quantity commute. This is seen in the following way. The probability to find the system in the fluxion space element $\mathrm{d}\varphi$ is given by $f\mathrm{d}\varphi$. This measure $f\mathrm{d}\varphi$ remains constant in time if one follows the region of fluxion space points in their natural motion in fluxion space. Therefore one has:

$$\frac{\partial}{\partial t}\langle a \rangle \equiv \frac{\partial}{\partial t}\int af\,\mathrm{d}\varphi = \int \frac{\mathrm{d}a}{\mathrm{d}t} f\,\mathrm{d}\varphi \equiv \left\langle \frac{\mathrm{d}a}{\mathrm{d}t} \right\rangle, \tag{2}$$

where $\mathrm{d}/\mathrm{d}t$ is the total time derivative in fluxion space $\partial/\partial t + \sum_{ki} \dot{\boldsymbol{R}}_{ki}\cdot\nabla_{ki} + \sum_{ki} \ddot{\boldsymbol{R}}_{ki}\cdot\dot{\nabla}_{ki}$ (where $\dot{\nabla}_{ki} \equiv \partial/\partial\dot{\boldsymbol{R}}_{ki}$ and where $\ddot{\boldsymbol{R}}_{ki}$, as a consequence of the equations of motion, depends on all $\boldsymbol{R}_{lj}$ and $\dot{\boldsymbol{R}}_{lj}$ and on time). Equation (2) shows that time differentiation and averaging commute.

Space differentiation and averaging commute trivially:

$$\nabla\langle a \rangle = \langle \nabla a \rangle, \tag{3}$$

because the fluxion space distribution function does not depend on the space coordinates.

Often the quantities $a$ are sums of functions which depend on the variables pertinent to one atom or to two atoms only. In such cases one may perform a number of integrations in the expression (1) for the average $A$. For instance, if $a = \sum_k a_k(\boldsymbol{R}_{ki}, \dot{\boldsymbol{R}}_{ki})$ (where $\boldsymbol{R}_{ki}$ stands for $\boldsymbol{R}_{k1}, \boldsymbol{R}_{k2}, \ldots$) or alternatively, with mass centre and internal coordinates and velocities, $a = \sum_k a_k(\boldsymbol{R}_k, \dot{\boldsymbol{R}}_k, \boldsymbol{r}_{ki}, \dot{\boldsymbol{r}}_{ki})$ one may write the average $A$ in terms of a 'one-point distribution function' $f_1(\boldsymbol{R}_1, \dot{\boldsymbol{R}}_1, \boldsymbol{r}_{1i}, \dot{\boldsymbol{r}}_{1i}; t)$ as

$$A = \int a_1(\boldsymbol{R}_1, \dot{\boldsymbol{R}}_1, \boldsymbol{r}_{1i}, \dot{\boldsymbol{r}}_{1i}) f_1(\boldsymbol{R}_1, \dot{\boldsymbol{R}}_1, \boldsymbol{r}_{1i}, \dot{\boldsymbol{r}}_{1i}; t) \mathrm{d}\boldsymbol{R}_1 \, \mathrm{d}\dot{\boldsymbol{R}}_1 \prod_i \mathrm{d}\boldsymbol{r}_{1i} \mathrm{d}\dot{\boldsymbol{r}}_{1i}, \tag{4}$$

or, in a shorter notation,

$$A = \int a_1(1) f_1(1; t) \mathrm{d}1, \tag{5}$$

where $f_1(1; t)\mathrm{d}1$ is the probability, normalized to $N$ (the number of atoms in the system), to find an atom with parameters in the range $\mathrm{d}\boldsymbol{R}_1 \, \mathrm{d}\dot{\boldsymbol{R}}_1 \prod_i \mathrm{d}\boldsymbol{r}_{1i} \, \mathrm{d}\dot{\boldsymbol{r}}_{1i}$ around the point $\boldsymbol{R}_1, \dot{\boldsymbol{R}}_1, \boldsymbol{r}_{1i}, \dot{\boldsymbol{r}}_{1i}$ $(i = 1, 2, \ldots)$ in fluxion space.

Likewise, if $a$ has the form $a = \sum_{k, l(k \neq l)} a_{kl}(k, l)$, one may write

$$A = \int a_{12}(1, 2) f_2(1, 2; t) \mathrm{d}1 \, \mathrm{d}2, \tag{6}$$

where $f_2(1, 2; t)\mathrm{d}1 \, \mathrm{d}2$ is the joint probability, normalized to $N(N-1)$, to find one atom in the range d1 and another in the range d2.

It will be convenient to introduce also a two-point correlation function defined as

$$c_2(1, 2; t) \equiv f_2(1, 2; t) - f_1(1; t) f_1(2; t). \tag{7}$$

This correlation function has the property to vanish rapidly with increasing atomic distances for systems without long range order, such as fluids. For crystals however this is not the case.

In the preceding the treatment was confined to one-component systems. The extension to mixtures of several components is straightforward. In that case one has to introduce distribution functions for each separate species. The one-point distribution function $f_1^a(1; t)$ now gets an extra label $a$ which indicates the species. Now $f_1^a(1; t)\mathrm{d}1$ is the probability, normalized to $N^a$ (the number of atoms of species $a$), to find an atom of species $a$ with parameters in the range d1 in fluxion space. Similarly the two-point distribution function is defined in such way that $f_2^{ab}(1, 2; t)\mathrm{d}1 \, \mathrm{d}2$ is the joint probability, normalized to $N^a N^b$ (if $a \neq b$) or $N^a(N^a - 1)$ (if $a = b$), to find an atom of species $a$ in the range d1 and an atom of species $b$ in the range d2. The correlation function $c_2^{ab}(1, 2; t)$ is now defined as the difference $f_2^{ab}(1, 2; t) - f_1^a(1; t) f_1^b(2; t)$.

## 3 *The Maxwell equations*

The Maxwell equations may be found from the atomic field equations (I.35) by the statistical procedure of the preceding section. Indeed one gets:

$$\begin{aligned}\langle \nabla\cdot \boldsymbol{e}\rangle &= \langle \rho^{\mathrm{e}}\rangle - \langle \nabla\cdot \boldsymbol{p}\rangle,\\ -\langle \partial_0\, \boldsymbol{e}\rangle + \langle \nabla\wedge \boldsymbol{b}\rangle &= \langle \boldsymbol{j}\rangle/c + \langle \partial_0\, \boldsymbol{p}\rangle + \langle \nabla\wedge \boldsymbol{m}\rangle,\\ \langle \nabla\cdot \boldsymbol{b}\rangle &= 0,\\ \langle \partial_0\, \boldsymbol{b}\rangle + \langle \nabla\wedge \boldsymbol{e}\rangle &= 0.\end{aligned} \tag{8}$$

With the help of the commutation rules (2) and (3) one obtains from (8) the set of equations

$$\begin{aligned}\nabla\cdot\langle \boldsymbol{e}\rangle &= \langle \rho^{\mathrm{e}}\rangle - \nabla\cdot\langle \boldsymbol{p}\rangle,\\ -\partial_0\langle \boldsymbol{e}\rangle + \nabla\wedge\langle \boldsymbol{b}\rangle &= \langle \boldsymbol{j}\rangle/c + \partial_0\langle \boldsymbol{p}\rangle + \nabla\wedge\langle \boldsymbol{m}\rangle,\\ \nabla\cdot\langle \boldsymbol{b}\rangle &= 0,\\ \partial_0\langle \boldsymbol{b}\rangle + \nabla\wedge\langle \boldsymbol{e}\rangle &= 0.\end{aligned} \tag{9}$$

With the notations (1) for the macroscopic quantities, i.e. macroscopic fields

$$\boldsymbol{E} = \langle \boldsymbol{e}\rangle, \qquad \boldsymbol{B} = \langle \boldsymbol{b}\rangle, \tag{10}$$

the macroscopic charge and current densities

$$\varrho^{\mathrm{e}} = \langle \rho^{\mathrm{e}}\rangle, \qquad \boldsymbol{J} = \langle \boldsymbol{j}\rangle, \tag{11}$$

and the macroscopic polarization vectors

$$\boldsymbol{P} = \langle \boldsymbol{p}\rangle, \qquad \boldsymbol{M} = \langle \boldsymbol{m}\rangle, \tag{12}$$

one may write for the set (9):

$$\begin{aligned}\nabla\cdot \boldsymbol{E} &= \varrho^{\mathrm{e}} - \nabla\cdot \boldsymbol{P},\\ -\partial_0\, \boldsymbol{E} + \nabla\wedge \boldsymbol{B} &= \boldsymbol{J}/c + \partial_0\, \boldsymbol{P} + \nabla\wedge \boldsymbol{M},\\ \nabla\cdot \boldsymbol{B} &= 0,\\ \partial_0\, \boldsymbol{B} + \nabla\wedge \boldsymbol{E} &= 0.\end{aligned} \tag{13}$$

These are the *Maxwell equations*. With the definitions of the displacement vectors

$$\boldsymbol{D} = \boldsymbol{E} + \boldsymbol{P}, \qquad \boldsymbol{H} = \boldsymbol{B} - \boldsymbol{M}, \tag{14}$$

they may be written alternatively as

$$
\begin{aligned}
\nabla \cdot \boldsymbol{D} &= \varrho^{\mathrm{e}}, \\
-\partial_0 \boldsymbol{D} + \nabla \wedge \boldsymbol{H} &= \boldsymbol{J}/c, \\
\nabla \cdot \boldsymbol{B} &= 0, \\
\partial_0 \boldsymbol{B} + \nabla \wedge \boldsymbol{E} &= 0.
\end{aligned} \tag{15}
$$

The latter could also have been obtained by averaging the atomic equations (I.37).

Maxwell's equations have thus been found from the atomic field equations, which in turn followed from Lorentz's microscopic field equations. This completes the derivation.

The macroscopic equation of conservation of charge

$$\partial \varrho^{\mathrm{e}}/\partial t = -\nabla \cdot \boldsymbol{J} \tag{16}$$

follows from the averaging of (I.38) with the help of the definitions (11) and the fact that averaging and differentiation commute.

Up to order $c^{-1}$ the solutions of the Maxwell equations (13) are

$$
\begin{aligned}
\boldsymbol{E}(\boldsymbol{R}, t) &= \boldsymbol{E}_{\mathrm{e}}(\boldsymbol{R}, t) - \nabla \int \{\varrho^{\mathrm{e}}(\boldsymbol{R}', t) - \nabla' \cdot \boldsymbol{P}(\boldsymbol{R}', t)\} \frac{1}{4\pi|\boldsymbol{R}-\boldsymbol{R}'|} \mathrm{d}\boldsymbol{R}', \\
\boldsymbol{B}(\boldsymbol{R}, t) &= \boldsymbol{B}_{\mathrm{e}}(\boldsymbol{R}, t) \\
&\quad + \nabla \wedge \int \left\{ c^{-1}\boldsymbol{J}(\boldsymbol{R}', t) + c^{-1} \frac{\partial \boldsymbol{P}(\boldsymbol{R}', t)}{\partial t} + \nabla' \wedge \boldsymbol{M}(\boldsymbol{R}', t) \right\} \frac{1}{4\pi|\boldsymbol{R}-\boldsymbol{R}'|} \mathrm{d}\boldsymbol{R}'.
\end{aligned} \tag{17}
$$

One may verify that these are the solutions of (13) by inserting them and using (16) (v. problem 1). The first terms at the right-hand sides are the external fields, which are solutions of the field equations without sources.

The macroscopic charge and current densities $\varrho^{\mathrm{e}}$ and $\boldsymbol{J}$ are the averages (11) of (I.33)

$$
\begin{aligned}
\varrho^{\mathrm{e}}(\boldsymbol{R}, t) &= \langle \rho^{\mathrm{e}} \rangle = \langle \sum_k e_k \, \delta(\boldsymbol{R}_k - \boldsymbol{R}) \rangle, \\
\boldsymbol{J}(\boldsymbol{R}, t) &= \langle \boldsymbol{j} \rangle = \langle \sum_k e_k \, \boldsymbol{v}_k \, \delta(\boldsymbol{R}_k - \boldsymbol{R}) \rangle,
\end{aligned} \tag{18}
$$

or, in terms of one-point distribution functions, as in (4):

$$
\begin{aligned}
\varrho^{\mathrm{e}}(\boldsymbol{R}, t) &= \sum_a e_a f_1^a(\boldsymbol{R}; t), \\
\boldsymbol{J}(\boldsymbol{R}, t) &= \sum_a \int e_a \boldsymbol{v}_1 f_1^a(\boldsymbol{R}, \boldsymbol{v}_1; t) \mathrm{d}\boldsymbol{v}_1,
\end{aligned} \tag{19}
$$

where the summation over $a$ is extended over the number of species in the system. These expressions include contributions from all charged 'atoms', such as ions and free conduction electrons.

Furthermore the macroscopic electric and magnetic polarizations are given by the averages (12) of (I.34):

$$\begin{aligned} \boldsymbol{P}(\boldsymbol{R}, t) &= \langle \boldsymbol{p} \rangle = \langle \sum_{n=1}^{\infty} (-1)^{n-1} \boldsymbol{\nabla}^{n-1} : \sum_k \overline{\boldsymbol{\mu}}_k^{(n)} \delta(\boldsymbol{R}_k - \boldsymbol{R}) \rangle, \\ \boldsymbol{M}(\boldsymbol{R}, t) &= \langle \boldsymbol{m} \rangle = \langle \sum_{n=1}^{\infty} (-1)^{n-1} \boldsymbol{\nabla}^{n-1} : \sum_k (\overline{\boldsymbol{\nu}}_k^{(n)} + c^{-1} \overline{\boldsymbol{\mu}}_k^{(n)} \wedge \boldsymbol{v}_k) \delta(\boldsymbol{R}_k - \boldsymbol{R}) \rangle, \end{aligned} \tag{20}$$

or in terms of one-point distribution functions:

$$\begin{aligned} \boldsymbol{P}(\boldsymbol{R}, t) &= \sum_a \sum_{n=1}^{\infty} (-1)^{n-1} \boldsymbol{\nabla}^{n-1} : \int \overline{\boldsymbol{\mu}}_1^{(n)} f_1^a(\boldsymbol{R}, 1; t) \mathrm{d}1, \\ \boldsymbol{M}(\boldsymbol{R}, t) &= \sum_a \sum_{n=1}^{\infty} (-1)^{n-1} \boldsymbol{\nabla}^{n-1} : \int (\overline{\boldsymbol{\nu}}_1^{(n)} + \overline{\boldsymbol{\mu}}_1^{(n)} \wedge \boldsymbol{\beta}_1) f_1^a(\boldsymbol{R}, 1; t) \mathrm{d}1, \end{aligned} \tag{21}$$

where the symbol 1 now indicates all atomic parameters except for the position i.e. $\overline{\boldsymbol{\mu}}_1^{(n)}$ in the first line and $\overline{\boldsymbol{\mu}}_1^{(n)}$, $\overline{\boldsymbol{\nu}}_1^{(n)}$ and $\boldsymbol{\beta}_1 \equiv \boldsymbol{v}_1/c$ in the second line. Just as the multipole moments may be permanent or induced or both, the polarization vectors $\boldsymbol{P}$ and $\boldsymbol{M}$ are the total polarizations, due to permanent or induced effects or both.

The quantities (18–21) are continuous functions of space and time coordinates; they contain the charge $e_k$ and the multipole moments $\overline{\boldsymbol{\mu}}_k^{(n)}$ and $\overline{\boldsymbol{\nu}}_k^{(n)}$ as atomic characteristics.

A few remarks may be made on the result. In the first place it is seen to be valid for completely arbitrary polarizations of the material, that is to say polarizations due to both permanent and induced (by means of external and internal fields) electromagnetic moments of the atoms. The derivation leads to polarization vectors $\boldsymbol{P}$ and $\boldsymbol{M}$ expressed in (20) and (21) as statistical averages involving the electromagnetic moments of the atoms. The derivation is therefore completely independent of the 'constitutive relations', by which connexions between the polarizations and the fields are given, usually in terms of electric and magnetic susceptibilities. In fact these connexions belong to the dynamics (or statics) of the system, not to its set of field equations.

In the second place the derivation shows the secondary character of the displacement vectors $\boldsymbol{D}$ and $\boldsymbol{H}$. They may be obtained from their definitions (14) together with (10) and (12) for the fields and polarizations. The set (15) is useful to formulate the boundary conditions which lead to the well-known operational definitions of the Maxwell fields $\boldsymbol{E}$, $\boldsymbol{B}$, $\boldsymbol{D}$ and $\boldsymbol{H}$ in certain cavities. But otherwise the set (13) is to be preferred since it shows better the microscopic origin of the equations, as is apparent from the derivation.

The field equations (13) may be shown to be covariant under Galilei transformations. These form a group consisting of spatial rotations and pure Galilei transformations. The latter have the form

$$\begin{aligned} \boldsymbol{R}' &= \boldsymbol{R}+\boldsymbol{V}t, \\ t' &= t, \end{aligned} \tag{22}$$

where $\boldsymbol{V}$ is the transformation velocity (independent of space and time). The covariance of the field equations under rotations is guaranteed by the fact that they have been written in vector notation. The covariance under pure Galilean transformations requires some further inspection. From (22) we have for the transformations of the partial derivations with respect to space and time

$$\begin{aligned} \boldsymbol{\nabla}' &= \boldsymbol{\nabla}, \\ \frac{\partial}{\partial t'} &= \frac{\partial}{\partial t} - \boldsymbol{V}\cdot\boldsymbol{\nabla}. \end{aligned} \tag{23}$$

Furthermore the distribution functions are invariant:

$$f_1^{a'}(1'; t') = f_1^a(1, t), \tag{24}$$

as a consequence of their probability interpretation. Here $1'$ denotes the transformed quantities of the atom 1, for instance $\boldsymbol{R}_1' = \boldsymbol{R}_1+\boldsymbol{V}t$, $\boldsymbol{v}_1' = \boldsymbol{v}_1+\boldsymbol{V}$, $\overline{\boldsymbol{\mu}}_1' = \overline{\boldsymbol{\mu}}_1$ and $\overline{\boldsymbol{\nu}}_1' = \overline{\boldsymbol{\nu}}_1$. With the help of these formulae one proves the transformation properties of the charge and current densities (19) and the electric and magnetic polarizations (21):

$$\begin{aligned} \varrho^{e\prime}(\boldsymbol{R}', t') &= \varrho^{e}(\boldsymbol{R}, t), \\ \boldsymbol{J}'(\boldsymbol{R}', t') &= \boldsymbol{J}(\boldsymbol{R}, t)+\boldsymbol{V}\varrho^{e}(\boldsymbol{R}, t), \end{aligned} \tag{25}$$

$$\begin{aligned} \boldsymbol{P}'(\boldsymbol{R}', t') &= \boldsymbol{P}(\boldsymbol{R}, t), \\ \boldsymbol{M}'(\boldsymbol{R}', t') &= \boldsymbol{M}(\boldsymbol{R}, t)-c^{-1}\boldsymbol{V}\wedge\boldsymbol{P}(\boldsymbol{R}, t). \end{aligned} \tag{26}$$

From (23) and (25) one proves the covariance of the charge conservation law (16). Furthermore from (23), (25) and (26) one finds that the field equations are covariant if one imposes the transformation formulae for the fields

$$\begin{aligned} \boldsymbol{E}'(\boldsymbol{R}', t') &= \boldsymbol{E}(\boldsymbol{R}, t)-c^{-1}\boldsymbol{V}\wedge\boldsymbol{B}(\boldsymbol{R}, t), \\ \boldsymbol{B}'(\boldsymbol{R}', t') &= \boldsymbol{B}(\boldsymbol{R}, t)+c^{-1}\boldsymbol{V}\wedge\boldsymbol{E}(\boldsymbol{R}, t). \end{aligned} \tag{27}$$

We note again that in the present non-relativistic theory only terms up to order $c^{-1}$ are included. As a consequence the transformation formula for the electric field contains in fact only the external magnetic field $\boldsymbol{B}_e$, since the magnetic field generated by the sources is itself of order $c^{-1}$, as follows from the solution (17).

In a fashion analogous to the definition of the macroscopic charge density, which was the average of the atomic charge density, one can define macroscopic electric and magnetic multipole densities as:

$$\begin{aligned} \bar{\mathscr{P}}^{(n)} &= \langle \sum_k \bar{\boldsymbol{\mu}}_k^{(n)} \delta(\boldsymbol{R}_k - \boldsymbol{R}) \rangle = \sum_a \int \bar{\boldsymbol{\mu}}_1^{(n)} f_1^a(\boldsymbol{R}, 1; t) \mathrm{d}1 \equiv \sum_a \bar{\mathscr{P}}_a^{(n)}, \\ \bar{\mathscr{M}}^{(n)} &= \langle \sum_k \bar{\boldsymbol{\nu}}_k^{(n)} \delta(\boldsymbol{R}_k - \boldsymbol{R}) \rangle = \sum_a \int \bar{\boldsymbol{\nu}}_1^{(n)} f_1^a(\boldsymbol{R}, 1; t) \mathrm{d}1 \equiv \sum_a \bar{\mathscr{M}}_a^{(n)}, \end{aligned} \tag{28}$$

where $n$ indicates the multipole order: $n = 1$ dipole, $n = 2$ quadrupole, etc. These macroscopic multipole densities are functions of space and time coordinates $\boldsymbol{R}$ and $t$.

The expression (20) or (21) for the electric polarization vector $\boldsymbol{P}$ can be written in terms of the electric multipole densities (28)

$$\boldsymbol{P} = \sum_{n=1}^{\infty} (-1)^{n-1} \nabla^{n-1} \vdots \bar{\mathscr{P}}^{(n)} = \bar{\mathscr{P}}^{(1)} - \nabla \cdot \bar{\mathscr{P}}^{(2)} + \nabla\nabla : \bar{\mathscr{P}}^{(3)} - \dots . \tag{29}$$

The right-hand side is a series expansion involving all multipole densities.

In the Maxwell equations (13) appear, besides $\varrho^e$ and $\boldsymbol{P}$, also the current density $\boldsymbol{J}$ and the magnetization vector $\boldsymbol{M}$. The latter quantity, which is given in (20) or (21), cannot be expressed in terms of the multipole densities (28) alone, just as the current density cannot be expressed in terms of the charge density. In both cases the reason is that the atomic velocities $\boldsymbol{v}_k$ appear in a particular way as the expressions (18–19) and (20–21) for $\boldsymbol{J}$ and $\boldsymbol{M}$ show. It is convenient for the physical discussion to resolve the atomic velocity $\boldsymbol{v}_k \equiv \boldsymbol{\beta}_k c$ into a local mean velocity $\boldsymbol{v} \equiv \boldsymbol{\beta} c$ and a velocity fluctuation $\hat{\boldsymbol{v}}_k \equiv \hat{\boldsymbol{\beta}}_k c$:

$$\boldsymbol{v}_k = \boldsymbol{v} + \hat{\boldsymbol{v}}_k . \tag{30}$$

(The local mean velocity is in general still a function of space and time coordinates $\boldsymbol{R}$ and $t$.) Then with the expression in (19) for the charge density one can write the current density of (19) as

$$\boldsymbol{J} = \varrho^e \boldsymbol{v} + \sum_a \int e_a \hat{\boldsymbol{v}}_1 f_1^a(\boldsymbol{R}, \boldsymbol{v}_1; t) \mathrm{d}\boldsymbol{v}_1 \equiv \varrho^e \boldsymbol{v} + \boldsymbol{I}. \tag{31}$$

In this way $\boldsymbol{J}$ has been resolved into a 'convection' current $\varrho^e \boldsymbol{v}$ and a 'conduction' current $\boldsymbol{I}$. The fact that the former produces a magnetic field as well has been demonstrated experimentally by Rowland[1].

Similarly (30) can be used in the expression (21) for the magnetization vector $\boldsymbol{M}$. Then, using the definitions of the multipole densities (28), one can write

$$\boldsymbol{M} = \sum_{n=1}^{\infty} (-1)^{n-1} \nabla^{n-1} \vdots \left\{ \overline{\mathscr{M}}^{(n)} + \overline{\mathscr{P}}^{(n)} \wedge \boldsymbol{\beta} + \sum_a \int \overline{\boldsymbol{\mu}}_1^{(n)} \wedge \hat{\boldsymbol{\beta}}_1 f_1^a(\boldsymbol{R}, 1; t) \mathrm{d}1 \right\}, \quad (32)$$

because differentiation and averaging commute.

Alternatively, with the help of expression (29) for the polarization vector, the magnetization vector reads

$$\boldsymbol{M} = \sum_{n=1}^{\infty} (-1)^{n-1} \nabla^{n-1} \vdots \overline{\mathscr{M}}^{(n)} + \boldsymbol{P} \wedge \boldsymbol{\beta} + \sum_{n=1}^{\infty} (-1)^{n-1} \nabla^{n-1} \vdots \sum_a \int \overline{\boldsymbol{\mu}}_1^{(n)} \wedge \hat{\boldsymbol{\beta}}_1 f_1^a(\boldsymbol{R}, 1; t) \mathrm{d}1. \quad (33)$$

The physical significance of special cases of these forms will be discussed in the next section, where practical examples are treated. But it may be remarked already here that $\boldsymbol{M}$ shows three contributions, 1st: a sum which contains all magnetic multipole densities, analogous in structure to the series in the electric polarization (29), 2nd: a convection term due to the convection motion $\boldsymbol{v} \equiv \boldsymbol{\beta} c$ of the total polarization vector $\boldsymbol{P}$, and 3rd: a fluctuation term, which contains the atomic electric multipole moments $\overline{\boldsymbol{\mu}}_k^{(n)}$ and the velocity fluctuations $\hat{\boldsymbol{\beta}}_k c$. The last term plays a role if the carriers of electric multipole moments $\overline{\boldsymbol{\mu}}_k^{(n)}$ do not all have the same velocity (i.e. if $\hat{\boldsymbol{\beta}}_k \neq 0$). Mazur and Nijboer[2] gave the first example of such a term. Expression (32) shows explicitly that $\boldsymbol{M}$ cannot be expressed in terms of the multipole densities (28) alone: the first and the second terms are functions of these multipole densities, but not the third.

Let us summarize the general results obtained so far. The Maxwell equations were found in the form of the set (13). It contains the macroscopic fields $\boldsymbol{E}$ and $\boldsymbol{B}$, and moreover the four macroscopic quantities $\varrho^e$, $\boldsymbol{J}$, $\boldsymbol{P}$ and $\boldsymbol{M}$, for which expressions were found:

(a) The charge density $\varrho^e$, given as the average of the atomic charge density in formula (18) or (19).

[1] H. A. Rowland, Am. J. Sci. **15**(1878)30.

[2] P. Mazur and B. R. A. Nijboer, Physica **19**(1953)971; cf. reviews by P. Mazur, Adv. Chem. Phys. **1**(1958)309 and S. R. de Groot, The Maxwell equations (North-Holland Publ. Co., Amsterdam 1969).

(b) The current density $\boldsymbol{J}$, given as the sum (31) of a convection current, due to the bulk motion of the charge density, and a conduction current, due to the fluctuations in the velocities of the atomic charges.
(c) The electric polarization vector $\boldsymbol{P}$, expressed as a series expansion (29) in the macroscopic electric multipole densities defined in (28).
(d) The magnetic polarization vector $\boldsymbol{M}$, given by (33). It contains in the first place a series in the macroscopic magnetic multipole densities of (28). Furthermore two terms describing the effects of moving electric multipoles occur: a convection term, due to the bulk motion of the electric polarization, and a fluctuation (or conduction) term, due to the fluctuations in the velocities of the atomic electric multipole moments.

## 4 *Applications*

### a. *The polarizations up to dipole moments*

To simplify the discussion of the various physical systems let us give some explicit formulae, containing lowest order multipoles. In fact in not too dense systems one can limit oneself to the consideration of atomic charges and dipole moments only. Then the polarizations (21) become

$$\begin{aligned} \boldsymbol{P} &= \sum_a \int \overline{\boldsymbol{\mu}}_1^{(1)} f_1^a(\boldsymbol{R}, 1; t)\mathrm{d}1, \\ \boldsymbol{M} &= \sum_a \int (\overline{\boldsymbol{v}}_1^{(1)} + \overline{\boldsymbol{\mu}}_1^{(1)} \wedge \boldsymbol{\beta}_1) f_1^a(\boldsymbol{R}, 1; t)\mathrm{d}1. \end{aligned} \tag{34}$$

In these expressions occur the macroscopic electric dipole density $\overline{\mathscr{P}}^{(1)}$ and the macroscopic magnetic dipole density $\overline{\mathscr{M}}^{(1)}$, defined in (28). With the use of these quantities, and the splitting of the atomic velocity in a local mean velocity $\boldsymbol{\beta}c$ and a deviation $\hat{\boldsymbol{\beta}}_k c$ from it, one can write the polarizations as

$$\begin{aligned} \boldsymbol{P} &= \overline{\mathscr{P}}^{(1)}, \\ \boldsymbol{M} &= \overline{\mathscr{M}}^{(1)} + \overline{\mathscr{P}}^{(1)} \wedge \boldsymbol{\beta} + \sum_a \int \overline{\boldsymbol{\mu}}_1^{(1)} \wedge \hat{\boldsymbol{\beta}}_1 f_1^a(\boldsymbol{R}, 1; t)\mathrm{d}1. \end{aligned} \tag{35}$$

The electric polarization vector $\boldsymbol{P}$ could be expressed in terms of the macroscopic electric dipole density alone. The magnetic polarization vector $\boldsymbol{M}$ contains the magnetic dipole density. Furthermore terms due to moving electric dipoles are present. First a convection term, due to the bulk velocity $\boldsymbol{v} = \boldsymbol{\beta}c$

of the electric dipole density, is present. Its curl, which occurs in the Maxwell equations, is called the Röntgen current. It has been observed experimentally[1]. But $\boldsymbol{M}$ contains also terms due to the fluctuations $\hat{\boldsymbol{\beta}}_k c$ in the velocity of the carriers of the electric dipole moments.

Only if the carriers of the electric dipole moments all have the same velocity $\boldsymbol{\beta}_k = \boldsymbol{\beta}$ do the fluctuation terms vanish. The polarizations then reduce to

$$\boldsymbol{P} = \overline{\mathscr{P}}^{(1)}, \qquad \boldsymbol{M} = \overline{\mathscr{M}}^{(1)} + \overline{\mathscr{P}}^{(1)} \wedge \boldsymbol{\beta}. \tag{36}$$

These expressions are the same as Lorentz's original results[2]. Lorentz's model did not include the possibility of the appearance of the fluctuation terms. It should be noted that in contrast with the general formulae the special expressions (36) are functions of the macroscopic dipole densities alone.

If the system is completely at rest, i.e. if all atoms have velocities $\boldsymbol{\beta}_k = 0$, then (36) further reduces to

$$\boldsymbol{P} = \overline{\mathscr{P}}^{(1)}, \qquad \boldsymbol{M} = \overline{\mathscr{M}}^{(1)}. \tag{37}$$

b. *The polarizations up to quadrupole moments*

In the dipole plus quadrupole approximation one retains the terms with $n = 1$ and 2 in (21):

$$\begin{aligned} \boldsymbol{P} &= \sum_a \int \bar{\boldsymbol{\mu}}_1^{(1)} f_1^a(\boldsymbol{R}, 1; t)\mathrm{d}1 - \sum_a \nabla \cdot \int \bar{\boldsymbol{\mu}}_1^{(2)} f_1^a(\boldsymbol{R}, 1; t)\mathrm{d}1, \\ \boldsymbol{M} &= \sum_a \int (\bar{\boldsymbol{\nu}}_1^{(1)} + \bar{\boldsymbol{\mu}}_1^{(1)} \wedge \boldsymbol{\beta}_1) f_1^a(\boldsymbol{R}, 1; t)\mathrm{d}1 \\ &\quad - \sum_a \nabla \cdot \int (\bar{\boldsymbol{\nu}}_1^{(2)} + \bar{\boldsymbol{\mu}}_1^{(2)} \wedge \boldsymbol{\beta}_1) f_1^a(\boldsymbol{R}, 1; t)\mathrm{d}1. \end{aligned} \tag{38}$$

Introducing the macroscopic dipole and quadrupole densities (28) with $n = 1$ and 2, and resolving the atomic velocity $\boldsymbol{\beta}_k c$ into a local mean velocity $\boldsymbol{\beta} c$ and a velocity fluctuation $\hat{\boldsymbol{\beta}}_k c$ one can write (38) as

$$\begin{aligned} \boldsymbol{P} &= \overline{\mathscr{P}}^{(1)} - \nabla \cdot \overline{\mathscr{P}}^{(2)}, \\ \boldsymbol{M} &= \overline{\mathscr{M}}^{(1)} - \nabla \cdot \overline{\mathscr{M}}^{(2)} + (\overline{\mathscr{P}}^{(1)} - \nabla \cdot \overline{\mathscr{P}}^{(2)}) \wedge \boldsymbol{\beta} + \sum_a \int \bar{\boldsymbol{\mu}}_1^{(1)} \wedge \hat{\boldsymbol{\beta}}_1 f_1^a(\boldsymbol{R}, 1; t)\mathrm{d}1 \\ &\quad - \sum_a \nabla \cdot \int \bar{\boldsymbol{\mu}}_1^{(2)} \wedge \hat{\boldsymbol{\beta}}_1 f_1^a(\boldsymbol{R}, 1; t)\mathrm{d}1. \end{aligned} \tag{39}$$

[1] W. C. Röntgen, Ann. Phys. Chem. **35**(1888)264, **40**(1890)93; A. Eichenwald, Ann. Phys. Chem. **11**(1903)1, 421.

[2] H. A. Lorentz, Proc. Roy. Acad. Amsterdam (1902)254; Enc. Math. Wiss. V 2, fasc. 1 (Teubner, Leipzig 1904) 200.

The electric polarization vector $\boldsymbol{P}$ is equal to the macroscopic electric dipole density minus the divergence of the electric quadrupole density. The magnetization contains three dipole contributions; the first is the macroscopic magnetic dipole density, the second is a convection term due to the bulk motion of the macroscopic electric dipole density $\overline{\mathscr{P}}^{(1)}$, and the third is a fluctuation term, due to the fluctuations in the velocities of the individual atomic dipole moments $\overline{\boldsymbol{\mu}}_k^{(1)}$. This last term plays a role in systems in which the electric dipoles do not all have the same velocity. (In the following sections some practical examples will be given.) In all terms the negative divergence of a quadrupole term is added to the corresponding dipole terms. In uniform systems the quadrupole terms will thus not play a role, but for instance boundaries will give quadrupole (and perhaps even higher multipole order) contributions.

If all electric multipoles have the same velocity ($\boldsymbol{\beta}_k = \boldsymbol{\beta}$, the fluctuation terms disappear from the expressions for the magnetization vector $\boldsymbol{M}$. Formulae (39) then simplify to

$$\begin{aligned} \boldsymbol{P} &= \overline{\mathscr{P}}^{(1)} - \nabla\cdot\overline{\mathscr{P}}^{(2)}, \\ \boldsymbol{M} &= \overline{\mathscr{M}}^{(1)} - \nabla\cdot\overline{\mathscr{M}}^{(2)} + (\overline{\mathscr{P}}^{(1)} - \nabla\cdot\overline{\mathscr{P}}^{(2)})\wedge\boldsymbol{\beta}. \end{aligned} \tag{40}$$

These expressions were originally found by Frenkel[1]. They include quadrupolar effects, but otherwise their validity is limited in the same way as Lorentz's, since the fluctuation terms of (39) are missing. Earlier Fokker[2] found these formulae, but without the magnetic quadrupole term.

For the still more special case of no motion at all ($\boldsymbol{\beta}_k = 0$), the expressions further reduce to

$$\boldsymbol{P} = \overline{\mathscr{P}}^{(1)} - \nabla\cdot\overline{\mathscr{P}}^{(2)}, \qquad \boldsymbol{M} = \overline{\mathscr{M}}^{(1)} - \nabla\cdot\overline{\mathscr{M}}^{(2)}. \tag{41}$$

Full symmetry between electric and magnetic terms is present only in this last static case. Rosenfeld[3] obtained these expressions but without the magnetic quadrupole term.

## c. *Examples*

In this subsection the Maxwell equations for specific physical systems will be discussed. The expressions for the material quantities which occur in the

[1] J. Frenkel, Lehrbuch der Elektrodynamik II (Springer, Berlin 1928), p. 26.

[2] A. D. Fokker, Phil. Mag. **39**(1920)404; Versl. Kon. Acad. Wet. Amsterdam **28**(1920) 1040; Relativiteitstheorie (Noordhoff, Groningen 1929).

[3] L. Rosenfeld, Theory of electrons (North-Holland Publ. Co., Amsterdam 1951); cf. J. Voisin, Physica **25**(1959)195.

Maxwell equations, namely the charge and current densities $\varrho^{\mathrm{e}}$ and $\boldsymbol{J}$, and the electric and magnetic polarization vectors $\boldsymbol{P}$ and $\boldsymbol{M}$, will depend on the characteristics of the particular physical model studied.

(i) *Metals.* A metal is supposed to consist of free electrons moving in a rigid lattice formed by positively charged ions. The 'stable groups' of point particles of the present theory are then those free electrons and ions. They will be labelled by an index $a = \mathrm{I}$ and $a = \mathrm{II}$ respectively. The model is then specified by assigning charges to the electrons, and charges and electric and magnetic dipole moments to the ions. Furthermore in the model one supposes that the metal can only move as a whole, with a velocity $\boldsymbol{v} \equiv \boldsymbol{\beta} c$. This means that all ions move with this velocity. The free electrons however have velocities $\boldsymbol{\beta}_1 c = \boldsymbol{\beta} c + \hat{\boldsymbol{\beta}}_1 c$. On the basis of these properties of the model we can now give the expressions for the material quantities $\varrho^{\mathrm{e}}$, $\boldsymbol{J}$, $\boldsymbol{P}$ and $\boldsymbol{M}$, which occur in the Maxwell equations. The charge density (19) becomes

$$\varrho^{\mathrm{e}} = e_{\mathrm{I}}\, f_1^{\mathrm{I}}(\boldsymbol{R}, t) + e_{\mathrm{II}}\, f_1^{\mathrm{II}}(\boldsymbol{R}, t). \tag{42}$$

The current density (31) is

$$\boldsymbol{J} = \varrho^{\mathrm{e}}\boldsymbol{v} + \boldsymbol{I}, \qquad \boldsymbol{I} = e_{\mathrm{I}} \int \hat{\boldsymbol{\beta}}_1 c f_1^{\mathrm{I}}(\boldsymbol{R}, 1; t)\mathrm{d}1. \tag{43}$$

The convection current $\varrho^{\mathrm{e}}\boldsymbol{v}$ contains contributions (42) from the free electrons and the ions. The conduction current $\boldsymbol{I}$ contains only contributions from the free electrons, since the ions have no velocity fluctuations. The polarization vectors $\boldsymbol{P}$ and $\boldsymbol{M}$ follow from the formulae (36) for the dipole case, with the macroscopic dipole densities (28), as

$$\boldsymbol{P} = \bar{\mathscr{P}}_{\mathrm{II}}^{(1)}, \qquad \boldsymbol{M} = \bar{\mathscr{M}}_{\mathrm{II}}^{(1)} + \bar{\mathscr{P}}_{\mathrm{II}}^{(1)} \wedge \boldsymbol{\beta}. \tag{44}$$

Only the ions contribute. Since the ions have no velocity fluctuations no fluctuation contribution arises in $\boldsymbol{M}$. The free electrons do not give rise to such a contribution either, because, although they do have velocity fluctuations, they possess no dipole moments. This is the reason why the expressions (44) turn out to be of the particular type (36). The latter were also Lorentz's results. So for the model of the metal – and Lorentz apparently had this model in mind – these results are justified from the general theory.

A possible influence of multipole moments of higher than dipole order might have been taken into consideration. However in a system in which the total charges of 'stable groups' play a role, their effects usually overshadow those due to the dipole moments. The corrections obtained by taking into account also quadrupole effects are then negligibly small.

(ii) *Insulators.* If the stable groups in insulators are positively and negatively charged ions of the kind I and II, then the model can be specified as follows. The ions possess charges as well as electric and magnetic multipole moments of order $n = 1, 2, \ldots$. The system is a rigid lattice moving with the velocity $\boldsymbol{v} = \boldsymbol{\beta} c$ as a whole. This means that the ions have this velocity, so that all velocity fluctuations vanish. Often the charges of the two kinds of ions just cancel so that the system as a whole is electrically neutral. Then we have the charge density $\varrho^{\mathrm{e}} = 0$, just as for an atomic or molecular lattice. The current density $\boldsymbol{J}$ (31) also vanishes because both the convection current $\varrho^{\mathrm{e}}\boldsymbol{v}$ and the conduction current $\boldsymbol{I}$ are zero: the first because $\varrho^{\mathrm{e}} = 0$, and the second because all velocity fluctuations vanish. Since the ion lattice is usually fairly closely packed, one may need to include terms of higher than dipole order into the polarizations. Because no fluctuation motion is present, the expressions (29) and (32) become

$$
\begin{aligned}
\boldsymbol{P} &= \sum_{n=1}^{\infty} (-1)^{n-1} \boldsymbol{\nabla}^{n-1} \vdots\, \bar{\mathscr{P}}^{(n)}, \\
\boldsymbol{M} &= \sum_{n=1}^{\infty} (-1)^{n-1} \boldsymbol{\nabla}^{n-1} \vdots\, (\bar{\mathscr{M}}^{(n)} + \bar{\mathscr{P}}^{(n)} \wedge \boldsymbol{\beta}),
\end{aligned}
\tag{45}
$$

with the complete multipole series. One has here for the macroscopic multipole densities:

$$
\bar{\mathscr{P}}^{(n)} = \bar{\mathscr{P}}^{(n)}_{\mathrm{I}} + \bar{\mathscr{P}}^{(n)}_{\mathrm{II}}, \qquad \bar{\mathscr{M}}^{(n)} = \bar{\mathscr{M}}^{(n)}_{\mathrm{I}} + \bar{\mathscr{M}}^{(n)}_{\mathrm{II}}. \tag{46}
$$

The number of space differentiations in (45) increases with the multipole order. The effects of the higher order multipole densities show up especially at boundaries between different media.

(iii) *Plasmas.* A plasma is a gas in which a sensible proportion of the atoms or molecules is ionized, so that virtually the properties of the system are completely determined by the effects of the charges of the ions and free electrons. In fact in practice one neglects completely the multipole moments of the atoms, molecules and ions. In such a model the charges of the ions and free electrons determine the value of the charge–current densities $\varrho^{\mathrm{e}}$ and $\boldsymbol{J}$, which are given by (18). If all multipole moments are neglected, the polarization vectors $\boldsymbol{P}$ and $\boldsymbol{M}$ vanish. The Maxwell equations (13) read then

$$
\begin{aligned}
\boldsymbol{\nabla}\cdot\boldsymbol{E} &= \varrho^{\mathrm{e}}, \\
-\partial_0 \boldsymbol{E} + \boldsymbol{\nabla} \wedge \boldsymbol{B} &= c^{-1}\boldsymbol{J}, \\
\boldsymbol{\nabla}\cdot\boldsymbol{B} &= 0, \\
\partial_0 \boldsymbol{B} + \boldsymbol{\nabla} \wedge \boldsymbol{E} &= 0.
\end{aligned}
\tag{47}
$$

These equations form indeed the starting point which is generally adopted in plasma theory. In this connexion it should be noted that sometimes electric polarization vectors are introduced that are not defined as the average over the microscopic electric multipole moments. As a matter of fact various 'effective polarization vectors' can be found in literature. One of these is the 'effective polarization vector' $\boldsymbol{P}^*$, which is related to the charge density $\varrho^{\mathrm{e}}$ as

$$\varrho^{\mathrm{e}} = -\nabla\cdot\boldsymbol{P}^*. \tag{48}$$

Then one introduces also an 'effective displacement vector' $\boldsymbol{D}^*$, defined as

$$\boldsymbol{D}^* = \boldsymbol{E}+\boldsymbol{P}^*. \tag{49}$$

Another formal 'effective displacement vector' which is sometimes used is the quantity $\boldsymbol{D}^{**}$, which satisfies

$$\partial_0 \boldsymbol{D}^{**} = \partial_0 \boldsymbol{E}+c^{-1}\boldsymbol{J}. \tag{50}$$

Then one can accordingly also introduce 'effective dielectric constants'. The dielectric constants $\boldsymbol{E}$ are normally defined as the proportionality constants between the fields $\boldsymbol{D}$ and $\boldsymbol{E}$. The 'effective constants' $\varepsilon^*$ or $\varepsilon^{**}$ are defined as the proportionality constants between the fields $\boldsymbol{D}^*$ or $\boldsymbol{D}^{**}$ and $\boldsymbol{E}$. Some of these 'effective' quantities may be useful abbreviations in certain cases, but one should not confuse them with the ordinary polarization and displacement vectors, which are directly connected to the multipole moments of the particles in the system.

In particular the model of a plasma is such that the refractive index $n$ is different from unity, whereas $\varepsilon = 1$, since $\boldsymbol{D} = \boldsymbol{E}$. In fact for a plasma $n^2$ is therefore not equal to $\varepsilon$. Again one can of course introduce an 'effective dielectric constant' equal to the square of the refractive index, but one should not confuse it with the ordinary dielectric constant $\varepsilon$.

(iv) *Fluids.* In a fluid which consists of neutral molecules the charge density $\varrho^{\mathrm{e}}$ and current density $\boldsymbol{J}$ both vanish. The polarizations $\boldsymbol{P}$ and $\boldsymbol{M}$ contain contributions from the various molecular multipole moments.

In a gas it is usual to limit oneself to the dipole contributions alone. The velocity $\boldsymbol{\beta}_k c$ of the molecules will again be written as the sum of a local mean velocity $\boldsymbol{\beta}c$ (which depends on space and time coordinates $\boldsymbol{R}$ and $t$) and a velocity fluctuation $\hat{\boldsymbol{\beta}}_k c$. Then the dipole approximation (35) for the polarization vectors applies. The electric polarization vector is simply equal to the electric dipole density $\bar{\mathscr{P}}^{(1)}$. The magnetic polarization vector contains three contributions. The first is the magnetic dipole density; the second is a 'convection term' and the third has a fluctuation character: even for a gas at rest this term will not disappear.

In a liquid, because of its high density, one may need to take into account the effects of higher multipole moments, for instance the quadrupole moments. In the latter case the polarizations are given by (39).

(v) *Electrolytes.* In an electrolyte one has positively and negatively charged ions, and usually also a neutral component. In this respect it is similar to a plasma. However the density of an electrolyte is much higher than the density of a plasma. The effects of dipoles (or even higher order multipoles) should therefore be taken into account. The model adopted here is a mixture of three components labelled by the indices I, II and III of which I and II are ions, with charges $e_{\mathrm{I}}$ and $e_{\mathrm{II}}$, and III neutral molecules; all three components are supposed to carry electric and magnetic dipole moments.

The charge density (19) gets the form

$$\varrho^{\mathrm{e}} = \sum_{a=\mathrm{I}}^{\mathrm{II}} e_a f_1^a(\boldsymbol{R}; t). \tag{51}$$

The velocities of the ions and molecules can be written as the sum of a local mean velocity of all ions and molecules and fluctuation velocities. Then the current density (31) is

$$\boldsymbol{J} = \varrho^{\mathrm{e}} \boldsymbol{v} + \boldsymbol{I}, \tag{52}$$

where $\varrho^{\mathrm{e}}\boldsymbol{v}$ is the convection current and $\boldsymbol{I}$ the conduction current:

$$\boldsymbol{I} = \sum_{a=\mathrm{I}}^{\mathrm{II}} \int e_a \hat{\boldsymbol{v}}_1 f_1^a(\boldsymbol{R}, \boldsymbol{v}_1; t) \mathrm{d}\boldsymbol{v}_1 . \tag{53}$$

With the macroscopic electric and magnetic dipole densities (21) for the ions and the molecules $\overline{\mathscr{P}}_a^{(1)}$ and $\overline{\mathscr{M}}_a^{(1)}$ ($a = \mathrm{I, II, III}$) one can write the polarization vectors (35) as

$$\begin{aligned} \boldsymbol{P} &= \sum_{a=\mathrm{I}}^{\mathrm{III}} \overline{\mathscr{P}}_a^{(1)}, \\ \boldsymbol{M} &= \sum_{a=\mathrm{I}}^{\mathrm{III}} \left\{ \overline{\mathscr{M}}_a^{(1)} + \overline{\mathscr{P}}_a^{(1)} \wedge \boldsymbol{\beta} + \int \overline{\boldsymbol{\mu}}_1^{(1)} \wedge \hat{\boldsymbol{\beta}}_1 f_1^a(\boldsymbol{R}, 1; t) \mathrm{d}1 \right\} . \end{aligned} \tag{54}$$

The polarization vector $\boldsymbol{P}$ contains the three electric dipole densities. The magnetization vector $\boldsymbol{M}$ consists of three kinds of terms. First the magnetic dipole densities due to the ions and molecules appear. Then the convection of the electric dipole densities with the bulk velocity $\boldsymbol{v} = \boldsymbol{\beta}c$ gives a contribution. Finally the electric dipoles also have fluctuations in their velocities around the bulk motion. These effects give rise to the last term. They do not occur in solids, but they can play a role in fluid systems, in which freely moving electric multipoles exist.

## 5 *The momentum and energy equations*

### a. *Introduction*

The motion of matter in bulk is described by the balance equations of momentum, energy and angular momentum. The derivation of the former two from the corresponding atomic laws by means of a statistical averaging procedure will be the subject of this section, while the latter will be discussed separately in the following section. As a result we shall find macroscopic laws that contain quantities, such as the pressure, the internal energy and the heat flow, which are given as statistical averages of atomic quantities.

In contrast to the field equations the material equations mentioned contain quantities that are two-point functions on the atomic level, so that they contain two-point distribution functions (or correlation functions) on the macroscopic level. As a consequence one will have to distinguish in the course of the treatment between physical systems for which these correlation functions show marked differences: systems in which the correlations have short range – such as gases, liquids, plasmas and amorphous or polycrystalline solids – and systems such as crystalline solids with correlation functions of long range character.

### b. *The mass conservation law*

In the course of the derivation of the momentum and energy laws, we shall need the macroscopic mass conservation law, which is an immediate consequence of the atomic conservation law. In fact, the atomic mass density is

$$\rho(\boldsymbol{R}, t) = \sum_k m_k \, \delta(\boldsymbol{R}_k - \boldsymbol{R}), \tag{55}$$

where $m_k$ is the mass of the (identical) atoms for a one-component system[1].

[1] For formal convenience we treat in the following subsections one-component systems. The generalization to mixtures is obvious (cf. subsection *g*).

Then the macroscopic mass density becomes:

$$\varrho(\boldsymbol{R}, t) = \langle \sum_k m_k\, \delta(\boldsymbol{R}_k - \boldsymbol{R})\rangle \equiv m f_1(\boldsymbol{R}; t), \tag{56}$$

where $f_1(\boldsymbol{R}; t)$ is a one-point distribution function, which depends only on the position $\boldsymbol{R}_1$ $(= \boldsymbol{R})$ and the time. The time derivative of the macroscopic mass density (56) is according to (2)

$$\frac{\partial \varrho}{\partial t} = \langle \sum_k m_k\, \boldsymbol{v}_k \cdot \nabla_k\, \delta(\boldsymbol{R}_k - \boldsymbol{R})\rangle, \tag{57}$$

since in this case $\mathrm{d}/\mathrm{d}t = \boldsymbol{v}_k \cdot \nabla_k$ (with $\boldsymbol{v}_k$ the atomic velocity $\dot{\boldsymbol{R}}_k$). Introducing the local barycentric velocity $\boldsymbol{v}(\boldsymbol{R}, t)$ by

$$\varrho(\boldsymbol{R}, t)\boldsymbol{v}(\boldsymbol{R}, t) = \langle \sum_k m_k\, \boldsymbol{v}_k\, \delta(\boldsymbol{R}_k - \boldsymbol{R})\rangle \equiv \int m \boldsymbol{v}_1\, f_1(\boldsymbol{R}, \boldsymbol{v}_1; t)\mathrm{d}\boldsymbol{v}_1\,, \tag{58}$$

we may write (57) in the form

$$\frac{\partial \varrho}{\partial t} = -\nabla \cdot (\varrho \boldsymbol{v}), \tag{59}$$

which is the macroscopic law of mass conservation.

### c. *The momentum balance*

The momentum law is obtained by taking the time derivative of (58). With (2) and the equation of motion (I.50) one gets

$$\frac{\partial \varrho \boldsymbol{v}}{\partial t} = -\nabla \cdot \langle \sum_k m_k\, \boldsymbol{v}_k\, \boldsymbol{v}_k\, \delta(\boldsymbol{R}_k - \boldsymbol{R})\rangle + \langle \sum_k (\boldsymbol{f}_k^{\mathrm{L}} + \boldsymbol{f}_k^{\mathrm{S}})\delta(\boldsymbol{R}_k - \boldsymbol{R})\rangle. \tag{60}$$

Introducing the velocity fluctuation $\hat{\boldsymbol{v}}_k$ as

$$\hat{\boldsymbol{v}}_k(\boldsymbol{R}, t) \equiv \boldsymbol{v}_k - \boldsymbol{v}(\boldsymbol{R}, t), \tag{61}$$

we obtain from (60) with (56) and (58) the momentum balance equation

$$\frac{\partial \varrho \boldsymbol{v}}{\partial t} = -\nabla \cdot (\varrho \boldsymbol{v} \boldsymbol{v} + \mathrm{P}^{\mathrm{K}}) + \boldsymbol{F}^{\mathrm{L}} + \boldsymbol{F}^{\mathrm{S}}, \tag{62}$$

where the kinetic pressure

$$\mathrm{P}^{\mathrm{K}} \equiv \int m \hat{\boldsymbol{v}}_1\, \hat{\boldsymbol{v}}_1\, f_1(\boldsymbol{R}, \boldsymbol{v}_1; t)\mathrm{d}\boldsymbol{v}_1 \tag{63}$$

is written with the help of an appropriate one-point distribution function. Furthermore we introduced the abbreviations $\boldsymbol{F}^{\mathrm{L}} \equiv \langle \sum_k \boldsymbol{f}_k^{\mathrm{L}} \delta(\boldsymbol{R}_k - \boldsymbol{R})\rangle$ and $\boldsymbol{F}^{\mathrm{S}} \equiv \langle \sum_k \boldsymbol{f}_k^{\mathrm{S}} \delta(\boldsymbol{R}_k - \boldsymbol{R})\rangle$ for the long and short range terms.

The *long range* term contains $\boldsymbol{f}_k^{\mathrm{L}}$, which has been specified in (I.54). We shall first treat the part with the external fields $(\boldsymbol{E}_{\mathrm{e}}, \boldsymbol{B}_{\mathrm{e}})$ (it will be called $\boldsymbol{F}_{\mathrm{e}}^{\mathrm{L}}$). It can be expressed in terms of the macroscopic charge and current densities (v. (18)):

$$\varrho^{\mathrm{e}}(\boldsymbol{R}, t) = \langle \sum_k e_k \delta(\boldsymbol{R}_k - \boldsymbol{R})\rangle, \qquad \boldsymbol{J}(\boldsymbol{R}, t) = \langle \sum_k e_k \boldsymbol{v}_k \delta(\boldsymbol{R}_k - \boldsymbol{R})\rangle, \tag{64}$$

(where $e_k = e$ is the charge of the (identical) atoms) and the macroscopic polarization and magnetization densities which read, if only dipoles contribute for the system under consideration (v. (20))[1]

$$\begin{aligned} \boldsymbol{P}(\boldsymbol{R}, t) &= \langle \sum_k \overline{\boldsymbol{\mu}}_k^{(1)} \delta(\boldsymbol{R}_k - \boldsymbol{R})\rangle, \\ \boldsymbol{M}(\boldsymbol{R}, t) &= \langle \sum_k (\overline{\boldsymbol{v}}_k^{(1)} + \overline{\boldsymbol{\mu}}_k^{(1)} \wedge \boldsymbol{v}_k/c) \delta(\boldsymbol{R}_k - \boldsymbol{R})\rangle. \end{aligned} \tag{65}$$

(Note that according to (I.30) the magnetization is of the order $c^{-1}$, while the other three densities are of the order $c^0$.) In this way we get the expression

$$\begin{aligned} \boldsymbol{F}_{\mathrm{e}}^{\mathrm{L}} = \varrho^{\mathrm{e}} \boldsymbol{E}_{\mathrm{e}} + c^{-1} \boldsymbol{J} \wedge \boldsymbol{B}_{\mathrm{e}} + (\nabla \boldsymbol{E}_{\mathrm{e}}) \cdot \boldsymbol{P} + (\nabla \boldsymbol{B}_{\mathrm{e}}) \cdot \boldsymbol{M} \\ + c^{-1} \langle \sum_k \frac{\mathrm{d}}{\mathrm{d}t} \{\overline{\boldsymbol{\mu}}_k^{(1)} \wedge \boldsymbol{B}_{\mathrm{e}}(\boldsymbol{R}_k, t)\} \delta(\boldsymbol{R}_k - \boldsymbol{R})\rangle. \end{aligned} \tag{66}$$

The last term becomes with (2)

$$c^{-1} \frac{\partial}{\partial t} \langle \sum_k \overline{\boldsymbol{\mu}}_k^{(1)} \wedge \boldsymbol{B}_{\mathrm{e}} \delta(\boldsymbol{R}_k - \boldsymbol{R})\rangle + c^{-1} \nabla \cdot \langle \sum_k \boldsymbol{v}_k \overline{\boldsymbol{\mu}}_k^{(1)} \wedge \boldsymbol{B}_{\mathrm{e}} \delta(\boldsymbol{R}_k - \boldsymbol{R})\rangle. \tag{67}$$

If (61) and (65) are used in this expression and the result is substituted into (66), one gets

$$\begin{aligned} \boldsymbol{F}_{\mathrm{e}}^{\mathrm{L}} &= \varrho^{\mathrm{e}} \boldsymbol{E}_{\mathrm{e}} + c^{-1} \boldsymbol{J} \wedge \boldsymbol{B}_{\mathrm{e}} + (\nabla \boldsymbol{E}_{\mathrm{e}}) \cdot \boldsymbol{P} + (\nabla \boldsymbol{B}_{\mathrm{e}}) \cdot \boldsymbol{M} \\ &\quad + c^{-1} \frac{\partial}{\partial t} (\boldsymbol{P} \wedge \boldsymbol{B}_{\mathrm{e}}) + c^{-1} \nabla \cdot (\boldsymbol{v} \boldsymbol{P} \wedge \boldsymbol{B}_{\mathrm{e}}) \\ &\quad + c^{-1} \nabla \cdot \int \hat{\boldsymbol{v}}_1 \overline{\boldsymbol{\mu}}_1^{(1)} \wedge \boldsymbol{B}_{\mathrm{e}} f_1(\boldsymbol{R}, \boldsymbol{v}_1, \overline{\boldsymbol{\mu}}_1^{(1)}; t) \mathrm{d}\boldsymbol{v}_1 \, \mathrm{d}\overline{\boldsymbol{\mu}}_1^{(1)}. \end{aligned} \tag{68}$$

[1] Higher multipoles could be included at this point, v. problem 3 for the case of quadrupoles. As a result it turns out that the macrosopic force density which for the dipole case will be given in (106) can no longer be expressed in terms of the Maxwell fields and the total polarizations alone. (Cf. also H. A. Haus, Ann. Physics **45**(1967)314.)

The remaining part of the long range term of (62), which is due to the interatomic interactions (v. (I.54)) may be written with the help of a two-point distribution function:

$$\boldsymbol{F}^{\mathrm{L}}-\boldsymbol{F}_{\mathrm{e}}^{\mathrm{L}} = -\int \sum_{n,m=0}^{\infty} \left(\overline{\boldsymbol{\mu}}_1^{(n)} : \nabla_1^n \overline{\boldsymbol{\mu}}_2^{(m)} : \nabla_2^m \nabla_1 \frac{1}{4\pi|\boldsymbol{R}_1-\boldsymbol{R}_2|}\right)$$
$$\delta(\boldsymbol{R}-\boldsymbol{R}_1) f_2(\boldsymbol{R}_1, 1, \boldsymbol{R}_2, 2; t)\mathrm{d}\boldsymbol{R}_1\, \mathrm{d}1\, \mathrm{d}\boldsymbol{R}_2\, \mathrm{d}2, \quad (69)$$

where 1 (and 2) denote the whole set of electric multipole moments $\overline{\boldsymbol{\mu}}_1^{(n)}$ (and $\overline{\boldsymbol{\mu}}_2^{(m)}$) with $n, m = 0, 1, \ldots$. Let us split the right-hand side into two parts with the help of (7). In the 'uncorrelated' part, which contains the product of one-point distribution functions, we add a factor $\delta(\boldsymbol{R}'-\boldsymbol{R}_2)$ and an integration over $\boldsymbol{R}'$. Furthermore we introduce the macroscopic charge and polarization densities (64–65), omitting here all multipole moments of order two and higher. The latter moments would give rise to terms containing the macroscopic multipole densities of order two and higher, which are assumed to be negligible in our system. In the 'correlated' part (which contains the correlation function) the integration over $\boldsymbol{R}_1$ may be performed. In this way we obtain:

$$\boldsymbol{F}^{\mathrm{L}}-\boldsymbol{F}_{\mathrm{e}}^{\mathrm{L}} = -\int \{\varrho^{\mathrm{e}}(\boldsymbol{R}, t)+\boldsymbol{P}(\boldsymbol{R}, t)\cdot\nabla\}\{\varrho^{\mathrm{e}}(\boldsymbol{R}', t)+\boldsymbol{P}(\boldsymbol{R}', t)\cdot\nabla'\}\nabla \frac{1}{4\pi|\boldsymbol{R}-\boldsymbol{R}'|}\mathrm{d}\boldsymbol{R}'$$
$$-\int \sum_{n,m=0}^{\infty} \left(\overline{\boldsymbol{\mu}}_1^{(n)} : \nabla^n \overline{\boldsymbol{\mu}}_2^{(m)} : \nabla'^m \nabla \frac{1}{4\pi|\boldsymbol{R}-\boldsymbol{R}'|}\right) c_2(\boldsymbol{R}, 1, \boldsymbol{R}', 2; t)\mathrm{d}\boldsymbol{R}'\, \mathrm{d}1\, \mathrm{d}2. \quad (70)$$

The total long range force density is now given by the sum of (68) and (70). Introducing the macroscopic electric and magnetic fields (17), which read up to order $c^{-1}$ and $c^0$ respectively:

$$\begin{aligned}\boldsymbol{E}(\boldsymbol{R}, t) &= \boldsymbol{E}_{\mathrm{e}}(\boldsymbol{R}, t)-\int \{\varrho^{\mathrm{e}}(\boldsymbol{R}', t)+\boldsymbol{P}(\boldsymbol{R}', t)\cdot\nabla'\}\nabla \frac{1}{4\pi|\boldsymbol{R}-\boldsymbol{R}'|}\mathrm{d}\boldsymbol{R}', \\ \boldsymbol{B}(\boldsymbol{R}, t) &= \boldsymbol{B}_{\mathrm{e}}(\boldsymbol{R}, t),\end{aligned} \quad (71)$$

(v. problem 1) we obtain thus up to order $c^{-1}$

$$\boldsymbol{F}^{\mathrm{L}} = \varrho^{\mathrm{e}}\boldsymbol{E}+c^{-1}\boldsymbol{J}\wedge\boldsymbol{B}+(\nabla\boldsymbol{E})\cdot\boldsymbol{P}+(\nabla\boldsymbol{B})\cdot\boldsymbol{M}$$
$$+c^{-1}\frac{\partial}{\partial t}(\boldsymbol{P}\wedge\boldsymbol{B})+c^{-1}\nabla\cdot(\boldsymbol{v}\boldsymbol{P}\wedge\boldsymbol{B})-\nabla\cdot\mathsf{P}^{\mathrm{F}}+\boldsymbol{F}^{\mathrm{C}}, \quad (72)$$

with

$$\mathsf{P}^{\mathrm{F}} \equiv -c^{-1}\int \hat{\boldsymbol{v}}_1 \overline{\boldsymbol{\mu}}_1^{(1)} \wedge \boldsymbol{B} f_1(\boldsymbol{R}, \boldsymbol{v}_1, \overline{\boldsymbol{\mu}}_1^{(1)}; t)\mathrm{d}\boldsymbol{v}_1\, \mathrm{d}\overline{\boldsymbol{\mu}}_1^{(1)}, \quad (73)$$

a contribution to the pressure tensor due to the action of the field $\boldsymbol{B}$ on the electric dipoles. Furthermore $\boldsymbol{F}^{\mathrm{C}}$ is the 'correlation contribution' given by the last term in (70). It reads written with $\boldsymbol{s}$ for the interatomic separation $\boldsymbol{R}-\boldsymbol{R}'$:

$$\boldsymbol{F}^{\mathrm{C}} \equiv -\int \sum_{n,m=0}^{\infty} \left\{(-1)^m \overline{\boldsymbol{\mu}}_1^{(n)} \vdots \nabla_s^n \overline{\boldsymbol{\mu}}_2^{(m)} \vdots \nabla_s^m \nabla_s \frac{1}{4\pi s}\right\} c_2(\boldsymbol{R}, 1, \boldsymbol{R}-\boldsymbol{s}, 2; t)\, \mathrm{d}\boldsymbol{s}\, \mathrm{d}1\, \mathrm{d}2. \quad (74)$$

The *short range* term $\boldsymbol{F}^{\mathrm{S}} \equiv \langle \sum_k \boldsymbol{f}_k^{\mathrm{S}} \delta(\boldsymbol{R}_k-\boldsymbol{R})\rangle$ at the right-hand side of (62) contains the force $\boldsymbol{f}_k^{\mathrm{S}}$ given in (I.52). We may write it with the help of an appropriate two-point distribution function. Again performing the integration over $\boldsymbol{R}_1$ and introducing the integration variable $\boldsymbol{s} = \boldsymbol{R}-\boldsymbol{R}_2$ we get with $\boldsymbol{r}_{ki} \equiv \boldsymbol{R}_{ki}-\boldsymbol{R}_k$:

$$\boldsymbol{F}^{\mathrm{S}} = \int \left\{-\sum_{i,j} \nabla_s \frac{e_i e_j}{4\pi|\boldsymbol{s}+\boldsymbol{r}_{1i}-\boldsymbol{r}_{2j}|} + \sum_{n,m=0}^{\infty} (-1)^m \overline{\boldsymbol{\mu}}_1^{(n)} \vdots \nabla_s^n \overline{\boldsymbol{\mu}}_2^{(m)} \vdots \nabla_s^m \nabla_s \frac{1}{4\pi s}\right\} f_2(\boldsymbol{R}, 1, \boldsymbol{R}-\boldsymbol{s}, 2; t) \mathrm{d}\boldsymbol{s}\, \mathrm{d}1\, \mathrm{d}2, \quad (75)$$

where $e_i$ and $e_j$ are the charges of the constituent particles $i$ and $j$ of the (identical) atoms.

The equation (62) with (63) and (72–75) constitutes the macroscopic balance of momentum. It will be studied further for specific systems in subsections *f*, *g* and *h*.

## d. *The energy balance*

The macroscopic energy law will be derived from the atomic energy equation (I.63). Let us first consider the macroscopic quantity

$$\left\langle \sum_k \left(\tfrac{1}{2}m_k \boldsymbol{v}_k^2 + \tfrac{1}{2}\sum_i m_{ki} \dot{\boldsymbol{r}}_{ki}^2 + \sum_{i,j(i\neq j)} \frac{e_{ki} e_{kj}}{8\pi|\boldsymbol{r}_{ki}-\boldsymbol{r}_{kj}|}\right) \delta(\boldsymbol{R}_k-\boldsymbol{R})\right\rangle \quad (76)$$

($\boldsymbol{v}_k$ is the atomic velocity $\dot{\boldsymbol{R}}_k$). Introducing the appropriate distribution functions, the notations $m$ for the mass $m_k$ of the (identical) atoms, and $m_i$ and $e_i$ for the mass $m_{ki}$ and charge $e_{ki}$ of their constituent particles $i$, one gets for this expression, with the help of (56), (58) and (61):

$$\tfrac{1}{2}\varrho \boldsymbol{v}^2 + \varrho u^{\mathrm{K}} \quad (77)$$

with an internal energy density

$$\varrho u^{\mathrm{K}} \equiv \int \left(\tfrac{1}{2}m \hat{\boldsymbol{v}}_1^2 + \tfrac{1}{2}\sum_i m_i \dot{\boldsymbol{r}}_{1i}^2 + \sum_{i,j(i\neq j)} \frac{e_i e_j}{8\pi|\boldsymbol{r}_{1i}-\boldsymbol{r}_{1j}|}\right) f_1(\boldsymbol{R}, 1; t) \mathrm{d}1, \quad (78)$$

due to the velocity fluctuations and the intra-atomic kinetic and potential energies.

The energy balance equation is obtained by taking the time derivative of (77) in its form (76) and by using (2) and (I.63):

$$\frac{\partial}{\partial t}(\tfrac{1}{2}\varrho \boldsymbol{v}^2+\varrho u^{\mathrm{K}})$$
$$= -\nabla\cdot\left\langle \sum_k \boldsymbol{v}_k\left(\tfrac{1}{2}m_k \boldsymbol{v}_k^2+\tfrac{1}{2}\sum_i m_{ki}\dot{\boldsymbol{r}}_{ki}^2+\sum_{i,\,j(i\neq j)}\frac{e_{ki}e_{kj}}{8\pi|\boldsymbol{r}_{ki}-\boldsymbol{r}_{kj}|}\right)\delta(\boldsymbol{R}_k-\boldsymbol{R})\right\rangle$$
$$+\langle \sum_k (\psi_k^{\mathrm{L}}+\psi_k^{\mathrm{S}})\delta(\boldsymbol{R}_k-\boldsymbol{R})\rangle. \quad (79)$$

Splitting the atomic velocities according to (61) one obtains with the help of appropriate distribution functions the energy balance equation:

$$\frac{\partial}{\partial t}(\tfrac{1}{2}\varrho \boldsymbol{v}^2+\varrho u^{\mathrm{K}}) = -\nabla\cdot\{\boldsymbol{v}(\tfrac{1}{2}\varrho\boldsymbol{v}^2+\varrho u^{\mathrm{K}})+\mathbf{P}^{\mathrm{K}}\cdot\boldsymbol{v}+\boldsymbol{J}_q^{\mathrm{K}}\}+\Psi^{\mathrm{L}}+\Psi^{\mathrm{S}}, \quad (80)$$

with $\mathbf{P}^{\mathrm{K}}$ the kinetic pressure given in (63) and

$$\boldsymbol{J}_q^{\mathrm{K}} \equiv \int \hat{\boldsymbol{v}}_1\left(\tfrac{1}{2}m\hat{\boldsymbol{v}}_1^2+\tfrac{1}{2}\sum_i m_i\dot{\boldsymbol{r}}_{1i}^2+\sum_{i,\,j(i\neq j)}\frac{e_ie_j}{8\pi|\boldsymbol{r}_{1i}-\boldsymbol{r}_{1j}|}\right)f_1(\boldsymbol{R},1;t)\mathrm{d}1, \quad (81)$$

a contribution to the heat flow. (In fact it is due to the transport, with the velocity fluctuation, of the atomic quantity occurring in (78).) Furthermore the last two terms of (80) represent long and short range contributions defined as $\Psi^{\mathrm{L,S}} = \langle \sum_k \psi_k^{\mathrm{L,S}}\delta(\boldsymbol{R}_k-\boldsymbol{R})\rangle$, with $\psi_k^{\mathrm{L,S}}$ as specified in (I.66) and (I.65). They will now be investigated in detail.

The *long range* part contains a contribution due to the external fields $(\boldsymbol{E}_{\mathrm{e}}, \boldsymbol{B}_{\mathrm{e}})$ which may be written in the form

$$\Psi_{\mathrm{e}}^{\mathrm{L}} = \boldsymbol{J}\cdot\boldsymbol{E}_{\mathrm{e}}-\boldsymbol{M}\cdot\frac{\partial \boldsymbol{B}_{\mathrm{e}}}{\partial t}+\langle\sum_k\{\boldsymbol{v}_k\cdot(\nabla_k\boldsymbol{E}_{\mathrm{e}})\cdot\overline{\boldsymbol{\mu}}_k^{(1)}+\dot{\boldsymbol{\mu}}_k^{(1)}\cdot\boldsymbol{E}_{\mathrm{e}}\}\delta(\boldsymbol{R}_k-\boldsymbol{R})\rangle, \quad (82)$$

where (64) and (65) have been used. With the help of (2) one may write

$$\langle\sum_k \dot{\overline{\boldsymbol{\mu}}}_k^{(1)}\delta(\boldsymbol{R}_k-\boldsymbol{R})\rangle = \frac{\partial}{\partial t}\langle\sum_k\overline{\boldsymbol{\mu}}_k^{(1)}\delta(\boldsymbol{R}_k-\boldsymbol{R})\rangle+\nabla\cdot\langle\sum_k \boldsymbol{v}_k\overline{\boldsymbol{\mu}}_k^{(1)}\delta(\boldsymbol{R}_k-\boldsymbol{R})\rangle. \quad (83)$$

Using this identity and (65) we get for (82):

$$\Psi_{\mathrm{e}}^{\mathrm{L}} = \boldsymbol{J}\cdot\boldsymbol{E}_{\mathrm{e}}+\frac{\partial \boldsymbol{P}}{\partial t}\cdot\boldsymbol{E}_{\mathrm{e}}-\boldsymbol{M}\cdot\frac{\partial \boldsymbol{B}_{\mathrm{e}}}{\partial t}+\nabla\cdot\{\langle\sum_k \boldsymbol{v}_k\overline{\boldsymbol{\mu}}_k^{(1)}\delta(\boldsymbol{R}_k-\boldsymbol{R})\rangle\cdot\boldsymbol{E}_{\mathrm{e}}\}. \quad (84)$$

Introducing velocity fluctuations (61) and the appropriate distribution functions, and using again (65), we obtain finally

$$\Psi_e^L = \boldsymbol{J}\cdot\boldsymbol{E}_e + \frac{\partial \boldsymbol{P}}{\partial t}\cdot\boldsymbol{E}_e + \nabla\cdot(\boldsymbol{v}\boldsymbol{P}\cdot\boldsymbol{E}_e) - \boldsymbol{M}\cdot\frac{\partial \boldsymbol{B}_e}{\partial t} + \nabla\cdot\left[\left\{\int \hat{\boldsymbol{v}}_1\,\overline{\boldsymbol{\mu}}_1^{(1)} f_1(\boldsymbol{R}, \boldsymbol{v}_1, \overline{\boldsymbol{\mu}}_1^{(1)}; t)\mathrm{d}\boldsymbol{v}_1\,\mathrm{d}\overline{\boldsymbol{\mu}}_1^{(1)}\right\}\cdot\boldsymbol{E}_e\right]. \quad (85)$$

The interatomic contribution to $\Psi^L$ (v. the first term of (I.66)) reads, written with distribution functions,

$$\Psi^L - \Psi_e^L = -\int\left\{\sum_{m=0}^{\infty}\left(\sum_{n=0}^{\infty}\overline{\boldsymbol{\mu}}_1^{(n)} \vdots \nabla_1^n\,\boldsymbol{v}_1\cdot\nabla_1 + \sum_{n=1}^{\infty}\dot{\overline{\boldsymbol{\mu}}}_1^{(n)} \vdots \nabla_1^n\right)\overline{\boldsymbol{\mu}}_2^{(m)} \vdots \nabla_2^m\,\frac{1}{4\pi|\boldsymbol{R}_1-\boldsymbol{R}_2|}\right\} \delta(\boldsymbol{R}-\boldsymbol{R}_1) f_2(\boldsymbol{R}_1, 1, \boldsymbol{R}_2, 2; t)\mathrm{d}\boldsymbol{R}_1\,\mathrm{d}1\,\mathrm{d}\boldsymbol{R}_2\,\mathrm{d}2. \quad (86)$$

With the help of (7) the right-hand side may be split into two parts. In the uncorrelated part we introduce the macroscopic charge and polarization densities (64–65) taking only dipole moments into account and using (61) and (83). Then we get:

$$\begin{aligned}\Psi^L - \Psi_e^L = &-\int\left\{\boldsymbol{J}(\boldsymbol{R}, t) + \frac{\partial \boldsymbol{P}(\boldsymbol{R}, t)}{\partial t}\right\}\cdot\nabla\{\varrho^e(\boldsymbol{R}', t) + \boldsymbol{P}(\boldsymbol{R}', t)\cdot\nabla'\}\frac{1}{4\pi|\boldsymbol{R}-\boldsymbol{R}'|}\mathrm{d}\boldsymbol{R}' \\ &-\nabla\cdot\int \boldsymbol{v}(\boldsymbol{R}, t)\boldsymbol{P}(\boldsymbol{R}, t)\cdot\nabla\{\varrho^e(\boldsymbol{R}', t) + \boldsymbol{P}(\boldsymbol{R}', t)\cdot\nabla'\}\frac{1}{4\pi|\boldsymbol{R}-\boldsymbol{R}'|}\mathrm{d}\boldsymbol{R}' \\ &-\nabla\cdot\int \hat{\boldsymbol{v}}_1\left[\overline{\boldsymbol{\mu}}_1^{(1)}\cdot\nabla\{\varrho^e(\boldsymbol{R}', t) + \boldsymbol{P}(\boldsymbol{R}', t)\cdot\nabla'\}\frac{1}{4\pi|\boldsymbol{R}-\boldsymbol{R}'|}\right] f_1(\boldsymbol{R}, \boldsymbol{v}_1, \overline{\boldsymbol{\mu}}_1^{(1)}; t)\mathrm{d}\boldsymbol{R}'\,\mathrm{d}\boldsymbol{v}_1\,\mathrm{d}\overline{\boldsymbol{\mu}}_1^{(1)} \\ &-\int\sum_{m=0}^{\infty}\left\{\left(\sum_{n=0}^{\infty}\overline{\boldsymbol{\mu}}_1^{(n)} \vdots \nabla_1^n\,\boldsymbol{v}_1\cdot\nabla_1 + \sum_{n=1}^{\infty}\dot{\overline{\boldsymbol{\mu}}}_1^{(n)} \vdots \nabla_1^n\right)\overline{\boldsymbol{\mu}}_2^{(m)} \vdots \nabla_2^m\,\frac{1}{4\pi|\boldsymbol{R}_1-\boldsymbol{R}_2|}\right\} \\ &\qquad \delta(\boldsymbol{R}-\boldsymbol{R}_1) c_2(\boldsymbol{R}_1, 1, \boldsymbol{R}_2, 2; t)\mathrm{d}\boldsymbol{R}_1\,\mathrm{d}1\,\mathrm{d}\boldsymbol{R}_2\,\mathrm{d}2. \quad (87)\end{aligned}$$

The total long range contribution is the sum of (85) and (87). It reads written with the Maxwell fields (71) and the pressure $\mathbf{P}^F$ (73)

$$\Psi^L = \boldsymbol{J}\cdot\boldsymbol{E} + \frac{\partial \boldsymbol{P}}{\partial t}\cdot\boldsymbol{E} + \nabla\cdot(\boldsymbol{v}\boldsymbol{P}\cdot\boldsymbol{E}) - \boldsymbol{M}\cdot\frac{\partial \boldsymbol{B}}{\partial t} - \nabla\cdot(\mathbf{P}^F\cdot\boldsymbol{v} + \boldsymbol{J}_q^F) + \Psi^C, \quad (88)$$

with

$$\boldsymbol{J}_q^F \equiv -\left\{\int \hat{\boldsymbol{v}}_1\,\overline{\boldsymbol{\mu}}_1^{(1)} f_1(\boldsymbol{R}, \boldsymbol{v}_1, \overline{\boldsymbol{\mu}}_1^{(1)}; t)\mathrm{d}\boldsymbol{v}_1\,\mathrm{d}\overline{\boldsymbol{\mu}}_1^{(1)}\right\}\cdot(\boldsymbol{E} + c^{-1}\boldsymbol{v}\wedge\boldsymbol{B}), \quad (89)$$

a contribution to the heat flow due to the interaction of the fields $(\boldsymbol{E}, \boldsymbol{B})$ and the electric dipoles. Furthermore the correlation contribution $\Psi^{\mathrm{C}}$ given by the last term of (87) may be written as

$$\Psi^{\mathrm{C}} \equiv -\int \sum_{m=0}^{\infty} (-1)^m \left\{ \left( \sum_{n=0}^{\infty} \overline{\boldsymbol{\mu}}_1^{(n)} : \nabla_s^n \boldsymbol{v}_1 \cdot \nabla_s + \sum_{n=1}^{\infty} \dot{\overline{\boldsymbol{\mu}}}_1^{(n)} : \nabla_s^n \right) \overline{\boldsymbol{\mu}}_2^{(m)} : \nabla_s^m \frac{1}{4\pi s} \right\}$$

$$c_2(\boldsymbol{R}, 1, \boldsymbol{R}-\boldsymbol{s}, 2; t) \mathrm{d}\boldsymbol{s} \, \mathrm{d}1 \, \mathrm{d}2, \quad (90)$$

where the integration over $\boldsymbol{R}_1$ has been performed and the variable $\boldsymbol{s} \equiv \boldsymbol{R}-\boldsymbol{R}_2$ introduced.

The *short range* term $\Psi^{\mathrm{S}} \equiv \langle \sum_k \psi_k^{\mathrm{S}} \delta(\boldsymbol{R}_k - \boldsymbol{R}) \rangle$, where $\psi_k^{\mathrm{S}}$ is given in (I.65), may be written in terms of a two-point distribution function:

$$\Psi^{\mathrm{S}} = -\int \left\{ \sum_{i,j} (\boldsymbol{v}_1 + \dot{\boldsymbol{r}}_{1i}) \cdot \nabla_s \frac{e_i e_j}{4\pi |\boldsymbol{s} + \boldsymbol{r}_{1i} - \boldsymbol{r}_{2j}|} \right.$$

$$\left. - \sum_{m=0}^{\infty} (-1)^m \left( \sum_{n=0}^{\infty} \overline{\boldsymbol{\mu}}_1^{(n)} : \nabla_s^n \boldsymbol{v}_1 \cdot \nabla_s + \sum_{n=1}^{\infty} \dot{\overline{\boldsymbol{\mu}}}_1^{(n)} : \nabla_s^n \right) \overline{\boldsymbol{\mu}}_2^{(m)} : \nabla_s^m \frac{1}{4\pi s} \right\}$$

$$f_2(\boldsymbol{R}, 1, \boldsymbol{R}-\boldsymbol{s}, 2; t) \mathrm{d}\boldsymbol{s} \, \mathrm{d}1 \, \mathrm{d}2. \quad (91)$$

The equation (80) with (78), (81), (88–91) is the balance equation of energy. The correlation and short range contributions (90) and (91) will be studied for special systems in the following subsections.

### e. *The short range terms in the momentum and energy laws*

We shall consider in this subsection the terms $\boldsymbol{F}^{\mathrm{S}}$ and $\Psi^{\mathrm{S}}$ occurring in the momentum and energy equations (62) and (80), and given explicitly in (75) and (91). Their short range character will allow us to write them in a convenient form.

Since the bracket expression in (75) vanishes if the atoms are outside each other the integral needs to be extended over small values of $\boldsymbol{s}$ only. In sufficiently homogeneous systems the two-point distribution function $f_2(\boldsymbol{R}, 1, \boldsymbol{R}-\boldsymbol{s}, 2; t)$ in (75) varies slowly as a function of $\boldsymbol{R}$, i.e. appreciably only over macroscopic distances whereas it varies rapidly as a function of the interatomic distance $\boldsymbol{s}$. Hence one may limit oneself in the integral to the first two terms in a Taylor expansion of $f_2$ as a function of $\boldsymbol{R}$ [1]:

$$f_2(\boldsymbol{R}, 1, \boldsymbol{R}-\boldsymbol{s}, 2; t)$$

$$= f_2(\boldsymbol{R}+\tfrac{1}{2}\boldsymbol{s}, 1, \boldsymbol{R}-\tfrac{1}{2}\boldsymbol{s}, 2; t) - \tfrac{1}{2}\boldsymbol{s} \cdot \nabla f_2(\boldsymbol{R}+\tfrac{1}{2}\boldsymbol{s}, 1, \boldsymbol{R}-\tfrac{1}{2}\boldsymbol{s}, 2; t), \quad (92)$$

[1] Cf. J. H. Irving and J. G. Kirkwood, J. Chem. Phys. **18**(1950)817.

(with $\nabla \equiv \partial/\partial \boldsymbol{R}$). In this way (75) becomes

$$\boldsymbol{F}^{\mathrm{S}} = -\nabla\cdot\mathbf{P}^{\mathrm{S}} \tag{93}$$

with the abbreviation

$$\mathbf{P}^{\mathrm{S}} \equiv -\int \left\{ \boldsymbol{s}\nabla_s \left( \sum_{i,j} \frac{e_i e_j}{8\pi|\boldsymbol{s}+\boldsymbol{r}_{1i}-\boldsymbol{r}_{2j}|} - \sum_{n,m=0}^{\infty} (-1)^m \overline{\boldsymbol{\mu}}_1^{(n)} \vdots \nabla_s^n \overline{\boldsymbol{\mu}}_2^{(m)} \vdots \nabla_s^m \frac{1}{8\pi s} \right) \right\}$$
$$f_2(\boldsymbol{R}+\tfrac{1}{2}\boldsymbol{s}, 1, \boldsymbol{R}-\tfrac{1}{2}\boldsymbol{s}, 2; t)\mathrm{d}\boldsymbol{s}\,\mathrm{d}1\,\mathrm{d}2. \tag{94}$$

The latter quantity will turn out to be a contribution to the pressure tensor. Owing to the (trivial) symmetry of $f_2$ with respect to an interchange of the first pair of variables with the second pair, the first term of the right-hand side of (92) does not give rise to a term in (93).

Let us now turn to the discussion of the short range term $\Psi^{\mathrm{S}}$ (91) of the energy equation. The expression between brackets in (91) vanishes if the atoms are outside each other. With the use of the expansion (92) we obtain now

$$\Psi^{\mathrm{S}} = \Psi^{\mathrm{S}*} - \nabla\cdot(\mathbf{P}^{\mathrm{S}}\cdot\boldsymbol{v} + \boldsymbol{J}_q^{\mathrm{S}'}). \tag{95}$$

The first term $\Psi^{\mathrm{S}*}$ reads like (91) but with $f_2(\boldsymbol{R}+\frac{1}{2}\boldsymbol{s}, 1, \boldsymbol{R}-\frac{1}{2}\boldsymbol{s}, 2; t)$ instead of $f_2(\boldsymbol{R}, 1, \boldsymbol{R}-\boldsymbol{s}, 2; t)$, while the second contains, apart from a term with the local velocity $\boldsymbol{v}$, a divergence of the vector

$$\boldsymbol{J}_q^{\mathrm{S}'} \equiv -\int \boldsymbol{s} \left\{ \sum_{i,j} (\hat{\boldsymbol{v}}_1 + \dot{\boldsymbol{r}}_{1i})\cdot\nabla_s \frac{e_i e_j}{8\pi|\boldsymbol{s}+\boldsymbol{r}_{1i}-\boldsymbol{r}_{2j}|} \right.$$
$$\left. - \sum_{m=0}^{\infty} (-1)^m \left( \sum_{n=0}^{\infty} \overline{\boldsymbol{\mu}}_1^{(n)} \vdots \nabla_s^n \hat{\boldsymbol{v}}_1\cdot\nabla_s + \sum_{n=1}^{\infty} \dot{\overline{\boldsymbol{\mu}}}_1^{(n)} \vdots \nabla_s^n \right) \overline{\boldsymbol{\mu}}_2^{(m)} \vdots \nabla_s^m \frac{1}{8\pi s} \right\}$$
$$f_2(\boldsymbol{R}+\tfrac{1}{2}\boldsymbol{s}, 1, \boldsymbol{R}-\tfrac{1}{2}\boldsymbol{s}, 2; t)\mathrm{d}\boldsymbol{s}\,\mathrm{d}1\,\mathrm{d}2. \tag{96}$$

The contribution $\Psi^{\mathrm{S}*}$ to (95) may be written in a simpler form. For that purpose we shall consider the time derivative of the quantity

$$\varrho u^{\mathrm{S}} \equiv \int \left( \sum_{i,j} \frac{e_i e_j}{8\pi|\boldsymbol{s}+\boldsymbol{r}_{1i}-\boldsymbol{r}_{2j}|} - \sum_{n,m=0}^{\infty} (-1)^m \overline{\boldsymbol{\mu}}_1^{(n)} \vdots \nabla_s^n \overline{\boldsymbol{\mu}}_2^{(m)} \vdots \nabla_s^m \frac{1}{8\pi s} \right)$$
$$f_2(\boldsymbol{R}+\tfrac{1}{2}\boldsymbol{s}, 1, \boldsymbol{R}-\tfrac{1}{2}\boldsymbol{s}, 2; t)\mathrm{d}\boldsymbol{s}\,\mathrm{d}1\,\mathrm{d}2, \tag{97}$$

which will turn out to be a contribution to the internal energy density. With the identity

$$\frac{\partial f_2(\boldsymbol{R}+\tfrac{1}{2}\boldsymbol{s},\, \boldsymbol{r}_{1i},\, \boldsymbol{R}-\tfrac{1}{2}\boldsymbol{s},\, \boldsymbol{r}_{2j};\, t)}{\partial t} = -\int \left\{ \frac{\boldsymbol{v}_1+\boldsymbol{v}_2}{2} \cdot \nabla + (\boldsymbol{v}_1-\boldsymbol{v}_2)\cdot \nabla_s \right.$$

$$\left. +\dot{\boldsymbol{r}}_{1i}\cdot\nabla_{r_{1i}} + \dot{\boldsymbol{r}}_{2j}\cdot\nabla_{r_{2j}} \right\} f_2(\boldsymbol{R}+\tfrac{1}{2}\boldsymbol{s},\, \boldsymbol{v}_1,\, \boldsymbol{r}_{1i},\, \dot{\boldsymbol{r}}_{1i},\, \boldsymbol{R}-\tfrac{1}{2}\boldsymbol{s},\, \boldsymbol{v}_2,\, \boldsymbol{r}_{2j},\, \dot{\boldsymbol{r}}_{2j};\, t)$$

$$\mathrm{d}\boldsymbol{v}_1\, \mathrm{d}\boldsymbol{v}_2 \prod_{i,j} \mathrm{d}\dot{\boldsymbol{r}}_{1i} \mathrm{d}\dot{\boldsymbol{r}}_{2j}, \qquad (98)$$

which follows from the conservation of the number of particles (or the probability in fluxion space), we get for the time derivative of (97) after partial integration and the use of the (trivial) symmetry $f_2(\boldsymbol{R}+\frac{1}{2}\boldsymbol{s}, 1, \boldsymbol{R}-\frac{1}{2}\boldsymbol{s}, 2; t) = f_2(\boldsymbol{R}-\frac{1}{2}\boldsymbol{s}, 2, \boldsymbol{R}+\frac{1}{2}\boldsymbol{s}, 1; t)$:

$$\frac{\partial \varrho u^{\mathrm{S}}}{\partial t} = -\nabla\cdot(\boldsymbol{v}\varrho u^{\mathrm{S}} + \boldsymbol{J}_q^{\mathrm{S}''}) - \Psi^{\mathrm{S}*} \qquad (99)$$

with the abbreviation

$$\boldsymbol{J}_q^{\mathrm{S}''} \equiv \int (\hat{\boldsymbol{v}}_1+\hat{\boldsymbol{v}}_2)\left( \sum_{i,j} \frac{e_i e_j}{16\pi|\boldsymbol{s}+\boldsymbol{r}_{1i}-\boldsymbol{r}_{2j}|} - \sum_{n,m=0}^{\infty} (-1)^m \overline{\boldsymbol{\mu}}_1^{(n)} : \nabla_s^n \overline{\boldsymbol{\mu}}_2^{(m)} : \nabla_s^m \frac{1}{16\pi s} \right)$$

$$f_2(\boldsymbol{R}+\tfrac{1}{2}\boldsymbol{s}, 1, \boldsymbol{R}-\tfrac{1}{2}\boldsymbol{s}, 2; t)\mathrm{d}\boldsymbol{s}\, \mathrm{d}1\, \mathrm{d}2. \qquad (100)$$

The divergence in (99) contains a convective part $\boldsymbol{v}\varrho u^{\mathrm{S}}$ and the conductive part (100).

With $\Psi^{\mathrm{S}*}$ from (99) we get for (95):

$$\Psi^{\mathrm{S}} = -\nabla\cdot(\boldsymbol{v}\varrho u^{\mathrm{S}} + \mathbf{P}^{\mathrm{S}}\cdot\boldsymbol{v} + \boldsymbol{J}_q^{\mathrm{S}}) - \frac{\partial \varrho u^{\mathrm{S}}}{\partial t}. \qquad (101)$$

Since $\Psi^{\mathrm{S}}$ occurred in the energy equation (80), it appears that $\varrho u^{\mathrm{S}}$ (97) is a contribution to the internal energy density, and that $\boldsymbol{J}_q^{\mathrm{S}}$, which is the sum of $\boldsymbol{J}_q^{\mathrm{S}'}$ (96) and $\boldsymbol{J}_q^{\mathrm{S}''}$ (100), is a contribution to the heat flow.

In this way, the short range quantities $\boldsymbol{F}^{\mathrm{S}}$ and $\Psi^{\mathrm{S}}$, which occur in the momentum and energy equations, have been found, in formulae (93) and (101), for sufficiently homogeneous systems. It may be noted that if the 'atoms' carry charges, but no multipole moments (as in plasmas, for instance), the quantities $\boldsymbol{F}^{\mathrm{S}}$ and $\Psi^{\mathrm{S}}$ simply vanish.

### f. *The momentum and energy equations for fluids*

In this subsection we want to consider the momentum and energy equations for systems in which the correlation function has short range i.e.

$c_2(\boldsymbol{R}, 1, \boldsymbol{R}-\boldsymbol{s}, 2; t)$ vanishes rapidly as the interatomic distance $\boldsymbol{s}$ increases. This is usually the case for one-component fluid systems, at least if the constituent atoms are electrically neutral. (Amorphous and polycrystalline solids of neutral atoms are other examples of systems with short-range correlation functions, to which the treatment of this subsection applies.) For such systems we shall cast the correlation terms $\boldsymbol{F}^{\mathrm{C}}$ (74) and $\Psi^{\mathrm{C}}$ (90) in the momentum and energy equations in a convenient form.

In normal fluids the correlation function becomes negligible for $\boldsymbol{s}$ greater than the so-called 'correlation length'. This correlation length is much smaller than the distances over which the macroscopic quantities change appreciably. Then the correlation function may be approximated by the first two terms of a Taylor expansion

$$c_2(\boldsymbol{R}, 1, \boldsymbol{R}-\boldsymbol{s}, 2; t) = c_2(\boldsymbol{R}+\tfrac{1}{2}\boldsymbol{s}, 1, \boldsymbol{R}-\tfrac{1}{2}\boldsymbol{s}, 2; t) \\ -\tfrac{1}{2}\boldsymbol{s}\cdot\nabla c_2(\boldsymbol{R}+\tfrac{1}{2}\boldsymbol{s}, 1, \boldsymbol{R}-\tfrac{1}{2}\boldsymbol{s}, 2; t), \quad (102)$$

the Irving–Kirkwood approximation[1].

For a fluid of neutral atoms the correlation term $\boldsymbol{F}^{\mathrm{C}}$ (74) in the *momentum equation* becomes then a divergence, because the first term at the right-hand side of (102) gives no contribution. In fact (74) with this first term vanishes owing to the trivial symmetry of $c_2(\boldsymbol{R}+\frac{1}{2}\boldsymbol{s}, 1, \boldsymbol{R}-\frac{1}{2}\boldsymbol{s}, 2; t)$ with respect to an interchange of the first and second pairs of arguments. In this way one gets

$$\boldsymbol{F}^{\mathrm{C}} = -\nabla\cdot\mathbf{P}^{\mathrm{C}} \quad (103)$$

with the correlation pressure given by

$$\mathbf{P}^{\mathrm{C}} \equiv -\int \sum_{n,m=1}^{\infty} (-1)^m \left(\boldsymbol{s}\nabla_{\boldsymbol{s}}\overline{\boldsymbol{\mu}}_1^{(n)} \vdots \nabla_{\boldsymbol{s}}^n\overline{\boldsymbol{\mu}}_2^{(m)} \vdots \nabla_{\boldsymbol{s}}^m \frac{1}{8\pi s}\right) \\ c_2(\boldsymbol{R}+\tfrac{1}{2}\boldsymbol{s}, 1, \boldsymbol{R}-\tfrac{1}{2}\boldsymbol{s}, 2; t)\mathrm{d}\boldsymbol{s}\,\mathrm{d}1\,\mathrm{d}2. \quad (104)$$

The momentum equation for a fluid of neutral atoms is obtained from (62) with (72), (93) and (103). It reads finally

$$\frac{\partial \varrho \boldsymbol{v}}{\partial t} = -\nabla\cdot(\varrho\boldsymbol{v}\boldsymbol{v}+\mathbf{P})+\boldsymbol{F}, \quad (105)$$

where the force density is

$$\boldsymbol{F} = (\nabla\boldsymbol{E})\cdot\boldsymbol{P}+(\nabla\boldsymbol{B})\cdot\boldsymbol{M}+c^{-1}\frac{\partial}{\partial t}(\boldsymbol{P}\wedge\boldsymbol{B})+c^{-1}\nabla\cdot(\boldsymbol{v}\boldsymbol{P}\wedge\boldsymbol{B}) \quad (106)$$

[1] J. H. Irving and J. G. Kirkwood, op. cit.

(for systems of neutral atoms $\varrho^e$ and $\boldsymbol{J}$ vanish) and the pressure tensor

$$\mathbf{P} = \mathbf{P}^K + \mathbf{P}^F + \mathbf{P}^S + \mathbf{P}^C. \tag{107}$$

The momentum equation (105) has the form of a *balance equation*, not of a conservation law. Indeed the momentum density $\varrho\boldsymbol{v}$ does not only change as a consequence of momentum flow $\varrho\boldsymbol{v}\boldsymbol{v}+\mathbf{P}$, but also as a consequence of momentum 'production' $\boldsymbol{F}$. The latter source term arises because the system is not closed. It vanishes if the electromagnetic fields are not present. (The expression (106) contains a time derivative and a divergence of quantities which might be grouped with the momentum density and the momentum flow respectively. This has not been done in order to keep together terms which depend exclusively on the Maxwell fields $\boldsymbol{E}$, $\boldsymbol{B}$, the polarizations $\boldsymbol{P}$, $\boldsymbol{M}$ and the velocity $\boldsymbol{v}$.)

The momentum flow of which (minus) the divergence appears in (105) consists of a convection part ($\varrho\boldsymbol{v}\boldsymbol{v}$) together with the pressure tensor ($\mathbf{P}$). The latter quantity contains the kinetic pressure tensor $\mathbf{P}^K$ (63), a term $\mathbf{P}^F$ (73) with the magnetic field $\boldsymbol{B}$ and the potential pressure tensor $\mathbf{P}^S+\mathbf{P}^C$, where $\mathbf{P}^S$ (94) (with the atomic charge $\bar{\mu}_1^{(0)} = \bar{\mu}_2^{(0)} = 0$) contains a short range interatomic interaction multiplied by a distribution function $f_2$ with a long range, whereas $\mathbf{P}^C$ (104) contains the long range part of the interatomic interaction multiplied by a correlation function $c_2$ of short range. (The second term in the short range pressure $\mathbf{P}^S$ (94) has the same structure as the correlation pressure $\mathbf{P}^C$ (104). Their sum might, according to (7), be written with a product of one-point distribution functions. However, the way in which the potential pressure tensor $\mathbf{P}^S+\mathbf{P}^C$ has been written here has the advantage, as stated above, that the short-range character, of both contributions separately, is explicitly apparent.)

Furthermore the momentum balance equation (105) contains as a source term the force density $\boldsymbol{F}$ (106) exerted by the field ($\boldsymbol{E}$, $\boldsymbol{B}$) on a medium with polarizations ($\boldsymbol{P}$, $\boldsymbol{M}$). It includes the Kelvin force $(\nabla\boldsymbol{E})\cdot\boldsymbol{P}$ on an electrically polarized medium and three force terms of magnetic origin: the first of these, $(\nabla\boldsymbol{B})\cdot\boldsymbol{M}$, is analogous to the Kelvin force while the other two describe a coupling of the magnetic field and the electric polarization[1].

The momentum equation (105) may be written in the form of a *conservation law* by using the identity

[1] Part of these results were obtained already by H. A. Lorentz, Enc. Math. Wiss. V 2, fasc. 1 (Teubner, Leipzig 1904)200; A. Einstein and J. Laub, Ann. Physik **26**(1908)541; W. Dällenbach, Phys. Z. **27**(1926)632; P. Mazur and S. R. de Groot, Physica **22**(1956) 657; A. N. Kaufman, Phys. Fluids **8**(1965)935.

$$(\nabla \boldsymbol{E})\cdot\boldsymbol{P}+(\nabla\boldsymbol{B})\cdot\boldsymbol{M} = \nabla\cdot\{\boldsymbol{DE}+\boldsymbol{BH}-(\tfrac{1}{2}E^2+\tfrac{1}{2}B^2-\boldsymbol{M}\cdot\boldsymbol{B})\mathrm{U}\}-c^{-1}\frac{\partial}{\partial t}(\boldsymbol{D}\wedge\boldsymbol{B}), \tag{108}$$

which follows from the Maxwell equations (13) with (14) for systems of neutral atoms (**U** is the unit tensor). Substituting this identity into (106) and the result into (105) one obtains

$$\frac{\partial(\varrho\boldsymbol{v}+c^{-1}\boldsymbol{E}\wedge\boldsymbol{B})}{\partial t} = -\nabla\cdot\{\varrho\boldsymbol{vv}+\mathbf{P}-\boldsymbol{DE}-\boldsymbol{BH}-c^{-1}\boldsymbol{vP}\wedge\boldsymbol{B}+(\tfrac{1}{2}E^2+\tfrac{1}{2}B^2-\boldsymbol{M}\cdot\boldsymbol{B})\mathrm{U}\}. \tag{109}$$

This equation forms the conservation law of total momentum for a fluid of neutral atoms in an electromagnetic field. Both the momentum density and the momentum flow consist of a material part and a field part.

The correlation term $\Psi^{\mathrm{C}}$ (90), which plays a role in the *energy equation,* may likewise be written in a special form for fluid systems of neutral atoms. Since for such systems the approximation (102), which has the same structure as (92), is valid, one may follow the same procedure as in subsection *e*. In this way one obtains (cf. (101))

$$\Psi^{\mathrm{C}} = -\nabla\cdot(\boldsymbol{v}\varrho u^{\mathrm{C}}+\mathbf{P}^{\mathrm{C}}\cdot\boldsymbol{v}+\boldsymbol{J}_q^{\mathrm{C}})-\frac{\partial\varrho u^{\mathrm{C}}}{\partial t}, \tag{110}$$

where

$$\varrho u^{\mathrm{C}} \equiv \int \sum_{n,m=1}^{\infty}(-1)^m\left(\overline{\boldsymbol{\mu}}_1^{(n)} : \nabla_s^n\overline{\boldsymbol{\mu}}_2^{(m)} : \nabla_s^m\frac{1}{8\pi s}\right) c_2(\boldsymbol{R}+\tfrac{1}{2}\boldsymbol{s}, 1, \boldsymbol{R}-\tfrac{1}{2}\boldsymbol{s}, 2; t)\mathrm{d}\boldsymbol{s}\,\mathrm{d}1\,\mathrm{d}2 \tag{111}$$

and

$$\boldsymbol{J}_q^{\mathrm{C}} \equiv -\int \sum_{n,m=1}^{\infty}(-1)^m\left[\{(\boldsymbol{s}\hat{\boldsymbol{v}}_1\cdot\nabla_s-\hat{\boldsymbol{v}}_1)\overline{\boldsymbol{\mu}}_1^{(n)} : \nabla_s^n+\boldsymbol{s}\dot{\overline{\boldsymbol{\mu}}}_1^{(n)} : \nabla_s^n\}\overline{\boldsymbol{\mu}}_2^{(m)} : \nabla_s^m\frac{1}{8\pi s}\right] c_2(\boldsymbol{R}+\tfrac{1}{2}\boldsymbol{s}, 1, \boldsymbol{R}-\tfrac{1}{2}\boldsymbol{s}, 2; t)\mathrm{d}\boldsymbol{s}\,\mathrm{d}1\,\mathrm{d}2 \tag{112}$$

are the correlation contributions to the internal energy density and the heat flow. The energy equation for a fluid of neutral atoms is obtained from (80) with (88), (101) and (110). It reads finally:

$$\frac{\partial}{\partial t}(\tfrac{1}{2}\varrho v^2+\varrho u) = -\nabla\cdot\{\boldsymbol{v}(\tfrac{1}{2}\varrho v^2+\varrho u)+\mathbf{P}\cdot\boldsymbol{v}+\boldsymbol{J}_q\}+\Psi, \tag{113}$$

where the 'power' density is ($\boldsymbol{J} = 0$ for systems of neutral atoms)

$$\Psi \equiv \frac{\partial \boldsymbol{P}}{\partial t}\cdot\boldsymbol{E}+\nabla\cdot(\boldsymbol{v}\boldsymbol{P}\cdot\boldsymbol{E})-\boldsymbol{M}\cdot\frac{\partial \boldsymbol{B}}{\partial t}, \tag{114}$$

the specific internal energy

$$u \equiv u^{\mathrm{K}}+u^{\mathrm{S}}+u^{\mathrm{C}}, \tag{115}$$

and the heat flow

$$\boldsymbol{J}_q \equiv \boldsymbol{J}_q^{\mathrm{K}}+\boldsymbol{J}_q^{\mathrm{F}}+\boldsymbol{J}_q^{\mathrm{S}}+\boldsymbol{J}_q^{\mathrm{C}}. \tag{116}$$

The energy equation (113) has the form of a balance equation. It shows that the sum of the bulk kinetic energy density $\frac{1}{2}\varrho \boldsymbol{v}^2$ and the internal energy density $\varrho u$ changes as a consequence of two causes: through the divergence of an energy flow and through a source term. The specific internal energy (115) consists of three parts. In the first place a contribution $u^{\mathrm{K}}$ (78) formed with the help of one-point distribution functions occurs. It consists of the energy due to atomic velocity fluctuations and the total intra-atomic energy. It will be referred to as the kinetic part of the specific internal energy. Furthermore two contributions $u^{\mathrm{S}}$ (97) (with the atomic charge $\bar{\mu}_1^{(0)} = \bar{\mu}_2^{(0)} = 0$) and $u^{\mathrm{C}}$ (111) with two-point distribution functions arise; these terms, which are due to interatomic forces, will be called together the potential part of the specific internal energy.

The energy flow in (113) contains besides convection terms with the local velocity $\boldsymbol{v}$, the heat flow $\boldsymbol{J}_q$ (116). The latter consists of a kinetic part $\boldsymbol{J}_q^{\mathrm{K}}$ (81), a part $\boldsymbol{J}_q^{\mathrm{F}}$ (89), due to the action of the fields on the electric dipoles, and two terms $\boldsymbol{J}_q^{\mathrm{S}}$ ($\equiv \boldsymbol{J}_q^{\mathrm{S}'}+\boldsymbol{J}_q^{\mathrm{S}''}$, v. (96) and (100) with $\bar{\mu}_1^{(0)} = \bar{\mu}_2^{(0)} = 0$) and $\boldsymbol{J}^{\mathrm{C}}$ (112), which form together the potential part of the heat flow. (Just as in the potential pressure tensor one could have combined here part of the short range internal energy with the correlation energy in such a way that a product of two one-point distribution functions occurs. The same remark applies to the potential part of the heat flow.)

The power density $\Psi$ (114) contains two terms with time derivatives, showing the same asymmetry as was present in the atomic energy equation. This asymmetry will play a role in the first law of thermodynamics, as will be shown in section 7. Furthermore the power density $\Psi$ contains a divergence of a vector which might be shifted to the energy flow. We have preferred to keep it together with the other terms containing the macroscopic Maxwell fields $\boldsymbol{E}$, $\boldsymbol{B}$, $\boldsymbol{P}$ and $\boldsymbol{M}$.

The balance equation (113) may be transformed into a conservation law. In fact from Maxwell's equations it follows for neutral and current-free systems that:

$$\frac{\partial \boldsymbol{P}}{\partial t}\cdot\boldsymbol{E}-\boldsymbol{M}\cdot\frac{\partial \boldsymbol{B}}{\partial t} = -c\nabla\cdot(\boldsymbol{E}\wedge\boldsymbol{H})-\frac{1}{2}\frac{\partial}{\partial t}(\boldsymbol{E}^2+\boldsymbol{B}^2). \tag{117}$$

If this is inserted in (113–114) one gets

$$\frac{\partial}{\partial t}(\tfrac{1}{2}\varrho\boldsymbol{v}^2+\varrho u+\tfrac{1}{2}\boldsymbol{E}^2+\tfrac{1}{2}\boldsymbol{B}^2)$$
$$= -\nabla\cdot\{\boldsymbol{v}(\tfrac{1}{2}\varrho\boldsymbol{v}^2+\varrho u)+\mathbf{P}\cdot\boldsymbol{v}+\boldsymbol{J}_q+c\boldsymbol{E}\wedge\boldsymbol{H}-\boldsymbol{v}\boldsymbol{P}\cdot\boldsymbol{E}\}, \tag{118}$$

which expresses the conservation of total energy for a fluid system of neutral atoms. Both the energy density and the energy flow consist of a material part and a field part. The latter includes the Poynting vector.

The right-hand sides of the conservation laws (109) and (118) of momentum and energy contain the total momentum flow and the total energy flow respectively. Since only the divergences of these quantities play a role, they are determined up to a divergence-free part. The expressions given are thus not uniquely fixed, although they appear to be the simplest ones[1].

## g. *Mixtures, in particular plasmas*

In the preceding the treatment was confined to one-component systems. The extension to mixtures of several components is straightforward. In that case one has to introduce distribution functions for each separate species. The one-point distribution function $f_1^a(1;t)$ carries an extra label $a$ which indicates the species. Now $f_1^a(1;t)\mathrm{d}1$ is the probability, normalized to $N^a$ (the number of atoms of species $a$), to find an atom of species $a$ with parameters in the range d1 in fluxion space. Similarly the two-point distribution function is defined in such a way that $f_2^{ab}(1,2;t)\mathrm{d}1\,\mathrm{d}2$ is the joint probability, normalized to $N^aN^b$ (if $a \neq b$) or $N^a(N^a-1)$ (if $a = b$), to find an atom of species $a$ in the range d1 and an atom of species $b$ in the range d2. The correlation function $c_2^{ab}(1,2;t)$ is now defined as the difference $f_2^{ab}(1,2;t) - f_1^a(1;t)f_1^b(2;t)$.

A case in which the use of this kind of distribution functions is essential is a plasma consisting of a mixture of oppositely charged ions and electrons, of which the internal structure is supposed to play no role. For such a plasma

[1] Much discussion has been devoted to this point, in particular with respect to the Poynting vector: v. G. H. Livens, Phil. Mag. **34**(1917)385; C. O. Hines, Canad. J. Phys. **30**(1952) 123; F. Bopp, Ann. Physik **11**(1963)35; E. M. Pugh and G. E. Pugh, Am. J. Phys. **35**(1967) 153; L. W. Zelby, Am. J. Phys. **35**(1967)1094; W. Shockley, Phys. Lett. **28A**(1968)185.

the macroscopic mass density is

$$\varrho(\boldsymbol{R}, t) = \sum_a m_a f_1^a(\boldsymbol{R}; t), \tag{119}$$

and the local barycentric velocity is defined by

$$\varrho(\boldsymbol{R}, t)\boldsymbol{v}(\boldsymbol{R}, t) = \sum_a \int m_a \boldsymbol{v}_1 f_1^a(\boldsymbol{R}, \boldsymbol{v}_1; t)\mathrm{d}\boldsymbol{v}_1 . \tag{120}$$

They satisfy the macroscopic law of mass conservation (59). The *momentum* law that follows from the atomic equation (I.50) reads (cf. (62)) in the case of the plasma

$$\frac{\partial \varrho \boldsymbol{v}}{\partial t} = -\nabla \cdot (\varrho \boldsymbol{v}\boldsymbol{v} + \mathbf{P}^{\mathrm{K}}) + \boldsymbol{F}^{\mathrm{L}}, \tag{121}$$

where the kinetic pressure is now (cf. (63))

$$\mathbf{P}^{\mathrm{K}} \equiv \sum_a \int m_a \hat{\boldsymbol{v}}_1 \hat{\boldsymbol{v}}_1 f_1^a(\boldsymbol{R}, \boldsymbol{v}_1; t)\mathrm{d}\boldsymbol{v}_1 \tag{122}$$

with the velocity fluctuation given by (61). Furthermore the force density is (cf. (72))

$$\boldsymbol{F}^{\mathrm{L}} = \varrho^{\mathrm{e}}\boldsymbol{E} + c^{-1}\boldsymbol{J} \wedge \boldsymbol{B} + \boldsymbol{F}^{\mathrm{C}}, \tag{123}$$

where the macroscopic charge and current densities are (18) or alternatively (19):

$$\varrho^{\mathrm{e}}(\boldsymbol{R}, t) = \sum_a e_a f_1^a(\boldsymbol{R}, t), \qquad \boldsymbol{J}(\boldsymbol{R}, t) = \sum_a \int e_a \boldsymbol{v}_1 f_1^a(\boldsymbol{R}, \boldsymbol{v}_1; t)\mathrm{d}\boldsymbol{v}_1 \tag{124}$$

and the correlation force density (cf. (74))

$$\boldsymbol{F}^{\mathrm{C}} = -\sum_{a,b} \int \left(\nabla_s \frac{e_a e_b}{4\pi s}\right) c_2^{ab}(\boldsymbol{R}, \boldsymbol{R}-\boldsymbol{s}; t)\mathrm{d}\boldsymbol{s}. \tag{125}$$

(The terms (73) and (75) are absent in (123) and (121) respectively because the internal structure of the charged particles has been ignored here.)

Let us now consider the correlation force $\boldsymbol{F}^{\mathrm{C}}$ for the special case of a plasma in which the correlation function vanishes rapidly if the interparticle distance becomes of macroscopic order. This is the case for plasmas without space charge and sufficiently near equilibrium, as a result of Debye shielding[1]. Then using the Irving–Kirkwood approximation (102) we may write (125)

[1] A. N. Kaufman, Phys. Fluids **6**(1963)1574, who gives a treatment similar to that of this subsection.

as a divergence:

$$\boldsymbol{F}^{\mathrm{C}} = -\nabla\cdot\mathbf{P}^{\mathrm{C}} \tag{126}$$

with the correlation pressure

$$\mathbf{P}^{\mathrm{C}} = -\sum_{a,b}\int\left(\boldsymbol{s}\nabla_s \frac{e_a e_b}{8\pi s}\right) c_2^{ab}(\boldsymbol{R}+\tfrac{1}{2}\boldsymbol{s}, \boldsymbol{R}-\tfrac{1}{2}\boldsymbol{s}; t)\mathrm{d}\boldsymbol{s}. \tag{127}$$

The momentum equation (121) may now be written as

$$\frac{\partial \varrho \boldsymbol{v}}{\partial t} = -\nabla\cdot(\varrho\boldsymbol{v}\boldsymbol{v}+\mathbf{P})+c^{-1}\boldsymbol{J}\wedge\boldsymbol{B}, \tag{128}$$

where the total pressure is the sum of the kinetic and the correlation pressure

$$\mathbf{P} = \mathbf{P}^{\mathrm{K}}+\mathbf{P}^{\mathrm{C}} \tag{129}$$

and where the last term of (128) is the Lorentz force in a neutral medium.

The *energy* law that follows for the plasma from the atomic equation (I.63) is (cf. (80))

$$\frac{\partial}{\partial t}(\tfrac{1}{2}\varrho v^2+\varrho u^{\mathrm{K}}) = -\nabla\cdot\{\boldsymbol{v}(\tfrac{1}{2}\varrho v^2+\varrho u^{\mathrm{K}})+\mathbf{P}^{\mathrm{K}}\cdot\boldsymbol{v}+\boldsymbol{J}_q^{\mathrm{K}}\}+\Psi^{\mathrm{L}}, \tag{130}$$

where the kinetic part of the internal energy density is given by (cf. (78))

$$\varrho u^{\mathrm{K}} = \sum_a \int \tfrac{1}{2}m_a \hat{v}_1^2 f_1^a(\boldsymbol{R}, \boldsymbol{v}_1; t)\mathrm{d}\boldsymbol{v}_1, \tag{131}$$

and the kinetic part of the heat flow by (cf. (81))

$$\boldsymbol{J}_q^{\mathrm{K}} = \sum_a \int \hat{\boldsymbol{v}}_1 \tfrac{1}{2}m_a \hat{v}_1^2 f_1^a(\boldsymbol{R}, \boldsymbol{v}_1; t)\mathrm{d}\boldsymbol{v}_1. \tag{132}$$

Furthermore the power density is here (cf. (88))

$$\Psi^{\mathrm{L}} = \boldsymbol{J}\cdot\boldsymbol{E}+\Psi^{\mathrm{C}} \tag{133}$$

with the current density $\boldsymbol{J}$ (124) and the correlation contribution (cf. (90))

$$\Psi^{\mathrm{C}} = -\sum_{a,b}\int\left(\boldsymbol{v}_1\cdot\nabla_s \frac{e_a e_b}{4\pi s}\right) c_2^{ab}(\boldsymbol{R}, \boldsymbol{v}_1, \boldsymbol{R}-\boldsymbol{s}; t)\mathrm{d}\boldsymbol{s}\,\mathrm{d}\boldsymbol{v}_1. \tag{134}$$

(The terms (89) and (91) are absent from (133) and (130) respectively because the charged particles have no internal structure.) For a sufficiently homogeneous plasma without space charge ($\varrho^{\mathrm{e}} = 0$) we apply the Irving–Kirkwood approximation (102) on the correlation power density (134):

$$\Psi^{\mathrm{C}} = -\sum_{a,b}\int \boldsymbol{v}_1\cdot\nabla_s \frac{e_a e_b}{4\pi s}\, c_2^{ab}(\boldsymbol{R}+\tfrac{1}{2}\boldsymbol{s},\, \boldsymbol{v}_1,\, \boldsymbol{R}-\tfrac{1}{2}\boldsymbol{s};\, t)\mathrm{d}\boldsymbol{s}\,\mathrm{d}\boldsymbol{v}_1$$

$$+\nabla\cdot\sum_{a,b}\int s\boldsymbol{v}_1\cdot\nabla_s \frac{e_a e_b}{8\pi s}\, c_2^{ab}(\boldsymbol{R}+\tfrac{1}{2}\boldsymbol{s},\, \boldsymbol{v}_1,\, \boldsymbol{R}-\tfrac{1}{2}\boldsymbol{s};\, t)\mathrm{d}\boldsymbol{s}\,\mathrm{d}\boldsymbol{v}_1\,. \tag{135}$$

The first term at the right-hand side may be transformed to the sum of a time derivative and a divergence, since as a consequence of the conservation of probability in fluxion space one has

$$\sum_{a,b}\frac{\partial}{\partial t}\int \frac{e_a e_b}{8\pi s}\, c_2^{ab}(\boldsymbol{R}+\tfrac{1}{2}\boldsymbol{s},\, \boldsymbol{R}-\tfrac{1}{2}\boldsymbol{s};\, t)\mathrm{d}\boldsymbol{s}$$

$$= -\nabla\cdot\sum_{a,b}\int (\boldsymbol{v}_1+\boldsymbol{v}_2)\frac{e_a e_b}{16\pi s}\, c_2^{ab}(\boldsymbol{R}+\tfrac{1}{2}\boldsymbol{s},\, \boldsymbol{v}_1,\, \boldsymbol{R}-\tfrac{1}{2}\boldsymbol{s},\, \boldsymbol{v}_2;\, t)\mathrm{d}\boldsymbol{s}\,\mathrm{d}\boldsymbol{v}_1\,\mathrm{d}\boldsymbol{v}_2$$

$$+\sum_{a,b}\int (\boldsymbol{v}_1-\boldsymbol{v}_2)\cdot\nabla_s \frac{e_a e_b}{8\pi s}\, c_2^{ab}(\boldsymbol{R}+\tfrac{1}{2}\boldsymbol{s},\, \boldsymbol{v}_1,\, \boldsymbol{R}-\tfrac{1}{2}\boldsymbol{s},\, \boldsymbol{v}_2;\, t)\mathrm{d}\boldsymbol{s}\,\mathrm{d}\boldsymbol{v}_1\,\mathrm{d}\boldsymbol{v}_2 \tag{136}$$

(where in the last term a partial integration has been performed). Owing to the symmetry of the correlation function the last term is equal to minus the first term at the right-hand side in (135). This allows us to write (135) in the form (cf. (110))

$$\Psi^{\mathrm{C}} = -\nabla\cdot(\boldsymbol{v}\varrho u^{\mathrm{C}}+\mathbf{P}^{\mathrm{C}}\cdot\boldsymbol{v}+\boldsymbol{J}_q^{\mathrm{C}})-\frac{\partial \varrho u^{\mathrm{C}}}{\partial t}\,, \tag{137}$$

where now the correlation part of the internal energy density is

$$\varrho u^{\mathrm{C}} = \sum_{a,b}\int \frac{e_a e_b}{8\pi s}\, c_2^{ab}(\boldsymbol{R}+\tfrac{1}{2}\boldsymbol{s},\, \boldsymbol{R}-\tfrac{1}{2}\boldsymbol{s};\, t)\mathrm{d}\boldsymbol{s}, \tag{138}$$

and the correlation part of the heat flow

$$\boldsymbol{J}_q^{\mathrm{C}} = \sum_{a,b}\int\left\{(\hat{\boldsymbol{v}}_1 - s\hat{\boldsymbol{v}}_1\cdot\nabla_s)\frac{e_a e_b}{8\pi s}\right\} c_2(\boldsymbol{R}+\tfrac{1}{2}\boldsymbol{s},\, \boldsymbol{v}_1,\, \boldsymbol{R}-\tfrac{1}{2}\boldsymbol{s};\, t)\mathrm{d}\boldsymbol{s}\,\mathrm{d}\boldsymbol{v}_1\,. \tag{139}$$

The energy equation follows if (133) with (137) is substituted into (130):

$$\frac{\partial}{\partial t}(\tfrac{1}{2}\varrho v^2+\varrho u) = -\nabla\cdot\{\boldsymbol{v}(\tfrac{1}{2}\varrho v^2+\varrho u)+\mathbf{P}\cdot\boldsymbol{v}+\boldsymbol{J}_q\}+\boldsymbol{J}\cdot\boldsymbol{E}, \tag{140}$$

where the total internal specific energy

$$u = u^{\mathrm{K}}+u^{\mathrm{C}} \tag{141}$$

is the sum of a kinetic part, given in (131), and a correlation part, given in

(138). Furthermore the total pressure **P** has been given in (129) and the total heat flow

$$J_q = J_q^{\mathrm{K}} + J_q^{\mathrm{C}} \tag{142}$$

is again the sum of a kinetic part (132) and a correlation part (139). The source term in (140) is the well-known electric power density.

One may write both the momentum law (128) and the energy law (140) in the form of conservation laws, if one uses the identities

$$c^{-1}\boldsymbol{J}\wedge\boldsymbol{B} = \nabla\cdot\{\boldsymbol{EE}+\boldsymbol{BB}-\tfrac{1}{2}(E^2+B^2)\mathrm{U}\}-c^{-1}\frac{\partial}{\partial t}(\boldsymbol{E}\wedge\boldsymbol{B}) \tag{143}$$

and

$$\boldsymbol{J}\cdot\boldsymbol{E} = -c\nabla\cdot(\boldsymbol{E}\wedge\boldsymbol{B})-\frac{1}{2}\frac{\partial}{\partial t}(E^2+B^2) \tag{144}$$

respectively, which both follow directly from Maxwell's equations for a neutral unpolarized medium ($\varrho^{\mathrm{e}} = 0$, $\boldsymbol{P} = 0$, $\boldsymbol{M} = 0$). One obtains in this way

$$\frac{\partial}{\partial t}(\varrho\boldsymbol{v}+c^{-1}\boldsymbol{E}\wedge\boldsymbol{B}) = -\nabla\cdot\{\varrho\boldsymbol{vv}+\mathbf{P}-\boldsymbol{EE}-\boldsymbol{BB}+\tfrac{1}{2}(E^2+B^2)\mathrm{U}\} \tag{145}$$

for the conservation of total momentum, and

$$\frac{\partial}{\partial t}(\tfrac{1}{2}\varrho v^2+\varrho u+\tfrac{1}{2}E^2+\tfrac{1}{2}B^2) = -\nabla\cdot\{\boldsymbol{v}(\tfrac{1}{2}\varrho v^2+\varrho u)+\mathbf{P}\cdot\boldsymbol{v}+\boldsymbol{J}_q+c\boldsymbol{E}\wedge\boldsymbol{B}\} \tag{146}$$

for the conservation of total energy.

### h. *Crystalline solids*

Up to subsection *e* the derivations were independent of the nature of the systems, provided that these were sufficiently uniform. The latter restriction was made in order to write the short range terms in the momentum and energy equations in convenient form. In subsections *f* and *g*, where the correlation terms were discussed, it was necessary to specify the system further: we confined ourselves to fluids of neutral atoms (and amorphous or polycrystalline solids) and to neutral plasmas, for which the correlation function is of short range. Then the Irving–Kirkwood approximation could be employed.

In this subsection we shall study systems with correlation functions of arbitrary range, such as crystalline solids. Even then it is possible to transform the correlation terms to a divergence, or to a divergence and a time derivative. This may be done with the help of the following artifice[1]. Let us write the following identity for the correlation function of a one-component system[2]

$$c_2(\boldsymbol{R}, 1, \boldsymbol{R}-\boldsymbol{s}, 2; t)-c_2(\boldsymbol{R}+\tfrac{1}{2}\boldsymbol{s}, 1, \boldsymbol{R}-\tfrac{1}{2}\boldsymbol{s}, 2; t)$$
$$= -\int_{-1}^{0} \frac{\partial}{\partial\lambda} c_2\{\boldsymbol{R}+\tfrac{1}{2}(\lambda+1)\boldsymbol{s}, 1, \boldsymbol{R}+\tfrac{1}{2}(\lambda-1)\boldsymbol{s}, 2; t\}\mathrm{d}\lambda. \quad (147)$$

Since the correlation function in the integrand depends on position coordinates which are combinations of $\boldsymbol{R}+\frac{1}{2}\lambda\boldsymbol{s}$ and $\frac{1}{2}\boldsymbol{s}$ one may replace the operator $\partial/\partial\lambda$ by $\frac{1}{2}\boldsymbol{s}\cdot\nabla$. In this way one obtains for (147):

$$c_2(\boldsymbol{R}, 1, \boldsymbol{R}-\boldsymbol{s}, 2; t)$$
$$= c_2(\boldsymbol{R}+\tfrac{1}{2}\boldsymbol{s}, 1, \boldsymbol{R}-\tfrac{1}{2}\boldsymbol{s}, 2; t)-\tfrac{1}{2}\boldsymbol{s}\cdot\nabla c_2^+(\boldsymbol{R}+\tfrac{1}{2}\boldsymbol{s}, 1, \boldsymbol{R}-\tfrac{1}{2}\boldsymbol{s}, 2; t), \quad (148)$$

with the 'mean correlation function':

$$c_2^+(\boldsymbol{R}+\tfrac{1}{2}\boldsymbol{s}, 1, \boldsymbol{R}-\tfrac{1}{2}\boldsymbol{s}, 2; t) \equiv \int_{-1}^{0} c_2\{\boldsymbol{R}+\tfrac{1}{2}(\lambda+1)\boldsymbol{s}, 1, \boldsymbol{R}+\tfrac{1}{2}(\lambda-1)\boldsymbol{s}, 2; t\}\mathrm{d}\lambda. \quad (149)$$

(If $c_2$ vanishes rapidly with increasing interatomic distance $|\boldsymbol{s}|$ so that a correlation length exists, and if moreover $c_2$ changes slowly if both positions are shifted over a distance of the order of the correlation length one may consider the integrand in (149) as a constant. Choosing its value at $\lambda = 0$ the 'mean correlation function' $c_2^+$ reduces then to the ordinary correlation function $c_2$. In this case (148) reduces to the Irving–Kirkwood approximation (102).)

Using (148) which has the same form as (102) we may find expressions for the correlation terms in the momentum and energy laws for systems with long range correlations. These are in form very similar to the expressions valid for fluids. In fact for the case of long range order the momentum balance becomes

$$\frac{\partial \varrho \boldsymbol{v}}{\partial t} = -\nabla\cdot(\varrho\boldsymbol{v}\boldsymbol{v}+\mathsf{P})+\varrho^{e}\boldsymbol{E}+c^{-1}\boldsymbol{J}\wedge\boldsymbol{B}+(\nabla\boldsymbol{E})\cdot\boldsymbol{P}+(\nabla\boldsymbol{B})\cdot\boldsymbol{M}$$
$$+c^{-1}\frac{\partial}{\partial t}(\boldsymbol{P}\wedge\boldsymbol{B})+c^{-1}\nabla\cdot(\boldsymbol{v}\boldsymbol{P}\wedge\boldsymbol{B}) \quad (150)$$

[1] J. H. Irving and J. G. Kirkwood, op. cit.

[2] If the system is a mixture one should add indices $a$ and $b$ to the distribution functions and the atomic parameters such as the charges and the masses. Furthermore summations over $a$ and $b$ are then to be added.

with a pressure tensor $\mathbf{P}$ that consists of four contributions (cf. (107)), viz $\mathbf{P}^{\mathrm{K}}$ (63), $\mathbf{P}^{\mathrm{F}}$ (73), $\mathbf{P}^{\mathrm{S}}$ (94) and the correlation pressure $\mathbf{P}^{\mathrm{C}}$ given by (cf. 104)):

$$\mathbf{P}^{\mathrm{C}} \equiv -\int \sum_{n,m=0}^{\infty} (-1)^m \left( s \nabla_s \overline{\boldsymbol{\mu}}_1^{(n)} : \nabla_s^n \overline{\boldsymbol{\mu}}_2^{(m)} : \nabla_s^m \frac{1}{8\pi s} \right) c_2^+(\boldsymbol{R}+\tfrac{1}{2}\boldsymbol{s}, 1, \boldsymbol{R}-\tfrac{1}{2}\boldsymbol{s}, 2; t) \mathrm{d}\boldsymbol{s}\, \mathrm{d}1\, \mathrm{d}2. \quad (151)$$

Likewise the energy balance equation for the case of long range order gets the form

$$\frac{\partial}{\partial t}(\tfrac{1}{2}\varrho \boldsymbol{v}^2 + \varrho u) = -\nabla \cdot \{\boldsymbol{v}(\tfrac{1}{2}\varrho \boldsymbol{v}^2 + \varrho u) + \mathbf{P}\cdot \boldsymbol{v} + \boldsymbol{J}_q\} + \boldsymbol{J}\cdot\boldsymbol{E} + \frac{\partial \boldsymbol{P}}{\partial t}\cdot\boldsymbol{E} + \nabla\cdot(\boldsymbol{v}\boldsymbol{P}\cdot\boldsymbol{E}) - \boldsymbol{M}\cdot\frac{\partial \boldsymbol{B}}{\partial t}, \quad (152)$$

with a specific internal energy $u$ that consists of three parts (cf. (115)), viz $u^{\mathrm{K}}$ (78), $u^{\mathrm{S}}$ (97) and the correlation contribution $u^{\mathrm{C}}$ given by (cf. (111))

$$\varrho u^{\mathrm{C}} \equiv \int \sum_{n,m=0}^{\infty} (-1)^m \left( \overline{\boldsymbol{\mu}}_1^{(n)} : \nabla_s^n \overline{\boldsymbol{\mu}}_2^{(m)} : \nabla_s^m \frac{1}{8\pi s} \right) c_2(\boldsymbol{R}+\tfrac{1}{2}\boldsymbol{s}, 1, \boldsymbol{R}-\tfrac{1}{2}\boldsymbol{s}, 2; t) \mathrm{d}\boldsymbol{s}\, \mathrm{d}1\, \mathrm{d}2. \quad (153)$$

Furthermore the heat flow consists of four parts (cf. (116)), viz $\boldsymbol{J}_q^{\mathrm{K}}$ (81), $\boldsymbol{J}_q^{\mathrm{F}}$ (89), $\boldsymbol{J}_q^{\mathrm{S}}$ (96), (100) and the correlation contribution given by (cf. (112))

$$\boldsymbol{J}_q^{\mathrm{C}} \equiv -\int s \sum_{n,m=0}^{\infty} (-1)^m \left\{ (\hat{\boldsymbol{v}}_1 \cdot \nabla_s \overline{\boldsymbol{\mu}}_1^{(n)} : \nabla_s^n + \dot{\underline{\boldsymbol{\mu}}}_1^{(n)} : \nabla_s^n) \overline{\boldsymbol{\mu}}_2^{(m)} : \nabla_s^m \frac{1}{8\pi s} \right\} c_2^+(\boldsymbol{R}+\tfrac{1}{2}\boldsymbol{s}, 1, \boldsymbol{R}-\tfrac{1}{2}\boldsymbol{s}, 2; t) \mathrm{d}\boldsymbol{s}\, \mathrm{d}1\, \mathrm{d}2$$
$$+ \int \hat{\boldsymbol{v}}_1 \sum_{n,m=0}^{\infty} (-1)^m \left( \overline{\boldsymbol{\mu}}_1^{(n)} : \nabla_s^n \overline{\boldsymbol{\mu}}_2^{(m)} : \nabla_s^m \frac{1}{8\pi s} \right) c_2(\boldsymbol{R}+\tfrac{1}{2}\boldsymbol{s}, 1, \boldsymbol{R}-\tfrac{1}{2}\boldsymbol{s}, 2; t) \mathrm{d}\boldsymbol{s}\, \mathrm{d}1\, \mathrm{d}2, \quad (154)$$

where only in the first part the 'mean correlation function' occurs.

Finally it should be remarked that the procedure employed to write the correlation terms in the form of a divergence (or a divergence and a time derivative) is not unique. However, the statistical expressions obtained have been preferred because of their formal resemblance to the corresponding expressions for fluid systems.

i. *Galilean invariance*

In the preceding non-relativistic balance equations of momentum and energy have been obtained. We shall investigate whether they are indeed covariant with respect to the Galilei group. The rotational invariance of all equations is manifest since they have been formulated in tensor notation. The transformation character with respect to a pure Galilean transformation (22) needs to be considered in some detail.

Let us first show that the conservation law of mass (59) possesses Galilean covariance. Indeed the mass density defined in (56) is an invariant

$$\varrho'(\boldsymbol{R}', t') = \varrho(\boldsymbol{R}, t), \tag{155}$$

as follows from the invariance of the one-point distribution function (cf. (24)):

$$f_1'(1'; t') = f_1(1; t), \tag{156}$$

where 1 stands for the position, velocity etc. of atom 1; primes indicate the corresponding transformed quantities. Furthermore it follows from (58) with (155) and (156) that

$$\boldsymbol{v}'(\boldsymbol{R}', t') = \boldsymbol{v}(\boldsymbol{R}, t) + \boldsymbol{V}. \tag{157}$$

Now from (23), (155) and (157) one proves immediately that (59) is a covariant equation, i.e. also valid with primes.

We now turn to the momentum balance (105–106) for fluid systems of neutral atoms. With the help of the mass conservation law (59) it may be brought into the form

$$\varrho \frac{\mathrm{d}\boldsymbol{v}}{\mathrm{d}t} = -\boldsymbol{\nabla}\cdot\mathbf{P} + (\boldsymbol{\nabla}\boldsymbol{E})\cdot\boldsymbol{P} + (\boldsymbol{\nabla}\boldsymbol{B})\cdot\boldsymbol{M} + c^{-1}\varrho \frac{\mathrm{d}}{\mathrm{d}t}(v\boldsymbol{P} \wedge \boldsymbol{B}), \tag{158}$$

where $v \equiv \varrho^{-1}$ is the specific volume and

$$\frac{\mathrm{d}}{\mathrm{d}t} \equiv \frac{\partial}{\partial t} + \boldsymbol{v}\cdot\boldsymbol{\nabla}, \tag{159}$$

the material time derivative. By inspection of the expression (107) with (63), (73), (94) and (104) it follows from (27), (61) and the invariance of the distribution function expressed by (156) and

$$\begin{aligned} f_2'(1', 2'; t') &= f_2(1, 2; t), \\ c_2'(1', 2'; t') &= c_2(1, 2; t), \end{aligned} \tag{160}$$

that the pressure **P** is invariant with respect to pure Galilean transformations:

$$\mathbf{P}'(\boldsymbol{R}', t') = \mathbf{P}(\boldsymbol{R}, t). \tag{161}$$

Hence, according to (23), the term $\nabla \cdot \mathbf{P}$ in the momentum equation (158) is invariant with respect to pure Galilean transformations. The other terms in (158), viz $\varrho \mathrm{d}\boldsymbol{v}/\mathrm{d}t$, $(\nabla \boldsymbol{E})\cdot \boldsymbol{P} + (\nabla \boldsymbol{B})\cdot \boldsymbol{M}$ and $c^{-1}\varrho(\mathrm{d}/\mathrm{d}t)(\boldsymbol{v}\boldsymbol{P} \wedge \boldsymbol{B})$ are also, separately, invariant under pure Galilean transformations as follows from (23), (26), (27), (155), (157) and (159), where terms of order $c^{-2}$ must of course be discarded. In this way the Galilean covariance of the momentum balance has been proved for the case of fluids systems of neutral atoms. The momentum balance for neutral plasmas and for systems with long range correlations (subsections *g* and *h*) also possess Galilean covariance as follows by a similar reasoning from the transformation formulae.

Let us discuss now the Galilean covariance of the energy balance equation (113) for fluids of neutral atoms. With the help of the mass conservation law (59) and the material time derivative (159) it may be written as

$$\varrho \frac{\mathrm{d}}{\mathrm{d}t}(\tfrac{1}{2}v^2 + u) = -\nabla \cdot (\mathbf{P} \cdot \boldsymbol{v} + \boldsymbol{J}_q)$$
$$+ \varrho \frac{\mathrm{d}(\boldsymbol{v}\boldsymbol{P})}{\mathrm{d}t} \cdot \boldsymbol{E} - \boldsymbol{M} \cdot \frac{\mathrm{d}\boldsymbol{B}}{\mathrm{d}t} + \boldsymbol{v} \cdot \{(\nabla \boldsymbol{E}) \cdot \boldsymbol{P} + (\nabla \boldsymbol{B}) \cdot \boldsymbol{M}\}. \tag{162}$$

If the first term is rewritten with the help of the (Galilei covariant) momentum equation (158) we obtain

$$\varrho \frac{\mathrm{d}u}{\mathrm{d}t} = -\nabla \cdot \boldsymbol{J}_q - \tilde{\mathbf{P}} : (\nabla \boldsymbol{v})$$
$$+ \varrho \frac{\mathrm{d}(\boldsymbol{v}\boldsymbol{P})}{\mathrm{d}t} \cdot (\boldsymbol{E} + c^{-1}\boldsymbol{v} \wedge \boldsymbol{B}) - (\boldsymbol{M} + c^{-1}\boldsymbol{v} \wedge \boldsymbol{P}) \cdot \frac{\mathrm{d}\boldsymbol{B}}{\mathrm{d}t}, \tag{163}$$

where $\tilde{\mathbf{P}}$ is the transposed of the tensor **P** (or $\tilde{P}_{ij} = P_{ji}$) and the double dot indicates a double contraction of two tensors ($\mathbf{A} : \mathbf{B} \equiv \sum_{i,j} A_{ij} B_{ji}$). With (61), (155), (156), (157) and (160) it follows by inspection that the contributions (78), (97) and (111) to the specific internal energy *u* (115) are each Galilei invariant, so that

$$u'(\boldsymbol{R}', t') = u(\boldsymbol{R}, t). \tag{164}$$

Consequently with (23), (155) and (159) it follows that the left-hand side of (163) is invariant. Furthermore with the help of (27), (61), (156), (157) and (160) it is seen that the contributions (81), (89), (96), (100) and (112) to the heat flow $\boldsymbol{J}_q$ (116) are each Galilei invariant, so that

$$\boldsymbol{J}'_q(\boldsymbol{R}', t') = \boldsymbol{J}_q(\boldsymbol{R}, t). \tag{165}$$

Hence the invariance of the first term at the right-hand side of (163) follows if (23) is used. The invariance of the second term at the right-hand side of (163) follows from (23), (157) and (161). Finally if one uses (23), (26), (27), (155) and (157) one may prove the invariance of each of the last two terms of (163). In this way the Galilean covariance of equation (163) has been established, and hence that of the energy equation (162). We have chosen to give this proof via equation (163) because, in contrast with (162), all its terms are separately invariant. For (neutral) plasmas and for systems with long range correlations the Galilean covariance of the energy equations is proved in an analogous way.

## 6 *The angular momentum equations*

### a. *The inner angular momentum balance*

On a par with the momentum and energy equations derived in the preceding section, macroscopic angular momentum laws will now be obtained by averaging the corresponding atomic equations. The macroscopic angular momentum density is defined as the average:

$$\boldsymbol{S}(\boldsymbol{R}, t) = \langle \sum_k \bar{\boldsymbol{s}}_k \, \delta(\boldsymbol{R}_k - \boldsymbol{R}) \rangle, \tag{166}$$

where $\bar{\boldsymbol{s}}_k$ is the inner angular momentum (I.68) of atom $k$. The time derivative of $\boldsymbol{S}(\boldsymbol{R}, t)$ is found with the help of the lemma (2) and (I.76):

$$\frac{\partial \boldsymbol{S}}{\partial t} = -\nabla \cdot \langle \sum_k \boldsymbol{v}_k \bar{\boldsymbol{s}}_k \, \delta(\boldsymbol{R}_k - \boldsymbol{R}) \rangle + \langle \sum_k (\boldsymbol{d}_k^{\mathrm{L}} + \boldsymbol{d}_k^{\mathrm{S}}) \delta(\boldsymbol{R}_k - \boldsymbol{R}) \rangle. \tag{167}$$

Introducing the velocity fluctuation $\hat{\boldsymbol{v}}_k$ (61) we get

$$\frac{\partial \boldsymbol{S}}{\partial t} = -\nabla \cdot (\boldsymbol{v}\boldsymbol{S} + \mathbf{J}_s^{\mathrm{K}}) + \boldsymbol{D}^{\mathrm{L}} + \boldsymbol{D}^{\mathrm{S}*} \tag{168}$$

with the kinetic flow of inner angular momentum

$$\mathbf{J}_s^{\mathrm{K}} = \int \hat{\boldsymbol{v}}_1 \bar{\boldsymbol{s}}_1 \, f_1(\boldsymbol{R}, \boldsymbol{v}_1, \bar{\boldsymbol{s}}_1; t) \mathrm{d}\boldsymbol{v}_1 \, \mathrm{d}\bar{\boldsymbol{s}}_1 \tag{169}$$

and the abbreviations $\boldsymbol{D}^{\mathrm{L},\mathrm{S}*} = \langle \sum_k \boldsymbol{d}_k^{\mathrm{L},\mathrm{S}} \delta(\boldsymbol{R}_k - \boldsymbol{R}) \rangle$ with $\boldsymbol{d}_k^{\mathrm{L},\mathrm{S}}$ given by (I.77–78).

Let us first consider the external field part $\boldsymbol{D}_{\mathrm{e}}^{\mathrm{L}}$ of $\boldsymbol{D}^{\mathrm{L}}$. It may be written with the help of the polarizations (65), if one uses the vector identity $\boldsymbol{a}\wedge(\boldsymbol{b}\wedge\boldsymbol{c})+\text{cycl.}=0$:

$$\boldsymbol{D}_{\mathrm{e}}^{\mathrm{L}} = \boldsymbol{P}\wedge\boldsymbol{E}_{\mathrm{e}}+\boldsymbol{M}\wedge\boldsymbol{B}_{\mathrm{e}}+c^{-1}\boldsymbol{v}\wedge(\boldsymbol{P}\wedge\boldsymbol{B}_{\mathrm{e}}) + c^{-1}\int\hat{\boldsymbol{v}}_1\wedge(\overline{\boldsymbol{\mu}}_1^{(1)}\wedge\boldsymbol{B}_{\mathrm{e}})f_1(\boldsymbol{R},\boldsymbol{v}_1,\overline{\boldsymbol{\mu}}_1^{(1)};t)\mathrm{d}\boldsymbol{v}_1\,\mathrm{d}\overline{\boldsymbol{\mu}}_1^{(1)}. \tag{170}$$

The part of the long range term in (168) that is due to the interatomic interactions is

$$\boldsymbol{D}^{\mathrm{L}}-\boldsymbol{D}_{\mathrm{e}}^{\mathrm{L}} = \int\left(\sum_{n,m=0}^{\infty} n\nabla_1\wedge\overline{\boldsymbol{\mu}}_1^{(n)}:\nabla_1^{n-1}\overline{\boldsymbol{\mu}}_2^{(m)}:\nabla_2^{m}\frac{1}{4\pi|\boldsymbol{R}_1-\boldsymbol{R}_2|}\right) \delta(\boldsymbol{R}-\boldsymbol{R}_1)f_2(\boldsymbol{R}_1,1,\boldsymbol{R}_2,2;t)\mathrm{d}\boldsymbol{R}_1\,\mathrm{d}\boldsymbol{R}_2\,\mathrm{d}1\,\mathrm{d}2, \tag{171}$$

where 1 and 2 indicate all electric multipole moments. We split this expression with the help of (7). In the uncorrelated part we introduce the macroscopic charge and polarization densities (64) and (65), omitting (just as in § 5*c*) higher multipole moments. In this way (171) becomes

$$\boldsymbol{D}^{\mathrm{L}}-\boldsymbol{D}_{\mathrm{e}}^{\mathrm{L}} = -\int\boldsymbol{P}(\boldsymbol{R},t)\wedge\nabla\{\varrho^{\mathrm{e}}(\boldsymbol{R}',t)+\boldsymbol{P}(\boldsymbol{R}',t)\cdot\nabla'\}\frac{1}{4\pi|\boldsymbol{R}-\boldsymbol{R}'|}\mathrm{d}\boldsymbol{R}' + \int\left(\sum_{n,m=0}^{\infty} n\nabla\wedge\overline{\boldsymbol{\mu}}_1^{(n)}:\nabla^{n-1}\overline{\boldsymbol{\mu}}_2^{(m)}:\nabla'^{m}\frac{1}{4\pi|\boldsymbol{R}-\boldsymbol{R}'|}\right) c_2(\boldsymbol{R},1,\boldsymbol{R}',2;t)\mathrm{d}\boldsymbol{R}'\,\mathrm{d}1\,\mathrm{d}2. \tag{172}$$

The total long range moment density which is given by the sum of (170) and (171) may be written with the help of (71) as

$$\boldsymbol{D}^{\mathrm{L}} = \boldsymbol{P}\wedge\boldsymbol{E}+\boldsymbol{M}\wedge\boldsymbol{B}+c^{-1}\boldsymbol{v}\wedge(\boldsymbol{P}\wedge\boldsymbol{B})+\boldsymbol{D}^{\mathrm{F}}+\boldsymbol{D}^{\mathrm{C}*}, \tag{173}$$

where the last two contributions are given by

$$\boldsymbol{D}^{\mathrm{F}} \equiv c^{-1}\int\hat{\boldsymbol{v}}_1\wedge(\overline{\boldsymbol{\mu}}_1^{(1)}\wedge\boldsymbol{B})f_1(\boldsymbol{R},\boldsymbol{v}_1,\overline{\boldsymbol{\mu}}_1^{(1)};t)\mathrm{d}\boldsymbol{v}_1\,\mathrm{d}\overline{\boldsymbol{\mu}}_1^{(1)}, \tag{174}$$

$$\boldsymbol{D}^{\mathrm{C}*} \equiv \int\boldsymbol{d}^{\mathrm{C}}(\boldsymbol{s},1,2)c_2(\boldsymbol{R},1,\boldsymbol{R}-\boldsymbol{s},2;t)\mathrm{d}\boldsymbol{s}\,\mathrm{d}1\,\mathrm{d}2, \tag{175}$$

with the abbreviation

$$\boldsymbol{d}^{\mathrm{C}}(\boldsymbol{s},1,2) \equiv \sum_{n,m=0}^{\infty}(-1)^m n\nabla_{\boldsymbol{s}}\wedge\overline{\boldsymbol{\mu}}_1^{(n)}:\nabla_{\boldsymbol{s}}^{n-1}\overline{\boldsymbol{\mu}}_2^{(m)}:\nabla_{\boldsymbol{s}}^{m}\frac{1}{4\pi s}. \tag{176}$$

Finally the short range moment in (168) reads

$$\boldsymbol{D}^{\mathrm{S}*} = \int \boldsymbol{d}^{\mathrm{S}}(\boldsymbol{s}, 1, 2) f_2(\boldsymbol{R}, 1, \boldsymbol{R}-\boldsymbol{s}, 2; t)\mathrm{d}\boldsymbol{s}\,\mathrm{d}1\,\mathrm{d}2 \tag{177}$$

with the abbreviation

$$\boldsymbol{d}^{\mathrm{S}}(\boldsymbol{s}, 1, 2) \equiv \int \left\{ -\sum_{i,j} \boldsymbol{r}_{1i} \wedge \nabla_{\boldsymbol{s}} \frac{e_i e_j}{4\pi|\boldsymbol{s}+\boldsymbol{r}_{1i}-\boldsymbol{r}_{2j}|} - \boldsymbol{d}^{\mathrm{C}}(\boldsymbol{s}, 1, 2) \right\} f_2(\boldsymbol{R}, 1, \boldsymbol{R}-\boldsymbol{s}, 2; t)\mathrm{d}\boldsymbol{s}\,\mathrm{d}1\,\mathrm{d}2. \tag{178}$$

The integrand in (177) vanishes if the atoms are outside each other. Since the two-point distribution function remains practically unchanged if both $\boldsymbol{R}$ and $\boldsymbol{R}-\boldsymbol{s}$ are shifted over a distance of the order of an atomic diameter, one may write the short range moment (177) with the help of (92) as

$$\boldsymbol{D}^{\mathrm{S}*} = -\nabla \cdot \mathbf{J}_s^{\mathrm{S}} + \boldsymbol{D}^{\mathrm{S}}, \tag{179}$$

with the short range contribution to the inner angular momentum flow

$$\mathbf{J}_s^{\mathrm{S}} \equiv \int \tfrac{1}{2}\boldsymbol{s}\boldsymbol{d}^{\mathrm{S}}(\boldsymbol{s}, 1, 2) f_2(\boldsymbol{R}+\tfrac{1}{2}\boldsymbol{s}, 1, \boldsymbol{R}-\tfrac{1}{2}\boldsymbol{s}, 2; t)\mathrm{d}\boldsymbol{s}\,\mathrm{d}1\,\mathrm{d}2, \tag{180}$$

and the source term:

$$\boldsymbol{D}^{\mathrm{S}} \equiv \int \boldsymbol{d}^{\mathrm{S}}(\boldsymbol{s}, 1, 2) f_2(\boldsymbol{R}+\tfrac{1}{2}\boldsymbol{s}, 1, \boldsymbol{R}-\tfrac{1}{2}\boldsymbol{s}, 2; t)\mathrm{d}\boldsymbol{s}\,\mathrm{d}1\,\mathrm{d}2. \tag{181}$$

The correlation contribution $\boldsymbol{D}^{\mathrm{C}*}$ will be specified for particular systems in the next subsections.

b. *Fluid systems*

For fluid systems of neutral atoms (and amorphous or polycrystalline solids) the correlation function has usually short range. In that case it may be expanded as in (102). In that way the correlation contribution $\boldsymbol{D}^{\mathrm{C}*}$ (175) gets the form

$$\boldsymbol{D}^{\mathrm{C}*} = -\nabla \cdot \mathbf{J}_s^{\mathrm{C}} + \boldsymbol{D}^{\mathrm{C}} \tag{182}$$

with the correlation contribution to the inner angular momentum flow:

$$\mathbf{J}_s^{\mathrm{C}} \equiv \int \tfrac{1}{2}\boldsymbol{s}\boldsymbol{d}^{\mathrm{C}}(\boldsymbol{s}, 1, 2) c_2(\boldsymbol{R}+\tfrac{1}{2}\boldsymbol{s}, 1, \boldsymbol{R}-\tfrac{1}{2}\boldsymbol{s}, 2; t)\mathrm{d}\boldsymbol{s}\,\mathrm{d}1\,\mathrm{d}2 \tag{183}$$

and the source term

$$\boldsymbol{D}^{\mathrm{C}} \equiv \int \boldsymbol{d}^{\mathrm{C}}(\boldsymbol{s}, 1, 2) c_2(\boldsymbol{R}+\tfrac{1}{2}\boldsymbol{s}, 1, \boldsymbol{R}-\tfrac{1}{2}\boldsymbol{s}, 2; t) \mathrm{d}\boldsymbol{s}\, \mathrm{d}1\, \mathrm{d}2, \tag{184}$$

where $\boldsymbol{d}^{\mathrm{C}}(\boldsymbol{s}, 1, 2)$ has been given by (176) (with the atomic charges $\bar{\mu}_1^{(0)} = \bar{\mu}_2^{(0)} = 0$).

The balance equation (168) of the inner angular momentum becomes, for fluid systems of neutral atoms, upon insertion of (173) and (179):

$$\frac{\partial \boldsymbol{S}}{\partial t} = -\nabla\cdot(\boldsymbol{v}\boldsymbol{S}+\mathsf{J}_s)+\boldsymbol{D}_s+\boldsymbol{P}\wedge\boldsymbol{E}+\boldsymbol{M}\wedge\boldsymbol{B}+c^{-1}\boldsymbol{v}\wedge(\boldsymbol{P}\wedge\boldsymbol{B}). \tag{185}$$

The conduction flow of inner angular momentum consists of three parts:

$$\mathsf{J}_s = \mathsf{J}_s^{\mathrm{K}}+\mathsf{J}_s^{\mathrm{S}}+\mathsf{J}_s^{\mathrm{C}}, \tag{186}$$

where the various contributions have been given in (169), (180) and (183) with (176) and (178) (with the atomic charges $\bar{\mu}_1^{(0)} = \bar{\mu}_2^{(0)} = 0$). Furthermore the source term contains a material part

$$\boldsymbol{D}_s = \boldsymbol{D}^{\mathrm{F}}+\boldsymbol{D}^{\mathrm{S}}+\boldsymbol{D}^{\mathrm{C}}, \tag{187}$$

where the three contributions have been given in (174), (181) and (184) with (176) and (178) (again with $\bar{\mu}_1^{(0)} = \bar{\mu}_2^{(0)} = 0$). The other source terms are the torque densities which the Maxwell fields exert on the polarization densities in the moving fluid.

The source terms with the Maxwell fields may be written in a simpler form if 'rest frame quantities' are introduced. The rest frame (denoted by primes) is related to the observer's frame by a pure Galilean transformation (22) with transformation velocity $\boldsymbol{V} = -\boldsymbol{v}$, such that $\boldsymbol{v}' = 0$ (cf. (157)). Then the Maxwell fields and polarizations transform according to (26) and (27) (up to order $c^{-1}$):

$$\begin{aligned} \boldsymbol{E}' &= \boldsymbol{E}+c^{-1}\boldsymbol{v}\wedge\boldsymbol{B}, \qquad & \boldsymbol{B}' &= \boldsymbol{B}-c^{-1}\boldsymbol{v}\wedge\boldsymbol{E}, \\ \boldsymbol{P}' &= \boldsymbol{P}, & \boldsymbol{M}' &= \boldsymbol{M}+c^{-1}\boldsymbol{v}\wedge\boldsymbol{P}. \end{aligned} \tag{188}$$

With the help of these formulae we get for the source term (up to order $c^{-1}$)

$$\boldsymbol{P}\wedge\boldsymbol{E}+\boldsymbol{M}\wedge\boldsymbol{B}+c^{-1}\boldsymbol{v}\wedge(\boldsymbol{P}\wedge\boldsymbol{B}) = \boldsymbol{P}'\wedge\boldsymbol{E}'+\boldsymbol{M}'\wedge\boldsymbol{B}'. \tag{189}$$

In fluid systems quite often the rest frame polarizations $\boldsymbol{P}'$ and $\boldsymbol{M}'$ are parallel to the rest frame fields $\boldsymbol{E}'$ and $\boldsymbol{B}'$ respectively. In that case the field source terms (189) vanish.

The equation (185) is a balance equation: the inner angular momentum is not conserved. A conserved quantity is obtained if the orbital angular momentum

$$\boldsymbol{L}(\boldsymbol{R}, t) \equiv \boldsymbol{R} \wedge (\varrho \boldsymbol{v} + c^{-1} \boldsymbol{E} \wedge \boldsymbol{B}) \tag{190}$$

is added to it. The time derivative of the latter quantity may be obtained with the help of (109):

$$\frac{\partial \boldsymbol{L}}{\partial t} = -\boldsymbol{\nabla}\cdot(\boldsymbol{v}\boldsymbol{L} + \mathbf{J}_l) + \boldsymbol{P}_{\mathrm{A}} - \boldsymbol{D} \wedge \boldsymbol{E} - \boldsymbol{B} \wedge \boldsymbol{H} - c^{-1} \boldsymbol{v} \wedge (\boldsymbol{P} \wedge \boldsymbol{B}). \tag{191}$$

Here the conduction flow of orbital angular momentum is

$$\mathbf{J}_l = -\mathbf{P} \wedge \boldsymbol{R} - \boldsymbol{D}(\boldsymbol{R} \wedge \boldsymbol{E}) - \boldsymbol{B}(\boldsymbol{R} \wedge \boldsymbol{H}) - c^{-1} \boldsymbol{v}\boldsymbol{R} \wedge (\boldsymbol{D} \wedge \boldsymbol{B}) + \boldsymbol{\epsilon}\cdot\boldsymbol{R}(\tfrac{1}{2}E^2 + \tfrac{1}{2}B^2 - \boldsymbol{M}\cdot\boldsymbol{B}). \tag{192}$$

Furthermore $\boldsymbol{P}_{\mathrm{A}} \equiv \boldsymbol{\epsilon} : \tilde{\mathbf{P}}$ is the antisymmetric part of the pressure tensor $\mathbf{P}$ ($\boldsymbol{\epsilon}$ is the Levi-Civita tensor with components $\varepsilon^{ijk}$ so that the components of $\boldsymbol{P}_{\mathrm{A}}$ are $P_{\mathrm{A}}^i = \varepsilon^{ijk} P_{jk}$).

If (185) and (191) are added one obtains a conservation law for the total angular momentum density $\boldsymbol{S}+\boldsymbol{L}$, since the source terms cancel, as we shall now show. Indeed the source terms with fields in (185) and (191) cancel immediately if one uses the definitions $\boldsymbol{D} = \boldsymbol{E}+\boldsymbol{P}$ and $\boldsymbol{H} = \boldsymbol{B}-\boldsymbol{M}$. Furthermore the antisymmetric part of the pressure tensor follows from (107) with (63), (73), (94) and (104):

$$\begin{aligned}\boldsymbol{P}_{\mathrm{A}} = {} & -c^{-1} \int \hat{\boldsymbol{v}}_1 \wedge (\overline{\boldsymbol{\mu}}_1^{(1)} \wedge \boldsymbol{B}) f_1(\boldsymbol{R}, 1; t)\mathrm{d}1 \\ & - \int \Big\{ \boldsymbol{s} \wedge \boldsymbol{\nabla}_{\boldsymbol{s}} \Big( \sum_{i,j} \frac{e_i e_j}{8\pi|\boldsymbol{s}+\boldsymbol{r}_{1i}-\boldsymbol{r}_{2j}|} - \sum_{n,m=1}^{\infty} (-1)^m \overline{\boldsymbol{\mu}}_1^{(n)} : \boldsymbol{\nabla}_{\boldsymbol{s}}^n \overline{\boldsymbol{\mu}}_2^{(m)} : \boldsymbol{\nabla}_{\boldsymbol{s}}^m \frac{1}{8\pi s} \Big) \Big\} \\ & \qquad f_2(\boldsymbol{R}+\tfrac{1}{2}\boldsymbol{s}, 1, \boldsymbol{R}-\tfrac{1}{2}\boldsymbol{s}, 2; t)\mathrm{d}\boldsymbol{s}\,\mathrm{d}1\,\mathrm{d}2 \\ & - \int \sum_{n,m=1}^{\infty} (-1)^m \Big( \boldsymbol{s} \wedge \boldsymbol{\nabla}_{\boldsymbol{s}} \overline{\boldsymbol{\mu}}_1^{(n)} : \boldsymbol{\nabla}_{\boldsymbol{s}}^n \overline{\boldsymbol{\mu}}_2^{(m)} : \boldsymbol{\nabla}_{\boldsymbol{s}}^m \frac{1}{8\pi s} \Big) \\ & \qquad c_2(\boldsymbol{R}+\tfrac{1}{2}\boldsymbol{s}, 1, \boldsymbol{R}-\tfrac{1}{2}\boldsymbol{s}, 2; t)\mathrm{d}\boldsymbol{s}\,\mathrm{d}1\,\mathrm{d}2. \end{aligned} \tag{193}$$

The first term at the right-hand side is equal to $-\boldsymbol{D}^{\mathrm{F}}$ (174). The second term may be transformed to

$$\begin{aligned} \int \Big\{ & \sum_{i,j} (\boldsymbol{r}_{1i} - \boldsymbol{r}_{2j}) \wedge \boldsymbol{\nabla}_{\boldsymbol{s}} \frac{e_i e_j}{8\pi|\boldsymbol{s}+\boldsymbol{r}_{1i}-\boldsymbol{r}_{2j}|} \\ & - \sum_{n,m=1}^{\infty} (-1)^m n (\overline{\boldsymbol{\mu}}_1^{(n)} : \boldsymbol{\nabla}_{\boldsymbol{s}}^{n-1}) \wedge \boldsymbol{\nabla}_{\boldsymbol{s}} \overline{\boldsymbol{\mu}}_2^{(m)} : \boldsymbol{\nabla}_{\boldsymbol{s}}^m \frac{1}{8\pi s} \\ & - \sum_{n,m=1}^{\infty} (-1)^m m \overline{\boldsymbol{\mu}}_1^{(n)} : \boldsymbol{\nabla}_{\boldsymbol{s}}^n (\overline{\boldsymbol{\mu}}_2^{(m)} : \boldsymbol{\nabla}_{\boldsymbol{s}}^{m-1}) \wedge \boldsymbol{\nabla}_{\boldsymbol{s}} \frac{1}{8\pi s} \Big\} \\ & \qquad f_2(\boldsymbol{R}+\tfrac{1}{2}\boldsymbol{s}, 1, \boldsymbol{R}-\tfrac{1}{2}\boldsymbol{s}, 2; t)\mathrm{d}\boldsymbol{s}\,\mathrm{d}1\,\mathrm{d}2. \end{aligned} \tag{194}$$

Using the symmetry of the distribution function $f_2$ with respect to the interchange of 1 and 2 together with $\boldsymbol{s}$ and $-\boldsymbol{s}$, we find that (194) is equal to $-\boldsymbol{D}^{\mathrm{S}}$ (181) with (178) and (176). In the same way one finds that the third term of (193) is equal to $-\boldsymbol{D}^{\mathrm{C}}$ (184) with (176). So finally the antisymmetric part (193) of the pressure tensor is equal to the material source term (187) of the inner angular momentum balance:

$$\boldsymbol{P}_{\mathrm{A}} = -\boldsymbol{D}_s. \tag{195}$$

This allows one to write (185) in the alternative form:

$$\frac{\partial \boldsymbol{S}}{\partial t} = -\nabla\cdot(\boldsymbol{v}\boldsymbol{S}+\mathbf{J}_s)-\boldsymbol{P}_{\mathrm{A}}+\boldsymbol{P}\wedge\boldsymbol{E}+\boldsymbol{M}\wedge\boldsymbol{B}+c^{-1}\boldsymbol{v}\wedge(\boldsymbol{P}\wedge\boldsymbol{B}). \tag{196}$$

If we add equations (191) and (196) we have now

$$\frac{\partial(\boldsymbol{L}+\boldsymbol{S})}{\partial t} = -\nabla\cdot\{\boldsymbol{v}(\boldsymbol{L}+\boldsymbol{S})+\mathbf{J}_l+\mathbf{J}_s\}, \tag{197}$$

which is the conservation law of total angular momentum $\boldsymbol{L}+\boldsymbol{S}$ for a fluid system of neutral atoms.

### c. *Plasmas*

In plasmas the internal structure of the ions is usually disregarded. Then the inner angular momentum does not occur either. The angular momentum is thus entirely of orbital origin

$$\boldsymbol{L}(\boldsymbol{R}, t) \equiv \boldsymbol{R}\wedge(\varrho\boldsymbol{v}+c^{-1}\boldsymbol{E}\wedge\boldsymbol{B}). \tag{198}$$

From the conservation law (145) of total momentum for plasmas follows:

$$\frac{\partial \boldsymbol{L}}{\partial t} = -\nabla\cdot(\boldsymbol{v}\boldsymbol{L}+\mathbf{J}_l), \tag{199}$$

where we used the symmetrical character of the material pressure tensor (129) with (122) and (127). Furthermore the flow of angular momentum is (cf. (192)):

$$\mathbf{J}_l = -\mathbf{P}\wedge\boldsymbol{R}-\boldsymbol{E}(\boldsymbol{R}\wedge\boldsymbol{E})-\boldsymbol{B}(\boldsymbol{R}\wedge\boldsymbol{B})-c^{-1}\boldsymbol{v}\boldsymbol{R}\wedge(\boldsymbol{E}\wedge\boldsymbol{B})+\boldsymbol{\epsilon}\cdot\boldsymbol{R}(\tfrac{1}{2}E^2+\tfrac{1}{2}B^2) \tag{200}$$

(with $\boldsymbol{\epsilon}$ the Levi-Civita tensor). Hence as (199) shows the angular momentum satisfies a local conservation law.

d. *Crystalline solids*

In the preceding two subsections we studied systems with short range correlations. Now we shall turn to the general case of arbitrary range correlations – as occur in crystalline solids – including the effects due to charged atoms. Then all results of subsection *a* remain valid[1].

Since the correlation function is no longer of short range, the Irving–Kirkwood approximation (102) is not applicable. Nevertheless by use of an artifice of the type (148) we may still obtain a formula (182), such that again (195) is valid. Indeed with the help of the identity

$$c_2(\boldsymbol{R}, 1, \boldsymbol{R}-\boldsymbol{s}, 2; t) = c_2^{\ddagger}(\boldsymbol{R}+\tfrac{1}{2}\boldsymbol{s}, 1, \boldsymbol{R}-\tfrac{1}{2}\boldsymbol{s}; t) - \tfrac{1}{2}\boldsymbol{s}\cdot\nabla c_2^{\ddagger\!\!\!\ddagger}(\boldsymbol{R}+\tfrac{1}{2}\boldsymbol{s}, 1, \boldsymbol{R}-\tfrac{1}{2}\boldsymbol{s}; t), \tag{201}$$

where

$$c_2^{\ddagger}(\boldsymbol{R}+\tfrac{1}{2}\boldsymbol{s}, 1, \boldsymbol{R}-\tfrac{1}{2}\boldsymbol{s}, 2; t) \equiv \frac{1}{2}\int_{-1}^{+1} c_2\{\boldsymbol{R}+\tfrac{1}{2}(\lambda+1)\boldsymbol{s}, 1, \boldsymbol{R}+\tfrac{1}{2}(\lambda-1)\boldsymbol{s}, 2; t\}\mathrm{d}\lambda \tag{202}$$

and

$$c_2^{\ddagger\!\!\!\ddagger}(\boldsymbol{R}+\tfrac{1}{2}\boldsymbol{s}, 1, \boldsymbol{R}-\tfrac{1}{2}\boldsymbol{s}, 2; t) \equiv \frac{1}{2}\int_{-1}^{+1}(1-\lambda) c_2\{\boldsymbol{R}+\tfrac{1}{2}(\lambda+1)\boldsymbol{s}, 1, \boldsymbol{R}+\tfrac{1}{2}(\lambda-1)\boldsymbol{s}, 2; t\}\mathrm{d}\lambda, \tag{203}$$

one may write (cf. (182))

$$\boldsymbol{D}^{\mathrm{C}*} = -\nabla\cdot\mathbf{J}_s^{\mathrm{C}} + \boldsymbol{D}^{\mathrm{C}}. \tag{204}$$

Here the correlation part of the inner angular momentum flow is

$$\mathbf{J}_s^{\mathrm{C}} = \int \tfrac{1}{2}\boldsymbol{s}\boldsymbol{d}^{\mathrm{C}}(\boldsymbol{s}, 1, 2) c_2^{\ddagger\!\!\!\ddagger}(\boldsymbol{R}+\tfrac{1}{2}\boldsymbol{s}, 1, \boldsymbol{R}-\tfrac{1}{2}\boldsymbol{s}, 2; t)\mathrm{d}\boldsymbol{s}\,\mathrm{d}1\,\mathrm{d}2, \tag{205}$$

while the source term is

$$\boldsymbol{D}^{\mathrm{C}} = \int \boldsymbol{d}^{\mathrm{C}}(\boldsymbol{s}, 1, 2) c_2^{\ddagger}(\boldsymbol{R}+\tfrac{1}{2}\boldsymbol{s}, 1, \boldsymbol{R}-\tfrac{1}{2}\boldsymbol{s}, 2; t)\mathrm{d}\boldsymbol{s}\,\mathrm{d}1\,\mathrm{d}2. \tag{206}$$

In these expressions $\boldsymbol{d}^{\mathrm{C}}(\boldsymbol{s}, 1, 2)$ is given by (176).

Furthermore one may prove that just as in subsection *b*

$$\boldsymbol{D}^{\mathrm{C}} = -\boldsymbol{P}_{\mathrm{A}}^{\mathrm{C}}, \tag{207}$$

where $\boldsymbol{P}_{\mathrm{A}}^{\mathrm{C}}$ is the antisymmetric part of the correlation pressure (151). Therefore also (195) is valid in the present case, and consequently (196) and (197).

[1] For mixtures an extra summation over the indices labelling the species should be added.

e. *Galilean invariance*

The inner angular momentum equation (196) is Galilei invariant, as may be shown if one rewrites this equation with the help of the law of mass conservation (59) and the relation (189) as:

$$\varrho \frac{\mathrm{d}(v\boldsymbol{S})}{\mathrm{d}t} = -\nabla\cdot\mathbf{J}_s - \boldsymbol{P}_\mathrm{A} + \boldsymbol{P}'\wedge\boldsymbol{E}' + \boldsymbol{M}'\wedge\boldsymbol{B}', \tag{208}$$

where $v = \varrho^{-1}$ is the specific volume, $\mathrm{d}/\mathrm{d}t$ is the material time derivative (159) and the primes denote rest frame quantities (188). From inspection of the various terms of (208) it may be proved that they are all separately Galilei invariant.

## 7 *The laws of thermodynamics*

a. *The first law*

The first law of thermodynamics for fluid systems with neutral atoms (and for amorphous or polycrystalline solids) will follow from the energy equation (113–114) which may be written as

$$\varrho \frac{\mathrm{d}}{\mathrm{d}t}(\tfrac{1}{2}\boldsymbol{v}^2 + u) = -\nabla\cdot(\mathbf{P}\cdot\boldsymbol{v} + \boldsymbol{J}_q) + \varrho \frac{\mathrm{d}(v\boldsymbol{P})}{\mathrm{d}t}\cdot\boldsymbol{E} - \boldsymbol{M}\cdot\frac{\mathrm{d}\boldsymbol{B}}{\mathrm{d}t} + \boldsymbol{v}\cdot\{(\nabla\boldsymbol{E})\cdot\boldsymbol{P} + (\nabla\boldsymbol{B})\cdot\boldsymbol{M}\}, \tag{209}$$

where (59) and (159) have been used and where $v = \varrho^{-1}$ is the specific volume. At the left-hand side the sum of the specific macroscopic kinetic and internal energies appears. A balance of internal energy alone is obtained if the momentum law (105–106) with (59) or (158) is used in (209):

$$\varrho \frac{\mathrm{d}u}{\mathrm{d}t} = -\nabla\cdot\boldsymbol{J}_q - \tilde{\mathbf{P}} : (\nabla\boldsymbol{v}) + \varrho \frac{\mathrm{d}(v\boldsymbol{P})}{\mathrm{d}t}\cdot(\boldsymbol{E} + c^{-1}\boldsymbol{v}\wedge\boldsymbol{B}) - (\boldsymbol{M} + c^{-1}\boldsymbol{v}\wedge\boldsymbol{P})\cdot\frac{\mathrm{d}\boldsymbol{B}}{\mathrm{d}t}, \tag{210}$$

where $\tilde{\mathbf{P}}$ is the transposed pressure tensor. Each of the terms of this balance equation is separately Galilei invariant (v. subsection 5*i*). In particular the rest frame fields and polarizations (188) (up to order $c^{-1}$) appear (v. (26–27)), so that we may write (210) as

$$\varrho \frac{\mathrm{d}u}{\mathrm{d}t} = -\nabla\cdot\boldsymbol{J}_q - \tilde{\mathbf{P}} : \nabla\boldsymbol{v} + \varrho \frac{\mathrm{d}(v\boldsymbol{P}')}{\mathrm{d}t}\cdot\boldsymbol{E}' - \boldsymbol{M}'\cdot\frac{\mathrm{d}\boldsymbol{B}'}{\mathrm{d}t} \tag{211}$$

(it should be remembered that $\boldsymbol{M}'$ is of order $c^{-1}$). Traditionally one writes the divergence of the heat flow $\boldsymbol{J}_q$ in terms of the 'supplied heat' $\mathrm{d}q/\mathrm{d}t$ per unit mass and time:

$$\nabla\cdot\boldsymbol{J}_q = -\varrho \frac{\mathrm{d}q}{\mathrm{d}t}. \tag{212}$$

Then (211) gets the form

$$\frac{\mathrm{d}q}{\mathrm{d}t} = \frac{\mathrm{d}u}{\mathrm{d}t} + v\tilde{\mathbf{P}} : \nabla\boldsymbol{v} - \frac{\mathrm{d}(v\boldsymbol{P}')}{\mathrm{d}t}\cdot\boldsymbol{E}' + v\boldsymbol{M}'\cdot\frac{\mathrm{d}\boldsymbol{B}'}{\mathrm{d}t}. \tag{213}$$

This is the *first law of thermodynamics* for fluids (and amorphous or polycrystalline solids) of neutral atoms in an electromagnetic field. If the pressure tensor is a scalar $p$ (times the unit tensor) the second term at the right-hand side reads $vp\nabla\cdot\boldsymbol{v}$ or, if one uses (59), $p\mathrm{d}v/\mathrm{d}t$, the usual form. All quantities of (213) are well-defined as statistical expressions in terms of atomic quantities (v. section 5*f*). In particular the specific internal energy $u$ has been given by (115) as the sum of the three contributions $u^{\mathrm{K}}$ (78), $u^{\mathrm{S}}$ (97) and $u^{\mathrm{C}}$ (111). It should be noted that the polarization terms in (213) show a special asymmetry, which is a direct consequence of the asymmetry present already on the atomic level (chapter I, section 5*b*). This asymmetry may of course be removed by means of a Legendre transformation of the internal energy[1]. For instance, with the transformation

$$\hat{u} = u + v\boldsymbol{M}'\cdot\boldsymbol{B}', \tag{214}$$

one gets instead of (213)

$$\frac{\mathrm{d}q}{\mathrm{d}t} = \frac{\mathrm{d}\hat{u}}{\mathrm{d}t} + v\tilde{\mathbf{P}} : \nabla\boldsymbol{v} - \frac{\mathrm{d}(v\boldsymbol{P}')}{\mathrm{d}t}\cdot\boldsymbol{E}' - \frac{\mathrm{d}(v\boldsymbol{M}')}{\mathrm{d}t}\cdot\boldsymbol{B}'. \tag{215}$$

However the introduction of the energy $\hat{u}$ is rather artificial: the energy $u$, in contrast to $\hat{u}$, has a clear-cut physical meaning from the microscopic point of view.

For a neutral plasma the first law of thermodynamics follows from the energy equation (140) with (59), (128), (159), (188) and (212) as

$$\frac{\mathrm{d}q}{\mathrm{d}t} = \frac{\mathrm{d}u}{\mathrm{d}t} + v\tilde{\mathbf{P}} : \nabla\boldsymbol{v} - v\boldsymbol{J}\cdot\boldsymbol{E}', \tag{216}$$

[1] For a discussion of various types of Legendre transformations in the first and second laws for magnetized media see for instance H. A. Leupold, Am. J. Phys. **37**(1969)1047. Compare also the microscopic considerations of A. N. Kaufman and T. Soda, J. Chem. Phys. **37**(1962)1988.

where $\boldsymbol{J}$ is the electric current density (purely conductive; the convective part $\varrho^e \boldsymbol{v}$ is not present in a neutral plasma). The 'supplied heat' which figures in the left-hand side and has been given by (212) is essentially the divergence of the heat flow and represents therefore the heat supplied through conduction by the surroundings. The last term at the right-hand side represents the Joule heat produced per unit mass and time.

Finally for systems with correlations of arbitrary range the first law follows from (152) with (59), (150), (159), (188) and (212):

$$\frac{\mathrm{d}q}{\mathrm{d}t} = \frac{\mathrm{d}u}{\mathrm{d}t} + v\tilde{\mathsf{P}} : \nabla \boldsymbol{v} - v\boldsymbol{J}'\cdot\boldsymbol{E}' - \frac{\mathrm{d}(v\boldsymbol{P}')}{\mathrm{d}t}\cdot\boldsymbol{E}' + v\boldsymbol{M}'\cdot\frac{\mathrm{d}\boldsymbol{B}'}{\mathrm{d}t}, \tag{217}$$

where $\boldsymbol{J}' = \boldsymbol{J} - \varrho^e \boldsymbol{v}$ is the rest frame (or conduction) electric current density.

b. *The second law for fluids*

The microscopic basis of the second law of thermodynamics has a character which is different from that of the laws established so far. The latter were all statistical averages of corresponding microscopic equations, whereas the second law contains a new quantity, the entropy, which is not the average of a microscopic quantity. Furthermore the system for which one wants to derive the second law has to be specified in more details as to its statistical properties: here we shall confine ourselves to systems in equilibrium described by a canonical ensemble.

In the present subsection we shall be concerned with the derivation of a second law for systems of neutral atoms in which only short range correlations are present, namely fluids and amorphous or polycrystalline solids.

The theory may be developed along two slightly different lines. In the first conception one considers a system at rest enclosed in a vessel and surrounded by a heat bath in a uniform and time-independent field. As a consequence of the fact that the polarizations are discontinuous at the surface it turns out then that the pressure and related thermodynamic quantities vary over the sample. In the other conception one avoids non-uniformities due to surface effects by dividing a large polarized system into nearly uniform cells, still containing many atoms. These cells are then described by a canonical (or grand) ensemble with their environments playing the role of a heat bath. As external fields the averages of the fields arising from the surroundings of the cell are employed, so that correlations between particles inside and outside the cell are neglected. This is the reason why such an approach is only applicable to systems with short range correlations. We shall employ it (in this subsection) for fluids and amorphous or polycrystalline solids.

The uniform Maxwell fields $\boldsymbol{E}$ and $\boldsymbol{B}$ in the cell are related to the fields $\boldsymbol{E}_e(\boldsymbol{R})$ and $\boldsymbol{B}_e(\boldsymbol{R})$, due to the surroundings of the cell, as

$$\begin{aligned} \boldsymbol{E} &= \boldsymbol{E}_e(\boldsymbol{R})+\boldsymbol{P}\cdot\int^V \nabla\nabla \frac{1}{4\pi|\boldsymbol{R}-\boldsymbol{R}'|}\,\mathrm{d}\boldsymbol{R}', \\ \boldsymbol{B} &= \boldsymbol{B}_e(\boldsymbol{R})+\boldsymbol{M}+\boldsymbol{M}\cdot\int^V \nabla\nabla \frac{1}{4\pi|\boldsymbol{R}-\boldsymbol{R}'|}\,\mathrm{d}\boldsymbol{R}' \end{aligned} \tag{218}$$

(cf. (17)), where $\boldsymbol{P}$ and $\boldsymbol{M}$ are the uniform polarizations in the cell and where the integrals are extended over its volume $V$. They are to be understood as the sum of a principal value and an integral over a small surface around $\boldsymbol{R}$ (v. problem 1). The equations (218) are satisfied by uniform external fields $\boldsymbol{E}_e$ and $\boldsymbol{B}_e$ if the integral

$$\int^V \nabla\nabla \frac{1}{4\pi|\boldsymbol{R}-\boldsymbol{R}'|}\,\mathrm{d}\boldsymbol{R}' \quad (\equiv -\mathsf{L}) \tag{219}$$

is independent of the position $\boldsymbol{R}$. This is the case if the sample has ellipsoidal shape (see appendix I). In that case the expressions (218) may be written as

$$\begin{aligned} \boldsymbol{E} &= \boldsymbol{E}_e-\mathsf{L}\cdot\boldsymbol{P}, \\ \boldsymbol{B} &= \boldsymbol{B}_e+\boldsymbol{M}-\mathsf{L}\cdot\boldsymbol{M}, \end{aligned} \tag{220}$$

where the tensor $\mathsf{L}$, which is equal to (minus) the integral (219), is called the 'depolarizing tensor'. (It depends on the shape of the boundary.)

The cell will be described with the help of the canonical ensemble

$$\mathrm{e}^{-F^*/kT} = C\int \mathrm{e}^{-H/kT}\,\mathrm{d}\boldsymbol{q}\,\mathrm{d}\boldsymbol{p}, \tag{221}$$

where $F^*$ is the free energy[1], $T$ the temperature and $C$ a constant (depending on the number of atoms in the system), while $H(\boldsymbol{q}, \boldsymbol{p})$ is the Hamiltonian for a dipole system (appendix II, formula (A32), with the atomic charges $e_k = 0$):

[1] An asterisk is written at the symbol for the free energy to distinguish it from $F = U-TS$ with $U$ the total internal energy. In fact we shall find that the latter will differ from the average Hamiltonian $\langle H\rangle$, which will be denoted by $U^*$ ($= F^*+TS$; v. (227) and (232)).

$$H(\boldsymbol{q}, \boldsymbol{p}) = \sum_k \left( \frac{\boldsymbol{P}_k^2}{2m_k} + \sum_{i=1}^{f-1} \frac{\boldsymbol{p}_{ki}^2}{2m_{ki}} - \sum_{i,j=1}^{f-1} \frac{\boldsymbol{p}_{ki} \cdot \boldsymbol{p}_{kj}}{2m_k} \right)$$

$$+ \sum_k \sum_{i,j=1(i \neq j)}^{f} \frac{e_{ki}\, e_{kj}}{8\pi |\boldsymbol{R}_{ki}(\boldsymbol{q}) - \boldsymbol{R}_{kj}(\boldsymbol{q})|} + \sum_{k,l(k \neq l)} \sum_{i,j=1}^{f} \frac{e_{ki}\, e_{lj}}{8\pi |\boldsymbol{R}_{ki}(\boldsymbol{q}) - \boldsymbol{R}_{lj}(\boldsymbol{q})|}$$

$$- \sum_k \left\{ \bar{\boldsymbol{\mu}}_k^{(1)} \cdot \boldsymbol{E}_e + \left( \bar{\boldsymbol{v}}_k^{(1)} + c^{-1} \bar{\boldsymbol{\mu}}_k^{(1)} \wedge \frac{\boldsymbol{P}_k}{m_k} \right) \cdot \boldsymbol{B}_e \right\}, \tag{222}$$

where $\bar{\boldsymbol{\mu}}_k^{(1)}$ and $\bar{\boldsymbol{v}}_k^{(1)}$ are the electric and magnetic dipole moments.

The free energy $F^*$ is a function of the external fields $\boldsymbol{E}_e$ and $\boldsymbol{B}_e$, the temperature $T$ and the position of the boundaries of the system. The partial derivatives of the free energy with respect to the external fields follow from (221) and (222):

$$\begin{aligned} \frac{\partial F^*}{\partial \boldsymbol{E}_e} &= -\langle \sum_k \bar{\boldsymbol{\mu}}_k^{(1)} \rangle = -V\boldsymbol{P}, \\ \frac{\partial F^*}{\partial \boldsymbol{B}_e} &= -\left\langle \sum_k \left( \bar{\boldsymbol{v}}_k^{(1)} + c^{-1} \bar{\boldsymbol{\mu}}_k^{(1)} \wedge \frac{\boldsymbol{P}_k}{m_k} \right) \right\rangle = -V\boldsymbol{M}, \end{aligned} \tag{223}$$

where (65) has been used. The brackets indicate canonical ensemble averages so that for a dynamical variable $a = a(\boldsymbol{q}, \boldsymbol{p})$ one has the average value $\langle a \rangle = C \int a \exp \{(F^* - H)/kT\} \mathrm{d}\boldsymbol{q}\, \mathrm{d}\boldsymbol{p}$. Furthermore the partial derivative with respect to the temperature gives the entropy $S$ of the system:

$$\frac{\partial F^*}{\partial T} = -S. \tag{224}$$

The free energy changes also if the boundary changes. We consider infinitesimal variations of the position vector $\boldsymbol{R}$ (choosing the centre of the cell as the origin of the coordinate system):

$$\delta \boldsymbol{R} = \delta \boldsymbol{\epsilon} \cdot \boldsymbol{R}, \tag{225}$$

with a uniform (infinitesimal) deformation tensor $\delta\boldsymbol{\epsilon}$. Then the ellipsoidal shape of the boundaries remains ellipsoidal. (The external fields $\boldsymbol{E}_e$ and $\boldsymbol{B}_e$ may be kept constant during such a deformation by adjusting the charges on condenser plates and currents in coils around the total system in the proper way.)

The total change of the free energy now follows from (223–225):

$$\delta F^* = -S\delta T - V\boldsymbol{P} \cdot \delta \boldsymbol{E}_e - V\boldsymbol{M} \cdot \delta \boldsymbol{B}_e + \mathsf{A} : \delta \boldsymbol{\epsilon}, \tag{226}$$

where the tensor $\mathbf{A}$, which is contracted twice with $\delta\boldsymbol{\epsilon}$ (i.e. $\mathbf{A}:\delta\boldsymbol{\epsilon} \equiv \sum_{i,j} A_{ij}\delta\varepsilon_{ji}$), has still to be determined.

The free energy is connected to the average $\langle H\rangle = U^*$ of the Hamiltonian $H$ by

$$F^* = U^* - TS, \tag{227}$$

where $H$ is given by (222) or by (A37) (with $e_k = 0$). The third term at the right-hand side of (A37) may be expanded as a series if the atoms are outside each other. For that reason we write (A37) in the form

$$\begin{aligned} H = K + \sum_k \sum_{i,j=1(i\neq j)}^{f} \frac{e_{ki}e_{kj}}{8\pi|\boldsymbol{R}_{ki}-\boldsymbol{R}_{kj}|} \\ + \sum_{k,l(k\neq l)} \sum_{n,m=1}^{\infty} \overline{\boldsymbol{\mu}}_k^{(n)} : \nabla_k^n \overline{\boldsymbol{\mu}}_l^{(m)} : \nabla_l^m \frac{1}{8\pi|\boldsymbol{R}_k-\boldsymbol{R}_l|} \\ + \sum_{k,l(k\neq l)} \left( \sum_{i,j=1}^{f} \frac{e_{ki}e_{lj}}{8\pi|\boldsymbol{R}_{ki}-\boldsymbol{R}_{lj}|} - \sum_{n,m=1}^{\infty} \overline{\boldsymbol{\mu}}_k^{(n)} : \nabla_k^n \overline{\boldsymbol{\mu}}_l^{(m)} : \nabla_l^m \frac{1}{8\pi|\boldsymbol{R}_k-\boldsymbol{R}_l|} \right) \\ - \sum_k \overline{\boldsymbol{\mu}}_k^{(1)} \cdot \boldsymbol{E}_e, \end{aligned} \tag{228}$$

with $K$ the kinetic energy and where the penultimate term vanishes if the atoms are outside each other. The canonical average of (228) is the total energy of the system. Using (I.41), (I.44), (61) with $\boldsymbol{v} = 0$ and appropriate (time-independent) one- and two-point distribution functions it gets the form

$$\begin{aligned} U^* \equiv \langle H\rangle = \int \left( \tfrac{1}{2}m\hat{\boldsymbol{v}}_1^2 + \tfrac{1}{2}\sum_i m_i \dot{\boldsymbol{r}}_{1i}^2 + \sum_{i,j(i\neq j)} \frac{e_i e_j}{8\pi|\boldsymbol{r}_{1i}-\boldsymbol{r}_{2j}|} \right) f_1(\boldsymbol{R}_1, 1)\mathrm{d}\boldsymbol{R}_1\,\mathrm{d}1 \\ + \int \sum_{n,m=1}^{\infty} \overline{\boldsymbol{\mu}}_1^{(n)} : \nabla_1^n \overline{\boldsymbol{\mu}}_2^{(m)} : \nabla_2^m \frac{1}{8\pi|\boldsymbol{R}_1-\boldsymbol{R}_2|} f_2(\boldsymbol{R}_1, 1, \boldsymbol{R}_2, 2)\mathrm{d}\boldsymbol{R}_1\,\mathrm{d}\boldsymbol{R}_2\,\mathrm{d}1\,\mathrm{d}2 \\ + \int \left( \sum_{i,j} \frac{e_i e_j}{8\pi|\boldsymbol{R}_{1i}-\boldsymbol{R}_{2j}|} - \sum_{n,m=1}^{\infty} \overline{\boldsymbol{\mu}}_1^{(n)} : \nabla_1^n \overline{\boldsymbol{\mu}}_2^{(m)} : \nabla_2^m \frac{1}{8\pi|\boldsymbol{R}_1-\boldsymbol{R}_2|} \right) \\ f_2(\boldsymbol{R}_1, 1, \boldsymbol{R}_2, 2)\mathrm{d}\boldsymbol{R}_1\,\mathrm{d}\boldsymbol{R}_2\,\mathrm{d}1\,\mathrm{d}2 \\ - \int \overline{\boldsymbol{\mu}}_1^{(1)} \cdot \boldsymbol{E}_e\, f_1(\boldsymbol{R}_1, \overline{\boldsymbol{\mu}}_1^{(1)})\mathrm{d}\boldsymbol{R}_1\,\mathrm{d}\overline{\boldsymbol{\mu}}_1^{(1)}, \end{aligned} \tag{229}$$

where the integrals are extended over the volume of the system. Since the system is uniform the integrals over $\boldsymbol{R}_1$ in the first and last terms may be performed. In the second term we split $f_2$ into a correlation function of the type (7) and a product of two one-point distribution functions. The latter give rise to an uncorrelated part, in which we omit all multipole moments of orders two and higher (as in subsections 5*c*, *d* such moments would yield terms containing macroscopic multipole densities of higher order, which are assumed to be negligible in our system). We then obtain

$$U^* = V\varrho u^{\mathrm{K}} + \boldsymbol{PP} : \int \nabla_1 \nabla_2 \frac{1}{8\pi|\boldsymbol{R}_1-\boldsymbol{R}_2|} \mathrm{d}\boldsymbol{R}_1 \mathrm{d}\boldsymbol{R}_2$$
$$+ \int \sum_{n,m=1}^{\infty} (-1)^m \overline{\boldsymbol{\mu}}_1^{(n)} \vdots \nabla_s^n \overline{\boldsymbol{\mu}}_2^{(m)} \vdots \nabla_s^m \frac{1}{8\pi s} c_2(\boldsymbol{R}+\tfrac{1}{2}\boldsymbol{s}, 1, \boldsymbol{R}-\tfrac{1}{2}\boldsymbol{s}, 2)\mathrm{d}\boldsymbol{R}\,\mathrm{d}\boldsymbol{s}\,\mathrm{d}1\,\mathrm{d}2$$
$$+ \int \left( \sum_{i,j} \frac{e_i e_j}{8\pi|\boldsymbol{s}+\boldsymbol{r}_{1i}-\boldsymbol{r}_{2j}|} - \sum_{n,m=1}^{\infty} (-1)^m \overline{\boldsymbol{\mu}}_1^{(n)} \vdots \nabla_s^n \overline{\boldsymbol{\mu}}_2^{(m)} \vdots \nabla_s^m \frac{1}{8\pi s} \right)$$
$$f_2(\boldsymbol{R}+\tfrac{1}{2}\boldsymbol{s}, 1, \boldsymbol{R}-\tfrac{1}{2}\boldsymbol{s}, 2)\mathrm{d}\boldsymbol{R}\,\mathrm{d}\boldsymbol{s}\,\mathrm{d}1\,\mathrm{d}2 - \boldsymbol{P}\cdot\boldsymbol{E}_{\mathrm{e}}, \quad (230)$$

where (65) and (78) have been used, and new integration variables $\boldsymbol{R}$ and $\boldsymbol{s}$ have been introduced in the third and fourth terms at the right-hand side. The limits of the integration over $\boldsymbol{s}$ depend on the value of the variable $\boldsymbol{R}$. However since both the correlation length and the dimension of the atoms (which is the range of the first factor in the integrand of the fourth term) are small compared to the dimension of the system, effectively the limits of the integration over $\boldsymbol{s}$ depend on $\boldsymbol{R}$ only in a small region near the surface. Neglecting these surface effects, and using the fact that the system is uniform, we perform the integrals over $\boldsymbol{R}$ in the third and fourth term. Furthermore the integral in the second term may be written as (see appendix I):

$$\int \nabla_1 \nabla_2 \frac{1}{4\pi|\boldsymbol{R}_1-\boldsymbol{R}_2|} \mathrm{d}\boldsymbol{R}_1 \mathrm{d}\boldsymbol{R}_2 = -V\int^V \nabla\nabla \frac{1}{4\pi|\boldsymbol{R}|} \mathrm{d}\boldsymbol{R} \equiv V\mathbf{L}, \quad (231)$$

where $\boldsymbol{R}$ is the position with respect to the centre of the ellipsoid and where $\mathbf{L}$ is the depolarizing tensor (which depends on the shape of the system). In this way we obtain as the average Hamiltonian, using the internal energy $u$ (115) with (97) (neutral atoms have $\bar{\mu}_k^{(0)} = 0$) and (111)

$$U^* = U + V(\tfrac{1}{2}\boldsymbol{PP} : \mathbf{L} - \boldsymbol{P}\cdot\boldsymbol{E}_{\mathrm{e}}), \quad (232)$$

with $U = V\varrho u$ the total internal energy.

Finally the tensor $\mathbf{A}$ occurring in (226) has to be found. In appendix III it is proved that $\mathbf{A}$ is the following average

$$\mathbf{A} = -\left\langle \sum_k \left( \frac{\partial H}{\partial \boldsymbol{P}_k} \boldsymbol{P}_k - \boldsymbol{R}_k \frac{\partial H}{\partial \boldsymbol{R}_k} \right) \right\rangle \quad (233)$$

with $\boldsymbol{R}_k$ and $\boldsymbol{P}_k$ the canonical centre of mass coordinates and momenta of atom $k$. We substitute into this expression the Hamiltonian (222), add and subtract the multipole series expansion of the third term and use the appropriate time-independent distribution functions. This gives

$$\begin{aligned}
\mathsf{A} = & -\int m\hat{\boldsymbol{v}}_1\,\hat{\boldsymbol{v}}_1\, f_1(\boldsymbol{R}_1, \boldsymbol{v}_1)\mathrm{d}\boldsymbol{R}_1\,\mathrm{d}\boldsymbol{v}_1 \\
& + \int \sum_{n,m=1}^{\infty} \left\{(\boldsymbol{R}_1 - \boldsymbol{R}_2)\nabla_1\, \bar{\mu}_1^{(n)} \vdots \nabla_1^n\, \bar{\mu}_2^{(m)} \vdots \nabla_2^m \frac{1}{8\pi|\boldsymbol{R}_1 - \boldsymbol{R}_2|}\right\} \\
& \qquad\qquad f_2(\boldsymbol{R}_1, 1, \boldsymbol{R}_2, 2)\mathrm{d}\boldsymbol{R}_1\,\mathrm{d}\boldsymbol{R}_2\,\mathrm{d}1\,\mathrm{d}2 \\
& + \int \left\{(\boldsymbol{R}_1 - \boldsymbol{R}_2)\nabla_1 \left(\sum_{i,j} \frac{e_i e_j}{8\pi|\boldsymbol{R}_{1i} - \boldsymbol{R}_{2j}|}\right.\right. \\
& \left.\left. - \sum_{n,m=1}^{\infty} \bar{\mu}_1^{(n)} \vdots \nabla_1^n\, \bar{\mu}_2^{(m)} \vdots \nabla_2^m \frac{1}{8\pi|\boldsymbol{R}_1 - \boldsymbol{R}_2|}\right)\right\} f_2(\boldsymbol{R}_1, 1, \boldsymbol{R}_2, 2)\mathrm{d}\boldsymbol{R}_1\,\mathrm{d}\boldsymbol{R}_2\,\mathrm{d}1\,\mathrm{d}2 \\
& + c^{-1} \int \hat{\boldsymbol{v}}_1\, \bar{\mu}_1^{(1)} \wedge \boldsymbol{B}_\mathrm{e}\, f_1(\boldsymbol{R}_1, \boldsymbol{v}_1, \bar{\mu}_1^{(1)})\mathrm{d}\boldsymbol{R}_1\,\mathrm{d}\boldsymbol{v}_1\,\mathrm{d}\bar{\mu}_1^{(1)},
\end{aligned} \tag{234}$$

where we used the Hamilton equation $\boldsymbol{v}_k \equiv \dot{\boldsymbol{R}}_k = \partial H/\partial \boldsymbol{P}_k$ and the expression (61) with $\boldsymbol{v} = 0$. Because of the uniformity of the system the space integrals in the first and last terms may be performed. Then they become, apart from a factor $-V$, equal to the sum of the kinetic pressure $\mathsf{P}^\mathrm{K}$ (63) and the field dependent part of the pressure $\mathsf{P}^\mathrm{F}$ (73). In the second term we introduce a correlation function with the help of (7). In the uncorrelated part we omit all multipole moments of order 2 and higher, as in (230). In this way we obtain for (234)

$$\begin{aligned}
\mathsf{A} = & -V(\mathsf{P}^\mathrm{K} + \mathsf{P}^\mathrm{F}) + \int (\boldsymbol{R}_1 - \boldsymbol{R}_2)\nabla_1\, \boldsymbol{P}\cdot\nabla_1\, \boldsymbol{P}\cdot\nabla_2 \frac{1}{8\pi|\boldsymbol{R}_1 - \boldsymbol{R}_2|}\mathrm{d}\boldsymbol{R}_1\,\mathrm{d}\boldsymbol{R}_2 \\
& + \int \sum_{n,m=1}^{\infty} (-1)^m \left(\boldsymbol{s}\nabla_s\, \bar{\mu}_1^{(n)} \vdots \nabla_s^n\, \bar{\mu}_2^{(m)} \vdots \nabla_s^m \frac{1}{8\pi s}\right) \\
& \qquad\qquad c_2(\boldsymbol{R} + \tfrac{1}{2}\boldsymbol{s}, 1, \boldsymbol{R} - \tfrac{1}{2}\boldsymbol{s}, 2)\mathrm{d}\boldsymbol{R}\,\mathrm{d}\boldsymbol{s}\,\mathrm{d}1\,\mathrm{d}2 \\
& + \int \left\{\boldsymbol{s}\nabla_s \left(\sum_{i,j} \frac{e_i e_j}{8\pi|\boldsymbol{s} + \boldsymbol{r}_{1i} - \boldsymbol{r}_{2j}|} - \sum_{n,m=1}^{\infty} (-1)^m \bar{\mu}_1^{(n)} \vdots \nabla_s^n\, \bar{\mu}_2^{(m)} \vdots \nabla_s^m \frac{1}{8\pi s}\right)\right\} \\
& \qquad\qquad f_2(\boldsymbol{R} + \tfrac{1}{2}\boldsymbol{s}, 1, \boldsymbol{R} - \tfrac{1}{2}\boldsymbol{s}, 2)\mathrm{d}\boldsymbol{R}\,\mathrm{d}\boldsymbol{s}\,\mathrm{d}1\,\mathrm{d}2,
\end{aligned} \tag{235}$$

where (65) has been used and new integration variables $\boldsymbol{R}$ and $\boldsymbol{s}$ have been introduced. Just as for the Hamiltonian, the integrations over $\boldsymbol{s}$ in the third and fourth term are effectively to be extended over a volume small compared to that of the system, so that the integration over $\boldsymbol{R}$ may be performed, the system being uniform. Furthermore the second term may be written as (see appendix I):

$$\int (\boldsymbol{R}_1-\boldsymbol{R}_2)\nabla_1\nabla_1\nabla_2 \frac{1}{8\pi|\boldsymbol{R}_1-\boldsymbol{R}_2|}\,\mathrm{d}\boldsymbol{R}_1\,\mathrm{d}\boldsymbol{R}_2 = -V\int \boldsymbol{R}\nabla\nabla\nabla \frac{1}{8\pi|\boldsymbol{R}|}\,\mathrm{d}\boldsymbol{R} \equiv -\tfrac{1}{2}V\mathbf{K}, \quad (236)$$

where $\boldsymbol{R}$ measures the position relative to the centre of the system and $\mathbf{K}$ is a tensor of the fourth rank, which depends on the shape of the boundary of the ellipsoidal system. In this way we obtain for (235), using (107) with (94) and (104),

$$\mathbf{A} = -V(\mathbf{P}+\tfrac{1}{2}\mathbf{K} : \boldsymbol{P}\boldsymbol{P}), \quad (237)$$

where $\mathbf{P}$ is the pressure tensor and the last indices of $\mathbf{K}$ are contracted with those of the two factors $\boldsymbol{P}$ (the electric polarization).

This result could have been found along different lines, namely by starting from the expression (A48) of appendix III according to which

$$\delta_\varepsilon F^* = -\int^S \boldsymbol{n}\cdot\mathbf{P}_{\text{out}}\cdot\delta\boldsymbol{\epsilon}\cdot\boldsymbol{R}\,\mathrm{d}S, \quad (238)$$

where $\mathbf{P}_{\text{out}}$ is the pressure exerted by a wall (supposed to be unpolarizable) which separates the cell from its surroundings and $\boldsymbol{n}$ is the normal to the wall. This pressure is not equal to the pressure $\mathbf{P}$, just inside the boundary. The reason for this difference is that the electromagnetic fields are discontinuous across the boundary. In fact it follows from momentum conservation in the form (109), applied to a thin volume element with surfaces on either side of the boundary between the separation wall and the cell, that for a system in equilibrium and at rest one has

$$\boldsymbol{n}\cdot(\mathbf{P}-\mathbf{P}_{\text{out}}) = \boldsymbol{n}\cdot\{\boldsymbol{DE}+\boldsymbol{BH}-(\tfrac{1}{2}E^2+\tfrac{1}{2}B^2-\boldsymbol{M}\cdot\boldsymbol{B})\mathrm{U}\} - \boldsymbol{n}\cdot\{\boldsymbol{EE}+\boldsymbol{BB}-(\tfrac{1}{2}E^2+\tfrac{1}{2}B^2)\mathrm{U}\}_{\text{out}}. \quad (239)$$

The fields just inside the cell and inside the separation wall are connected by relations which are consequences of the Maxwell equations:

$$\begin{aligned} \boldsymbol{n}\cdot\boldsymbol{D} &= \boldsymbol{n}\cdot\boldsymbol{E}_{\text{out}}, & \boldsymbol{E}-\boldsymbol{nn}\cdot\boldsymbol{E} &= \boldsymbol{E}_{\text{out}}-\boldsymbol{nn}\cdot\boldsymbol{E}_{\text{out}}, \\ \boldsymbol{n}\cdot\boldsymbol{B} &= \boldsymbol{n}\cdot\boldsymbol{B}_{\text{out}}, & \boldsymbol{H}-\boldsymbol{nn}\cdot\boldsymbol{H} &= \boldsymbol{B}_{\text{out}}-\boldsymbol{nn}\cdot\boldsymbol{B}_{\text{out}}. \end{aligned} \quad (240)$$

Insertion of these formulae into (239) leads to

$$\boldsymbol{n}\cdot(\mathbf{P}-\mathbf{P}_{\text{out}}) = -\tfrac{1}{2}\boldsymbol{n}(\boldsymbol{P}\cdot\boldsymbol{n})^2, \quad (241)$$

up to order $c^{-1}$. (Terms quadratic in the magnetization are of order $c^{-2}$ and hence not considered in the non-relativistic theory.) The right-hand side of equation (241) is usually called the Liénard pressure[1]. (Incidentally this formula shows how one could measure the pressure $\mathbf{P}$ just inside a polarizable system: $\mathbf{P}_{\text{out}}$ may be measured by means of a manometer made of unpolarizable material, while a value of the polarization $\boldsymbol{P}$ may be obtained independently.) If one introduces (241) into (238) and uses the assumed uniformity of $\mathbf{P}$ and $\delta\boldsymbol{\epsilon}$ one finds

$$\delta_{\varepsilon} F^* = -V\mathbf{P} : \delta\boldsymbol{\epsilon} - \tfrac{1}{2}\int^{S} \boldsymbol{n}\cdot\delta\boldsymbol{\epsilon}\cdot\boldsymbol{R}(\boldsymbol{P}\cdot\boldsymbol{n})^2 \,\mathrm{d}S. \tag{242}$$

This expression is indeed equal to $\mathbf{A} : \delta\boldsymbol{\epsilon}$ with $\mathbf{A}$ given by (237) as follows with the help of the identity valid for the tensor $\mathbf{K}$ (defined in (236)) of a volume of ellipsoidal shape (v. problem 5)

$$\mathbf{K} = \frac{1}{V}\int^{S} \boldsymbol{Rnnn}\,\mathrm{d}S. \tag{243}$$

Collecting the results (223), (224) and (237) and substituting them into (226) with (227), we have found now for the change of the entropy

$$T\delta S = \delta U^* + V(\mathbf{P} + \tfrac{1}{2}\mathbf{K} : \boldsymbol{PP}) : \delta\boldsymbol{\epsilon} + V\boldsymbol{P}\cdot\delta\boldsymbol{E}_{\mathrm{e}} + V\boldsymbol{M}\cdot\delta\boldsymbol{B}_{\mathrm{e}}. \tag{244}$$

If one inserts moreover (232) one obtains, dividing the result by the total (constant) mass $M$ of the system:

$$T\delta s = \delta(u + \tfrac{1}{2}v\boldsymbol{PP} : \mathbf{L}) + v(\mathbf{P} + \tfrac{1}{2}\mathbf{K} : \boldsymbol{PP}) : \delta\boldsymbol{\epsilon} - \boldsymbol{E}_{\mathrm{e}}\cdot\delta(v\boldsymbol{P}) + v\boldsymbol{M}\cdot\delta\boldsymbol{B}_{\mathrm{e}}, \tag{245}$$

where $s = S/M$ is the specific entropy and $v = V/M$ the specific volume. In this relation the external fields occur, *not* the Maxwell fields. We may introduce the latter instead of the former, by using (220) and also the relation (proved in appendix I) that gives the change of the depolarizing tensor if the shape of the boundary is changed:

$$\delta(v^{-1}\mathbf{L}) = -v^{-1}\delta\boldsymbol{\epsilon} : \mathbf{K}. \tag{246}$$

We then obtain, up to order $c^{-1}$ (noting that $\boldsymbol{M}$ is of order $c^{-1}$ already) the entropy law

$$T\delta s = \delta u + v\mathbf{P} : \delta\boldsymbol{\epsilon} - \boldsymbol{E}\cdot\delta(v\boldsymbol{P}) + v\boldsymbol{M}\cdot\delta\boldsymbol{B}. \tag{247}$$

This law will be further studied, first for fluids, then for amorphous or polycrystalline substances.

[1] A. Liénard, Ann. Physique **20**(1923)249.

A *fluid* system at rest is isotropic in the absence of polarizations and fields. Then the entropy depends only on the internal energy $u$ and the specific volume $v$, not on the shape of the boundary. Therefore in this case only the scalar part of the tensor $\delta\boldsymbol{\epsilon}$ should contribute to the second term and hence $\mathbf{P}$ reduces to a scalar $p$ (times the unit tensor $\mathbf{U}$). In this way, since $v\mathbf{U} : \delta\boldsymbol{\epsilon}$ is the change of volume $\delta v$, the entropy law (247) becomes then

$$T\mathrm{d}s = \mathrm{d}u + p\,\mathrm{d}v, \tag{248}$$

where we have written differentials, because now $s$ is a function of $u$ and $v$.

If fields and polarizations are present, it is not immediately clear that again $\mathbf{P}$ is diagonal, since now the isotropy of the system is perturbed. However, if the polarization vectors $\boldsymbol{P}$, $\boldsymbol{M}$ are assumed to depend only on the specific volume $v$, the specific entropy $s$ (or temperature) and the fields, i.e.

$$\boldsymbol{P} = \boldsymbol{P}(v, s, \boldsymbol{E}, \boldsymbol{B}), \qquad \boldsymbol{M} = \boldsymbol{M}(v, s, \boldsymbol{E}, \boldsymbol{B}), \tag{249}$$

the entropy law (247) may be integrated at constant $\boldsymbol{\epsilon}$ and $s$ with the result

$$u = u_0 + \Delta u, \tag{250}$$

where $u_0$ is the specific energy at zero polarizations and fields, which depends only on $v$ and $s$. Furthermore $\Delta u$ is a function of $v$, $s$, $\boldsymbol{E}$ and $\boldsymbol{B}$ or (with (249)) of $v$, $s$, $v\boldsymbol{P}$ and $\boldsymbol{B}$. Therefore $u$ depends on these variables so that $\delta u$ contains only the trace of $\delta\boldsymbol{\epsilon}$ which is equal to $v^{-1}\delta v$. Hence from (247) it follows that in equilibrium the tensor $\mathbf{P}$ reduces to a scalar pressure $p\mathbf{U}$ for the fluid systems studied. So finally the non-relativistic second law (or 'Gibbs relation') becomes for a (one-component) fluid of neutral atoms

$$T\mathrm{d}s = \mathrm{d}u + p\,\mathrm{d}v - \boldsymbol{E}\cdot\mathrm{d}(v\boldsymbol{P}) + v\boldsymbol{M}\cdot\mathrm{d}\boldsymbol{B}. \tag{251}$$

(It should be kept in mind that all quantities have been defined for a system at rest. In particular the fields and polarizations are therefore the same as the primed quantities of the preceding subsection.) The field terms in the second law (251) show the same asymmetry as has been discussed in connexion with those appearing in the first law (213).

For *amorphous* or *polycrystalline* substances it is not possible to reduce the entropy law (247) to the simple form (251): the pressure tensor does not reduce to a multiple of the unit tensor. In order to obtain a Gibbs relation from (247) we start by expressing the infinitesimal deformation $\delta\boldsymbol{\epsilon}$ in terms of state variables; $\delta\boldsymbol{\epsilon}$ itself is not a stable variable as its definition (225) shows, because $\boldsymbol{R}$ is the position of a point in the deformed state. Let us introduce, as state variables characterizing the position $\boldsymbol{R}$ of a point of the substance in terms of its position $\boldsymbol{R}^\circ$ in a fixed reference state (denoted by the symbol °), a tensor $\hat{\boldsymbol{\eta}}$ by means of the relation

$$\boldsymbol{R} = \hat{\boldsymbol{\eta}}\cdot\boldsymbol{R}^{\circ}. \tag{252}$$

Here $\hat{\boldsymbol{\eta}}$ is independent of $\boldsymbol{R}$ since only uniform deformations of the (uniform) cell are considered. Comparison of (252) with (225) shows that one has for $\delta\boldsymbol{\epsilon}$:

$$\delta\boldsymbol{\epsilon} = \delta\hat{\boldsymbol{\eta}}\cdot\hat{\boldsymbol{\eta}}^{-1}. \tag{253}$$

In this way $\delta\boldsymbol{\epsilon}$ is expressed in variations of the state variable $\hat{\boldsymbol{\eta}}$.

It is convenient to write the deformation tensor $\hat{\boldsymbol{\eta}}$ as a product of a symmetric 'dilatation' tensor $\boldsymbol{\eta}$ ($= \tilde{\boldsymbol{\eta}}$, the transposed tensor), and an orthogonal 'rotation' tensor $\boldsymbol{\eta}_{\mathrm{A}}$ (so that $\tilde{\boldsymbol{\eta}}_{\mathrm{A}} = \boldsymbol{\eta}_{\mathrm{A}}^{-1}$):

$$\hat{\boldsymbol{\eta}} = \boldsymbol{\eta}_{\mathrm{A}}\cdot\boldsymbol{\eta}. \tag{254}$$

If the sample is only slightly deformable the tensor $\boldsymbol{\eta}$ is nearly equal to the unit tensor $\mathbf{U}$. Furthermore the tensor $\boldsymbol{\eta}_{\mathrm{A}}$ might be parametrized in terms of three angles, for instance the Eulerian angles. Introducing (254) into (253) we get for the variation

$$\delta\boldsymbol{\epsilon} = (\delta\boldsymbol{\eta}_{\mathrm{A}}\cdot\boldsymbol{\eta}+\boldsymbol{\eta}_{\mathrm{A}}\cdot\delta\boldsymbol{\eta})\cdot\boldsymbol{\eta}^{-1}\cdot\boldsymbol{\eta}_{\mathrm{A}}^{-1} \simeq \delta\boldsymbol{\eta}_{\mathrm{A}}\cdot\tilde{\boldsymbol{\eta}}_{\mathrm{A}}+\boldsymbol{\eta}_{\mathrm{A}}\cdot\delta\boldsymbol{\eta}\cdot\tilde{\boldsymbol{\eta}}_{\mathrm{A}}, \tag{255}$$

where in the second expression $\boldsymbol{\eta}$ could be replaced by the unit tensor. The first term in the last member of (255) is antisymmetric as follows from $\boldsymbol{\eta}_{\mathrm{A}}\cdot\tilde{\boldsymbol{\eta}}_{\mathrm{A}} = \mathbf{U}$, while the second term is symmetric, since $\delta\boldsymbol{\eta}$ is symmetric and $\boldsymbol{\eta}_{\mathrm{A}}$ orthogonal. With the help of (255) we obtain as the second law for an amorphous or polycrystalline solid:

$$T\mathrm{d}s = \mathrm{d}u+v(\boldsymbol{\eta}_{\mathrm{A}}\cdot\mathbf{P}\cdot\boldsymbol{\eta}_{\mathrm{A}}) : \mathrm{d}\boldsymbol{\eta}+v\mathbf{P} : (\mathrm{d}\boldsymbol{\eta}_{\mathrm{A}}\cdot\tilde{\boldsymbol{\eta}}_{\mathrm{A}})-\boldsymbol{E}\cdot\mathrm{d}(v\boldsymbol{P})+v\boldsymbol{M}\cdot\mathrm{d}\boldsymbol{B}, \tag{256}$$

which is written with differentials since now all quantities are state variables. In the second term at the right-hand side only the symmetrical part $\mathbf{P}_{\mathrm{S}} = \frac{1}{2}(\mathbf{P}+\tilde{\mathbf{P}})$ of the pressure contributes, since $\mathrm{d}\boldsymbol{\eta}$ is symmetric and $\boldsymbol{\eta}_{\mathrm{A}}$ is orthogonal. In the third term only the antisymmetrical part $\frac{1}{2}(\mathbf{P}-\tilde{\mathbf{P}})$ remains, because the bracket expression is antisymmetric. From the angular momentum balance equation (196) it follows by employing the uniformity of the cell that

$$\boldsymbol{P}_{\mathrm{A}} = \boldsymbol{P}\wedge\boldsymbol{E}+\boldsymbol{M}\wedge\boldsymbol{B}, \tag{257}$$

or alternatively,

$$\tfrac{1}{2}(\mathbf{P}-\tilde{\mathbf{P}}) = \tfrac{1}{2}(\boldsymbol{PE}-\boldsymbol{EP}+\boldsymbol{MB}-\boldsymbol{BM}). \tag{258}$$

Then the second law (256) becomes

$$T\mathrm{d}s = \mathrm{d}u+v(\boldsymbol{\eta}_{\mathrm{A}}\cdot\mathbf{P}_{\mathrm{S}}\cdot\boldsymbol{\eta}_{\mathrm{A}}) : \mathrm{d}\boldsymbol{\eta}+v(\boldsymbol{PE}+\boldsymbol{MB}) : (\mathrm{d}\boldsymbol{\eta}_{\mathrm{A}}\cdot\boldsymbol{\eta}_{\mathrm{A}})-\boldsymbol{E}\cdot\mathrm{d}(v\boldsymbol{P})+v\boldsymbol{M}\cdot\mathrm{d}\boldsymbol{B}. \tag{259}$$

This form shows that the entropy changes not only through a change of the internal energy and through symmetric deformations, but also due to effects of the electromagnetic fields as is apparent from the last three terms. The first of the latter in particular shows the effect of rotation of the ellipsoidal cell as a whole in the external field. (If desired so, the tensor $\boldsymbol{\eta}_A$ may be expressed in terms of the three Euler angles with respect to a fixed reference state.)

Often the polarizations are parallel to the fields for amorphous and polycrystalline solids. Then (257), (258) and the third term at the right-hand side of (259) vanish, so that we are left with the second law

$$T\mathrm{d}s = \mathrm{d}u + v(\tilde{\boldsymbol{\eta}}_A\cdot\mathbf{P}_S\cdot\boldsymbol{\eta}_A) : \mathrm{d}\boldsymbol{\eta} - \boldsymbol{E}\cdot\mathrm{d}(v\boldsymbol{P}) + v\boldsymbol{M}\cdot\mathrm{d}\boldsymbol{B}. \tag{260}$$

In this way the Gibbs relations for uniform fluids and amorphous or polycrystalline solids have been found as the laws (251) and (260). The only difference between these two cases consists in the occurrence of a pressure tensor in (260) and a scalar pressure in (251).

## c. *The second law for plasmas*

The method used to derive a second law for fluids will be employed in this subsection to find that for plasmas.

Let us consider a uniform cell as a subsystem of a neutral plasma at rest in a uniform and constant field. The plasma is a mixture of charged particles of which the internal structure is disregarded. In such a system the uniform Maxwell fields are connected to the fields $\boldsymbol{E}_e(\boldsymbol{R})$ and $\boldsymbol{B}_e(\boldsymbol{R})$ from outside the cell as (v. (17))

$$\begin{aligned} \boldsymbol{E} &= \boldsymbol{E}_e(\boldsymbol{R}) - \varrho^e \nabla \int \frac{1}{4\pi|\boldsymbol{R}-\boldsymbol{R}'|}\,\mathrm{d}\boldsymbol{R}', \\ \boldsymbol{B} &= \boldsymbol{B}_e(\boldsymbol{R}) - c^{-1}\boldsymbol{J}\wedge\nabla \int \frac{1}{4\pi|\boldsymbol{R}-\boldsymbol{R}'|}\,\mathrm{d}\boldsymbol{R}', \end{aligned} \tag{261}$$

where $\varrho^e$ and $\boldsymbol{J}$ are the uniform charge and current densities. Hence the external fields are uniform only if the macroscopic charge and current densities $\varrho^e$ and $\boldsymbol{J}$ vanish. For that reason we only consider neutral plasmas without currents. Then the Maxwell fields are equal to the external fields.

The Hamiltonian for the plasma is (cf. (A26) or (A32) of appendix II)

$$H = \sum_k \frac{\boldsymbol{P}_k^2}{2m_k} + \sum_{k,l(k\neq l)} \frac{e_k e_l}{8\pi|\boldsymbol{R}_k-\boldsymbol{R}_l|} + \sum_k e_k \left\{\varphi_e(\boldsymbol{R}_k) - c^{-1}\frac{\boldsymbol{P}_k}{m_k}\cdot\boldsymbol{A}_e(\boldsymbol{R}_k)\right\}. \tag{262}$$

The potentials describe a constant and uniform external field and may thus be chosen as

$$\varphi_e(\boldsymbol{R}_k) = -\boldsymbol{R}_k \cdot \boldsymbol{E}_e, \quad \boldsymbol{A}_e(\boldsymbol{R}_k) = \tfrac{1}{2}\boldsymbol{B}_e \wedge \boldsymbol{R}_k. \tag{263}$$

(If a different gauge would have been used, the resulting Hamiltonian might be transformed to (262) with (263) by means of a canonical transformation.)

The free energy[1] $F$ is again a function of the external fields $\boldsymbol{E}_e$, $\boldsymbol{B}_e$, the temperature $T$ and the position of the boundary of the system. The partial derivatives with respect to the external fields and the temperature are up to order $c^{-1}$:

$$\frac{\partial F}{\partial \boldsymbol{E}_e} = -\int \boldsymbol{R}\varrho^e(\boldsymbol{R})\mathrm{d}\boldsymbol{R} = 0, \quad \frac{\partial F}{\partial \boldsymbol{B}_e} = -\tfrac{1}{2}c^{-1}\int \boldsymbol{R}\wedge \boldsymbol{J}(\boldsymbol{R})\mathrm{d}\boldsymbol{R} = 0, \tag{264}$$

with the charge and current densities (124), and

$$\frac{\partial F}{\partial T} = -S. \tag{265}$$

In this way we get for the change of the free energy

$$\delta F = -S\delta T + \mathbf{A} : \delta\boldsymbol{\epsilon}, \tag{266}$$

where $\mathbf{A}$ has still to be determined.

The free energy $F = \langle H\rangle - TS$ follows from the average of the Hamiltonian (cf. (A37)) for the plasma

$$H = K + \sum_{k,l(k\neq l)} \frac{e_k e_l}{8\pi|\boldsymbol{R}_k - \boldsymbol{R}_l|} + \sum_k e_k \varphi_e(\boldsymbol{R}_k). \tag{267}$$

The average of this expression is

$$\langle H\rangle = \sum_a \int \tfrac{1}{2}m_a \hat{\boldsymbol{v}}_1^2 f_1^a(\boldsymbol{R}_1, \boldsymbol{v}_1)\mathrm{d}\boldsymbol{R}_1\,\mathrm{d}\boldsymbol{v}_1 + \sum_{a,b}\int \frac{e_a e_b}{8\pi|\boldsymbol{R}_1 - \boldsymbol{R}_2|} f_2^{ab}(\boldsymbol{R}_1, \boldsymbol{R}_2)\mathrm{d}\boldsymbol{R}_1\,\mathrm{d}\boldsymbol{R}_2, \tag{268}$$

where (263) and the vanishing of the charge density have been taken into account. In the first term the integration over $\boldsymbol{R}_1$ may be performed as a

[1] Here the free energy is denoted by a symbol $F$ without asterisk, since it will turn out that the average Hamiltonian $\langle H\rangle$ is equal to the total internal energy (v. (270)).

consequence of the uniformity of the system. In the second term we write $f_2^{ab}$ as the sum of a correlation function $c_2^{ab}$ and the product $f_1^a f_1^b$. The latter, uncorrelated part gives no contribution, since the charge density vanishes, so that we obtain now, using also (131)

$$\langle H \rangle = V\varrho u^{\mathrm{K}} + \sum_{a,b} \int \frac{e_a e_b}{8\pi s} c_2^{ab}(\boldsymbol{R}+\tfrac{1}{2}\boldsymbol{s}, \boldsymbol{R}-\tfrac{1}{2}\boldsymbol{s}) \mathrm{d}\boldsymbol{R}\, \mathrm{d}\boldsymbol{s}. \tag{269}$$

For neutral plasmas in equilibrium the correlation function has short range. Then the integral over $\boldsymbol{R}$ may be performed, since the system is uniform. In this way we get with (138) and (141)

$$\langle H \rangle = V\varrho u. \tag{270}$$

The tensor $\mathbf{A}$, which occurs in (266), follows from (A55) of appendix III with (262):

$$\begin{aligned}\mathbf{A} = &- \sum_a \int m_a \hat{\boldsymbol{v}}_1 \hat{\boldsymbol{v}}_1 f_1^a(\boldsymbol{R}_1, \boldsymbol{v}_1) \mathrm{d}\boldsymbol{R}_1 \, \mathrm{d}\boldsymbol{v}_1 - \sum_a e_a \int \boldsymbol{R}_1 \boldsymbol{E}_\mathrm{e} f_1^a(\boldsymbol{R}_1) \mathrm{d}\boldsymbol{R}_1 \\ &+ \tfrac{1}{2} c^{-1} \sum_a \int e_a \{\hat{\boldsymbol{v}}_1(\boldsymbol{R}_1 \wedge \boldsymbol{B}_\mathrm{e}) - \boldsymbol{R}_1(\hat{\boldsymbol{v}}_1 \wedge \boldsymbol{B}_\mathrm{e})\} f_1^a(\boldsymbol{R}_1, \boldsymbol{v}_1) \mathrm{d}\boldsymbol{R}_1 \, \mathrm{d}\boldsymbol{v}_1 \\ &+ \sum_{a,b} \int (\boldsymbol{R}_1 - \boldsymbol{R}_2) \boldsymbol{\nabla}_1 \frac{e_a e_b}{8\pi |\boldsymbol{R}_1 - \boldsymbol{R}_2|} f_2^{ab}(\boldsymbol{R}_1, \boldsymbol{R}_2) \mathrm{d}\boldsymbol{R}_1 \, \mathrm{d}\boldsymbol{R}_2,\end{aligned} \tag{271}$$

where we have used the Hamilton equation $\partial H/\partial \boldsymbol{P}_k = \dot{\boldsymbol{R}}_k \equiv \boldsymbol{v}_k$. In the first term the integration over $\boldsymbol{R}_1$ may be performed. In the second and third terms one recognizes the charge and current densities (124), which vanish in the plasma studied. In the fourth term we split $f_2^{ab}$ into $c_2^{ab}$ and $f_1^a f_1^b$. Again the latter part gives no contribution because the charge density is zero. In this way (271) becomes, with the help of (122),

$$\mathbf{A} = -V\mathbf{P}^{\mathrm{K}} + \sum_{a,b} \int \boldsymbol{s} \boldsymbol{\nabla}_s \frac{e_a e_b}{8\pi s} c_2^{ab}(\boldsymbol{R}+\tfrac{1}{2}\boldsymbol{s}, \boldsymbol{R}-\tfrac{1}{2}\boldsymbol{s}) \mathrm{d}\boldsymbol{R}\, \mathrm{d}\boldsymbol{s}. \tag{272}$$

The integration over $\boldsymbol{R}$ may be performed owing to the short range character of the correlation function. Then we get with (127) and (129):

$$\mathbf{A} = -V\mathbf{P}. \tag{273}$$

So finally we obtain from (266), with (264), (265), (270) and (273), dividing by the total mass $M$,

$$T\delta s = \delta u + v\mathbf{P} : \delta \boldsymbol{\epsilon}, \tag{274}$$

where $s = S/M$ and $v = V/M$ are the specific entropy and volume. From

(274) it follows that the specific energy at constant $s$ and $\boldsymbol{\epsilon}$ does not change if fields are switched on:

$$u = u_0, \tag{275}$$

where $u_0$ is the specific energy at zero fields, which depends only on $u$ and $s$, since the plasma is isotropic in the absence of fields. So in (274) only the trace of $\delta\boldsymbol{\epsilon}$ (which is equal to $v^{-1}\delta v$) may occur. Therefore the second law for a neutral and current-free plasma becomes

$$T\mathrm{d}s = \mathrm{d}u + p\,\mathrm{d}v, \tag{276}$$

showing that the pressure tensor in a plasma reduces to a scalar at equilibrium.

## d. *The second law for crystalline solids*

For the systems with short range correlations treated so far we derived the second law of thermodynamics by considering a uniform cell as a subsystem of a larger system, which played the role of a heat bath. For the case of systems with long range correlations – as crystalline solids – such a division of the system into cells is no longer feasible. One has to consider in that case the system as a whole. As a consequence one can no longer suppose that the system is uniform: non-uniformities will enter the system through boundary effects (even if simple shapes are chosen for this boundary).

We use again the canonical ensemble to describe the system in uniform external fields in a heat bath, limiting ourselves to systems without space charge. Then the existence of a thermodynamic limit has been proved[1]. At equilibrium the electric current density in the system at rest will vanish (since then both the conduction and convection currents are zero). The Hamiltonian is given by (A32) with the potentials (263) for the uniform external fields. The partial derivatives of the free energy with respect to the external fields are, up to order $c^{-1}$,

$$\frac{\partial F^*}{\partial \boldsymbol{E}_\mathrm{e}} = -\langle \sum_k (e_k \boldsymbol{R}_k + \overline{\boldsymbol{\mu}}_k^{(1)})\rangle = -\int \boldsymbol{P}(\boldsymbol{R})\mathrm{d}\boldsymbol{R},$$

$$\frac{\partial F^*}{\partial \boldsymbol{B}_\mathrm{e}} = -\left\langle \sum_k \left(\tfrac{1}{2}c^{-1} e_k \boldsymbol{R}_k \wedge \frac{\boldsymbol{P}_k}{m_k} + \overline{\boldsymbol{v}}_k^{(1)} + c^{-1}\overline{\boldsymbol{\mu}}_k^{(1)} \wedge \frac{\boldsymbol{P}_k}{m_k}\right)\right\rangle = -\int \boldsymbol{M}(\boldsymbol{R})\mathrm{d}\boldsymbol{R}, \tag{277}$$

where the expressions (64–65) have been used and the fact that the charge and current densities vanish. Furthermore the partial derivative of the free

[1] J. L. Lebowitz and E. H. Lieb, Phys. Rev. Lett. **22**(1969)631.

energy with respect to the temperature is

$$\frac{\partial F^*}{\partial T} = -S. \tag{278}$$

The total change of the free energy is now

$$\delta F^* = -S\delta T - V\bar{\boldsymbol{P}}\cdot\delta\boldsymbol{E}_e - V\overline{\boldsymbol{M}}\cdot\delta\boldsymbol{B}_e + \delta_\varepsilon F^*, \tag{279}$$

where we have introduced the notation $\overline{X}$ for space averages of a quantity $X(\boldsymbol{R})$:

$$\overline{X} \equiv \frac{1}{V}\int X(\boldsymbol{R})\mathrm{d}\boldsymbol{R}, \tag{280}$$

and where $\delta_\varepsilon F^*$ is the change of the free energy through deformations. The canonical average $\langle H\rangle \equiv U^*$ of the Hamiltonian (A37) with (263) is (cf. (232))

$$U^* = U - \int \boldsymbol{P}(\boldsymbol{R}_1)\boldsymbol{P}(\boldsymbol{R}_2) : \nabla_1\nabla_1 \frac{1}{8\pi|\boldsymbol{R}_1-\boldsymbol{R}_2|}\mathrm{d}\boldsymbol{R}_1\,\mathrm{d}\boldsymbol{R}_2 - V\bar{\boldsymbol{P}}\cdot\boldsymbol{E}_e, \tag{281}$$

where (115) with (78), (97) and (153) have been used and where $U$ is the total internal energy $V\overline{\varrho u}$ of the system.

The change of the free energy $\delta_\varepsilon F^*$ under deformations follows from (A53) of appendix III, with the Hamiltonian (A32). The term with the deformation gradient tensor $\delta\mathsf{e}$ in (A53) gives

$$-\int \{\mathsf{P}^{\mathrm{K}}(\boldsymbol{R}) + \mathsf{P}^{\mathrm{F}}(\boldsymbol{R})\} : \delta\mathsf{e}(\boldsymbol{R})\mathrm{d}\boldsymbol{R}, \tag{282}$$

as follows from the explicit forms (63) and (73) of the kinetic and 'field' part of the pressure tensor. The term with $\delta\boldsymbol{\epsilon}$ in (A53) becomes upon introduction of (A32)

$$\left\langle \sum_{k,l(k\neq l)} \sum_{i,j} \{\boldsymbol{R}_k\cdot\delta\tilde{\boldsymbol{\epsilon}}(\boldsymbol{R}_k) - \boldsymbol{R}_l\cdot\delta\tilde{\boldsymbol{\epsilon}}(\boldsymbol{R}_l)\}\cdot\nabla_k \frac{e_{ki}\,e_{lj}}{8\pi|\boldsymbol{R}_{ki}-\boldsymbol{R}_{lj}|}\right\rangle. \tag{283}$$

If this expression is split into a long range and a short range part by making a multipole expansion and if appropriate two-point distribution and correlation functions are introduced, one finds

$$-\int \{\boldsymbol{R}_1\cdot\delta\tilde{\boldsymbol{\epsilon}}(\boldsymbol{R}_1) - \boldsymbol{R}_2\cdot\delta\tilde{\boldsymbol{\epsilon}}(\boldsymbol{R}_2)\}\boldsymbol{P}(\boldsymbol{R}_1)\boldsymbol{P}(\boldsymbol{R}_2) \,\vdots\, \nabla_1\nabla_1\nabla_1 \frac{1}{8\pi|\boldsymbol{R}_1-\boldsymbol{R}_2|}\mathrm{d}\boldsymbol{R}_1\,\mathrm{d}\boldsymbol{R}_2$$

$$+\int\Big[\{(\boldsymbol{R}+\tfrac{1}{2}\boldsymbol{s})\cdot\delta\tilde{\boldsymbol{\epsilon}}(\boldsymbol{R}+\tfrac{1}{2}\boldsymbol{s}) - (\boldsymbol{R}-\tfrac{1}{2}\boldsymbol{s})\cdot\delta\tilde{\boldsymbol{\epsilon}}(\boldsymbol{R}-\tfrac{1}{2}\boldsymbol{s})\}\cdot\nabla_s \sum_{n,m=0}^{\infty}(-1)^m\overline{\boldsymbol{\mu}}_1^{(n)}$$

$$\vdots\, \nabla_s^n\overline{\boldsymbol{\mu}}_2^{(m)} \,\vdots\, \nabla_s^m \frac{1}{8\pi s}\Big]\, c_2(\boldsymbol{R}+\tfrac{1}{2}\boldsymbol{s}, 1, \boldsymbol{R}-\tfrac{1}{2}\boldsymbol{s}, 2)\mathrm{d}\boldsymbol{R}\,\mathrm{d}\boldsymbol{s}\,\mathrm{d}1\,\mathrm{d}2$$

$$+\int\left[\{(\boldsymbol{R}+\tfrac{1}{2}\boldsymbol{s})\cdot\delta\tilde{\boldsymbol{\epsilon}}(\boldsymbol{R}+\tfrac{1}{2}\boldsymbol{s})-(\boldsymbol{R}-\tfrac{1}{2}\boldsymbol{s})\cdot\delta\tilde{\boldsymbol{\epsilon}}(\boldsymbol{R}-\tfrac{1}{2}\boldsymbol{s})\}\cdot\nabla_{s}\left(\sum_{i,j}\frac{e_{1i}e_{2j}}{8\pi|\boldsymbol{s}+\boldsymbol{r}_{1i}-\boldsymbol{r}_{2j}|}\right.\right.$$

$$\left.\left.-\sum_{n,m=0}^{\infty}(-1)^m\overline{\boldsymbol{\mu}}_1^{(n)}:\nabla_s^n\overline{\boldsymbol{\mu}}_2^{(m)}:\nabla_s^m\frac{1}{8\pi s}\right)\right]$$

$$f_2(\boldsymbol{R}+\tfrac{1}{2}\boldsymbol{s},1,\boldsymbol{R}-\tfrac{1}{2}\boldsymbol{s},2)\mathrm{d}\boldsymbol{R}\,\mathrm{d}\boldsymbol{s}\,\mathrm{d}1\,\mathrm{d}2, \quad (284)$$

where in the second and third term we introduced the variables $\boldsymbol{R}$ and $\boldsymbol{s}$, connected with $\boldsymbol{R}_1$ and $\boldsymbol{R}_2$ by $\boldsymbol{R}_1 = \boldsymbol{R}+\frac{1}{2}\boldsymbol{s}$ and $\boldsymbol{R}_2 = \boldsymbol{R}-\frac{1}{2}\boldsymbol{s}$. In the last term the integration over $\boldsymbol{s}$ is extended over small values of $s$ only, so that one may expand the factors $(\boldsymbol{R}\pm\frac{1}{2}\boldsymbol{s})\cdot\delta\tilde{\boldsymbol{\epsilon}}(\boldsymbol{R}\pm\frac{1}{2}\boldsymbol{s})$ around $\boldsymbol{R}\cdot\delta\tilde{\boldsymbol{\epsilon}}(\boldsymbol{R})$ and break off after the second term. Then the last term of (284) becomes

$$-\int \mathbf{P}^{\mathrm{S}}(\boldsymbol{R}):\delta\mathbf{e}(\boldsymbol{R})\mathrm{d}\boldsymbol{R}, \quad (285)$$

where we introduced the short range pressure tensor (94) and the tensor $\delta\mathbf{e}$ (A51). The second term in (284) may be written in the form

$$-\int \mathbf{P}^{\mathrm{C}}(\boldsymbol{R}):\delta\mathbf{e}(\boldsymbol{R})\mathrm{d}\boldsymbol{R}, \quad (286)$$

as we shall now prove. To that end we introduce the correlation pressure (151) into (286); owing to the symmetry of the integrand of (151) one may employ $c_2^{\ddagger}$ (202) instead of $c_2^{+}$ (149). Then one gets

$$-\int \mathbf{P}^{\mathrm{C}}(\boldsymbol{R}):\delta\mathbf{e}(\boldsymbol{R})\mathrm{d}\boldsymbol{R} = \int_{-1}^{1}\mathrm{d}\lambda\int_{-\infty}^{\infty}\mathrm{d}\boldsymbol{R}'\int_{-\infty}^{\infty}\mathrm{d}\boldsymbol{s}\int\mathrm{d}1\,\mathrm{d}2\sum_{n,m=0}^{\infty}(-1)^m$$

$$\left\{\boldsymbol{s}\cdot\delta\tilde{\mathbf{e}}(\boldsymbol{R}'-\tfrac{1}{2}\lambda\boldsymbol{s})\cdot\nabla_s\overline{\boldsymbol{\mu}}_1^{(n)}:\nabla_s^n\overline{\boldsymbol{\mu}}_2^{(m)}:\nabla_s^m\frac{1}{16\pi s}\right\}c_2(\boldsymbol{R}'+\tfrac{1}{2}\boldsymbol{s},1,\boldsymbol{R}'-\tfrac{1}{2}\boldsymbol{s},2), \quad (287)$$

where we introduced the integration variables $\boldsymbol{R}' = \boldsymbol{R}+\frac{1}{2}\lambda\boldsymbol{s}$. Effectively the integrations over $\boldsymbol{R}'$ and $\boldsymbol{s}$ just as those over $\boldsymbol{R}$ and $\boldsymbol{s}$ are extended over those values for which the arguments of the correlation function are inside the volume of the system. Since the correlation function vanishes if these arguments indicate positions outside the volume, we may for convenience write $-\infty$ and $\infty$ as integration limits. The first factor in the integrand in (287) may be written in an alternative form

$$\boldsymbol{s}\cdot\delta\tilde{\mathbf{e}}(\boldsymbol{R}'-\tfrac{1}{2}\lambda\boldsymbol{s})\cdot\nabla_s = -2\frac{\partial}{\partial\lambda}\{(\boldsymbol{R}'-\tfrac{1}{2}\lambda\boldsymbol{s})\cdot\delta\tilde{\boldsymbol{\epsilon}}(\boldsymbol{R}'-\tfrac{1}{2}\lambda\boldsymbol{s})\}\cdot\nabla_s, \quad (288)$$

if we use the definition (A51) of $\delta\mathbf{e}$. If this expression is inserted into (287) the integration over $\lambda$ may be performed. Then one recovers indeed the second term of (284).

We have found now for the change of the free energy under deformations

$$\delta_\varepsilon F^* = -\int \mathbf{P}(\boldsymbol{R}) : \delta \mathbf{e}(\boldsymbol{R}) \mathrm{d}\boldsymbol{R}$$
$$-\int \{\boldsymbol{R}_1 \cdot \delta\tilde{\boldsymbol{\epsilon}}(\boldsymbol{R}_1) - \boldsymbol{R}_2 \cdot \delta\tilde{\boldsymbol{\epsilon}}(\boldsymbol{R}_2)\} \boldsymbol{P}(\boldsymbol{R}_1)\boldsymbol{P}(\boldsymbol{R}_2) \vdots \nabla_1 \nabla_1 \nabla_1 \frac{1}{8\pi|\boldsymbol{R}_1 - \boldsymbol{R}_2|} \mathrm{d}\boldsymbol{R}_1 \, \mathrm{d}\boldsymbol{R}_2 . \quad (289)$$

This form for the change of the free energy seems to be dependent on the deformation tensor $\delta\boldsymbol{\epsilon}$ (or $\delta\mathbf{e}$ (A51)) throughout the volume. We may however obtain an alternative form of $\delta_\varepsilon F^*$ which shows explicitly that only the deformation at the surface comes into play. Such an expression follows if one employs (A48) of the third appendix. Together with (241) one then finds

$$\delta_\varepsilon F^* = -\int^S \boldsymbol{n} \cdot \mathbf{P} \cdot \delta\boldsymbol{\epsilon} \cdot \boldsymbol{R} \, \mathrm{d}S - \tfrac{1}{2} \int^S \boldsymbol{n} \cdot \delta\boldsymbol{\epsilon} \cdot \boldsymbol{R} (\boldsymbol{P} \cdot \boldsymbol{n})^2 \mathrm{d}S. \quad (290)$$

(For uniform pressure and deformation tensors this formula reduces to (242).) One may prove the equivalence of (289) and (290) by performing a partial integration in the first term of (289), using the definition (A51) of $\delta\mathbf{e}$, and employing the identity

$$\int (\nabla \cdot \mathbf{P}) \cdot \delta\boldsymbol{\epsilon} \cdot \boldsymbol{R} \, \mathrm{d}\boldsymbol{R} + \tfrac{1}{2} \int^S \boldsymbol{n} \cdot \delta\boldsymbol{\epsilon} \cdot \boldsymbol{R} (\boldsymbol{P} \cdot \boldsymbol{n})^2 \mathrm{d}S$$
$$= \int \{\boldsymbol{R}_1 \cdot \delta\tilde{\boldsymbol{\epsilon}}(\boldsymbol{R}_1) - \boldsymbol{R}_2 \cdot \delta\tilde{\boldsymbol{\epsilon}}(\boldsymbol{R}_2)\} \boldsymbol{P}(\boldsymbol{R}_1)\boldsymbol{P}(\boldsymbol{R}_2) \vdots \nabla_1 \nabla_1 \nabla_1 \frac{1}{8\pi|\boldsymbol{R}_1 - \boldsymbol{R}_2|} \mathrm{d}\boldsymbol{R}_1 \, \mathrm{d}\boldsymbol{R}_2 , \quad (291)$$

which holds for a polarized system in equilibrium. The proof of this relation (v. problem 5) makes use of the equation of motion (150), which for the present system in equilibrium and at rest reads

$$\nabla \cdot \mathbf{P} = (\nabla \boldsymbol{E}) \cdot \boldsymbol{P} + (\nabla \boldsymbol{B}) \cdot \boldsymbol{M}. \quad (292)$$

In the second term at the right-hand side only the external magnetic field appears (v. (71)). This field is uniform, so that (292) reduces to

$$\nabla \cdot \mathbf{P} = (\nabla \boldsymbol{E}) \cdot \boldsymbol{P}. \quad (293)$$

(For uniform pressure, deformation tensor and polarization the relation (291) reduces to (243), as follows by employing (236).)

The change of entropy is found now if one substitutes (227) and (289) or (290) into (279). In the latter case, i.e. with (290), we get the entropy law

$$T\delta S = \delta U^* + \int^S \boldsymbol{n}\cdot\mathbf{P}\cdot\delta\boldsymbol{\epsilon}\cdot\boldsymbol{R}\,\mathrm{d}S + \tfrac{1}{2}\int^S \boldsymbol{n}\cdot\delta\boldsymbol{\epsilon}\cdot\boldsymbol{R}(\boldsymbol{P}\cdot\boldsymbol{n})^2\mathrm{d}S + V\bar{\boldsymbol{P}}\cdot\delta\boldsymbol{E}_\mathrm{e} + V\overline{\boldsymbol{M}}\cdot\delta\boldsymbol{B}_\mathrm{e}, \tag{294}$$

while in the first case, i.e. with (289), the result is

$$T\delta S = \delta U^* + \int \mathbf{P}(\boldsymbol{R}) : \delta\mathbf{e}(\boldsymbol{R})\mathrm{d}\boldsymbol{R} + \int \{\boldsymbol{R}_1\cdot\delta\tilde{\boldsymbol{\epsilon}}(\boldsymbol{R}_1) - \boldsymbol{R}_2\cdot\delta\tilde{\boldsymbol{\epsilon}}(\boldsymbol{R}_2)\}\boldsymbol{P}(\boldsymbol{R}_1)\boldsymbol{P}(\boldsymbol{R}_2)$$

$$\vdots\, \boldsymbol{\nabla}_1\boldsymbol{\nabla}_1\boldsymbol{\nabla}_1 \frac{1}{8\pi|\boldsymbol{R}_1-\boldsymbol{R}_2|}\mathrm{d}\boldsymbol{R}_1\,\mathrm{d}\boldsymbol{R}_2 + V\bar{\boldsymbol{P}}\cdot\delta\boldsymbol{E}_\mathrm{e} + V\overline{\boldsymbol{M}}\cdot\delta\boldsymbol{B}_\mathrm{e}. \tag{295}$$

The form (294) shows that the change of entropy depends only on the deformation tensor at the boundary, while (295) has a form that reduces for uniform pressure, deformation and polarizations to (244) of subsection *b*. If one inserts moreover (281) the relation (295) becomes

$$T\delta S = \delta U + \int \mathbf{P}(\boldsymbol{R}) : \delta\mathbf{e}(\boldsymbol{R})\mathrm{d}\boldsymbol{R} - \boldsymbol{E}_\mathrm{e}\cdot\delta(V\bar{\boldsymbol{P}})$$

$$-\delta\left\{\int \boldsymbol{P}(\boldsymbol{R}_1)\boldsymbol{P}(\boldsymbol{R}_2) : \boldsymbol{\nabla}_1\boldsymbol{\nabla}_1 \frac{1}{8\pi|\boldsymbol{R}_1-\boldsymbol{R}_2|}\mathrm{d}\boldsymbol{R}_1\,\mathrm{d}\boldsymbol{R}_2\right\}$$

$$+\int \{\boldsymbol{R}_1\cdot\delta\tilde{\boldsymbol{\epsilon}}(\boldsymbol{R}_1) - \boldsymbol{R}_2\cdot\delta\tilde{\boldsymbol{\epsilon}}(\boldsymbol{R}_2)\}\boldsymbol{P}(\boldsymbol{R}_1)\boldsymbol{P}(\boldsymbol{R}_2) \,\vdots\, \boldsymbol{\nabla}_1\boldsymbol{\nabla}_1\boldsymbol{\nabla}_1 \frac{1}{8\pi|\boldsymbol{R}_1-\boldsymbol{R}_2|}$$

$$\mathrm{d}\boldsymbol{R}_1\,\mathrm{d}\boldsymbol{R}_2 + V\overline{\boldsymbol{M}}\cdot\delta\boldsymbol{B}_\mathrm{e}. \tag{296}$$

We may cast this law in a form which contains the Maxwell fields instead of the external fields. Let us consider the fourth term of the right-hand side separately. As a consequence of the variation the polarization changes and also the boundary of the integral, so that it may be written as

$$-\int [\{\delta_0\boldsymbol{P}(\boldsymbol{R}_1)\}\boldsymbol{P}(\boldsymbol{R}_2) + \boldsymbol{P}(\boldsymbol{R}_1)\delta_0\boldsymbol{P}(\boldsymbol{R}_2)] : \boldsymbol{\nabla}_1\boldsymbol{\nabla}_1 \frac{1}{8\pi|\boldsymbol{R}_1-\boldsymbol{R}_2|}\mathrm{d}\boldsymbol{R}_1\,\mathrm{d}\boldsymbol{R}_2$$

$$-\int^V \mathrm{d}S_1 \int^S \mathrm{d}\boldsymbol{R}_2\,\delta\boldsymbol{R}_1\cdot\boldsymbol{n}\boldsymbol{P}(\boldsymbol{R}_1)\boldsymbol{P}(\boldsymbol{R}_2) : \boldsymbol{\nabla}_1\boldsymbol{\nabla}_1 \frac{1}{8\pi|\boldsymbol{R}_1-\boldsymbol{R}_2|}$$

$$-\int^S \mathrm{d}S_2 \int^V \mathrm{d}\boldsymbol{R}_1\,\delta\boldsymbol{R}_2\cdot\boldsymbol{n}\boldsymbol{P}(\boldsymbol{R}_1)\boldsymbol{P}(\boldsymbol{R}_2) : \boldsymbol{\nabla}_1\boldsymbol{\nabla}_1 \frac{1}{8\pi|\boldsymbol{R}_1-\boldsymbol{R}_2|}, \tag{297}$$

where in the first term $\delta_0\boldsymbol{P}(\boldsymbol{R}_i)$ is the 'syntopic' variation (i.e. $\boldsymbol{P}'(\boldsymbol{R}_i)-\boldsymbol{P}(\boldsymbol{R}_i)$) of the polarization. In the second term the integration with respect to $\boldsymbol{R}_1$ is extended over the surface $S$ of the sample with surface element $\mathrm{d}S_1$ and normal $\boldsymbol{n}$, while the integration with respect to $\boldsymbol{R}_2$ is performed over the volume $V$. Furthermore $\delta\boldsymbol{R}_1$ is the variation $\delta\boldsymbol{\epsilon}(\boldsymbol{R}_1)\cdot\boldsymbol{R}_1$. Similar remarks apply to the third term. If the last two terms are transformed with Gauss's theorem we obtain for (297):

$$-\int [\{\delta\boldsymbol{P}(\boldsymbol{R}_1)\boldsymbol{P}(\boldsymbol{R}_2)+\boldsymbol{P}(\boldsymbol{R}_1)\delta\boldsymbol{P}(\boldsymbol{R}_2)\} : \nabla_1\nabla_1$$
$$+\{\boldsymbol{R}_1\cdot\delta\tilde{\boldsymbol{\epsilon}}(\boldsymbol{R}_1)-\boldsymbol{R}_2\cdot\delta\tilde{\boldsymbol{\epsilon}}(\boldsymbol{R}_2)\}\boldsymbol{P}(\boldsymbol{R}_1)\boldsymbol{P}(\boldsymbol{R}_2) \,\vdots\, \nabla_1\nabla_1\nabla_1$$
$$+\{\mathrm{Tr}\ \delta\mathbf{e}(\boldsymbol{R}_1)+\mathrm{Tr}\ \delta\mathbf{e}(\boldsymbol{R}_2)\}\boldsymbol{P}(\boldsymbol{R}_1)\boldsymbol{P}(\boldsymbol{R}_2) : \nabla_1\nabla_1]\,\frac{1}{8\pi|\boldsymbol{R}_1-\boldsymbol{R}_2|}\,\mathrm{d}\boldsymbol{R}_1\,\mathrm{d}\boldsymbol{R}_2\,, \tag{298}$$

where the 'asyntopic' variations of the polarizations ($i = 1, 2$) are

$$\delta\boldsymbol{P}(\boldsymbol{R}_i) \equiv \boldsymbol{P}'(\boldsymbol{R}_i')-\boldsymbol{P}(\boldsymbol{R}_i) = \delta_0\,\boldsymbol{P}(\boldsymbol{R}_i)+\boldsymbol{R}_i\cdot\delta\tilde{\boldsymbol{\epsilon}}(\boldsymbol{R}_i)\cdot\nabla_i\,\boldsymbol{P}(\boldsymbol{R}_i). \tag{299}$$

The third term at the right-hand side of (296) may likewise be written in terms of the asyntopic variation (299) since

$$\delta(V\bar{\boldsymbol{P}}) \equiv \delta\int \boldsymbol{P}(\boldsymbol{R})\mathrm{d}\boldsymbol{R} = \int \delta\boldsymbol{P}(\boldsymbol{R})\mathrm{d}\boldsymbol{R}+\int \{\mathrm{Tr}\ \delta\mathbf{e}(\boldsymbol{R})\}\boldsymbol{P}(\boldsymbol{R})\mathrm{d}\boldsymbol{R}. \tag{300}$$

Substituting (298) and (300) into (296), we obtain with the expressions (71) for the Maxwell fields (with vanishing charge density) the non-relativistic entropy law

$$T\delta S = \delta U+\int [\mathbf{P}(\boldsymbol{R}) : \delta\mathbf{e}(\boldsymbol{R})-\boldsymbol{E}(\boldsymbol{R})\cdot\delta\boldsymbol{P}(\boldsymbol{R})$$
$$-\{\mathrm{Tr}\ \delta\mathbf{e}(\boldsymbol{R})\}\boldsymbol{P}(\boldsymbol{R})\cdot\boldsymbol{E}(\boldsymbol{R})+\boldsymbol{M}(\boldsymbol{R})\cdot\delta\boldsymbol{B}]\mathrm{d}\boldsymbol{R}, \tag{301}$$

or with the bar notation for volume averages

$$T\delta S = \delta U+V\overline{\mathbf{P} : \delta\mathbf{e}}-V\overline{\boldsymbol{E}\cdot\delta\boldsymbol{P}}-V\overline{(\mathrm{Tr}\ \delta\mathbf{e})\boldsymbol{P}\cdot\boldsymbol{E}}+V\overline{\boldsymbol{M}\cdot\delta\boldsymbol{B}}. \tag{302}$$

This form of the entropy law is closely analogous to (247) if the latter is multiplied by the total (constant) mass $M$ of the system (so that $s$, $u$ and $v$ are replaced by $S$, $U$ and $V$). The difference between these formulae is that (302) contains the tensor $\delta\mathbf{e}$ instead of $\delta\boldsymbol{\epsilon}$ and volume averages instead of uniform quantities.

The entropy law (301) or (302) contains the tensor $\delta\mathbf{e}(\boldsymbol{R})$, which characterizes the deformation throughout the volume. However, one may show by a transformation of the right-hand side of (301) that effectively only values of the deformation tensor at the surface come in, just as in (289–290).

With the help of (A51), (299) and partial integrations one may write (301) as

$$T\delta S = \delta U + \int^{S} \boldsymbol{n}\cdot(\mathbf{P}\cdot\delta\boldsymbol{\epsilon}\cdot\boldsymbol{R} - \delta\boldsymbol{\epsilon}\cdot\boldsymbol{R}\boldsymbol{P}\cdot\boldsymbol{E})\mathrm{d}S$$
$$- \int \{(\nabla\cdot\mathbf{P})\cdot\delta\boldsymbol{\epsilon}\cdot\boldsymbol{R} + \boldsymbol{E}\cdot\delta_0\boldsymbol{P} - \boldsymbol{R}\cdot\delta\tilde{\boldsymbol{\epsilon}}\cdot(\nabla\boldsymbol{E})\cdot\boldsymbol{P} - \boldsymbol{M}\cdot\delta_0\boldsymbol{B}\}\mathrm{d}\boldsymbol{R}, \quad (303)$$

where $\boldsymbol{n}$ is the outward normal to the surface element $\mathrm{d}S$. We now substitute the equation of motion (293), with the result

$$T\delta S = \delta U + \int^{S} \boldsymbol{n}\cdot(\mathbf{P}\cdot\delta\boldsymbol{\epsilon}\cdot\boldsymbol{R} - \delta\boldsymbol{\epsilon}\cdot\boldsymbol{R}\boldsymbol{P}\cdot\boldsymbol{E})\mathrm{d}S - \int (\boldsymbol{E}\cdot\delta_0\boldsymbol{P} - \boldsymbol{M}\cdot\delta_0\boldsymbol{B})\mathrm{d}\boldsymbol{R}, \quad (304)$$

which may also be obtained directly from (294). This formula shows that the change of entropy depends only on the value at the surface of the deformation tensor $\delta\boldsymbol{\epsilon}$. It depends moreover on the variation of the total energy, of the polarization and of the magnetic field throughout the system.

It should be noted that the infinitesimal $\delta\boldsymbol{\epsilon}$ is defined by (225) and is thus not the variation of a state variable. Just as in subsection *b* let us introduce as state variables characterizing the deformation of the boundary a tensor $\hat{\boldsymbol{\eta}}$, which gives the transformation of the position $\boldsymbol{R}$ of the boundary from a fixed position $\boldsymbol{R}^\circ$ in a reference state denoted by the symbol $^\circ$ to a deformed state

$$\boldsymbol{R} = \hat{\boldsymbol{\eta}}(\boldsymbol{R}^\circ)\cdot\boldsymbol{R}^\circ. \quad (305)$$

(In contrast to the case of subsection *b* the tensor $\hat{\boldsymbol{\eta}}$ is now a function of $\boldsymbol{R}^\circ$.) For an infinitesimal deformation we have then

$$\delta\boldsymbol{R} = \delta\hat{\boldsymbol{\eta}}(\boldsymbol{R}^\circ)\cdot\boldsymbol{R}^\circ = \delta\hat{\boldsymbol{\eta}}(\boldsymbol{R}^\circ)\cdot\hat{\boldsymbol{\eta}}^{-1}(\boldsymbol{R}^\circ)\cdot\boldsymbol{R}, \quad (306)$$

so that the variation $\delta\boldsymbol{\epsilon}(\boldsymbol{R})$ is

$$\delta\boldsymbol{\epsilon}(\boldsymbol{R}) = \delta\hat{\boldsymbol{\eta}}(\boldsymbol{R}^\circ)\cdot\hat{\boldsymbol{\eta}}^{-1}(\boldsymbol{R}^\circ). \quad (307)$$

We want to write the tensor $\hat{\boldsymbol{\eta}}(\boldsymbol{R}^\circ)$ as the product of an orthogonal tensor $\boldsymbol{\eta}_{\mathrm{A}}$ (independent of $\boldsymbol{R}^\circ$) and a tensor $\boldsymbol{\eta}(\boldsymbol{R}^\circ)$. To fix $\boldsymbol{\eta}_{\mathrm{A}}$ we factorize the surface integral

$$\int^{S^\circ} \hat{\boldsymbol{\eta}}(\boldsymbol{R}^\circ)\cdot\boldsymbol{R}^\circ\boldsymbol{n}^\circ\mathrm{d}S^\circ \quad (308)$$

(where $\boldsymbol{n}^\circ$ is the normal to the surface element $\mathrm{d}S^\circ$ of the surface $S^\circ$ in the reference state) into a product of the orthogonal tensor $\boldsymbol{\eta}_{\mathrm{A}}$ and a symmetric tensor. This condition fixes the tensor $\boldsymbol{\eta}_{\mathrm{A}}$. With the help of $\boldsymbol{\eta}_{\mathrm{A}}$ we now define $\boldsymbol{\eta}(\boldsymbol{R}^\circ)$ by means of the relation

$$\hat{\boldsymbol{\eta}}(\boldsymbol{R}^\circ) = \boldsymbol{\eta}_{\mathrm{A}}\cdot\boldsymbol{\eta}(\boldsymbol{R}^\circ). \quad (309)$$

In view of (308) the tensor $\boldsymbol{\eta}(\boldsymbol{R}^\circ)$ has the property that

$$\int^{S^\circ} \boldsymbol{\eta}(\boldsymbol{R}^\circ)\cdot\boldsymbol{R}^\circ\boldsymbol{n}^\circ \mathrm{d}S^\circ \tag{310}$$

is symmetric.

The reason for factorizing the tensor $\hat{\boldsymbol{\eta}}$ in the particular way described above is that a rotation of the body as a whole without deformations is described by a change of $\boldsymbol{\eta}_\mathrm{A}$, leaving $\boldsymbol{\eta}(\boldsymbol{R}^\circ)$ invariant. (From (310) it is apparent that if $\boldsymbol{\eta}(\boldsymbol{R}^\circ)$ is uniform over the surface, then this tensor $\boldsymbol{\eta}$ is itself symmetric.)

If the sample is only slightly deformable the tensor $\boldsymbol{\eta}(\boldsymbol{R}^\circ)$ is nearly equal to the unit tensor $\mathbf{U}$. Introducing (309) into (307) we get for the variation $\delta\boldsymbol{\epsilon}$:

$$\delta\boldsymbol{\epsilon}(\boldsymbol{R}) = \{\delta\boldsymbol{\eta}_\mathrm{A}\cdot\boldsymbol{\eta}(\boldsymbol{R}^\circ)+\boldsymbol{\eta}_\mathrm{A}\cdot\delta\boldsymbol{\eta}(\boldsymbol{R}^\circ)\}\cdot\boldsymbol{\eta}^{-1}(\boldsymbol{R}^\circ)\cdot\boldsymbol{\eta}_\mathrm{A}^{-1} \simeq \delta\boldsymbol{\eta}_\mathrm{A}\cdot\tilde{\boldsymbol{\eta}}_\mathrm{A}+\boldsymbol{\eta}_\mathrm{A}\cdot\delta\boldsymbol{\eta}(\tilde{\boldsymbol{\eta}}_\mathrm{A}\cdot\boldsymbol{R})\cdot\tilde{\boldsymbol{\eta}}_\mathrm{A}, \tag{311}$$

where in the second expression $\boldsymbol{\eta}(\boldsymbol{R}^\circ)$ could be replaced by the unit tensor and $\delta\boldsymbol{\eta}(\boldsymbol{R}^\circ)$ by $\delta\boldsymbol{\eta}(\tilde{\boldsymbol{\eta}}_\mathrm{A}\cdot\boldsymbol{R})$ as a consequence of the fact that the deformations are small. The first term in the last member of (311) is antisymmetric as follows from the orthogonality of $\boldsymbol{\eta}_\mathrm{A}$. With the help of (304) we obtain for the change of entropy

$$T\delta S = \delta U + \int^S \boldsymbol{n}\cdot(\mathbf{P}-\boldsymbol{P}\cdot\boldsymbol{E}\mathbf{U})\cdot\delta\boldsymbol{\eta}_\mathrm{A}\cdot\tilde{\boldsymbol{\eta}}_\mathrm{A}\cdot\boldsymbol{R}\,\mathrm{d}S$$
$$+\int^S \boldsymbol{n}\cdot(\mathbf{P}-\boldsymbol{P}\cdot\boldsymbol{E}\mathbf{U})\cdot\boldsymbol{\eta}_\mathrm{A}\cdot\delta\boldsymbol{\eta}(\tilde{\boldsymbol{\eta}}_\mathrm{A}\cdot\boldsymbol{R})\cdot\tilde{\boldsymbol{\eta}}_\mathrm{A}\cdot\boldsymbol{R}\,\mathrm{d}S - \int(\boldsymbol{E}\cdot\delta_0\boldsymbol{P}-\boldsymbol{M}\cdot\delta_0\boldsymbol{B})\mathrm{d}\boldsymbol{R}, \tag{312}$$

where the quantities $\mathbf{P}$, $\boldsymbol{E}$, $\boldsymbol{B}$, $\boldsymbol{P}$ and $\boldsymbol{M}$ all depend on the space coordinates $\boldsymbol{R}$. With the use of Gauss's theorem and (293) the first integral at the right-hand side may be transformed, so that one gets

$$T\delta S = \delta U + \int \mathbf{P} : (\delta\boldsymbol{\eta}_\mathrm{A}\cdot\tilde{\boldsymbol{\eta}}_\mathrm{A})\mathrm{d}\boldsymbol{R}$$
$$+\int^S \boldsymbol{n}\cdot(\mathbf{P}-\boldsymbol{P}\cdot\boldsymbol{E}\mathbf{U})\cdot\boldsymbol{\eta}_\mathrm{A}\cdot\delta\boldsymbol{\eta}(\tilde{\boldsymbol{\eta}}_\mathrm{A}\cdot\boldsymbol{R})\cdot\tilde{\boldsymbol{\eta}}_\mathrm{A}\cdot\boldsymbol{R}\,\mathrm{d}S - \int(\boldsymbol{E}\cdot\delta_\mathrm{A}\boldsymbol{P}-\boldsymbol{M}\cdot\delta_\mathrm{A}\boldsymbol{B})\mathrm{d}\boldsymbol{R}, \tag{313}$$

where we used the fact that the trace of $\delta\boldsymbol{\eta}_\mathrm{A}\cdot\tilde{\boldsymbol{\eta}}_\mathrm{A}$ vanishes, and the notation

$$\delta_\mathrm{A} \equiv \delta_0 + \boldsymbol{R}\cdot\boldsymbol{\eta}_\mathrm{A}\cdot\delta\tilde{\boldsymbol{\eta}}_\mathrm{A}\cdot\boldsymbol{\nabla}. \tag{314}$$

In the second term at the right-hand side of (313) only the antisymmetrical part $\frac{1}{2}(\mathbf{P}-\tilde{\mathbf{P}})$ of the pressure tensor contributes, because the other factor is antisymmetric. It follows from the angular balance equation (196) (cf. subsection 6*d*) by integration over the volume that

$$\int \boldsymbol{P}_{\mathrm{A}} \mathrm{d}\boldsymbol{R} = \int (\boldsymbol{P} \wedge \boldsymbol{E} + \boldsymbol{M} \wedge \boldsymbol{B}) \mathrm{d}\boldsymbol{R} \tag{315}$$

as a consequence of the fact that the system is in equilibrium and at rest. Alternatively one may write

$$\int \tfrac{1}{2}(\mathbf{P} - \tilde{\mathbf{P}}) \mathrm{d}\boldsymbol{R} = \int \tfrac{1}{2}(\boldsymbol{PE} - \boldsymbol{EP} + \boldsymbol{MB} - \boldsymbol{BM}) \mathrm{d}\boldsymbol{R}. \tag{316}$$

Substituting this relation into (313) we get the entropy law

$$T\delta S = \delta U + \int^{S} \boldsymbol{n}\cdot(\mathbf{P} - \boldsymbol{P}\cdot\boldsymbol{E}\mathbf{U})\cdot\boldsymbol{\eta}_{\mathrm{A}}\cdot\delta\boldsymbol{\eta}(\tilde{\boldsymbol{\eta}}_{\mathrm{A}}\cdot\boldsymbol{R})\cdot\tilde{\boldsymbol{\eta}}_{\mathrm{A}}\cdot\boldsymbol{R}\,\mathrm{d}S$$
$$- \int \{\boldsymbol{E}\cdot\delta_{\mathrm{A}}\boldsymbol{P} - \boldsymbol{M}\cdot\delta_{\mathrm{A}}\boldsymbol{B} - (\boldsymbol{PE} + \boldsymbol{MB}) : (\delta\boldsymbol{\eta}_{\mathrm{A}}\cdot\tilde{\boldsymbol{\eta}}_{\mathrm{A}})\}\mathrm{d}\boldsymbol{R}, \tag{317}$$

which gives the entropy in its dependence on the change of the total internal energy, of the electric polarization, the magnetic field and the state variables $\boldsymbol{\eta}_{\mathrm{A}}$ and $\boldsymbol{\eta}(\tilde{\boldsymbol{\eta}}_{\mathrm{A}}\cdot\boldsymbol{R})$ at the surface.

In the case that the solid is not rotated but only deformed, the entropy law (317) reduces to

$$T\delta S = \delta U + \int^{S} \boldsymbol{n}\cdot(\mathbf{P} - \boldsymbol{P}\cdot\boldsymbol{E}\mathbf{U})\cdot\delta\boldsymbol{\eta}(\boldsymbol{R})\cdot\boldsymbol{R}\,\mathrm{d}S - \int (\boldsymbol{E}\cdot\delta_0\boldsymbol{P} - \boldsymbol{M}\cdot\delta_0\boldsymbol{B})\mathrm{d}\boldsymbol{R}, \tag{318}$$

since then $\boldsymbol{\eta}_{\mathrm{A}} = \mathbf{U}$.

In certain cases the polarizations in a system with long range correlations in a uniform external field are approximately uniform if the sample has ellipsoidal shape. This is indeed only an approximation, since not all physical quantities are uniform (the pressure tensor, for instance, will in general vary over the sample, cf. section 8*b*). In that case the entropy law (317) becomes

$$T\mathrm{d}S = \mathrm{d}U + \int^{S} \boldsymbol{n}\cdot\mathbf{P}(\boldsymbol{R})\cdot\boldsymbol{\eta}_{\mathrm{A}}\cdot\mathrm{d}\boldsymbol{\eta}(\tilde{\boldsymbol{\eta}}_{\mathrm{A}}\cdot\boldsymbol{R})\cdot\tilde{\boldsymbol{\eta}}_{\mathrm{A}}\cdot\boldsymbol{R}\,\mathrm{d}S + V(\boldsymbol{PE} + \boldsymbol{MB}) : (\mathrm{d}\boldsymbol{\eta}_{\mathrm{A}}\cdot\tilde{\boldsymbol{\eta}}_{\mathrm{A}})$$
$$- \boldsymbol{P}\cdot\boldsymbol{E}\int^{S} \boldsymbol{n}\cdot\boldsymbol{\eta}_{\mathrm{A}}\cdot\mathrm{d}\boldsymbol{\eta}(\tilde{\boldsymbol{\eta}}_{\mathrm{A}}\cdot\boldsymbol{R})\cdot\tilde{\boldsymbol{\eta}}_{\mathrm{A}}\cdot\boldsymbol{R}\,\mathrm{d}S - V\boldsymbol{E}\cdot\mathrm{d}\boldsymbol{P} + V\boldsymbol{M}\cdot\mathrm{d}\boldsymbol{B}, \tag{319}$$

which has been written with differentials since now all quantities are state variables. The fourth term at the right-hand side is equal to $-\boldsymbol{P}\cdot\boldsymbol{E}\,\mathrm{d}V$ as follows from (225), (311), Gauss's theorem and the fact that $\mathrm{d}\boldsymbol{\eta}_{\mathrm{A}}\cdot\tilde{\boldsymbol{\eta}}_{\mathrm{A}}$ is traceless. Therefore (319) becomes

$$T\mathrm{d}S = \mathrm{d}U + \int^S \boldsymbol{n}\cdot\mathbf{P}(\boldsymbol{R})\cdot\boldsymbol{\eta}_{\mathrm{A}}\cdot\mathrm{d}\boldsymbol{\eta}(\tilde{\boldsymbol{\eta}}_{\mathrm{A}}\cdot\boldsymbol{R})\cdot\tilde{\boldsymbol{\eta}}_{\mathrm{A}}\cdot\boldsymbol{R}\,\mathrm{d}S + V(\boldsymbol{PE}+\boldsymbol{MB}) : (\mathrm{d}\boldsymbol{\eta}_{\mathrm{A}}\cdot\tilde{\boldsymbol{\eta}}_{\mathrm{A}}) - \boldsymbol{E}\cdot\mathrm{d}(V\boldsymbol{P}) + V\boldsymbol{M}\cdot\mathrm{d}\boldsymbol{B}. \quad (320)$$

In the case that no rotations of the solid are considered this relation reduces to

$$T\mathrm{d}S = \mathrm{d}U + \int^S \boldsymbol{n}\cdot\mathbf{P}(\boldsymbol{R})\cdot\mathrm{d}\boldsymbol{\eta}(\boldsymbol{R})\cdot\boldsymbol{R}\,\mathrm{d}S - \boldsymbol{E}\cdot\mathrm{d}(V\boldsymbol{P}) + V\boldsymbol{M}\cdot\mathrm{d}\boldsymbol{B}, \quad (321)$$

since then $\boldsymbol{\eta}_{\mathrm{A}} = \mathbf{U}$ (it also follows from (318) as a special case)[1]. The set of integrability conditions includes the relations between polarizations and fields ($i, j = 1, 2, 3$)

$$\frac{\partial P_i}{\partial E_j} = \frac{\partial P_j}{\partial E_i}, \qquad \frac{\partial P_i}{\partial B_j} = \frac{\partial M_j}{\partial E_i}, \qquad \frac{\partial M_i}{\partial B_j} = \frac{\partial M_j}{\partial B_i}, \quad (322)$$

where the polarizations have been considered as functions of $\boldsymbol{E}$, $\boldsymbol{B}$, $T$ or $S$, and $\boldsymbol{\eta}$. (These relations are trivially valid for isotropic substances.)

The Gibbs relation (320–321), which is an approximation to the entropy law (317–318), is the final result for crystalline solids. It shows how the total entropy is a function of the total energy, the polarization, the magnetic field and the deformation tensor at the boundary. The law is the counterpart of the Gibbs relations (251) for fluids and (260) for amorphous or polycrystalline solids. In contrast to these local laws the result just found has the form of a global law: it makes no sense to subdivide a crystalline solid (in which long range correlations are present) into nearly uniform cells for which local laws may be derived.

### e. *The entropy balance equation*

For fluid systems of neutral atoms we found a first law of the form (211) and a second law of the form (251). The latter equation has been derived for a system at rest so that the fields and polarizations are measured in the rest

[1] In (321) the complete pressure tensor $\mathbf{P}$ appears. If the tensor $\mathrm{d}\boldsymbol{\eta}(\boldsymbol{R})$ is uniform over the surface, it follows from (310) that it is symmetric. With the help of Gauss's theorem, (293) and the uniformity of the electric field $\boldsymbol{E}$ one may write then the second term at the right-hand side of (321) as $(\int \mathbf{P}(\boldsymbol{R})\mathrm{d}\boldsymbol{R}) : \mathrm{d}\boldsymbol{\eta}$. Since now $\mathrm{d}\boldsymbol{\eta}$ is symmetric only the symmetric part of the pressure tensor comes into play. The same statement may be made for another special case, namely that of a pressure tensor which is uniform over the surface. In that case it follows directly from (310) that the second term at the right-hand side of (321) contains only the symmetric part of this uniform pressure tensor. Similar remarks apply to the corresponding terms in (320).

frame. They are therefore the same as those occurring in (211). The second law, which we may write as

$$\varrho \frac{\mathrm{d}s}{\mathrm{d}t} = \frac{\varrho}{T} \frac{\mathrm{d}u}{\mathrm{d}t} + \frac{\varrho}{T} p \frac{\mathrm{d}v}{\mathrm{d}t} - \frac{\varrho}{T} \boldsymbol{E}' \cdot \frac{\mathrm{d}(v\boldsymbol{P}')}{\mathrm{d}t} + \frac{1}{T} \boldsymbol{M}' \cdot \frac{\mathrm{d}\boldsymbol{B}'}{\mathrm{d}t}, \tag{323}$$

is supposed to be valid also for fluid systems which are not too far from equilibrium. If we substitute the first law (211) into this equation we obtain the balance of entropy

$$\varrho \frac{\mathrm{d}s}{\mathrm{d}t} = -\nabla \cdot \left(\frac{\boldsymbol{J}_q}{T}\right) - \frac{\boldsymbol{J}_q}{T^2} \cdot \nabla T - \frac{1}{T} (\tilde{\mathsf{P}} - p\mathsf{U}) : \nabla \boldsymbol{v}, \tag{324}$$

where we used mass conservation in the form $\varrho \mathrm{d}v/\mathrm{d}t = \nabla \cdot \boldsymbol{v}$, as follows from (59) and (159). This balance equation shows that the entropy changes as the result of the divergence of an entropy (conduction) flow $\boldsymbol{J}_q/T$ and an entropy source strength arising from heat conduction and viscous phenomena. In equilibrium the source term vanishes since the temperature and velocity fields are then uniform. (Moreover simultaneously the heat flow $\boldsymbol{J}_q$ and the viscous pressure $\mathsf{P} - p\mathsf{U}$ also vanish then.) Outside equilibrium the entropy source strength is positive, as may be shown if the distribution function is known to satisfy particular equations, like Boltzmann's.

In the preceding we assumed that the quantities $\boldsymbol{E}'$ and $\boldsymbol{M}'$ in (211) were equal to the equilibrium values $\boldsymbol{E}'_{\mathrm{eq}}$ and $\boldsymbol{M}'_{\mathrm{eq}}$ of these quantities occurring in (323), so that in the entropy source strength no electromagnetic contributions appear. If however $\boldsymbol{E}'$ and $\boldsymbol{M}'$ are supposed different from their equilibrium values we obtain instead of (324) as the balance of entropy:

$$\begin{aligned} \varrho \frac{\mathrm{d}s}{\mathrm{d}t} = & -\nabla \cdot \left(\frac{\boldsymbol{J}_q}{T}\right) - \frac{\boldsymbol{J}_q}{T^2} \cdot \nabla T - \frac{1}{T} (\tilde{\mathsf{P}} - p\mathsf{U}) : \nabla \boldsymbol{v} \\ & + \frac{\varrho}{T} (\boldsymbol{E}' - \boldsymbol{E}'_{\mathrm{eq}}) \cdot \frac{\mathrm{d}(v\boldsymbol{P}')}{\mathrm{d}t} - \frac{1}{T} (\boldsymbol{M}' - \boldsymbol{M}'_{\mathrm{eq}}) \cdot \frac{\mathrm{d}\boldsymbol{B}'}{\mathrm{d}t}. \end{aligned} \tag{325}$$

The last two terms show which contributions to the entropy production arise from electromagnetic phenomena. They represent the entropy source strength due to electric and magnetic relaxation.

For amorphous and polycrystalline solids the first law (211) may be written in a slightly different form if one uses the relation

$$\nabla \boldsymbol{v} = \nabla \left(\frac{\mathrm{d}\boldsymbol{\epsilon}}{\mathrm{d}t} \cdot \boldsymbol{R}\right) = \frac{\mathrm{d}\tilde{\boldsymbol{\epsilon}}}{\mathrm{d}t}, \tag{326}$$

as follows from the definition (225) of the uniform deformation tensor $\delta\boldsymbol{\epsilon}$.

Inserting this expression we then find:

$$\varrho \frac{du}{dt} = -\nabla \cdot \boldsymbol{J}_q - \mathbf{P} : \frac{d\boldsymbol{\epsilon}}{dt} + \varrho \frac{dv\boldsymbol{P}'}{dt} \cdot \boldsymbol{E}' - \boldsymbol{M}' \cdot \frac{d\boldsymbol{B}'}{dt}. \tag{327}$$

If this relation is combined with the entropy law (260) (which will be assumed to be valid in the neighbourhood of equilibrium) written as

$$\rho \frac{ds}{dt} = \frac{\varrho}{T} \frac{du}{dt} + \frac{1}{T} \mathbf{P}_{eq} : \frac{d\boldsymbol{\epsilon}}{dt} - \frac{\varrho}{T} \boldsymbol{E}' \cdot \frac{d(v\boldsymbol{P}')}{dt} + \frac{1}{T} \boldsymbol{M}' \cdot \frac{d\boldsymbol{B}'}{dt}, \tag{328}$$

with a symmetric equilibrium pressure, one gets an entropy balance which has the same form as (325) but for the third term at the right-hand side, which reads now

$$-\frac{1}{T} (\mathbf{P} - \mathbf{P}_{eq}) : \frac{d\boldsymbol{\epsilon}}{dt}. \tag{329}$$

For a neutral plasma the first law of thermodynamics has been given in (216) and the second law in (276). For a plasma not too far from equilibrium one obtains the entropy balance equation

$$\varrho \frac{ds}{dt} = -\nabla \cdot \left( \frac{\boldsymbol{J}_q}{T} \right) - \frac{\boldsymbol{J}_q}{T^2} \cdot \nabla T - \frac{1}{T} (\tilde{\mathbf{P}} - p\mathbf{U}) : \nabla \boldsymbol{v} + \frac{1}{T} \boldsymbol{J} \cdot \boldsymbol{E}', \tag{330}$$

where the last term represents the entropy source due to Joule heat produced in the plasma.

Since for systems with long range correlations only a global entropy law has been derived, it is not possible to find a local entropy balance equation in the same way as above. The global entropy production law will follow by combining the first and second laws, both in their global forms. The global form of the first law is a direct consequence of (217) with (212). In fact, integrating (217) over the mass of the system, one finds

$$\frac{dQ}{dt} = \frac{dU}{dt} + \int \left\{ \tilde{\mathbf{P}} : \nabla \boldsymbol{v} - \boldsymbol{J}' \cdot \boldsymbol{E}' - \varrho \frac{d(v\boldsymbol{P}')}{dt} \cdot \boldsymbol{E}' + \boldsymbol{M}' \cdot \frac{d\boldsymbol{B}'}{dt} \right\} d\boldsymbol{R} \tag{331}$$

with the amount of heat added to the system per unit of time

$$\frac{dQ}{dt} = -\int \boldsymbol{J}_q \cdot \boldsymbol{n} \, dS. \tag{332}$$

Now one has for a non-uniform solid system (cf. (326)):

$$\nabla \boldsymbol{v} = \nabla \left( \frac{d\boldsymbol{\epsilon}}{dt} \cdot \boldsymbol{R} \right) = \frac{d\tilde{\boldsymbol{\epsilon}}}{dt}, \tag{333}$$

as follows from (225) and (A51). With this relation (331) becomes

$$\frac{\mathrm{d}Q}{\mathrm{d}t} = \frac{\mathrm{d}U}{\mathrm{d}t} + \int \left\{ \mathbf{P} : \frac{\mathrm{d}\mathbf{e}}{\mathrm{d}t} - \boldsymbol{J}' \cdot \boldsymbol{E}' - \varrho \frac{\mathrm{d}(v\boldsymbol{P}')}{\mathrm{d}t} \cdot \boldsymbol{E}' + \boldsymbol{M}' \cdot \frac{\mathrm{d}\boldsymbol{B}'}{\mathrm{d}t} \right\} \mathrm{d}\boldsymbol{R}. \tag{334}$$

This form of the first law is to be compared with the entropy law (301). In the latter we divide by $\delta t$. Then it becomes

$$T\frac{\mathrm{d}S}{\mathrm{d}t} = \frac{\mathrm{d}U}{\mathrm{d}t} + \int \left\{ \mathbf{P} : \frac{\mathrm{d}\mathbf{e}}{\mathrm{d}t} - \boldsymbol{E}' \cdot \frac{\mathrm{d}\boldsymbol{P}'}{\mathrm{d}t} - \left( \mathrm{Tr}\, \frac{\mathrm{d}\mathbf{e}}{\mathrm{d}t} \right) \boldsymbol{P}' \cdot \boldsymbol{E}' + \boldsymbol{M}' \cdot \frac{\mathrm{d}\boldsymbol{B}'}{\mathrm{d}t} \right\} \mathrm{d}\boldsymbol{R}. \tag{335}$$

At the right-hand side we added primes to indicate that the quantities are taken in the rest frame (the second law has been derived for a system at rest). Using the fact that, as a consequence of (333) Tr $(\mathrm{d}\mathbf{e}/\mathrm{d}t)$ is equal to $\boldsymbol{\nabla} \cdot \boldsymbol{v}$ or to $\varrho \mathrm{d}v/\mathrm{d}t$ with $v = \varrho^{-1}$ the specific volume (as follows from the conservation of mass), one may write (335) in the form:

$$T\frac{\mathrm{d}S}{\mathrm{d}t} = \frac{\mathrm{d}U}{\mathrm{d}t} + \int \left\{ \mathbf{P} : \frac{\mathrm{d}\mathbf{e}}{\mathrm{d}t} - \varrho \frac{\mathrm{d}(v\boldsymbol{P}')}{\mathrm{d}t} \cdot \boldsymbol{E}' + \boldsymbol{M}' \cdot \frac{\mathrm{d}\boldsymbol{B}'}{\mathrm{d}t} \right\} \mathrm{d}\boldsymbol{R}. \tag{336}$$

Again we assume that this law remains valid if the system is near equilibrium. Then one finds, by combining (334) and (336), for the global entropy balance equation

$$T\frac{\mathrm{d}S}{\mathrm{d}t} = \frac{\mathrm{d}Q}{\mathrm{d}t} - \int \left\{ (\mathbf{P} - \mathbf{P}_{\mathrm{eq}}) : \frac{\mathrm{d}\mathbf{e}}{\mathrm{d}t} - \varrho \frac{\mathrm{d}(v\boldsymbol{P}')}{\mathrm{d}t} \cdot (\boldsymbol{E}' - \boldsymbol{E}'_{\mathrm{eq}}) + (\boldsymbol{M}' - \boldsymbol{M}'_{\mathrm{eq}}) \cdot \frac{\mathrm{d}\boldsymbol{B}'}{\mathrm{d}t} \right\} \mathrm{d}\boldsymbol{R}, \tag{337}$$

where we added some indices eq to distinguish the equilibrium values, occurring in the second law, from the non-equilibrium values occurring in the first law. Note that the temperature has been supposed to be uniform, so that no term with the gradient of the temperature appears in (337), in contrast with what was the case in (325). At the right-hand side of (337) appears, apart from a term with the supplied heat, a volume integral which contains elastic, electric and magnetic relaxation terms.

A particular case, which arises for ferromagnetic materials, is that of a system in which magnetic hysteresis occurs. If one considers a cyclic process, in which no elastic after-effects occur, we have, if no heat is added,

$$T\Delta S = -\int \oint (\boldsymbol{M}' - \boldsymbol{M}'_{\mathrm{eq}}) \cdot \mathrm{d}\boldsymbol{B}' \, \mathrm{d}\boldsymbol{R} \tag{338}$$

for the entropy production $\Delta S$ per cycle.

# 8 *Helmholtz and Kelvin forces*

## a. *Fluids*

Let us consider a fluid of neutral atoms in which constitutive relations exist between the polarizations and fields:

$$\begin{aligned} \boldsymbol{P}' &= \kappa(v, T)\boldsymbol{E}', \\ \boldsymbol{M}' &= \chi(v, T)\boldsymbol{B}'. \end{aligned} \tag{339}$$

Primes have been added to indicate that the quantities are counted in the rest frame. The electric and magnetic susceptibilities $\kappa$ and $\chi$ depend on the specific volume $v$ and the temperature $T$. The second law for such a fluid in local equilibrium has been given in (251) and may be written in the form

$$\mathrm{d}f = -p\,\mathrm{d}v - s\,\mathrm{d}T + \boldsymbol{E}'\cdot\mathrm{d}(v\boldsymbol{P}') - v\boldsymbol{M}'\cdot\mathrm{d}\boldsymbol{B}' \tag{340}$$

with the specific free energy

$$f = u - Ts. \tag{341}$$

The differential expression (340) may be integrated at constant specific volume and temperature. Then one finds for the difference of the specific free energy in the presence and that in the absence of fields:

$$f - f_0 = v\int(\boldsymbol{E}'\cdot\mathrm{d}\boldsymbol{P}' - \boldsymbol{M}'\cdot\mathrm{d}\boldsymbol{B}'). \tag{342}$$

With (339) this relation becomes

$$f - f_0 = \tfrac{1}{2}v(\boldsymbol{P}'\cdot\boldsymbol{E}' - \boldsymbol{M}'\cdot\boldsymbol{B}'). \tag{343}$$

The scalar equilibrium pressure follows from the specific free energy by differentiation with respect to the specific volume at constant temperature, specific polarization $v\boldsymbol{P}'$ and magnetic field $\boldsymbol{B}'$, as (340) shows. Hence the pressure $p = -\partial f/\partial v$ is connected to the pressure $p_0 = -\partial f_0/\partial v$ for the same values of $v$ and $T$, but with switched-off fields by a relation[1] which follows from (343):

$$p - p_0 = \frac{1}{2}\left(\boldsymbol{P}'\cdot\boldsymbol{E}' + \boldsymbol{M}'\cdot\boldsymbol{B}' + v\frac{\partial\kappa}{\partial v}\boldsymbol{E}'^2 + v\frac{\partial\chi}{\partial v}\boldsymbol{B}'^2\right). \tag{344}$$

The specific entropy follows from the specific free energy by differentiation with respect to temperature $T$ at constant $v$, $v\boldsymbol{P}'$ and $\boldsymbol{B}'$. From (343) one has

$$s - s_0 = -\frac{1}{2}\left(v\frac{\partial\kappa}{\partial T}\boldsymbol{E}'^2 + v\frac{\partial\chi}{\partial T}\boldsymbol{B}'^2\right), \tag{345}$$

[1] P. Mazur and I. Prigogine, Mém. Acad. Roy. Belg. (Cl. Sc.) **28**(1953)fasc. 1; cf. W. F. Brown jr., Am. J. Phys. **19**(1951)290, 333.

so that the difference of the specific energies is

$$u-u_0 = \tfrac{1}{2}v\left(\boldsymbol{P}'\cdot\boldsymbol{E}'-\boldsymbol{M}'\cdot\boldsymbol{B}'+T\frac{\partial\kappa}{\partial T}\boldsymbol{E}'^2+T\frac{\partial\chi}{\partial T}\boldsymbol{B}'^2\right), \tag{346}$$

as follows from (343) with (341).

The relation (344) shows how the pressure changes if the fields are switched on. The pressure $p$ is the equilibrium value of the pressure $\mathbf{P}$ used so far. We shall call it the Kelvin pressure, to distinguish it from the pressure $p_0$, defined at equilibrium and with switched-off fields. The latter will be called the Helmholtz pressure at that specific volume and temperature.

In the equation of motion for a fluid of neutral atoms, which has been given in (105) with (106) or in (158), we introduce the Helmholtz pressure instead of the Kelvin pressure by using (344). Then we obtain

$$\varrho\frac{\mathrm{d}\boldsymbol{v}}{\mathrm{d}t} = -\boldsymbol{\nabla}p_0-\boldsymbol{\nabla}\cdot\boldsymbol{\Pi}+(\boldsymbol{\nabla}\boldsymbol{E})\cdot\boldsymbol{P}+(\boldsymbol{\nabla}\boldsymbol{B})\cdot\boldsymbol{M}$$
$$+c^{-1}\varrho\frac{\mathrm{d}}{\mathrm{d}t}(v\boldsymbol{P}\wedge\boldsymbol{B})-\tfrac{1}{2}\boldsymbol{\nabla}\left(\boldsymbol{P}'\cdot\boldsymbol{E}'+\boldsymbol{M}'\cdot\boldsymbol{B}'+v\frac{\partial\kappa}{\partial v}\boldsymbol{E}'^2+v\frac{\partial\chi}{\partial v}\boldsymbol{B}'^2\right), \tag{347}$$

where the viscous pressure tensor

$$\boldsymbol{\Pi} \equiv \mathbf{P}-p\mathbf{U} \tag{348}$$

has been introduced. Alternatively, introducing rest frame quantities with the help of (26) and (27), we have for the equation of motion, using also (339)

$$\varrho\frac{\mathrm{d}\boldsymbol{v}}{\mathrm{d}t} = -\boldsymbol{\nabla}p_0-\boldsymbol{\nabla}\cdot\boldsymbol{\Pi}+\mathscr{F}, \tag{349}$$

where the 'Helmholtz' force density is:

$$\mathscr{F} \equiv -\frac{1}{2}\left\{\boldsymbol{E}'^2\boldsymbol{\nabla}\kappa+\boldsymbol{B}'^2\boldsymbol{\nabla}\chi+\boldsymbol{\nabla}\left(v\frac{\partial\kappa}{\partial v}\boldsymbol{E}'^2+v\frac{\partial\chi}{\partial v}\boldsymbol{B}'^2\right)\right\}+c^{-1}(\boldsymbol{\nabla}\boldsymbol{v})\cdot(\boldsymbol{P}'\wedge\boldsymbol{B}')$$
$$+c^{-1}\varrho\frac{\mathrm{d}}{\mathrm{d}t}(v\boldsymbol{P}'\wedge\boldsymbol{B}')+(\boldsymbol{\nabla}\boldsymbol{E}')\cdot(\boldsymbol{P}'-\boldsymbol{P}'_{\mathrm{eq}})+(\boldsymbol{\nabla}\boldsymbol{B}')\cdot(\boldsymbol{M}'-\boldsymbol{M}'_{\mathrm{eq}}). \tag{350}$$

Here $\boldsymbol{P}'_{\mathrm{eq}}$ and $\boldsymbol{M}'_{\mathrm{eq}}$ represent the equilibrium values (339) in the rest frame.

The Helmholtz force $\mathscr{F}$ has a simpler form in the important special case of fluids in equilibrium and at rest in time-independent fields. Then the expression (350) reduces to

$$\mathscr{F} = -\frac{1}{2}\left\{\boldsymbol{E}'^2\boldsymbol{\nabla}\kappa+\boldsymbol{B}'^2\boldsymbol{\nabla}\chi+\boldsymbol{\nabla}\left(v\frac{\partial\kappa}{\partial v}\boldsymbol{E}'^2+v\frac{\partial\chi}{\partial v}\boldsymbol{B}'^2\right)\right\}. \tag{351}$$

The equation of motion reduces under these circumstances to

$$\mathscr{F} = \nabla p_0 . \tag{352}$$

The expression (351) contains the Helmholtz terms found on the basis of energy considerations[1]. The expression for static electric dipole systems has been derived already in a statistical treatment[2]. Earlier the connexion between Kelvin and Helmholtz forces and pressures had been found from thermodynamics[3].

Often one employs a magnetic susceptibility $\tilde{\chi}$ defined by

$$\boldsymbol{M}' = \tilde{\chi}(v, T)\boldsymbol{H}' \tag{353}$$

instead of the second line of (339). The connexion between the two susceptibilities is then

$$\tilde{\chi} = \frac{\chi}{1-\chi} . \tag{354}$$

With the help of this relation one may eliminate $\chi$ in favour of $\tilde{\chi}$ in the relations (344–346) and in (347), (350) and (351). The latter becomes in particular

$$\mathscr{F} = -\frac{1}{2}\left\{\boldsymbol{E}'^2\nabla\kappa + \boldsymbol{H}'^2\nabla\tilde{\chi} + \nabla\left(v\frac{\partial\kappa}{\partial v}\boldsymbol{E}'^2 + v\frac{\partial\tilde{\chi}}{\partial v}\boldsymbol{H}'^2\right)\right\} . \tag{355}$$

For practical applications one may alternatively use the equation of motion (105–106), which contains the Kelvin pressure and the Kelvin force, or the equation of motion (349–350), which has been written in terms of the Helmholtz pressure and the Helmholtz force. However the latter has a more limited validity, since it may only be employed if the system is characterized by linear constitutive relations. In its form (351–352) it may be applied only to equilibrium situations.

From the equation of motion (352) with (351) for a fluid in equilibrium and at rest, one may obtain the density distribution that arises if a static electromagnetic field is switched on. In fact since the Helmholtz pressure $p_0$ is a function of the density $\varrho = v^{-1}$ and the temperature $T$, one may write (352)

[1] D. J. Korteweg, Ann. Phys. Chem. **9**(1880)48; H. von Helmholtz, Ann. Phys. Chem. **13**(1881)385.

[2] P. Mazur and S. R. de Groot, Physica **22**(1956)657.

[3] P. Mazur and I. Prigogine, op. cit. For a review and applications see S. R. de Groot and P. Mazur, Non-equilibrium thermodynamics (North-Holland Publ. Co., Amsterdam 1962); A. Sanfeld, Introduction to the thermodynamics of charged and polarized layers (Wiley, London 1968).

with (351) as

$$\nabla p_0 = \tfrac{1}{2}\varrho\nabla\left(\frac{\partial\kappa}{\partial\varrho}\boldsymbol{E}'^2+\frac{\partial\chi}{\partial\varrho}\boldsymbol{B}'^2\right) \tag{356}$$

for constant $T$. If one defines now a function $\varphi$ as

$$\varphi(p_0, T) = \int_{\bar{p}_0}^{p_0} \varrho^{-1}(\hat{p}_0, T)\mathrm{d}\hat{p}_0, \tag{357}$$

which follows from the equation of state $\varrho = \varrho(p_0, T)$ (the lower limit is an arbitrary, but fixed constant), one may write (356) as

$$\varphi(p_0, T)-\frac{1}{2}\frac{\partial\kappa}{\partial\varrho}\boldsymbol{E}'^2-\frac{1}{2}\frac{\partial\chi}{\partial\varrho}\boldsymbol{B}'^2 = \text{const.}, \tag{358}$$

i.e. independent of the position in the fluid.

For an incompressible liquid at uniform temperature the function $\varphi(p_0, T)$ is equal to $v(p_0-\bar{p}_0)$ with constant $v(T)$. Then one finds from (358)

$$p_0(\boldsymbol{R}) = p_0(\boldsymbol{R}_0)+\tfrac{1}{2}\varrho\left\{\frac{\partial\kappa}{\partial\varrho}\boldsymbol{E}'^2(\boldsymbol{R})+\frac{\partial\chi}{\partial\varrho}\boldsymbol{B}'^2(\boldsymbol{R})\right\}, \tag{359}$$

where $\boldsymbol{R}_0$ denotes a position in the liquid where the fields vanish. Combining this result with the relation (344) between the Kelvin and Helmholtz pressures, one finds for the Kelvin pressure in an incompressible liquid at constant temperature

$$p(\boldsymbol{R}) = p_0(\boldsymbol{R}_0)+\tfrac{1}{2}\{\boldsymbol{P}'(\boldsymbol{R})\cdot\boldsymbol{E}'(\boldsymbol{R})+\boldsymbol{M}'(\boldsymbol{R})\cdot\boldsymbol{B}'(\boldsymbol{R})\}. \tag{360}$$

An alternative way to derive this formula starts from the equation of motion (105–106), which may be written for the present case of a fluid in equilibrium and at rest

$$\nabla p = \tfrac{1}{2}\nabla(\boldsymbol{P}'\cdot\boldsymbol{E}'+\boldsymbol{M}'\cdot\boldsymbol{B}')+c^{-1}\frac{\partial}{\partial t}(\boldsymbol{P}'\wedge\boldsymbol{B}')-\tfrac{1}{2}(\boldsymbol{E}'^2\nabla\kappa+\boldsymbol{B}'^2\nabla\chi). \tag{361}$$

For an incompressible liquid at constant temperature the last terms vanish, so that one recovers for the static case (360).

If the dependence of the susceptibilities on the density is given by the Clausius–Mossotti laws

$$\frac{\kappa}{\kappa+3} \sim \varrho, \qquad \frac{\chi}{3-2\chi} \sim \varrho, \tag{362}$$

one has for the partial derivatives

$$\varrho \frac{\partial \kappa}{\partial \varrho} = \kappa + \tfrac{1}{3}\kappa^2,$$
$$\varrho \frac{\partial \chi}{\partial \varrho} = \chi - \tfrac{2}{3}\chi^2. \tag{363}$$

If one inserts these relations into (359) one obtains

$$p_0(\boldsymbol{R}) = p_0(\boldsymbol{R}_0) + \tfrac{1}{2}\{\boldsymbol{P}'(\boldsymbol{R})\cdot\boldsymbol{E}'_{\mathrm{L}}(\boldsymbol{R}) + \boldsymbol{M}'(\boldsymbol{R})\cdot\boldsymbol{B}'_{\mathrm{L}}(\boldsymbol{R})\} \tag{364}$$

with the Lorentz cavity fields

$$\boldsymbol{E}'_{\mathrm{L}} \equiv \boldsymbol{E}' + \tfrac{1}{3}\boldsymbol{P}', \qquad \boldsymbol{B}'_{\mathrm{L}} \equiv \boldsymbol{B}' - \tfrac{2}{3}\boldsymbol{M}'. \tag{365}$$

For the electric case the relation (364) has been checked experimentally by measuring the index of refraction of a liquid placed between the plates of a condensor, which gives the pressures $p_0(\boldsymbol{R})$ and $p_0(\boldsymbol{R}_0)$ [1].

For an ideal gas the equation of state has the Boyle–Gay-Lussac form

$$\varrho = mp_0/kT \tag{366}$$

with $m$ the mass of the molecules and $k$ Boltzmann's constant. Then the function $\varphi$ becomes

$$\varphi(p_0, T) = \frac{kT}{m}\log\frac{p_0}{\bar{p}_0}. \tag{367}$$

Inserting this into (358) one finds

$$\frac{\varrho(\boldsymbol{R})}{\varrho(\boldsymbol{R}_0)} = \frac{p_0(\boldsymbol{R})}{p_0(\boldsymbol{R}_0)} = \exp\left\{\frac{m}{2kT}\left(\frac{\partial\kappa}{\partial\varrho}\boldsymbol{E}'^2 + \frac{\partial\chi}{\partial\varrho}\boldsymbol{B}'^2\right)\right\}, \tag{368}$$

where at the right-hand side the quantities depend on $\boldsymbol{R}$. With (363) and (365) this relation reduces to

$$\frac{\varrho(\boldsymbol{R})}{\varrho(\boldsymbol{R}_0)} = \exp\left\{\frac{m}{2\varrho kT}(\boldsymbol{P}'\cdot\boldsymbol{E}'_{\mathrm{L}} + \boldsymbol{M}'\cdot\boldsymbol{B}'_{\mathrm{L}})\right\}. \tag{369}$$

This formula shows that the quantity $-(m/2\varrho)(\boldsymbol{P}'\cdot\boldsymbol{E}'_{\mathrm{L}} + \boldsymbol{M}'\cdot\boldsymbol{B}'_{\mathrm{L}})$ may be looked upon as the energy of a particle with an electric and a magnetic dipole moment in a field.

A useful application of the expression (360) arises if one considers a solid body at rest immersed in an incompressible liquid at uniform tempera-

[1] S. S. Hakim and J. B. Higham, Proc. Phys. Soc. **80**(1962)190.

ture. In equilibrium the equation of motion (150) with (108) gives upon integration over a volume with a boundary that lies just outside the solid:

$$M\frac{\mathrm{d}\boldsymbol{v}}{\mathrm{d}t} = -\int\{p\mathsf{U}-\boldsymbol{E}'\boldsymbol{D}'-\boldsymbol{H}'\boldsymbol{B}'+(\tfrac{1}{2}E'^2+\tfrac{1}{2}B'^2-\boldsymbol{M}'\cdot\boldsymbol{B}')\mathsf{U}\}\cdot\boldsymbol{n}\,\mathrm{d}S, \qquad (370)$$

where $M$ is the total mass of the solid. If one employs now (360) one obtains the equation

$$M\frac{\mathrm{d}\boldsymbol{v}}{\mathrm{d}t} = \int\{\boldsymbol{E}'\boldsymbol{D}'+\boldsymbol{H}'\boldsymbol{B}'-\tfrac{1}{2}(\boldsymbol{E}'\cdot\boldsymbol{D}'+\boldsymbol{B}'\cdot\boldsymbol{H}')\mathsf{U}\}\cdot\boldsymbol{n}\,\mathrm{d}S, \qquad (371)$$

where (14) has been used. The right-hand side contains the field pressure tensor of Maxwell and Heaviside. The derivation shows that it corresponds to a material pressure which is the pressure $p_0(\boldsymbol{R}_0)$ at a point $\boldsymbol{R}_0$ in the liquid where the electromagnetic fields are zero. It will depend on the experimental situation whether such a pressure is accessible to measurement.

A second application of the expression (360) for the Kelvin pressure in an incompressible liquid consists in the evaluation of the radiation pressure on a metallic surface immersed in a liquid. Consider a plane electromagnetic wave

$$\begin{aligned} \boldsymbol{E}_\mathrm{i} &= \boldsymbol{E}_0\cos(\omega t-\boldsymbol{k}\cdot\boldsymbol{R}),\\ \boldsymbol{B}_\mathrm{i} &= n(\boldsymbol{n}\wedge\boldsymbol{E}_0)\cos(\omega t-\boldsymbol{k}\cdot\boldsymbol{R}), \end{aligned} \qquad (372)$$

with $\omega$ the circular frequency, $\boldsymbol{k}$ the wave vector, $n$ the refractive index and $\boldsymbol{n}\equiv\boldsymbol{k}/k$ the direction of propagation, hitting a plane metallic surface perpendicularly. The wave is assumed to be totally reflected, so that its reflected part has the form

$$\begin{aligned} \boldsymbol{E}_\mathrm{r} &= -\boldsymbol{E}_0\cos(\omega t+\boldsymbol{k}\cdot\boldsymbol{R}),\\ \boldsymbol{B}_\mathrm{r} &= n(\boldsymbol{n}\wedge\boldsymbol{E}_0)\cos(\omega t+\boldsymbol{k}\cdot\boldsymbol{R}) \end{aligned} \qquad (373)$$

(the metallic surface passes through the origin of coordinates). If one averages the law (150) or (109) over a period $2\pi\omega^{-1}$, one finds, taking the fluid to be at rest, that

$$\nabla\cdot\{\overline{\mathsf{P}}-\overline{\boldsymbol{D}\boldsymbol{E}}-\overline{\boldsymbol{B}\boldsymbol{H}}+(\tfrac{1}{2}\overline{E^2}+\tfrac{1}{2}\overline{B^2}-\overline{\boldsymbol{M}\cdot\boldsymbol{B}})\mathsf{U}\} = 0, \qquad (374)$$

where the bars indicate time averages. Applying this formula to a cylinder with unit cross-section and its axis parallel to $\boldsymbol{n}$, lying half in the fluid and half in the metal, one obtains upon using Gauss's theorem

$$\boldsymbol{n}\cdot\overline{\mathsf{P}}_{\mathrm{fluid}}-\boldsymbol{n}\cdot\overline{\mathsf{P}}_{\mathrm{metal}}+(\tfrac{1}{2}\overline{E^2}+\tfrac{1}{2}\overline{B^2}-\overline{\boldsymbol{M}\cdot\boldsymbol{B}})\boldsymbol{n} = 0. \qquad (375)$$

If the fluid is assumed to be in equilibrium, its time averaged pressure is diagonal and follows from (360). (The time derivative $c^{-1}(\partial/\partial t)(\boldsymbol{P}\wedge\boldsymbol{B})$ which is present in (361) drops out if one employs it for time averaged quantities.) Then one finds:

$$\boldsymbol{n}\cdot\overline{\mathbf{P}}_{\text{metal}} = \bar{p}_{\text{fluid},0}(\boldsymbol{R}_0)\boldsymbol{n}+\tfrac{1}{2}\{\overline{\boldsymbol{D}(\boldsymbol{R})\cdot\boldsymbol{E}(\boldsymbol{R})}+\overline{\boldsymbol{B}(\boldsymbol{R})\cdot\boldsymbol{H}(\boldsymbol{R})}\}\boldsymbol{n}, \tag{376}$$

where $\boldsymbol{R}$ is a position in the light beam and $\boldsymbol{R}_0$ a position outside of it. The average values occurring in the second term at the right-hand side follow from (372–373). One gets $2\varepsilon\boldsymbol{E}_0^2\boldsymbol{n}$, because the refractive index $n$ is equal to $(\varepsilon\mu)^{\frac{1}{2}}$. Therefore one obtains as the radiation pressure, which is the difference of the left-hand side and the first term at the right-hand side of (376)

$$p_{\text{rad}} = 2\varepsilon\boldsymbol{E}_0^2. \tag{377}$$

Introducing the amplitude of the absolute value of the Poynting vector $\boldsymbol{S} = c\boldsymbol{E}\wedge\boldsymbol{H}$ of the incident wave, which is

$$|\boldsymbol{S}| = c\sqrt{\frac{\varepsilon}{\mu}}\,\boldsymbol{E}_0^2, \tag{378}$$

one finds for the radiation pressure

$$p_{\text{rad}} = 2c^{-1}n|\boldsymbol{S}|. \tag{379}$$

For the case of vacuum ($n = 1$) this result has been found experimentally[1]. If one compares radiation pressures in different media, keeping $|\boldsymbol{S}|$ constant, one finds from (379) that the radiation pressures are proportional to the refractive index, a second result which has been checked experimentally[2]. The derivation shows that due to the time averaging the terms with time derivatives in the momentum law drop out: in other words neither the material nor the field momentum density play a role in the discussion of radiation pressure.

### b. *Crystalline solids*

Amorphous and polycrystalline solids may be discussed along similar lines as above. In contrast with these, crystalline solid systems cannot be described by thermodynamics in local formulation: only global laws could be derived in the preceding section. Yet it is possible, at least in principle, to find the

[1] P. Lebedew, Ann. Physik **6**(1901)433; E. F. Nichols and G. F. Hull, Phys. Rev. **13**(1901) 307, **17**(1903)26.

[2] R. V. Jones, Nature **167**(1951)439; R. V. Jones and J. C. S. Richards, Proc. Roy. Soc. **221A**(1954)480.

deformation of a solid system from thermodynamical considerations. To that end one may start from the entropy law (294). The infinitesimal deformation tensor $\delta\boldsymbol{\epsilon}$ may be expressed in terms of state variables by means of relation (311). If the solid does not rotate, (i.e. if $\boldsymbol{\eta}_A = 1$) and if only small deformations are considered, one finds from (294) for the change of free energy $F^* \equiv U^* - TS$ for a solid at rest in a uniform external field:

$$\mathrm{d}F^* = -S\,\mathrm{d}T - \int^S \boldsymbol{n}\cdot\mathbf{P}(\boldsymbol{R})\cdot\mathrm{d}\boldsymbol{\eta}(\boldsymbol{R})\cdot\boldsymbol{R}\,\mathrm{d}S - \tfrac{1}{2}\int^S \boldsymbol{n}\cdot\mathrm{d}\boldsymbol{\eta}(\boldsymbol{R})\cdot\boldsymbol{R}\{\boldsymbol{P}(\boldsymbol{R})\cdot\boldsymbol{n}\}^2\mathrm{d}S - V\overline{\boldsymbol{P}}\cdot\mathrm{d}\boldsymbol{E}_e - V\overline{\boldsymbol{M}}\cdot\mathrm{d}\boldsymbol{B}_e. \quad (380)$$

(In contrast to the preceding subsection no primes were added although again rest frame quantities are meant.) From this relation one may find the difference of the free energies in the presence ($F^*$) and absence ($F_0^*$) of external fields, at constant surface deformation $\boldsymbol{\eta}(\boldsymbol{R})$ and temperature $T$:

$$F^* - F_0^* = -V\int(\overline{\boldsymbol{P}}\cdot\mathrm{d}\boldsymbol{E}_e + \overline{\boldsymbol{M}}\cdot\mathrm{d}\boldsymbol{B}_e). \quad (381)$$

If in particular the polarizations are proportional to the fields one finds simply

$$F^* - F_0^* = -\tfrac{1}{2}V(\overline{\boldsymbol{P}}\cdot\boldsymbol{E}_e + \overline{\boldsymbol{M}}\cdot\boldsymbol{B}_e). \quad (382)$$

The relation (380) then shows that the difference of the pressure tensor at the surface (contracted with the normal on the surface) in the presence and in the absence of external fields follows by taking a functional derivative of (381) (or (382)) with respect to $\boldsymbol{\eta}(\boldsymbol{R})$, the deformation tensor at the surface. Since the (normal component of the) pressure tensor in the presence of external fields follows directly from the Liénard expression (241), we find in this way the normal pressure at the surface in the absence of external fields, but with the same values of the temperature and of the deformation tensor. Hence the problem to find the deformation $\boldsymbol{\eta}(\boldsymbol{R})$ at the surface under the influence of external fields has been reduced now to a problem of ordinary (field-free) elasticity theory.

The programme as sketched above is not feasible in general since the determination of the functional derivative of (381) presents difficulties in practical cases. One of these is the way in which the shape of the sample enters through the occurrence of the external fields instead of the Maxwell fields. A way to avoid this difficulty is to start from a second law which contains the Maxwell fields rather than the external fields, namely relation (321) (valid for an ellipsoidal sample), which entails a free energy change

$$\mathrm{d}F = -S\,\mathrm{d}T - \int^{S} \boldsymbol{n}\cdot\mathbf{P}(\boldsymbol{R})\cdot\mathrm{d}\boldsymbol{\eta}(\boldsymbol{R})\cdot\boldsymbol{R}\,\mathrm{d}S + \boldsymbol{E}\cdot\mathrm{d}(V\boldsymbol{P}) - V\boldsymbol{M}\cdot\mathrm{d}\boldsymbol{B}. \tag{383}$$

Again we assumed that the solid does not rotate and that only small deformations occur. The sample has been chosen ellipsoidal, so that the polarizations and fields are approximately uniform. The difference $F-F_0$ of the free energy in the presence and the absence of external fields, but with the same surface deformation $\boldsymbol{\eta}(\boldsymbol{R})$ and temperature $T$ may now be found. We assume that the polarizations depend on the Maxwell fields through linear relations of the form

$$\begin{aligned} \boldsymbol{P} &= \boldsymbol{\varkappa}(\boldsymbol{\eta}, T)\cdot\boldsymbol{E},\\ \boldsymbol{M} &= \boldsymbol{\chi}(\boldsymbol{\eta}, T)\cdot\boldsymbol{B}, \end{aligned} \tag{384}$$

with symmetrical susceptibility tensors $\boldsymbol{\varkappa}$ and $\boldsymbol{\chi}$ (v. (322)). In these relations $\boldsymbol{\eta}$ stands for the whole set of deformation tensors everywhere at the surface. Therefore one may write for small deformations the following expressions

$$\begin{aligned} \boldsymbol{\varkappa}^{-1}(\boldsymbol{\eta}, T) &= \boldsymbol{\varkappa}_0^{-1}(T) + \frac{1}{V}\int^{S} \boldsymbol{\varkappa}_1^{-1}(\boldsymbol{R}, T)\cdot\{\boldsymbol{\eta}(\boldsymbol{R}) - \mathbf{U}\}\cdot\boldsymbol{R}\,\mathrm{d}S,\\ \boldsymbol{\chi}(\boldsymbol{\eta}, T) &= \boldsymbol{\chi}_0(T) + \frac{1}{V}\int^{S} \boldsymbol{\chi}_1(\boldsymbol{R}, T)\cdot\{\boldsymbol{\eta}(\boldsymbol{R}) - \mathbf{U}\}\cdot\boldsymbol{R}\,\mathrm{d}S, \end{aligned} \tag{385}$$

for the inverse electric and the magnetic susceptibilities. ($\boldsymbol{\varkappa}_0^{-1}$ and $\boldsymbol{\chi}_0$ are tensors of the second rank, while $\boldsymbol{\varkappa}_1^{-1}$ and $\boldsymbol{\chi}_1$ have three indices.) From (383) with (384) it follows that one has

$$F - F_0 = \tfrac{1}{2}V(\boldsymbol{P}\cdot\boldsymbol{E} - \boldsymbol{M}\cdot\boldsymbol{B}). \tag{386}$$

By differentiating this relation functionally with respect to $\boldsymbol{\eta}(\boldsymbol{R})$ at constant temperature $T$, total electric polarization $V\boldsymbol{P}$ and magnetic field $\boldsymbol{B}$, one finds, according to (383), the normal component of the difference between the pressure tensors at the surface in the presence and in the absence of fields:

$$\boldsymbol{n}\cdot\{\mathbf{P}(\boldsymbol{R}) - \mathbf{P}_0(\boldsymbol{R})\} = -\tfrac{1}{2}\boldsymbol{PP} : \boldsymbol{\varkappa}_1^{-1}(\boldsymbol{R}, T) + \tfrac{1}{2}\boldsymbol{BB} : \boldsymbol{\chi}_1(\boldsymbol{R}, T) + \tfrac{1}{2}(\boldsymbol{P}\cdot\boldsymbol{E} + \boldsymbol{M}\cdot\boldsymbol{B})\boldsymbol{n}. \tag{387}$$

Here we used the fact that the volume change that accompanies an infinitesimal change of a deformation $\boldsymbol{\eta}$ is given by

$$\mathrm{d}V = \int \boldsymbol{n}\cdot\mathrm{d}\boldsymbol{\eta}(\boldsymbol{R})\cdot\boldsymbol{R}\,\mathrm{d}S, \tag{388}$$

as follows from the definition of $\boldsymbol{\eta}$. (For a fluid it follows from the isotropy of the system and the fact that the susceptibilities depend only on the total volume of the system, that

$$\varkappa_0^{-1}(T) = \mathbf{U}\kappa^{-1}(T),$$
$$\varkappa_1^{-1}(T) = \mathbf{U}\boldsymbol{n}v\,\frac{\partial\kappa^{-1}(v, T)}{\partial v} \tag{389}$$

and similarly for the magnetic susceptibilities[1]. Indeed one finds back now (344) from (387).)

With the help of the Liénard expression (241) we obtain now for the normal component of the pressure tensor in the absence of fields, but with the same deformation at the surface and the same temperature:

$$\boldsymbol{n}\cdot\mathbf{P}_0(\boldsymbol{R}) = \boldsymbol{n}\cdot\mathbf{P}_{\text{out}}(\boldsymbol{R}) - \tfrac{1}{2}\boldsymbol{n}(\boldsymbol{P}\cdot\boldsymbol{n})^2 + \tfrac{1}{2}\boldsymbol{PP} : \varkappa_1^{-1}(\boldsymbol{R}, T)$$
$$-\tfrac{1}{2}\boldsymbol{BB} : \chi_1(\boldsymbol{R}, T) - \tfrac{1}{2}(\boldsymbol{P}\cdot\boldsymbol{E} + \boldsymbol{M}\cdot\boldsymbol{B})\boldsymbol{n}, \tag{390}$$

with $\mathbf{P}_{\text{out}}$ the pressure outside the system (in the presence of fields). The expression (390) may be looked upon as a boundary condition for an ordinary (field-free) elasticity problem. It shows that the boundary value $\boldsymbol{n}\cdot\mathbf{P}_0$ consists of two parts, 1st: two terms that represent the effect of the outward and Liénard pressures (the corresponding deformation is called the electrostrictive form effect), and 2nd: three terms which contain the constants that characterize the material and which form a generalization to solids of (minus) the right-hand side of (344) (the corresponding deformation is called the electro- and magnetostriction effect *sensu stricto*).

For a uniform scalar outward pressure – as the atmospheric pressure – one may evaluate, with the usual methods of elasticity, the deformation at the surface of the sample due to the form effect, at least for spherical shapes cut from substances with simple crystal symmetries (as for instance the cubic symmetries)[2]. One finds in this way a non-uniform deformation at the surface. Comparison with the experimental data showed that the total deformation at the surface has the same non-uniform character in the sense that there is a uniform difference. This means that the proper electro-magnetostriction gives rise to a uniform deformation over the surface. One may

[1] Note that the specific volume $v$ enters as a parameter at the right-hand side. This corresponds to the parametric dependence in (384) of the quantities $\kappa$ and $\chi$ on the reference state with respect to which $\eta$ is defined.

[2] R. Gersdorf, Physica **26**(1960)553 for the magnetostrictive case; v. also R. R. Birss and S. R. Adamson, Brit. J. Appl. Phys. **1**(1968)631, R. R. Birss and B. C. Hegarty, Brit. J. Appl. Phys. **1**(1968)789 for the calculation of the form effect for prolate spheroids.

conclude from this fact that the tensors $\boldsymbol{\varkappa}_1^{-1}$ and $\boldsymbol{\chi}_1$, which occur in (390), have the following dependence on the position $\boldsymbol{R}$

$$
\begin{aligned}
\{\boldsymbol{\varkappa}_1^{-1}(\boldsymbol{R}, T)\}^{ijk} &= n_l\{\hat{\boldsymbol{\varkappa}}_1^{-1}(T)\}^{ijlk}, \\
\{\boldsymbol{\chi}_1(\boldsymbol{R}, T)\}^{ijk} &= n_l\{\hat{\boldsymbol{\chi}}_1(T)\}^{ijlk},
\end{aligned}
\tag{391}
$$

where the fourth-rank symbols (symmetric in $i$ and $j$) are independent of the position at the surface, the only dependence on $\boldsymbol{R}$ being represented by the outward normal unit vector $\boldsymbol{n}$ (which occurs also in the left-hand side of (390)). The theory of the proper electro-magnetostriction is concerned now with the study of the quantities $\hat{\boldsymbol{\varkappa}}_1^{-1}(T)$ and $\hat{\boldsymbol{\chi}}_1(T)$. (A microscopic theory of electro-magnetostriction will be given for a simple model of a magnetic material in chapter X, § 6.) Phenomenologically one may employ the crystal symmetry to reduce the number of independent electro-magnetostriction constants, which occur in the two tensors of the fourth rank in (391). For instance for an isotropic or polycrystalline solid the number of independent electrostriction components $\hat{\boldsymbol{\varkappa}}_1^{-1}(T)$ reduces from 54 to 2 since one has then

$$
(\hat{\boldsymbol{\varkappa}}_1^{-1})^{ijkl} = a\delta^{ij}\delta^{kl} + b(\delta^{ik}\delta^{jl} + \delta^{il}\delta^{jk}),
\tag{392}
$$

because the Kronecker deltas are the only invariant quantities with respect to rotations. (For fluid systems it follows from (389) that $b$ vanishes while $a$ is equal to $v\partial\kappa^{-1}/\partial v$.) Similar remarks apply to the magnetostriction constants.

APPENDIX I

# On the depolarizing tensor

In section 7*b* occurs the integral (219), which in general depends on the position $\boldsymbol{R}$. We want to prove first that for an ellipsoidal volume the integral is in fact independent of $\boldsymbol{R}$ (if $\boldsymbol{R}$ is inside the volume) so that we then have

$$\int^{V} \nabla'\nabla' \frac{1}{4\pi|\boldsymbol{R}-\boldsymbol{R}'|}\,\mathrm{d}\boldsymbol{R}' = \int^{V} \nabla'\nabla' \frac{1}{4\pi|\boldsymbol{R}'|}\,\mathrm{d}\boldsymbol{R}' \equiv -\mathbf{L}, \tag{A1}$$

where the centre of the ellipsoid has been chosen as the origin of the coordinate system. The quantity $\mathbf{L}$ is called the 'depolarizing tensor'.

The second integral in (A1) depends only on the shape of the ellipsoidal volume and not on its scale, so that we may replace it by an integral over a small volume around the origin and of the same shape as the ellipsoid. This means that it is sufficient to prove instead of (A1) the vanishing of the integral

$$\mathscr{P}\int^{V} \nabla'\nabla' \frac{1}{4\pi|\boldsymbol{R}-\boldsymbol{R}'|}\,\mathrm{d}\boldsymbol{R}', \tag{A2}$$

where the principal value sign indicates that an infinitesimal ellipsoid of the same shape as the large one with centre $\boldsymbol{R}$ has to be excluded from the integration over $\boldsymbol{R}'$. By a conveniently chosen linear transformation of coordinates

$$\hat{\boldsymbol{R}} = \mathbf{A}\cdot\boldsymbol{R}, \tag{A3}$$

it is possible to transform the ellipsoid to a sphere. Then it becomes sufficient to prove the vanishing of the integral (omitting the circumflexes of $\hat{\boldsymbol{R}}$ and $\hat{\boldsymbol{R}}'$):

$$\mathscr{P}_0\int^{V_0} \nabla'\cdot\mathbf{A}\nabla'\cdot\mathbf{A} \frac{1}{4\pi|\mathbf{A}^{-1}\cdot(\boldsymbol{R}-\boldsymbol{R}')|}(\det\mathbf{A})^{-1}\mathrm{d}\boldsymbol{R}', \tag{A4}$$

where the integration is now extended over a spherical volume $V_0$ and where the principal value sign indicates the exclusion of an infinitesimal sphere around $\boldsymbol{R}$ from the integration over $\boldsymbol{R}'$. The denominator may be written as $4\pi|(\boldsymbol{R}-\boldsymbol{R}')+(\mathbf{A}^{-1}-\mathbf{U})\cdot(\boldsymbol{R}-\boldsymbol{R}')|$. Then, if the reciprocal of this expression is Taylor expanded, it follows that it is sufficient to prove that for all integer $n \geqslant 0$ the integral

$$\mathscr{P}_0 \int^{V_0} \boldsymbol{\nabla}'\boldsymbol{\nabla}'(\boldsymbol{R}-\boldsymbol{R}')^n \boldsymbol{\nabla}'^n \frac{1}{4\pi|\boldsymbol{R}-\boldsymbol{R}'|} \, \mathrm{d}\boldsymbol{R}' \tag{A5}$$

vanishes.

Let us first consider these integrals for $\boldsymbol{R}=0$. With Gauss's theorem, (A5) becomes for this case, apart from a factor $(-1)^n$,

$$\int^{S_0} \boldsymbol{n}'\boldsymbol{\nabla}'\boldsymbol{R}'^n \boldsymbol{\nabla}'^n \frac{1}{4\pi|\boldsymbol{R}'|} \, \mathrm{d}S' - \int^{s_0} \boldsymbol{n}'\boldsymbol{\nabla}'\boldsymbol{R}'^n \boldsymbol{\nabla}'^n \frac{1}{4\pi|\boldsymbol{R}'|} \, \mathrm{d}S', \tag{A6}$$

where $S_0$ is the surface of the large sphere, and $s_0$ of the infinitesimal one; $\boldsymbol{n}'$ is the unit vector normal to the integration surfaces. Each of the integrals is independent of the scale of the sphere so that they are equal. Hence the expression (A6) vanishes.

Since now the vanishing of (A5) is proved for the case $\boldsymbol{R} = 0$, it is sufficient to prove that (A5) is independent of $\boldsymbol{R}$ in order to ensure its vanishing everywhere. The derivative of (A5) with respect to $\boldsymbol{R}$ is

$$-\int^{S_0} \boldsymbol{n}'\boldsymbol{\nabla}'\boldsymbol{\nabla}'(\boldsymbol{R}-\boldsymbol{R}')^n \boldsymbol{\nabla}'^n \frac{1}{4\pi|\boldsymbol{R}-\boldsymbol{R}'|} \, \mathrm{d}S'. \tag{A7}$$

To prove the vanishing of this derivative we expand the integrand with respect to $\boldsymbol{R}$. We find then that it is sufficient to prove the vanishing of the expression

$$\int^{S_0} \boldsymbol{n}'\boldsymbol{\nabla}'^{m+2} \left( \boldsymbol{R}'^n \boldsymbol{\nabla}'^n \frac{1}{4\pi|\boldsymbol{R}'|} \right) \mathrm{d}S' \tag{A8}$$

for all $m, n \geqslant 0$, or alternatively the vanishing of

$$\int^{S_0} \boldsymbol{n}'\boldsymbol{R}'^n \boldsymbol{\nabla}'^{n+m+2} \frac{1}{4\pi|\boldsymbol{R}'|} \, \mathrm{d}S'. \tag{A9}$$

This integral is an *invariant* tensor of rank $2n+m+3$, which is symmetric in the second up to and including the $(n+1)$th Cartesian index, and symmetric in the last $n+m+2$ indices. Moreover the traces taken with a pair of indices from the last $n+m+2$ vanish, since $\Delta'(1/4\pi|\boldsymbol{R}'|)$ is zero for $\boldsymbol{R}' \neq 0$. Therefore (A9) is an element of the direct product space of tensors of rank 1, symmetric tensors of rank $n$ and symmetric, traceless tensors of rank $n+m+2$. Symmetric tensors of rank $n$ form a reducible representation of the rotation group which contains irreducible representations of dimension $2n+1$ and lower, whereas symmetric, traceless tensors of rank $n+m+2$

form an irreducible representation of dimension $2n+2m+5$. The direct product of tensors of these two types contains only irreducible representations of a dimensionality $2m+5$ and higher, so that the direct product space mentioned above, of which (A9) is an element, contains irreducible representations of dimensionality higher than 1. Since (A9) is an *invariant* tensor in this direct product space it must vanish identically. Thus, retracing the chain of reasoning, it is now proved that the first member of (A1) is indeed independent of $\boldsymbol{R}$, and hence equal to minus the depolarizing tensor $\mathbf{L}$.

As a corollary of (A1) one finds by integrating over $\boldsymbol{R}$:

$$\iint^V \nabla'\nabla' \frac{1}{4\pi|\boldsymbol{R}-\boldsymbol{R}'|} \, \mathrm{d}\boldsymbol{R}\, \mathrm{d}\boldsymbol{R}' = -V\mathbf{L}, \tag{A10}$$

a formula which has been used in (231).

In a way analogous to the proof given above one may show that the integral

$$\int^V (\boldsymbol{R}-\boldsymbol{R}')\nabla'\nabla'\nabla' \frac{1}{8\pi|\boldsymbol{R}-\boldsymbol{R}'|} \, \mathrm{d}\boldsymbol{R}' \tag{A11}$$

is independent of $\boldsymbol{R}$, so that we may write

$$\int^V (\boldsymbol{R}-\boldsymbol{R}')\nabla'\nabla'\nabla' \frac{1}{8\pi|\boldsymbol{R}-\boldsymbol{R}'|} \, \mathrm{d}\boldsymbol{R}' = \int^V (\boldsymbol{R}-\boldsymbol{R}')\nabla'\nabla'\nabla' \frac{1}{8\pi|\boldsymbol{R}'|} \, \mathrm{d}\boldsymbol{R}' \equiv -\tfrac{1}{2}\mathbf{K}, \tag{A12}$$

where $\mathbf{K}$ is a tensor with four Cartesian indices. As a corollary it follows that

$$\iint^V (\boldsymbol{R}-\boldsymbol{R}')\nabla'\nabla'\nabla' \frac{1}{8\pi|\boldsymbol{R}-\boldsymbol{R}'|} \, \mathrm{d}\boldsymbol{R}\, \mathrm{d}\boldsymbol{R}' = -\tfrac{1}{2}V\mathbf{K}. \tag{A13}$$

Finally we want to prove the identity

$$\delta\mathbf{L} = (\mathrm{Tr}\, \delta\boldsymbol{\epsilon})\mathbf{L} - \delta\boldsymbol{\epsilon} : \mathbf{K}, \tag{A14}$$

which gives the variation of the tensor $\mathbf{L}$ when the boundary of the ellipsoidal volume is deformed according to

$$\delta\boldsymbol{R} = \delta\boldsymbol{\epsilon}\cdot\boldsymbol{R}, \tag{A15}$$

where $\delta\boldsymbol{\epsilon}$ is the (uniform) deformation tensor and where the centre of the ellipsoid is the origin of the coordinate system. From the definition (A1) of $\mathbf{L}$ one has

$$\delta\mathbf{L} = -\int^S \boldsymbol{n}\cdot\delta\boldsymbol{R}\nabla\nabla \frac{1}{4\pi|\boldsymbol{R}|} \, \mathrm{d}S, \tag{A16}$$

where the integration is extended over the surface of the ellipsoid with normal $\boldsymbol{n}$. With Gauss's theorem and (A15) this becomes

$$\delta \mathbf{L} = -\int^{V} \nabla \cdot \left( \delta \boldsymbol{\epsilon} \cdot \boldsymbol{R} \nabla \nabla \frac{1}{4\pi |\boldsymbol{R}|} \right) \mathrm{d}\boldsymbol{R}. \tag{A17}$$

Performing the differentiation one gets

$$\delta \mathbf{L} = -(\mathrm{Tr}\, \delta \boldsymbol{\epsilon}) \int^{V} \nabla \nabla \frac{1}{4\pi |\boldsymbol{R}|} \mathrm{d}\boldsymbol{R} - \int^{V} \delta \boldsymbol{\epsilon} : \boldsymbol{R} \nabla \nabla \nabla \frac{1}{4\pi |\boldsymbol{R}|} \mathrm{d}\boldsymbol{R}, \tag{A18}$$

which in view of the definitions of $\mathbf{L}$ and $\mathbf{K}$ is indeed (A14).

A corollary is obtained by noting that $\mathrm{Tr}\, \delta \boldsymbol{\epsilon}$ is $V^{-1}\delta V$ with $V$ the volume of the ellipsoid:

$$\delta(V^{-1}\mathbf{L}) = -V^{-1}\delta \boldsymbol{\epsilon} : \mathbf{K}. \tag{A19}$$

It has been employed in subsection 7*b* (246).

As an example let us derive the tensors $\mathbf{L}$ and $\mathbf{K}$ for a sphere. From (A1) it follows for a sphere that $\mathbf{L}$ is an invariant tensor with two indices and hence a multiple of the unit tensor. The factor is determined by calculating the trace of $\mathbf{L}$:

$$\mathrm{Tr}\, \mathbf{L} = -\int^{V} \Delta \frac{1}{4\pi |\boldsymbol{R}|} \mathrm{d}\boldsymbol{R} = 1, \tag{A20}$$

so that for the sphere

$$\mathbf{L} = \tfrac{1}{3}\mathbf{U}. \tag{A21}$$

Furthermore it follows from (A12) for a sphere that $\mathbf{K}$ is an invariant tensor with four indices and hence of the form

$$K_{ijkl} = \alpha \delta_{ij}\delta_{kl} + \beta \delta_{ik}\delta_{jl} + \gamma \delta_{il}\delta_{jk}. \tag{A22}$$

From the symmetry of $\mathbf{K}$ in its last three indices (see the definition (A12)) it follows that $\alpha$, $\beta$ and $\gamma$ are equal. Furthermore through contraction of the last pair of indices one has:

$$\mathbf{K} : \mathbf{U} = -\int \boldsymbol{R} \nabla \delta(\boldsymbol{R}) \mathrm{d}\boldsymbol{R} = \mathbf{U} \tag{A23}$$

as follows from partial integration. In this way we have obtained for a sphere:

$$K_{ijkl} = \tfrac{1}{5}(\delta_{ij}\delta_{kl} + \delta_{ik}\delta_{jl} + \delta_{il}\delta_{jk}). \tag{A24}$$

The identity (A14) becomes for small deformations of the sphere

$$\delta \mathbf{L} = \tfrac{2}{15}(\mathrm{Tr}\, \delta \boldsymbol{\epsilon})\mathbf{U} - \tfrac{1}{5}\delta \boldsymbol{\epsilon} - \tfrac{1}{5}\delta \tilde{\boldsymbol{\epsilon}}, \tag{A25}$$

where the results (A21) and (A24) have been utilized.

# The Hamiltonian for a system of composite particles in an external field

The Hamiltonian of a system of particles $ki$ (grouped in atoms or other stable entities numbered by $k$, while the constituent particles are labelled by $i = 1, 2, \ldots, f$) with charges $e_{ki}$, positions $\boldsymbol{R}_{ki}$ and momenta $\boldsymbol{P}_{ki}$, which move in an external field with scalar and vector potentials $\varphi_e$ and $\boldsymbol{A}_e$, reads up to order $c^{-1}$ (v. (I.16)):

$$H(\boldsymbol{R}_{ki}, \boldsymbol{P}_{ki}, t) = \sum_{k,i} \frac{\boldsymbol{P}_{ki}^2}{2m_{ki}} + \sum_{k} \sum_{i,j(i\neq j)} \frac{e_{ki}e_{kj}}{8\pi|\boldsymbol{R}_{ki}-\boldsymbol{R}_{kj}|}$$

$$+ \sum_{k,l(k\neq l)} \sum_{i,j} \frac{e_{ki}e_{lj}}{8\pi|\boldsymbol{R}_{ki}-\boldsymbol{R}_{lj}|} + \sum_{k,i} e_{ki}\left\{\varphi_e(\boldsymbol{R}_{ki}, t) - c^{-1}\frac{\boldsymbol{P}_{ki}}{m_{ki}}\cdot\boldsymbol{A}_e(\boldsymbol{R}_{ki}, t)\right\}. \quad \text{(A26)}$$

Let us now introduce new canonical coordinates $\boldsymbol{q}_{ki}$ and momenta $\hat{\boldsymbol{p}}_{ki}$, such that the $\boldsymbol{q}_{ki}$ are the centre of mass of atom $k$ and (independent) relative coordinates of the constituent particles with respect to this centre:

$$\begin{aligned}
\boldsymbol{q}_{ki} &= \boldsymbol{R}_{ki} - \boldsymbol{R}_k \equiv \boldsymbol{R}_{ki} - \sum_{j=1}^{f}(m_{kj}/m_k)\boldsymbol{R}_{kj}, \quad (i = 1, \ldots, f-1),\\
\boldsymbol{q}_{kf} &= \boldsymbol{R}_k \equiv \sum_{j=1}^{f}(m_{kj}/m_k)\boldsymbol{R}_{kj},\\
\hat{\boldsymbol{p}}_{ki} &= \boldsymbol{P}_{ki} - (m_{ki}/m_{kf})\boldsymbol{P}_{kf}, \quad (i = 1, \ldots, f-1),\\
\hat{\boldsymbol{p}}_{kf} &= \sum_{i=1}^{f}\boldsymbol{P}_{ki}.
\end{aligned} \quad \text{(A27)}$$

This is a canonical transformation, as may be checked by evaluating the Poisson brackets. Inversion of (A27) gives

$$\begin{aligned}
\boldsymbol{R}_{ki} &= \boldsymbol{q}_{kf} + (1-\delta_{if})\boldsymbol{q}_{ki} - \delta_{if}\sum_{j=1}^{f-1}(m_{kj}/m_{kf})\boldsymbol{q}_{kj},\\
\boldsymbol{P}_{ki} &= (m_{ki}/m_k)\hat{\boldsymbol{p}}_{kf} + (1-\delta_{if})\hat{\boldsymbol{p}}_{ki} - (m_{ki}/m_k)\sum_{j=1}^{f-1}\hat{\boldsymbol{p}}_{kj}.
\end{aligned} \quad \text{(A28)}$$

We substitute these relations into the Hamiltonian (A26) and expand the potentials around the centres of mass $\boldsymbol{R}_k$, retaining only terms up to first derivatives of the potentials:

$$
\begin{aligned}
H(\boldsymbol{q}, \hat{\boldsymbol{p}}, t) = {} & \sum_k \left( \frac{\hat{\boldsymbol{p}}_{kf}^2}{2m_k} + \sum_{i=1}^{f-1} \frac{\hat{\boldsymbol{p}}_{ki}^2}{2m_{ki}} - \sum_{i,j=1}^{f-1} \frac{\hat{\boldsymbol{p}}_{ki}\cdot\hat{\boldsymbol{p}}_{kj}}{2m_k} \right) \\
& + \sum_k \sum_{i,j=1(i\neq j)}^{f} \frac{e_{ki}\, e_{kj}}{8\pi|\boldsymbol{R}_{ki}(\boldsymbol{q}) - \boldsymbol{R}_{kj}(\boldsymbol{q})|} + \sum_{k,l(k\neq l)} \sum_{i,j=1}^{f} \frac{e_{ki}\, e_{lj}}{8\pi|\boldsymbol{R}_{ki}(\boldsymbol{q}) - \boldsymbol{R}_{lj}(\boldsymbol{q})|} \\
& + \sum_k e_k \left\{ \varphi_{\mathrm{e}}(\boldsymbol{R}_k, t) - c^{-1} \frac{\hat{\boldsymbol{p}}_{kf}}{m_k} \cdot \boldsymbol{A}_{\mathrm{e}}(\boldsymbol{R}_k, t) \right\} \\
& + \sum_k \sum_{i=1}^{f} e_{ki} \Bigg[ \{\boldsymbol{R}_{ki}(\boldsymbol{q}) - \boldsymbol{q}_{kf}\}\cdot\nabla_k \left\{ \varphi_{\mathrm{e}}(\boldsymbol{R}_k, t) - c^{-1} \frac{\hat{\boldsymbol{p}}_{kf}}{m_k} \cdot \boldsymbol{A}_{\mathrm{e}}(\boldsymbol{R}_k, t) \right\} \\
& - c^{-1} \left\{ \frac{\boldsymbol{P}_{ki}(\hat{\boldsymbol{p}})}{m_{ki}} - \frac{\hat{\boldsymbol{p}}_{kf}}{m_k} \right\} \cdot \boldsymbol{A}_{\mathrm{e}}(\boldsymbol{R}_k, t) \\
& - c^{-1} \{\boldsymbol{R}_{ki}(\boldsymbol{q}) - \boldsymbol{q}_{kf}\}\cdot\nabla_k \boldsymbol{A}_{\mathrm{e}}(\boldsymbol{R}_k, t) \cdot \left\{ \frac{\boldsymbol{P}_{ki}(\hat{\boldsymbol{p}})}{m_{ki}} - \frac{\hat{\boldsymbol{p}}_{kf}}{m_k} \right\} \Bigg],
\end{aligned}
\tag{A29}
$$

where $\boldsymbol{R}_{ki}(\boldsymbol{q})$ and $\boldsymbol{P}_{ki}(\hat{\boldsymbol{p}})$ stand for the right-hand sides of (A28). This expression may be transformed by means of a second canonical transformation in order to cast the last three terms in gauge invariant form. This transformation is given by the generating function:

$$
\begin{aligned}
F(\boldsymbol{q}, \boldsymbol{p}, t) = {} & \sum_k \boldsymbol{R}_k\cdot\boldsymbol{P}_k + \sum_k \sum_{i=1}^{f-1} \boldsymbol{q}_{ki}\cdot\boldsymbol{p}_{ki} \\
& + c^{-1} \sum_k \sum_{i=1}^{f} e_{ki} \boldsymbol{A}_{\mathrm{e}}(\boldsymbol{R}_k, t)\cdot\{\boldsymbol{R}_{ki}(\boldsymbol{q}) - \boldsymbol{q}_{kf}\} \\
& + \tfrac{1}{2} c^{-1} \sum_k \sum_{i=1}^{f} e_{ki} \{\boldsymbol{R}_{ki}(\boldsymbol{q}) - \boldsymbol{q}_{kf}\}\cdot\nabla_k \boldsymbol{A}_{\mathrm{e}}(\boldsymbol{R}_k, t)\cdot\{\boldsymbol{R}_{ki}(\boldsymbol{q}) - \boldsymbol{q}_{kf}\},
\end{aligned}
\tag{A30}
$$

depending on the old coordinates $\boldsymbol{q}_{ki}$, $\boldsymbol{q}_{kf} \equiv \boldsymbol{R}_k$, the new momenta $\boldsymbol{p}_{ki}$, $\boldsymbol{p}_{kf} \equiv \boldsymbol{P}_k$ and time. With the help of the transformation formulae

$$
\begin{aligned}
\boldsymbol{q}_{ki} &= \frac{\partial F}{\partial \boldsymbol{p}_{ki}} \quad (i = 1, \ldots, f-1), & \boldsymbol{R}_k &= \frac{\partial F}{\partial \boldsymbol{P}_k}, \\
\hat{\boldsymbol{p}}_{ki} &= \frac{\partial F}{\partial \boldsymbol{q}_{ki}} \quad (i = 1, \ldots, f-1), & \hat{\boldsymbol{p}}_{kf} &= \frac{\partial F}{\partial \boldsymbol{R}_k},
\end{aligned}
\tag{A31}
$$

it follows that the coordinates do not change. Furthermore the new Hamiltonian gets an extra term $\partial F/\partial t$. If we substitute (A31) with (A30) into (A29), neglecting again second derivatives of the potentials, and using the vector identity $(\boldsymbol{a} \wedge \boldsymbol{b})\cdot(\boldsymbol{c} \wedge \boldsymbol{d}) = (\boldsymbol{a}\cdot\boldsymbol{c})(\boldsymbol{b}\cdot\boldsymbol{d}) - (\boldsymbol{a}\cdot\boldsymbol{d})(\boldsymbol{b}\cdot\boldsymbol{c})$, we get as the new Hamiltonian (up to $c^{-1}$):

$$
\begin{aligned}
H(\boldsymbol{q}, \boldsymbol{p}, t) = {} & \sum_k \left( \frac{\boldsymbol{P}_k^2}{2m_k} + \sum_{i=1}^{f-1} \frac{\boldsymbol{p}_{ki}^2}{2m_{ki}} - \sum_{i,j=1}^{f-1} \frac{\boldsymbol{p}_{ki}\cdot\boldsymbol{p}_{kj}}{2m_k} \right) \\
& + \sum_k \sum_{i,j=1(i\neq j)}^{f} \frac{e_{ki}\, e_{kj}}{8\pi|\boldsymbol{R}_{ki}(\boldsymbol{q}) - \boldsymbol{R}_{kj}(\boldsymbol{q})|} + \sum_{k,l(k\neq l)} \sum_{i,j=1}^{f} \frac{e_{ki}\, e_{lj}}{8\pi|\boldsymbol{R}_{ki}(\boldsymbol{q}) - \boldsymbol{R}_{lj}(\boldsymbol{q})|} \\
& + \sum_k e_k \left\{ \varphi_{\mathrm{e}}(\boldsymbol{R}_k, t) - c^{-1} \frac{\boldsymbol{P}_k}{m_k} \cdot \boldsymbol{A}_{\mathrm{e}}(\boldsymbol{R}_k, t) \right\} \\
& - \sum_k \left\{ \overline{\boldsymbol{\mu}}_k^{(1)} \cdot \boldsymbol{E}_{\mathrm{e}}(\boldsymbol{R}_k, t) + \left( \overline{\boldsymbol{\nu}}_k^{(1)} + c^{-1} \overline{\boldsymbol{\mu}}_k^{(1)} \wedge \frac{\boldsymbol{P}_k}{m_k} \right) \cdot \boldsymbol{B}_{\mathrm{e}}(\boldsymbol{R}_k, t) \right\}, \qquad \text{(A32)}
\end{aligned}
$$

with the abbreviations

$$
\begin{aligned}
\overline{\boldsymbol{\mu}}_k^{(1)}(\boldsymbol{q}) &\equiv \sum_{i=1}^{f} e_{ki} \{ \boldsymbol{R}_{ki}(\boldsymbol{q}) - \boldsymbol{q}_{kf} \}, \\
\overline{\boldsymbol{\nu}}_k^{(1)}(\boldsymbol{q}, \boldsymbol{p}) &\equiv \tfrac{1}{2} c^{-1} \sum_{i=1}^{f} e_{ki} \{ \boldsymbol{R}_{ki}(\boldsymbol{q}) - \boldsymbol{q}_{kf} \} \wedge \left\{ (1-\delta_{if}) \frac{\boldsymbol{p}_{ki}}{m_{ki}} - \sum_{j=1}^{f-1} \frac{\boldsymbol{p}_{kj}}{m_k} \right\}. \qquad \text{(A33)}
\end{aligned}
$$

These quantities are the electric and magnetic dipole moments, written in terms of the canonical coordinates and momenta. For the electric dipole moment this is obvious since $\boldsymbol{q}_{kf} \equiv \boldsymbol{R}_k$, the position of the centre of mass. For the magnetic dipole moment it follows because the Hamilton equations yield up to $c^0$:

$$
(1-\delta_{if}) \frac{\boldsymbol{p}_{ki}}{m_{ki}} - \sum_{j=1}^{f-1} \frac{\boldsymbol{p}_{kj}}{m_k} = \dot{\boldsymbol{R}}_{ki}(\boldsymbol{q}) - \dot{\boldsymbol{q}}_{kf}. \qquad \text{(A34)}
$$

(We note in passing that the Hamiltonian equations for the coordinates $\boldsymbol{R}_k$ and momenta $\boldsymbol{P}_k$ lead indeed to the equations of motion (I.50).)

We shall need also an expression for the kinetic energy of the system:

$$
\begin{aligned}
K(\boldsymbol{q}, \dot{\boldsymbol{q}}, t) \ & \left(\equiv \sum_k \sum_{i=1}^{f} \tfrac{1}{2} m_{ki} \dot{\boldsymbol{R}}_{ki}^2 \right) \\
& = \sum_k \tfrac{1}{2} m_k \dot{\boldsymbol{R}}_k^2 + \sum_k \left( \sum_{i=1}^{f-1} \tfrac{1}{2} m_{ki} \dot{\boldsymbol{q}}_{ki}^2 + \sum_{i,j=1}^{f-1} \frac{m_{ki}\, m_{kj}}{2m_{kf}} \dot{\boldsymbol{q}}_{ki}\cdot\dot{\boldsymbol{q}}_{kj} \right), \qquad \text{(A35)}
\end{aligned}
$$

where (A28) has been used. We may express it in terms of the coordinates and momenta, by using the Hamilton equations $\dot{\boldsymbol{q}}_{ki} = \partial H/\partial \boldsymbol{p}_{ki}$ with (A32–

33). Then we get up to $c^{-1}$

$$K(\boldsymbol{q}, \boldsymbol{p}, t) = \sum_k \left\{ \frac{\boldsymbol{P}_k^2}{2m_k} - c^{-1} e_k \frac{\boldsymbol{P}_k}{m_k} \cdot \boldsymbol{A}_e(\boldsymbol{R}_k, t) - c^{-1} \left( \bar{\boldsymbol{\mu}}_k^{(1)} \wedge \frac{\boldsymbol{P}_k}{m_k} \right) \cdot \boldsymbol{B}_e(\boldsymbol{R}_k, t) \right\}$$

$$+ \sum_k \left\{ \sum_{i=1}^{f-1} \frac{\boldsymbol{p}_{ki}^2}{2m_{ki}} - \sum_{i,j=1}^{f-1} \frac{\boldsymbol{p}_{ki} \cdot \boldsymbol{p}_{kj}}{2m_k} - \bar{\boldsymbol{\nu}}_k^{(1)} \cdot \boldsymbol{B}_e(\boldsymbol{R}_k, t) \right\}. \tag{A36}$$

Therefore the Hamiltonian (A32) may be written in the form

$$H(\boldsymbol{q}, \boldsymbol{p}, t) = K(\boldsymbol{q}, \boldsymbol{p}, t) + \sum_k \sum_{i,j=1 (i \neq j)}^{f} \frac{e_{ki} e_{kj}}{8\pi |\boldsymbol{R}_{ki}(\boldsymbol{q}) - \boldsymbol{R}_{kj}(\boldsymbol{q})|}$$

$$+ \sum_{k,l (k \neq l)} \sum_{i,j=1}^{f} \frac{e_{ki} e_{lj}}{8\pi |\boldsymbol{R}_{ki}(\boldsymbol{q}) - \boldsymbol{R}_{lj}(\boldsymbol{q})|} + \sum_k e_k \varphi_e(\boldsymbol{R}_k, t) - \sum_k \bar{\boldsymbol{\mu}}_k^{(1)} \cdot \boldsymbol{E}_e(\boldsymbol{R}_k, t). \tag{A37}$$

It should be noted that the terms with the vector potential and the magnetic field appear explicitly in (A32) but are hidden in the kinetic energy in (A37).

# Deformations and free energy

In this appendix we want to study the change of the free energy $F^*$ with the change of the boundaries of the system.

The system considered consists of atoms, carrying charges and dipole moments, in a uniform and time-independent external field $\boldsymbol{E}_e$, $\boldsymbol{B}_e$. It is then described by the Hamiltonian (A32) of appendix II. At equilibrium it is represented by the canonical ensemble:

$$e^{-F^*/kT} = C\int^V e^{-H/kT}\mathrm{d}\boldsymbol{q}\,\mathrm{d}\boldsymbol{p}, \tag{A38}$$

where $V$ is the volume and $T$ the temperature. (The constant $C$ depends only on the particle number.) The free energy is thus a function of $T$, $\boldsymbol{E}_e$, $\boldsymbol{B}_e$ and the boundary of the volume $V$. The integrations over the coordinates $q$ may be extended to infinity if a wall potential (which is infinite if one of the particles of the system is outside the volume $V$) is included in the Hamiltonian. We take as the wall potential

$$U^W = \sum_k U_k^W(\boldsymbol{R}_k) \tag{A39}$$

with the functions $U_k^W$ zero for $\boldsymbol{R}_k$ inside the volume and infinite for $\boldsymbol{R}_k$ outside the volume. Here $\boldsymbol{R}_k$ is the centre of mass of the atom $k$. In the first instance one might be inclined to write as the wall potential $\sum_{k,i} U_{ki}^W(\boldsymbol{R}_{ki})$, where the sum is extended over all constituent particles. However, since the dimensions of the stable atoms are small compared to the volume the use of (A39) instead of this expression is justified if surface effects are neglected. Thus we write instead of (A38)

$$e^{-F^*/kT} = C\int^\infty e^{-(H+U^W)/kT}\mathrm{d}\boldsymbol{q}\,\mathrm{d}\boldsymbol{p}. \tag{A40}$$

Let us change the positions $\boldsymbol{R}^W$ of the walls by means of an infinitesimal transformation

$$\boldsymbol{R}^{W'} = \boldsymbol{R}^W + \delta\boldsymbol{R}^W = \{\mathsf{U} + \delta\boldsymbol{\epsilon}(\boldsymbol{R}^W)\}\cdot\boldsymbol{R}^W \tag{A41}$$

with an infinitesimal tensor $\delta\boldsymbol{\epsilon}$ that depends on the position of the wall $\boldsymbol{R}^{\mathrm{W}}$. (The unit tensor is denoted by $\mathbf{U}$.) Then the wall potential $U^{\mathrm{W}}$ is transformed to $U^{\mathrm{W}'} = \sum_k U_k^{\mathrm{W}'}(\boldsymbol{R}_k)$, such that $U_k^{\mathrm{W}'}$ is infinite if the position $\boldsymbol{R}_k$ is outside the new boundary, and zero inside. Alternatively one may say that $U_k^{\mathrm{W}'}$ is infinite if $\{\mathbf{U}-\delta\boldsymbol{\epsilon}(\boldsymbol{R}_k)\}\cdot\boldsymbol{R}_k$ is outside the old boundary; in other words

$$U^{\mathrm{W}'} = \sum_k U_k^{\mathrm{W}}[\{\mathbf{U}-\delta\boldsymbol{\epsilon}(\boldsymbol{R}_k)\}\cdot\boldsymbol{R}_k]. \tag{A42}$$

From (A40) we find now the change $\delta_\varepsilon F^*$ of $F^*$ with $\delta\boldsymbol{\epsilon}$ at constant $T$, $\boldsymbol{E}_{\mathrm{e}}$ and $\boldsymbol{B}_{\mathrm{e}}$:

$$\delta_\varepsilon F^* = C\,\mathrm{e}^{F^*/kT}\int^{\infty} \delta U^{\mathrm{W}} \mathrm{e}^{-(H+U^{\mathrm{W}})/kT}\mathrm{d}\boldsymbol{q}\,\mathrm{d}\boldsymbol{p}, \tag{A43}$$

where $\delta U^{\mathrm{W}} \equiv U^{\mathrm{W}'} - U^{\mathrm{W}}$. With (A42) this becomes, up to terms linear in $\delta\boldsymbol{\epsilon}(\boldsymbol{R}_k)$

$$\delta_\varepsilon F^* = -\langle \sum_k \boldsymbol{R}_k\cdot\delta\tilde{\boldsymbol{\epsilon}}(\boldsymbol{R}_k)\cdot\nabla_k\, U_k^{\mathrm{W}}(\boldsymbol{R}_k)\rangle, \tag{A44}$$

where the brackets denote the canonical ensemble average and where $\delta\tilde{\boldsymbol{\epsilon}}$ is the transposed matrix of $\delta\boldsymbol{\epsilon}$. The right-hand side contains the force $-\nabla_k U_k^{\mathrm{W}}(\boldsymbol{R}_k)$ exerted by the wall on atom $k$. This expression is only different from zero if $\boldsymbol{R}_k$ is situated at the wall so that one may write it as a sum of contributions due to the various surface elements $\mathrm{d}S$ of the wall:

$$-\nabla_k\, U_k^{\mathrm{W}}(\boldsymbol{R}_k) = \int^{S} \boldsymbol{f}_k^{\mathrm{W}}(\boldsymbol{R}_k)\delta(\boldsymbol{R}_k-\boldsymbol{R})\mathrm{d}S. \tag{A45}$$

Inserting this expression into (A44), one finds

$$\delta_\varepsilon F^* = \int^{S} \boldsymbol{R}\cdot\delta\tilde{\boldsymbol{\epsilon}}(\boldsymbol{R})\cdot\boldsymbol{f}^{\mathrm{W}}(\boldsymbol{R})\mathrm{d}S, \tag{A46}$$

where

$$\boldsymbol{f}^{\mathrm{W}}(\boldsymbol{R}) \equiv \langle \sum_k \boldsymbol{f}_k^{\mathrm{W}}\delta(\boldsymbol{R}_k-\boldsymbol{R})\rangle \tag{A47}$$

is the average force per unit surface exerted by the wall. The latter is conventionally written as $-\boldsymbol{n}\cdot\mathbf{P}_{\mathrm{out}}(\boldsymbol{R})$, where $\boldsymbol{n}$ is the outward normal to the boundary and $\mathbf{P}_{\mathrm{out}}(\boldsymbol{R})$ the pressure tensor in the wall. Therefore (A46) has the form

$$\delta_\varepsilon F^* = -\int^{S} \boldsymbol{n}\cdot\mathbf{P}_{\mathrm{out}}\cdot\delta\boldsymbol{\epsilon}\cdot\boldsymbol{R}\,\mathrm{d}S. \tag{A48}$$

An alternative expression for $\delta_\varepsilon F^*$ is obtained if one uses the virial theorem in the form

$$\left\langle \frac{\mathrm{d}}{\mathrm{d}t} \{ \sum_k \boldsymbol{P}_k \cdot \delta\boldsymbol{\epsilon}(\boldsymbol{R}_k) \cdot \boldsymbol{R}_k \} \right\rangle = 0, \tag{A49}$$

which follows from the fact that for a stationary ensemble the average of a total time derivative of a dynamical quantity vanishes. In (A49) the total time derivative has to be read as the Poisson bracket with the total Hamiltonian $H + U^{\mathrm{W}}$, which includes the wall potential. Hence (A49) may be written as

$$\left\langle \sum_k \left\{ \frac{\partial U^{\mathrm{W}}}{\partial \boldsymbol{R}_k} \cdot \delta\boldsymbol{\epsilon}(\boldsymbol{R}_k) \cdot \boldsymbol{R}_k + \frac{\partial H}{\partial \boldsymbol{R}_k} \cdot \delta\boldsymbol{\epsilon}(\boldsymbol{R}_k) \cdot \boldsymbol{R}_k - \boldsymbol{P}_k \cdot \delta\mathbf{e}(\boldsymbol{R}_k) \cdot \frac{\partial H}{\partial \boldsymbol{P}_k} \right\} \right\rangle = 0, \tag{A50}$$

where we introduced the infinitesimal deformation gradient tensor $\delta\mathbf{e}$ which is defined by a relation involving the partial derivatives of $\delta\boldsymbol{\epsilon}$:

$$\delta\tilde{\mathbf{e}}(\boldsymbol{R}) \equiv \frac{\partial}{\partial \boldsymbol{R}} \{ \delta\boldsymbol{\epsilon}(\boldsymbol{R}) \cdot \boldsymbol{R} \}. \tag{A51}$$

In particular if $\delta\boldsymbol{\epsilon}$ represents a homogeneous deformation, i.e. is independent of $\boldsymbol{R}$, we have

$$\delta\mathbf{e} = \delta\boldsymbol{\epsilon}. \tag{A52}$$

With the help of the identity (A50), we find for (A44)

$$\delta_\varepsilon F^* = - \left\langle \sum_k \left\{ \boldsymbol{P}_k \cdot \delta\mathbf{e}(\boldsymbol{R}_k) \cdot \frac{\partial H}{\partial \boldsymbol{P}_k} - \frac{\partial H}{\partial \boldsymbol{R}_k} \cdot \delta\boldsymbol{\epsilon}(\boldsymbol{R}_k) \cdot \boldsymbol{R}_k \right\} \right\rangle. \tag{A53}$$

In particular if $\delta\boldsymbol{\epsilon}$ is independent of the position, one finds for the change of the free energy

$$\delta_\varepsilon F^* = \mathbf{A} : \delta\boldsymbol{\epsilon} \tag{A54}$$

with the tensor $\mathbf{A}$ defined as

$$\mathbf{A} \equiv - \left\langle \sum_k \left( \frac{\partial H}{\partial \boldsymbol{P}_k} \boldsymbol{P}_k - \boldsymbol{R}_k \frac{\partial H}{\partial \boldsymbol{R}_k} \right) \right\rangle. \tag{A55}$$

The results (A53) and (A54–55) are used in the main text.

# PROBLEMS

**1.** Show that the solution of the first Maxwell equation $\nabla \cdot \boldsymbol{E} = \varrho^{\mathrm{e}} - \nabla \cdot \boldsymbol{P}$ for the electrostatic case has the form given in the first line of (17). Check this by noting that one may replace the integral in the solution mentioned by one in which a small volume around $\boldsymbol{R}$ is excluded from the integration over $\boldsymbol{R}'$:

$$-\mathscr{P}\int \{\varrho^{\mathrm{e}}(\boldsymbol{R}', t) - \nabla' \cdot \boldsymbol{P}(\boldsymbol{R}', t)\} \nabla \frac{1}{4\pi|\boldsymbol{R}-\boldsymbol{R}'|} \mathrm{d}\boldsymbol{R}',$$

since the integral is convergent. The advantage of the latter way of writing the integral becomes apparent if one takes the divergence: the divergence of the integrand does not give a contribution now.

If in the expression given above a partial integration is performed in the second term, one finds for the electrostatic field an expression with convergent integrals:

$$\boldsymbol{E}(\boldsymbol{R}, t) = \boldsymbol{E}_{\mathrm{e}}(\boldsymbol{R}, t) - \mathscr{P}\int \{\varrho^{\mathrm{e}}(\boldsymbol{R}', t) + \boldsymbol{P}(\boldsymbol{R}', t) \cdot \nabla'\} \nabla \frac{1}{4\pi|\boldsymbol{R}-\boldsymbol{R}'|} \mathrm{d}\boldsymbol{R}' - \int^{s'} \boldsymbol{n}' \cdot \boldsymbol{P}(\boldsymbol{R}', t) \nabla \frac{1}{4\pi|\boldsymbol{R}-\boldsymbol{R}'|} \mathrm{d}S',$$

where $s'$ is the surface of the small volume that is excluded from the integration in the first integral, while $\boldsymbol{n}'$ is the normal pointing in the direction away from $\boldsymbol{R}$. The latter expression for $\boldsymbol{E}(\boldsymbol{R}, t)$ is conventionally written in the form of the first line of (71), which contains a semi-convergent integral.

An alternative form for the electric field may be obtained by starting again from (17), leaving the nabla operator outside the integral and performing a partial integration. Check that one obtains then

$$\boldsymbol{E}(\boldsymbol{R}, t) = \boldsymbol{E}_{\mathrm{e}}(\boldsymbol{R}, t) - \nabla \int \{\varrho^{\mathrm{e}}(\boldsymbol{R}', t) + \boldsymbol{P}(\boldsymbol{R}', t) \cdot \nabla'\} \frac{1}{4\pi|\boldsymbol{R}-\boldsymbol{R}'|} \mathrm{d}\boldsymbol{R}',$$

which contains again convergent integrals. Going still one step further one may write all nabla operators before the integral

$$\boldsymbol{E}(\boldsymbol{R}, t) = \boldsymbol{E}_{\mathrm{e}}(\boldsymbol{R}, t) - \nabla \int \varrho^{\mathrm{e}}(\boldsymbol{R}', t) \frac{1}{4\pi|\boldsymbol{R}-\boldsymbol{R}'|} \mathrm{d}\boldsymbol{R}' + \nabla\nabla \cdot \int \boldsymbol{P}(\boldsymbol{R}', t) \frac{1}{4\pi|\boldsymbol{R}-\boldsymbol{R}'|} \mathrm{d}\boldsymbol{R}'$$

(with convergent integrals).

The second expression of this problem may also be obtained directly from the last formula.

**2.** Show by choosing for the excluded volume a small sphere with centre $\boldsymbol{R}$, that one may write the electric field as

$$\boldsymbol{E}(\boldsymbol{R}, t) = \boldsymbol{E}_e(\boldsymbol{R}, t) - \mathscr{P}_{\text{sph}} \int \{\varrho^e(\boldsymbol{R}', t) + \boldsymbol{P}(\boldsymbol{R}', t)\cdot\nabla'\}\nabla \frac{1}{4\pi|\boldsymbol{R}-\boldsymbol{R}'|} \mathrm{d}\boldsymbol{R}' - \tfrac{1}{3}\boldsymbol{P}(\boldsymbol{R}, t).$$

(The combination $\boldsymbol{E}+\frac{1}{3}\boldsymbol{P}$ is called the Lorentz cavity field.) Comparing this result with (71) one may write symbolically

$$\nabla_s \nabla_s \frac{1}{4\pi s} = \mathscr{P}_{\text{sph}} \nabla_s \nabla_s \frac{1}{4\pi s} - \tfrac{1}{3}\mathbf{U}\delta(\boldsymbol{s})$$

with $\mathbf{U}$ the unit three-tensor. Taking the trace one finds

$$\Delta_s \frac{1}{4\pi s} = -\delta(\boldsymbol{s}).$$

Prove along similar lines the symbolic relation

$$s^i \nabla_s^j \nabla_s^k \nabla_s^l \frac{1}{4\pi s} = \mathscr{P}_{\text{sph}}\, s^i \nabla_s^j \nabla_s^k \nabla_s^l \frac{1}{4\pi s} + \tfrac{1}{5}(\delta^{ij}\delta^{kl} + \delta^{ik}\delta^{jl} + \delta^{il}\delta^{jk})\delta(\boldsymbol{s})$$

(cf. (A24)). Contracting the indices $i$ and $j$, one finds

$$\boldsymbol{s}\cdot\nabla_s \nabla_s \nabla_s \frac{1}{4\pi s} = \mathscr{P}_{\text{sph}}\, \boldsymbol{s}\cdot\nabla_s \nabla_s \nabla_s \frac{1}{4\pi s} + \mathbf{U}\delta(\boldsymbol{s}).$$

Comparing with the second relation of this problem, one obtains the identity

$$\boldsymbol{s}\cdot\nabla_s \nabla_s \nabla_s \frac{1}{4\pi s} = -3\nabla_s \nabla_s \frac{1}{4\pi s}.$$

(The identity is trivial for $\boldsymbol{s} \neq 0$, since then it follows immediately by differentiation.)

**3.** Show by employing the equation of motion (found in problem 2 of chapter I) for a particle with dipoles and quadrupoles that the equation of motion for a fluid of neutral atoms with dipoles and quadrupoles has the form (105) with the force density

$$\boldsymbol{F} = (\nabla \boldsymbol{E})\cdot\boldsymbol{P} + (\nabla \boldsymbol{B})\cdot\boldsymbol{M} + c^{-1}\frac{\partial}{\partial t}(\boldsymbol{P}\wedge\boldsymbol{B}) + c^{-1}\nabla\cdot(\boldsymbol{v}\boldsymbol{P}\wedge\boldsymbol{B}) + \Delta\boldsymbol{F},$$

where $\boldsymbol{P}$ and $\boldsymbol{M}$ are the complete polarizations (39), including dipoles and quadrupoles. Furthermore the components of $\Delta\boldsymbol{F}$ read

$$\Delta F^i = \nabla_j \Big\{(\nabla^i E_k)\bar{\mathscr{P}}^{(2)jk} + (\nabla^i B_k)(\bar{\mathscr{M}}^{(2)jk} + c^{-1}\varepsilon^{klm}\bar{\mathscr{P}}^{(2)j}{}_{.l}\, v_m) + c^{-1}\varepsilon^{ikl}\frac{\partial}{\partial t}(\bar{\mathscr{P}}^{(2)j}{}_{.k}\, B_l) + c^{-1}\varepsilon^{ikl}\nabla_m(v^m \bar{\mathscr{P}}^{(2)j}{}_{.k}\, B_l)\Big\}$$

($\boldsymbol{\epsilon}$ is the Levi-Civita tensor).

The 'field' part $\mathbf{P}^{\mathrm{F}}$ of the material pressure, occurring in the equation of motion (105), reads in the present case instead of (73)

$$P^{\mathrm{F}ij} = -c^{-1}\int \{\hat{v}_1^i\, \varepsilon^{jkl}(\bar{\mu}_{1k}^{(1)} + \bar{\mu}_{1km}^{(2)}\nabla^m)B_l + \varepsilon_{klm}\bar{\mu}_1^{(2)ik}\hat{v}_1^l\, \nabla^j B^m\} f_1(\boldsymbol{R}, 1; t)\mathrm{d}1.$$

The expression for $\boldsymbol{F}$ shows that the inclusion of quadrupoles has the effect that the force density can no longer be expressed in terms of the Maxwell fields and the complete polarizations: extra terms with the quadrupole densities $\bar{\mathscr{P}}^{(2)}$ and $\bar{\mathscr{M}}^{(2)}$ occur in $\Delta\boldsymbol{F}$.

**4.** Consider the double integral

$$\int^{V_1} \boldsymbol{A}(\boldsymbol{R}_1)\cdot\nabla_1\nabla_1\cdot\left(\int^{V_2}\boldsymbol{B}(\boldsymbol{R}_2)\frac{1}{4\pi|\boldsymbol{R}_1-\boldsymbol{R}_2|}\mathrm{d}\boldsymbol{R}_2\right)\mathrm{d}\boldsymbol{R}_1. \qquad \text{(P1)}$$

In particular we want to study the limit of this expression for $V_1$ and $V_2$ both tending to a volume $V$, always keeping $V_1$ smaller than $V_2$. Note that the integral may be written also as

$$\int^{V_1} \boldsymbol{A}(\boldsymbol{R}_1)\cdot\nabla_1\left\{\mathscr{P}\int^{V_2}\boldsymbol{B}(\boldsymbol{R}_2)\cdot\nabla_1\frac{1}{4\pi|\boldsymbol{R}_1-\boldsymbol{R}_2|}\mathrm{d}\boldsymbol{R}_2\right\}\mathrm{d}\boldsymbol{R}_1,$$

where the principal value excludes a small volume around $\boldsymbol{R}_1$ from the integration over $\boldsymbol{R}_2$. Show that this may be transformed to

$$\mathscr{P}\int^{V_1}\int^{V_2}\boldsymbol{A}(\boldsymbol{R}_1)\boldsymbol{B}(\boldsymbol{R}_2):\nabla_1\nabla_1\frac{1}{4\pi|\boldsymbol{R}_1-\boldsymbol{R}_2|}\mathrm{d}\boldsymbol{R}_1\,\mathrm{d}\boldsymbol{R}_2 - \int^{V_1}\left\{\int^{S_2}\boldsymbol{A}(\boldsymbol{R}_1)\cdot\nabla_1\,\boldsymbol{n}_2\cdot\boldsymbol{B}(\boldsymbol{R}_2)\frac{1}{4\pi|\boldsymbol{R}_1-\boldsymbol{R}_2|}\mathrm{d}S_2\right\}\mathrm{d}\boldsymbol{R}_1,$$

where $s_2$ is the surface of the small volume excluded in the first integral over $\boldsymbol{R}_2$. Prove by means of a partial integration and Gauss's theorem that one may write this as

$$-\int^{V_1}\left\{\int^{S_2} \boldsymbol{A}(\boldsymbol{R}_1)\cdot\nabla_1\, \boldsymbol{n}_2\cdot\boldsymbol{B}(\boldsymbol{R}_2)\frac{1}{4\pi|\boldsymbol{R}_1-\boldsymbol{R}_2|}\,\mathrm{d}S_2\right\}\mathrm{d}\boldsymbol{R}_1$$

$$+\int^{V_1}\int^{V_2}\left[\boldsymbol{A}(\boldsymbol{R}_1)\cdot\nabla_1\{\nabla_2\cdot\boldsymbol{B}(\boldsymbol{R}_2)\}\frac{1}{4\pi|\boldsymbol{R}_1-\boldsymbol{R}_2|}\,\mathrm{d}\boldsymbol{R}_2\right]\mathrm{d}\boldsymbol{R}_1,$$

where $S_2$ is the surface of $V_2$. Show by another partial integration and application of Gauss's theorem that this expression is equal to:

$$-\int^{S_1}\int^{S_2}\boldsymbol{n}_1\cdot\boldsymbol{A}(\boldsymbol{R}_1)\boldsymbol{n}_2\cdot\boldsymbol{B}(\boldsymbol{R}_2)\frac{1}{4\pi|\boldsymbol{R}_1-\boldsymbol{R}_2|}\,\mathrm{d}S_1\,\mathrm{d}S_2$$

$$+\int^{V_1}\int^{S_2}\{\nabla_1\cdot\boldsymbol{A}(\boldsymbol{R}_1)\}\boldsymbol{n}_2\cdot\boldsymbol{B}(\boldsymbol{R}_2)\frac{1}{4\pi|\boldsymbol{R}_1-\boldsymbol{R}_2|}\,\mathrm{d}\boldsymbol{R}_1\,\mathrm{d}S_2$$

$$+\int^{S_1}\int^{V_2}\boldsymbol{n}_1\cdot\boldsymbol{A}(\boldsymbol{R}_1)\{\nabla_2\cdot\boldsymbol{B}(\boldsymbol{R}_2)\}\frac{1}{4\pi|\boldsymbol{R}_1-\boldsymbol{R}_2|}\,\mathrm{d}S_1\,\mathrm{d}\boldsymbol{R}_2$$

$$-\int^{V_1}\int^{V_2}\{\nabla_1\cdot\boldsymbol{A}(\boldsymbol{R}_1)\}\{\nabla_2\cdot\boldsymbol{B}(\boldsymbol{R}_2)\}\frac{1}{4\pi|\boldsymbol{R}_1-\boldsymbol{R}_2|}\,\mathrm{d}\boldsymbol{R}_1\,\mathrm{d}\boldsymbol{R}_2, \tag{P2}$$

where (as in all preceding formulae of this problem) the surface and volume of the first integration are smaller than those of the second.

If we had started from the double integral

$$\int^{V_2}\boldsymbol{B}(\boldsymbol{R}_2)\cdot\nabla_2\,\nabla_2\cdot\left(\int^{V_1}\boldsymbol{A}(\boldsymbol{R}_1)\frac{1}{4\pi|\boldsymbol{R}_1-\boldsymbol{R}_2|}\,\mathrm{d}\boldsymbol{R}_1\right)\mathrm{d}\boldsymbol{R}_2 \tag{P3}$$

with $V_1$ greater than $V_2$, we would have found the same result, but with surface and volume of the first integration greater than those of the second. Prove now that in the limit of $V_1$ and $V_2$ both tending to $V$ the limit prescriptions – $V_1$ greater than or smaller than $V_2$ – give the same result. Hence it is proved now that the limit of the integrals (P1) with $V_1 < V_2$ and (P3) with $V_1 > V_2$ are equal. For that reason this limit is conventionally written as

$$\iint^{V}\boldsymbol{A}(\boldsymbol{R}_1)\boldsymbol{B}(\boldsymbol{R}_2):\nabla_1\nabla_1\frac{1}{4\pi|\boldsymbol{R}_1-\boldsymbol{R}_2|}\,\mathrm{d}\boldsymbol{R}_1\,\mathrm{d}\boldsymbol{R}_2\,.$$

The same situation occurs in connexion with integrals of the type

$$\iint^{V_1}\boldsymbol{B}(\boldsymbol{R}_1)\cdot\nabla_1\nabla_1\nabla_1:\left[\int^{V_2}\{\boldsymbol{A}(\boldsymbol{R}_1')-\boldsymbol{A}(\boldsymbol{R}_2)\}\boldsymbol{C}(\boldsymbol{R}_2)\frac{1}{4\pi|\boldsymbol{R}_1-\boldsymbol{R}_2|}\,\mathrm{d}\boldsymbol{R}_2\right]$$
$$\delta(\boldsymbol{R}_1-\boldsymbol{R}_1')\mathrm{d}\boldsymbol{R}_1\,\mathrm{d}\boldsymbol{R}_1'$$

with $V_1 < V_2$ and

$$-\iint^{V_2} \boldsymbol{C}(\boldsymbol{R}_2)\cdot\nabla_2\nabla_2\nabla_2 : \left[\int^{V_1} \{\boldsymbol{A}(\boldsymbol{R}_1)-\boldsymbol{A}(\boldsymbol{R}_2')\}\boldsymbol{B}(\boldsymbol{R}_1)\frac{1}{4\pi|\boldsymbol{R}_1-\boldsymbol{R}_2|}\mathrm{d}\boldsymbol{R}_1\right] \delta(\boldsymbol{R}_2-\boldsymbol{R}_2')\mathrm{d}\boldsymbol{R}_2\,\mathrm{d}\boldsymbol{R}_2'$$

with $V_1 > V_2$ (the reason for the occurrence of the delta function and the extra integration variable being that the functions $\boldsymbol{A}$ are not to be differentiated). These two integrals are equal in the limit $V_1$, $V_2 \to V$ and are conventionally written as

$$\iint^{V} \{\boldsymbol{A}(\boldsymbol{R}_1)-\boldsymbol{A}(\boldsymbol{R}_2)\}\boldsymbol{B}(\boldsymbol{R}_1)\boldsymbol{C}(\boldsymbol{R}_2) : \nabla_1\nabla_1\nabla_1\frac{1}{4\pi|\boldsymbol{R}_1-\boldsymbol{R}_2|}\mathrm{d}\boldsymbol{R}_1\,\mathrm{d}\boldsymbol{R}_2,$$

as may be proved along similar lines.

In the main text examples of integrals like the two mentioned here occur frequently. They are always written in the conventional symbolic way, but they must be understood in the sense described above.

However integrals may be considered where the order of integration does matter. An example is given in the next problem.

**5.** Consider the integral

$$\int^{V_1} \boldsymbol{A}(\boldsymbol{R}_1)\cdot\nabla_1\nabla_1\nabla_1\cdot\left(\int^{V_2} \boldsymbol{B}(\boldsymbol{R}_2)\frac{1}{4\pi|\boldsymbol{R}_1-\boldsymbol{R}_2|}\mathrm{d}\boldsymbol{R}_2\right)\mathrm{d}\boldsymbol{R}_1 \tag{P4}$$

with $V_1 < V_2$. Prove along the same lines as in the preceding problem that this integral may be written as

$$\int^{V_1} A_{1i}\nabla_1\left(\mathscr{P}\int^{V_2} \boldsymbol{B}_2\cdot\nabla_1\nabla_1^i\frac{1}{4\pi|\boldsymbol{R}_1-\boldsymbol{R}_2|}\mathrm{d}\boldsymbol{R}_2\right)\mathrm{d}\boldsymbol{R}_1 - \int^{V_1} A_{1i}\nabla_1\left(\int^{S_2} \boldsymbol{n}_2\cdot\boldsymbol{B}_2\nabla_1^i\frac{1}{4\pi|\boldsymbol{R}_1-\boldsymbol{R}_2|}\mathrm{d}S_2\right)\mathrm{d}\boldsymbol{R}_1,$$

where the vector notation could no longer be maintained completely because of the order of the differentiations. The symbols $\boldsymbol{A}_1$ and $\boldsymbol{B}_2$ are abbreviations for $\boldsymbol{A}(\boldsymbol{R}_1)$ and $\boldsymbol{B}(\boldsymbol{R}_2)$. Furthermore $S_2$ is the surface of the small volume around $\boldsymbol{R}_1$. Partial integration and Gauss's theorem lead to a form

$$-\int^{V_1}\int^{S_2} \boldsymbol{A}_1\cdot\nabla_1\boldsymbol{n}_2\cdot\boldsymbol{B}_2\nabla_1\frac{1}{4\pi|\boldsymbol{R}_1-\boldsymbol{R}_2|}\mathrm{d}\boldsymbol{R}_1\,\mathrm{d}S_2 + \int^{V_1} A_{1i}\nabla_1\left(\mathscr{P}\int^{V_2} (\nabla_2\cdot\boldsymbol{B}_2)\nabla_1^i\frac{1}{4\pi|\boldsymbol{R}_1-\boldsymbol{R}_2|}\mathrm{d}\boldsymbol{R}_2\right)\mathrm{d}\boldsymbol{R}_1.$$

Show that by taking once more the same steps in the second integral, one arrives at

$$-\int^{S_1}\int^{S_2} \boldsymbol{n}_1\cdot\boldsymbol{A}_1\, \boldsymbol{n}_2\cdot\boldsymbol{B}_2\, \nabla_1 \frac{1}{4\pi|\boldsymbol{R}_1-\boldsymbol{R}_2|}\, \mathrm{d}S_1\, \mathrm{d}S_2$$

$$+\int^{V_1}\int^{S_2} (\nabla_1\cdot\boldsymbol{A}_1)\boldsymbol{n}_2\cdot\boldsymbol{B}_2\, \nabla_1 \frac{1}{4\pi|\boldsymbol{R}_1-\boldsymbol{R}_2|}\, \mathrm{d}\boldsymbol{R}_1\, \mathrm{d}S_2$$

$$-\int^{V_1}\int^{S_2} (\nabla_2\cdot\boldsymbol{B}_2)\boldsymbol{n}_2\, \boldsymbol{A}_1\cdot\nabla_1 \frac{1}{4\pi|\boldsymbol{R}_1-\boldsymbol{R}_2|}\, \mathrm{d}\boldsymbol{R}_1\, \mathrm{d}S_2$$

$$+\int^{V_1}\int^{V_2} \{\nabla_2(\nabla_2\cdot\boldsymbol{B}_2)\}\boldsymbol{A}_1\cdot\nabla_1 \frac{1}{4\pi|\boldsymbol{R}_1-\boldsymbol{R}_2|}\, \mathrm{d}\boldsymbol{R}_1\, \mathrm{d}\boldsymbol{R}_2 \qquad \text{(P5)}$$

with $V_1$ and $S_1$ smaller than $V_2$ and $S_2$.

Check that the integral

$$-\int^{V_2} \boldsymbol{B}(\boldsymbol{R}_2)\cdot\nabla_2\, \nabla_2\, \nabla_2\cdot \left(\int^{V_1} \boldsymbol{A}(\boldsymbol{R}_1) \frac{1}{4\pi|\boldsymbol{R}_1-\boldsymbol{R}_2|}\, \mathrm{d}\boldsymbol{R}_1\right) \mathrm{d}\boldsymbol{R}_2 \qquad \text{(P6)}$$

with $V_1 > V_2$ may likewise be written in a form similar to that given in (P5) but with the replacements

$$\boldsymbol{n}_1 \leftrightarrow -\boldsymbol{n}_2, \qquad \nabla_1 \leftrightarrow -\nabla_2, \qquad \boldsymbol{A}_1 \leftrightarrow \boldsymbol{B}_2.$$

Show, by performing repeated partial integrations and employing Gauss's theorem, that one has for the difference of the two integrals (P4) and (P6) in the limit $V_1, V_2 \to V$:

$$\lim_{V_1,V_2\to V} \left\{\int^{V_1(<V_2)} \boldsymbol{A}(\boldsymbol{R}_1)\cdot\nabla_1\, \nabla_1\, \nabla_1\cdot \left(\int^{V_2} \boldsymbol{B}(\boldsymbol{R}_2) \frac{1}{4\pi|\boldsymbol{R}_1-\boldsymbol{R}_2|}\, \mathrm{d}\boldsymbol{R}_2\right) \mathrm{d}\boldsymbol{R}_1\right.$$

$$\left. + \int^{V_2(<V_1)} \boldsymbol{B}(\boldsymbol{R}_2)\cdot\nabla_2\, \nabla_2\, \nabla_2\cdot \left(\int^{V_1} \boldsymbol{A}(\boldsymbol{R}_1) \frac{1}{4\pi|\boldsymbol{R}_1-\boldsymbol{R}_2|}\, \mathrm{d}\boldsymbol{R}_1\right) \mathrm{d}\boldsymbol{R}_2\right\}$$

$$= -\lim_{S_1,S_2\to S} \left(\overset{S_1<S_2}{\iint} - \overset{S_1>S_2}{\iint}\right) \boldsymbol{n}_1\cdot\boldsymbol{A}_1\, \boldsymbol{n}_2\cdot\boldsymbol{B}_2\, \nabla_1 \frac{1}{4\pi|\boldsymbol{R}_1-\boldsymbol{R}_2|}\, \mathrm{d}S_1\, \mathrm{d}S_2. \qquad \text{(P7)}$$

Show that the right-hand side of this relation is equal to

$$-\int^{S} \boldsymbol{n}\cdot\boldsymbol{A}\, \boldsymbol{n}\cdot\boldsymbol{B}\boldsymbol{n}\, \mathrm{d}S \qquad \text{(P8)}$$

by proving first that one has for the integral over an infinite plane $S$

$$\int^{S} \nabla \frac{1}{4\pi|\boldsymbol{R}-\boldsymbol{R}'|}\, \mathrm{d}S = -\tfrac{1}{2}\boldsymbol{n},$$

where $\boldsymbol{R}'$ is a point outside the plane and $\boldsymbol{n}$ is the normal to the plane pointing in the direction away from $\boldsymbol{R}'$.

With (P7) and (P8) one may prove now (291) of the main text. Employ to that end in the first term of the left-hand side of (291) the relation (293) together with the expression for the Maxwell field given at the end of problem 1.

Prove finally (243), valid for an ellipsoidal volume, as a particular case of (291). (The pressure $\mathbf{P}$, the polarization $\boldsymbol{P}$ and the deformation $\delta\boldsymbol{\epsilon}$ are all uniform in this case and may be taken outside the integrals.)

**6.** Show from the integrability relations

$$\frac{\partial P_i}{\partial E_{ej}} = \frac{\partial P_j}{\partial E_{ei}} \qquad (i, j = 1, 2, 3)$$

(v. (380)) and the connexion (220) between the external and Maxwell fields

$$\boldsymbol{E}_e = \boldsymbol{E} + \mathbf{L}\cdot\boldsymbol{P}$$

with the symmetrical depolarizing tensor $\mathbf{L}$ (219) that one has

$$\frac{\partial P_i}{\partial E_j} = \frac{\partial P_j}{\partial E_i}$$

(v. (383)).

**7.** Show by introduction of the new integration variables $\check{\boldsymbol{R}} = \tilde{\boldsymbol{\eta}}_A\cdot\boldsymbol{R}$ and the introduction of the abbreviations:

$$\begin{aligned} &\check{\boldsymbol{P}}(\check{\boldsymbol{R}}) \equiv \tilde{\boldsymbol{\eta}}_A\cdot\boldsymbol{P}(\boldsymbol{R}), \qquad \check{\boldsymbol{M}}(\check{\boldsymbol{R}}) \equiv \tilde{\boldsymbol{\eta}}_A\cdot\boldsymbol{M}(\boldsymbol{R}) \\ &\check{\boldsymbol{E}}(\check{\boldsymbol{R}}) \equiv \tilde{\boldsymbol{\eta}}_A\cdot\boldsymbol{E}(\boldsymbol{R}), \qquad \check{\boldsymbol{B}}(\check{\boldsymbol{R}}) \equiv \tilde{\boldsymbol{\eta}}_A\cdot\boldsymbol{B}(\boldsymbol{R}) \\ &\check{\mathbf{P}}(\check{\boldsymbol{R}}) \equiv \tilde{\boldsymbol{\eta}}_A\cdot\mathbf{P}(\boldsymbol{R})\cdot\boldsymbol{\eta}_A, \end{aligned}$$

that the entropy law (317) may be written in the form:

$$T\delta S = \delta U + \int^{\check{S}} \boldsymbol{n}\cdot(\check{\mathbf{P}} - \check{\boldsymbol{P}}\cdot\check{\boldsymbol{E}}\mathbf{U})\cdot\delta\boldsymbol{\eta}(\boldsymbol{R})\cdot\boldsymbol{R}\,\mathrm{d}S - \int (\check{\boldsymbol{E}}\cdot\delta_0\check{\boldsymbol{P}} - \check{\boldsymbol{M}}\cdot\delta_0\check{\boldsymbol{B}})\mathrm{d}\boldsymbol{R}$$

(where $\boldsymbol{R}$ has been written instead of $\check{\boldsymbol{R}}$ and also $\delta_0$ instead of $\check{\delta}_0$, since they are merely integration variables). This entropy law has the same form as (318) as is to be expected from (305) with (309).

**8.** In Quincke's experimental arrangement a U-shaped tube with one of its legs inside a condenser is filled with an incompressible, electrically polarizable liquid. Show with the help of the formulae (360) and (241) that the difference $\Delta h$ in height between the two liquid columns is given by

$$\varrho g \Delta h = \tfrac{1}{2}\kappa \boldsymbol{E}^2$$

with $\varrho$ the density, $g$ the acceleration of gravity, $\kappa$ the electric susceptibility and $\boldsymbol{E}$ the electric field at the surface of the liquid. The liquid inside the condenser has a level higher than on the other side.

**9.** Prove that the volume average of the pressure tensor in a uniformly polarized solid of ellipsoidal shape is given by

$$\overline{\mathbf{P}} = \mathbf{P}_{\text{out}} - \tfrac{1}{2}\mathbf{K} : \boldsymbol{PP},$$

if the outward pressure tensor is uniform. In order to prove this, write the components $\bar{P}^{ij}$ of the left-hand side as

$$\bar{P}^{ij} \equiv \frac{1}{V}\int^{V} P^{kj}\nabla^{k}R^{i}\,\mathrm{d}\boldsymbol{R}$$

and apply then a partial integration, Gauss's theorem and the equation of motion $\boldsymbol{\nabla}\cdot\mathbf{P} = 0$ (v. (293)). The use of the relations (241) and (243) leads then to the right-hand side.

(The form effect of electrostriction described by the first two terms of (390) leads to a non-uniform deformation. The so-called 'uniform form effect' is obtained if one calculates instead the deformation corresponding to the average normal pressure $\boldsymbol{n}\cdot\overline{\mathbf{P}}$ given above. This calculation is much simpler than that of the non-uniform effect, in particular for more complicated geometries. It gives an estimate of the order of magnitude only.)

# PART B

## *Covariant formulation of classical electrodynamics*

# CHAPTER III

# Charged point particles

## 1 Introduction

Classical electrodynamics in the non-relativistic approximation formed the subject of the preceding two chapters. Since the field equations are covariant with respect to Lorentz transformations one wants to give the complete classical theory of electrodynamics in the framework of special relativity. The first step of this programme will be concerned with the study of the fields and equations of motion of charged point particles. The results will serve as a basis for the derivation of the laws of electrodynamics for composite particles and for matter in bulk, which will be treated in the following pair of chapters.

In this chapter expressions of the fields generated by charged point particles are derived. It will be useful to give them not only in their covariant form, but also as series expansions in powers of $c^{-1}$. Subsequently the equation of motion for charged particles with the inclusion of radiation damping terms will be discussed. An important ingredient for their derivation is the evaluation of the self-fields of the particles at their own position.

## 2 The field equations

### a. *Covariant formulation*

The microscopic Lorentz equations for the electromagnetic fields $\boldsymbol{e}$ and $\boldsymbol{b}$, produced at a time $t$ and a position $\boldsymbol{R}$ by a set of point particles with charges $e_j$ ($j = 1, 2, \ldots$), positions $\boldsymbol{R}_j(t)$ and velocities $\mathrm{d}\boldsymbol{R}_j(t)/\mathrm{d}t \equiv c\boldsymbol{\beta}_j(t)$ read:

$$\begin{aligned}\nabla\cdot\boldsymbol{e} &= \sum_j e_j\,\delta(\boldsymbol{R}_j-\boldsymbol{R}),\\ -\partial_0\,\boldsymbol{e}+\nabla\wedge\boldsymbol{b} &= \sum_j e_j\,\boldsymbol{\beta}_j\,\delta(\boldsymbol{R}_j-\boldsymbol{R}),\\ \nabla\cdot\boldsymbol{b} &= 0,\\ \partial_0\,\boldsymbol{b}+\nabla\wedge\boldsymbol{e} &= 0,\end{aligned} \tag{1}$$

where $\partial_0$ and $\nabla$ indicate differentiations with respect to $ct$ and $\boldsymbol{R}$. We may put these equations into a covariant form by introducing the notations $R^\alpha$ ($\alpha = 0, 1, 2, 3$) for $(ct, \boldsymbol{R})$, $R_j^\alpha$ for $(ct_j, \boldsymbol{R}_j)$, $\partial_\alpha$ for $(\partial_0, \nabla)$, $f^{\alpha\beta}$ ($\alpha, \beta = 0, 1, 2, 3$) for the antisymmetric field tensor with components $(f^{01}, f^{02}, f^{03}) = \boldsymbol{e}$ and $(f^{23}, f^{31}, f^{12}) = \boldsymbol{b}$. (We use the metric $g^{00} = -1$, $g^{ii} = 1$ if $i = 1, 2, 3$, $g^{\alpha\beta} = 0$ if $\alpha \neq \beta$. The inner product $a_\alpha b^\alpha$ of two four-vectors $a^\alpha$ and $b^\alpha$ will sometimes be denoted as $a{\cdot}b$.) In this way the first two equations of (1) are the cases $\alpha = 0$ and $\alpha = 1, 2, 3$ of

$$\partial_\beta f^{\alpha\beta} = \sum_j e_j \int \frac{\mathrm{d}R_j^\alpha}{\mathrm{d}R_j^0} \delta(\boldsymbol{R}_j - \boldsymbol{R})\delta(R_j^0 - R^0)\mathrm{d}R_j^0. \tag{2}$$

Considering $R_j^0$ as a function $R_j^0(s_j)$ of an arbitrary parameter $s_j$ for each particle $j$, we can write (2) as

$$\partial_\beta f^{\alpha\beta} = \sum_j e_j \int \frac{\mathrm{d}R_j^\alpha(s_j)}{\mathrm{d}s_j} \delta^{(4)}\{R_j(s_j) - R\}\mathrm{d}s_j, \tag{3}$$

where $\delta^{(4)}\{R_j(s_j) - R\}$ is the four-dimensional delta function. The parameters $s_j$, which are integration variables, may be chosen independently for each trajectory ($j = 1, 2, \ldots$). For convenience we shall choose for $s_j$ a monotonically increasing function of the time $c^{-1}R_j^0$. We shall write equation (3) as

$$\partial_\beta f^{\alpha\beta} = c^{-1} j^\alpha, \tag{4}$$

with the four-current $j^\alpha$ given by:

$$c^{-1} j^\alpha(R) \equiv \sum_j e_j \int u_j^\alpha(s_j)\delta^{(4)}\{R_j(s_j) - R\}\mathrm{d}s_j, \tag{5}$$

where we introduced the abbreviation $u_j^\alpha$ for $\mathrm{d}R_j^\alpha(s_j)/\mathrm{d}s_j$; it represents the four-velocity if $s_j$ is the proper time. The components $c^{-1}j^\alpha = (\rho, c^{-1}\boldsymbol{j})$ are the sources of the first two equations of (1). From this expression the conservation of charge

$$\partial_\alpha j^\alpha = 0 \tag{6}$$

follows immediately, since $u_j^\alpha \partial_\alpha$ acting on the delta function is equal to $-\mathrm{d}/\mathrm{d}s_j$ acting on it.

The last two field equations of (1) may be written in covariant form:

$$\partial_\alpha f_{\beta\gamma} + \partial_\beta f_{\gamma\alpha} + \partial_\gamma f_{\alpha\beta} = 0. \tag{7}$$

With the help of the completely antisymmetric Levi-Civita tensor $\varepsilon_{\alpha\beta\gamma\delta}$ (with $\varepsilon_{\alpha\beta\gamma\delta} = \pm 1$ if $\alpha$, $\beta$, $\gamma$, $\delta$ is an even/odd permutation of 0, 1, 2, 3,

$\varepsilon_{\alpha\beta\gamma\delta} = 0$ if two indices are equal), the dual field tensor $f^*_{\alpha\beta} = \frac{1}{2}\varepsilon_{\alpha\beta\gamma\delta} f^{\gamma\delta}$ can be introduced, with components $(f^*_{01}, f^*_{02}, f^*_{03}) = \boldsymbol{b}$ and $(f^*_{23}, f^*_{31}, f^*_{12}) = \boldsymbol{e}$. In terms of this field tensor we can write (7) as:

$$\partial_\beta f^{*\alpha\beta} = 0. \tag{8}$$

The covariant equations (4) and (7) give the fields as measured in the space-time reference frame $(ct, \boldsymbol{R})$.

## b. *The solutions of the field equations*

In order to solve the equations (4) and (8), we note first that the general form of the solution of (8) is

$$f^{*\alpha\beta} = \varepsilon^{\alpha\beta\gamma\delta} \partial_\gamma a_\delta, \tag{9}$$

where $a^\alpha$ is an arbitrary four-vector[1]. The field tensor $f^{\alpha\beta} = -\frac{1}{2}\varepsilon^{\alpha\beta\gamma\delta} f^*_{\gamma\delta}$ becomes thus

$$f^{\alpha\beta} = \partial^\alpha a^\beta - \partial^\beta a^\alpha. \tag{10}$$

The vector $a^\alpha$ is called the four-potential, with components $(\varphi, \boldsymbol{a})$. Substitution into (4) gives:

$$\Box\, a^\alpha - \partial^\alpha \partial_\beta a^\beta = -c^{-1} j^\alpha, \tag{11}$$

where the d'Alembertian $\Box$ is the operator $\partial_\alpha \partial^\alpha$. From the form of (10) it follows that the potentials $a^\alpha$ are not uniquely determined by the fields $f^{\alpha\beta}$: a gauge transformation (I.5) $a'^\alpha = a^\alpha + \partial^\alpha \psi$ with arbitrary $\psi$ yields the same fields. As a consequence one may choose the potentials such that the Lorentz condition (I.6)

$$\partial_\alpha a^\alpha = 0 \tag{12}$$

is fulfilled. Then one finds the wave equation for the four-potential

$$\Box\, a^\alpha = -c^{-1} j^\alpha, \tag{13}$$

of which the solutions will be needed in the following. The general solution of this inhomogeneous linear differential equation may be written as the sum of one (arbitrary) particular solution and the general solution of the corre-

[1] In the first instance one finds from (8) that $f^{*\alpha\beta}$ has the form $\partial_\gamma(\lambda^{\alpha\beta\gamma\delta}\mu_\delta)$ with $\lambda^{\alpha\beta\gamma\delta}$ a tensor, antisymmetric only in its first three indices, and $\mu_\delta$ an arbitrary four-vector. The tensor $\lambda^{\alpha\beta\gamma\delta}\mu_\delta$ has four independent components. Instead one may introduce the four-vector $a^\alpha \equiv \frac{1}{6}\varepsilon^{\alpha\beta\gamma\delta}\lambda_{\beta\gamma\delta\zeta}\mu^\zeta$ with the inverse $\lambda^{\alpha\beta\gamma\delta}\mu_\delta = \varepsilon^{\alpha\beta\gamma\delta}a_\delta$. (Strictly spoken the quantity $a^\alpha$ in (9) does not have to be a four-vector, if only its non-covariant part is such that it drops out in $f^{*\alpha\beta}$. One may however confine oneself to a four-vector $a^\alpha$ without impairing the generality of $f^{*\alpha\beta}$.)

sponding homogeneous equation. (This general theorem remains valid if the subsidiary condition (12) is added, since the latter is linear and homogeneous.)

We may fix the arbitrary particular solution of (13) by imposing certain requirements. In the first place we want to confine ourselves to solutions linear in $j^\alpha$ (such linear solutions exist since (13) is linear itself). Then the general form of the particular solution sought is

$$a^\alpha(R) = c^{-1} \int G^{\alpha\beta}(R, R') j_\beta(R') \mathrm{d}^4 R' \tag{14}$$

with the Green function $G^{\alpha\beta}(R, R')$, which must satisfy the equation

$$\Box\, G^{\alpha\beta}(R, R') = -g^{\alpha\beta}\delta^{(4)}(R-R') \tag{15}$$

with $g^{\alpha\beta}$ the metric tensor.

In the second place we require the invariance of the Green function with respect to Poincaré transformations without inversions, i.e. translations in space and time and Lorentz transformations without inversions. In other words we require for the Poincaré transformation

$$\hat{R}^\alpha = \Lambda^\alpha_{.\beta} R^\beta + c^\alpha \tag{16}$$

(with $\Lambda^{\alpha\beta}\Lambda_{\alpha\gamma} = g^\beta_\gamma$, as follows from the defining relation $\hat{R}^\alpha \hat{R}_\alpha = R^\alpha R_\alpha$ for $c^\alpha = 0$) that the Poincaré transform

$$\hat{a}^\alpha(\hat{R}) = \Lambda^\alpha_{.\beta} a^\beta(R) \tag{17}$$

of the four-potential follows from the Poincaré transformed four-current

$$\hat{j}^\alpha(\hat{R}) = \Lambda^\alpha_{.\beta} j^\beta(R) \tag{18}$$

by the relation

$$\hat{a}^\alpha(R) = c^{-1} \int G^{\alpha\beta}(R, R') \hat{j}_\beta(R') \mathrm{d}^4 R' \tag{19}$$

with the *same* Green function as in (14). With $\hat{R}$ and $\hat{R}'$ instead of $R$ and $R'$ one has for (19) with (17) and (18)

$$\Lambda^\alpha_{.\beta} a^\beta(R) = c^{-1} \int G^{\alpha\beta}(\hat{R}, \hat{R}') \Lambda_{\beta\gamma} j^\gamma(R') \mathrm{d}^4 \hat{R}'. \tag{20}$$

Inserting (14) in the left-hand side, and introducing the new integration variable $R'$ instead of $\hat{R}'$ at the right-hand side (the Jacobian of this transformation is unity) gives, since $j^\alpha$ is arbitrary:

$$G^{\alpha\beta}(\hat{R}, \hat{R}') = \Lambda^\alpha_{.\gamma} \Lambda^\beta_{.\delta} G^{\gamma\delta}(R, R'), \tag{21}$$

where we used the property $\Lambda^{\alpha\beta}\Lambda_{\alpha\gamma} = g_\gamma^\beta$ or alternatively $\Lambda^{\alpha\beta} = (\Lambda^{-1})^{\beta\alpha}$. In particular for pure translations one finds

$$G^{\alpha\beta}(R+c, R'+c) = G^{\alpha\beta}(R, R'), \tag{22}$$

so that $G^{\alpha\beta}(R, R')$ depends only on the difference $R-R'$:

$$G^{\alpha\beta}(R, R') = G^{\alpha\beta}(R-R'). \tag{23}$$

For Lorentz transformations the condition (21), with (23), reads

$$G^{\alpha\beta}(\Lambda\cdot R) = \Lambda^{\alpha}_{\cdot\gamma}\Lambda^{\beta}_{\cdot\delta}G^{\gamma\delta}(R). \tag{24}$$

To solve equation (15) with (23) we take its Fourier transform

$$k^2\tilde{G}^{\alpha\beta}(k) = g^{\alpha\beta}, \tag{25}$$

where we employed the Fourier transforms $\tilde{G}^{\alpha\beta}(k)$ and 1 of the Green function and the delta function according to

$$G^{\alpha\beta}(R-R') = \frac{1}{(2\pi)^4}\int \tilde{G}^{\alpha\beta}(k)\mathrm{e}^{ik\cdot(R-R')}\mathrm{d}^4k, \tag{26}$$

$$\delta^{(4)}(R-R') = \frac{1}{(2\pi)^4}\int \mathrm{e}^{ik\cdot(R-R')}\mathrm{d}^4k. \tag{27}$$

The Fourier transform of the relation (24) is

$$\tilde{G}^{\alpha\beta}(\Lambda\cdot k) = \Lambda^{\alpha}_{\cdot\gamma}\Lambda^{\beta}_{\cdot\delta}\tilde{G}^{\gamma\delta}(k). \tag{28}$$

Hence the Fourier transform $\tilde{G}^{\alpha\beta}(k)$ is a tensor which depends only on the vector $k^\alpha$, on invariant tensors (as $g^{\alpha\beta}$) and scalars. This means that the solution of (25) has the form

$$\tilde{G}^{\alpha\beta}(k) = \frac{1}{k^2}g^{\alpha\beta} + \{\lambda^+\theta(k)+\lambda^-\theta(-k)\}g^{\alpha\beta}\delta(k^2) + \{\mu^+\theta(k)+\mu^-\theta(-k)\}k^\alpha k^\beta\delta(k^2), \tag{29}$$

where $\theta(k) = 1$ if $k^0 \geqslant 0$ and 0 if $k^0 < 0$ and where $\lambda^+$, $\lambda^-$, $\mu^+$ and $\mu^-$ are arbitrary constants. (Since the Lorentz transformation $\Lambda^{\alpha\beta}$ in (28) did not contain inversions we could not exclude the possibility that $\tilde{G}^{\alpha\beta}$ has different values on the positive and negative parts of the light cone in $k$-space ($k^2 = 0$).) The Green function $G^{\alpha\beta}(R-R')$ follows now by substitution of (29) into (26). The corresponding four-potential is then obtained with (14) and (23), and the field with (10). Then the contribution due to the 'longitudinal' term with $k^\alpha k^\beta$ in (29) drops out. For that reason it may be suppressed from now on, and we write the Green function as

$$G^{\alpha\beta}(R-R') = g^{\alpha\beta}G(R-R') \tag{30}$$

with the abbreviation

$$G(R) = \frac{1}{(2\pi)^4}\int \left[\frac{1}{k^2} + \{\lambda^+\theta(k)+\lambda^-\theta(-k)\}\delta(k^2)\right] e^{ik\cdot R}d^4k. \tag{31}$$

With (23), (30) and (31) inserted into (14), we have obtained the general form of the particular solution that satisfies the requirements of linearity and covariance (without the longitudinal terms). It contains two arbitrary constants, which may be chosen at will.

Since the first term of the integrand of (31) has poles if $k^2 = 0$, we need a prescription for the treatment of these poles. Since the prescription must be invariant ($G(R)$ itself has to be invariant) all poles on the positive part of the light cone ($k^0 = |\boldsymbol{k}|$) should be treated in the same way, and likewise all poles on the negative part ($k^0 = -|\boldsymbol{k}|$). If various (invariant) prescriptions for the integration of the first term are used, one obtains results which differ from each other by a multiple of the residues. Exactly such a multiple of residues is obtained if one performs the integration over $k^0$ in the second and third parts of $G(R)$. For that reason taking all possible (invariant) prescriptions for the integration and omitting the $\lambda^+$- and $\lambda^-$-terms is equivalent to taking one invariant prescription, but maintaining the $\lambda^+$- and $\lambda^-$-terms with arbitrary values for these parameters. One may thus write instead of (31):

$$G(R) = \frac{1}{(2\pi)^4}\int_C \frac{1}{k^2} e^{ik\cdot R}d^4k \tag{32}$$

with an arbitrary invariant integration contour C in the complex $k$-space.

A prescription (of which we shall show that it is indeed invariant) consists in replacing $k^2$ by $(k+i\varepsilon)^2$ or by $(k-i\varepsilon)^2$, where $\varepsilon^\alpha$ is a time-like infinitesimal four-vector with positive time-component $\varepsilon^0$, and integrating along the real axes $k^0$, $k^1$, $k^2$ and $k^3$. The resulting Green functions will be labelled by the indices r and a respectively:

$$G_{\mathrm{r,a}}(R) = \frac{1}{(2\pi)^4}\int \frac{1}{(k\pm i\varepsilon)^2} e^{ik\cdot R}d^4k. \tag{33}$$

For both signs the integral is real (since it is seen to be equal to its complex conjugate).

The poles of the integrand lie at

$$k^0 = |\boldsymbol{k}| \mp i\left(\varepsilon^0 - \frac{\boldsymbol{\varepsilon}\cdot\boldsymbol{k}}{|\boldsymbol{k}|}\right) \tag{34}$$

and at

$$k^0 = -|\boldsymbol{k}| \mp i\left(\varepsilon^0 + \frac{\boldsymbol{\varepsilon}\cdot\boldsymbol{k}}{|\boldsymbol{k}|}\right). \tag{35}$$

Since $\varepsilon^\alpha$ is a time-like vector with $\varepsilon^0 > 0$, the factors between brackets are both infinitesimally positive, independent of the precise values of the components of $\varepsilon^\alpha$ and $\boldsymbol{k}$.

The integral (33) may now be calculated by performing first the integration over $k^0$. For the upper sign, which we shall consider first, the poles (34) and (35) lie just below the real $k^0$-axis in the neighbourhood of $|\boldsymbol{k}|$ and $-|\boldsymbol{k}|$. For $R^0 < 0$ one may close the contour by a semi-circle in the upper part of the complex $k^0$-plane. Since then no poles are surrounded by the contour the result is zero, i.e.

$$G_r(R) = 0 \qquad (\text{if } R^0 < 0). \tag{36}$$

For $R^0 > 0$ one closes the contour by a semi-circle in the lower part of the complex $k^0$-plane. Cauchy's theorem then gives

$$G_r(R) = -\frac{i}{2(2\pi)^3}\int \mathrm{d}\boldsymbol{k}\, \mathrm{e}^{i\boldsymbol{k}\cdot\boldsymbol{R}}\, \frac{\mathrm{e}^{i|\boldsymbol{k}|R^0} - \mathrm{e}^{-i|\boldsymbol{k}|R^0}}{|\boldsymbol{k}|} \qquad (\text{if } R^0 > 0). \tag{37}$$

Performing the integration (first over the angles and then over the absolute value of $\boldsymbol{k}$) one gets thus

$$G_r(R) = \frac{1}{4\pi|\boldsymbol{R}|}\,\delta(R^0 - |\boldsymbol{R}|) \qquad (\text{if } R^0 > 0). \tag{38}$$

Combining (36) and (38) we have as the Green function $G_r$ for all $R^0$:

$$G_r(R) = \frac{1}{4\pi|\boldsymbol{R}|}\,\delta(R^0 - |\boldsymbol{R}|). \tag{39}$$

From the derivation it is apparent that the precise position of the poles in the complex $k^0$-plane is irrelevant, provided that they lie below the real axis and infinitesimally near to $k^0 = +|\boldsymbol{k}|$ and $-|\boldsymbol{k}|$. From (34) and (35) it then follows that the Green function $G_r(R)$ given by (33) has the same value for all infinitesimal time-like $\varepsilon^\alpha$ with $\varepsilon^0 > 0$. Hence the prescription for the integration was indeed an invariant prescription, since all such $\varepsilon^\alpha$ transform into each other under Lorentz transformations.

From (14) with (23) and (30) it follows that with the choice of the Green function (39) the potential only depends on the charge-current at time-space points which are earlier than the observer's time ($R'^0 \equiv ct' < ct \equiv R^0$). For this reason the Green function $G_r$ in question is said to have retarded character.

The Green function $G_a(R)$ (33) is found in a similar way (lower signs in (33–35)):

$$G_a(R) = \frac{1}{4\pi|\boldsymbol{R}|}\,\delta(R^0+|\boldsymbol{R}|). \tag{40}$$

It has the property to lead to advanced potentials, since it vanishes if $R^0 > 0$.

From the property for delta functions

$$\delta\{f(x)\} = \sum_n \frac{1}{|\partial f/\partial x|}\,\delta(x-x_n), \tag{41}$$

with $x_n$ the (simple) roots of $f(x) = 0$, it follows that

$$\delta(R^2) = \frac{1}{2|\boldsymbol{R}|}\,\{\delta(R^0-|\boldsymbol{R}|)+\delta(R^0+|\boldsymbol{R}|)\}, \tag{42}$$

and hence, with $\theta(R) = 1$ for $R^0 \geqslant 0$ and $\theta(R) = 0$ for $R^0 < 0$,

$$\delta(R^2)\theta(\pm R) = \frac{1}{2|\boldsymbol{R}|}\,\delta(R^0 \mp |\boldsymbol{R}|). \tag{43}$$

Then the retarded and advanced Green functions (39) and (40) may be written in the form

$$G_{r,a}(R) = \frac{1}{2\pi}\,\delta(R^2)\theta(\pm R), \tag{44}$$

which shows explicitly their invariant character.

The integration prescription, contained in (33), is such that the contour passes either below both poles, or above both poles in the complex $k^0$-plane, as (34) and (35) show. The only other independent invariant integration prescription consists in letting the contour pass below one of the poles and above the other one. This is achieved if in (32) one writes $k^2 - i\varepsilon$ for $k^2$:

$$G_f(R) = \frac{1}{(2\pi)^4}\int \frac{1}{k^2-i\varepsilon}\,e^{ik\cdot R}\mathrm{d}^4k. \tag{45}$$

The choice of $+i\varepsilon$ in the denominator leads to a Green function

$$G_{\mathrm{af}}(R) = \frac{1}{(2\pi)^4}\int \frac{1}{k^2+i\varepsilon}\,\mathrm{e}^{ik\cdot R}\mathrm{d}^4k, \tag{46}$$

which depends on the other three. In fact from (33), (45) and (46) it may be proved that the following linear relation between the four Green functions exists:

$$G_{\mathrm{r}}+G_{\mathrm{a}} = G_{\mathrm{f}}+G_{\mathrm{af}}. \tag{47}$$

(Inspection of the various contours shows immediately the validity of this connexion.)

The evaluation of the Green functions (45) and (46) proceeds in the same fashion as before, by closing the contours (distinguishing between $R^0 > 0$ and $R^0 < 0$) and applying Cauchy's theorem. This leads to the results

$$G_{\mathrm{f}}(R) = \frac{i}{4\pi^2}\,\frac{1}{R^2+i\varepsilon}, \tag{48}$$

$$G_{\mathrm{af}}(R) = -\frac{i}{4\pi^2}\,\frac{1}{R^2-i\varepsilon}. \tag{49}$$

The latter function is the complex conjugate of the former, as is also visible in (45) and (46).

The most general invariant Green function is a linear combination of the independent functions $G_{\mathrm{r}}$, $G_{\mathrm{a}}$ and $G_{\mathrm{f}}$, i.e.

$$G = \alpha G_{\mathrm{r}}+\beta G_{\mathrm{a}}+(1-\alpha-\beta)G_{\mathrm{f}}, \tag{50}$$

where $\alpha$ and $\beta$ are complex constants. The sum of the three coefficients has to be equal to 1 in order to get a solution of the inhomogeneous equation (15) with (23) and (30).

Since the vector potential and the four-current are both real functions, we want to confine ourselves from now on to invariant Green functions which are real. This imposes conditions on the complex constants in (50). In fact one finds from the reality condition $G = G^*$ and the relations $G_{\mathrm{r}}^* = G_{\mathrm{r}}$, $G_{\mathrm{a}}^* = G_{\mathrm{a}}$ and $G_{\mathrm{f}}^* = G_{\mathrm{af}} = G_{\mathrm{r}}+G_{\mathrm{a}}-G_{\mathrm{f}}$ that

$$\mathrm{re}\,\alpha+\mathrm{re}\,\beta = 1, \qquad \mathrm{im}\,\alpha-\mathrm{im}\,\beta = 0. \tag{51}$$

The general form of the real invariant Green function is hence (with the notations $\xi \equiv \mathrm{re}\,\alpha$ and $\eta \equiv 2\,\mathrm{im}\,\alpha$)

$$G = G^* = \xi G_{\mathrm{r}}+(1-\xi)G_{\mathrm{a}}+\eta\,\mathrm{im}\,G_{\mathrm{f}}, \tag{52}$$

where $\xi$ and $\eta$ are arbitrary real constants. The function im $G_f$ follows from (48) with the identity (for real $x$)

$$\frac{1}{x \pm i\varepsilon} = \mathscr{P}\frac{1}{x} \mp i\pi\delta(x) \tag{53}$$

($\mathscr{P}$ indicates the principal value). One finds then

$$\text{im } G_f = \frac{1}{4\pi^2}\mathscr{P}\frac{1}{R^2}. \tag{54}$$

With this expression and (44) the general form (52) of the real invariant Green function becomes

$$G(R) = G^*(R) = \frac{1}{2\pi}\,\delta(R^2)\{\xi\theta(R)+(1-\xi)\theta(-R)\}+\eta\mathscr{P}\frac{1}{R^2}, \tag{55}$$

(a factor $(4\pi^2)^{-1}$ has been absorbed into the coefficient $\eta$).

The particular solutions of the (inhomogeneous) wave equation (13) that follow from (14) with (23), (30) and (55) read

$$a^\alpha(R) = c^{-1}\int\left[\frac{1}{2\pi}\,\delta\{(R-R')^2\}\{\xi\theta(R-R')+(1-\xi)\theta(-R+R')\} + \eta\mathscr{P}\frac{1}{(R-R')^2}\right] j^\alpha(R')\mathrm{d}^4R'. \tag{56}$$

The general solution of the wave equation may be obtained by adding the general solution of the homogeneous equation to this expression with an arbitrary but fixed choice for $\xi$ and $\eta$. As to the expression (56), it represents, as we have seen, the general solution of the inhomogeneous equation subject to the conditions that it be linear in the source $j^\alpha$ and connected with the latter by means of an invariant, real Green function. In the following we shall be concerned with these solutions only. They still contain two parameters, which we shall now fix with the help of a further requirement.

In the three terms of (56) the space-time points $R'$ of the source and $R$ of the observer are related in three different ways as a consequence of the different properties of the three Green functions. In fact the first term contains the retarded Green function $G_r(R-R')$ (44) which ensures that the signal from the source travels with the speed of light and reaches the observer at a later time. The second term, with the advanced Green function $G_a(R-R')$ (44), gives a contribution to the potential at an observer's time earlier than the source term. The third term is an integral over the whole of four-space, except for the light cone; it contains even source points $R'$ which are at a space-like distance from the observer's point $R$. If one wants to exclude the acausal effects described by the last two terms, one has to choose the parameters $\xi = 1$ and $\eta = 0$. Then one arrives at the retarded linear solution, with an invariant and real Green function, of the inhomogeneous wave equation:

$$a_{\mathrm{r}}^{\alpha}(R) = \frac{c^{-1}}{2\pi}\int \delta\{(R-R')^2\}\theta(R-R')j^{\alpha}(R')\mathrm{d}^4R'. \tag{57}$$

Inserting the expression (5) for the four-current we obtain

$$a_{\mathrm{r}}^{\alpha}(R) = \frac{1}{2\pi}\sum_j e_j\int u_j^{\alpha}(s_j)\delta[\{R-R_j(s_j)\}^2]\theta\{(R-R_j(s_j)\}\mathrm{d}s_j, \tag{58}$$

with the abbreviation $u_j^{\alpha}(s_j) \equiv \mathrm{d}R_j^{\alpha}(s_j)/\mathrm{d}s_j$.

Up to now the Lorentz condition has not been imposed explicitly on the solution. However one may verify that the solution (58) as it stands satisfies the Lorentz condition (12).

The retarded fields follow by insertion of (58) into (10):

$$f_{\mathrm{r}}^{\alpha\beta}(R) = \frac{1}{2\pi}\sum_j e_j\int \{u_j^{\beta}(s_j)\partial^{\alpha}-u_j^{\alpha}(s_j)\partial^{\beta}\}\delta[\{R-R_j(s_j)\}^2]\theta\{R-R_j(s_j)\}\mathrm{d}s_j. \tag{59}$$

For future use in calculations we shall also need the advanced potentials and fields. They follow from (57–59) by replacing $\theta(x)$ by $\theta(-x)$. Half the sum and half the difference of the retarded and advanced fields are conventionally called the 'plus' and 'minus' fields. (The former is a solution of the field equations with sources, whereas the latter is a solution of the source-free field equations.) They are obtained by replacing $\theta(x)$ in (57–59) by $\frac{1}{2}\{\theta(x)\pm\theta(-x)\}$, i.e. by $\frac{1}{2}$ and $\frac{1}{2}\varepsilon(x)$ respectively ($\varepsilon(x) = 1$ for $x^0 \geqslant 0$, $\varepsilon(x) = -1$ for $x^0 < 0$). One should note in this connexion that although we shall often treat the retarded and advanced fields on the same footing, only the first will represent the physical fields, while the latter (and also the plus and minus fields) will only serve as mathematical ancillaries.

### c. *Expansion of the retarded and advanced potentials and fields into powers of* $c^{-1}$

The retarded potential (58) and the corresponding advanced one may be written in an alternative form, if we use the identity (43):

$$a_{\mathrm{r,a}}^{\alpha}(\boldsymbol{R}, t) = \sum_j e_j\int u_j^{\alpha}(s_j)\,\frac{\delta\{R^0-R_j^0(s_j)\mp|\boldsymbol{R}-\boldsymbol{R}_j(s_j)|\}}{4\pi|\boldsymbol{R}-\boldsymbol{R}_j(s_j)|}\,\mathrm{d}s_j \tag{60}$$

with $u_j^\alpha(s_j) \equiv \mathrm{d}R_j^\alpha/\mathrm{d}s_j$. Choosing in particular as the parametrization of the world lines their time $t_j \equiv c^{-1}R_j^0$, we obtain for the retarded and advanced scalar and vector potentials $(\varphi, \boldsymbol{a}) = a^\alpha$:

$$\begin{aligned}\varphi_{\mathrm{r,a}}(\boldsymbol{R}, t) &= \sum_j e_j \int \frac{\delta\{t-t_j \mp c^{-1}|\boldsymbol{R}-\boldsymbol{R}_j(t_j)|\}}{4\pi|\boldsymbol{R}-\boldsymbol{R}_j(t_j)|}\,\mathrm{d}t_j,\\ \boldsymbol{a}_{\mathrm{r,a}}(\boldsymbol{R}, t) &= \sum_j e_j \int \boldsymbol{\beta}_j(t_j)\frac{\delta\{t-t_j \mp c^{-1}|\boldsymbol{R}-\boldsymbol{R}_j(t_j)|\}}{4\pi|\boldsymbol{R}-\boldsymbol{R}_j(t_j)|}\,\mathrm{d}t_j\end{aligned} \tag{61}$$

due to a set of point sources. Here $c\boldsymbol{\beta}_j(t_j)$ stands for the velocity $\mathrm{d}\boldsymbol{R}_j(t_j)/\mathrm{d}t_j$.

Let us consider in the following the potentials due to a single particle, of which for convenience we denote the charge as $e$ instead of $e_j$ and the velocity as $c\boldsymbol{\beta}$ instead of $c\boldsymbol{\beta}_j$. If we furthermore employ the abbreviations $\boldsymbol{r} = \boldsymbol{R}-\boldsymbol{R}_j$ and $r = |\boldsymbol{r}|$ and write $t'$ instead of $t_j$ for the integration variable, we have the potentials

$$\begin{aligned}\varphi_{\mathrm{r,a}}(\boldsymbol{R}, t) &= e\int \frac{1}{4\pi r(t')}\,\delta\left\{t-t'\mp\frac{r(t')}{c}\right\}\mathrm{d}t',\\ \boldsymbol{a}_{\mathrm{r,a}}(\boldsymbol{R}, t) &= e\int \frac{\boldsymbol{\beta}(t')}{4\pi r(t')}\,\delta\left\{t-t'\mp\frac{r(t')}{c}\right\}\mathrm{d}t'.\end{aligned} \tag{62}$$

The delta function may be expanded in a Taylor series

$$\delta\left\{t-t'\mp\frac{r(t')}{c}\right\} = \sum_{n=0}^{\infty}\frac{\{\pm r(t')\}^n}{c^n n!}\frac{\partial^n\delta(t-t')}{\partial t'^n}. \tag{63}$$

Substitution into (62) and partial integration over $t'$ yield the expansions:

$$\varphi(\boldsymbol{R}, t) = \sum_{n=0}^{\infty}\varphi^{(n)}(\boldsymbol{R}, t), \qquad \boldsymbol{a}(\boldsymbol{R}, t) = \sum_{n=1}^{\infty}\boldsymbol{a}^{(n)}(\boldsymbol{R}, t), \tag{64}$$

with the partial potentials $\varphi^{(n)}$ and $\boldsymbol{a}^{(n)}$ of order $c^{-n}$ given by:

$$\begin{aligned}\varphi_{\mathrm{r,a}}^{(n)}(\boldsymbol{R}, t) &= \frac{e}{4\pi}\frac{(\mp 1)^n}{c^n n!}\frac{\partial^n r^{n-1}}{\partial t^n}, \quad (n = 0, 1, \ldots),\\ \boldsymbol{a}_{\mathrm{r,a}}^{(n+1)}(\boldsymbol{R}, t) &= \frac{e}{4\pi}\frac{(\mp 1)^n}{c^n n!}\frac{\partial^n(r^{n-1}\boldsymbol{\beta})}{\partial t^n}, \quad (n = 0, 1, \ldots).\end{aligned} \tag{65}$$

In this way 'synchronous' expressions for the potentials in the form of power series in $c^{-1}$ have been obtained. From these expressions (in which $\partial/\partial t$ denotes the sum of an explicit time derivative and an implicit one of the form $-c\boldsymbol{\beta}\cdot\nabla$) it follows that the power series in $c^{-1}$ are in fact expansions with respect to a set of dimensionless parameters, namely the 'retardation time'

$r/c \equiv t_r$ multiplied by $c\beta/r$, by $\dot{\beta}/\beta$, by $\ddot{\beta}/\dot{\beta}$, etc. These parameters are of two types: the first is simply $\beta$ and thus independent of the retardation time, while the others have the general form of the retardation time divided by a characteristic time of the (accelerated) motion of the source. Thus the series may be broken off if the velocity of the source and moreover the distance between source and observer are not too large. (Sometimes it may be useful to consider separately expansions with respect to one of the parameters only, then the others need not be limited in magnitude.)

The potentials satisfy the Lorentz gauge condition, which reads for the partial potentials

$$\partial_0 \varphi^{(n)} + \nabla \cdot \boldsymbol{a}^{(n+1)} = 0, \qquad (n = 0, 1, \dots) \tag{66}$$

and for the total potentials

$$\partial_0 \varphi + \nabla \cdot \boldsymbol{a} = 0. \tag{67}$$

If the lowest orders of (65) are evaluated one finds the expressions:

$$\begin{aligned}
\varphi^{(0)}_{\mathrm{r,a}} &= \frac{e}{4\pi r}, \\
\varphi^{(1)}_{\mathrm{r,a}} &= 0, \\
\varphi^{(2)}_{\mathrm{r,a}} &= \frac{e}{4\pi r}\left\{\tfrac{1}{2}\beta^2 - \tfrac{1}{2}(\boldsymbol{n}\cdot\boldsymbol{\beta})^2 - \frac{r}{2c}(\boldsymbol{n}\cdot\dot{\boldsymbol{\beta}})\right\}, \\
\varphi^{(3)}_{\mathrm{r,a}} &= \pm\frac{e}{4\pi r}\left\{-\frac{r}{c}(\boldsymbol{\beta}\cdot\dot{\boldsymbol{\beta}}) + \frac{r^2}{3c^2}(\boldsymbol{n}\cdot\ddot{\boldsymbol{\beta}})\right\}, \\
\boldsymbol{a}^{(1)}_{\mathrm{r,a}} &= \frac{e\boldsymbol{\beta}}{4\pi r}, \\
\boldsymbol{a}^{(2)}_{\mathrm{r,a}} &= \mp\frac{e\dot{\boldsymbol{\beta}}}{4\pi c}, \\
\boldsymbol{a}^{(3)}_{\mathrm{r,a}} &= \frac{e\boldsymbol{\beta}}{4\pi r}\left\{\tfrac{1}{2}\beta^2 - \tfrac{1}{2}(\boldsymbol{n}\cdot\boldsymbol{\beta})^2 - \frac{r}{2c}(\boldsymbol{n}\cdot\dot{\boldsymbol{\beta}})\right\} - \frac{e\dot{\boldsymbol{\beta}}(\boldsymbol{n}\cdot\boldsymbol{\beta})}{4\pi c} + \frac{er\ddot{\boldsymbol{\beta}}}{8\pi c^2}.
\end{aligned} \tag{68}$$

The fields follow from these potentials. We write the result as

$$\boldsymbol{e}(\boldsymbol{R}, t) = \sum_{n=0}^{\infty} \boldsymbol{e}^{(n)}(\boldsymbol{R}, t), \qquad \boldsymbol{b}(\boldsymbol{R}, t) = \sum_{n=0}^{\infty} \boldsymbol{b}^{(n)}(\boldsymbol{R}, t) \tag{69}$$

with partial fields $\boldsymbol{e}^{(n)}$ and $\boldsymbol{b}^{(n)}$ of order $c^{-n}$ given by the following connexions with the partial potentials:

$$\boldsymbol{e}^{(n)} = -\nabla\varphi^{(n)} - \partial_0 \boldsymbol{a}^{(n-1)}, \qquad \boldsymbol{b}^{(n)} = \nabla\wedge\boldsymbol{a}^{(n)}, \qquad (n = 0, 1, \ldots), \tag{70}$$

where, by definition, $\boldsymbol{a}^{(-1)}$ and $\boldsymbol{a}^{(0)}$ vanish. They read with (65) inserted (for $n = 0, 1, \ldots$)

$$\begin{aligned} \boldsymbol{e}^{(n)}_{\mathrm{r,a}} &= -\frac{e}{4\pi}\frac{(\mp 1)^n}{c^n n!}\nabla\frac{\partial^n r^{n-1}}{\partial t^n} - \frac{e}{4\pi}\frac{(\mp 1)^n}{c^{n-1}(n-2)!}\frac{\partial^{n-2}(r^{n-3}\boldsymbol{\beta})}{\partial t^{n-2}}, \\ \boldsymbol{b}^{(n)}_{\mathrm{r,a}} &= \frac{e}{4\pi}\frac{(\mp 1)^{n-1}}{c^{n-1}(n-1)!}\nabla\wedge\frac{\partial^{n-1}(r^{n-2}\boldsymbol{\beta})}{\partial t^{n-1}}. \end{aligned} \tag{71}$$

The lowest orders are explicitly

$$\begin{aligned} \boldsymbol{e}^{(0)}_{\mathrm{r,a}} &= \frac{e\boldsymbol{n}}{4\pi r^2}, \\ \boldsymbol{e}^{(1)}_{\mathrm{r,a}} &= 0, \\ \boldsymbol{e}^{(2)}_{\mathrm{r,a}} &= \frac{e\boldsymbol{n}\boldsymbol{\beta}^2}{8\pi r^2} - \frac{3e\boldsymbol{n}(\boldsymbol{n}\cdot\boldsymbol{\beta})^2}{8\pi r^2} - \frac{e\boldsymbol{n}(\boldsymbol{n}\cdot\dot{\boldsymbol{\beta}})}{8\pi rc} - \frac{e\dot{\boldsymbol{\beta}}}{8\pi rc}, \\ \boldsymbol{e}^{(3)}_{\mathrm{r,a}} &= \pm\frac{e\ddot{\boldsymbol{\beta}}}{6\pi c^2}, \\ \boldsymbol{b}^{(0)}_{\mathrm{r,a}} &= 0, \\ \boldsymbol{b}^{(1)}_{\mathrm{r,a}} &= \frac{e\boldsymbol{\beta}\wedge\boldsymbol{n}}{4\pi r^2}, \\ \boldsymbol{b}^{(2)}_{\mathrm{r,a}} &= 0, \\ \boldsymbol{b}^{(3)}_{\mathrm{r,a}} &= \frac{e\boldsymbol{\beta}\wedge\boldsymbol{n}}{8\pi r^2}\left\{\boldsymbol{\beta}^2 - 3(\boldsymbol{n}\cdot\boldsymbol{\beta})^2 - \frac{r}{c}(\boldsymbol{n}\cdot\dot{\boldsymbol{\beta}})\right\} - \frac{e\dot{\boldsymbol{\beta}}\wedge\boldsymbol{n}(\boldsymbol{n}\cdot\boldsymbol{\beta})}{4\pi rc} + \frac{e\dot{\boldsymbol{\beta}}\wedge\boldsymbol{\beta}}{8\pi rc} - \frac{e\ddot{\boldsymbol{\beta}}\wedge\boldsymbol{n}}{8\pi c^2}. \end{aligned} \tag{72}$$

The partial advanced fields are related to the partial retarded fields by:

$$\boldsymbol{e}^{(n)}_{\mathrm{a}} = (-1)^n \boldsymbol{e}^{(n)}_{\mathrm{r}}, \qquad \boldsymbol{b}^{(n)}_{\mathrm{a}} = (-1)^{n+1}\boldsymbol{b}^{(n)}_{\mathrm{r}}. \tag{73}$$

Since $\boldsymbol{e}^{(1)}$, $\boldsymbol{b}^{(0)}$ and $\boldsymbol{b}^{(2)}$ vanish, the lowest orders in which retarded and advanced fields differ are $n = 3$ for the electric field and $n = 4$ for the magnetic field.

The plus and minus fields, defined as half the sum and difference of the retarded and advanced fields, may also be written as series in $c^{-1}$:

$$\boldsymbol{e}_{\pm} = \sum_{n=0}^{\infty}\boldsymbol{e}^{(n)}_{\pm}, \qquad \boldsymbol{b}_{\pm} = \sum_{n=0}^{\infty}\boldsymbol{b}^{(n)}_{\pm} \tag{74}$$

with the partial fields

$$\begin{aligned} \boldsymbol{e}_{\pm}^{(n)} &= \tfrac{1}{2}\{1 \pm (-1)^n\}\boldsymbol{e}_{\mathrm{r}}^{(n)}, \\ \boldsymbol{b}_{\pm}^{(n)} &= \tfrac{1}{2}\{1 \pm (-1)^{n+1}\}\boldsymbol{b}_{\mathrm{r}}^{(n)}. \end{aligned} \tag{75}$$

In particular it follows thus that the lowest order minus fields different from zero are $n = 3$ for the electric field and $n = 4$ for the magnetic field.

In the non-relativistic theory, when only terms up to order $c^{-1}$ are taken into account, only the partial fields $\boldsymbol{e}^{(0)}$ and $\boldsymbol{b}^{(1)}$ occur. Then no retardation effects are included, as is also manifest from the fact that the retarded and advanced fields are the same in this order (or, in other words, the minus field vanishes then).

In the preceding we employed potentials which satisfy the Lorentz condition (66). The field may be found alternatively from potentials $\varphi'$ and $\boldsymbol{a}'$ in a different gauge. The partial potentials are then related to the Lorentz partial potentials (65) as

$$\begin{aligned} \varphi'^{(n)} &= \varphi^{(n)} - \partial_0 \psi^{(n-1)}, \\ \boldsymbol{a}'^{(n)} &= \boldsymbol{a}^{(n)} + \nabla\psi^{(n)}, \end{aligned} \tag{76}$$

with an arbitrary gauge function

$$\psi = \sum_n \psi^{(n)}. \tag{77}$$

The potentials (76) give the same fields as before, as follows from (70).

In particular one may require for the vector potential: $\nabla\cdot\boldsymbol{a}' = 0$, or

$$\nabla\cdot\boldsymbol{a}'^{(n)} = 0. \tag{78}$$

This defines the 'Coulomb gauge'. Quantities in this gauge will be indicated by an index (C). From the second line of (76) and (78) we get the differential equations

$$\Delta\psi^{(n)} = -\nabla\cdot\boldsymbol{a}^{(n)} \tag{79}$$

or with the partial potentials (65) inserted:

$$\Delta\psi_{\mathrm{r,a}}^{(n)} = \frac{e}{4\pi}\frac{(\mp 1)^{n+1}}{c^n(n-1)!}\frac{\partial^n r^{n-2}}{\partial t^n}. \tag{80}$$

This equation has as a solution

$$\psi_{\mathrm{r,a}}^{(n)} = \frac{e}{4\pi}\frac{(\mp 1)^{n+1}}{c^n(n+1)!}\frac{\partial^n r^n}{\partial t^n}, \qquad (n = 0, 1, \dots) \tag{81}$$

(for negative values of $n$ the function vanishes by definition). The Coulomb gauge potentials now follow by insertion of (65) and (81) into (76):

$$\varphi^{(0)}_{(\mathrm{C})\mathrm{r,a}} = \frac{e}{4\pi r}, \qquad \varphi^{(n)}_{(\mathrm{C})\mathrm{r,a}} = 0, \qquad (n = 1, 2, \ldots),$$

$$\boldsymbol{a}^{(n+1)}_{(\mathrm{C})\mathrm{r,a}} = \frac{e}{4\pi}\left\{\frac{(\mp 1)^n}{c^n n!}\frac{\partial^n(r^{n-1}\boldsymbol{\beta})}{\partial t^n} + \nabla\frac{(\mp 1)^n}{c^{n+1}(n+2)!}\frac{\partial^{n+1} r^{n+1}}{\partial t^{n+1}}\right\}, \quad (n = 0, 1, 2, \ldots). \tag{82}$$

The partial potentials of lowest order read explicitly

$$\begin{aligned}
\varphi^{(0)}_{(\mathrm{C})\mathrm{r,a}} &= \frac{e}{4\pi r},\\
\varphi^{(n)}_{(\mathrm{C})\mathrm{r,a}} &= 0, \quad (n = 1, 2, \ldots),\\
\boldsymbol{a}^{(1)}_{(\mathrm{C})\mathrm{r,a}} &= \frac{e\{\boldsymbol{\beta} + \boldsymbol{n}(\boldsymbol{n}\cdot\boldsymbol{\beta})\}}{8\pi r},\\
\boldsymbol{a}^{(2)}_{(\mathrm{C})\mathrm{r,a}} &= \mp\frac{e\dot{\boldsymbol{\beta}}}{6\pi c},\\
\boldsymbol{a}^{(3)}_{(\mathrm{C})\mathrm{r,a}} &= -\frac{3}{8}\frac{e\boldsymbol{n}(\boldsymbol{n}\cdot\boldsymbol{\beta})^3}{4\pi r} - \frac{1}{8}\frac{e\boldsymbol{\beta}(\boldsymbol{n}\cdot\boldsymbol{\beta})^2}{4\pi r} + \frac{3}{8}\frac{e\boldsymbol{n}(\boldsymbol{n}\cdot\boldsymbol{\beta})\boldsymbol{\beta}^2}{4\pi r} + \frac{1}{8}\frac{e\boldsymbol{\beta}\boldsymbol{\beta}^2}{4\pi r}\\
&\quad - \frac{3}{8}\frac{e\boldsymbol{n}(\boldsymbol{n}\cdot\boldsymbol{\beta})(\boldsymbol{n}\cdot\dot{\boldsymbol{\beta}})}{4\pi c} - \frac{1}{8}\frac{e\boldsymbol{\beta}(\boldsymbol{n}\cdot\dot{\boldsymbol{\beta}})}{4\pi c} + \frac{3}{8}\frac{e\boldsymbol{n}(\boldsymbol{\beta}\cdot\dot{\boldsymbol{\beta}})}{4\pi c} - \frac{1}{8}\frac{e\boldsymbol{n}(\boldsymbol{n}\cdot\ddot{\boldsymbol{\beta}})r}{4\pi c^2}\\
&\quad - \frac{5}{8}\frac{e\dot{\boldsymbol{\beta}}(\boldsymbol{n}\cdot\boldsymbol{\beta})}{4\pi c} + \frac{3}{8}\frac{e\ddot{\boldsymbol{\beta}}r}{4\pi c^2}.
\end{aligned} \tag{83}$$

The potentials used in the well-known *Darwin Lagrangian* are the same as $\varphi^{(0)}_{(\mathrm{C})}$ and $\boldsymbol{a}^{(1)}_{(\mathrm{C})}$ given above. From (82) it follows now that, since $\varphi^{(1)}_{(\mathrm{C})}$ and $\varphi^{(2)}_{(\mathrm{C})}$ vanish and $\boldsymbol{a}^{(2)}_{(\mathrm{C})}$ is independent of $\boldsymbol{R}$, one finds from the 'Darwin potentials' the fields with $n = 0, 1, 2$. In other words *the fields* $\boldsymbol{e}$ *and* $\boldsymbol{b}$ *which play a role in the Darwin approximation are correct up to order* $c^{-2}$ (v. problem 6).

## d. *The Liénard–Wiechert potentials and fields*

The retarded and advanced four-potentials of a single particle with charge $e$, four-position $R^{\alpha}_1(s)$ (with $s$ an arbitrary parameter along the world line of the particle, which increases as a monotonic function with time) and the derivative $u^{\alpha}(s) = \mathrm{d}R^{\alpha}_1/\mathrm{d}s$ are (cf. 58)):

$$a^{\alpha}_{\mathrm{r,a}}(R) = \frac{1}{2\pi}e\int u^{\alpha}(s)\delta[\{R - R_1(s)\}^2]\theta[\pm\{R - R_1(s)\}]\mathrm{d}s. \tag{84}$$

This integral may be calculated if the delta function is written in an alternative form by making use of the property (41). Denoting the two roots of the light cone equation

$$\{R-R_1(s)\}^2 = 0 \tag{85}$$

by $s_r$ and $s_a$ (where $R^0 - R_1^0(s_{r,a}) \gtrless 0$), we have for $R \neq R_1(s_{r,a})$:

$$\delta[\{R-R_1(s)\}^2] = \frac{1}{2|u(s_r)\cdot\{R-R_1(s_r)\}|}\delta(s-s_r) + \frac{1}{2|u(s_a)\cdot\{R-R_1(s_a)\}|}\delta(s-s_a). \tag{86}$$

Hence we have

$$\delta[\{R-R_1(s)\}^2]\theta[\pm\{R-R_1(s)\}] = \frac{1}{2|u(s_{r,a})\cdot\{R-R_1(s_{r,a})\}|}\delta(s-s_{r,a}). \tag{87}$$

The potentials (84) then get the form

$$a^{\alpha}_{r,a}(R) = \frac{eu^{\alpha}(s_{r,a})}{4\pi|u(s_{r,a})\cdot\{R-R_1(s_{r,a})\}|}. \tag{88}$$

The expression between the bars is negative and positive for the retarded and advanced solution respectively, since $u^{\alpha}$ is a time-like vector with a positive zero-component. Therefore we may write (88) alternatively, using moreover the abbreviation $r^{\alpha} = R^{\alpha} - R_1^{\alpha}$, as:

$$a^{\alpha}_{r,a} = \mp \left.\frac{eu^{\alpha}}{4\pi u\cdot r}\right|_{r,a}, \tag{89}$$

where the bar with the suffixes r, a indicates that one should take the dynamical quantities $r^{\alpha}$ and $u^{\alpha}$ at $s = s_r$ and $s = s_a$ respectively.

The formula (89) shows again, explicitly, that the parametrization of the world line may be arbitrarily chosen without changing the result, since both the numerator and the denominator contain one differentiation with respect to $s$. If one chooses in particular the time $R_1^0/c$ of the particle as the parameter one has

$$u^{\alpha} = c(1, \boldsymbol{\beta}), \qquad (\boldsymbol{\beta} \equiv \mathrm{d}\boldsymbol{R}_1/\mathrm{d}R_1^0) \tag{90}$$

and, since $r^0_{r,a} = \pm|\boldsymbol{r}| \equiv \pm r$ according to (85),

$$u\cdot r = \mp c\kappa_{r,a} r, \qquad (\kappa_{r,a} \equiv 1 \mp \boldsymbol{\beta}\cdot\boldsymbol{n}, \quad \boldsymbol{n} \equiv \boldsymbol{r}/r). \tag{91}$$

Then one obtains for the scalar and vector potentials $(\varphi, \boldsymbol{a}) \equiv a^{\alpha}$:

$$\varphi_{r,a} = \left.\frac{e}{4\pi\kappa_{r,a} r}\right|_{r,a}, \qquad \boldsymbol{a}_{r,a} = \left.\frac{e\boldsymbol{\beta}}{4\pi\kappa_{r,a} r}\right|_{r,a}, \tag{92}$$

which are the expressions of Liénard and Wiechert[1].

[1] A. Liénard, L'éclairage électrique **16**(1898)5, 53, 106; E. Wiechert, Archives néerlandaises **5**(1900)549.

The retarded and advanced fields follow by differentiation of the potentials. From (89) one has

$$\partial^\alpha a^\beta_{\mathrm{r,a}} = \left[ \pm \frac{eu^\alpha u^\beta}{4\pi(u\cdot r)^2} \mp \frac{e}{4\pi} \left\{ \frac{\mathrm{d}}{\mathrm{d}s} \left( \frac{u^\beta}{u\cdot r} \right) \right\} \partial^\alpha s \right] \Bigg|_{\mathrm{r,a}} . \tag{93}$$

Here the partial derivative of the parameter $s$ follows from differentiation of the light cone equation (85):

$$\partial^\alpha s = \frac{r^\alpha}{u\cdot r} . \tag{94}$$

With (93) and (94) we find for the fields (10):

$$f^{\alpha\beta}_{\mathrm{r,a}} = \mp \frac{e}{4\pi u\cdot r} \frac{\mathrm{d}}{\mathrm{d}s} \left( \frac{r^\alpha u^\beta - r^\beta u^\alpha}{u\cdot r} \right) \Bigg|_{\mathrm{r,a}} , \tag{95}$$

or, if the differentiation is carried out,

$$f^{\alpha\beta}_{\mathrm{r,a}} = \left\{ \pm \frac{e}{4\pi(u\cdot r)^3} (a\cdot r - u^2)(r^\alpha u^\beta - r^\beta u^\alpha) \mp \frac{e}{4\pi(u\cdot r)^2} (r^\alpha a^\beta - r^\beta a^\alpha) \right\} \Bigg|_{\mathrm{r,a}} , \tag{96}$$

where $a^\alpha \equiv \mathrm{d}u^\alpha/\mathrm{d}s$ (not to be confused with the four-potential).

Choosing for the parameter $s$ the time component $R^0_1/c$ one obtains, with (90), (91) and

$$a^\alpha = c(0, \dot{\boldsymbol{\beta}}), \qquad (\dot{\boldsymbol{\beta}} \equiv c\,\mathrm{d}^2\boldsymbol{R}_1/\mathrm{d}R^{02}_1), \tag{97}$$

for the fields:

$$\begin{aligned} \boldsymbol{e}_{\mathrm{r,a}} &= \frac{e}{4\pi} \left\{ \frac{\boldsymbol{n}\mp\boldsymbol{\beta}}{\gamma^2 \kappa^3_{\mathrm{r,a}} r^2} + \frac{(\boldsymbol{n}\mp\boldsymbol{\beta})\dot{\boldsymbol{\beta}}\cdot\boldsymbol{n}}{c\kappa^3_{\mathrm{r,a}} r} - \frac{\dot{\boldsymbol{\beta}}}{c\kappa^2_{\mathrm{r,a}} r} \right\} \Bigg|_{\mathrm{r,a}} , \\ \boldsymbol{b}_{\mathrm{r,a}} &= \frac{e}{4\pi} \left( \frac{\boldsymbol{\beta}\wedge\boldsymbol{n}}{\gamma^2 \kappa^3_{\mathrm{r,a}} r^2} + \frac{\boldsymbol{\beta}\wedge\boldsymbol{n}\,\dot{\boldsymbol{\beta}}\cdot\boldsymbol{n}}{c\kappa^3_{\mathrm{r,a}} r} \pm \frac{\dot{\boldsymbol{\beta}}\wedge\boldsymbol{n}}{c\kappa^2_{\mathrm{r,a}} r} \right) \Bigg|_{\mathrm{r,a}} , \end{aligned} \tag{98}$$

where $\gamma \equiv (1-\beta^2)^{-\frac{1}{2}}$. These expressions show that the fields consist of two types of terms: one without the acceleration, proportional to $|\boldsymbol{r}|^{-2}$, and another with acceleration, proportional to $|\boldsymbol{r}|^{-1}$ (all taken at the retarded or the advanced times).

## e. *The self-field of a charged particle*

In the following we shall need the field due to a point particle with charge $e$ at the position of the particle itself[1]. To that purpose we shall start from ex-

[1] P. A. M. Dirac, Proc. Roy. Soc. A **167**(1938)148.

pression (96) for the retarded and the advanced fields, which reads

$$f^{\alpha\beta}_{\mathrm{r,a}} = \left\{ \pm \frac{e}{4\pi(u\cdot r)^3}(a\cdot r + c^2)(r^\alpha u^\beta - r^\beta u^\alpha) \mp \frac{e}{4\pi(u\cdot r)^2}(r^\alpha a^\beta - r^\beta a^\alpha) \right\}\Bigg|_{\mathrm{r,a}}, \tag{99}$$

where we have chosen for the arbitrary parameter $s$ of (96) the proper time along the world line. Furthermore we have $r^\alpha(s) \equiv R^\alpha - R^\alpha_1(s)$, with $R^\alpha_1(s)$ the four-position of the particle and $R^\alpha$ the four-position of the observer, $u^\alpha = \mathrm{d}R^\alpha_1(s)/\mathrm{d}s$ the four-velocity ($u^2 = -c^2$) and $a^\alpha = \mathrm{d}u^\alpha(s)/\mathrm{d}s$ the four-acceleration. The indices r and a denote that one has to take for $s$ the retarded and advanced values $s_\mathrm{r}$ and $s_\mathrm{a}$ respectively, which satisfy

$$\{R - R_1(s_{\mathrm{r,a}})\}^2 = 0, \qquad R^0 - R^0_1(s_{\mathrm{r,a}}) \gtrless 0. \tag{100}$$

The field (99) will be considered here for positions of the observer

$$R^\alpha = R^\alpha_1(s_1) + \varepsilon n^\alpha, \tag{101}$$

with fixed $s_1$ and space-like unit vector $n^\alpha$ orthogonal to $u^\alpha(s_1)$:

$$n^2 = 1, \qquad u(s_1)\cdot n = 0. \tag{102}$$

If the parameter $\varepsilon$ ($> 0$) tends to zero, one gets the expression for the field at the position $R^\alpha_1(s_1)$ of the particle. If (101) is substituted into (100) one finds upon a Taylor expansion of $R^\alpha_1(s_{\mathrm{r,a}})$ around $R^\alpha_1(s_1)$ that $s_\mathrm{r}$ and $s_\mathrm{a}$ are the two roots of the following equation in $s$:

$$\{\varepsilon n^\alpha - (s-s_1)u^\alpha(s_1) - \tfrac{1}{2}(s-s_1)^2 a^\alpha(s_1) - \tfrac{1}{6}(s-s_1)^3 \dot{a}^\alpha(s_1) + \ldots\}^2 = 0. \tag{103}$$

(The dot indicates a differentiation with respect to $s_1$.) Solving for $s$ in terms of $\varepsilon$ we find the roots $s_\mathrm{r}$ and $s_\mathrm{a}$:

$$s_{\mathrm{r,a}} - s_1 = \mp c^{-1}\varepsilon[1 - \tfrac{1}{2}c^{-2}a\cdot n\varepsilon + \{\tfrac{3}{8}c^{-4}(a\cdot n)^2 - \tfrac{1}{24}c^{-4}a^2 \pm \tfrac{1}{6}c^{-3}\dot{a}\cdot n\}\varepsilon^2 + \ldots], \tag{104}$$

where at the right-hand side the quantities $a$ and $\dot{a}$ depend on $s_1$. By Taylor expansions of those quantities in (99) that depend on $s_\mathrm{r}$ or $s_\mathrm{a}$ around their values at $s_1$ and introduction of (104) we get expansions in powers of the parameter $\varepsilon$. In this way one finds, using also (101), the auxiliary formulae

$$u(s)\cdot r(s) = \mp c\varepsilon[1 + \tfrac{1}{2}c^{-2}a\cdot n\varepsilon + \{-\tfrac{1}{8}c^{-4}(a\cdot n)^2 + \tfrac{1}{8}c^{-4}a^2 \mp \tfrac{1}{3}c^{-3}\dot{a}\cdot n\}\varepsilon^2 + \ldots], \tag{105}$$

$$a(s)\cdot r(s) = \varepsilon\{a\cdot n \mp c^{-1}(\dot{a}\cdot n \mp \tfrac{1}{2}c^{-1}a^2)\varepsilon + \ldots\}, \tag{106}$$

$$r^\alpha(s)u^\beta(s)-r^\beta(s)u^\alpha(s) = \varepsilon\{n^\alpha u^\beta+(\mp c^{-1}n^\alpha a^\beta-\tfrac{1}{2}c^{-2}u^\alpha a^\beta)\varepsilon$$
$$+(\pm\tfrac{1}{2}c^{-3}a{\cdot}nn^\alpha a^\beta+\tfrac{1}{2}c^{-2}n^\alpha\dot{a}^\beta+\tfrac{1}{2}c^{-4}a{\cdot}nu^\alpha a^\beta\pm\tfrac{1}{3}c^{-3}u^\alpha\dot{a}^\beta)\varepsilon^2+\ldots\}$$
$$-(\alpha,\beta), \quad (107)$$

$$r^\alpha(s)a^\beta(s)-r^\beta(s)a^\alpha(s) = \varepsilon\{n^\alpha a^\beta\pm c^{-1}u^\alpha a^\beta$$
$$+(\mp c^{-1}n^\alpha\dot{a}^\beta\mp\tfrac{1}{2}c^{-3}a{\cdot}nu^\alpha a^\beta-c^{-2}u^\alpha\dot{a}^\beta)\varepsilon+\ldots\}-(\alpha,\beta), \quad (108)$$

where $(\alpha, \beta)$ stands for the preceding terms with $\alpha$ and $\beta$ interchanged.

With the use of these expressions we obtain for the field (99) up to order $\varepsilon^0$:

$$f^{\alpha\beta}_{\mathrm{r,a}} = -\frac{e}{4\pi}c^{-1}\varepsilon^{-2}[n^\alpha u^\beta+\tfrac{1}{2}c^{-2}(u^\alpha a^\beta-a{\cdot}nn^\alpha u^\beta)\varepsilon$$
$$+c^{-2}\{\tfrac{1}{8}c^{-2}a^2n^\alpha u^\beta-\tfrac{3}{4}c^{-2}a{\cdot}nu^\alpha a^\beta+\tfrac{3}{8}c^{-2}(a{\cdot}n)^2n^\alpha u^\beta$$
$$-\tfrac{1}{2}n^\alpha\dot{a}^\beta\mp\tfrac{2}{3}c^{-1}u^\alpha\dot{a}^\beta\}\varepsilon^2+\ldots]-(\alpha,\beta). \quad (109)$$

It is useful to write also separately half the sum and half the difference of the retarded and the advanced fields, i.e. the plus and minus fields. From (109) one has then immediately

$$f^{\alpha\beta}_{+} = -\frac{e}{4\pi}c^{-1}\varepsilon^{-2}[n^\alpha u^\beta+\tfrac{1}{2}c^{-2}(u^\alpha a^\beta-a{\cdot}nn^\alpha u^\beta)\varepsilon$$
$$+c^{-2}\{\tfrac{1}{8}c^{-2}a^2n^\alpha u^\beta-\tfrac{3}{4}c^{-2}a{\cdot}nu^\alpha a^\beta+\tfrac{3}{8}c^{-2}(a{\cdot}n)^2n^\alpha u^\beta$$
$$-\tfrac{1}{2}n^\alpha\dot{a}^\beta\}\varepsilon^2+\ldots]-(\alpha,\beta), \quad (110)$$

$$f^{\alpha\beta}_{-} = \frac{e}{4\pi}\tfrac{2}{3}c^{-4}(u^\alpha\dot{a}^\beta-u^\beta\dot{a}^\alpha)+\ldots. \quad (111)$$

While the plus field diverges at the world line, the minus field is finite in the neighbourhood of the world line. The minus field at the world line is the part of the self-field that will give rise to the radiation damping force in the equation of motion.

An alternative way to obtain the minus part of the self-field starts from the expressions (69) with (71) for the fields developed in powers of $c^{-1}$. In the series of problems 7–12 it is shown how the expression (111) may be obtained in this way.

## 3 *The equation of motion*

### a. *A single particle in a field*

If a particle with charge $e$ and mass $m$ is moving in an external field $F^{\alpha\beta}$, it will be subject to a Lorentz force. As a direct generalization of the non-relativistic law one might write

$$ma^{\alpha} = c^{-1}eF^{\alpha\beta}u_{\beta}, \tag{112}$$

where $u^{\alpha}$ and $a^{\alpha}$ are the four-velocity $\mathrm{d}R_1^{\alpha}(s)/\mathrm{d}s$ and the four-acceleration $\mathrm{d}^2R_1^{\alpha}(s)/\mathrm{d}s^2$ with $R_1^{\alpha}(s)$ the four-position and $s$ the proper time. The field has to be taken at the four-position $R_1^{\alpha}(s)$ of the particle.

The equation of motion (112) is certainly not complete, since it does not contain the damping force due to the fact that the particle emits radiation and hence is subject to a recoil force. Such a recoil force may be added *ad hoc*. It is however more illuminating to obtain it starting from the equation

$$m_0\, a^{\alpha} = c^{-1}e(F^{\alpha\beta} + f_{\mathrm{r}}^{\alpha\beta})u_{\beta}. \tag{113}$$

At the left-hand side we have written a constant $m_0$ – the 'bare mass'. Its relation to the experimental mass $m$, which has been written in (112), will become apparent from the following. Furthermore at the right-hand side figures the total field which is the sum of the external field $F^{\alpha\beta}$ and the retarded field $f_{\mathrm{r}}^{\alpha\beta}$, generated by the particle itself. The fields have to be taken at the four-position of the particle. One should thus have to substitute for $f_{\mathrm{r}}^{\alpha\beta}$ the expression (109), or the sum of (110) and (111), taken with $\varepsilon \to 0$. The minus field $f_{-}^{\alpha\beta}$ presents no difficulties but the plus field $f_{+}^{\alpha\beta}$ diverges at the world line. Therefore the equation has to be handled with caution.

To begin with, (113) will be written in the form of a conservation law of energy and momentum. To that purpose we first write it in the form of a local equation by multiplying it by a four-dimensional delta function and adding an integration over proper times:

$$cm_0 \int a^{\alpha}(s)\delta^{(4)}\{R_1(s) - R\}\mathrm{d}s$$
$$= \int e[F^{\alpha\beta}\{R_1(s)\} + f_{\mathrm{r}}^{\alpha\beta}\{R_1(s)\}]u_{\beta}(s)\delta^{(4)}\{R_1(s) - R\}\mathrm{d}s. \tag{114}$$

The right-hand side of (114) may be transformed with the help of the field equation (4) with (5) for a single particle. In the left-hand side we may perform a partial integration. In this way the equation becomes

$$c\partial_{\beta}\, m_0 \int u^{\alpha}u^{\beta}\delta^{(4)}\{R_1(s) - R\}\mathrm{d}s = \{F^{\alpha\beta}(R) + f_{\mathrm{r}}^{\alpha\beta}(R)\}\partial_{\gamma} f_{\mathrm{r}\beta}^{\cdot\gamma}(R). \tag{115}$$

With the use of the homogeneous equations (7) for $f_{\mathrm{r}}^{\alpha\beta}$ and the equations of the form (4) and (7), but without sources, for the external field $F^{\alpha\beta}$ one may cast the right-hand side in the form of a divergence. Thus (115) is then indeed a conservation law of the form

$$\partial_\beta t^{\alpha\beta}_{\text{tot}} = 0 \tag{116}$$

with an energy–momentum tensor defined as

$$\begin{aligned} t^{\alpha\beta}_{\text{tot}}(R) = cm_0 \int u^\alpha u^\beta \delta^{(4)}\{R_1(s)-R\}\mathrm{d}s + F^{\alpha\gamma}(R) f^{\beta}_{\mathrm{r}.\gamma}(R) + f^{\alpha\gamma}_{\mathrm{r}}(R) F^{\beta}_{.\gamma}(R) \\ + f^{\alpha\gamma}_{\mathrm{r}}(R) f^{\beta}_{\mathrm{r}.\gamma}(R) - \{\tfrac{1}{2} f^{\gamma\varepsilon}_{\mathrm{r}}(R) F_{\gamma\varepsilon}(R) + \tfrac{1}{4} f^{\gamma\varepsilon}_{\mathrm{r}}(R) f_{\mathrm{r}\gamma\varepsilon}(R)\} g^{\alpha\beta}. \end{aligned} \tag{117}$$

We note that instead of (117) one might alternatively use in (116) a tensor with a field part of the Maxwell–Heaviside type

$$-(F+f_{\mathrm{r}})^{\alpha\gamma}(F+f_{\mathrm{r}})^{\cdot\beta}_{\gamma} - \tfrac{1}{4}(F+f_{\mathrm{r}})^{\gamma\varepsilon}(F+f_{\mathrm{r}})_{\gamma\varepsilon} g^{\alpha\beta}, \tag{118}$$

since it differs from the field part in (117) only by a divergence-free contribution.

Let us, following Dirac[1], integrate (116) over a narrow tube around the world line. For each value of the proper time $s$ the tube section is chosen to be spherical with constant radius $\varepsilon$ in that Lorentz frame in which the particle is momentarily at rest. Furthermore the tube extends from proper time $s_1$ to proper time $s_2$ and is closed by plane surfaces through $R^\alpha_1(s_1)$ and $R^\alpha_1(s_2)$ with normals $c^{-1}u^\alpha(s_1)$ and $c^{-1}u^\alpha(s_2)$ respectively. A convenient starting point for this integration is the following hybrid of equations (114) and (116–117), viz.

$$\begin{aligned} cm_0 \int a^\alpha \delta^{(4)}(R_1 - R)\mathrm{d}s = \int e(F^{\alpha\beta} + f^{\alpha\beta}_{-}) u_\beta \delta^{(4)}(R_1 - R)\mathrm{d}s \\ + \partial_\beta (f^{\alpha\gamma}_{+} f^{\cdot\beta}_{+\gamma} + \tfrac{1}{4} f^{\gamma\varepsilon}_{+} f_{+\gamma\varepsilon} g^{\alpha\beta}), \end{aligned} \tag{119}$$

where the retarded field has been split into a plus and a minus part according to

$$f^{\alpha\beta}_{\mathrm{r}} = \tfrac{1}{2}(f^{\alpha\beta}_{\mathrm{r}} + f^{\alpha\beta}_{\mathrm{a}}) + \tfrac{1}{2}(f^{\alpha\beta}_{\mathrm{r}} - f^{\alpha\beta}_{\mathrm{a}}) = f^{\alpha\beta}_{+} + f^{\alpha\beta}_{-} \tag{120}$$

with $f^{\alpha\beta}_{\mathrm{a}}$ the advanced field. The reason for a different treatment of the plus and minus fields is that the minus field, in contrast to the plus field, is finite on the world line[2]. Integration of (119) over the tube and application of

[1] P. A. M. Dirac, op. cit.

[2] Alternatively one may refrain from splitting the retarded field, or split it in a different way, v. C. Teitelboim, Phys. Rev. **D 1**(1970)1572, **D 3**(1971)297, to obtain the same final equation of motion.

Gauss's theorem give now

$$cm_0 \int_{s_1}^{s_2} a^\alpha \mathrm{d}s = \int_{s_1}^{s_2} e(F^{\alpha\beta} + f_-^{\alpha\beta}) u_\beta \, \mathrm{d}s$$
$$+ \int_{\Sigma_{\text{lat}}} (f_+^{\alpha\gamma} f_{+\gamma}^{\cdot\beta} + \tfrac{1}{4} f_+^{\gamma\varepsilon} f_{+\gamma\varepsilon} g^{\alpha\beta}) n_\beta \, \mathrm{d}^3\Sigma + \Phi^\alpha(s_2) - \Phi^\alpha(s_1). \qquad (121)$$

Here $\Sigma_{\text{lat}}$ indicates the lateral part of the surface of the tube, while $\Phi^\alpha(s_1)$ and $\Phi^\alpha(s_2)$ are integrals over the closing plane surfaces of the tube; the definition of $\Phi^\alpha(s)$ is:

$$\Phi^\alpha(s) \equiv -c^{-1} \int_{\Sigma_{\text{sect}}} (f_+^{\alpha\gamma} f_{+\gamma}^{\cdot\beta} + \tfrac{1}{4} f_+^{\gamma\varepsilon} f_{+\gamma\varepsilon} g^{\alpha\beta}) u_\beta \, \mathrm{d}^3\Sigma. \qquad (122)$$

The index at the integration sign means that the integral is to be extended over that part of the plane surface $\Sigma(s)$ (which passes through $R_1^\alpha(s)$ and has normal $c^{-1}u^\alpha(s)$) that lies within the tube, i.e. over a sphere with radius $\varepsilon$ around the world line in the plane surface $\Sigma(s)$.

In the first term at the right-hand side of (121) one may substitute (111) for the minus field and in the second term (110) for the plus field. Then the integrand of the first term becomes:

$$eF^{\alpha\beta}u_\beta + \frac{e^2c^{-2}}{6\pi} \Delta_\beta^\alpha(u)\dot{a}^\beta, \qquad (123)$$

where the tensor $\Delta_\beta^\alpha(u)$ is defined as $\delta_\beta^\alpha + c^{-2}u^\alpha u_\beta$. The integrand of the second term becomes:

$$(f_+^{\alpha\gamma} f_{+\gamma}^{\cdot\beta} + \tfrac{1}{4} f_+^{\gamma\varepsilon} f_{+\gamma\varepsilon} g^{\alpha\beta}) n_\beta$$
$$= \frac{e^2}{16\pi^2} \varepsilon^{-4} [\tfrac{1}{2} n^\alpha - \tfrac{1}{2} c^{-2}(a^\alpha + n^\alpha a \cdot n)\varepsilon + \{\tfrac{1}{2} c^{-4}(a \cdot n)^2 n^\alpha$$
$$- \tfrac{1}{2} c^{-4} a^2 n^\alpha + \tfrac{5}{4} c^{-4} a \cdot n a^\alpha\}\varepsilon^2 + \ldots]. \qquad (124)$$

This expression has to be integrated over the lateral part $\Sigma_{\text{lat}}$ of the surface of the tube. We now need an expression for the surface element $\mathrm{d}^3\Sigma$. Let us consider to that end a surface element at the position

$$R^\alpha = R_1^\alpha(s) + \varepsilon n^\alpha(s), \qquad (125)$$

namely a (three-dimensional) strip with edges parallel to the velocity $u^\alpha(s)$, with basis situated in $\Sigma(s)$ and top situated in $\Sigma(s+\mathrm{d}s)$. The height $h$ of the strip is given by the condition that $R^\alpha + hu^\alpha/c$ lie in the plane $\Sigma(s+\mathrm{d}s)$, i.e. by the condition

$$\{R^\alpha + hu^\alpha(s)/c - R_1^\alpha(s+\mathrm{d}s)\} \cdot u_\alpha(s+\mathrm{d}s) = 0. \qquad (126)$$

After expansion in powers of d$s$, insertion of (125) and the use of the relation (v. (102)):

$$n(s)\cdot u(s) = 0 \tag{127}$$

one finds

$$h = c\mathrm{d}s(1+c^{-2}\varepsilon n\cdot a). \tag{128}$$

The surface element $\mathrm{d}^3\Sigma$ of $\Sigma_{\mathrm{lat}}$ at the position $R^\alpha$ may now be written as

$$\mathrm{d}^3\Sigma = c\varepsilon^2\mathrm{d}^2\Omega\,\mathrm{d}s(1+c^{-2}\varepsilon n\cdot a), \tag{129}$$

where $\mathrm{d}^2\Omega$ is the differential of the solid angle which parametrizes the direction of the unit vector $n^\alpha$ in the frame in which $u^\alpha = (c, 0, 0, 0)$ [1].

If one inserts (124) and (129) into the second term at the right-hand side of (121), one finds, up to terms of order $\varepsilon^0$

$$\frac{e^2c}{16\pi^2}\varepsilon^{-2}\int_{s_1}^{s_2}\{\tfrac{1}{2}n^\alpha-\tfrac{1}{2}c^{-2}a^\alpha\varepsilon+(\tfrac{3}{4}c^{-4}a\cdot na^\alpha-\tfrac{1}{2}c^{-4}a^2n^\alpha)\varepsilon^2\}\mathrm{d}^2\Omega\,\mathrm{d}s. \tag{130}$$

After integration over the solid angle we obtain

$$-\frac{e^2c^{-1}}{8\pi\varepsilon}\int_{s_1}^{s_2}a^\alpha\mathrm{d}s, \tag{131}$$

since the integration of the unit vector $n^\alpha$ yields a vanishing result.

With (123) and (131) we obtain from (121), since $s_1$ and $s_2$ are arbitrary,

$$m_0\,a^\alpha = c^{-1}eF^{\alpha\beta}u_\beta+\frac{e^2}{6\pi}c^{-3}\Delta^\alpha_\beta(u)\dot{a}^\beta-\frac{e^2c^{-2}a^\alpha}{8\pi\varepsilon}+c^{-1}\frac{\mathrm{d}\Phi^\alpha}{\mathrm{d}s}. \tag{132}$$

We are left with the task to calculate the four-vector $\Phi^\alpha$ (122). In the frame in which $u^\alpha$ is $(c, 0, 0, 0)$ the components of $\Phi^\alpha$ are

$$\begin{aligned}\Phi^0 &= -\tfrac{1}{2}\int_0^\varepsilon r^2\mathrm{d}r\int^{4\pi}\mathrm{d}^2\Omega(e_+^2+b_+^2),\\ \boldsymbol{\Phi} &= -\int_0^\varepsilon r^2\mathrm{d}r\int^{4\pi}\mathrm{d}^2\Omega(\boldsymbol{e}_+\wedge\boldsymbol{b}_+),\end{aligned} \tag{133}$$

where we have written the integration element $\mathrm{d}^3\Sigma$ as $r^2\,\mathrm{d}r\,\mathrm{d}^2\Omega$. We may insert here the expressions (74) with (75) and (71) for $\boldsymbol{e}_+$ and $\boldsymbol{b}_+$. They show that the partial fields $\boldsymbol{e}_+^{(n)}$ (and $\boldsymbol{b}_+^{(n)}$) are given by (71) for $n$ even (odd) and equal

[1] From (129) one finds an expression for the volume of a parallelepiped with basis $\mathrm{d}^3\Sigma$ in $\Sigma(s)$ at the position $R^\alpha$, with edges parallel to $u^\alpha$ and top in $\Sigma(s+\mathrm{d}s)$, namely $\mathrm{d}^4V = c\mathrm{d}^3\Sigma\mathrm{d}s[1+c^{-2}\{R-R_1(s)\}\cdot a(s)]$. An expression of this type will be needed in section 3$c$ of the following chapter.

to zero for $n$ odd (even). In the expression (71) the time derivative stands for the sum of an explicit time derivative (which we shall denote here as $\partial/\partial t$) and an implicit time derivative $-c\boldsymbol{\beta}\cdot\nabla$. In this way we have

$$\begin{aligned}
\boldsymbol{e}_+^{(n)} &= \frac{-e}{4\pi c^n n!}\nabla\left(\frac{\partial}{\partial t}-c\boldsymbol{\beta}\cdot\nabla\right)^n r^{n-1}-\frac{e}{4\pi c^{n-1}(n-2)!}\left(\frac{\partial}{\partial t}-c\boldsymbol{\beta}\cdot\nabla\right)^{n-2}(r^{n-3}\boldsymbol{\beta}), \\
&\qquad\qquad (n \text{ even}), \\
\boldsymbol{e}_+^{(n)} &= 0, \qquad (n \text{ odd}), \\
\boldsymbol{b}_+^{(n)} &= \frac{e}{4\pi c^{n-1}(n-1)!}\nabla\wedge\left(\frac{\partial}{\partial t}-c\boldsymbol{\beta}\cdot\nabla\right)^{n-1}(r^{n-2}\boldsymbol{\beta}), \qquad (n \text{ odd}), \\
\boldsymbol{b}_+^{(n)} &= 0, \qquad (n \text{ even}).
\end{aligned} \tag{134}$$

We used the frame in which $u^\alpha = (c, 0, 0, 0)$ and hence $\boldsymbol{\beta} = 0$. To find the consequences for the expressions given, one has to work out the powers of the operator $\partial/\partial t - c\boldsymbol{\beta}\cdot\nabla$. This gives in the first term of the expression for $\boldsymbol{e}_+^{(n)}$ (for $n$ even) a sum of terms which contains $p$ times the operator $\partial/\partial t$ and $n-p$ times the operator $-c\boldsymbol{\beta}\cdot\nabla$ in all possible arrangements with $p = 0, \ldots, n$. Since $\boldsymbol{\beta} = 0$ the contribution to $\boldsymbol{e}_+^{(n)}$ will vanish if the number ($p$) of times that the operator $\partial/\partial t$ occurs is smaller than the number ($n-p$) of times that the velocity $c\boldsymbol{\beta}$ is present. Hence only the terms with $p \geqslant \frac{1}{2}n$ contribute. These terms contain less than $\frac{1}{2}n+1$ times the operator $\nabla$; since they operate on $r^{n-1}$ the lowest power of $r$ which occurs in the first term of $\boldsymbol{e}_+^{(n)}$ for $n$ even is $r^{\frac{1}{2}n-2}$. With an analogous reasoning one finds that the lowest power in the second term of $\boldsymbol{e}_+^{(n)}$ for $n$ even ($\geqslant 4$) has the exponent $\frac{1}{2}n-1$ and the lowest power in $\boldsymbol{b}_+^{(n)}$ for $n$ odd ($\geqslant 3$) has exponent $\frac{1}{2}n-\frac{3}{2}$. In other words the partial electric and magnetic fields for $\boldsymbol{\beta} = 0$ have the form

$$\begin{aligned}
\boldsymbol{e}_+^{(n)} &= \mathrm{O}(r^{\frac{1}{2}n-2}), && (n = 0, 2, 4, \ldots), \\
\boldsymbol{e}_+^{(n)} &= 0, && (n = 1, 3, 5, \ldots), \\
\boldsymbol{b}_+^{(n)} &= \mathrm{O}(r^{\frac{1}{2}n-\frac{3}{2}}), && (n = 3, 5, 7, \ldots), \\
\boldsymbol{b}_+^{(n)} &= 0, && (n = 1; n = 0, 2, 4, \ldots).
\end{aligned} \tag{135}$$

The expressions (133) for $\Phi^0$ and $\boldsymbol{\Phi}$ may be written in terms of the partial fields:

$$\begin{aligned}
\Phi^0 &= -\sum_{n=0(\text{even})}^{\infty}\sum_{m=0(\text{even})}^{n}\frac{1}{2}\int_0^\varepsilon r^2\mathrm{d}r\int^{4\pi}\mathrm{d}^2\Omega\,\boldsymbol{e}_+^{(m)}\cdot\boldsymbol{e}_+^{(n-m)} \\
&\quad -\sum_{n=6(\text{even})}^{\infty}\sum_{m=3(\text{odd})}^{n-3}\frac{1}{2}\int_0^\varepsilon r^2\mathrm{d}r\int^{4\pi}\mathrm{d}^2\Omega\,\boldsymbol{b}_+^{(m)}\cdot\boldsymbol{b}_+^{(n-m)}, \\
\boldsymbol{\Phi} &= -\sum_{n=3(\text{odd})}^{\infty}\sum_{m=0(\text{even})}^{n-3}\int_0^\varepsilon r^2\mathrm{d}r\int^{4\pi}\mathrm{d}^2\Omega\,\boldsymbol{e}_+^{(m)}\wedge\boldsymbol{b}_+^{(n-m)}.
\end{aligned} \tag{136}$$

From (135) it follows that

$$\boldsymbol{e}_+^{(m)} \wedge \boldsymbol{b}_+^{(n-m)} = \mathrm{O}(r^{n-\frac{7}{2}}), \qquad (n = 3, 5, \dots; m = 0, 2, \dots, n-3). \tag{137}$$

This means that the integrand for the expression $\boldsymbol{\Phi}$ contains powers of $r$ with exponents greater than 0, so that the integrand vanishes in the limit in which $\varepsilon$ tends to zero:

$$\boldsymbol{\Phi} = 0. \tag{138}$$

Furthermore it follows from (135) that

$$\begin{aligned} \boldsymbol{e}_+^{(m)} \cdot \boldsymbol{e}_+^{(n-m)} &= \mathrm{O}(r^{\frac{1}{2}n-4}), \qquad (n = 0, 2, \dots; m = 0, 2, \dots, n), \\ \boldsymbol{b}_+^{(m)} \cdot \boldsymbol{b}_+^{(n-m)} &= \mathrm{O}(r^{\frac{1}{2}n-3}), \qquad (n = 6, 8, \dots; m = 3, 5, \dots, n-3). \end{aligned} \tag{139}$$

Hence one finds that in the limit $\varepsilon \to 0$ the expression for $\Phi^0$ becomes

$$\Phi^0 = -\int_0^\varepsilon r^2 \mathrm{d}r \int^{4\pi} \mathrm{d}^2\Omega (\tfrac{1}{2}\boldsymbol{e}_+^{(0)2} + \boldsymbol{e}_+^{(0)} \cdot \boldsymbol{e}_+^{(2)}). \tag{140}$$

With the expressions (72) (with $\boldsymbol{\beta} = 0$) we get

$$\Phi^0 = -\int_0^\varepsilon \mathrm{d}r \int^{4\pi} \mathrm{d}^2\Omega \, \frac{e^2}{32\pi^2 r} \left( \frac{1}{r} - 2\, \frac{\boldsymbol{\beta} \cdot \boldsymbol{n}}{c} \right) \tag{141}$$

and thus, after integration over the angles,

$$\Phi^0 = -\frac{e^2}{8\pi} \int_0^\varepsilon \frac{1}{r^2} \, \mathrm{d}r. \tag{142}$$

The expressions (138) and (142), valid in the rest frame, show that in an arbitrary frame the expression for the four-vector $\Phi^\alpha$ is, up to terms of order $\varepsilon^0$,

$$\Phi^\alpha = -\frac{e^2 u^\alpha}{8\pi c} \int_0^\varepsilon \frac{1}{r^2} \, \mathrm{d}r. \tag{143}$$

Substituting this result into the equation of motion (132) and using the identity

$$\frac{1}{\varepsilon} = \int_\varepsilon^\infty \frac{1}{r^2} \, \mathrm{d}r, \tag{144}$$

one finds an equation of motion, which is independent of $\varepsilon$:

$$\left( m_0 + c^{-2} \int_0^\infty \frac{e^2}{8\pi r^2} \, \mathrm{d}r \right) a^\alpha = c^{-1} e F^{\alpha\beta} u_\beta + \frac{e^2}{6\pi} \, c^{-3} \Delta^\alpha_\beta(u) \dot{a}^\beta. \tag{145}$$

This equation has a pathological character in so far that it contains a divergent integral at the left-hand side, which represents the Coulomb energy of the charged point particle. This difficulty may be veiled by means of a 'renormalization' procedure, namely by writing

$$m_0 + c^{-2}\int_0^\infty \frac{e^2}{8\pi r^2}\,\mathrm{d}r = m \tag{146}$$

and taking $m$ to be the finite (experimental) mass of the point particle. ($m_0$ turns out to be negative infinite.) With this artifice the equation of motion (145) gets its final form:

$$ma^\alpha = c^{-1}eF^{\alpha\beta}u_\beta + \frac{e^2}{6\pi}c^{-3}\Delta^\alpha_\beta(u)\dot{a}^\beta. \tag{147}$$

This is Dirac's equation of motion. Dirac did not calculate an expression for $\Phi^\alpha$, but instead assumed that it is proportional to the four-velocity $u^\alpha$.

The general form (147) of the equation of motion may be cast into the form of a local law, in a way analogous to that leading from (113) to (116) with (117). First we write it as

$$ma^\alpha = c^{-1}e(F^{\alpha\beta} + f_-^{\alpha\beta})u_\beta, \tag{148}$$

where (111) has been used. We then multiply by the four-dimensional delta function $\delta^{(4)}\{R_1(s) - R\}$ (with $R_1(s)$ the four-position, depending on the proper time $s$) and integrate over $s$ (cf. (114)):

$$m\int a^\alpha \delta^{(4)}\{R_1(s) - R\}\mathrm{d}s = c^{-1}e\int (F^{\alpha\beta} + f_-^{\alpha\beta})u_\beta\,\delta^{(4)}\{R_1(s) - R\}\mathrm{d}s. \tag{149}$$

This equation may be written as a conservation law or as a balance equation. An equation of the latter type is obtained by bringing the left-hand side in the form of a divergence by means of a partial integration and by using at the right-hand side the field equation for the plus and minus parts of the field $f^{\alpha\beta}$:

$$f^{\alpha\beta} \equiv f_+^{\alpha\beta} + f_-^{\alpha\beta}. \tag{150}$$

These equations, which follow from (4) and (7) with (5), read

$$\begin{aligned} &\partial_\beta f_+^{\alpha\beta} = e\int u^\alpha(s)\delta^{(4)}\{R_1(s) - R\}\mathrm{d}s, \\ &\partial_\beta f_-^{\alpha\beta} = 0, \\ &\partial^\alpha f_\pm^{\beta\gamma} + \partial^\beta f_\pm^{\gamma\alpha} + \partial^\gamma f_\pm^{\alpha\beta} = 0. \end{aligned} \tag{151}$$

Then one obtains for (149) the balance equation:

$$\partial_\beta t^{\alpha\beta} = f^\alpha, \tag{152}$$

where we introduced the energy–momentum tensor of the charged particle (cf. (117)):

$$t^{\alpha\beta}(R) = cm \int u^\alpha u^\beta \delta^{(4)}\{R_1(s)-R\}\mathrm{d}s + f_+^{\alpha\gamma}(R) f_{-\cdot\gamma}^{\beta}(R) + f_-^{\alpha\gamma}(R) f_{+\cdot\gamma}^{\beta}(R) - \tfrac{1}{2} f_+^{\gamma\varepsilon}(R) f_{-\gamma\varepsilon}(R) g^{\alpha\beta} \tag{153}$$

(which is seen to be symmetric) and the Lorentz force density

$$f^\alpha(R) = c^{-1} F^{\alpha\beta}(R) j_\beta(R). \tag{154}$$

Here $j^\alpha(R)$ is the four-current, given by (5):

$$c^{-1} j^\alpha(R) = e \int u^\alpha(s) \delta^{(4)}\{R_1(s)-R\}\mathrm{d}s. \tag{155}$$

The balance equation (152) with (153–154) represents the local law, corresponding to the equation of motion (147) or (148).

The energy–momentum tensor contains as components $t^{00}$ the energy density, $ct^{0i}$ the energy flow, $c^{-1}t^{i0}$ the momentum density and $t^{ij}$ the momentum flow. The total energy and the total momentum follow by integrating $t^{00}$ and $c^{-1}t^{i0}$ respectively over the whole of space for a fixed time $t$. From the expressions (110) and (111) for the plus and minus fields one notices that these integrals converge in the neighbourhood of the world line. However for large distances from the world line it follows from the expressions (96) or (98) that the fields diminish inversely proportionally to the distance if the particle suffers accelerations in the remote past and future. As a consequence the integrals for the total energy and the total momentum of the charged particle diverge in that case. However if one imposes the subsidiary condition that in the remote past and future the particle is not accelerated, one is left – according to (96) or (98) – with fields that diminish inversely proportionally to the square of the distance. Thus the integrals mentioned converge under those circumstances. This means that one is led to the conclusion that only those solutions of the equation of motion (147) or (152) make sense, for which the subsidiary condition about the asymptotic behaviour of the particle is satisfied[1]. This is indeed a necessary condition since the equation (147) as it stands would allow runaway solutions of self-accelerating particles.

[1] R. Haag, Z. Naturf. **10A**(1955)752; F. Rohrlich, Ann. Physics **13**(1961)93.

Often the derivative $\dot{a}^\alpha$ of the acceleration which appears in the last term of (147) is much smaller than the acceleration itself divided by the characteristic time $e^2/mc^3$. Then one may limit oneself to the truncated equation (112). In non-relativistic approximation the equations (112) and (147) reduce both to the form that has been used in the previous chapters.

### b. *A set of particles in a field*

A particle $i$ of a set moves in the combined field due to the other particles $j$ ($\neq i$) and to sources outside the system considered:

$$\sum_{j(\neq i)} f_j^{\alpha\beta}(R) + F^{\alpha\beta}(R). \tag{156}$$

Hence the equation of motion of particle $i$ becomes (cf. (147))

$$m_i a_i^\alpha = c^{-1} e_i \Big( \sum_{j(\neq i)} f_j^{\alpha\beta} + F^{\alpha\beta} \Big) u_{i\beta} + \frac{e_i^2}{6\pi} c^{-3} \Delta_\beta^\alpha(u_i) \dot{a}_i^\beta, \tag{157}$$

where $m_i$ is the mass of the particle, $a_i^\alpha$ its four-acceleration, $e_i$ its charge and $u_i^\alpha$ its four-velocity. The fields depend on the four-position $R_i^\alpha$.

A local law follows from (157) in a similar way as discussed for a single particle. One then finds the balance

$$\partial_\beta t^{\alpha\beta} = f^\alpha, \tag{158}$$

with the (symmetric) energy–momentum tensor of the set of charged particles defined as:

$$\begin{aligned} t^{\alpha\beta}(R) = {} & c \sum_i m_i \int u_i^\alpha(s_i) u_i^\beta(s_i) \delta^{(4)}\{R_i(s_i) - R\} \mathrm{d}s_i \\ & + \sum_{i,j(i\neq j)} \{ f_i^{\alpha\gamma}(R) f_{j.\gamma}^{\beta}(R) - \tfrac{1}{4} f_i^{\gamma\varepsilon}(R) f_{j\gamma\varepsilon}(R) g^{\alpha\beta} \} \\ & + \sum_i \{ f_{+i}^{\alpha\gamma}(R) f_{-i.\gamma}^{\beta}(R) + f_{-i}^{\alpha\gamma}(R) f_{+i.\gamma}^{\beta}(R) - \tfrac{1}{2} f_{+i}^{\gamma\varepsilon}(R) f_{-i\gamma\varepsilon}(R) g^{\alpha\beta} \} \end{aligned} \tag{159}$$

and the Lorentz force density

$$f^\alpha(R) = c^{-1} F^{\alpha\beta}(R) j_\beta(R), \tag{160}$$

that contains the four-current

$$c^{-1} j^\alpha(R) = \sum_i e_i \int u_i^\alpha(s_i) \delta^{(4)}\{R_i(s_i) - R\} \mathrm{d}s_i . \tag{161}$$

The balance equation (158) has the property that it reduces to a conservation law if no external electromagnetic fields $F^{\alpha\beta}$ are present, as (160) shows. This is a reflection of the fact that (159) is the complete energy–momentum tensor of the set of charged particles and the fields generated by them.

In the preceding subsection it has been remarked that often the self-force term in the equation of motion may be neglected. Then one must also omit the last sum of the energy–momentum tensor (159).

The balance equation (158) with (159) and (160) is not the only way to write the equation of motion (157) in a local form. An alternative way is discussed in the appendix. In the next chapter it will turn out that the form given here has certain advantages over the alternative.

In this survey of the classical covariant theory of charged particles and their fields we considered the field equations with their solutions and the equations of motion for the particles in each other's fields. These results of the microscopic theory will form the basis for the theory of atoms, considered as groups of point particles, and of matter in bulk.

APPENDIX

# On an energy-momentum tensor with 'local' character

In the main text we have written the equation of motion for a set of charged particles in an external field in the form of a balance equation. The way to arrive at such a balance equation is by no means unique. It depends on the choice of what is called the force density acting on the particles and what is called the energy–momentum tensor of the particles. An alternative way will be discussed in this appendix.

The energy–momentum tensor (159) that figures in the balance equation (158) contains the fields generated by all particles of the system. These fields may be split into plus and minus fields that have a different behaviour in the neighbourhood of the world line, as has been shown in section 2*e*. A special different form of the balance equation may now be obtained by writing the contributions of the minus fields generated by all particles at the right-hand side of (158), so that one gets

$$\partial_\beta t_+^{\alpha\beta} = f^{*\alpha} \tag{A1}$$

with the energy–momentum tensor

$$t_+^{\alpha\beta}(R) = c\sum_i m_i \int u_i^\alpha(s_i)u_i^\beta(s_i)\delta^{(4)}\{R_i(s_i)-R\}\mathrm{d}s_i + \sum_{i,j(i\neq j)} \{f_{+i}^{\alpha\gamma}(R)f_{+j.\gamma}^{\beta}(R)-\tfrac{1}{4}f_{+i}^{\gamma\varepsilon}(R)f_{+j\gamma\varepsilon}(R)g^{\alpha\beta}\}, \tag{A2}$$

(which is also symmetric) and the force density

$$f^{*\alpha}(R) = f^\alpha(R)+\sum_{i,j} e_i \int f_{-j}^{\alpha\beta}(R)u_{i\beta}\,\delta^{(4)}\{R_i(s_i)-R\}\mathrm{d}s_i, \tag{A3}$$

where $f^\alpha(R)$ has been given in (160–161).

The equation (A1) is not modified if a divergenceless tensor is added to $t_+^{\alpha\beta}$. This freedom will be used to introduce instead of $t_+^{\alpha\beta}$ a different tensor $t^{*\alpha\beta}$, which is 'local' in a sense that will be explained below.

To obtain such a tensor we shall start from the expression which Wheeler and Feynman[1] have given for the total energy–momentum of a set of charged

[1] J. A. Wheeler and R. P. Feynman, Rev. Mod. Phys. **21**(1949)425; J. W. Dettmann and A. Schild, Phys. Rev. **95**(1954)1057; cf. footnote on page 160.

particles which interact via plus fields. This total momentum $p^\alpha(\Sigma)$ associated to a plane space-like surface $\Sigma$ with normal $n^\alpha$ (and thus given by the equation

$$n{\cdot}R + c\tau = 0 \tag{A4}$$

with fixed $\tau$) reads as follows

$$\begin{aligned} p^\alpha(\Sigma) = {} & \sum_i m_i u_i^\alpha(s_i^0) \\ & - \frac{1}{4\pi c} \sum_{i,j(i\neq j)} e_i e_j \int \{u_i{\cdot}u_j(R_i - R_j)^\alpha - (R_i - R_j){\cdot}u_i u_j^\alpha - (R_i - R_j){\cdot}u_j u_i^\alpha\} \\ & \{\theta(n{\cdot}R_i + c\tau)\theta(-n{\cdot}R_j - c\tau) - \theta(-n{\cdot}R_i - c\tau)\theta(n{\cdot}R_j + c\tau)\} \\ & \delta'\{(R_i - R_j)^2\} \mathrm{d}s_i \, \mathrm{d}s_j, \end{aligned} \tag{A5}$$

where $s_i^0$ is the proper time of the intersection of world line $i$ with $\Sigma$, i.e. given by:

$$n{\cdot}R_i(s_i^0) + c\tau = 0. \tag{A6}$$

The first term of $p^\alpha(\Sigma)$ is the contribution of the material momentum of the particles, while the second is the momentum carried by the plus fields. The unit step functions $\theta$ ensure that in (A5) only those parts of the world lines of $i$ and $j$ contribute for which $R_i^\alpha$ is on one side of the surface $\Sigma$ and $R_j^\alpha$ on the other. One may write the difference of products of step functions alternatively as

$$\theta(-s_i + s_i^0)\theta(s_j - s_j^0) - \theta(s_i - s_i^0)\theta(-s_j + s_j^0). \tag{A7}$$

The prime at the delta function in (A5) indicates a differentiation with respect to its argument. From the form of $p^\alpha(\Sigma)$ it is apparent that only finite parts of the world lines of the particles $i$ contribute to the field momentum, namely those parts that lie outside the light cones having their top at the intersections of the world lines $j$ ($\neq i$) with the plane $\Sigma$.

From the expression (A5) for $p^\alpha(\Sigma)$ one may infer the energy–momentum tensor $t_{\mp}^{\alpha\beta}$ which gives back $p^\alpha(\Sigma)$ by integration over the plane $\Sigma$, i.e.

$$-c^{-1} \int_\Sigma t_{\mp}^{\alpha\beta} \mathrm{d}^3\Sigma_\beta = p^\alpha(\Sigma). \tag{A8}$$

(The tensor $t_{\mp}^{\alpha\beta}$ will lead to the tensor $t^{*\alpha\beta}$, which we are trying to find.) To that end we write a tensor with the plus field momentum localized on the line that joins the positions $R_i^\alpha$ and $R_j^\alpha$ and with the particle momentum localized – of course – at the positions $R_i^\alpha$ themselves

$$t_{\ddagger}^{\alpha\beta}(R) = c\sum_i m_i \int u_i^\alpha u_i^\beta \delta^{(4)}(R_i - R)\mathrm{d}s_i$$
$$+\frac{1}{4\pi}\sum_{i,j(i\neq j)} e_i e_j \int\int_{\lambda=0}^{1} \{u_i\cdot u_j(R_i-R_j)^\alpha-(R_i-R_j)\cdot u_i u_j^\alpha-(R_i-R_j)\cdot u_j u_i^\alpha\}$$
$$(R_i-R_j)^\beta \delta'\{(R_i-R_j)^2\}\delta^{(4)}\{R_i+\lambda(R_j-R_i)-R\}\mathrm{d}s_i\,\mathrm{d}s_j\,\mathrm{d}\lambda. \quad \text{(A9)}$$

We have to show whether this tensor satisfies (A8) and moreover whether it differs from $t_{+}^{\alpha\beta}$ (A2) by a divergenceless part so that

$$\partial_\beta t_{\ddagger}^{\alpha\beta} = \partial_\beta t_{+}^{\alpha\beta}. \quad \text{(A10)}$$

To begin with we check the validity of (A8) by integrating $t_{\ddagger}^{\alpha\beta}$ over a space-like plane $\Sigma$. One has then to employ the identity

$$|(R_i-R_j)\cdot n|\int_\Sigma\int_{\lambda=0}^{1} \delta^{(4)}\{R_i+\lambda(R_j-R_i)-R\}\mathrm{d}\lambda\,\mathrm{d}^3\Sigma$$
$$= \theta(n\cdot R_i+c\tau)\theta(-n\cdot R_j-c\tau)+\theta(-n\cdot R_i-c\tau)\theta(n\cdot R_j+c\tau), \quad \text{(A11)}$$

which may be proved by evaluating the left-hand side in a Lorentz frame in which $\Sigma$ is purely space-like (i.e. in which $n^\alpha = (1, 0, 0, 0)$) and using the relation (41). Then one obtains for the integral of $t_{\ddagger}^{\alpha\beta}$:

$$-c^{-1}\int_\Sigma t_{\ddagger}^{\alpha\beta}\mathrm{d}\Sigma_\beta = -m_i\int u_i^\alpha u_i\cdot n\delta^{(4)}(R_i-R)\mathrm{d}s_i\,\mathrm{d}^3\Sigma$$
$$-\frac{1}{4\pi c}\sum_{i,j(i\neq j)} e_i e_j\int \{u_i\cdot u_j(R_i-R_j)^\alpha-(R_i-R_j)\cdot u_i u_j^\alpha-(R_i-R_j)\cdot u_j u_i^\alpha\}$$
$$\varepsilon\{(R_i-R_j)\cdot n\}\{\theta(n\cdot R_i+c\tau)\theta(-n\cdot R_j-c\tau)+\theta(-n\cdot R_i-c\tau)\theta(n\cdot R_j+c\tau)\}$$
$$\delta'\{(R_i-R_j)^2\}\mathrm{d}s_i\,\mathrm{d}s_j. \quad \text{(A12)}$$

In the first term at the right-hand side the integrations over $s_i$ and $\Sigma$ may be performed if use is made of (41). As a result one finds the first term at the right-hand side of (A5). In the second term at the right-hand side of (A12) the product of the $\varepsilon$-function and the sum of products of the $\theta$-functions is equal to the difference of the products of the $\theta$-functions so that also the second term at the right-hand side of (A5) is recovered.

The use of $t_{\ddagger}^{\alpha\beta}$ instead of $t_{+}^{\alpha\beta}$ in (A1) is justified if we succeed in proving (A10). With the explicit form (A9) of $t_{\ddagger}^{\alpha\beta}$ one finds, by performing a partial integration in the first term and using the chain rule of differentiation, for its divergence

$$\partial_\beta t_{\mp}^{\alpha\beta} = c \sum_i m_i \int a_i^\alpha \delta^{(4)}(R_i - R)\mathrm{d}s_i$$
$$+ \frac{1}{4\pi} \sum_{i,j(i\neq j)} e_i e_j \int\int_{\lambda=0}^{1} \{u_i \cdot u_j (R_i - R_j)^\alpha - (R_i - R_j)\cdot u_i u_j^\alpha - (R_i - R_j)\cdot u_j u_i^\alpha\}$$
$$\delta'\{(R_i - R_j)^2\} \frac{\partial}{\partial \lambda} \delta^{(4)}\{R_i + \lambda(R_j - R_i) - R\}\mathrm{d}s_i \mathrm{d}s_j \mathrm{d}\lambda. \quad \text{(A13)}$$

The integration over $\lambda$ may be performed so that the difference of the two delta functions $\delta^{(4)}(R_j - R)$ and $\delta^{(4)}(R_i - R)$ arises. By using the antisymmetry – in $i$ and $j$ – of the remaining part of the integrand the last term of (A13) becomes

$$-\frac{1}{2\pi} \sum_{i,j(i\neq j)} e_i e_j \int \{u_i \cdot u_j (R_i - R_j)^\alpha - (R_i - R_j)\cdot u_i u_j^\alpha - (R_i - R_j)\cdot u_j u_i^\alpha\}$$
$$\delta'\{(R_i - R_j)^2\}\delta^{(4)}(R_i - R)\mathrm{d}s_i \mathrm{d}s_j. \quad \text{(A14)}$$

The last term between the brackets gives a vanishing contribution since it leads to an integrand which contains the factor $(\mathrm{d}/\mathrm{d}s_j)\delta\{(R_i - R_j)^2\}$. The integral of this term over $s_j$ may then be performed, with a vanishing result. The other terms may be handled similarly, with the result

$$-\frac{1}{4\pi} \sum_{i,j(i\neq j)} e_i e_j \int [(\partial_i^\alpha u_j^\beta - \partial_i^\beta u_j^\alpha) u_{i\beta}\, \delta\{(R_i - R_j)^2\}]\delta^{(4)}(R_i - R)\mathrm{d}s_i \mathrm{d}s_j. \quad \text{(A15)}$$

In this expression one recognizes the plus field due to particle $j$ at the position $i$:

$$f_{+j}^{\alpha\beta}(R_i) = \frac{1}{4\pi} e_j \int (u_j^\beta \partial_i^\alpha - u_j^\alpha \partial_i^\beta)\delta\{(R_i - R_j)^2\}\mathrm{d}s_j, \quad \text{(A16)}$$

as follows from (59) and the subsequent remarks. Collecting the results (A13) with (A15) we have obtained

$$\partial_\beta t_{\mp}^{\alpha\beta} = c \sum_i \int \{m_i a_i^\alpha - c^{-1} e_i \sum_{j(\neq i)} f_{+j}^{\alpha\beta}(R_i) u_{i\beta}\}\delta^{(4)}(R_i - R)\mathrm{d}s_i. \quad \text{(A17)}$$

From the expression (A2) for $t_{+}^{\alpha\beta}$ it now follows that we have found here the relation (A10), which shows that the tensor $t_{\mp}^{\alpha\beta}$ (A9) may be used as well as $t_{+}^{\alpha\beta}$ (A2) in (A1)[1].

[1] The reader may note that the Wheeler–Feynman expression (A5) has played a heuristic role, since it was only used to infer the tensor $t_{\mp}^{\alpha\beta}$ (A9) from it. The main result is the equivalence of $t_{\mp}^{\alpha\beta}$ and $t_{+}^{\alpha\beta}$ in the sense of (A10). The relation (A8) clarifies the connexion between $t_{\mp}^{\alpha\beta}$ and the expression (A5).

In contrast to $t_{+}^{\alpha\beta}$ the tensor $t_{\mp}^{\alpha\beta}$ has 'local' character in so far that it vanishes for positions outside the 'envelope' of the world lines.

The tensor $t_{\mp}^{\alpha\beta}$ is asymmetric with respect to the indices $\alpha$ and $\beta$. The symmetry may be restored, without loosing the locality, by adding another divergenceless tensor, as we shall now show. The first two terms of the right-hand side of (A9) are already symmetric in $\alpha$ and $\beta$ so that we shall focus our attention on the third and fourth term. In the third term we may employ the identity

$$(R_i-R_j)\cdot u_i\,\delta'\{(R_i-R_j)^2\} = \frac{1}{2}\frac{\mathrm{d}}{\mathrm{d}s_i}\,\delta\{(R_i-R_j)^2\} \tag{A18}$$

and perform a partial integration with respect to $s_i$. If the fourth term is treated in a similar way and use is made of the symmetry in $i$ and $j$ the third and fourth terms of $t_{\mp}^{\alpha\beta}$ become

$$\frac{1}{4\pi}\sum_{i,j(i\neq j)} e_i e_j \iint_{\lambda=0}^{1} \{u_i^\alpha u_j^\beta + \lambda u_i^\alpha (R_i-R_j)^\beta u_j^\gamma \partial_\gamma\}$$
$$\delta\{(R_i-R_j)^2\}\delta^{(4)}\{R_i+\lambda(R_j-R_i)-R\}\mathrm{d}s_i\,\mathrm{d}s_j\,\mathrm{d}\lambda. \tag{A19}$$

By adding a divergenceless term to this expression one obtains

$$\frac{1}{4\pi}\sum_{i,j(i\neq j)} e_i e_j \iint_{\lambda=0}^{1} \{u_i^\alpha u_j^\beta + \lambda u_i^\alpha u_j^\beta (R_i-R_j)^\gamma \partial_\gamma\}$$
$$\delta\{(R_i-R_j)^2\}\delta^{(4)}\{R_i+\lambda(R_j-R_i)-R\}\mathrm{d}s_i\,\mathrm{d}s_j\,\mathrm{d}\lambda. \tag{A20}$$

If we use once more the chain rule of differentiation we may write this as

$$\frac{1}{4\pi}\sum_{i,j(i\neq j)} e_i e_j \iint_{\lambda=0}^{1} \frac{\partial}{\partial\lambda}\,[\lambda u_i^\alpha u_j^\beta \delta\{(R_i-R_j)^2\}$$
$$\delta^{(4)}\{R_i+\lambda(R_j-R_i)-R\}]\mathrm{d}s_i\,\mathrm{d}s_j\,\mathrm{d}\lambda, \tag{A21}$$

so that the integration over $\lambda$ may be performed. Only the value at the boundary $\lambda = 1$ contributes:

$$\frac{1}{4\pi}\sum_{i,j(i\neq j)} e_i e_j \int u_i^\alpha u_j^\beta \delta\{(R_i-R_j)^2\}\delta^{(4)}(R_j-R)\mathrm{d}s_i\,\mathrm{d}s_j\,. \tag{A22}$$

This is still not symmetric, since $j$ alone occurs in the second delta function. Let us write (A22) as half the sum plus half the difference of that expression and its transposed:

$$\frac{1}{8\pi}\sum_{i,j(i\neq j)} e_i e_j \int (u_i^\alpha u_j^\beta + u_i^\beta u_j^\alpha)\delta\{(R_i-R_j)^2\}\delta^{(4)}(R_i-R)\mathrm{d}s_i\,\mathrm{d}s_j$$
$$+\frac{1}{8\pi}\sum_{i,j(i\neq j)} e_i e_j \iint_{\lambda=0}^{1} u_i^\alpha u_j^\beta (R_i-R_j)^\gamma \partial_\gamma\,\delta\{(R_i-R_j)^2\}$$
$$\delta^{(4)}\{R_j+\lambda(R_i-R_j)-R\}\mathrm{d}s_i\,\mathrm{d}s_j\,\mathrm{d}\lambda. \tag{A23}$$

Adding a divergenceless part to the second term, we get

$$\frac{1}{8\pi}\sum_{i,j(i\neq j)} e_i e_j \int (u_i^\alpha u_j^\beta + u_i^\beta u_j^\alpha)\delta\{(R_i-R_j)^2\}\delta^{(4)}(R_i-R)\mathrm{d}s_i\,\mathrm{d}s_j$$

$$+\frac{1}{8\pi}\sum_{i,j(i\neq j)} e_i e_j \int\int_{\lambda=0}^{1} \{u_i^\alpha(R_i-R_j)^\beta + u_i^\beta(R_i-R_j)^\alpha\}u_j^\gamma \partial_\gamma \delta\{(R_i-R_j)^2\}$$

$$\delta^{(4)}\{R_j+\lambda(R_i-R_j)-R\}\mathrm{d}s_i\,\mathrm{d}s_j\,\mathrm{d}\lambda. \quad \text{(A24)}$$

Indeed the second part of the second term has vanishing divergence in view of the antisymmetry in $i$ and $j$ (together with $\lambda \leftrightarrow 1-\lambda$); the first part of the second term has the same divergence as the second term of (A23).

Since the expression (A24) is symmetric with respect to an interchange of $\alpha$ and $\beta$, we have found now a symmetric energy–momentum tensor with the same divergence as $t_{\ddagger}^{\alpha\beta}$ (and hence as $t_{+}^{\alpha\beta}$). It is found by adding (A24) to the first two terms at the right-hand side of (A9):

$$t^{*\alpha\beta} = c\sum_i m_i \int u_i^\alpha u_i^\beta \delta^{(4)}(R_i-R)\mathrm{d}s_i$$

$$+\frac{1}{8\pi}\sum_{i,j(i\neq j)} e_i e_j \int (u_i^\alpha u_j^\beta + u_i^\beta u_j^\alpha)\delta\{(R_i-R_j)^2\}\delta^{(4)}(R_i-R)\mathrm{d}s_i\,\mathrm{d}s_j$$

$$+\frac{1}{8\pi}\sum_{i,j(i\neq j)} e_i e_j \int\int_{\lambda=0}^{1} [2u_i\cdot u_j(R_i-R_j)^\alpha(R_i-R_j)^\beta\delta'\{(R_i-R_j)^2\}$$

$$+\{u_i^\alpha(R_i-R_j)^\beta + u_i^\beta(R_i-R_j)^\alpha\}u_j^\gamma\partial_\gamma\delta\{(R_i-R_j)^2\}]$$

$$\delta^{(4)}\{R_j+\lambda(R_i-R_j)-R\}\mathrm{d}s_i\,\mathrm{d}s_j\,\mathrm{d}\lambda. \quad \text{(A25)}$$

This symmetric tensor[1], which still has 'local' character, satisfies the energy–momentum balance

$$\partial_\beta t^{*\alpha\beta} = f^{*\alpha} \quad \text{(A26)}$$

with the force density (A3).

The two balance equations (158) and (A26) are both equivalent with the equation (157). Each of them may be used for the description of the system of charged particles. As remarked in the main text the typical feature of the balance equation (158) with (159–160) is that it reduces to a conservation

[1] An energy–momentum tensor of the same type has been given by S. Emid and J. Vlieger (Physica **52**(1971)329) in the form of a series expansion, valid for a continuous charge–current distribution. If that formula is applied to a set of charged point particles, one finds a result that is nearly the same as the expression that is obtained from (A25) if the integrand of the last term there is developed into powers of $\lambda$ and the integration over $\lambda$ is performed – the sole difference being that the summations in (A25) exclude the contributions $i=j$, thus avoiding infinite self energy contributions.

law if no external fields are present. The typical feature of (A26) with (A3) and (A25) is that the contributions of the minus fields (which, just as the external fields, are finite at the world lines of the particles) are written at the right-hand side just as has been done in (148). The energy–momentum tensor (A25) has local character in the sense that it vanishes outside the region in which the world lines are situated, in contrast to the energy–momentum tensor (159) which also diminishes with increasing distances in space-like directions, but which does not possess a finite support in such directions.

Often the effect of the minus fields on the equation of motion is small (as may be discussed if they are explicitly evaluated: see the appendix of the following chapter). If they are neglected the force density (A3) reduces to (160).

In the next chapter we shall employ (158) extensively, but we shall also have occasion to show, in an appendix, the consequences of the use of (A26).

# PROBLEMS

**1.** Show that the expressions (44), (48) and (49) for the Green functions $G_r$, $G_a$, $G_f$ and $G_{af}$ satisfy the relation (47).

Hint: use the identity (53).

**2.** Prove that the retarded and advanced potentials due to a source $j^\alpha(\boldsymbol{R}, t)$ may be written as

$$a^\alpha_{r,a}(\boldsymbol{R}, t) = \frac{c^{-1}}{4\pi}\int j^\alpha\left(\boldsymbol{R}', t \mp \frac{|\boldsymbol{R}-\boldsymbol{R}'|}{c}\right)\frac{1}{|\boldsymbol{R}-\boldsymbol{R}'|}\,\mathrm{d}\boldsymbol{R}'.$$

This may be proved from (14) with (23), (30), (39) and (40), or alternatively from (57) with (43).

**3.** Prove – by insertion – that the potentials (61) satisfy the equations

$$\Box\,\varphi(\boldsymbol{R}, t) = -\rho^e(\boldsymbol{R}, t) = -\sum_j e_j\,\delta\{\boldsymbol{R}-\boldsymbol{R}_j(t)\},$$

$$\Box\,\boldsymbol{a}(\boldsymbol{R}, t) = -c^{-1}\boldsymbol{j}(\boldsymbol{R}, t) = -\sum_j e_j\,\boldsymbol{\beta}_j(t)\delta\{\boldsymbol{R}-\boldsymbol{R}_j(t)\},$$

which are the equations (13) with (5) in three-dimensional notation (check that first).

**4.** Show that the expressions (61) for the potentials may be written alternatively as the following integrals over the space-coordinates

$$\varphi_{r,a}(\boldsymbol{R}, t) = \sum_j e_j\int\frac{\delta\{\boldsymbol{R}_j(t\mp c^{-1}|\boldsymbol{R}-\boldsymbol{R}'|)-\boldsymbol{R}\}}{4\pi|\boldsymbol{R}-\boldsymbol{R}'|}\,\mathrm{d}\boldsymbol{R}',$$

$$\boldsymbol{a}_{r,a}(\boldsymbol{R}, t) = \sum_j e_j\int\boldsymbol{\beta}_j(t\mp c^{-1}|\boldsymbol{R}-\boldsymbol{R}'|)\frac{\delta\{\boldsymbol{R}_j(t\mp c^{-1}|\boldsymbol{R}-\boldsymbol{R}'|)-\boldsymbol{R}\}}{4\pi|\boldsymbol{R}-\boldsymbol{R}'|}\,\mathrm{d}\boldsymbol{R}'.$$

Hint: add a factor $\delta\{\boldsymbol{R}_j(t_j)-\boldsymbol{R}'\}$ and an integration over $\boldsymbol{R}'$ to (61). Then replace in the integrand $\boldsymbol{R}_j(t_j)$ by $\boldsymbol{R}'$ and integrate over $t_j$.

**5.** Prove that for the field $f^{\alpha\beta}$ (95) or (96) one has the identity

$$\varepsilon_{\alpha\beta\gamma\delta}\,f^{\alpha\beta}r^\gamma|_{r,a} = 0,$$

where $f^{\alpha\beta}$ has to be understood as the quantity 'inside' the bar symbol, as explained after (89).

Show that this identity reads in three-dimensional notation

$$\boldsymbol{b}|_{\mathrm{r,a}} = \pm(\boldsymbol{n}\wedge\boldsymbol{e})|_{\mathrm{r,a}}$$

and

$$\boldsymbol{b}\cdot\boldsymbol{n}|_{\mathrm{r,a}} = 0$$

with the same convention for the fields $\boldsymbol{e}$ and $\boldsymbol{b}$.

**6.** Prove that the Hamilton equations that follow from the so-called Darwin Hamiltonian (C. G. Darwin, Phil. Mag. **39**(1920)537) for two particles with charges $e_1$, $e_2$, masses $m_1$, $m_2$, positions $\boldsymbol{R}_1$, $\boldsymbol{R}_2$, momenta $\boldsymbol{P}_1$, $\boldsymbol{P}_2$, moving in each others fields

$$H(\boldsymbol{P}_1, \boldsymbol{R}_1, \boldsymbol{P}_2, \boldsymbol{R}_2) = \sum_{i=1}^{2}\left(\frac{\boldsymbol{P}_i^2}{2m_i} - \frac{\boldsymbol{P}_i^4}{8m_i^3c^2}\right) + \left\{1 - \frac{\boldsymbol{P}_1\cdot\mathsf{T}(\boldsymbol{R}_1-\boldsymbol{R}_2)\cdot\boldsymbol{P}_2}{2m_1 m_2 c^2}\right\}\frac{e_1 e_2}{4\pi|\boldsymbol{R}_1-\boldsymbol{R}_2|},$$

with $\mathsf{T}(\boldsymbol{R}_1-\boldsymbol{R}_2) \equiv \mathsf{U}+(\boldsymbol{R}_1-\boldsymbol{R}_2)(\boldsymbol{R}_1-\boldsymbol{R}_2)/|\boldsymbol{R}_1-\boldsymbol{R}_2|^2$, lead to equations of motion for the particles that are, up to order $c^{-2}$, equal to the $c^{-2}$-approximations of the complete relativistic equations (157).

Hint: Use the relevant expressions from (72) for the retarded fields.

**7.** In the following series of problems the minus part (111) of the self-field at the world line of the particle will be obtained from the expressions (69) with (71) for the fields in terms of series in powers of $c^{-1}$.

Prove that in the limit $\boldsymbol{r}\to 0$ the expressions (75) with (71) for the partial minus fields of order $n$ (i.e. $n$ odd for the electric and $n$ even for the magnetic field) lead to:

$$\begin{aligned}
\boldsymbol{e}_-^{(n)} = {} & -\frac{ec^{-2}}{4\pi n!}\nabla\sum_{\substack{k,l=0\\k+l\leq n-2}}^{n-2}(n-2-k-l)\{\ddot{\boldsymbol{\beta}}\cdot\nabla(\boldsymbol{\beta}\cdot\nabla)^{n-3} \\
& \qquad +(n-3-k)(\dot{\boldsymbol{\beta}}\cdot\nabla)^2(\boldsymbol{\beta}\cdot\nabla)^{n-4}\}r^{n-1} \\
& +\frac{ec^{-2}}{4\pi(n-2)!}\sum_{\substack{k,l=0\\k+l\leq n-3}}^{n-3}[\ddot{\boldsymbol{\beta}}(\boldsymbol{\beta}\cdot\nabla)^{n-3}+(n-3-k)\dot{\boldsymbol{\beta}}\dot{\boldsymbol{\beta}}\cdot\nabla(\boldsymbol{\beta}\cdot\nabla)^{n-4} \\
& +(n-3-k-l)\{\dot{\boldsymbol{\beta}}\dot{\boldsymbol{\beta}}\cdot\nabla(\boldsymbol{\beta}\cdot\nabla)^{n-4}+\boldsymbol{\beta}\ddot{\boldsymbol{\beta}}\cdot\nabla(\boldsymbol{\beta}\cdot\nabla)^{n-4} \\
& \qquad +(n-4-k)\boldsymbol{\beta}(\dot{\boldsymbol{\beta}}\cdot\nabla)^2(\boldsymbol{\beta}\cdot\nabla)^{n-5}\}]r^{n-3},
\end{aligned}$$

and to a similar expression for $\boldsymbol{b}_-^{(n)}$. (The dots indicate time derivatives.)

Hints: remember that $\partial/\partial t$ in (71) contains an explicit and an implicit differentiation, i.e. $\partial/\partial t = (\partial/\partial t)_{\text{expl}} - c\boldsymbol{\beta}\cdot\boldsymbol{\nabla}$. Furthermore note that in the limit $\boldsymbol{r} \to 0$ only terms with $\boldsymbol{\nabla}^m r^m$, or in other words with two explicit time differentiations, subsist.

**8.** Prove the relations

$$\sum_{k,l=0(k+l\leqq n)}^{n} (n-k) = \tfrac{1}{3}n(n+1)(n+2),$$

$$\sum_{k,l=0(k+l\leqq n)}^{n} (n-k-l) = \tfrac{1}{6}n(n+1)(n+2),$$

$$\sum_{k,l=0(k+l\leqq n)}^{n} (n-k-l)(n-1-k) = \tfrac{1}{8}(n-1)n(n+1)(n+2).$$

Hint: use the sums $\sum_{k=0}^{n} k = \frac{1}{2}n(n+1)$, $\sum_{k=0}^{n} k^2 = \frac{1}{6}n(n+1)(2n+1)$ and $\sum_{k=0}^{n} k^3 = \frac{1}{4}n^2(n+1)^2$.

**9.** Prove the relations ($n$ odd)

$$(\boldsymbol{\beta}\cdot\boldsymbol{\nabla})^{n-2}r^{n-1} = (n-1)!\beta^{n-3}\boldsymbol{\beta}\cdot\boldsymbol{r},$$

$$(\boldsymbol{\beta}\cdot\boldsymbol{\nabla})^{n-3}r^{n-1} = \tfrac{1}{2}(n-1)(n-3)!\beta^{n-3}r^2 + \tfrac{1}{2}(n-1)(n-3)(n-3)!\beta^{n-5}(\boldsymbol{\beta}\cdot\boldsymbol{r})^2,$$

$$(\boldsymbol{\beta}\cdot\boldsymbol{\nabla})^{n-4}r^{n-1} = \tfrac{1}{2}(n-1)(n-3)(n-4)!\beta^{n-5}\boldsymbol{\beta}\cdot\boldsymbol{r}r^2 + \tfrac{1}{6}(n-1)(n-3)(n-5)(n-4)!\beta^{n-7}(\boldsymbol{\beta}\cdot\boldsymbol{r})^3.$$

**10.** Prove from the results of the preceding three problems that

$$\boldsymbol{e}_-^{(n)}(\boldsymbol{r}=0) = \frac{e}{4\pi}c^{-2}\{\tfrac{1}{3}(n-1)\ddot{\boldsymbol{\beta}}\beta^{n-3} + \tfrac{1}{4}(n-1)(n-3)\dot{\boldsymbol{\beta}}\boldsymbol{\beta}\cdot\dot{\boldsymbol{\beta}}\beta^{n-5}\}, \quad (n \text{ odd}),$$

$$\boldsymbol{b}_-^{(n)}(\boldsymbol{r}=0) = \frac{e}{4\pi}c^{-2}\{\tfrac{1}{3}(n-2)\boldsymbol{\beta}\wedge\ddot{\boldsymbol{\beta}}\beta^{n-4} + \tfrac{1}{4}(n-2)(n-4)\boldsymbol{\beta}\wedge\dot{\boldsymbol{\beta}}\boldsymbol{\beta}\cdot\dot{\boldsymbol{\beta}}\beta^{n-6}\}, \quad (n \text{ even}).$$

**11.** Prove from the preceding result that the total minus field at the position of the particle is given by

$$\boldsymbol{e}_-(\boldsymbol{r}=0) = \frac{e}{4\pi}c^{-2}(\tfrac{2}{3}\gamma^4\ddot{\boldsymbol{\beta}} + 2\gamma^6\dot{\boldsymbol{\beta}}\boldsymbol{\beta}\cdot\dot{\boldsymbol{\beta}}),$$

$$\boldsymbol{b}_-(\boldsymbol{r}=0) = \frac{e}{4\pi}c^{-2}(\tfrac{2}{3}\gamma^4\boldsymbol{\beta}\wedge\ddot{\boldsymbol{\beta}} + 2\gamma^6\boldsymbol{\beta}\wedge\dot{\boldsymbol{\beta}}\boldsymbol{\beta}\cdot\dot{\boldsymbol{\beta}}).$$

**12.** Show that the results given in the preceding problem are the components of the tensor (111) (at the world line):

$$f^{\alpha\beta} = \frac{e}{6\pi} c^{-4}(u^\alpha \dot{a}^\beta - u^\beta \dot{a}^\alpha),$$

where $u^\alpha$ and $a^\alpha$ are the four-velocity and four-acceleration and the dot now stands for a differentiation with respect to proper time.

**13.** Show by use of (A18) and a partial integration of the second and third terms in the first bracket expression of the integrand in (A5) that one may write this formula alternatively as

$$p^\alpha(\Sigma) = \sum_i m_i u_i^\alpha(s_i^0) - \frac{1}{4\pi c} \sum_{i,j(i\neq j)} e_i e_j \int u_i \cdot u_j (R_i - R_j)^\alpha$$

$$\{\theta(n\cdot R_i + c\tau)\theta(-n\cdot R_j - c\tau) - \theta(-n\cdot R_i - c\tau)\theta(n\cdot R_j + c\tau)\}\delta'\{(R_i - R_j)^2\}\mathrm{d}s_i\,\mathrm{d}s_j$$

$$- \frac{1}{4\pi c} \sum_{i,j(i\neq j)} e_i e_j \int u_j^\alpha n\cdot u_i\, \delta(n\cdot R_i + c\tau)\delta\{(R_i - R_j)^2\}\mathrm{d}s_i\,\mathrm{d}s_j .$$

Prove the ancillary formula (cf. (A11))

$$(R_i - R_j)\cdot n \int_\Sigma \int_{\lambda=0}^{1} u_j \cdot \partial \delta^{(4)}\{R_j + \lambda(R_i - R_j) - R\}\mathrm{d}^3\Sigma\, \mathrm{d}\lambda$$

$$= -u_j \cdot n\{\delta(n\cdot R_i + c\tau) - \delta(n\cdot R_j + c\tau)\}.$$

Hint: evaluate the left-hand side in a Lorentz frame in which $\Sigma$ is purely space-like; then one may apply Gauss's theorem to prove that only the time differentiation (i.e. $u_j^0 \partial_0$) gives a contribution.

Show with the help of the latter formula that the total momentum (in a plane $\Sigma$) which corresponds to the tensor $t^{*\alpha\beta}$ (A25) is equal to that which corresponds to $t_\ddagger^{\alpha\beta}$ (A9), i.e. that

$$p^{*\alpha}(\Sigma) = -c^{-1} \int t^{*\alpha\beta} \mathrm{d}^3\Sigma_\beta = -c^{-1} \int t_\ddagger^{\alpha\beta} \mathrm{d}^3\Sigma_\beta = p^\alpha(\Sigma),$$

where the last member is given in the first formula of this problem (v. also (A8)).

CHAPTER IV

# Composite particles

## 1 *Introduction*

The covariant laws of electrodynamics of composite particles will be derived in this chapter on the basis of the equations that govern the fields and motion of charged point particles.

The equations for the fields generated by composite particles will be found by introducing covariantly defined multipole moments, through which the inner structure of the composite particles will be characterized. As a consequence the field equations will contain in their sources the effects of the motion of these multipoles. It will turn out that the electric and magnetic properties then appear in a more symmetric fashion than in the non-relativistic approximation.

The central problem in the derivation of equations of motion of composite particles in an electromagnetic field consists in defining a covariant centre of energy. It will be shown how such a definition can be obtained and how it leads to covariant equations of motion for particles endowed with multipole moments, at least in the case of weak fields.

## 2 *The field equations*

### a. *The atomic series expansion*

Let us consider a system in which the charged point particles (electrons and nuclei) are grouped into stable entities (such as atoms, ions, molecules, free electrons), which will be called 'atoms' here. The particles will be labelled by double indices $ki$, where $k$ numbers the atoms and $i$ their constituent particles. Then the field equations (III.4–5) become:

$$\partial_\beta f^{\alpha\beta} = \sum_{k,i} e_{ki} \int u_{ki}^{\alpha}(s_{ki})\delta^{(4)}\{R_{ki}(s_{ki})-R\}\mathrm{d}s_{ki}. \qquad (1)$$

Here $s_{ki}$ is the proper time along the world line of particle $ki$, $R^{\alpha}_{ki}(s_{ki})$ its four-position, $u^{\alpha}_{ki}(s_{ki})$ its four-velocity, and $e_{ki}$ its charge. From the form of the integral it follows that it is not necessary to choose the proper time as the parameter along the world line. A different parametrization will indeed be introduced.

Let us now choose a (at the moment arbitrary) privileged world line $R^{\alpha}_{k}$ that describes the motion of atom $k$ as a whole, with the proper time $s_k$. This parametrization can be carried over to the world lines of the constituent particles $ki$ by defining the point $R^{\alpha}_{ki}(s_k)$ on the world line of $ki$ with the help of the relation:

$$R_{ki}(s_k)\cdot n_k(s_k) = R_k(s_k)\cdot n_k(s_k), \tag{2}$$

where $n^{\alpha}_{k}(s_k)$ is an arbitrary, fixed time-like unit vector depending on $s_k$. (Later on we shall make various choices for this unit vector.) Internal atomic parameters $r^{\alpha}_{ki}(s_k)$ are now introduced by

$$r^{\alpha}_{ki}(s_k) = R^{\alpha}_{ki}(s_k) - R^{\alpha}_{k}(s_k). \tag{3}$$

Introducing this definition into (1) and expanding the delta function in powers of $r^{\alpha}_{ki}$, one obtains

$$\partial_\beta f^{\alpha\beta} = \sum_{k,i} e_{ki} \int \left( u^{\alpha}_{k}(s_k) + \frac{\mathrm{d}r^{\alpha}_{ki}}{\mathrm{d}s_k} \right) \sum_{n=0}^{\infty} \frac{(-1)^n}{n!} (r_{ki}\cdot\partial)^n \delta^{(4)}\{R_k(s_k) - R\}\mathrm{d}s_k, \tag{4}$$

where $u^{\alpha}_{k}(s_k) \equiv \mathrm{d}R^{\alpha}_{k}/\mathrm{d}s_k$ and where the fact has been used that $\partial/\partial R^{\alpha}_{k}$ acting on the delta function is the same as $-\partial_\alpha$ acting on it. The first term on the right-hand side with the term $n = 0$ of the series expansion is

$$c^{-1} j^{\alpha}(R) = \sum_{k} e_k \int u^{\alpha}_{k}(s_k) \delta^{(4)}\{R_k(s_k) - R\}\mathrm{d}s_k \tag{5}$$

(with $e_k$ the total charge $\sum_i e_{ki}$ of atom $k$). This four-vector has components $\alpha = 0$ and $\alpha = 1, 2, 3$, which are equal to the charge density and the current density (divided by $c$). Equation (4) can now be written as

$$\partial_\beta f^{\alpha\beta}(R) = c^{-1} j^{\alpha}(R) + \partial_\beta \sum_{k,i} e_{ki} \left\{ \sum_{n=1}^{\infty} \frac{(-1)^n}{n!} \int u^{\alpha}_{k} r^{\beta}_{ki} (r_{ki}\cdot\partial)^{n-1} \right.$$
$$\left. + \sum_{n=0}^{\infty} \frac{(-1)^n}{n!} \int \frac{\mathrm{d}r^{\alpha}_{ki}}{\mathrm{d}s_k} r^{\beta}_{ki} (r_{ki}\cdot\partial)^{n-1} \right\} \delta^{(4)}(R_k - R)\mathrm{d}s_k. \tag{6}$$

One can complete the second term at the right-hand side to the divergence of an antisymmetric tensor by subtracting a term of similar structure but with $r^{\alpha}_{ki}\mathrm{d}R^{\beta}_{k}/\mathrm{d}s_k$ instead of $(\mathrm{d}R^{\alpha}_{k}/\mathrm{d}s_k)r^{\beta}_{ki}$. If this extra term is partially integrated and then added to the last term, one obtains

$$\partial_\beta f^{\alpha\beta} = c^{-1} j^\alpha + \partial_\beta m^{\alpha\beta}. \tag{7}$$

This equation has already the form of Maxwell's inhomogeneous equations. It contains the antisymmetric tensor

$$m^{\alpha\beta} = \sum_{k,i} e_{ki} \sum_{n=1}^{\infty} \frac{(-1)^{n+1}}{n!} \int \left\{ r_{ki}^\alpha u_k^\beta - u_k^\alpha r_{ki}^\beta + \frac{n}{n+1} \left( r_{ki}^\alpha \frac{\mathrm{d}r_{ki}^\beta}{\mathrm{d}s_k} - \frac{\mathrm{d}r_{ki}^\alpha}{\mathrm{d}s_k} r_{ki}^\beta \right) \right\} (r_{ki} \cdot \partial)^{n-1} \delta^{(4)}(R_k - R) \mathrm{d}s_k, \tag{8}$$

which will be called the atomic polarization tensor.

The internal coordinates $r_{ki}^\alpha(s_k)$ enter this expression in certain combinations which we shall call the 'covariant electric and magnetic multipole moments', defined as

$$\mu_{(n)k}^{\alpha_1 \ldots \alpha_n} = \frac{1}{n!} \sum_i e_{ki} r_{ki}^{\alpha_1} \ldots r_{ki}^{\alpha_n}, \qquad (n = 1, 2, \ldots),$$

$$\nu_{(n)k}^{\alpha_1 \ldots \alpha_{n+1}} = c^{-1} \frac{n}{(n+1)!} \sum_i e_{ki} r_{ki}^{\alpha_1} \ldots r_{ki}^{\alpha_{n-1}} \left( r_{ki}^{\alpha_n} \frac{\mathrm{d}r_{ki}^{\alpha_{n+1}}}{\mathrm{d}s_k} - r_{ki}^{\alpha_{n+1}} \frac{\mathrm{d}r_{ki}^{\alpha_n}}{\mathrm{d}s_k} \right), \qquad (n = 1, 2, \ldots). \tag{9}$$

Their dependence on the time-like unit vector $n_k^\alpha(s_k)$ is indicated by the index $(n)$. With the help of these quantities the polarization tensor becomes

$$m^{\alpha\beta} = \sum_k \sum_{n=1}^{\infty} (-1)^{n+1} \int (\mu_{(n)k}^{\alpha_1 \ldots \alpha_{n-1}\alpha} u_k^\beta - \mu_{(n)k}^{\alpha_1 \ldots \alpha_{n-1}\beta} u_k^\alpha + c\nu_{(n)k}^{\alpha_1 \ldots \alpha_{n-1}\alpha\beta}) \partial_{\alpha_1 \ldots \alpha_{n-1}} \delta^{(4)}(R_k - R) \mathrm{d}s_k, \tag{10}$$

where we have written $\partial_{\alpha_1 \ldots \alpha_n}$ for $\partial_{\alpha_1} \partial_{\alpha_2} \ldots \partial_{\alpha_n}$. The combination of electric and magnetic multipole moments which occurs here will be employed frequently, and will be denoted as

$$\mathfrak{m}_{(n)k}^{\alpha_1 \ldots \alpha_{n+1}} \equiv c^{-1} (\mu_{(n)k}^{\alpha_1 \ldots \alpha_n} u_k^{\alpha_{n+1}} - \mu_{(n)k}^{\alpha_1 \ldots \alpha_{n-1}\alpha_{n+1}} u_k^{\alpha_n}) + \nu_{(n)k}^{\alpha_1 \ldots \alpha_{n+1}}. \tag{11}$$

In this way $m^{\alpha\beta}$ may be written in the compact form

$$m^{\alpha\beta} = c \sum_k \sum_{n=1}^{\infty} (-1)^{n+1} \int \mathfrak{m}_{(n)k}^{\alpha_1 \ldots \alpha_{n-1}\alpha\beta} \partial_{\alpha_1 \ldots \alpha_{n-1}} \delta^{(4)}(R_k - R) \mathrm{d}s_k. \tag{12}$$

The inhomogeneous equation (7) with (5) and (12), together with the homogeneous one

$$\partial^\alpha f^{\beta\gamma} + \partial^\beta f^{\gamma\alpha} + \partial^\gamma f^{\alpha\beta} = 0 \tag{13}$$

form the atomic equations[1] for the fields generated by atoms with charges $e_k$ and multipole moments $\mu_{(n)k}^{\alpha_1 \dots \alpha_n}$ and $\nu_{(n)k}^{\alpha_1 \dots \alpha_n}$ $(n = 1, 2, \dots)$. The retarded solution of (7) and (13) may be written in terms of the Green function (III.44), derived in the preceding chapter. One finds (cf. (III.59)):

$$f^{\alpha\beta} = \sum_k \int \hat{f}_k^{\alpha\beta} \mathrm{d}s_k = \sum_k \int (\hat{f}_{k(e)}^{\alpha\beta} + \hat{f}_{k(m)}^{\alpha\beta}) \mathrm{d}s_k , \tag{14}$$

where the contributions of the charges and multipoles are:

$$\hat{f}_{k(e)}^{\alpha\beta} \equiv -\frac{e_k}{2\pi} (u_k^\alpha \partial^\beta - u_k^\beta \partial^\alpha) \delta\{(R-R_k)^2\} \theta(R-R_k), \tag{15}$$

$$\hat{f}_{k(m)}^{\alpha\beta} \equiv -\sum_{n=1}^{\infty} \frac{(-1)^n c}{2\pi} \left\{ \mathrm{m}_{(n)k}^{\alpha_1 \dots \alpha_n \alpha} \partial^\beta_{.\alpha_1 \dots \alpha_n} - c^{-1} \left( \frac{\mathrm{d}}{\mathrm{d}s_k} - u_k \cdot \partial \right) \mu_k^{\alpha_1 \dots \alpha_{n-1} \alpha} \partial^\beta_{.\alpha_1 \dots \alpha_{n-1}} \right\}$$
$$\delta\{(R-R_k)^2\} \theta(R-R_k) - (\alpha, \beta). \tag{16}$$

Here $\mathrm{d}/\mathrm{d}s_k$ acts only on the electric multipole moment $\mu_k$. Furthermore $\theta(R-R_k)$ is the unit step function of $R^0 - R_k^0$. The symbol $(\alpha, \beta)$ indicates the preceding terms with $\alpha$ and $\beta$ interchanged. If (16) is inserted into (14), the second term gives no contribution since the integral over $s_k$ may be performed. The reason for retaining it is that (15) and (16) as they stand are together the multipole expanded form of the integrand of (III.59) (v. problem 4). This property will be useful later on.

It will be convenient to split the retarded field in the plus and minus fields (which are half the sum and half the difference of the retarded and advanced solutions respectively). The plus fields will satisfy an equation of the form (7) with (5) and (12), whereas the equation for the minus field will contain no sources. Since the advanced field has the same form as (14) but with $\theta(R_k - R)$ instead of $\theta(R-R_k)$, one finds then for the plus and minus fields

$$f_\pm^{\alpha\beta} = \sum_k \int \hat{f}_{\pm k}^{\alpha\beta} \mathrm{d}s_k = \sum_k \int (\hat{f}_{\pm k(e)}^{\alpha\beta} + \hat{f}_{\pm k(m)}^{\alpha\beta}) \mathrm{d}s_k , \tag{17}$$

where the contributions of the charges and multipoles are for the plus field

$$\hat{f}_{+k(e)}^{\alpha\beta} \equiv -\frac{e_k}{4\pi} (u_k^\alpha \partial^\beta - u_k^\beta \partial^\alpha) \delta\{(R-R_k)^2\}, \tag{18}$$

[1] A. N. Kaufman, Ann. Physics **18**(1962)264; H. Bacry, Ann. Physique **8**(1963)197; S. R. de Groot and L. G. Suttorp, Physica **31**(1965)1713; L. G. Suttorp, On the covariant derivation of macroscopic electrodynamics from electron theory, thesis, Amsterdam (1968).

$$\hat{f}^{\alpha\beta}_{+k(\mathrm{m})} \equiv -\sum_{n=1}^{\infty} \frac{(-1)^n c}{4\pi} \left\{ \mathfrak{m}^{\alpha_1 \ldots \alpha_n \alpha}_{(n)k} \partial^{\beta}_{.\alpha_1 \ldots \alpha_n} - c^{-1} \left( \frac{\mathrm{d}}{\mathrm{d}s_k} - u_k \cdot \partial \right) \mu_k^{\alpha_1 \ldots \alpha_{n-1} \alpha} \partial^{\beta}_{.\alpha_1 \ldots \alpha_{n-1}} \right\} \delta\{(R-R_k)^2\} - (\alpha, \beta), \quad (19)$$

while the minus fields contain an extra factor $\varepsilon(R-R_k) \equiv \theta(R-R_k) - \theta(R_k-R)$ after the delta functions. The partial fields fulfil the equations:

$$\partial_\beta \hat{f}^{\alpha\beta}_{+k(e)} = e_k u_k^\alpha \delta^{(4)}(R_k - R) + \frac{e_k}{4\pi} \partial^\alpha u_k \cdot \partial \delta\{(R-R_k)^2\}, \quad (20)$$

$$\partial_\beta \hat{f}^{\alpha\beta}_{+k(\mathrm{m})} = \sum_{n=1}^{\infty} (-1)^n c \left\{ \mathfrak{m}^{\alpha_1 \ldots \alpha_n \alpha}_{(n)k} \partial_{\alpha_1 \ldots \alpha_n} - c^{-1} \left( \frac{\mathrm{d}}{\mathrm{d}s_k} - u_k \cdot \partial \right) \mu_k^{\alpha_1 \ldots \alpha_{n-1} \alpha} \partial_{\alpha_1 \ldots \alpha_{n-1}} \right\} \delta^{(4)}(R_k - R)$$
$$+ \sum_{n=1}^{\infty} \frac{(-1)^{n-1}}{4\pi} \left( \frac{\mathrm{d}}{\mathrm{d}s_k} - u_k \cdot \partial \right) \mu_k^{\alpha_1 \ldots \alpha_n} \partial^{\alpha}_{.\alpha_1 \ldots \alpha_n} \delta\{(R-R_k)^2\}, \quad (21)$$

$$\partial_\beta \hat{f}^{\alpha\beta}_{-k(e)} = \frac{e_k}{4\pi} \partial^\alpha u_k \cdot \partial \delta\{(R-R_k)^2\} \varepsilon(R-R_k), \quad (22)$$

$$\partial_\beta \hat{f}^{\alpha\beta}_{-k(\mathrm{m})} = \sum_{n=1}^{\infty} \frac{(-1)^{n-1}}{4\pi} \left( \frac{\mathrm{d}}{\mathrm{d}s_k} - u_k \cdot \partial \right) \mu_k^{\alpha_1 \ldots \alpha_n} \partial^{\alpha}_{.\alpha_1 \ldots \alpha_n} \delta\{(R-R_k)^2\} \varepsilon(R-R_k). \quad (23)$$

If one integrates these four equations over $s_k$ only the first terms of the right-hand sides of (20) and (21) give contributions (v. (7) with (5) and (12)). The inhomogeneous atomic field equations may be written in an alternative form by introducing the atomic 'displacement tensor'

$$h^{\alpha\beta} \equiv f^{\alpha\beta} - m^{\alpha\beta}. \quad (24)$$

Then equation (7) becomes

$$\partial_\beta h^{\alpha\beta} = c^{-1} j^\alpha, \quad (25)$$

which has the same form as Maxwell's inhomogeneous equation.

Owing to the antisymmetry of $h^{\alpha\beta}$ equation (25) is consistent with the law of conservation of charge

$$\partial_\alpha j^\alpha = 0, \quad (26)$$

which may be proved from expression (5) for the four-current.

b. *Multipole moments*

The polarization tensor (10) contains the covariant multipole moments (9) which depend on the atomic internal coordinates $r_{ki}^{\alpha}$, measured in the observer's $(ct, \boldsymbol{R})$-frame. These covariant multipole moments have different values in different observer's Lorentz frames. They are therefore not constant properties which characterize the atoms. It is more convenient to characterize the internal electromagnetic structure of the atoms by means of parameters that are independent of the velocity of the atoms. This can be achieved with the help of parameters defined in a Lorentz frame in which the atom as a whole is at rest. Such an 'atomic frame' must have a (constant) velocity $\boldsymbol{v}$ equal to $(\mathrm{d}\boldsymbol{R}_k/\mathrm{d}t)_{t=t_0}$ (at a moment $t = t_0$) with respect to the reference frame $(ct, \boldsymbol{R})$ of the observer. Space-time coordinates of the reference frame $(ct, \boldsymbol{R})$ and of the atomic frame $(ct^{(0)}, \boldsymbol{R}^{(0)})$ are connected by a Lorentz transformation.

In the atomic frame the atom may be characterized by internal parameters $\boldsymbol{r}_{ki}^{(0)}$, which at the moment $t^{(0)} = t_0^{(0)}$ (corresponding to $t = t_0$ in the reference frame) are purely spatial vectors, i.e. $r_{ki}^{(0)0} = 0$. This corresponds to the choice $c^{-1}u_k^{\alpha}$ voor $n_k^{\alpha}$ as follows from (2) with (3). Since the atom suffers accelerations the atomic frame is just a momentary rest frame: only for $t = t_0$ does the atomic velocity $\mathrm{d}\boldsymbol{R}_k^{(0)}/\mathrm{d}t^{(0)}$ vanish. Hence one needs for a description in which the atomic parameters are independent of the velocity all the time a succession of momentary rest frames. This succession of Lorentz frames which is not a Lorentz frame itself will be called the permanent atomic rest frame (denoted by a prime). It coincides at time $t_0$ with the momentary atomic rest frame which has been denoted by (0).

The proper atomic multipole moments are certain useful combinations of the atomic internal parameters $\boldsymbol{r}'_{ki}$, defined in the permanent atomic frame[1]. The electric atomic $2^n$-pole moment is defined as[2]

$$\boldsymbol{\mu}_k^{(n)} = \frac{1}{n!}\sum_i e_{ki}(\boldsymbol{r}'_{ki})^n, \qquad (n = 1, 2, \ldots) \tag{27}$$

and the magnetic atomic $2^n$-pole moment as:

$$\boldsymbol{\nu}_k^{(n)} = \frac{n}{(n+1)!}\sum_i e_{ki}(\boldsymbol{r}'_{ki})^n \wedge \frac{\dot{\boldsymbol{r}}'_{ki}}{c}, \qquad (n = 1, 2, \ldots) \tag{28}$$

[1] A. N. Kaufman, op. cit. (dipole case), S. R. de Groot and J. Vlieger, Physica **31**(1965)125 (quadrupole case); S. R. de Groot and L. G. Suttorp, op. cit.; L. G. Suttorp, op. cit.

[2] For atoms in uniform motion these atomic multipole moments coincide with the space-space and space-time components of the covariant multipole moments (11) – with the unit vector $n^{\alpha}$ chosen as $u_k^{\alpha}/c$ – in the rest frame. If the atom suffers accelerations the relation between the covariant and atomic multipole moments is slightly more complicated (v. appendix II).

(the powers indicate polyads of three-vectors). The dot indicates a time derivative defined in a special way as explained in appendix II.

The polarization tensor, which in (10) is given in terms of the covariant multipole moments, may be written now as a function of the atomic multipole moments. The connexion between the two kinds of multipole moments follows from a Lorentz transformation (v. appendix II). With the help of the relations (A6), (A41) and (A43) one obtains, performing the integral over $s_k$, for the components $m^{i0} = p^i$ of the polarization tensor:

$$\begin{aligned}
\boldsymbol{p} = \sum_k \sum_{n=1}^{\infty} (-1)^{n+1} \Bigg[ & \sum_{p=0}^{n-1} \binom{n-1}{p} \partial_0^p \{ \boldsymbol{\Omega}_k (\gamma_k \boldsymbol{\beta}_k)^p (\nabla \cdot \boldsymbol{\Omega}_k^{-1})^{n-p-1} \vdots \boldsymbol{\mu}_k^{(n)} \delta(\boldsymbol{R}_k - \boldsymbol{R}) \} \\
& - \sum_{p=0}^{n-1} \binom{n-1}{p} \partial_0^p \{ (\gamma_k \boldsymbol{\beta}_k)^p (\nabla \cdot \boldsymbol{\Omega}_k^{-1})^{n-p-1} \vdots \boldsymbol{\nu}_k^{(n)} \wedge \boldsymbol{\beta}_k \delta(\boldsymbol{R}_k - \boldsymbol{R}) \} \\
& + \sum_{p=0}^{n-2} \binom{n-2}{p} \frac{(n-1)(n+1)}{n} \partial_0^p \{ (\gamma_k \boldsymbol{\beta}_k)^p (\nabla \cdot \boldsymbol{\Omega}_k^{-1})^{n-p-2} \gamma_k \partial_0 (\gamma_k \boldsymbol{\beta}_k) \cdot \boldsymbol{\Omega}_k \\
& \qquad \vdots \boldsymbol{\nu}_k^{(n)} \wedge \boldsymbol{\beta}_k \delta(\boldsymbol{R}_k - \boldsymbol{R}) \} \\
& - \sum_{p=0}^{n-2} \binom{n-2}{p} (n-1) \partial_0^p \{ \boldsymbol{\Omega}_k (\gamma_k \boldsymbol{\beta}_k)^p (\nabla \cdot \boldsymbol{\Omega}_k^{-1})^{n-p-2} \gamma_k \partial_0 (\gamma_k \boldsymbol{\beta}_k) \cdot \boldsymbol{\Omega}_k \\
& \qquad \vdots \boldsymbol{\mu}_k^{(n)} \delta(\boldsymbol{R}_k - \boldsymbol{R}) \} \Bigg], \qquad (29)
\end{aligned}$$

where the triple dot stands for an $n$- or $(n-1)$-fold contraction. This expression contains as external variables the position vector $\boldsymbol{R}_k$ of the atoms (entering only in the delta function), the velocity $\mathrm{d}\boldsymbol{R}_k/\mathrm{d}t \equiv \boldsymbol{\beta}_k c$ and higher time derivatives. Furthermore we used the abbreviation $\gamma_k = (1-\beta_k^2)^{-\frac{1}{2}}$ and the three-tensor ($\mathbf{U}$ is the unit three-tensor)

$$\boldsymbol{\Omega}_k = \mathbf{U} + (\gamma_k^{-1} - 1) \frac{\boldsymbol{\beta}_k \boldsymbol{\beta}_k}{\beta_k^2}, \qquad (30)$$

as well as its inverse

$$\boldsymbol{\Omega}_k^{-1} = \mathbf{U} + (\gamma_k - 1) \frac{\boldsymbol{\beta}_k \boldsymbol{\beta}_k}{\beta_k^2}. \qquad (31)$$

The three-tensor $\boldsymbol{\Omega}_k$ can be interpreted in terms of a Lorentz contraction, since for every three-vector $\boldsymbol{a} = \boldsymbol{a}_{//} + \boldsymbol{a}_{\perp}$ (split into a part parallel and a part perpendicular to the velocity $\boldsymbol{\beta}_k c$) one has according to definition (30)

$$\boldsymbol{\Omega}_k \cdot \boldsymbol{a} = \boldsymbol{a}_{\perp} + \sqrt{1-\beta_k^2}\, \boldsymbol{a}_{//}. \qquad (32)$$

This shows that the longitudinal component of the vector is subjected to a Lorentz contraction. In view of this property of $\boldsymbol{\Omega}_k$ we eliminate $\boldsymbol{\Omega}_k^{-1}$ from

(29) by means of the identity

$$\boldsymbol{\Omega}_k^{-1} = \boldsymbol{\Omega}_k + \gamma_k \boldsymbol{\beta}_k \boldsymbol{\beta}_k, \tag{33}$$

which is a consequence of (30) and (31). In the result for the polarization the atomic multipole moments then occur contracted with the tensor $\boldsymbol{\Omega}_k$. These quantities will be denoted with underlined symbols:

$$\underline{\boldsymbol{\mu}}_k^{(n)} \equiv \frac{1}{n!} \sum_i e_{ki} (\boldsymbol{\Omega}_k \cdot \boldsymbol{r}'_{ki})^n, \qquad (n = 1, 2, \ldots), \tag{34}$$

$$\underline{\boldsymbol{\nu}}_k^{(n)} \equiv \frac{n}{(n+1)!} \sum_i e_{ki} (\boldsymbol{\Omega}_k \cdot \boldsymbol{r}'_{ki})^{n-1} \boldsymbol{\Omega}_k \cdot \left( \boldsymbol{r}'_{ki} \wedge \frac{\dot{\boldsymbol{r}}'_{ki}}{c} \right), \qquad (n = 1, 2, \ldots). \tag{35}$$

According to (32) they represent atomic electromagnetic multipole moments of which the longitudinal components are submitted to a Lorentz contraction. With the help of (34) and (35) the electric polarization (29) becomes finally:

$$\begin{aligned} \boldsymbol{p} = \sum_k \sum_{n=1}^{\infty} \sum_{p=0}^{n-1} \sum_{q=0}^{p} \frac{(-1)^{n-1}(n-1)!}{(n-1-p)!(p-q)!} \\ \boldsymbol{\nabla}^{n-1-p} \partial_0^{p-q-1} [(\gamma_k \boldsymbol{\beta}_k)^{p-q} \partial_0 \{\gamma_k^p (\gamma_k \boldsymbol{\beta}_k)^q \vdots (\underline{\boldsymbol{\mu}}_k^{(n)} - \underline{\boldsymbol{\nu}}_k^{(n)} \wedge \boldsymbol{\beta}_k)\}] D_q \delta(\boldsymbol{R}_k - \boldsymbol{R}) \\ + \sum_k \sum_{n=2}^{\infty} \sum_{p=1}^{n-1} \sum_{q=0}^{p-1} \frac{(-1)^{n-1}(n-1)!}{n(n-1-p)!(p-q)!} \\ \boldsymbol{\nabla}^{n-1-p} \partial_0^{p-q-1} [\{\partial_0 (\gamma_k \boldsymbol{\beta}_k)^{p-q}\} \{\gamma_k^p (\gamma_k \boldsymbol{\beta}_k)^q \vdots (\underline{\boldsymbol{\nu}}_k^{(n)} \wedge \boldsymbol{\beta}_k)\}] D_q \delta(\boldsymbol{R}_k - \boldsymbol{R}). \end{aligned} \tag{36}$$

The nabla operator differentiates $\boldsymbol{R}$, which occurs only in the delta function. Furthermore the product $\partial_0^{-1}\partial_0$, which occurs if $p = q$ in the first sum, is to be considered as unity. The triple dot stands for an $(n-1)$-fold contraction. The symbol $D_q$ is defined by:

$$D_q \delta(\boldsymbol{R}_k - \boldsymbol{R}) \equiv \sum_{m=0}^{q} \frac{1}{m!(q-m)!} \partial_0^{q-m} \{(\boldsymbol{\beta}_k \cdot \boldsymbol{\nabla})^m \delta(\boldsymbol{R}_k - \boldsymbol{R})\}. \tag{37}$$

The magnetization $m^i = m^{jk}$ $(i, j, k = 1, 2, 3$ cycl.) follows by substitution of (A41) and (A42) into (10). Then one obtains an expression which is similar to (36) but with the replacements

$$\begin{aligned} \underline{\boldsymbol{\mu}}_k^{(n)} &\to \underline{\boldsymbol{\mu}}_k^{(n)} \wedge \boldsymbol{\beta}_k, \\ \underline{\boldsymbol{\nu}}_k^{(n)} \wedge \boldsymbol{\beta}_k &\to -\underline{\boldsymbol{\nu}}_k^{(n)}. \end{aligned} \tag{38}$$

This shows a symmetry between the components $\boldsymbol{p}$ and $\boldsymbol{m}$ of the polarization tensor $m^{\alpha\beta}$.

The polarization tensor obtained so far contains time derivatives of the delta function as is apparent from (37). They can be expressed in terms of spatial derivatives, because

$$\partial_0 \delta\{\boldsymbol{R}_k(t)-\boldsymbol{R}\} = -\boldsymbol{\beta}_k \cdot \nabla \delta\{\boldsymbol{R}_k(t)-\boldsymbol{R}\}. \tag{39}$$

By means of the chain rule of differentiation one then obtains a sum of spatial derivatives of the delta function, where each term has a factor in front of it, which may contain $\boldsymbol{\beta}_k$ and time derivatives of $\boldsymbol{\beta}_k$. For the values $q = 0$ and 1 one finds directly from (37)

$$D_0 = 1, \qquad D_1 = 0. \tag{40}$$

For dipoles ($n = 1$) and quadrupoles ($n = 2$) one only needs these values.

c. *The field equations*

The atomic field equations (7) and (13) read in three-dimensional notation

$$\begin{aligned} \nabla \cdot \boldsymbol{e} &= \rho^{\mathrm{e}} - \nabla \cdot \boldsymbol{p}, \\ -\partial_0 \boldsymbol{e} + \nabla \wedge \boldsymbol{b} &= \boldsymbol{j}/c + \partial_0 \boldsymbol{p} + \nabla \wedge \boldsymbol{m}, \\ \nabla \cdot \boldsymbol{b} &= 0, \\ \partial_0 \boldsymbol{b} + \nabla \wedge \boldsymbol{e} &= 0. \end{aligned} \tag{41}$$

Here $e^i = f^{0i}$, $b^i = f^{jk}$, $\rho^{\mathrm{e}} = j^0/c$, $j^i = j^i$, $p^i = -m^{0i}$ and $m^i = m^{jk}$ ($i, j, k = 1, 2, 3$ cycl.). These atomic field equations have the same form as Maxwell's equations. Their source terms include the atomic charge density

$$\rho^{\mathrm{e}} = \sum_k e_k \delta(\boldsymbol{R}_k - \boldsymbol{R}) \tag{42}$$

and the atomic current density

$$\boldsymbol{j} = \sum_k e_k \boldsymbol{\beta}_k c \delta(\boldsymbol{R}_k - \boldsymbol{R}), \tag{43}$$

as follows from (5). They satisfy the conservation law of charge (26) which reads in three-dimensional notation:

$$\frac{\partial \rho^{\mathrm{e}}}{\partial t} + \nabla \cdot \boldsymbol{j} = 0. \tag{44}$$

Furthermore the polarization vector is given in (36), while the magnetization follows with the replacement rules (38). They are sums of multipole contributions of various order $n$:

$$\boldsymbol{p} = \sum_{n=1}^{\infty} \boldsymbol{p}^{(n)}, \qquad \boldsymbol{m} = \sum_{n=1}^{\infty} \boldsymbol{m}^{(n)}, \tag{45}$$

of which the lowest orders will be given and discussed in the next subsection.

The sources $\rho^e$, $\boldsymbol{j}$, $\boldsymbol{p}$ and $\boldsymbol{m}$ which are functions of $\boldsymbol{R}$ and $t$ are all sums over the separate atoms $k$. They contain the internal atomic quantities $e_k$ (which are scalars), $\boldsymbol{\mu}_k^{(n)}$ and $\boldsymbol{\nu}_k^{(n)}$ (which are defined in the atomic frames). They also depend on the external quantities of the atoms, which are defined in the observer's $(ct, \boldsymbol{R})$-frame. The external quantities include the atomic positions, velocities and higher time derivatives of these. Time derivatives of the internal quantities $\boldsymbol{\mu}_k^{(n)}$ and $\boldsymbol{\nu}_k^{(n)}$ also occur. The atomic field equations are still equations for the microscopic fields $\boldsymbol{e}$ and $\boldsymbol{b}$, in which however the existence of stable groups of point particles is taken into account. They contain internal and external quantities, all referring to single atoms. Thus the atomic field equations can be said to be valid at the so-called *kinetic level* of the theory.

An alternative form of the atomic field equations follows by introducing the displacement vectors $d^i = h^{0i}$ and $h^i = h^{jk}$ $(i, j, k = 1, 2, 3$ cycl.) that form part of the displacement tensor (24). The latter definition reads for the components:

$$\boldsymbol{d} = \boldsymbol{e}+\boldsymbol{p}, \qquad \boldsymbol{h} = \boldsymbol{b}-\boldsymbol{m}, \tag{46}$$

so that one has for (41):

$$\begin{aligned} \nabla\cdot\boldsymbol{d} &= \rho^e, \\ -\partial_0\boldsymbol{d}+\nabla\wedge\boldsymbol{h} &= \boldsymbol{j}/c, \\ \nabla\cdot\boldsymbol{b} &= 0, \\ \partial_0\boldsymbol{b}+\nabla\wedge\boldsymbol{e} &= 0. \end{aligned} \tag{47}$$

These alternative atomic field equations have the form in which the macroscopic Maxwell equations are usually written.

### d. *Explicit expressions for the polarization tensor*

In this subsection explicit expressions for a few special cases of the polarization tensor will be written down. The contributions from the electric and magnetic dipole moments ($n = 1$) that follow from (36–40) are:

$$\boldsymbol{p}^{(1)} = \sum_k (\underline{\boldsymbol{\mu}}_k^{(1)}-\underline{\boldsymbol{\nu}}_k^{(1)}\wedge\boldsymbol{\beta}_k)\delta(\boldsymbol{R}_k-\boldsymbol{R}), \tag{48}$$

$$\boldsymbol{m}^{(1)} = \sum_k (\underline{\boldsymbol{\nu}}_k^{(1)}+\underline{\boldsymbol{\mu}}_k^{(1)}\wedge\boldsymbol{\beta}_k)\delta(\boldsymbol{R}_k-\boldsymbol{R}). \tag{49}$$

Here the motion of the atom gives rise to two separate effects. In the first place the electric polarization (48) contains a term due to the magnetic dipole moment and similarly the magnetic polarization (49) a term due to the electric dipole moment. Furthermore the dipoles are subject to a Lorentz contraction, since $\underline{\boldsymbol{\mu}}_k^{(1)}$ and $\underline{\boldsymbol{\nu}}_k^{(1)}$ can be rewritten, if use is made of the property (32) of the $\boldsymbol{\Omega}_k$ tensor:

$$\underline{\boldsymbol{\mu}}_k^{(1)} \equiv \boldsymbol{\Omega}_k \cdot \boldsymbol{\mu}_k^{(1)} = \boldsymbol{\mu}_{k,\perp}^{(1)} + \sqrt{1-\beta_k^2}\, \boldsymbol{\mu}_{k,//}^{(1)}, \tag{50}$$

$$\underline{\boldsymbol{\nu}}_k^{(1)} \equiv \boldsymbol{\Omega}_k \cdot \boldsymbol{\nu}_k^{(1)} = \boldsymbol{\nu}_{k,\perp}^{(1)} + \sqrt{1-\beta_k^2}\, \boldsymbol{\nu}_{k,//}^{(1)}, \tag{51}$$

where $\boldsymbol{\mu}_k^{(1)}$ and $\boldsymbol{\nu}_k^{(1)}$ are split into a part parallel with and a part perpendicular to the velocity $c\boldsymbol{\beta}_k$.

The terms $\boldsymbol{p}^{(2)}$ and $\boldsymbol{m}^{(2)}$ with electric and magnetic quadrupoles ($n = 2$) read:

$$\boldsymbol{p}^{(2)} = \sum_k [-\nabla \cdot (\underline{\boldsymbol{\mu}}_k^{(2)} - \underline{\boldsymbol{\nu}}_k^{(2)} \wedge \boldsymbol{\beta}_k) - \gamma_k \boldsymbol{\beta}_k \cdot \partial_0 \{\gamma_k (\underline{\boldsymbol{\mu}}_k^{(2)} - \underline{\boldsymbol{\nu}}_k^{(2)} \wedge \boldsymbol{\beta}_k)\} \\ - \tfrac{1}{2}\gamma_k\, \partial_0(\gamma_k \boldsymbol{\beta}_k) \cdot \underline{\boldsymbol{\nu}}_k^{(2)} \wedge \boldsymbol{\beta}_k]\, \delta(\boldsymbol{R}_k - \boldsymbol{R}), \tag{52}$$

$$\boldsymbol{m}^{(2)} = \sum_k [-\nabla \cdot (\underline{\boldsymbol{\nu}}_k^{(2)} + \underline{\boldsymbol{\mu}}_k^{(2)} \wedge \boldsymbol{\beta}_k) - \gamma_k \boldsymbol{\beta}_k \cdot \partial_0 \{\gamma_k (\underline{\boldsymbol{\nu}}_k^{(2)} + \underline{\boldsymbol{\mu}}_k^{(2)} \wedge \boldsymbol{\beta}_k)\} \\ + \tfrac{1}{2}\gamma_k\, \partial_0(\gamma_k \boldsymbol{\beta}_k) \cdot \underline{\boldsymbol{\nu}}_k^{(2)}]\, \delta(\boldsymbol{R}_k - \boldsymbol{R}). \tag{53}$$

Due to the motion of the atoms the electric and magnetic quadrupoles are subject to a Lorentz contraction since we can write:

$$\underline{\boldsymbol{\mu}}_k^{(2)} \equiv \boldsymbol{\Omega}_k \cdot \boldsymbol{\mu}_k^{(2)} \cdot \boldsymbol{\Omega}_k = \boldsymbol{\mu}_{k,\perp\perp}^{(2)} + \sqrt{1-\beta_k^2}\,(\boldsymbol{\mu}_{k,\perp//}^{(2)} + \boldsymbol{\mu}_{k,//\perp}^{(2)}) + (1-\beta_k^2)\boldsymbol{\mu}_{k,////}^{(2)}, \tag{54}$$

and an analogous equation for $\underline{\boldsymbol{\nu}}_k^{(2)}$. The leading terms of (52) and (53) contain the divergence of the electric and magnetic quadrupole densities. The second and third terms are relativistic corrections. They contain time derivatives of the atomic quadrupole moments which means that changing quadrupole moments contribute in a special way to the polarization tensor. This effect, which has a relativistic character, may be called the 'multipole fluxion effect'. Accelerated atoms carrying quadrupole moments also give rise to a special type of terms in the polarization tensor. This effect, which again does not exist in a non-relativistic theory may be called the 'acceleration effect'. The order of magnitude of these effects as compared to the main contributions in the polarization tensor will be discussed in the chapter on macroscopic theory.

Higher order multipole contributions may also be derived from (36–39). We shall not give these expressions explicitly.

The polarization tensor $(\boldsymbol{p}_k, \boldsymbol{m}_k)$ of a single atom $k$ to all multipole orders takes a simple form in the momentary atomic rest frame $(ct^{(0)}, \boldsymbol{R}^{(0)})$. In fact $\boldsymbol{p}_k^{(0)}$ and $\boldsymbol{m}_k^{(0)}$ follow from (36) with (37) and (39) (without the summation over $k$) and the corresponding expression obtained with (38), if one puts $\boldsymbol{\beta}_k^{(0)} = 0$. Then only $p = 0$ and $q = 0$ subsist in the leading term. The other term in (36) disappears altogether, whereas in $\boldsymbol{m}_k$ only its part with $q = 0$ is left over. Thus one gets

$$\boldsymbol{p}_k^{(0)} = \sum_{n=1}^{\infty} (-1)^{n-1} (\nabla^{(0)})^{n-1} : \boldsymbol{\mu}_k^{(n)} \delta(\boldsymbol{R}_k^{(0)} - \boldsymbol{R}^{(0)}), \tag{55}$$

$$\boldsymbol{m}_k^{(0)} = \sum_{n=1}^{\infty} (-1)^{n-1} (\nabla^{(0)})^{n-1} : \boldsymbol{\nu}_k^{(n)} \delta(\boldsymbol{R}_k^{(0)} - \boldsymbol{R}^{(0)})$$
$$- \sum_{n=2}^{\infty} \sum_{p=1}^{n-1} \frac{(-1)^{n-1}(n-1)!}{n(n-1-p)!} (\nabla^{(0)})^{n-1-p} (\partial_0^{(0)} \boldsymbol{\beta}_k^{(0)})^p : \boldsymbol{\nu}_k^{(n)} \delta(\boldsymbol{R}_k^{(0)} - \boldsymbol{R}^{(0)}). \tag{56}$$

Apart from the leading terms, which are divergences of multipole densities, a term representing the acceleration effect (with quadrupoles and higher moments) appears in the magnetization vector for an atom in its momentary atomic rest frame.

### e. *The non-relativistic and semi-relativistic limits*

In the so-called semi-relativistic approximation one retains terms of order $c^{-1}$, treating the atomic multipole moments as parameters characterizing the atom, without considering whether they contain a factor $c^{-1}$. Then, if terms of order $c^{-2}$ and higher are neglected in (36) with (37–39) so that $\gamma_k \simeq 1$ and $\boldsymbol{\Omega}_k \simeq \mathbf{U}$, we find:

$$\boldsymbol{p}_k(\boldsymbol{R}, t) \simeq \sum_{n=1}^{\infty} (-1)^{n-1} \nabla^{n-1} : (\boldsymbol{\mu}_k^{(n)} - \boldsymbol{\nu}_k^{(n)} \wedge \boldsymbol{\beta}_k) \delta(\boldsymbol{R}_k - \boldsymbol{R}), \tag{57}$$

$$\boldsymbol{m}_k(\boldsymbol{R}, t) \simeq \sum_{n=1}^{\infty} (-1)^{n-1} \nabla^{n-1} : (\boldsymbol{\nu}_k^{(n)} + \boldsymbol{\mu}_k^{(n)} \wedge \boldsymbol{\beta}_k) \delta(\boldsymbol{R}_k - \boldsymbol{R}). \tag{58}$$

These formulae show a symmetry in the sense that $\boldsymbol{p}_k$ contains terms due to moving magnetic multipoles $\boldsymbol{\nu}_k^{(n)}$, just as $\boldsymbol{m}_k$ contains contributions of the same type from moving electric multipoles $\boldsymbol{\mu}_k^{(n)}$.

The non-relativistic limiting case is obtained if one takes into account the fact that the magnetic multipoles contain a factor $c^{-1}$. Then, up to terms of order $c^{-1}$, one is left with (cf. (I.34)):

$$\boldsymbol{p}_k(\boldsymbol{R}, t) \simeq \sum_{n=1}^{\infty} (-1)^{n-1} \nabla^{n-1} : \boldsymbol{\mu}_k^{(n)} \delta(\boldsymbol{R}_k - \boldsymbol{R}), \tag{59}$$

$$\boldsymbol{m}_k(\boldsymbol{R}, t) \simeq \sum_{n=1}^{\infty} (-1)^{n-1} \nabla^{n-1} : (\boldsymbol{\nu}_k^{(n)} + \boldsymbol{\mu}_k^{(n)} \wedge \boldsymbol{\beta}_k) \delta(\boldsymbol{R}_k - \boldsymbol{R}). \tag{60}$$

The symmetry of (57) and (58) is lost now, since $\boldsymbol{p}_k$ (59) contains no terms with moving magnetic multipoles.

## 3 *The equations of motion of a composite particle in a field*

### a. *Introduction*

To obtain equations of motion of composite particles we shall have to start from the corresponding equations for the constituent particles. In the preceding we have written the equations of motion for a set of point particles in the form (III.158):

$$\partial_\beta t^{\alpha\beta} = f^\alpha. \tag{61}$$

Here $t^{\alpha\beta}$ is the (symmetric) energy–momentum tensor, which is the sum of a material contribution and a field contribution. The latter contains the fields generated by the constituent particles. It has been given in (III.159) as

$$t^{\alpha\beta} = c\sum_i m_i \int u_i^\alpha u_i^\beta \delta^{(4)}(R_i - R)\mathrm{d}s_i + \sum_{i,j(i\neq j)} (f_i^{\alpha\gamma} f_{j.\gamma}^{\beta} - \tfrac{1}{4} f_i^{\gamma\varepsilon} f_{j\gamma\varepsilon}\, g^{\alpha\beta})$$
$$+ \sum_i (f_{+i}^{\alpha\gamma} f_{-i.\gamma}^{\beta} + f_{-i}^{\alpha\gamma} f_{+i.\gamma}^{\beta} - \tfrac{1}{2} f_{+i}^{\gamma\varepsilon} f_{-i\gamma\varepsilon}\, g^{\alpha\beta}). \tag{62}$$

At the right-hand side appears the force density (III.160), which is the sum of the Lorentz forces acting on the individual particles:

$$f^\alpha = \sum_i e_i \int F^{\alpha\beta} u_{i\beta}\, \delta^{(4)}(R_i - R)\mathrm{d}s_i\,. \tag{63}$$

Starting from the energy–momentum balance given above we shall derive the equations of motion of a composite particle in an electromagnetic field. For that purpose we shall first have to define a covariant centre of energy. Subsequently the equations of motion will be obtained for charged composite particles which carry electromagnetic multipoles[1].

[1] The treatment will follow the derivation given in L. G. Suttorp and S. R. de Groot, N. Cim. **65A**(1970)245; v. also W. G. Dixon, N. Cim. **34**(1964)317; **38**(1965)1616. A treatment with a different definition of the centre of energy from the one used in the following was given in S.R. de Groot and L.G. Suttorp, Physica **37**(1967)284, 297; **39**(1968)84; L.G. Suttorp, On the covariant derivation of macroscopic electrodynamics from electron theory, thesis, Amsterdam (1968). Often equations have been postulated, i.e., either obtained from variational principles *ad hoc* or generalized from non-relativistic theory, e.g. J. Frenkel, Z. Phys. **37**(1926)243; V. Bargmann, L. Michel and V. L. Telegdi, Phys. Rev. Lett. **2**(1959) 435.

Part of this programme may also be performed with the use of the alternative energy–momentum balance discussed in the appendix of the preceding chapter. This will be done in an appendix to the present chapter.

### b. *Definition of a covariant centre of energy*

In this subsection we shall be concerned with the general definition of a centre of energy for systems described by a symmetric energy–momentum tensor $t^{\alpha\beta}$, that fulfils a balance equation:

$$\partial_\beta t^{\alpha\beta} = f^\alpha, \tag{64}$$

with $f^\alpha$ a force density that has a finite support in space-like directions. The general scheme will subsequently be applied for the tensor $t^{\alpha\beta}$ of the preceding subsection.

Let us consider systems for which the total momentum $p^\alpha$ over a plane space-like surface $\Sigma$ with normal $n^\alpha$

$$p^\alpha = -c^{-1}\int_\Sigma t^{\alpha\beta} n_\beta \,\mathrm{d}^3\Sigma \tag{65}$$

is a finite time-like vector, with a positive time-component (then the total energy of the system is positive). The tensor $t^{\alpha\beta}$ will be supposed to diminish in space-like directions with increasing distances in such a way that this integral (and those which we shall need later) are (semi-)convergent. A covariant centre of energy may then be defined[1] by considering those plane surfaces $\Sigma$ of which the normal $n^\alpha$ is parallel to $p^\alpha$. In these surfaces one then determines the centre of energy

$$X^\alpha = \frac{\int_\Sigma R^\alpha n_\beta t^{\beta\gamma} n_\gamma \,\mathrm{d}^3\Sigma}{\int_\Sigma n_\varepsilon t^{\varepsilon\zeta} n_\zeta \,\mathrm{d}^3\Sigma}. \tag{66}$$

(In the rest frame of $p^\alpha$ this formula reads indeed $\boldsymbol{X} = \int \boldsymbol{R} t^{00}\mathrm{d}\boldsymbol{R}/\int t^{00}\mathrm{d}\boldsymbol{R}$.) These centres of energy then satisfy the relation

$$p_\alpha s^{\alpha\beta} = 0, \tag{67}$$

where the inner angular momentum is

$$s^{\alpha\beta} = -c^{-1}\int_\Sigma \{(R-X)^\alpha t^{\beta\gamma} - (R-X)^\beta t^{\alpha\gamma}\} n_\gamma \,\mathrm{d}^3\Sigma, \tag{68}$$

assumed to be a finite quantity for the system under consideration.

[1] T. Nakano, Progr. Theor. Phys. **15**(1956)333; W. Tulczyjew, Acta Phys. Polon. **18**(1959) 393; W. G. Dixon, N. Cim. **34**(1964)317.

We shall prove now that the set of centres of energy defined in this way forms one single world line (or several discrete ones). Consider such a point $X^\alpha$ determined in a plane surface $\Sigma$ with normal parallel to $p^\alpha$. One may ask oneself now if there exists a point $X^\alpha + \delta X^\alpha$ (with $p_\alpha \, \delta X^\alpha = 0$) in the infinitesimal neighbourhood of $X^\alpha$, which is likewise a centre of energy, this time in a plane surface $\Sigma'$ with normal parallel to the corresponding momentum $p^\alpha + \delta p^\alpha$. In the proper frame of $p^\alpha$ one has from (65), (67) and (68)

$$\int_\Sigma t^{i0}(R^0, \boldsymbol{R})\mathrm{d}\boldsymbol{R} = 0 \qquad (R^0 = X^0), \tag{69}$$

$$\int_\Sigma (\boldsymbol{R}-\boldsymbol{X})t^{00}(R^0, \boldsymbol{R})\mathrm{d}\boldsymbol{R} = 0. \tag{70}$$

The proper frame of $p^\alpha + \delta p^\alpha$ is connected to the proper frame of $p^\alpha$ by an infinitesimal pure Lorentz transformation:

$$R^{0\prime} = R^0 + \boldsymbol{\varepsilon}\cdot\boldsymbol{R}, \qquad \boldsymbol{R}' = \boldsymbol{R} + \boldsymbol{\varepsilon}R^0 \tag{71}$$

for a certain value of $\boldsymbol{\varepsilon}$. The time-space point with coordinates $(R^0, \boldsymbol{X} + \delta\boldsymbol{X})$ has in the new frame the coordinates $(\hat{R}^{0\prime}, \boldsymbol{X}' + \delta\boldsymbol{X}')$ which are given by

$$\hat{R}^{0\prime} = R^0 + \boldsymbol{\varepsilon}\cdot\boldsymbol{X}, \tag{72}$$

$$\boldsymbol{X}' + \delta\boldsymbol{X}' = \boldsymbol{X} + \delta\boldsymbol{X} + \boldsymbol{\varepsilon}R^0 \tag{73}$$

up to terms linear in $\boldsymbol{\varepsilon}$ and $\delta\boldsymbol{X}$. Furthermore the space coordinates of an arbitrary point in $\Sigma'$ read in the new frame

$$\hat{\boldsymbol{R}}' = \hat{\boldsymbol{R}} + \boldsymbol{\varepsilon}\hat{R}^0, \tag{74}$$

where $\hat{\boldsymbol{R}}$ and $\hat{R}^0$ are connected by

$$\hat{R}^{0\prime} = \hat{R}^0 + \boldsymbol{\varepsilon}\cdot\hat{\boldsymbol{R}}. \tag{75}$$

From (72), (74) and (75) it follows now that

$$\hat{R}^0 = R^0 + \boldsymbol{\varepsilon}\cdot(\boldsymbol{X} - \hat{\boldsymbol{R}}'), \tag{76}$$

$$\hat{\boldsymbol{R}} = \hat{\boldsymbol{R}}' - \boldsymbol{\varepsilon}R^0. \tag{77}$$

In the proper frame of $p^\alpha + \delta p^\alpha$ the space components of the total momentum vanish (cf. (69)):

$$\int_{\Sigma'} t^{i0\prime}(\hat{R}^{0\prime}, \hat{\boldsymbol{R}}')\mathrm{d}\hat{\boldsymbol{R}}' = 0. \tag{78}$$

Using the transformation properties of a tensor one gets

$$\int_{\Sigma'} \{t^{i0}(\hat{R}^0, \hat{\boldsymbol{R}}) + \varepsilon^i t^{00}(\hat{R}^0, \hat{\boldsymbol{R}}) + \varepsilon_j t^{ij}(\hat{R}^0, \hat{\boldsymbol{R}})\} \mathrm{d}\hat{\boldsymbol{R}}' = 0. \tag{79}$$

Introducing (76) and (77) we obtain ($\partial_0 \equiv \partial/\partial \boldsymbol{R}^0$):

$$\int \{t^{i0}(R^0, \hat{\boldsymbol{R}}') - R^0 \boldsymbol{\varepsilon}\cdot(\partial/\partial\hat{\boldsymbol{R}}') t^{i0}(R^0, \hat{\boldsymbol{R}}') + \boldsymbol{\varepsilon}\cdot(\boldsymbol{X} - \hat{\boldsymbol{R}}')\partial_0 t^{i0}(R^0, \hat{\boldsymbol{R}}') + \varepsilon^i t^{00}(R^0, \hat{\boldsymbol{R}}') + \varepsilon_j t^{ij}(R^0, \hat{\boldsymbol{R}}')\} \mathrm{d}\hat{\boldsymbol{R}}' = 0. \tag{80}$$

With (69) it follows that the first term vanishes. From (72) and the fact that $\hat{R}^{0\prime}$ is constant in the integration it follows that $R^0$ is constant so that the second term gives no contribution either. Therefore we have:

$$\int \{\boldsymbol{\varepsilon}\cdot(\boldsymbol{X} - \hat{\boldsymbol{R}}')\partial_0 t^{i0}(R^0, \hat{\boldsymbol{R}}') + \varepsilon^i t^{00}(R^0, \hat{\boldsymbol{R}}') + \varepsilon_j t^{ij}(R^0, \hat{\boldsymbol{R}}')\} \mathrm{d}\hat{\boldsymbol{R}}' = 0. \tag{81}$$

With the equation of motion (64) and a partial integration this becomes

$$\int \{\boldsymbol{\varepsilon}\cdot(\boldsymbol{X} - \boldsymbol{R})\boldsymbol{f}(R^0, \boldsymbol{R}) + \boldsymbol{\varepsilon} t^{00}(R^0, \boldsymbol{R})\} \mathrm{d}\boldsymbol{R} = 0. \tag{82}$$

The first integral is in fact extended over a finite support since $f^\alpha$ has a finite support. Hence for sufficiently small force densities one has

$$\left| \boldsymbol{\varepsilon}\cdot \int (\boldsymbol{X} - \boldsymbol{R})\boldsymbol{f}(R^0, \boldsymbol{R})\mathrm{d}\boldsymbol{R} \right| < \left| \boldsymbol{\varepsilon} \int t^{00} \mathrm{d}\boldsymbol{R} \right|, \tag{83}$$

so that (82) cannot be satisfied. The conclusion is that no point exists in the infinitesimal neighbourhood of $X^\alpha$, which is also a centre of energy; in other words the set of centres of energy determines a discrete number of world lines.

The condition (67) thus leads to a situation completely different from that following from the condition $u_\alpha s^{\alpha\beta} = 0$ (with $u^\alpha = \mathrm{d}X^\alpha/\mathrm{d}s$ where $s$ is the proper time). As a matter of fact Møller proved[1] that the latter condition does not suffice to determine a world line. Moreover he showed that it cannot be supplemented by the requirement that $p^\alpha$ be parallel to $u^\alpha$ (in the

[1] C. Møller, Ann. Inst. H. Poincaré **11**(1949)251.

general case $f^{\alpha} \neq 0$)[1]. Nevertheless this condition is sometimes used[2], although it leads to peculiar solutions: even in the force-free case a special type of helical motions is possible. (In order to avoid this difficulty the form of the derived equations is sometimes[3] changed *ad hoc* by means of a so-called 'iteration process'.)

The general prescription for the construction of a centre of energy as given above will be applied now to the case of a composite particle (consisting of charged point particles) that moves in an external electromagnetic field. As we saw in the preceding subsection such a system may be described by means of the energy–momentum tensor (62). In order to be able to apply the construction of the energy centre given above, we must check whether all assumptions used there are justified. To begin with, the energy–momentum tensor $t^{\alpha\beta}$ (62) is indeed symmetric and the force density (63) has indeed a finite support in space-like directions. The next point to discuss is the convergence of the integrals, in particular of (65) and (68) for $p^{\alpha}$ and $s^{\alpha\beta}$. (The convergence of all other integrals occurring is determined by the latter two.) The integrals contain $t^{\alpha\beta}$, which is a quadratic function of the fields $f^{\alpha\beta}$ of which the behaviour for small and for large space-like distances follows from (III.110,111) and (III.96,98). The first two formulae show that the convergence at short distances presents no difficulties (note the presence of the condition $i \neq j$ in the second term). The latter formulae indicate that for large space-like distances the fields diminish inversely proportionally to those distances, so that the energy–momentum tensor diminishes only with the square of the distances. As a consequence the integral $p^{\alpha}$ (65) would diverge if no subsidiary conditions on the fields are imposed. This is a reflection of the fact that the total energy stored in the electromagnetic field would be infinitely great if the particles have been suffering accelerations from infinitely past to infinitely future times. Here we hit a well-known difficulty of classical theory. In such a theory a composite particle is not

[1] For the free composite particle centres of energy have been defined and discussed already by A. D. Fokker, Relativiteitstheorie (Noordhoff, Groningen 1929) 170; M. H. L. Pryce, Proc. Roy. Soc. **A195**(1949)62.

[2] H. Hönl and A. Papapetrou, Z. Phys. **112**(1939)512; **116**(1940)153; M. Mathisson, Proc. Cambr. Phil. Soc. **36**(1940)331; **38**(1942)40; H. J. Bhabha and H. C. Corben, Proc. Roy. Soc. A **178**(1941)273; J. Weyssenhoff and A. Raabe, Acta Phys. Polon. **9**(1947)7, 19, 26, 34, 46; H. C. Corben, N. Cim. **20**(1961)529; Phys. Rev. **121**(1961)1833; A. Białas, Acta Phys. Polon. **22**(1962)499; P. Nyborg, N. Cim. **31**(1964)1209; **32**(1964)1131; W. G. Dixon, J. Math. Phys. **8**(1967)1591; J. Vlieger, Physica **37**(1967)165.

[3] E. Plahte, Suppl. N. Cim. **4**(1966)291; J. Vlieger and S. Emid, Physica **41**(1969)368; S. Emid and J. Vlieger, Physica **52**(1971)329.

stable against radiation: as a consequence of the emitted radiation energy and momentum is lost unrestrictedly. This is a paradox if the total energy content in the initial state is finite. To obtain nevertheless classical equations of motion one imposes the subsidiary condition that in the remote past and future the particles are not accelerated (just as in the treatment of the foregoing chapter for a single particle). In this way the effects of radiation in the remote past and future are suppressed.

With the subsidiary condition it follows that for the discussion of the convergence of the integrals at large space-like distances only the velocity fields in (III.96) or (III.99) have to be taken into account. These retarded and advanced velocity fields due to particle $i$ are

$$f^{\alpha\beta}_{\mathrm{r,a},i} = \pm \frac{e_i c^2}{4\pi(u_i \cdot r_i)^3} (r^\alpha_i u^\beta_i - r^\beta_i u^\alpha_i)|_{\mathrm{r,a}}, \tag{84}$$

where $r^\alpha_i \equiv R^\alpha - R^\alpha_i$. If these fields are introduced into $t^{\alpha\beta}$ (62) it follows that the total momentum (65) over a space-like plane is indeed a finite four-vector (which is assumed to be time-like). However the integral (68) for the inner angular momentum is only conditionally convergent (as follows by counting the powers of $R$), so that a (Lorentz-invariant) prescription must be given for its evaluation. Since the integral is independent of the choice of the origin of coordinates (because only coordinate differences are involved) we may choose as origin a point lying in the plane $\Sigma$. The conditionally convergent integral splits into a convergent part with integrand $X^\alpha t^{\beta\gamma} - X^\beta t^{\alpha\gamma}$ and a semi-convergent part with integrand $R^\alpha t^{\beta\gamma} - R^\beta t^{\alpha\gamma}$. We shall confine our attention to the latter part. A prescription for the evaluation of this part is obtained by considering a three-sphere with radius $\rho$ around a point in the plane $\Sigma$. We then evaluate the integral $\int (R^\alpha t^{\beta\gamma} - R^\beta t^{\alpha\gamma}) \mathrm{d}^3\Sigma_\gamma$. We shall prove that this integral tends to a finite limiting value if the radius $\rho$ tends to infinity. Furthermore we shall show that it is independent of the precise location of the centre of the sphere.

To start with the latter let us consider a sphere of radius $\rho$ around the centre $C^\alpha_1$ and another sphere of the same radius around $C^\alpha_2$. The difference between the integrals extended over the two spheres has an integrand which is of the order of $\rho^{-3}$ (since for sufficiently large $\rho$ only the velocity fields, which are proportional to $\rho^{-2}$, come into play) and is extended over a volume of the order of $\rho^2$. Hence this difference tends to zero if $\rho$ grows indefinitely. Since the limiting value of the integral is now proved to be independent of the location of the sphere's centre, we shall choose this centre as the origin.

In order to prove the existence of the limit of the integral $\int (R^\alpha t^{\beta\gamma} - R^\beta t^{\alpha\gamma})$

$\times\, d^3\Sigma_\gamma$ over the sphere for $\rho \to \infty$ we have to show that the contribution of a spherical shell, lying between two such spheres, tends to zero if the smallest radius tends to infinity. To that end we substitute $t^{\alpha\beta}$ (62) with $f_{r,a,i}$ (84) into the integrand so that we get for the field dependent part:

$$\sum_{i,j(i\neq j)} e_i e_j c^4 \int R^\alpha \left\{ \frac{(r_i^\beta u_i^\varepsilon - r_i^\varepsilon u_i^\beta)(r_j^\gamma u_{j\varepsilon} - r_{j\varepsilon} u_j^\gamma)}{16\pi^2 (u_i\cdot r_i)^3 (u_j\cdot r_j)^3} - \frac{(r_i^\varepsilon u_i^\zeta - r_i^\zeta u_i^\varepsilon) r_{j\varepsilon} u_{j\zeta} g^{\beta\gamma}}{32\pi^2 (u_i\cdot r_i)^3 (u_j\cdot r_j)^3} \right\} \Bigg|_r d^3\Sigma_\gamma$$

$$+ \sum_i e_i^2 c^4 \int R^\alpha \left\{ \frac{(r_i^\beta u_i^\varepsilon - r_i^\varepsilon u_i^\beta)(r_i^\gamma u_{i\varepsilon} - r_{i\varepsilon} u_i^\gamma)}{16\pi^2 (u_i\cdot r_i)^6} - \frac{(r_i^\varepsilon u_i^\zeta - r_i^\zeta u_i^\varepsilon) r_{i\varepsilon} u_{i\zeta} g^{\beta\gamma}}{32\pi^2 (u_i\cdot r_i)^6} \right\} \Bigg|_- d^3\Sigma_\gamma - (\alpha, \beta), \quad (85)$$

where we introduced the symbol $-$ at the bar to indicate half the difference of the retarded and the advanced contribution. The symbol $(\alpha, \beta)$ indicates the preceding expression with $\alpha$ and $\beta$ interchanged. Since we have taken into account only velocity fields it is consistent to assume that the retarded position four-vector $R_i^\alpha|_r$ that occurs here is parallel to the retarded four-velocity $u_i^\alpha|_r$, at least for sufficiently large $\rho$, i.e. we write

$$R_i^\alpha|_r = -c^{-2} u_i\cdot R_i\, u_i^\alpha|_r + \xi_{ir}^\alpha, \quad (86)$$

where the factor in front of $u_i^\alpha$ is chosen such that the four-vector $\xi_{ir}^\alpha$ is orthogonal to the velocity ($\xi_{ir}\cdot u_i|_r = 0$). If the radii of the spheres tend to infinity, both the left-hand side and the first term at the right-hand side blow up, while the last term remains finite. Similar remarks apply for the connexion between the advanced position $R_i^\alpha|_a$ and the advanced velocity $u_i^\alpha|_a$:

$$R_i^\alpha|_a = -c^{-2} u_i\cdot R_i\, u_i^\alpha|_a + \xi_{ia}^\alpha. \quad (87)$$

If we substitute (86) and (87) into (85) (using the definition $r_i^\alpha = R^\alpha - R_i^\alpha$), we find terms which are independent of $\xi^\alpha$ and terms that contain $\xi^\alpha$. By counting the powers in $R$ one notices that the latter terms give vanishing contributions if $\rho$ tends to infinity. The remaining terms of the right-hand side of (85) read

$$\sum_{i,j(i\neq j)} e_i e_j c^4 \int \frac{R^\alpha\{u_i^\beta(R^2 u_j\cdot n - R\cdot u_j\, R\cdot n) - \frac{1}{2} n^\beta (R^2 u_i\cdot u_j - R\cdot u_i\, R\cdot u_j)\}}{16\pi^2 \{u_i\cdot(R-R_i)\}^3 \{u_j\cdot(R-R_j)\}^3} \Bigg|_r d^3\Sigma$$

$$+ \sum_i e_i^2 c^4 \int \frac{R^\alpha[u_i^\beta(R^2 u_i\cdot n - R\cdot u_i\, R\cdot n) - \frac{1}{2} n^\beta\{-R^2 c^2 - (R\cdot u_i)^2\}]}{16\pi^2 \{u_i\cdot(R-R_i)\}^6} \Bigg|_- d^3\Sigma - (\alpha, \beta), \quad (88)$$

where we employed the definition $d^3\Sigma_\gamma = n_\gamma d^3\Sigma$ (with $n^\alpha$ the normal to the surface $\Sigma$). Since $R$ as well as the origin are situated on the surface $\Sigma$ one has $R \cdot n = 0$. Furthermore we may use the light-cone equation $r_i^2 = 0$ to eliminate $u_i \cdot R_i$. Indeed if (86) or (87) is substituted into the light-cone equation written as $(R_i - R)^2 = 0$ one gets

$$u_i \cdot R_i|_{\mathrm{r,a}} = \{u_i \cdot R \pm \sqrt{(u_i \cdot R)^2 + R^2 c^2}\}|_{\mathrm{r,a}}, \tag{89}$$

where terms containing $\xi$ have been suppressed at the right-hand side. Then the denominators of (88) get the form

$$\begin{aligned} &16\pi^2\{(u_i \cdot R)^2 + R^2c^2\}^{\frac{3}{2}}\{(u_j \cdot R)^2 + R^2c^2\}^{\frac{3}{2}}, \\ &16\pi^2\{(u_i \cdot R)^2 + R^2c^2\}^3. \end{aligned} \tag{90}$$

If these denominators are used one finds immediately that the integral extended over the three-space between two spheres around the origin vanishes on grounds of symmetry (the integrand changes sign if $R$ is replaced by $-R$). Thus we have proved now that the semi-convergent integral for the inner angular momentum tends to a definite limit if spheres of increasing radii are chosen as integration domain.

In the course of the proof on the uniqueness of the energy centre as given in the first part of this subsection we made use in an essential way of the assumption that the external force density $f^\alpha$ can be made arbitrarily small. In the present case where $f^\alpha$ is given by (63) in terms of the external fields, this assumption is certainly justified since these fields can be made arbitrarily small. This remark completes the discussion of the validity of the application of the general centre of energy construction to the special case of a system described by an energy–momentum tensor $t^{\alpha\beta}$ (62) and acted upon by a force density $f^\alpha$ (63). The asymptotic conditions employed are just what one has to expect in a classical theory in which radiative collapse of bound states would occur if the particles would be allowed to suffer accelerations in the remote past (and future).

### c. *Charged dipole particles*

For a composite particle which satisfies the energy–momentum law (61) with the energy–momentum tensor $t^{\alpha\beta}$ (62) and the force density (63) the derivative of $p^\alpha$ with respect to the proper time $s$ of the world line $X^\alpha(s)$ is given by [1]:

[1] One may prove from (69) and (70) that for sufficiently small fields the world line is time-like so that a proper time $s$ along the world line may be introduced (see problem 2).

$$\frac{\mathrm{d}p^\alpha}{\mathrm{d}s} = -\frac{c^{-1}}{\mathrm{d}s}\left\{\int_{\Sigma(s+\mathrm{d}s)} t^{\alpha\beta}(R)n_\beta(s+\mathrm{d}s)\mathrm{d}^3\Sigma - \int_{\Sigma(s)} t^{\alpha\beta}(R)n_\beta(s)\mathrm{d}^3\Sigma\right\}, \quad (91)$$

where the right-hand side is given by the difference of two integrals of the form (65) over the surfaces $\Sigma(s+\mathrm{d}s)$ and $\Sigma(s)$, divided by $\mathrm{d}s$. To be able to apply Gauss's theorem here we must discuss the contribution from the surface $\Sigma_\infty(s, \mathrm{d}s)$ at infinity which closes the volume between the surfaces $\Sigma(s+\mathrm{d}s)$ and $\Sigma(s)$. Since in the remote past and future the particles suffer no accelerations, only the velocity fields have to be inserted in the integral

$$\int_{\Sigma_\infty(s,\mathrm{d}s)} t^{\alpha\beta}(R)n_{\infty,\beta}(R)\mathrm{d}^3\Sigma \quad (92)$$

(with $n^\alpha_\infty(R)$ the outward pointing normal on the surface $\Sigma_\infty(s, \mathrm{d}s)$). By employing the expressions (84) and counting the powers in $R$ one finds that this integral tends to zero, if the surface tends to infinity. Therefore one may apply Gauss's theorem to (91), with the result

$$\frac{\mathrm{d}p^\alpha}{\mathrm{d}s} = \frac{c^{-1}}{\mathrm{d}s}\int_{\Sigma(s)}^{\Sigma(s+\mathrm{d}s)} f^\alpha \mathrm{d}^4 V, \quad (93)$$

where the integral is extended over the volume bounded by the surfaces $\Sigma(s)$ and $\Sigma(s+\mathrm{d}s)$. The volume element $\mathrm{d}^4V$ may be written as

$$\mathrm{d}^4 V = J(R, s)\mathrm{d}s\,\mathrm{d}^3\Sigma, \quad (94)$$

where the Jacobian $J(R, s)$ is:

$$J(R, s) = -\frac{u_\alpha p^\alpha}{\sqrt{-p^2}}\left\{1 - \frac{\dot{p}_\beta(R-X)^\beta}{u_\gamma p^\gamma}\right\} \quad (95)$$

(v. problem 3), with $X^\alpha$, $u^\alpha \equiv \mathrm{d}X^\alpha/\mathrm{d}s$ and $p^\alpha$ functions of the proper time $s$. The equation of motion (93) becomes with (94)

$$\frac{\mathrm{d}p^\alpha}{\mathrm{d}s} = \mathfrak{f}^\alpha, \quad (96)$$

where $\mathfrak{f}^\alpha$ is the total four-force expressed in terms of the force density $f^\alpha(R)$:

$$\mathfrak{f}^\alpha(s) \equiv c^{-1}\int_{\Sigma(s)} f^\alpha(R)J(R, s)\mathrm{d}^3\Sigma. \quad (97)$$

For the derivative of the inner angular momentum (68) with respect to the proper time $s$ one finds

$$\frac{\mathrm{d}s^{\alpha\beta}}{\mathrm{d}s} = -\frac{c^{-1}}{\mathrm{d}s}\left\{\int_{\Sigma(s+\mathrm{d}s)} \{R^\alpha - X^\alpha(s+\mathrm{d}s)\}t^{\beta\gamma}(R)n_\gamma(s+\mathrm{d}s)\mathrm{d}^3\Sigma \right.$$
$$\left. -\int_{\Sigma(s)} \{R^\alpha - X^\alpha(s)\}t^{\beta\gamma}(R)n_\gamma(s)\mathrm{d}^3\Sigma\right\} -(\alpha,\beta) \quad (98)$$

again with $n^\alpha = p^\alpha/\sqrt{-p^2}$, depending on $s+\mathrm{d}s$ in the first integral and on $s$ in the second. The integrals over $\Sigma(s)$ and $\Sigma(s+\mathrm{d}s)$ are to be read as the limits of integrals over three-spheres with increasing radii in the planes $\Sigma(s)$ and $\Sigma(s+\mathrm{d}s)$. In order to apply Gauss's theorem we close the region between the two surfaces $\Sigma(s)$ and $\Sigma(s+\mathrm{d}s)$ by a four-sphere of large radius, of which the centre is the origin of coordinates. (The intersections of this four-sphere with the surfaces $\Sigma(s)$ and $\Sigma(s+\mathrm{d}s)$ consist of two three-spheres.) With the use of Gauss's theorem one may write then

$$\int_{\Sigma(s)}^{\Sigma(s+\mathrm{d}s)} \partial_\gamma[\{R^\alpha - X^\alpha(s)\}t^{\beta\gamma} - \{R^\beta - X^\beta(s)\}t^{\alpha\gamma}]\mathrm{d}^4V$$
$$= -\int_{\Sigma(s+\mathrm{d}s)} \{R^\alpha - X^\alpha(s)\}t^{\beta\gamma}(R)n_\gamma(s+\mathrm{d}s)\mathrm{d}^3\Sigma$$
$$+\int_{\Sigma(s)} \{R^\alpha - X^\alpha(s)\}t^{\beta\gamma}(R)n_\gamma(s)\mathrm{d}^3\Sigma$$
$$+\int_{\Sigma(s,\mathrm{d}s)} \{R^\alpha - X^\alpha(s)\}t^{\beta\gamma}(R)n_{\infty,\gamma}(R)\mathrm{d}^3\Sigma - (\alpha,\beta), \quad (99)$$

where the normal $n^\alpha_\infty$ is equal to $R^\alpha/\sqrt{R^2}$. The first two terms at the right-hand side may be written as

$$c\,\mathrm{d}s\left(\frac{\mathrm{d}s^{\alpha\beta}}{\mathrm{d}s} + u^\alpha p^\beta - u^\beta p^\alpha\right), \quad (100)$$

as follows by comparison with (98) and (65). The last term, which extends over the large four-sphere, vanishes with increasing radius, as may be seen in the following way. The velocity fields which are to be used in $t^{\alpha\beta}$ (62) decrease with the square of the inverse radius. As a consequence the part with $X^\alpha(s)$ goes to zero as counting of the powers in $R$ shows. As to the term with $R^\alpha$, it has the same form as (88) with $n^\alpha$ replaced by $n^\alpha_\infty = R^\alpha/\sqrt{R^2}$. Thus it vanishes. In this way we have found the change of inner angular momentum

$$\frac{\mathrm{d}s^{\alpha\beta}}{\mathrm{d}s} = \frac{c^{-1}}{\mathrm{d}s}\int_{\Sigma(s)}^{\Sigma(s+\mathrm{d}s)} \partial_\gamma[\{R^\alpha - X^\alpha(s)\}t^{\beta\gamma} - \{R^\beta - X^\beta(s)\}t^{\alpha\gamma}]\mathrm{d}^4V - (u^\alpha p^\beta - u^\beta p^\alpha).$$
$$(101)$$

If one uses the symmetry of the energy–momentum tensor $t^{\alpha\beta}$ (62), the energy–momentum law (61) and the expression (94) for the volume element one finds the inner angular momentum law

$$\frac{\mathrm{d}s^{\alpha\beta}}{\mathrm{d}s} = \mathfrak{d}^{\alpha\beta} - (u^{\alpha}p^{\beta} - u^{\beta}p^{\alpha}) \tag{102}$$

with the total torque

$$\mathfrak{d}^{\alpha\beta}(s) = c^{-1}\int_{\Sigma(s)} [\{R^{\alpha} - X^{\alpha}(s)\}f^{\beta}(R) - \{R^{\beta} - X^{\beta}(s)\}f^{\alpha}(R)]J(R, s)\mathrm{d}^3\Sigma, \tag{103}$$

containing the force density $f^{\alpha}(R)$.

With the explicit form (63) for the force density $f^{\alpha}(R)$ the total force (97) and the total torque (103) are completely specified:

$$\mathfrak{f}^{\alpha}(s) = c^{-1}\iint_{\Sigma(s)} F^{\alpha\beta}(R)\sum_i e_i \frac{\mathrm{d}R_{i\beta}}{\mathrm{d}s'}\,\delta^{(4)}\{R_i(s') - R\}J(R, s)\mathrm{d}^3\Sigma\,\mathrm{d}s', \tag{104}$$

$$\mathfrak{d}^{\alpha\beta}(s) = c^{-1}\iint_{\Sigma(s)} [\{R^{\alpha} - X^{\alpha}(s)\}F^{\beta\gamma}(R) - \{R^{\beta} - X^{\beta}(s)\}F^{\alpha\gamma}(R)] \sum_i e_i \frac{\mathrm{d}R_{i\gamma}}{\mathrm{d}s'}\,\delta^{(4)}\{R_i(s') - R\}J(R, s)\mathrm{d}^3\Sigma\,\mathrm{d}s'. \tag{105}$$

(For convenience's sake we choose a different parametrization of the world lines of the constituent particles, namely the parameter $s'$, which may be induced with the help of the surfaces $\Sigma(s')$ starting from the parametrization of the world line of the centre of energy.)

Because of the occurrence of the four-dimensional delta functions only the intersection points of $\Sigma(s)$ with the world lines of the constituent particles contribute to the integrals over $\Sigma(s)$. Therefore we may perform the integrations in (104) and (105). One obtains then

$$\mathfrak{f}^{\alpha}(s) = c^{-1}\sum_i e_i F^{\alpha\beta}\{R_i(s)\}\frac{\mathrm{d}R_{i\beta}}{\mathrm{d}s}, \tag{106}$$

$$\mathfrak{d}^{\alpha\beta}(s) = c^{-1}\sum_i e_i[\{R_i^{\alpha}(s) - X^{\alpha}(s)\}F^{\beta\gamma}\{R_i(s)\} - \{R_i^{\beta}(s) - X^{\beta}(s)\}F^{\alpha\gamma}\{R_i(s)\}]\frac{\mathrm{d}R_{i\gamma}}{\mathrm{d}s}. \tag{107}$$

The external fields $F^{\alpha\beta}$ in these expressions depend on the positions $R_i^{\alpha}$ of the constituent particles. We may expand them in Taylor series around the centre of energy $X^{\alpha}(s)$ of the composite particle. To that purpose let us introduce the relative positions

$$r_i^\alpha(s) \equiv R_i^\alpha(s) - X^\alpha(s) \tag{108}$$

of the constituent particles inside the composite particle. They fulfil the orthogonality relation (this follows from the construction of the centre of energy, described in the preceding subsection):

$$r_{i\alpha}(s)p^\alpha(s) = 0. \tag{109}$$

In the expanded expressions for (106) and (107) the internal coordinates $r_i^\alpha(s)$ may be grouped in such a way that only the covariant multipole moments (9) occur, with $p^\alpha/\sqrt{-p^2}$ as the time-like unit vector $n^\alpha$. (In the following we shall omit the index $(n) = (p/\sqrt{-p^2})$ of the multipole moments.) If we limit ourselves (in this subsection) to the contributions of the covariant electric and magnetic dipoles ($n = 1$):

$$\begin{aligned} \mu^\alpha &= \sum_i e_i r_i^\alpha, \\ \nu^{\alpha\beta} &= \tfrac{1}{2}c^{-1} \sum_i e_i \left( r_i^\alpha \frac{\mathrm{d}r_i^\beta}{\mathrm{d}s} - r_i^\beta \frac{\mathrm{d}r_i^\alpha}{\mathrm{d}s} \right), \end{aligned} \tag{110}$$

i.e. to slowly varying external fields, we find from (106) and (107)

$$\begin{aligned} \mathfrak{f}^\alpha(s) = {} & c^{-1}eF^{\alpha\beta}\{X(s)\}u_\beta(s) \\ & + \tfrac{1}{2}\partial^\alpha F^{\beta\gamma}\{X(s)\}[c^{-1}\{\mu_\beta(s)u_\gamma(s) - \mu_\gamma(s)u_\beta(s)\} + \nu_{\beta\gamma}(s)] \\ & + c^{-1}\frac{\mathrm{d}}{\mathrm{d}s}[F^{\alpha\beta}\{X(s)\}\mu_\beta(s)], \end{aligned} \tag{111}$$

$$\begin{aligned} \mathfrak{d}^{\alpha\beta}(s) = {} & F^{\alpha\gamma}\{X(s)\}\{-c^{-1}u_\gamma(s)\mu^\beta(s) + \nu_\gamma^{\cdot\beta}(s)\} \\ & - F^{\beta\gamma}\{X(s)\}\{-c^{-1}u_\gamma(s)\mu^\alpha(s) + \nu_\gamma^{\cdot\alpha}(s)\}, \end{aligned} \tag{112}$$

where $e = \sum_i e_i$ is the charge of the composite particle and where the homogeneous field equations have been used in the second term at the right-hand side of (111). It will be convenient to introduce the electromagnetic dipole moment tensor ((11) with two indices $\alpha_1, \alpha_2 = \alpha, \beta$):

$$\mathfrak{m}^{\alpha\beta}(s) = c^{-1}\{\mu^\alpha(s)u^\beta(s) - \mu^\beta(s)u^\alpha(s)\} + \nu^{\alpha\beta}(s). \tag{113}$$

The covariant electric dipole moment may be expressed in terms of this tensor. If one uses the orthogonality relation (109) and the fact that the covariant electric quadrupole moment $\frac{1}{2}\sum_i e_i r_i^\alpha r_i^\beta$ is neglected in this subsection, one gets:

$$\mu^\alpha(s) = c\mathfrak{m}^{\alpha\beta}(s)p_\beta(s)/u_\gamma(s)p^\gamma(s). \tag{114}$$

If the definition (113) and the relation (114) are used in the expressions (111) and (112) for the force and the torque, we obtain:

$$\mathfrak{f}^{\alpha} = c^{-1}eF^{\alpha\beta}(X)u_{\beta}+\tfrac{1}{2}\{\partial^{\alpha}F^{\beta\gamma}(X)\}\mathfrak{m}_{\beta\gamma}+\frac{\mathrm{d}}{\mathrm{d}s}\left\{F^{\alpha\beta}(X)\mathfrak{m}_{\beta\gamma}\frac{p^{\gamma}}{u_{\varepsilon}p^{\varepsilon}}\right\}, \tag{115}$$

$$\mathfrak{d}^{\alpha\beta} = F^{\alpha\gamma}(X)\mathfrak{m}_{\gamma}^{\cdot\beta}-F^{\beta\gamma}(X)\mathfrak{m}_{\gamma}^{\cdot\alpha}+\{u^{\alpha}F^{\beta\gamma}(X)-u^{\beta}F^{\alpha\gamma}(X)\}\mathfrak{m}_{\gamma\varepsilon}\frac{p^{\varepsilon}}{u_{\zeta}p^{\zeta}}. \tag{116}$$

These expressions are to be inserted in (96) and (102). Together with the supplementary condition (67) one has obtained then the equation of motion and of inner angular momentum of a composite particle with charge and dipole moments in an external field.

In order to discuss them we first consider the *field-free* case. Then the equations reduce to

$$\frac{\mathrm{d}p^{\alpha}}{\mathrm{d}s} = 0, \qquad \frac{\mathrm{d}s^{\alpha\beta}}{\mathrm{d}s} = p^{\alpha}u^{\beta}-p^{\beta}u^{\alpha}. \tag{117}$$

By differentiating the condition (67) one finds from these equations

$$p_{\alpha}p^{\alpha}u^{\beta}-p_{\alpha}u^{\alpha}p^{\beta} = 0. \tag{118}$$

Hence the four-vectors $p^{\alpha}$ and $u^{\alpha}$ are parallel, so that

$$p^{\alpha} = mu^{\alpha}, \tag{119}$$

where $m$ is defined as:

$$m = -c^{-2}p_{\alpha}u^{\alpha}. \tag{120}$$

Now (117) reduces to

$$\frac{\mathrm{d}u^{\alpha}}{\mathrm{d}s} = 0, \qquad \frac{\mathrm{d}s^{\alpha\beta}}{\mathrm{d}s} = 0, \tag{121}$$

since $\mathrm{d}m/\mathrm{d}s$ vanishes as follows from (119) and the first equation of (117).

In the case *with fields* differentiation of (67) and substitution of (96) and (102) leads to

$$p^{\alpha}u_{\beta}p^{\beta} = u^{\alpha}p_{\beta}p^{\beta}-s^{\alpha\beta}\mathfrak{f}_{\beta}-\mathfrak{d}^{\alpha\beta}p_{\beta}. \tag{122}$$

Hence now $p^{\alpha}$ is not parallel to $u^{\alpha}$. If this relation is multiplied by $u_{\alpha}$ one obtains the equality

$$p_{\alpha}p^{\alpha} = -c^{-2}(u_{\alpha}p^{\alpha})^{2}-c^{-2}u_{\alpha}s^{\alpha\beta}\mathfrak{f}_{\beta}-c^{-2}u_{\alpha}\mathfrak{d}^{\alpha\beta}p_{\beta}. \tag{123}$$

According to (119) and the condition (67) all terms on the right-hand side but the first are at least of second order in the fields, since the leading terms of $\mathfrak{f}^\alpha$ and $\mathfrak{d}^{\alpha\beta}$ are linear in the fields. Hence if one wants to confine oneself to terms linear in the fields the equality (122) may be written as

$$p^\alpha = mu^\alpha + \frac{1}{mc^2} s^{\alpha\beta}\mathfrak{f}_\beta + c^{-2}\mathfrak{d}^{\alpha\beta}u_\beta , \tag{124}$$

so that now the total momentum $p^\alpha$ is expressed in terms of $u^\alpha$, $m$, which is again defined as in (120), $s^{\alpha\beta}$, $\mathfrak{f}^\alpha$ and $\mathfrak{d}^{\alpha\beta}$. With this equality the equations of motion and of spin (96) and (102) become[1]:

$$\frac{\mathrm{d}(mu^\alpha)}{\mathrm{d}s} = \mathfrak{f}^\alpha - c^{-2}\frac{\mathrm{d}}{\mathrm{d}s}(s^{\alpha\beta}\mathfrak{f}_\beta/m + \mathfrak{d}^{\alpha\beta}u_\beta), \tag{125}$$

$$\frac{\mathrm{d}s^{\alpha\beta}}{\mathrm{d}s} = \Delta^\alpha_\gamma \Delta^\beta_\varepsilon \mathfrak{d}^{\gamma\varepsilon} + c^{-2}(s^{\alpha\gamma}u^\beta - s^{\beta\gamma}u^\alpha)\mathfrak{f}_\gamma/m. \tag{126}$$

with the tensor:

$$\Delta^{\alpha\beta} \equiv \Delta^{\alpha\beta}(u) \equiv g^{\alpha\beta} + c^{-2}u^\alpha u^\beta. \tag{127}$$

The force and torque have been given by (115), (116) or, if again only terms linear in the fields are retained, by:

$$\mathfrak{f}^\alpha = c^{-1}eF^{\alpha\beta}u_\beta + \tfrac{1}{2}(\partial^\alpha F^{\beta\gamma})\mathfrak{m}_{\beta\gamma} - c^{-2}\frac{\mathrm{d}}{\mathrm{d}s}(F^{\alpha\beta}\mathfrak{m}_{\beta\gamma}u^\gamma), \tag{128}$$

$$\mathfrak{d}^{\alpha\beta} = F^{\alpha\gamma}\mathfrak{m}_\gamma^{\cdot\varepsilon}\Delta^\beta_\varepsilon - F^{\beta\gamma}\mathfrak{m}_\gamma^{\cdot\varepsilon}\Delta^\alpha_\varepsilon . \tag{129}$$

Introducing these expressions into (125) and (126) we obtain the equations

[1] From (126) it follows that $s^{\alpha\beta}\mathfrak{f}_\beta/m = -(\mathrm{d}s^{\alpha\beta}/\mathrm{d}s)u_\beta$ up to terms linear in the fields. If one inserts this equality into (125) and (126) one gets (with the explicit expressions (128) and (129) for $\mathfrak{f}^\alpha$ and $\mathfrak{d}^{\alpha\beta}$) equations which have been discussed earlier[2] in connexion with the condition $u_\alpha s^{\alpha\beta} = 0$. Owing to the use of this different subsidiary condition it is then not possible to go back to (125) and (126); this fact is connected with the appearance of unwanted helical solutions, even in the field-free case. (In ref. 3, equations of the type just described were derived on the basis of an explicit construction of a central point and with the use of the Darwin approximation for the intra-atomic fields. Helical motions of macroscopic dimensions are then excluded.)

[2] C. Møller, op. cit.; cf. papers mentioned in footnote 2 on page 184.

[3] S. R. de Groot and L. G. Suttorp, Physica **37**(1967)284, 297; **39**(1968)84; L. G. Suttorp, On the covariant derivation of macroscopic electrodynamics from electron theory, thesis Amsterdam (1968).

of motion and of spin:

$$\frac{\mathrm{d}(mu^\alpha)}{\mathrm{d}s} = c^{-1}eF^{\alpha\beta}u_\beta + \tfrac{1}{2}(\partial^\alpha F^{\beta\gamma})\mathfrak{m}_{\beta\gamma} + c^{-2}\frac{\mathrm{d}}{\mathrm{d}s}\left[\Delta^\alpha_\beta\,\mathfrak{m}^{\beta\gamma}F_{\gamma\varepsilon}u^\varepsilon - F^{\alpha\beta}\mathfrak{m}_{\beta\gamma}u^\gamma\right.$$

$$\left. - \frac{s^{\alpha\beta}}{m}\left\{c^{-1}eF_{\beta\gamma}u^\gamma + \tfrac{1}{2}(\partial_\beta F_{\gamma\varepsilon})\mathfrak{m}^{\gamma\varepsilon} - c^{-2}\frac{\mathrm{d}}{\mathrm{d}s}(F_{\beta\gamma}\mathfrak{m}^{\gamma\varepsilon}u_\varepsilon)\right\}\right], \tag{130}$$

$$\frac{\mathrm{d}s^{\alpha\beta}}{\mathrm{d}s} = \Delta^\alpha_\gamma\Delta^\beta_\varepsilon(F^{\gamma\zeta}\mathfrak{m}_\zeta^{\cdot\varepsilon} - \mathfrak{m}^{\gamma\zeta}F_\zeta^{\cdot\varepsilon}) + c^{-2}\frac{1}{m}(s^{\alpha\gamma}u^\beta - s^{\beta\gamma}u^\alpha)$$

$$\left\{c^{-1}eF_{\gamma\varepsilon}u^\varepsilon + \tfrac{1}{2}(\partial_\gamma F_{\varepsilon\zeta})\mathfrak{m}^{\varepsilon\zeta} - c^{-2}\frac{\mathrm{d}}{\mathrm{d}s}(F_{\gamma\varepsilon}\mathfrak{m}^{\varepsilon\zeta}u_\zeta)\right\}. \tag{131}$$

The tensor $\mathfrak{m}^{\alpha\beta}(s)$ (113) contains the covariant electric and magnetic dipole moments $\mu^\alpha$ and $\nu^{\alpha\beta}$, which are defined with respect to the time-like unit vector $n^\alpha = p^\alpha/\sqrt{-p^2}$. Since we neglected quadratic field terms and since $\mathfrak{m}^{\alpha\beta}$ is always multiplied by the field, we may replace them by the covariant dipole moments, defined with respect to $c^{-1}u^\alpha$. (They will be denoted by the same symbols.) The latter multipole moments have been studied in detail in section 2.

The space parts of (130) and (131) will be written in three-dimensional notation. The four-velocity $u^\alpha$ is $(\gamma c, \gamma \boldsymbol{v})$ with $\gamma = (1-\boldsymbol{\beta}^2)^{-\frac{1}{2}}$ and $\boldsymbol{\beta} = \boldsymbol{v}/c$. Furthermore the space-space components of $s^{\alpha\beta}$ will be denoted by the vector $\boldsymbol{s}$ with components $s^i = \frac{1}{2}\varepsilon^{ijk}s_{jk}$. As far as they occur at the right-hand side of (130) and (131), the space-time components $s^{i0}$ of $s^{\alpha\beta}$ may be written as $(\boldsymbol{\beta}\wedge\boldsymbol{s})^i$ by the use of (67) with (124) together with the fact that we neglected quadratic field terms throughout. The field $F^{\alpha\beta}$ has the components $F^{ij} = \varepsilon^{ijk}B_k$ and $F^{0i} = E^i$. The covariant multipole moments occurring in $\mathfrak{m}^{\alpha\beta}$ may be expressed in terms of the atomic multipole moments, which are independent of the atomic velocities. One finds from (A41–43)

$$\mu^0 = \gamma\boldsymbol{\beta}\cdot\boldsymbol{\mu}^{(1)}, \qquad \mu^i = (\boldsymbol{\Omega}^{-1}\cdot\boldsymbol{\mu}^{(1)})^i, \qquad (i = 1,2,3),$$

$$\nu^{ij} = \gamma(\boldsymbol{\Omega}\cdot\boldsymbol{\nu}^{(1)})^k, \quad (i,j,k = 1,2,3 \text{ cycl.}), \qquad \nu^{i0} = -\gamma(\boldsymbol{\nu}^{(1)}\wedge\boldsymbol{\beta})^i, \quad (i = 1,2,3), \tag{132}$$

if only the atomic dipole moments are retained. The $\boldsymbol{\Omega}$-tensor is defined as

$$\boldsymbol{\Omega} = \mathbf{U} - \frac{\gamma}{\gamma+1}\boldsymbol{\beta}\boldsymbol{\beta}. \tag{133}$$

If these formulae are substituted into the definition (113) of $\mathfrak{m}^{\alpha\beta}$ one obtains with (A6)[1]

$$\begin{aligned} \mathfrak{m}^{i0} &\equiv \mathfrak{p}^i = \gamma(\boldsymbol{\Omega}\cdot\boldsymbol{\mu}^{(1)} + \boldsymbol{\beta}\wedge\boldsymbol{v}^{(1)})^i, \qquad (i = 1, 2, 3), \\ \mathfrak{m}^{ij} &\equiv \mathfrak{m}^k = \gamma(\boldsymbol{\Omega}\cdot\boldsymbol{v}^{(1)} - \boldsymbol{\beta}\wedge\boldsymbol{\mu}^{(1)})^k, \qquad (i, j, k = 1, 2, 3 \text{ cycl.}). \end{aligned} \tag{134}$$

We may now write the equations (130) and (131) in three-dimensional notation:

$$\begin{aligned} \frac{\mathrm{d}}{\mathrm{d}t}(\gamma m\boldsymbol{v}) = {} & e(\boldsymbol{E}+\boldsymbol{\beta}\wedge\boldsymbol{B}) + \gamma^{-1}\{(\boldsymbol{\nabla}\boldsymbol{E})\cdot\mathfrak{p} + (\boldsymbol{\nabla}\boldsymbol{B})\cdot\mathfrak{m}\} \\ & + c^{-1}\frac{\mathrm{d}}{\mathrm{d}t}[\gamma\{\mathfrak{p}\wedge\boldsymbol{B} - \mathfrak{m}\wedge\boldsymbol{E}) - \boldsymbol{\beta}\wedge(\mathfrak{p}\wedge\boldsymbol{E} + \mathfrak{m}\wedge\boldsymbol{B})\}] \\ & + c^{-1}\frac{\mathrm{d}}{\mathrm{d}t}\{\gamma^3\boldsymbol{\beta}(\mathfrak{p} - \boldsymbol{\beta}\wedge\mathfrak{m})\cdot\boldsymbol{\Omega}^2\cdot(\boldsymbol{E}+\boldsymbol{\beta}\wedge\boldsymbol{B})\} + c^{-1}\frac{\mathrm{d}\boldsymbol{\Phi}_s}{\mathrm{d}t}, \end{aligned} \tag{135}$$

$$\frac{\mathrm{d}\boldsymbol{s}}{\mathrm{d}t} = \gamma\boldsymbol{\Omega}^2\cdot(\mathfrak{p}\wedge\boldsymbol{E} + \mathfrak{m}\wedge\boldsymbol{B}) + \gamma\boldsymbol{\beta}\wedge(\mathfrak{p}\wedge\boldsymbol{B} - \mathfrak{m}\wedge\boldsymbol{E}) + \boldsymbol{\beta}\wedge\boldsymbol{\Phi}_s, \tag{136}$$

with the inner angular momentum terms $\boldsymbol{\Phi}_s$ given by:

$$\begin{aligned} \boldsymbol{\Phi}_s \equiv {} & \frac{e}{mc}\gamma\{\boldsymbol{s}\wedge(\boldsymbol{E}+\boldsymbol{\beta}\wedge\boldsymbol{B}) - \boldsymbol{s}\wedge\boldsymbol{\beta}\boldsymbol{\beta}\cdot\boldsymbol{E}\} \\ & + \frac{1}{mc}[\boldsymbol{s}\wedge\{(\boldsymbol{\nabla}\boldsymbol{E})\cdot\mathfrak{p} + (\boldsymbol{\nabla}\boldsymbol{B})\cdot\mathfrak{m}\} + \boldsymbol{s}\wedge\boldsymbol{\beta}\{(\partial_0\boldsymbol{E})\cdot\mathfrak{p} + (\partial_0\boldsymbol{B})\cdot\mathfrak{m}\}] \\ & + \frac{1}{mc^2}\gamma\boldsymbol{s}\wedge\frac{\mathrm{d}}{\mathrm{d}t}\{\gamma(\mathfrak{p} - \boldsymbol{\beta}\wedge\mathfrak{m})\wedge\boldsymbol{B} + \gamma\boldsymbol{E}\boldsymbol{\beta}\cdot\mathfrak{p}\} \\ & - \frac{1}{mc^2}\gamma\boldsymbol{s}\wedge\boldsymbol{\beta}\frac{\mathrm{d}}{\mathrm{d}t}\{\gamma(\mathfrak{p} - \boldsymbol{\beta}\wedge\mathfrak{m})\cdot\boldsymbol{E}\}. \end{aligned} \tag{137}$$

(For convenience the expressions (134) for $\mathfrak{p}$ and $\mathfrak{m}$ in terms of the atomic dipole moments have not been inserted in these formulae.)

The equation of motion (135) contains at its right-hand side the Lorentz force on a charge $e$, the 'Kelvin forces' on the electric and magnetic dipole moments $\boldsymbol{\mu}^{(1)}$ and $\boldsymbol{v}^{(1)}$, and three terms which are time derivatives of quantities of which the leading terms are

$$c^{-1}\left\{\boldsymbol{\mu}^{(1)}\wedge\boldsymbol{B} - \left(\boldsymbol{v}^{(1)} - \frac{e}{mc}\boldsymbol{s}\right)\wedge\boldsymbol{E}\right\}, \tag{138}$$

[1] The vector $\mathfrak{p}$ which is defined here should not be confused with the space part of $p^\alpha$ (65).

as follows with (134). The time derivative of the first term is the electrodynamic effect found already in the non-relativistic theory (v. (I.155)). The time derivative of the other terms (which are sometimes called the 'hidden momentum' of a magnetic dipole particle with inner angular momentum) is an analogous magnetodynamic effect, which contains the vector product of the electric field and a combination of the magnetic moment and the inner angular momentum. Such a magnetodynamic effect, which occurs already in Frenkel's work, was discussed extensively[1] in recent years on the basis of various *ad hoc* arguments, such as the assumed equality of action and reaction for the forces exerted by a charge and a magnetic dipole moment on each other. One finds then only the magnetic dipole term, not the inner angular momentum term of (138). (The latter is for an atom or a molecule small as compared to the former since the mass $m$ of the composite particle as a whole is much greater than the electronic masses which contribute to $\boldsymbol{s}$). Often the question of the equivalence of a magnetic dipole consisting of charged particles (or a current loop) and one consisting of two 'magnetic charges' plays a role in these discussions[2].

Since the terms given in (138) are of order $c^{-2}$ the mechanism which leads to this contribution to the force can be studied already in a theory which gives all terms up to order $c^{-2}$. Such a treatment is presented in the fifth appendix of this chapter[3].

The equation of inner angular momentum (136) shows at the right-hand side the change of $\boldsymbol{s}$ due to a torque of which the leading term is $\boldsymbol{\mu}^{(1)}\wedge\boldsymbol{E}+\boldsymbol{\nu}^{(1)}\wedge\boldsymbol{B}$, as follows from (134).

When one wants to compare the results (135–137) with the non-relativistic and the so-called 'semi-relativistic' ones (see appendix V), it should in the first place be borne in mind that the expression (137) is of order $c^{-1}$, so that it contributes neither in the non-relativistic nor in the semi-relativistic approximation of the equations (135) and (136). Furthermore one should also remember that in the non-relativistic and semi-relativistic limit the magnetic moment $\boldsymbol{\nu}^{(1)}$ is considered to be of order $c^{-1}$ and $c^{0}$ respectively. Then if one limits oneself to terms of order $c^{-1}$ one finds indeed the equa-

[1] P. Penfield jr. and H. A. Haus, The electrodynamics of moving media (M.I.T. Press, Cambridge, Mass. 1967) p. 215; Phys. Lett. **26A**(1968)412; Physica **42**(1969)447; O. Costa de Beauregard, Compt. Rend. **263B**(1966)1007, **264B**(1967)565, 731, **266B**(1968) 364, 1181; Phys. Lett. **24A**(1967)177, **25A**(1967)95, **26A**(1967)48; Cah. Physique **206**(1967) 373; N. Cim. **63B**(1969)611; W. Shockley and R. P. James, Phys. Rev. Lett. **18**(1967)876.

[2] v. B. D. H. Tellegen, Am. J. Phys. **30**(1962)650. From the present results it follows that the force on a magnetic dipole consisting of charged particles is exactly the same as that which is assumed to be valid for a 'magnetic charge' dipole.

[3] S. Coleman and J. H. Van Vleck, Phys. Rev. **171**(1968)1370 followed a similar approach.

tions (I.55) and (I.79) of the non-relativistic theory and (A118) and (A136) of the semi-relativistic theory (cf. also problems 6–8).

d. *Charged particles with magnetic dipole moment proportional to their inner angular momentum*

Let us consider the special case of a charged composite particle without electric dipole moment and with a magnetic dipole moment proportional to the inner angular momentum; then

$$\mathfrak{m}^{\alpha\beta} = \kappa s^{\alpha\beta}. \tag{139}$$

The equations of motion and of spin have been given by (96), (102), (124) with (67):

$$\frac{dp^\alpha}{ds} = \mathfrak{f}^\alpha, \tag{140}$$

$$\frac{ds^{\alpha\beta}}{ds} = \mathfrak{d}^{\alpha\beta} - (u^\alpha p^\beta - u^\beta p^\alpha), \tag{141}$$

$$p^\alpha = mu^\alpha + \frac{1}{mc^2} s^{\alpha\beta}\mathfrak{f}_\beta + c^{-2}\mathfrak{d}^{\alpha\beta}u_\beta, \tag{142}$$

$$p_\alpha s^{\alpha\beta} = 0. \tag{143}$$

The force and torque that follow from (128) and (129) with (139) are in this case:

$$\mathfrak{f}^\alpha = c^{-1}eF^{\alpha\beta}u_\beta + \tfrac{1}{2}(\partial^\alpha F^{\beta\gamma})\mathfrak{m}_{\beta\gamma}, \tag{144}$$

$$\mathfrak{d}^{\alpha\beta} = F^{\alpha\gamma}\mathfrak{m}_{\gamma}^{\cdot\beta} - F^{\beta\gamma}\mathfrak{m}_{\gamma}^{\cdot\alpha}, \tag{145}$$

where only terms linear in the field have been retained. If these expressions are inserted into (142) we obtain:

$$p^\alpha = mu^\alpha - c^{-2}\mathfrak{m}^{\alpha\beta}F_{\beta\gamma}u^\gamma + \frac{e}{mc^3} s^{\alpha\beta}F_{\beta\gamma}u^\gamma + \frac{1}{2mc^2} s^{\alpha\beta}(\partial_\beta F_{\gamma\varepsilon})\mathfrak{m}^{\gamma\varepsilon}. \tag{146}$$

From these equations one may prove that the square of the inner angular momentum $s_{\alpha\beta}s^{\alpha\beta}$ and the quantity

$$m^* = m + \tfrac{1}{2}c^{-2}F_{\alpha\beta}\mathfrak{m}^{\alpha\beta} \tag{147}$$

are conserved. In fact, multiplying (141) by $s_{\alpha\beta}$ and using (143) and (145), one gets

$$\frac{d}{ds}(s_{\alpha\beta}s^{\alpha\beta}) = 4s_{\alpha\beta}F^{\alpha\gamma}\mathfrak{m}_{\gamma}^{\cdot\beta}, \tag{148}$$

which vanishes, as follows if (139) is introduced. Furthermore the time derivative of $m$ (120) becomes with (140), (144) and (146)

$$\frac{\mathrm{d}m}{\mathrm{d}s} = -\tfrac{1}{2}c^{-2}\frac{\mathrm{d}F^{\alpha\beta}}{\mathrm{d}s}\mathfrak{m}_{\alpha\beta}+c^{-4}\frac{\mathrm{d}u_\alpha}{\mathrm{d}s}$$

$$\left\{\mathfrak{m}^{\alpha\beta}F_{\beta\gamma}u^\gamma - \frac{e}{mc}s^{\alpha\beta}F_{\beta\gamma}u^\gamma - \frac{1}{2m}s^{\alpha\beta}(\partial_\beta F_{\gamma\varepsilon})\mathfrak{m}^{\gamma\varepsilon}\right\}. \quad (149)$$

Since $\mathrm{d}u^\alpha/\mathrm{d}s$ vanishes in the field-free case, this equality becomes up to first order in the fields

$$\frac{\mathrm{d}m}{\mathrm{d}s}+\tfrac{1}{2}c^{-2}\frac{\mathrm{d}F^{\alpha\beta}}{\mathrm{d}s}\mathfrak{m}_{\alpha\beta} = 0. \quad (150)$$

Finally, if (141) is multiplied by $F_{\alpha\beta}$ one finds with (139), (145) and (146)

$$F_{\alpha\beta}\frac{\mathrm{d}\mathfrak{m}^{\alpha\beta}}{\mathrm{d}s} = 0, \quad (151)$$

if again only linear field terms are retained. From (150) and (151) it follows indeed that $m^*$ (147) is conserved.

Since $m^*$ is conserved we shall use it instead of $m$ in (146). With (147) this expression becomes up to terms linear in the fields:

$$p^\alpha = m^*u^\alpha - \tfrac{1}{2}c^{-2}F_{\beta\gamma}\mathfrak{m}^{\beta\gamma}u^\alpha - c^{-2}\mathfrak{m}^{\alpha\beta}F_{\beta\gamma}u^\gamma + \frac{e}{m^*c^3}s^{\alpha\beta}F_{\beta\gamma}u^\gamma$$
$$+\frac{1}{2m^*c^2}s^{\alpha\beta}(\partial_\beta F_{\gamma\varepsilon})\mathfrak{m}^{\gamma\varepsilon}. \quad (152)$$

In this expression appears the inner angular momentum $s^{\alpha\beta}$ together with the factor $e/m^*c$ containing the total charge of the composite particle. We shall call this combination the 'normal magnetic moment'

$$\mathfrak{m}^{\alpha\beta}_{(\mathrm{n})} = \frac{e}{m^*c}s^{\alpha\beta}. \quad (153)$$

The total magnetic moment (139) is the sum of this normal part and an 'anomalous magnetic moment' $\mathfrak{m}^{\alpha\beta}_{(\mathrm{a})}$:

$$\mathfrak{m}^{\alpha\beta} = \mathfrak{m}^{\alpha\beta}_{(\mathrm{n})}+\mathfrak{m}^{\alpha\beta}_{(\mathrm{a})}, \quad (154)$$

$$\mathfrak{m}^{\alpha\beta}_{(\mathrm{a})} \equiv \mathfrak{m}^{\alpha\beta} - \frac{e}{m^*c}s^{\alpha\beta} = \left(\kappa - \frac{e}{m^*c}\right)s^{\alpha\beta}. \quad (155)$$

For convenience one could introduce (for a charged composite particle) the 'gyromagnetic factor' $g$ by means of

$$\kappa \equiv \frac{ge}{2m^*c}. \tag{156}$$

Then we may write the anomalous magnetic moment alternatively in the form

$$\mathfrak{m}^{\alpha\beta}_{(a)} = \frac{(g-2)e}{2m^*c}\, s^{\alpha\beta}. \tag{157}$$

With (152) and (155) the equations of motion and spin (140) and (141) with (144) and (145) become:

$$m^* \frac{\mathrm{d}u^\alpha}{\mathrm{d}s} = c^{-1}eF^{\alpha\beta}u_\beta + \tfrac{1}{2}(\partial^\alpha F^{\beta\gamma})\mathfrak{m}_{\beta\gamma} + \tfrac{1}{2}c^{-2}\frac{\mathrm{d}F^{\beta\gamma}}{\mathrm{d}s}\mathfrak{m}_{\beta\gamma}u^\alpha + c^{-2}\mathfrak{m}^{\alpha\beta}_{(a)}\frac{\mathrm{d}F_{\beta\gamma}}{\mathrm{d}s}u^\gamma - \frac{1}{2m^*c^2}s^{\alpha\beta}\left(\partial_\beta \frac{\mathrm{d}F_{\gamma\varepsilon}}{\mathrm{d}s}\right)\mathfrak{m}^{\gamma\varepsilon}, \tag{158}$$

$$\frac{\mathrm{d}s^{\alpha\beta}}{\mathrm{d}s} = F^{\alpha\gamma}\mathfrak{m}_\gamma^{\cdot\beta} - F^{\beta\gamma}\mathfrak{m}_\gamma^{\cdot\alpha} + c^{-2}(u^\alpha \mathfrak{m}^{\beta\gamma}_{(a)} - u^\beta \mathfrak{m}^{\alpha\gamma}_{(a)})F_{\gamma\varepsilon}u^\varepsilon + \frac{1}{2m^*c^2}\{s^{\alpha\gamma}(\partial_\gamma F_{\varepsilon\zeta})\mathfrak{m}^{\varepsilon\zeta}u^\beta - s^{\beta\gamma}(\partial_\gamma F_{\varepsilon\zeta})\mathfrak{m}^{\varepsilon\zeta}u^\alpha\}. \tag{159}$$

In the last three terms at the right-hand side of equation (158) only the fields $F^{\alpha\beta}$ have been differentiated with respect to $s$, not the polarization tensors $\mathfrak{m}^{\alpha\beta}$, $\mathfrak{m}^{\alpha\beta}_{(a)}$, the inner angular momentum tensor $s^{\alpha\beta}$ and the four-velocity $u^\alpha$. The reason for this is that differentiation of the latter would have given rise to terms quadratic in the fields. (To see this use must be made of the proportionality of the polarization tensors $\mathfrak{m}^{\alpha\beta}$, $\mathfrak{m}^{\alpha\beta}_{(a)}$ and the inner angular momentum tensor $s^{\alpha\beta}$ together with the equations (121) for the field-free case.)

In the right-hand sides of (158) and (159) the leading terms with the polarization tensors contain the first derivatives of the field and the field itself respectively. Hence if the fields are sufficiently homogeneous the last terms in (158) and (159) may be discarded, so that we then have the simpler equations:

$$m^* \frac{\mathrm{d}u^\alpha}{\mathrm{d}s} = c^{-1}eF^{\alpha\beta}u_\beta + \tfrac{1}{2}(\partial^\alpha F^{\beta\gamma})\mathfrak{m}_{\beta\gamma} + \tfrac{1}{2}c^{-2}\frac{\mathrm{d}F^{\beta\gamma}}{\mathrm{d}s}\mathfrak{m}_{\beta\gamma}u^\alpha + c^{-2}\mathfrak{m}^{\alpha\beta}_{(a)}\frac{\mathrm{d}F_{\beta\gamma}}{\mathrm{d}s}u^\gamma, \tag{160}$$

$$\frac{\mathrm{d}s^{\alpha\beta}}{\mathrm{d}s} = F^{\alpha\gamma}\mathfrak{m}_\gamma^{\cdot\beta} - F^{\beta\gamma}\mathfrak{m}_\gamma^{\cdot\alpha} + c^{-2}(u^\alpha \mathfrak{m}^{\beta\gamma}_{(a)} - u^\beta \mathfrak{m}^{\alpha\gamma}_{(a)})F_{\gamma\varepsilon}u^\varepsilon. \tag{161}$$

These are the equations of motion and of spin for a composite particle with charge and magnetic dipole moment (proportional to the inner angular momentum) in sufficiently homogeneous external fields. The covariant magnetic dipole moment occurring in $\mathfrak{m}^{\alpha\beta}$ was defined with respect to the normal unit vector $p^{\alpha}/\sqrt{-p^2}$. As in the preceding subsection one might use as well the covariant magnetic dipole moment defined with respect to $c^{-1}u^{\alpha}$, because quadratic field terms have been neglected.

As in the preceding subsection we may write the covariant equations (160) and (161) in three-dimensional notation. The four-velocity $u^{\alpha}$ has components $(\gamma c, \gamma c\boldsymbol{\beta})$ with $\boldsymbol{\beta} = \boldsymbol{v}/c$, while the field components are $F^{0i} = E^i$, $F^{ij} = B^k$ $(i, j, k = 1, 2, 3$ cycl.). Since the electric dipole moment vanishes here, one may write the components $\mathfrak{m}^{i0} = \mathfrak{p}^i$ and $\mathfrak{m}^{ij} = \mathfrak{m}^k$ $(i, j, k = 1, 2, 3$ cycl.) of the magnetic moment tensor according to (134) as: $\boldsymbol{\mathfrak{p}} = \boldsymbol{\beta} \wedge \boldsymbol{\mathfrak{m}}$ and $\boldsymbol{\mathfrak{m}} = \gamma \boldsymbol{\Omega}\cdot\boldsymbol{v}^{(1)}$. From the proportionality of the anomalous and total polarization tensor (v. (139), (156) and (157)) it follows that the space-space component $\mathfrak{m}^i_{(a)} = \mathfrak{m}^{jk}_{(a)}$ $(i, j, k = 1, 2, 3$ cycl.) is $\mathfrak{m}^i_{(a)} = \{(g-2)/g\}\mathfrak{m}^i$, while the space-time component $\mathfrak{m}^{i0}_{(a)}$ is equal to $(\boldsymbol{\beta} \wedge \boldsymbol{\mathfrak{m}}_{(a)})^i$. In this way the equations of motion and of spin (160) and (161) get the form:

$$m^* \frac{\mathrm{d}(\gamma \boldsymbol{v})}{\mathrm{d}t} = e(\boldsymbol{E}+\boldsymbol{\beta}\wedge\boldsymbol{B})+\gamma^{-1}(\boldsymbol{\nabla}\boldsymbol{E})\cdot(\boldsymbol{\beta}\wedge\boldsymbol{\mathfrak{m}})+\gamma^{-1}(\boldsymbol{\nabla}\boldsymbol{B})\cdot\boldsymbol{\mathfrak{m}}$$

$$+c^{-1}\gamma\frac{\mathrm{d}}{\mathrm{d}t}\{\boldsymbol{\beta}\boldsymbol{\mathfrak{m}}\cdot(\boldsymbol{B}-\boldsymbol{\beta}\wedge\boldsymbol{E})-\boldsymbol{\mathfrak{m}}_{(a)}\wedge(\boldsymbol{E}+\boldsymbol{\beta}\wedge\boldsymbol{B})+\boldsymbol{\mathfrak{m}}_{(a)}\wedge\boldsymbol{\beta}\boldsymbol{\beta}\cdot\boldsymbol{E}\}, \qquad (162)$$

$$\frac{\mathrm{d}\boldsymbol{s}}{\mathrm{d}t} = \gamma^{-1}\boldsymbol{\mathfrak{m}}\wedge\boldsymbol{B}+\gamma^{-1}(\boldsymbol{\beta}\wedge\boldsymbol{\mathfrak{m}})\wedge\boldsymbol{E}+\gamma\boldsymbol{\beta}\cdot\boldsymbol{\mathfrak{m}}_{(a)}(\boldsymbol{E}+\boldsymbol{\beta}\wedge\boldsymbol{B})-\gamma^{-1}\boldsymbol{\mathfrak{m}}_{(a)}\,\boldsymbol{\beta}\cdot\boldsymbol{E}$$

$$-\gamma\boldsymbol{\beta}\boldsymbol{\beta}\cdot\boldsymbol{\mathfrak{m}}_{(a)}\,\boldsymbol{\beta}\cdot\boldsymbol{E}. \qquad (163)$$

(Time and space differentiations in the right-hand sides operate only on the fields, just as before.)

If the composite particle is momentarily at rest it is described by equations (162) and (163) with $\boldsymbol{\beta} = 0$. Then these equations reduce to:

$$m^* \frac{\mathrm{d}\boldsymbol{v}}{\mathrm{d}t} = e\boldsymbol{E}+(\boldsymbol{\nabla}\boldsymbol{B})\cdot\boldsymbol{\mathfrak{m}}-c^{-1}\boldsymbol{\mathfrak{m}}_{(a)}\wedge\frac{\mathrm{d}\boldsymbol{E}}{\mathrm{d}t}, \qquad (164)$$

$$\frac{\mathrm{d}\boldsymbol{s}}{\mathrm{d}t} = \boldsymbol{\mathfrak{m}}\wedge\boldsymbol{B}. \qquad (165)$$

Equation (164) or (162) contains the 'magnetodynamic effect', i.e. $c^{-1}\boldsymbol{\mathfrak{m}}_{(a)} \wedge \mathrm{d}\boldsymbol{E}/\mathrm{d}t$ or alternatively $c^{-1}(\mathrm{d}/\mathrm{d}t)(\boldsymbol{\mathfrak{m}}_{(a)} \wedge \boldsymbol{E})$ (the difference between these two expressions is of second order in the fields). Thus this time derivative of the so-called 'hidden momentum' $c^{-1}\mathfrak{m}_{(a)}\wedge \boldsymbol{E}$ contains only the *anomalous* magnetic moment.

## e. *Composite particles in an arbitrarily varying electromagnetic field*

In subsection *c* we derived the equations of motion (125) and (126) with the expressions (106) and (107) for the total force and torque. By making a Taylor expansion of the fields and confining ourselves to terms with the charge and electromagnetic dipole moments we found the expressions (115) and (116) (or (128–129)) for the total force and torque. If the fields in which the composite particle moves change rapidly, the limitation to dipole terms in the Taylor expansion is no longer justified. We must then consider the complete Taylor expansion of the total force (106) around the centre of energy $X^\alpha$

$$\mathfrak{f}^\alpha = c^{-1}\sum_{n=0}^{\infty}\frac{1}{n!}\sum_i e_i(r_i\cdot\partial)^n F^{\alpha\beta}(X)\left(\frac{\mathrm{d}X_\beta}{\mathrm{d}s}+\frac{\mathrm{d}r_{i\beta}}{\mathrm{d}s}\right), \tag{166}$$

where the relative positions $r_i^\alpha$ are defined by (108). This expression may be written in the form

$$\begin{aligned}\mathfrak{f}^\alpha &= c^{-1}eF^{\alpha\beta}(X)u_\beta\\ &+c^{-1}\sum_{n=1}^{\infty}\frac{1}{n!}\sum_i e_i(r_i\cdot\partial)^{n-1}(r_i^\gamma u_\beta - r_{i\beta}u^\gamma)\partial_\gamma F^{\alpha\beta}(X)\\ &+c^{-1}\sum_{n=1}^{\infty}\frac{1}{n!}\frac{\mathrm{d}}{\mathrm{d}s}\{\sum_i e_i(r_i\cdot\partial)^{n-1}F^{\alpha\beta}(X)r_{i\beta}\}\\ &+c^{-1}\sum_{n=1}^{\infty}\frac{n}{(n+1)!}\sum_i e_i(r_i\cdot\partial)^{n-1}\left(\frac{\mathrm{d}r_{i\beta}}{\mathrm{d}s}r_i^\gamma-\frac{\mathrm{d}r_i^\gamma}{\mathrm{d}s}r_{i\beta}\right)\partial_\gamma F^{\alpha\beta}(X),\end{aligned} \tag{167}$$

where $e=\sum_i e_i$ is the total charge of the composite particle. At the right-hand side one recognizes the covariant electric and magnetic multipole moments (9) with $p^\alpha/\sqrt{-p^2}$ as the time-like unit vector $n^\alpha$. Therefore by using the homogeneous field equations (13) and omitting the subscript $(n) = (p/\sqrt{-p^2})$ of the covariant multipoles, one may write the total force on the composite particle as:

$$\begin{aligned}\mathfrak{f}^\alpha &= c^{-1}eF^{\alpha\beta}u_\beta\\ &+\tfrac{1}{2}\sum_{n=1}^{\infty}(c^{-1}\mu^{\alpha_1\ldots\alpha_n}u^{\alpha_{n+1}}-c^{-1}\mu^{\alpha_1\ldots\alpha_{n-1}\alpha_{n+1}}u^{\alpha_n}+\nu^{\alpha_1\ldots\alpha_{n+1}})\partial^\alpha\partial_{\alpha_1}\ldots\partial_{\alpha_{n-1}}F_{\alpha_n,\alpha_{n+1}}\\ &\qquad+c^{-1}\sum_{n=1}^{\infty}\frac{\mathrm{d}}{\mathrm{d}s}(\mu^{\alpha_1\ldots\alpha_n}\partial_{\alpha_1}\ldots\partial_{\alpha_{n-1}}F^\alpha_{.\alpha_n})\end{aligned} \tag{168}$$

(with $\partial_{\alpha_1} \dots \partial_{\alpha_{n-1}} \equiv 1$ for $n = 1$). This result is the generalization of (111) to the case of arbitrarily varying fields, where all multipoles are needed.

The multipoles employed here were defined with respect to the normal unit vector $p^\alpha/\sqrt{-p^2}$. However, since quadratic field terms are neglected throughout again, one might as well use the multipole moments defined with respect to the normal unit vector $c^{-1}u^\alpha$, as follows from (124). The expression (168) may be written in compact form by introducing the abbreviation (11):

$$\mathfrak{m}^{\alpha_1 \dots \alpha_{n+1}} \equiv c^{-1}(\mu^{\alpha_1 \dots \alpha_n} u^{\alpha_{n+1}} - \mu^{\alpha_1 \dots \alpha_{n-1}\alpha_{n+1}} u^{\alpha_n}) + \nu^{\alpha_1 \dots \alpha_{n+1}}. \tag{169}$$

The electric multipole moment occurring in the last term of (168) may be expressed in terms of this quantity. If only terms of zero order in the fields are considered, one has

$$\mathfrak{m}^{\alpha_1 \dots \alpha_{n+1}} u_{\alpha_{n+1}} = -c\mu^{\alpha_1 \dots \alpha_n}, \tag{170}$$

as follows from (9) with (109), (121) and (124). The expression (168) becomes in this way

$$\mathfrak{f}^\alpha = c^{-1}eF^{\alpha\beta}u_\beta + \tfrac{1}{2}\sum_{n=1}^{\infty} (\partial^\alpha_{.\alpha_1 \dots \alpha_{n-1}} F_{\alpha_n, \alpha_{n+1}})\mathfrak{m}^{\alpha_1 \dots \alpha_{n+1}}$$
$$-c^{-2}\sum_{n=1}^{\infty} \frac{\mathrm{d}}{\mathrm{d}s}\{(\partial_{\alpha_1 \dots \alpha_{n-1}} F^\alpha_{.\alpha_n})\mathfrak{m}^{\alpha_1 \dots \alpha_{n+1}} u_{\alpha_{n+1}}\} \tag{171}$$

(with $\partial_{\alpha_1 \dots \alpha_n} \equiv \partial_{\alpha_1} \dots \partial_{\alpha_n}$), which is the generalization of (128) to all multipole orders.

In a similar way one may find an expression for the total torque (107), starting from its Taylor expansion

$$\mathfrak{d}^{\alpha\beta} = c^{-1}\sum_{n=0}^{\infty} \frac{1}{n!}\sum_i e_i(r_i \cdot \partial)^n (r_i^\alpha F^{\beta\gamma} - r_i^\beta F^{\alpha\gamma})\left(\frac{\mathrm{d}X_\gamma}{\mathrm{d}s} + \frac{\mathrm{d}r_{i\gamma}}{\mathrm{d}s}\right). \tag{172}$$

We find then, as a generalization of (112),

$$\mathfrak{d}^{\alpha\beta} = c^{-1}\sum_{n=1}^{\infty} n\mu^{\alpha_1 \dots \alpha_{n-1}\alpha}\partial_{\alpha_1 \dots \alpha_{n-1}} F^{\beta\gamma}u_\gamma$$
$$+c^{-1}\sum_{n=2}^{\infty}(n-1)\frac{\mathrm{d}}{\mathrm{d}s}(\mu^{\alpha_1 \dots \alpha_{n-2}\alpha\gamma})\partial_{\alpha_1 \dots \alpha_{n-2}} F^\beta_{.\gamma} + \sum_{n=1}^{\infty} \nu^{\alpha_1 \dots \alpha_{n-1}\alpha\gamma}\partial_{\alpha_1 \dots \alpha_{n-1}} F^\beta_{.\gamma}$$
$$-\sum_{n=2}^{\infty}(n-1)\nu^{\alpha_1 \dots \alpha_{n-2}\alpha\gamma\alpha_{n-1}}\partial_{\alpha_1 \dots \alpha_{n-1}} F^\beta_{.\gamma} - (\alpha, \beta). \tag{173}$$

The symbol $(\alpha, \beta)$ indicates terms of the same structure as written down explicitly, but with the indices $\alpha$ and $\beta$ interchanged. Instead of the multipole moments defined with respect to $p^\alpha/\sqrt{-p^2}$, we may use those defined with respect to $c^{-1}u^\alpha$. If one employs (169–170) and the homogeneous field equations one gets, limiting oneself to terms linear in the fields:

$$\mathfrak{d}^{\alpha\beta} = \sum_{n=1}^{\infty} (\partial_{\alpha_1 \dots \alpha_{n-1}} F^{\alpha}_{.\alpha_n}) \mathfrak{m}^{\alpha_1 \dots \alpha_{n+1}} \Delta^{\beta}_{\alpha_{n+1}} - \sum_{n=2}^{\infty} (n-1)(\partial_{\alpha_1 \dots \alpha_{n-1}} F^{\alpha}_{.\alpha_n}) \mathfrak{m}^{\beta\alpha_1 \dots \alpha_n}$$
$$+ c^{-2} \sum_{n=2}^{\infty} (n-1) \frac{\mathrm{d}}{\mathrm{d}s} \{(\partial_{\alpha_1 \dots \alpha_{n-2}} F^{\alpha}_{.\alpha_{n-1}}) \mathfrak{m}^{\beta\alpha_1 \dots \alpha_n} u_{\alpha_n}\} - (\alpha, \beta), \qquad (174)$$

which is the generalization of (129) to all multipole orders.

The equations of motion (125) and (126) are now specified by the expressions (168) and (173), or (171) and (174), for the total force and torque exerted on a composite particle in an arbitrarily varying external field.

### f. *A set of composite particles in a field*

In this subsection the equations governing the behaviour of a system of composite particles in an external field will be derived. The composite particles will be labelled by an index $k$, their constituent particles by the double index $ki$.

It is convenient to consider the law of mass conservation before studying the equation of motion. This conserved mass will be the *rest* mass of the composite particle, not the quantity $m$ (120), since the latter is not conserved in general. (These two quantities are indeed different since $m$ includes contributions from the intra-atomic field, as follows from its definition (120) with (65) and (62).) The rest mass flow density of the system of composite particles is defined as (cf. the analogous definition for the electric four-current density (5))

$$c \sum_k m_k^{\text{rest}} \int u_k^\alpha \delta^{(4)}(X_k - R) \mathrm{d}s_k, \qquad (175)$$

where the rest mass of composite particle $k$ is $m_k^{\text{rest}} \equiv \sum_i m_{ki}$. It obeys the conservation law

$$\partial_\alpha \left\{ c \sum_k m_k^{\text{rest}} \int u_k^\alpha \delta^{(4)}(X_k - R) \mathrm{d}s_k \right\} = 0. \qquad (176)$$

The equations of motion for particle $k$ have the same form as (125) and (126). From these equations one may obtain local balance equations of energy–momentum and angular momentum by multiplying them with the four-dimensional delta function $\delta^{(4)}\{X_k(s_k) - R\}$, integrating over $s_k$ and summing over $k$. Then one obtains, after a partial integration

$$c\partial_\beta \left\{ \sum_k \int m_k u_k^\alpha u_k^\beta \delta^{(4)}(X_k - R) \mathrm{d}s_k \right\} = c \sum_k \int \mathfrak{f}_k^\alpha \delta^{(4)}(X_k - R) \mathrm{d}s_k$$
$$- c^{-1} \partial_\beta \left\{ \sum_k \int (s_k^{\alpha\gamma} \mathfrak{f}_{k\gamma}/m_k + \mathfrak{d}_k^{\alpha\gamma} u_{k\gamma}) u_k^\beta \delta^{(4)}(X_k - R) \mathrm{d}s_k \right\}, \tag{177}$$

$$c\partial_\gamma \left\{ \sum_k \int s_k^{\alpha\beta} u_k^\gamma \delta^{(4)}(X_k - R) \mathrm{d}s_k \right\}$$
$$= c \sum_k \int \{ \Delta_{k\gamma}^\alpha \Delta_{k\varepsilon}^\beta \mathfrak{d}_k^{\gamma\varepsilon} + c^{-2} (s_k^{\alpha\gamma} u_k^\beta - s_k^{\beta\gamma} u_k^\alpha) \mathfrak{f}_{k\gamma}/m_k \} \delta^{(4)}(X_k - R) \mathrm{d}s_k. \tag{178}$$

The total force and torque $\mathfrak{f}_k^\alpha$ and $\mathfrak{d}_k^{\alpha\beta}$ have the forms (106) and (107). The field occurring in these expressions is now the combined field of the other particles $l$ ($\neq k$) and the external field:

$$f^{\alpha\beta} + F_e^{\alpha\beta} = \sum_{l(\neq k)} f_l^{\alpha\beta} + F_e^{\alpha\beta}, \tag{179}$$

so that one has

$$\mathfrak{f}_k^\alpha = c^{-1} \sum_i e_{ki} F_e^{\alpha\beta}(R_{ki}) u_{ki\beta} + c^{-1} \sum_i \sum_{l(\neq k)} e_{ki} f_l^{\alpha\beta}(R_{ki}) u_{ki\beta}, \tag{180}$$

$$\mathfrak{d}_k^{\alpha\beta} = c^{-1} \sum_i e_{ki} r_{ki}^\alpha F_e^{\beta\gamma}(R_{ki}) u_{ki\gamma} + c^{-1} \sum_i \sum_{l(\neq k)} e_{ki} r_{ki}^\alpha f_l^{\beta\gamma}(R_{ki}) u_{ki\gamma} - (\alpha, \beta), \tag{181}$$

where (108) and the notation $u_{ki}^\alpha$ for $\mathrm{d}R_{ki}^\alpha/\mathrm{d}s_k$ have been employed (since $s_k$ is the proper time of the central point of $k$ and not of the particle $ki$ one should note that $u_{ki}^\alpha$ is *not* the four-velocity of particle $ki$.)

The external field $F_e^{\alpha\beta}$ changes slowly over the dimension of a particle, whereas the fields $f_l^{\alpha\beta}$ may change rapidly at the position of particle $k$. Therefore we may expand the terms with the external fields in (180) and (181) and retain only the charge and electric and magnetic dipole moments, just as in subsection c. We then obtain as the external field contributions $\mathfrak{f}_{ke}^\alpha$ and $\mathfrak{d}_{ke}^{\alpha\beta}$ to the total force and torque (cf. (128–129)):

$$\mathfrak{f}_{ke}^\alpha = c^{-1} e_k F_e^{\alpha\beta}(X_k) u_{k\beta} + \tfrac{1}{2} \{ \partial^\alpha F_e^{\beta\gamma}(X_k) \} \mathfrak{m}_{k\beta\gamma} - c^{-2} \frac{\mathrm{d}}{\mathrm{d}s_k} \{ F_e^{\alpha\beta}(X_k) \mathfrak{m}_{k\beta\gamma} u_k^\gamma \}, \tag{182}$$

$$\mathfrak{d}_{ke}^{\alpha\beta} = F_e^{\alpha\gamma}(X_k) \mathfrak{m}_{k\gamma}^{\;\;\varepsilon} \Delta_{k\varepsilon}^\beta - (\alpha, \beta), \tag{183}$$

where $\mathfrak{m}_k^{\alpha\beta}$ was given in terms of the dipole moments $\mu_k^\alpha$ and $\nu_k^{\alpha\beta}$ by (11). (These dipole moments may be defined with respect to the unit vector $p^\alpha/\sqrt{-p^2}$ or $c^{-1}u^\alpha$; the difference between these two cases consist in terms quadratic in the fields.)

We now turn to the contributions in (180) and (181) with the fields $f_l^{\alpha\beta}$ due to the other composite particles. These fields have been found in the preceding chapter. They read for particle $l$:

$$f_l^{\alpha\beta}(R) = \int \hat{f}_l^{\alpha\beta}(R)\mathrm{d}s_l = \int \{\hat{f}_{+l}^{\alpha\beta}(R)+\hat{f}_{-l}^{\alpha\beta}(R)\}\mathrm{d}s_l, \tag{184}$$

where the partial 'plus field' is

$$\hat{f}_{+l}^{\alpha\beta}(R) \equiv -\sum_j \frac{e_{lj}}{4\pi}(u_{lj}^{\alpha}\partial^{\beta}-u_{lj}^{\beta}\partial^{\alpha})\delta\{(R-R_{lj})^2\} \tag{185}$$

and where the partial 'minus field' $\hat{f}_{-l}^{\alpha\beta}$ has the same form except for an extra factor $\varepsilon(R-R_{lj})$. Since we chose the proper time $s_l$ as the parameter along the world line $lj$, the four-vector $u_{lj}^{\alpha}$ is equal to $\mathrm{d}R_{lj}^{\alpha}/\mathrm{d}s_l$ (not the four-velocity of particle $lj$, as explained below formula (181).)

If the observer's point $R^{\alpha}$ is sufficiently far away from the sources we may make a multipole expansion of the field (184). One then obtains for the fields (v. (14–19))

$$f_l^{\alpha\beta} = f_{l(e)}^{\alpha\beta}+f_{l(m)}^{\alpha\beta}, \tag{186}$$

where

$$f_{l(e)}^{\alpha\beta} = \int (\hat{f}_{+l(e)}^{\alpha\beta}+\hat{f}_{-l(e)}^{\alpha\beta})\mathrm{d}s_l, \tag{187}$$

$$f_{l(m)}^{\alpha\beta} = \int (\hat{f}_{+l(m)}^{\alpha\beta}+\hat{f}_{-l(m)}^{\alpha\beta})\mathrm{d}s_l. \tag{188}$$

The partial plus fields are here

$$\hat{f}_{+l(e)}^{\alpha\beta} = -\frac{e_l}{4\pi}(u_l^{\alpha}\partial^{\beta}-u_l^{\beta}\partial^{\alpha})\delta\{(R-X_l)^2\}, \tag{189}$$

$$\hat{f}_{+l(m)}^{\alpha\beta} = -\sum_{n=1}^{\infty}\frac{(-1)^n c}{4\pi}\Big\{\mathfrak{m}_l^{\alpha_1\ldots\alpha_n\alpha}\partial^{\beta}_{.\alpha_1\ldots\alpha_n}$$
$$-c^{-2}\left(\frac{\mathrm{d}}{\mathrm{d}s_l}-u_l\cdot\partial\right)\mathfrak{m}_l^{\alpha_1\ldots\alpha_n\alpha}u_{l\alpha_n}\partial^{\beta}_{.\alpha_1\ldots\alpha_{n-1}}\Big\}\delta\{(R-X_l)^2\}-(\alpha,\beta), \tag{190}$$

where (170) has been used and where the differential operator $\mathrm{d}/\mathrm{d}s_l$ acts on $\mathfrak{m}_l$ and $u_l$. Similar expressions (i.e. with an additional factor $\varepsilon(R-X_l)$) may be written for the partial minus fields.

The contributions of the interatomic fields to the total force and torque (180) and (181) acting on composite particle $k$ is specified if one inserts the interatomic fields (184) with (185). If the atoms are sufficiently far apart one may perform a double multipole expansion, i.e. an expansion of the form (171) and (174) and moreover the expansion (186) with (187–190) of the interatomic field in terms of multipoles of its sources. If the atoms are near each other such a double expansion is not justified. Therefore we write in the general case the force and torque as sums of long range and short range contributions indicated by L and S:

$$\mathfrak{f}^{\alpha}_{k} = \mathfrak{f}^{\mathrm{L}\alpha}_{k} + \mathfrak{f}^{\mathrm{S}\alpha}_{k}, \tag{191}$$

$$\mathfrak{d}^{\alpha\beta}_{k} = \mathfrak{d}^{\mathrm{L}\alpha\beta}_{k} + \mathfrak{d}^{\mathrm{S}\alpha\beta}_{k}. \tag{192}$$

Here the long range contributions are

$$\mathfrak{f}^{\mathrm{L}\alpha}_{k} = \mathfrak{f}^{\alpha}_{ke} + \mathfrak{f}^{\mathrm{L}\alpha}_{k(ee)} + \mathfrak{f}^{\mathrm{L}\alpha}_{k(em)} + \mathfrak{f}^{\mathrm{L}\alpha}_{k(\mathrm{mm})}, \tag{193}$$

$$\mathfrak{d}^{\mathrm{L}\alpha\beta}_{k} = \mathfrak{d}^{\alpha\beta}_{ke} + \mathfrak{d}^{\mathrm{L}\alpha\beta}_{k(ee)} + \mathfrak{d}^{\mathrm{L}\alpha\beta}_{k(em)} + \mathfrak{d}^{\mathrm{L}\alpha\beta}_{k(\mathrm{mm})}. \tag{194}$$

The external field terms $\mathfrak{f}^{\alpha}_{ke}$ and $\mathfrak{d}^{\alpha}_{ke}$ were given already in (182) and (183). The next terms contain the contributions from the charges of the various composite particles. They get the form

$$\mathfrak{f}^{\mathrm{L}\alpha}_{k(ee)} = \sum_{l(\neq k)} \int \hat{\mathfrak{f}}^{\mathrm{L}\alpha}_{k;l(ee)} \, \mathrm{d}s_l, \tag{195}$$

$$\mathfrak{d}^{\mathrm{L}\alpha\beta}_{k(ee)} = 0, \tag{196}$$

with

$$\hat{\mathfrak{f}}^{\mathrm{L}\alpha}_{k;l(ee)} = c^{-1} e_k \hat{f}^{\alpha\beta}_{l(e)}(X_k) u_{k\beta}. \tag{197}$$

The plus field contribution in

$$\hat{\mathfrak{f}}^{\mathrm{L}\alpha}_{k;l(ee)} = \hat{\mathfrak{f}}^{\mathrm{L}\alpha}_{+k;l(ee)} + \hat{\mathfrak{f}}^{\mathrm{L}\alpha}_{-k;l(ee)} \tag{198}$$

follows by inserting in (197) the partial plus field (189). One finds

$$\hat{\mathfrak{f}}^{\mathrm{L}\alpha}_{+k;l(ee)} = -\frac{c^{-1}}{4\pi} e_k e_l (u^{\alpha}_{l} \partial^{\beta}_{k} - u^{\beta}_{l} \partial^{\alpha}_{k}) u_{k\beta} \, \delta(X^{2}_{kl}), \tag{199}$$

where the abbreviations $X^{\alpha}_{kl} \equiv X^{\alpha}_{k} - X^{\alpha}_{l}$ and $\partial_{k\alpha} \equiv \partial/\partial X^{\alpha}_{k}$ have been employed.

In the same way one finds for the last terms in (193) and (194):

$$\mathfrak{f}^{\mathrm{L}\alpha}_{k(\mathrm{mm})} = \sum_{l(\neq k)} \int \hat{\mathfrak{f}}^{\mathrm{L}\alpha}_{k;l(\mathrm{mm})} \, \mathrm{d}s_l, \tag{200}$$

$$\mathfrak{d}^{\mathrm{L}\alpha\beta}_{k(\mathrm{mm})} = \sum_{l(\neq k)} \int \hat{\mathfrak{d}}^{\mathrm{L}\alpha\beta}_{k;l(\mathrm{mm})} \, \mathrm{d}s_l, \tag{201}$$

with – in view of (171) and (174) – the partial forces and torques:

$$\hat{\mathfrak{f}}^{L\alpha}_{k;l(\mathrm{mm})} = \tfrac{1}{2}\sum_{n=1}^{\infty}\{\partial^{\alpha}_{k.\alpha_1\ldots\alpha_{n-1}}\hat{f}_{l(\mathrm{m})\alpha_n,\alpha_{n+1}}(X_k)\}\mathfrak{m}_k^{\alpha_1\ldots\alpha_{n+1}}$$
$$-c^{-2}\sum_{n=1}^{\infty}\frac{\mathrm{d}}{\mathrm{d}s_k}[\{\partial_{k\alpha_1\ldots\alpha_{n-1}}\hat{f}^{\alpha}_{l(\mathrm{m}).\alpha_n}(X_k)\}\mathfrak{m}_k^{\alpha_1\ldots\alpha_{n+1}}u_{k\alpha_{n+1}}], \quad (202)$$

$$\hat{\mathfrak{d}}^{L\alpha\beta}_{k;l(\mathrm{mm})} = \sum_{n=1}^{\infty}\{\partial_{k\alpha_1\ldots\alpha_{n-1}}\hat{f}^{\alpha}_{l(\mathrm{m}).\alpha_n}(X_k)\}\mathfrak{m}_k^{\alpha_1\ldots\alpha_{n+1}}\Delta^{\beta}_{k\alpha_{n+1}}$$
$$-\sum_{n=2}^{\infty}(n-1)\{\partial_{k\alpha_1\ldots\alpha_{n-1}}\hat{f}^{\alpha}_{l(\mathrm{m}).\alpha_n}(X_k)\}\mathfrak{m}_k^{\beta\alpha_1\ldots\alpha_n}$$
$$+c^{-2}\sum_{n=2}^{\infty}(n-1)\frac{\mathrm{d}}{\mathrm{d}s_k}[\{\partial_{k\alpha_1\ldots\alpha_{n-2}}\hat{f}^{\alpha}_{l(\mathrm{m}).\alpha_{n-1}}(X_k)\}\mathfrak{m}_k^{\beta\alpha_1\ldots\alpha_n}u_{k\alpha_n}]-(\alpha,\beta) \quad (203)$$

(with $\partial_{k\alpha_1\ldots\alpha_{n-1}} \equiv 1$ for $n = 1$). The plus field contributions in particular follow by inserting (190):

$$\hat{\mathfrak{f}}^{L\alpha}_{+k;l(\mathrm{mm})} = \frac{c}{4\pi}\sum_{n,m=1}^{\infty}(-1)^m\left\{\mathfrak{m}_k^{\alpha_1\ldots\alpha_{n+1}}\partial_{k\alpha_n}+c^{-2}\left(\frac{\mathrm{d}}{\mathrm{d}s_k}+u_k\cdot\partial_k\right)\mathfrak{m}_k^{\alpha_1\ldots\alpha_{n+1}}u_{k\alpha_n}\right\}$$
$$\left\{\mathfrak{m}_l^{\beta_1\ldots\beta_{m+1}}\partial_{k\beta_m}-c^{-2}\left(\frac{\mathrm{d}}{\mathrm{d}s_l}+u_l\cdot\partial_l\right)\mathfrak{m}_l^{\beta_1\ldots\beta_{m+1}}u_{l\beta_m}\right\}$$
$$\partial_{k\alpha_1\ldots\alpha_{n-1}\beta_1\ldots\beta_{m-1}}(\partial^{\alpha}_{k}g_{\alpha_{n+1}\beta_{m+1}}-\partial_{k\alpha_{n+1}}g^{\alpha}_{\beta_{m+1}})\delta(X^2_{kl}), \quad (204)$$

$$\hat{\mathfrak{d}}^{L\alpha\beta}_{+k;l(\mathrm{mm})} = \frac{c}{4\pi}\sum_{n,m=1}^{\infty}(-1)^{m+1}\left\{\Delta^{\alpha}_{k\alpha_{n+1}}\mathfrak{m}_k^{\alpha_1\ldots\alpha_{n+1}}\partial_{k\alpha_{n-1}}-(n-1)\mathfrak{m}_k^{\alpha\alpha_1\ldots\alpha_n}\partial_{k\alpha_{n-1}}\right.$$
$$\left.-c^{-2}(n-1)\left(\frac{\mathrm{d}}{\mathrm{d}s_k}+u_k\cdot\partial_k\right)\mathfrak{m}_k^{\alpha\alpha_1\ldots\alpha_n}u_{k\alpha_{n-1}}\right\}$$
$$\left\{\mathfrak{m}_l^{\beta_1\ldots\beta_{m+1}}\partial_{k\beta_m}-c^{-2}\left(\frac{\mathrm{d}}{\mathrm{d}s_l}+u_l\cdot\partial_l\right)\mathfrak{m}_l^{\beta_1\ldots\beta_{m+1}}u_{l\beta_m}\right\}$$
$$\partial_{k\alpha_1\ldots\alpha_{n-2}\beta_1\ldots\beta_{m-1}}(\partial^{\beta}_{k}g_{\alpha_n\beta_{m+1}}-g^{\beta}_{\beta_{m+1}}\partial_{k\alpha_n})\delta(X^2_{kl})-(\alpha,\beta). \quad (205)$$

The cross-terms $(e\mathfrak{m})$ have similar form.

The short range terms in (191) and (192) follow from (180) and (181). One finds:

$$\mathfrak{f}^{S\alpha}_{k} = \sum_{l(\neq k)}\int\hat{\mathfrak{f}}^{S\alpha}_{k;l}\,\mathrm{d}s_l, \quad (206)$$

$$\mathfrak{d}^{S\alpha\beta}_{k} = \sum_{l(\neq k)}\int\hat{\mathfrak{d}}^{S\alpha\beta}_{k;l}\mathrm{d}s_l, \quad (207)$$

with

$$\hat{\mathfrak{f}}^{S\alpha}_{k;l} = c^{-1}\sum_i e_{ki}\hat{f}_l^{\alpha\beta}(R_{ki})u_{ki\beta} - \hat{\mathfrak{f}}^{L\alpha}_{k;l(ee)} - \mathfrak{f}^{L\alpha}_{k;l(em)} - \hat{\mathfrak{f}}^{L\alpha}_{k;l(mm)}, \tag{208}$$

$$\hat{\mathfrak{d}}^{S\alpha\beta}_{k;l} = c^{-1}\sum_i e_{ki}\{r^{\alpha}_{ki}\hat{f}_l^{\beta\gamma}(R_{ki}) - r^{\beta}_{ki}\hat{f}_l^{\alpha\gamma}(R_{ki})\}u_{ki\gamma} - \hat{\mathfrak{d}}^{L\alpha\beta}_{k;l(em)} - \hat{\mathfrak{d}}^{L\alpha\beta}_{k;l(mm)}. \tag{209}$$

Explicitly the plus field parts become with (184) and (185):

$$\mathfrak{f}^{S\alpha}_{+k;l} = \frac{c^{-1}}{4\pi}\sum_{i,j} e_{ki}e_{lj}(u_{ki}\cdot u_{lj}\,\partial^{\alpha}_{ki} - u^{\alpha}_{lj}\,u_{ki}\cdot\partial_{ki})\delta(R^2_{ki,lj})$$
$$- \hat{\mathfrak{f}}^{L\alpha}_{+k;l(ee)} - \mathfrak{f}^{L\alpha}_{+k;l(em)} - \mathfrak{f}^{L\alpha}_{+k;l(mm)}, \tag{210}$$

$$\hat{\mathfrak{d}}^{S\alpha\beta}_{+k;l} = \frac{c^{-1}}{4\pi}\sum_{i,j} e_{ki}e_{lj}\,r^{\alpha}_{ki}(u_{ki}\cdot u_{lj}\,\partial^{\beta}_{ki} - u^{\beta}_{lj}\,u_{ki}\cdot\partial_{ki})\delta(R^2_{ki,lj}) - (\alpha,\beta)$$
$$- \hat{\mathfrak{d}}^{L\alpha\beta}_{+k;l(em)} - \hat{\mathfrak{d}}^{L\alpha\beta}_{+k;l(mm)}. \tag{211}$$

In this way expressions have been found for the force and torque that occur in the balance equations of energy–momentum and angular momentum (177) and (178) for a system of composite particles in an external field. They will form the starting point for the statistical considerations of the next chapter.

## APPENDIX I

# Some properties of the tensor $\boldsymbol{\Omega}$

In this appendix we collect a number of properties of the tensor $\boldsymbol{\Omega}(\boldsymbol{\beta})$, which is defined as

$$\boldsymbol{\Omega}(\boldsymbol{\beta}) = \mathbf{U}+\frac{\gamma^{-1}-1}{\beta^2}\boldsymbol{\beta}\boldsymbol{\beta} = \mathbf{U}-\frac{\gamma}{\gamma+1}\boldsymbol{\beta}\boldsymbol{\beta}, \tag{A1}$$

where $\gamma = (1-\beta^2)^{-\frac{1}{2}}$ and $\mathbf{U}$ is the unit tensor. Its frequent use in this chapter stems from the fact that it describes the effect of a Lorentz contraction. In fact, if $\boldsymbol{a}$ is an arbitrary vector one has from (A1):

$$\boldsymbol{\Omega}\cdot\boldsymbol{a} = \boldsymbol{a}_{\perp}+\sqrt{1-\beta^2}\boldsymbol{a}_{//}, \tag{A2}$$

where $\boldsymbol{a}_{//} = \boldsymbol{\beta}\boldsymbol{\beta}\cdot\boldsymbol{a}/\beta^2$ and $\boldsymbol{a}_{\perp} = \boldsymbol{a}-\boldsymbol{a}_{//}$ are the components of $\boldsymbol{a}$ orthogonal and parallel to $\boldsymbol{\beta}$ respectively. The longitudinal component of the vector $\boldsymbol{a}$ is thus seen to be subjected to a Lorentz contraction.

The inverse tensor $\boldsymbol{\Omega}^{-1}(\boldsymbol{\beta})$, which obeys the relation that its product with $\boldsymbol{\Omega}(\boldsymbol{\beta})$ is the unit tensor, is:

$$\boldsymbol{\Omega}^{-1}(\boldsymbol{\beta}) = \mathbf{U}+\frac{\gamma-1}{\beta^2}\boldsymbol{\beta}\boldsymbol{\beta} = \mathbf{U}+\frac{\gamma^2}{\gamma+1}\boldsymbol{\beta}\boldsymbol{\beta}, \tag{A3}$$

as may be checked directly. The tensor $\boldsymbol{\Omega}^{-1}$ occurs in the formulae for the Lorentz transformation (with transformation velocity $c\boldsymbol{\beta}$) connecting the frames $(ct_1, \boldsymbol{R}_1)$ and $(ct_2, \boldsymbol{R}_2)$:

$$\begin{aligned} ct_2 &= \gamma ct_1+\gamma\boldsymbol{\beta}\cdot\boldsymbol{R}_1, \\ \boldsymbol{R}_2 &= \boldsymbol{\Omega}^{-1}\cdot\boldsymbol{R}_1+\gamma\boldsymbol{\beta}ct_1. \end{aligned} \tag{A4}$$

On the contrary the transformation formulae for an antisymmetric tensor $\mathbf{A}$ with components $\boldsymbol{X} = (A^{01}, A^{02}, A^{03})$ and $\boldsymbol{Y} = (A^{23}, A^{31}, A^{12})$ contain the tensor $\boldsymbol{\Omega}$. They read

$$\begin{aligned} \boldsymbol{X}_2 &= \gamma(\boldsymbol{\Omega}\cdot\boldsymbol{X}_1-\boldsymbol{\beta}\wedge\boldsymbol{Y}_1), \\ \boldsymbol{Y}_2 &= \gamma(\boldsymbol{\Omega}\cdot\boldsymbol{Y}_1+\boldsymbol{\beta}\wedge\boldsymbol{X}_1). \end{aligned} \tag{A5}$$

From (A1) and (A3) one derives the properties

$$\boldsymbol{\Omega}^{-1} = \boldsymbol{\Omega}+\gamma\boldsymbol{\beta}\boldsymbol{\beta}, \tag{A6}$$

$$\boldsymbol{\Omega}\cdot\boldsymbol{\beta} = \gamma^{-1}\boldsymbol{\beta}, \tag{A7}$$

$$\boldsymbol{\Omega}^{-1}\cdot\boldsymbol{\beta} = \gamma\boldsymbol{\beta}, \tag{A8}$$

$$\boldsymbol{\beta}\wedge\boldsymbol{\Omega}\cdot\boldsymbol{a} = \boldsymbol{\beta}\wedge\boldsymbol{a}, \tag{A9}$$

$$\boldsymbol{\Omega}\cdot(\boldsymbol{a}\wedge\boldsymbol{b}) = \gamma^{-1}(\boldsymbol{\Omega}^{-1}\cdot\boldsymbol{a})\wedge(\boldsymbol{\Omega}^{-1}\cdot\boldsymbol{b}), \tag{A10}$$

$$\boldsymbol{\Omega}^{-1}\cdot(\boldsymbol{a}\wedge\boldsymbol{b}) = \gamma(\boldsymbol{\Omega}\cdot\boldsymbol{a})\wedge(\boldsymbol{\Omega}\cdot\boldsymbol{b}), \tag{A11}$$

where $\boldsymbol{a}$ and $\boldsymbol{b}$ are two arbitrary vectors. The squares of $\boldsymbol{\Omega}$ and $\boldsymbol{\Omega}^{-1}$ have the forms

$$\boldsymbol{\Omega}^2 = \mathbf{U}-\boldsymbol{\beta}\boldsymbol{\beta}, \tag{A12}$$

$$\boldsymbol{\Omega}^{-2} = \mathbf{U}+\gamma^2\boldsymbol{\beta}\boldsymbol{\beta}. \tag{A13}$$

The tensor $\boldsymbol{\Omega}^2$ occurs in the relation

$$\boldsymbol{\Omega}^2\cdot\partial_0(\boldsymbol{\beta}\gamma) = \gamma\partial_0\boldsymbol{\beta}. \tag{A14}$$

It may be proved from (A12) if use is made of the identity

$$\boldsymbol{\beta}\cdot\partial_0\boldsymbol{\beta} = \gamma^{-3}\partial_0\gamma, \tag{A15}$$

which follows from the definition of $\gamma$. Finally the determinant values of $\boldsymbol{\Omega}$ and $\boldsymbol{\Omega}^{-1}$ are

$$|\boldsymbol{\Omega}| = \gamma^{-1}, \qquad |\boldsymbol{\Omega}^{-1}| = \gamma, \tag{A16}$$

as follows from (A1) and (A3).

APPENDIX II

# The connexion between the covariant and the atomic multipole moments

The covariant multipole moments, depending on a certain time-like unit vector and defined in (9), contain quantities in the observer's frame $(ct, \boldsymbol{R})$. They still depend on the velocities of the atoms. Therefore we introduced atomic multipole moments (27, 28), defined in momentary rest frames; they are thus independent of the atomic velocities. The connexion between these atomic multipole moments and the covariant multipole moments (with the four-velocity chosen as the time-like unit vector) will be obtained in this appendix by studying the Lorentz transformation between the observer's frame and the momentary rest frames.

The covariant multipole moments contain the internal parameters $r_{ki}^{\alpha}$ (3) that fulfil the relation (2) with the four-velocity $c^{-1}u_k^{\alpha}(s_k)$ chosen as the unit vector $n_k^{\alpha}(s_k)$ i.e.:

$$r_{ki\alpha}(s_k)u_k^{\alpha}(s_k) \equiv -r_{ki}^0(s_k)u_k^0(s_k) + \boldsymbol{r}_{ki}(s_k)\cdot\boldsymbol{u}_k(s_k) = 0. \tag{A17}$$

This covariant condition means that in the atomic rest frame (where $\boldsymbol{u}_k^{(0)}(s_k) = 0$ at the moment $t = t_0$, or correspondingly for $s_k = s_{k0}$) the vectors $\boldsymbol{r}_{ki}^{(0)}$ become purely spatial ($r_{ki}^{(0)0}$ then vanishes according to the condition) and thus constitute at that moment the atomic parameters, which we want to employ for the characterization of the atoms.

The frame in which the atom as a whole is momentarily at rest must have a velocity $\boldsymbol{v}$ equal to $(\mathrm{d}\boldsymbol{R}_k/\mathrm{d}t)_{t=t_0}$ at the moment $t = t_0$ with respect to the reference frame $(ct, \boldsymbol{R})$ of the observer. Time–space coordinates of the reference frame $(ct, \boldsymbol{R})$ and the atomic rest frame $(ct^{(0)}, \boldsymbol{R}^{(0)})$ are connected by the Lorentz transformation (A4):

$$\begin{aligned} ct &= \gamma ct^{(0)} + \gamma\boldsymbol{\beta}\cdot\boldsymbol{R}^{(0)}, \\ \boldsymbol{R} &= \boldsymbol{\Omega}^{-1}\cdot\boldsymbol{R}^{(0)} + \gamma\boldsymbol{\beta}ct^{(0)}, \end{aligned} \tag{A18}$$

where the tensor (A3) has been used.

Since the atom suffers accelerations, at every moment $t_0$ one needs a different atomic rest frame. Every atomic frame is therefore only a *momentary* rest frame: only for $t = t_0$ does the atomic velocity $\mathrm{d}\boldsymbol{R}_k^{(0)}/\mathrm{d}t^{(0)}$ vanish. The transformation which connects the reference frame to the momentary atomic rest frame (in which the atom is at rest at time $t_0$) is therefore determined by the transformation velocity:

$$\boldsymbol{\beta}_k = \frac{\boldsymbol{v}_k}{c} = \frac{1}{c}\left(\frac{\mathrm{d}\boldsymbol{R}_k}{\mathrm{d}t}\right)_{t=t_0} = \left(\frac{\mathrm{d}\boldsymbol{R}_k/\mathrm{d}s_k}{\mathrm{d}R_k^0/\mathrm{d}s_k}\right)_{s_k=s_{k0}}, \tag{A19}$$

with $s_{k0}$ corresponding to $t_0$. For the atomic position four-vector $R_k^\alpha$ this Lorentz transformation reads

$$\begin{aligned} R_k^0(s_k) &= \gamma_k R_k^{(0)0}(s_k) + \gamma_k \boldsymbol{\beta}_k \cdot \boldsymbol{R}_k^{(0)}(s_k), \\ \boldsymbol{R}_k(s_k) &= \boldsymbol{\Omega}_k^{-1} \cdot \boldsymbol{R}_k^{(0)}(s_k) + \gamma_k \boldsymbol{\beta}_k R_k^{(0)0}(s_k). \end{aligned} \tag{A20}$$

Differentiation with respect to $s$ gives

$$\begin{aligned} \frac{\mathrm{d}R_k^0}{\mathrm{d}s_k} &= \gamma_k \frac{\mathrm{d}R_k^{(0)0}}{\mathrm{d}s_k} + \gamma_k \boldsymbol{\beta}_k \cdot \frac{\mathrm{d}\boldsymbol{R}_k^{(0)}}{\mathrm{d}s_k}, \\ \frac{\mathrm{d}\boldsymbol{R}_k}{\mathrm{d}s_k} &= \boldsymbol{\Omega}_k^{-1} \cdot \frac{\mathrm{d}\boldsymbol{R}_k^{(0)}}{\mathrm{d}s_k} + \gamma_k \boldsymbol{\beta}_k \frac{\mathrm{d}R_k^{(0)0}}{\mathrm{d}s_k}. \end{aligned} \tag{A21}$$

Similarly for the internal quantities one has the Lorentz transformation

$$\begin{aligned} r_{ki}^0(s_k) &= \gamma_k r_{ki}^{(0)0}(s_k) + \gamma_k \boldsymbol{\beta}_k \cdot \boldsymbol{r}_{ki}^{(0)}(s_k), \\ \boldsymbol{r}_{ki}(s_k) &= \boldsymbol{\Omega}_k^{-1} \cdot \boldsymbol{r}_{ki}^{(0)}(s_k) + \gamma_k \boldsymbol{\beta}_k r_{ki}^{(0)0}(s_k), \end{aligned} \tag{A22}$$

and thus for the derivatives

$$\begin{aligned} \frac{\mathrm{d}r_{ki}^0}{\mathrm{d}s_k} &= \gamma_k \frac{\mathrm{d}r_{ki}^{(0)0}}{\mathrm{d}s_k} + \gamma_k \boldsymbol{\beta}_k \cdot \frac{\mathrm{d}\boldsymbol{r}_{ki}^{(0)}}{\mathrm{d}s_k}, \\ \frac{\mathrm{d}\boldsymbol{r}_{ki}}{\mathrm{d}s_k} &= \boldsymbol{\Omega}_k^{-1} \cdot \frac{\mathrm{d}\boldsymbol{r}_{ki}^{(0)}}{\mathrm{d}s_k} + \gamma_k \boldsymbol{\beta}_k \frac{\mathrm{d}r_{ki}^{(0)0}}{\mathrm{d}s_k}. \end{aligned} \tag{A23}$$

In the momentary atomic rest frame the atomic velocity vanishes: $\mathrm{d}\boldsymbol{R}_k^{(0)}/\mathrm{d}s_k = 0$ for $s_k = s_{k0}$. Furthermore, as a consequence of the orthogonality relation (A17) the internal parameter $r_{ki}^\alpha$ becomes purely space-like: $r_{ki}^{(0)\alpha} = (0, \boldsymbol{r}_{ki}^{(0)})$ for $s_k = s_{k0}$. These relations can be formulated conveniently with the help of a coordinate frame in which the atom is at rest all the time. This frame, which will be called the *permanent atomic rest frame* (denoted by a prime) is a succession of Lorentz frames, not a Lorentz frame itself; it coincides at time $t_0$ with the momentary atomic rest frame (denoted by (0)). The permanent rest frame is connected with the reference frame by a Lorentz transformation with velocity

$$\boldsymbol{\beta}_k(s_k) = \frac{\mathrm{d}\boldsymbol{R}_k/\mathrm{d}s_k}{\mathrm{d}R_k^0/\mathrm{d}s_k}. \tag{A24}$$

With the help of $\boldsymbol{\beta}_k(s_k)$ we define $\boldsymbol{\Omega}_k(s_k)$ analogous to (A1) with $\gamma_k(s_k) = \{1-\beta_k^2(s_k)\}^{-\frac{1}{2}}$. In the permanent rest frame the atomic velocity vanishes identically: $(\mathrm{d}\boldsymbol{R}_k/\mathrm{d}s_k)' = 0$ for all $s_k$; moreover $(r_{ki}^0)' = 0$ for all $s_k$. From (A24) it follows that

$$\frac{\mathrm{d}R_k^0}{\mathrm{d}s_k} = c\gamma_k(s_k), \qquad \frac{\mathrm{d}\boldsymbol{R}_k}{\mathrm{d}s_k} = c\gamma_k(s_k)\boldsymbol{\beta}_k(s_k). \tag{A25}$$

Furthermore, (A22) may be written with the help of quantities in the permanent rest frame:

$$r_{ki}^0(s_k) = \gamma_k(s_k)\boldsymbol{\beta}_k(s_k)\cdot\boldsymbol{r}'_{ki}(s_k), \qquad \boldsymbol{r}_{ki}(s_k) = \boldsymbol{\Omega}_k^{-1}(s_k)\cdot\boldsymbol{r}'_{ki}(s_k). \tag{A26}$$

Finally (A23) becomes

$$\frac{\mathrm{d}r_{ki}^0}{\mathrm{d}s_k} = \gamma_k(s_k)\left(\frac{\mathrm{d}r_{ki}^0}{\mathrm{d}s_k}\right)' + \gamma_k(s_k)\boldsymbol{\beta}_k(s_k)\cdot\left(\frac{\mathrm{d}\boldsymbol{r}_{ki}}{\mathrm{d}s_k}\right)', \qquad \frac{\mathrm{d}\boldsymbol{r}_{ki}}{\mathrm{d}s_k} = \boldsymbol{\Omega}_k^{-1}(s_k)\cdot\left(\frac{\mathrm{d}\boldsymbol{r}_{ki}}{\mathrm{d}s_k}\right)' + \gamma_k(s_k)\boldsymbol{\beta}_k(s_k)\left(\frac{\mathrm{d}r_{ki}^0}{\mathrm{d}s_k}\right)'. \tag{A27}$$

Since internal quantities are to be defined in the atomic rest frame, but external quantities (atomic positions, velocities etc.) in the reference frame, one needs a few more consequences of the preceding transformation formulae. In the first place we want an expression for the second derivative of $\boldsymbol{R}_k^{(0)}(s_k)$ with respect to $s_k$. According to the Lorentz transformation one has

$$\left(\frac{\mathrm{d}^2\boldsymbol{R}_k}{\mathrm{d}s_k^2}\right)' = \boldsymbol{\Omega}_k^{-1}(s_k)\cdot\frac{\mathrm{d}^2\boldsymbol{R}_k}{\mathrm{d}s_k^2} - \gamma_k(s_k)\boldsymbol{\beta}_k(s_k)\frac{\mathrm{d}^2R_k^0}{\mathrm{d}s_k^2}. \tag{A28}$$

The second derivatives at the right-hand side follow from (A25):

$$\frac{\mathrm{d}^2\ {}_k}{\mathrm{d}s_k^2} = c^2\gamma_k\partial_0(\gamma_k\boldsymbol{\beta}_k), \qquad \frac{\mathrm{d}^2R_k^0}{\mathrm{d}s_k^2} = c^2\gamma_k\partial_0\gamma_k. \tag{A29}$$

Substituting these expressions and using (A6) and (A15) one finds:

$$\left(\frac{\mathrm{d}^2\boldsymbol{R}_k}{\mathrm{d}s_k^2}\right)' = c^2\gamma_k\,\boldsymbol{\Omega}_k\cdot\partial_0(\gamma_k\,\boldsymbol{\beta}_k). \tag{A30}$$

A second result can be obtained from the invariant condition (A17), which reads in the momentary atomic rest frame

$$r_{ki}^{(0)0}\,\frac{\mathrm{d}R_k^{(0)0}}{\mathrm{d}s_k} - \boldsymbol{r}_{ki}^{(0)}\cdot\frac{\mathrm{d}\boldsymbol{R}_k^{(0)}}{\mathrm{d}s_k} = 0. \tag{A31}$$

Differentiating this relation with respect to $s$ and taking into account $r_{ki}^{0\prime} = 0$ and $(\mathrm{d}\boldsymbol{R}_k/\mathrm{d}s_k)' = 0$, one finds in the permanent atomic rest frame

$$\left(\frac{\mathrm{d}r_{ki}^0}{\mathrm{d}s_k}\right)' = c^{-1}\boldsymbol{r}_{ki}'(s_k)\cdot(\mathrm{d}^2\boldsymbol{R}_k/\mathrm{d}s_k^2)', \tag{A32}$$

which, with (A30), becomes finally

$$\left(\frac{\mathrm{d}r_{ki}^0}{\mathrm{d}s_k}\right)' = c\gamma_k\,\boldsymbol{r}_{ki}'\cdot\boldsymbol{\Omega}_k\cdot\partial_0(\gamma_k\,\boldsymbol{\beta}_k). \tag{A33}$$

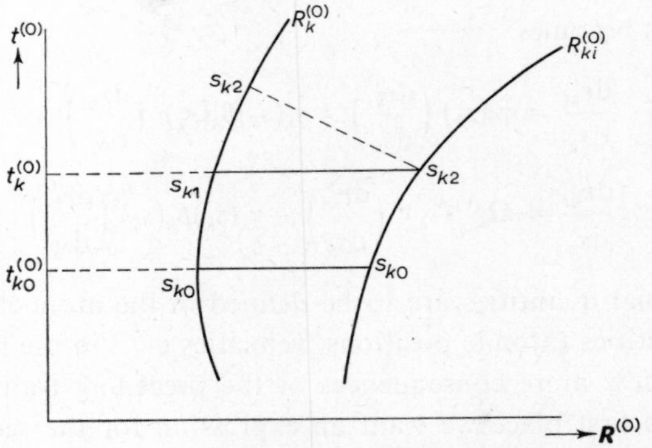


**Fig. 1. World lines of atom $k$ and constituent particle $ki$ in the momentary atomic rest frame.**

In the definition (28) of the magnetic multipole moments a time derivative of the internal coordinates occurs which is the limit of the difference of two purely spatial vectors divided by the corresponding time difference (cf. fig. 1):

$$\dot{\boldsymbol{r}}_{ki}' = \lim_{t_k{}^{(0)}\to t_{k0}{}^{(0)}} \frac{\boldsymbol{R}_{ki}^{(0)}(s_{k2}) - \boldsymbol{R}_k^{(0)}(s_{k1}) - \boldsymbol{r}_{ki}^{(0)}(s_{k0})}{t_k^{(0)} - t_{k0}^{(0)}}$$

$$= \lim_{s_{k1}\to s_{k0}} \frac{\boldsymbol{R}_{ki}^{(0)}(s_{k2}) - \boldsymbol{R}_k^{(0)}(s_{k1}) - \boldsymbol{r}_{ki}^{(0)}(s_{k0})}{s_{k1} - s_{k0}}. \tag{A34}$$

The values $s_{k1}$ and $s_{k2}$ of the parameter $s_k$ are related by (cf. fig. 1)

$$R_{ki}^{(0)0}(s_{k2}) = R_k^{(0)0}(s_{k1}), \tag{A35}$$

or, with the splitting (3)

$$R_k^{(0)0}(s_{k2})+r_{ki}^{(0)0}(s_{k2}) = R_k^{(0)0}(s_{k1}). \tag{A36}$$

A Taylor expansion of the left- and right-hand sides with respect to $s_{k2}-s_{k0}$ and $s_{k1}-s_{k0}$ respectively gives

$$s_{k2}-s_{k0} = \left(1+c^{-1}\frac{\mathrm{d}r_{ki}^{(0)0}}{\mathrm{d}s_k}\right)^{-1}(s_{k1}-s_{k0})+\dots. \tag{A37}$$

With the help of this relation, expression (A34) becomes after expansion of the numerator around $s_{k0}$:

$$\dot{\boldsymbol{r}}'_{ki} = \frac{\mathrm{d}\boldsymbol{r}_{ki}^{(0)}}{\mathrm{d}s_k}\left(1+c^{-1}\frac{\mathrm{d}r_{ki}^{(0)0}}{\mathrm{d}s_k}\right)^{-1}. \tag{A38}$$

This expression is derived for $s_k = s_{k0}$. Introducing quantities measured in the permanent rest frame we get a relation valid for all $s_k$:

$$\dot{\boldsymbol{r}}'_{ki} = \left(\frac{\mathrm{d}\boldsymbol{r}_{ki}}{\mathrm{d}s_k}\right)'\left\{1+c^{-1}\left(\frac{\mathrm{d}r_{ki}^{0}}{\mathrm{d}s_k}\right)'\right\}^{-1}. \tag{A39}$$

With the use of (A33) this can be written in the form

$$\left(\frac{\mathrm{d}\boldsymbol{r}_{ki}}{\mathrm{d}s_k}\right)' = \dot{\boldsymbol{r}}'_{ki}\{1+\gamma_k\,\boldsymbol{r}'_{ki}\cdot\boldsymbol{\Omega}_k\cdot\partial_0(\gamma_k\,\boldsymbol{\beta}_k)\}, \tag{A40}$$

which gives $(\mathrm{d}\boldsymbol{r}_{ki}/\mathrm{d}s_k)'$ in terms of the internal quantities $\boldsymbol{r}'_{ki}$ and $\dot{\boldsymbol{r}}'_{ki}$ occurring in the multipole moments which characterize the atomic structure.

With the help of the results that followed from the Lorentz transformation, we now write the connexion between the covariant multipole moments (9) with the choice $c^{-1}\mathrm{d}R_k^{\alpha}/\mathrm{d}s_k$ for $n^{\alpha}$ and the atomic multipole moments (27–28). To that end we substitute into (9) the transformation formulae (A25–A27), using also (A33) and (A40). Then one obtains (omitting the index $(n) = (c^{-1}\mathrm{d}R_k/\mathrm{d}s_k)$) for the covariant electric multipole moment

$$\mu_k^{0\dots0 i_{m+1}\dots i_n} = \{(\boldsymbol{\Omega}_k^{-1})^{n-m}(\gamma_k\,\boldsymbol{\beta}_k)^m : \boldsymbol{\mu}_k^{(n)}\}^{i_{m+1}\dots i_n},$$

$$(n = 1, 2, \dots;\ m = 0, 1, \dots, n), \tag{A41}$$

where the indices $\alpha_1, \dots, \alpha_m$ have been chosen as zero and the indices $\alpha_{m+1}, \dots, \alpha_n$ as $i_{m+1} \dots i_n$, which can take the values 1, 2, 3. The symmetrical character of the covariant electric multipole moment could be used to write the zeros first. The covariant magnetic multipole moment has components

$$\nu_k^{0\ldots 0 i_{m+1}\ldots i_{n-1} ij} = \Big\{ \gamma_k (\boldsymbol{\Omega}_k^{-1})^{n-m-1} (\gamma_k \boldsymbol{\beta}_k)^m : \boldsymbol{\nu}_k^{(n)} \cdot \boldsymbol{\Omega}_k$$

$$+ \frac{n(n+2)}{n+1} \gamma_k^2 (\boldsymbol{\Omega}_k^{-1})^{n-m-1} (\gamma_k \boldsymbol{\beta}_k)^m \partial_0 (\gamma_k \boldsymbol{\beta}_k) \cdot \boldsymbol{\Omega}_k : \boldsymbol{\nu}_k^{(n+1)} \cdot \boldsymbol{\Omega}_k$$

$$+ n\gamma_k^2 (\boldsymbol{\Omega}_k^{-1})^{n-m-1} (\gamma_k \boldsymbol{\beta}_k)^m \partial_0 (\gamma_k \boldsymbol{\beta}_k) \cdot \boldsymbol{\Omega}_k : \boldsymbol{\mu}_k^{(n+1)} \wedge \boldsymbol{\beta}_k \Big\}^{i_{m+1}\ldots i_{n-1} p},$$

$$(n = 1, 2, \ldots;\ m = 0, \ldots, n-1;\ i, j, p = 1, 2, 3 \text{ cycl.}), \quad \text{(A42)}$$

$$\nu_k^{0\ldots 0 i_{m+1}\ldots i_{n-1} i0} = \Big\{ -\gamma_k (\boldsymbol{\Omega}_k^{-1})^{n-m-1} (\gamma_k \boldsymbol{\beta}_k)^m \boldsymbol{\nu}_k^{(n)} \wedge \boldsymbol{\beta}_k$$

$$- \frac{n(n+2)}{n+1} \gamma_k^2 (\boldsymbol{\Omega}_k^{-1})^{n-m-1} (\gamma_k \boldsymbol{\beta}_k)^m \partial_0 (\gamma_k \boldsymbol{\beta}_k) \cdot \boldsymbol{\Omega}_k : \boldsymbol{\nu}_k^{(n+1)} \wedge \boldsymbol{\beta}_k$$

$$+ n\gamma_k^2 (\boldsymbol{\Omega}_k^{-1})^{n-m-1} (\gamma_k \boldsymbol{\beta}_k)^m \partial_0 (\gamma_k \boldsymbol{\beta}_k) \cdot \boldsymbol{\Omega}_k : \boldsymbol{\mu}_k^{(n+1)} \cdot \boldsymbol{\Omega}_k \Big\}^{i_{m+1}\ldots i_{n-1} i},$$

$$(n = 1, 2, \ldots;\ m = 0, \ldots, n-1), \quad \text{(A43)}$$

where we used (A6–A10) and the fact that the covariant magnetic multipole moment is symmetric in its first $n-1$ indices and antisymmetric in its last two indices.

From these connexionsit is apparent that in the absence of acceleration the purely space-like components of the covariant electromagnetic multipole moments (the case $m = 0$) in the rest frame coincide with the atomic electromagnetic moments; the mixed space-time components vanish in that frame. If accelerations are present this needs no longer be the case.

APPENDIX III

# On equations of motion with explicit radiation damping

In the main text we carried out a programme to obtain equations of motion for a composite particle in an external field. The basis of these equations consisted in the microscopic balance equation (61) with (62–63). Essential steps were the definition of a centre of energy and the derivation of the equations of motion and spin in the linear field approximation.

Part of this programme – but indeed only part of it – may be accomplished as well on the basis of a different form, namely (III.A26), of the balance equation, derived in the appendix of the preceding chapter:

$$\partial_\beta t^{*\alpha\beta} = f^{*\alpha} \tag{A44}$$

with the (symmetric) energy–momentum tensor $t^{*\alpha\beta}$ (III.A25), which consists of a material part and a field part that accounts for the interactions due to the plus fields of the particles:

$$\begin{aligned} t^{*\alpha\beta} &= c\sum_i m_i \int u_i^\alpha u_i^\beta \delta^{(4)}(R_i - R)\mathrm{d}s_i \\ &+ \frac{1}{8\pi}\sum_{i,j(i\neq j)} e_i e_j \int (u_i^\alpha u_j^\beta + u_i^\beta u_j^\alpha)\delta\{(R_i-R_j)^2\}\delta^{(4)}(R_i-R)\mathrm{d}s_i\,\mathrm{d}s_j \\ &+ \frac{1}{8\pi}\sum_{i,j(i\neq j)} e_i e_j \int\int_{\lambda=0}^{1} [2u_i\cdot u_j (R_i-R_j)^\alpha (R_i-R_j)^\beta \delta'\{(R_i-R_j)^2\} \\ &+ \{u_i^\alpha (R_i-R_j)^\beta + u_i^\beta (R_i-R_j)^\alpha\} u_j^\gamma \partial_\gamma \delta\{(R_i-R_j)^2\}] \\ &\qquad \delta^{(4)}\{R_j + \lambda(R_i-R_j) - R\}\mathrm{d}s_i\,\mathrm{d}s_j\,\mathrm{d}\lambda. \end{aligned} \tag{A45}$$

This tensor has the special property that it vanishes outside the domain enclosed by the world lines. The force density $f^{*\alpha}$ was found to be

$$f^{*\alpha} = f^\alpha + \sum_{i,j}\int f_{-j}^{\alpha\beta} u_{i\beta}\,\delta^{(4)}(R_i - R)\mathrm{d}s_i \tag{A46}$$

(v. (III.A3)). It contains, just as (63), the Lorentz forces due to the external field, and moreover those due to the minus fields, generated by the constituent particles $j$ (including the particle $i$ itself).

The first step in the derivation is the application of the general definition of the centre of energy as given in section 3*b* to the particular case of a composite particle described by the energy–momentum tensor $t^{*\alpha\beta}$ (A45) and acted upon by the force density $f^{*\alpha}$ (A46). To that end we have to check whether all assumptions used there are indeed justified for the present case.

To begin with, we notice that the tensor $t^{*\alpha\beta}$ is symmetric and that the force $f^{*\alpha}$ has finite support. Furthermore the integrals

$$p^{*\alpha} \equiv -c^{-1}\int t^{*\alpha\beta} n_\beta \,\mathrm{d}^3\Sigma \tag{A47}$$

for the total momentum (which is assumed to be time-like with positive time-component) and

$$s^{*\alpha\beta} \equiv -c^{-1}\int \{(R^\alpha - X^{*\alpha})t^{*\beta\gamma} - (R^\beta - X^{*\beta})t^{*\alpha\gamma}\} n_\gamma \,\mathrm{d}^3\Sigma \tag{A48}$$

for the inner angular momentum are both convergent *sensu stricto* because of the finiteness of the support of the tensor $t^{*\alpha\beta}$. The proof on the uniqueness of the centre of energy construction is now complicated by the difficulty that the force density $f^{*\alpha}$ (A46) cannot be made arbitrarily small by varying the external fields, due to the occurrence of the minus fields. In this case one has to *assume* that the centre of energy construction leads to a unique central point.

By making use of the balance equation (A44) one may derive by applying Gauss's theorem (in a straightforward manner since $t^{*\alpha\beta}$ has 'local character') the equations of motion and spin, which are analogous to (96) and (102):

$$\frac{\mathrm{d}p^{*\alpha}}{\mathrm{d}s^*} = \mathfrak{f}^{*\alpha}, \tag{A49}$$

$$\frac{\mathrm{d}s^{*\alpha\beta}}{\mathrm{d}s^*} = \mathfrak{d}^{*\alpha\beta} - (u^{*\alpha}p^{*\beta} - u^{*\beta}p^{*\alpha}). \tag{A50}$$

The total force and the total torque occurring here have the same form as (106) and (107) but with the sum of the external field and the total minus field instead of the external field *tout court*:

$$\mathfrak{f}^{*\alpha}(s^*) = c^{-1}\sum_i e_i[F^{\alpha\beta}\{R_i(s^*)\} + \sum_j f^{\alpha\beta}_{-j}\{R_i(s^*)\}]\,\frac{\mathrm{d}R_{i\beta}}{\mathrm{d}s^*}, \tag{A51}$$

$$\mathfrak{d}^{*\alpha\beta}(s^*) = c^{-1}\sum e_i\{R_i^\alpha(s_i^*) - X^{*\alpha}(s_i^*)\}[F^{\beta\gamma}\{R_i(s^*)\} + \sum_j f^{\beta\gamma}_{-j}\{R_i(s^*)\}]\,\frac{\mathrm{d}R_{i\gamma}}{\mathrm{d}s^*} - (\alpha, \beta). \tag{A52}$$

Since the minus fields are finite at the world lines of the particles (v. (III.111)) we may develop them into a multipole series just as the external field. If one retains the charge and dipole contributions one finds for the total force (and torque) quantities that are the sum of expressions $\mathfrak{f}^{\alpha}(s^*)$ (115) (and $\mathfrak{d}^{\alpha\beta}(s^*)$ (116)) with $X^{*\alpha}$, $u^{*\alpha}$, $\mathfrak{m}^{*\alpha\beta}$ and $p^{*\alpha}$ depending on $s^*$ instead of $X^{\alpha}$, $u^{\alpha}$, $\mathfrak{m}^{\alpha\beta}$ and $p^{\alpha}$ depending on $s$, and minus field contributions of similar form:

$$\mathfrak{f}^{*\alpha}(s^*) = \mathfrak{f}^{\alpha}(s^*)+\mathfrak{f}^{\alpha}_{-}(s^*), \tag{A53}$$

$$\mathfrak{d}^{*\alpha\beta}(s^*) = \mathfrak{d}^{\alpha\beta}(s^*)+\mathfrak{d}^{\alpha\beta}_{-}(s^*), \tag{A54}$$

with

$$\mathfrak{f}^{\alpha}_{-}(s^*) = c^{-1}ef^{\alpha\beta}_{-}(X^*)u^*_{\beta}+\tfrac{1}{2}\{\partial^{\alpha}f^{\beta\gamma}_{-}(X^*)\}\mathfrak{m}^*_{\beta\gamma}+\frac{\mathrm{d}}{\mathrm{d}s^*}\left\{f^{\alpha\beta}_{-}(X^*)\mathfrak{m}^*_{\beta\gamma}\frac{p^{*\gamma}}{u^*_{\varepsilon}p^{*\varepsilon}}\right\}, \tag{A55}$$

$$\mathfrak{d}^{\alpha\beta}_{-}(s^*) = f^{\alpha\gamma}_{-}(X^*)\mathfrak{m}^{*\cdot\beta}_{\gamma}+u^{*\alpha}f^{\beta\gamma}_{-}(X^*)\mathfrak{m}^*_{\gamma\varepsilon}\frac{p^{*\varepsilon}}{u^*_{\zeta}p^{*\zeta}}-(\alpha,\beta), \tag{A56}$$

where $f^{\alpha\beta}_{-}$ is the total minus field $\sum_j f^{\alpha\beta}_{-j}$. This total minus field may be developed into a multipole series in terms of its sources by using the formulae (17) with (18–19). Retaining only the charges and dipoles, we then get (suppressing the asterisks for convenience from now on):

$$f^{\alpha\beta}_{-}(R) = -\frac{e}{4\pi}\int(u^{\alpha}\partial^{\beta}-u^{\beta}\partial^{\alpha})\delta\{(R-X)^2\}\varepsilon(R-X)\mathrm{d}s$$
$$+\frac{c}{4\pi}(\mathfrak{m}^{\gamma\alpha}\partial^{\beta}\partial_{\gamma}-\mathfrak{m}^{\gamma\beta}\partial^{\alpha}\partial_{\gamma})\delta\{(R-X)^2\}\varepsilon(R-X)\mathrm{d}s. \tag{A57}$$

In order to evaluate (A55) and (A56) we have to calculate this minus field and its derivative at the position of the world line. The minus field due to the charge at the position of the world line has been found already in (III.111) of the preceding chapter:

$$f^{\alpha\beta}_{-(e)} = \frac{e}{6\pi}c^{-4}(u^{\alpha}\dot{a}^{\beta}-u^{\beta}\dot{a}^{\alpha}). \tag{A58}$$

where $a^{\alpha}$ is the four-acceleration $\mathrm{d}u^{\alpha}/\mathrm{d}s$. Its derivative at the position of the world line will be calculated in the next appendix. The result is (cf. (A77)):

$$\partial^{\gamma}f^{\alpha\beta}_{-(e)} = \frac{e}{12\pi}c^{-4}\{\ddot{a}^{\alpha}g^{\beta\gamma}+c^{-2}a\cdot\dot{a}g^{\alpha\gamma}u^{\beta}+c^{-2}a^2g^{\alpha\gamma}a^{\beta}-2c^{-2}u^{\alpha}a^{\beta}\dot{a}^{\gamma}$$
$$-4c^{-2}u^{\alpha}\dot{a}^{\beta}a^{\gamma}+u^{\gamma}(3c^{-4}a^2u^{\alpha}a^{\beta}-3c^{-2}u^{\alpha}\ddot{a}^{\beta}-2c^{-2}a^{\alpha}\dot{a}^{\beta})\}-(\alpha,\beta). \tag{A59}$$

It is possible to derive in a similar way expressions for the dipole minus field and its derivative. Since the results are rather lengthy[1] we shall give here only those terms which are independent of the first and higher derivatives of the velocity. These terms (which will be derived in the next appendix) read

$$f_{-(\mathrm{m})}^{\alpha\beta} = \frac{c^{-3}}{6\pi}(\overset{(3)}{\mathfrak{m}}{}^{\alpha\beta} + 2c^{-2}\overset{(3)}{\mathfrak{m}}{}^{\alpha\gamma}u_\gamma u^\beta - 2c^{-2}\overset{(3)}{\mathfrak{m}}{}^{\beta\gamma}u_\gamma u^\alpha), \tag{A60}$$

$$\partial^\gamma f_{-(\mathrm{m})}^{\alpha\beta} = -\frac{c^{-5}}{12\pi}(\overset{(4)}{\mathfrak{m}}{}^{\alpha\varepsilon}u_\varepsilon g^{\beta\gamma} + \overset{(4)}{\mathfrak{m}}{}^{\alpha\gamma}u^\beta + \overset{(4)}{\mathfrak{m}}{}^{\alpha\beta}u^\gamma + 6c^{-2}\overset{(4)}{\mathfrak{m}}{}^{\alpha\varepsilon}u_\varepsilon u^\beta u^\gamma) - (\alpha, \beta), \tag{A61}$$

where numbers above the symbols indicate the number of differentiations with respect to the proper time. If the expressions (A58–A61) are introduced into (A55) and (A56), one obtains the total minus field contribution of the force and torque exerted on the composite particle. If again only terms independent of the first and higher derivatives of the velocity are written down, one gets

$$\begin{aligned}\mathfrak{f}_-^\alpha = {} & -\frac{ec^{-4}}{6\pi}\overset{(3)}{\mathfrak{m}}{}^{\alpha\beta}u_\beta + \frac{c^{-5}}{12\pi}(\overset{(4)}{\mathfrak{m}}{}^{\alpha\beta}\mathfrak{m}_{\beta\gamma}u^\gamma \\ & + \mathfrak{m}^{\alpha\beta}\overset{(4)}{\mathfrak{m}}{}_{\beta\gamma}u^\gamma - u^\alpha\overset{(4)}{\mathfrak{m}}{}^{\beta\gamma}\mathfrak{m}_{\beta\gamma} - 6c^{-2}u^\alpha\overset{(4)}{\mathfrak{m}}{}^{\beta\gamma}u_\gamma u^\varepsilon\mathfrak{m}_{\beta\varepsilon}) \\ & + \frac{c^{-3}}{6\pi}\frac{\mathrm{d}}{\mathrm{d}s}\left\{(\overset{(3)}{\mathfrak{m}}{}^{\alpha\beta} + 2c^{-2}\overset{(3)}{\mathfrak{m}}{}^{\alpha\varepsilon}u_\varepsilon u^\beta - 2c^{-2}u^\alpha\overset{(3)}{\mathfrak{m}}{}^{\beta\varepsilon}u_\varepsilon)\mathfrak{m}_{\beta\gamma}\frac{p^\gamma}{u_\zeta p^\zeta}\right\},\end{aligned} \tag{A62}$$

$$\begin{aligned}\mathfrak{d}_-^{\alpha\beta} = \frac{c^{-3}}{6\pi}\Big[ & \overset{(3)}{\mathfrak{m}}{}^{\alpha\gamma}\dot{\mathfrak{m}}_\gamma{}^\beta + 2c^{-2}\overset{(3)}{\mathfrak{m}}{}^{\alpha\varepsilon}u_\varepsilon u^\gamma\dot{\mathfrak{m}}_\gamma{}^\beta \\ & + u^\alpha\left\{(\overset{(3)}{\mathfrak{m}}{}^{\beta\gamma} + 2c^{-2}\overset{(3)}{\mathfrak{m}}{}^{\beta\varepsilon}u_\varepsilon u^\gamma - 2c^{-2}u^\beta\overset{(3)}{\mathfrak{m}}{}^{\gamma\varepsilon}u_\varepsilon)\mathfrak{m}_{\gamma\zeta}\frac{p^\zeta}{u_\vartheta p^\vartheta} - 2c^{-2}\overset{(3)}{\mathfrak{m}}{}^{\gamma\varepsilon}u_\varepsilon\dot{\mathfrak{m}}_\gamma{}^\beta\right\}\Big] \\ & - (\alpha, \beta).\end{aligned} \tag{A63}$$

The equations of motion (A49) and (A50) are now completely specified if one substitutes (A53) and (A54) with (A62) and (A63). They contain explicitly terms that describe the radiation damping. Owing to this fact these equations do not reduce, for the field-free case, to such simple forms as derived in the main text. The problem of their solution requires the consideration of runaway solutions (just as in chapter III).

[1] S. Emid and J. Vlieger, Physica **52**(1971)329.

# The minus field of a charged dipole particle

In this appendix expressions will be derived for the minus field of a charged dipole particle at the position of the particle itself in a way similar to that of section 2*e* of the preceding chapter. We start from formulae (14–16) which give the general expressions for the retarded field. Combining it with the expression for the advanced field, one gets

$$f^{\alpha\beta}_{\mathrm{r,a}} \equiv f^{\alpha\beta}_{\mathrm{r,a(e)}} + f^{\alpha\beta}_{\mathrm{r,a(m)}} = -\frac{e}{2\pi}\int (u^\alpha\partial^\beta - u^\beta\partial^\alpha)\delta\{(R-X)^2\}\theta\{\pm(R-X)\}\mathrm{d}s$$

$$+\frac{c}{2\pi}\int (\mathfrak{m}^{\gamma\alpha}\partial^\beta\partial_\gamma - \mathfrak{m}^{\gamma\beta}\partial^\alpha\partial_\gamma)\delta\{(R-X)^2\}\theta\{\pm(R-X)\}\mathrm{d}s, \qquad \text{(A64)}$$

with $e$ the charge and $\mathfrak{m}^{\alpha\beta}$ the covariant dipole moment (11) of the particle. Performing the integration, one finds

$$f^{\alpha\beta}_{\mathrm{r,a(e)}} = \mp\frac{e}{4\pi}\partial^\alpha\left\{\left(\frac{u^\beta}{u\cdot r}\right)\Bigg|_{\mathrm{r,a}}\right\} - (\alpha, \beta), \qquad \text{(A65)}$$

$$f^{\alpha\beta}_{\mathrm{r,a(m)}} = \pm\frac{c}{4\pi}\partial^\alpha\partial_\gamma\left\{\left(\frac{\mathfrak{m}^{\gamma\beta}}{u\cdot r}\right)\Bigg|_{\mathrm{r,a}}\right\} - (\alpha, \beta), \qquad \text{(A66)}$$

with $r^\alpha = R^\alpha - X^\alpha$, where the suffixes r and a at the bar indicate that one should take the retarded and advanced expressions. The differentiations with respect to $R$ may be performed by making use of the equation (III.94) that follows from the light-cone equation.

The charge minus field that may be calculated from (A65) has been obtained already in section 2*e* of the preceding chapter. There we have found for the minus field at the world line (v. (III.111)):

$$f^{\alpha\beta}_{-(\mathrm{e})} = \frac{e}{6\pi}c^{-4}(u^\alpha\dot{a}^\beta - u^\beta\dot{a}^\alpha). \qquad \text{(A67)}$$

We are also interested (in view of the equation of motion) in the space–time derivative of the minus field, i.e. in $\partial^\gamma f^{\alpha\beta}_{-(\mathrm{e})}$. By making use of the projection operator $\Delta^{\alpha\beta} \equiv g^{\alpha\beta} + c^{-2}u^\alpha u^\beta$, one may split this derivative into two parts:

$$\partial^\gamma f^{\alpha\beta}_{-(\mathrm{e})} = \Delta^\gamma_\varepsilon\partial^\varepsilon f^{\alpha\beta}_{-(\mathrm{e})} - c^{-2}u^\gamma u^\varepsilon\partial_\varepsilon f^{\alpha\beta}_{-(\mathrm{e})}, \qquad \text{(A68)}$$

namely into a part which gives the derivative in a direction orthogonal to the four-velocity and a part that specifies the derivative in a direction parallel to $u^\alpha$. The latter part may be calculated directly from (A67), since

$$u^\varepsilon \partial_\varepsilon f^{\alpha\beta}_{-(e)} = \frac{\mathrm{d}}{\mathrm{d}s} f^{\alpha\beta}_{-(e)} = \frac{e}{6\pi} c^{-4}(a^\alpha \dot{a}^\beta + u^\alpha \ddot{a}^\beta - a^\beta \dot{a}^\alpha - u^\beta \ddot{a}^\alpha). \tag{A69}$$

To evaluate the former part one has to consider the minus field at a position $R^\alpha = X^\alpha(s_1) + \varepsilon n^\alpha$ (v. (III.101)) with fixed $s_1$ and space-like unit vector $n^\alpha$ orthogonal to $u^\alpha(s_1)$ and then take the derivative with respect to $\varepsilon n^\alpha$. To be able to perform this programme one has to push the series expansions in $\varepsilon$ as given in (III.103–109) one step further. The extension of (III.103) with one more term leads to an extension of (III.104) of the form:

$$\begin{aligned} s_{r,a} - s_1 = {} & \mp c^{-1}\varepsilon[1 - \tfrac{1}{2}c^{-2} a\cdot n\varepsilon + \{\tfrac{3}{8}c^{-4}(a\cdot n)^2 - \tfrac{1}{24}c^{-4}a^2 \pm \tfrac{1}{6}c^{-3}\dot{a}\cdot n\}\varepsilon^2 \\ & + \{-\tfrac{5}{16}c^{-6}(a\cdot n)^3 + \tfrac{5}{48}c^{-6} a\cdot n a^2 \mp \tfrac{1}{3}c^{-5} a\cdot n \dot{a}\cdot n \\ & - \tfrac{1}{24}c^{-4}\ddot{a}\cdot n \pm \tfrac{1}{24}c^{-5} a\cdot\dot{a}\}\varepsilon^3 + \ldots]. \end{aligned} \tag{A70}$$

Furthermore one has to calculate the extensions of (III.105–108). This leads to the series expansions:

$$\begin{aligned} u(s)\cdot r(s) = {} & \mp c\varepsilon[1 + \tfrac{1}{2}c^{-2} a\cdot n\varepsilon + \{-\tfrac{1}{8}c^{-4}(a\cdot n)^2 + \tfrac{1}{8}c^{-4}a^2 \mp \tfrac{1}{3}c^{-3}\dot{a}\cdot n\}\varepsilon^2 \\ & + \{\tfrac{1}{16}c^{-6}(a\cdot n)^3 - \tfrac{3}{16}c^{-6} a\cdot n a^2 \pm \tfrac{1}{3}c^{-5} a\cdot n \dot{a}\cdot n \\ & + \tfrac{1}{8}c^{-4}\ddot{a}\cdot n \mp \tfrac{1}{6}c^{-5} a\cdot\dot{a}\}\varepsilon^3 + \ldots], \end{aligned} \tag{A71}$$

$$\begin{aligned} a(s)\cdot r(s) = {} & \varepsilon\{a\cdot n \mp c^{-1}(\dot{a}\cdot n \mp \tfrac{1}{2}c^{-1}a^2)\varepsilon \\ & + (\mp\tfrac{5}{6}c^{-3} a\cdot\dot{a} - \tfrac{1}{2}c^{-4}a^2 a\cdot n + \tfrac{1}{2}c^{-2}\ddot{a}\cdot n \pm \tfrac{1}{2}c^{-3} a\cdot n \dot{a}\cdot n)\varepsilon^2 + \ldots\}, \end{aligned} \tag{A72}$$

$$\begin{aligned} r^\alpha(s)u^\beta(s) - r^\beta(s)u^\alpha(s) = {} & \varepsilon(n^\alpha u^\beta + (\mp c^{-1} n^\alpha a^\beta - \tfrac{1}{2}c^{-2}u^\alpha a^\beta)\varepsilon \\ & + (\pm\tfrac{1}{2}c^{-3} a\cdot n n^\alpha a^\beta + \tfrac{1}{2}c^{-2} n^\alpha \dot{a}^\beta + \tfrac{1}{2}c^{-4} a\cdot n u^\alpha a^\beta \pm \tfrac{1}{3}c^{-3}u^\alpha \dot{a}^\beta)\varepsilon^2 \\ & + [\{\mp\tfrac{3}{8}c^{-5}(a\cdot n)^2 \pm \tfrac{1}{24}c^{-5}a^2 - \tfrac{1}{6}c^{-4}\dot{a}\cdot n\} n^\alpha a^\beta - \tfrac{1}{2}c^{-4} a\cdot n n^\alpha \dot{a}^\beta \\ & \mp \tfrac{1}{6}c^{-3} n^\alpha \ddot{a}^\beta + \{-\tfrac{1}{2}c^{-6}(a\cdot n)^2 + \tfrac{1}{24}c^{-6}a^2 \mp \tfrac{1}{6}c^{-5}\dot{a}\cdot n\} u^\alpha a^\beta \\ & \mp \tfrac{1}{2}c^{-5} a\cdot n u^\alpha \dot{a}^\beta - \tfrac{1}{8}c^{-4}u^\alpha \ddot{a}^\beta - \tfrac{1}{12}c^{-4}a^\alpha \dot{a}^\beta]\varepsilon^3 + \ldots) - (\alpha, \beta), \end{aligned} \tag{A73}$$

$$\begin{aligned} r^\alpha(s)a^\beta(s) - r^\beta(s)a^\alpha(s) = {} & \varepsilon(n^\alpha a^\beta \pm c^{-1}u^\alpha a^\beta + (\mp c^{-1} n^\alpha \dot{a}^\beta \mp \tfrac{1}{2}c^{-3} a\cdot n u^\alpha a^\beta \\ & - c^{-2}u^\alpha \dot{a}^\beta)\varepsilon + [\pm\tfrac{1}{2}c^{-3} a\cdot n n^\alpha \dot{a}^\beta + \tfrac{1}{2}c^{-2} n^\alpha \ddot{a}^\beta + c^{-4} a\cdot n u^\alpha \dot{a}^\beta \\ & \pm \tfrac{1}{2}c^{-3}u^\alpha \ddot{a}^\beta \pm \tfrac{1}{2}c^{-3}a^\alpha \dot{a}^\beta + \{\pm\tfrac{3}{8}c^{-5}(a\cdot n)^2 \mp \tfrac{1}{24}c^{-5}a^2 \\ & + \tfrac{1}{6}c^{-4}\dot{a}\cdot n\} u^\alpha a^\beta]\varepsilon^2 + \ldots) - (\alpha, \beta). \end{aligned} \tag{A74}$$

If these expressions are substituted into (A65) and half the difference of the retarded and advanced field is taken, one finds for the minus field in the neighbourhood of the world line

$$f^{\alpha\beta}_{-(e)} = \frac{ec^{-4}}{12\pi}\{2u^\alpha\dot{a}^\beta+(c^{-2}a\cdot\dot{a}n^\alpha u^\beta+c^{-2}a^2n^\alpha a^\beta$$
$$-2c^{-2}\dot{a}\cdot nu^\alpha a^\beta-4c^{-2}a\cdot nu^\alpha\dot{a}^\beta-n^\alpha\ddot{a}^\beta)\varepsilon+\ldots\}-(\alpha,\beta). \quad (A75)$$

(Indeed the limit $\varepsilon \to 0$ gives back (A67).) The derivative of the minus field in a direction orthogonal to the four-velocity follows by taking the derivative with respect to $\varepsilon n^\alpha$:

$$\Delta^\gamma_\delta\,\partial^\delta f^{\alpha\beta}_{-(e)} = \lim_{\varepsilon\to 0}\frac{1}{\varepsilon}\,\Delta^\gamma_\delta\frac{\partial}{\partial n_\delta}f^{\alpha\beta}_{-(e)} = \frac{ec^{-4}}{12\pi}\,\Delta^\gamma_\delta(c^{-2}a\cdot\dot{a}g^{\alpha\delta}u^\beta+c^{-2}a^2g^{\alpha\delta}a^\beta$$
$$-2c^{-2}\dot{a}^\delta u^\alpha a^\beta-4c^{-2}a^\delta u^\alpha\dot{a}^\beta-g^{\alpha\delta}\ddot{a}^\beta)-(\alpha,\beta). \quad (A76)$$

Combining this relation with (A69) according to (A68) one finds

$$\partial^\gamma f^{\alpha\beta}_{-(e)} = \frac{e}{12\pi}\,c^{-4}\{\ddot{a}^\alpha g^{\beta\gamma}+c^{-2}a\cdot\dot{a}g^{\alpha\gamma}u^\beta+c^{-2}a^2g^{\alpha\gamma}a^\beta-2c^{-2}u^\alpha a^\beta\dot{a}^\gamma$$
$$-4c^{-2}u^\alpha\dot{a}^\beta a^\gamma+u^\gamma(3c^{-4}a^2u^\alpha a^\beta-3c^{-2}u^\alpha\ddot{a}^\beta-2c^{-2}a^\alpha\dot{a}^\beta)\}-(\alpha,\beta), \quad (A77)$$

which is the expression (A59). This formula shows that the derivative of the minus field, just as the minus field itself at the world line of the particle, disappears if no accelerations are present, i.e. if the particle moves uniformly.

We now turn to the dipole field given by (A66). It is possible to calculate (in an analogous fashion as used for the charge field) the general expression for the field and its derivative at the world line. The results are lengthy[1] and will not be reproduced here. Much more simple results are obtained if one retains right from the beginning only those terms which are independent of the first and higher derivatives of the velocity. Then formula (A66) becomes upon using (III.94)

$$f^{\alpha\beta}_{\mathrm{r,a(m)}} = \pm\frac{c}{4\pi}\frac{1}{(u\cdot r)^2}\left\{\frac{\dot{\mathfrak{m}}^{\alpha\gamma}r_\gamma r^\beta}{u\cdot r}-\frac{2\dot{\mathfrak{m}}^{\alpha\gamma}u_\gamma r^\beta}{u\cdot r}-\frac{2\dot{\mathfrak{m}}^{\alpha\gamma}r_\gamma u^\beta}{u\cdot r}-\frac{3\mathfrak{m}^{\alpha\gamma}r_\gamma r^\beta c^2}{(u\cdot r)^2}\right.$$
$$+\dot{\mathfrak{m}}^{\alpha\beta}+\frac{2\mathfrak{m}^{\alpha\gamma}u_\gamma u^\beta}{u\cdot r}+\frac{3\mathfrak{m}^{\alpha\gamma}u_\gamma r^\beta c^2}{(u\cdot r)^2}+\frac{3\mathfrak{m}^{\alpha\gamma}r_\gamma u^\beta c^2}{(u\cdot r)^2}+\frac{3\mathfrak{m}^{\alpha\gamma}r_\gamma r^\beta c^4}{(u\cdot r)^3}$$
$$\left.-\frac{\mathfrak{m}^{\alpha\beta}c^2}{u\cdot r}\right\}\Bigg|_{\mathrm{r,a}}-(\alpha,\beta). \quad (A78)$$

[1] Harish-Chandra, Proc. Roy. Soc. **A185**(1946)269; J. R. Ellis, J. Math. Phys. **7**(1966) 1185; S. Emid and J. Vlieger, Physica **52**(1971)329.

Its value near the world line, i.e. at a position $R^\alpha = X^\alpha(s)+\varepsilon n^\alpha$ follows by making use of formulae of the type (A70–A74). In particular, since we limit ourselves here to terms without accelerations, the equation (A70) now reduces to

$$s_{\mathrm{r,a}}-s_1 = \mp c^{-1}\varepsilon. \tag{A79}$$

Furthermore one finds then

$$u(s)\cdot r(s) = \mp c\varepsilon, \tag{A80}$$

$$r^\alpha(s) = \varepsilon n^\alpha \pm c^{-1}\varepsilon u^\alpha \tag{A81}$$

and expressions for $\mathfrak{m}^{\alpha\beta}(s)$ and its derivatives that follow immediately from the Taylor expansion around $s_1$. If these expressions are inserted into (A78) one obtains for the minus field in the neighbourhood of the world line up to order $\varepsilon$:

$$f^{\alpha\beta}_{-(\mathrm{m})} = \frac{c^{-3}}{6\pi}(\overset{(3)}{\mathfrak{m}}{}^{\alpha\beta}+2c^{-2}\overset{(3)}{\mathfrak{m}}{}^{\alpha\gamma}u_\gamma u^\beta-2c^{-2}\overset{(3)}{\mathfrak{m}}{}^{\beta\gamma}u_\gamma u^\alpha)$$
$$-\frac{\varepsilon}{12\pi}c^{-5}(\overset{(4)}{\mathfrak{m}}{}^{\alpha\gamma}u_\gamma n^\beta+\overset{(4)}{\mathfrak{m}}{}^{\alpha\gamma}n_\gamma u^\beta-\overset{(4)}{\mathfrak{m}}{}^{\beta\gamma}u_\gamma n^\alpha-\overset{(4)}{\mathfrak{m}}{}^{\beta\gamma}n_\gamma u^\alpha). \tag{A82}$$

From this formula one finds directly, by letting tend $\varepsilon$ to zero, for the field at the position of the particle

$$f^{\alpha\beta}_{-(\mathrm{m})} = \frac{c^{-3}}{6\pi}(\overset{(3)}{\mathfrak{m}}{}^{\alpha\beta}+2c^{-2}\overset{(3)}{\mathfrak{m}}{}^{\alpha\gamma}u_\gamma u^\beta-2c^{-2}\overset{(3)}{\mathfrak{m}}{}^{\beta\gamma}u_\gamma u^\alpha) \tag{A83}$$

and for the derivative of the field in the direction orthogonal to the four-velocity,

$$\Delta^\gamma_\delta \partial^\delta f^{\alpha\beta}_{-(\mathrm{m})} = -\frac{c^{-5}}{12\pi}\Delta^\gamma_\delta(\overset{(4)}{\mathfrak{m}}{}^{\alpha\varepsilon}u_\varepsilon g^{\beta\delta}+\overset{(4)}{\mathfrak{m}}{}^{\alpha\delta}u^\beta-\overset{(4)}{\mathfrak{m}}{}^{\beta\varepsilon}u_\varepsilon g^{\alpha\delta}-\overset{(4)}{\mathfrak{m}}{}^{\beta\delta}u^\alpha). \tag{A84}$$

Furthermore it follows from (A83) that the derivative in the direction of the four-velocity is

$$u^\varepsilon\partial_\varepsilon f^{\alpha\beta}_{-(\mathrm{m})} = \frac{c^{-3}}{6\pi}(\overset{(4)}{\mathfrak{m}}{}^{\alpha\beta}+2c^{-2}\overset{(4)}{\mathfrak{m}}{}^{\alpha\gamma}u_\gamma u^\beta-2c^{-2}\overset{(4)}{\mathfrak{m}}{}^{\beta\gamma}u_\gamma u^\alpha), \tag{A85}$$

where again accelerations have been neglected. Then, according to (A68), we find from (A84) and (A85)

$$\partial^\gamma f^{\alpha\beta}_{-(\mathrm{m})} = -\frac{c^{-5}}{12\pi}(\overset{(4)}{\mathfrak{m}}{}^{\alpha\varepsilon}u_\varepsilon g^{\beta\gamma}+\overset{(4)}{\mathfrak{m}}{}^{\alpha\gamma}u^\beta+\overset{(4)}{\mathfrak{m}}{}^{\alpha\beta}u^\gamma+6c^{-2}\overset{(4)}{\mathfrak{m}}{}^{\alpha\varepsilon}u_\varepsilon u^\beta u^\gamma)-(\alpha,\beta), \tag{A86}$$

which is the final result for the derivative of the dipole minus field at the position of the particle (at least for the terms without accelerations).

APPENDIX V

# Semi-relativistic equations of motion for a composite particle

### *1. The semi-relativistic approximation*

In the relativistic treatment of a composite particle in an external field equations of motion and of angular momentum have been found. They contain, as compared to the corresponding non-relativistic treatment, a number of relativistic effects. Amongst these figure in particular terms of order $c^{-2}$, which contain an explicit factor $c^{-1}$ and the magnetic dipole moment, which is itself of order $c^{-1}$. Hence these are the terms which would survive in the so-called *semi-relativistic* approximation (v. section 2*e*) in which one considers the magnetic dipole moment to be an atomic parameter of order $c^0$ and subsequently retains only terms up to order $c^{-1}$. It is possible to give a theory in which all these semi-relativistic effects are included, without going through the complete relativistic treatment. In this way some insight is gained about the origin of these terms, in particular about the so-called magnetodynamic effect, that contains the vector product of the magnetic dipole moment and the electric field. In the following we shall carry out this programme by first developing the theory up to order $c^{-2}$ (which includes the use of some notions of relativity) and then taking the semi-relativistic limit of the resulting equations.

### *2. The momentum and energy equations*

#### a. *The equation of motion*

In non-relativistic theory the equation of motion for a particle in an electromagnetic field contains the time derivative of mass times velocity. In a theory which includes all effects up to order $c^{-2}$, again the time derivative of the momentum appears. However, the momentum now contains not only the ordinary rest mass, but also a mass which is equivalent to the kinetic energy. The equation of motion for constituent particle $i$ of the composite particle reads then

$$\frac{\mathrm{d}}{\mathrm{d}t}\{m_i(1+\tfrac{1}{2}c^{-2}\dot{\boldsymbol{R}}_i^2)\dot{\boldsymbol{R}}_i\} = e_i\{\boldsymbol{e}_t(\boldsymbol{R}_i, t)+c^{-1}\dot{\boldsymbol{R}}_i \wedge \boldsymbol{b}_t(\boldsymbol{R}_i, t)\}, \qquad \text{(A87)}$$

where $e_i$ is the charge, $m_i$ the mass, $\boldsymbol{R}_i(t)$ the position and $\dot{\boldsymbol{R}}_i(t)$ the velocity of the particle $i$. The right-hand side is the Lorentz force, just as in non-relativistic theory. The total electromagnetic fields $\boldsymbol{e}_t$ and $\boldsymbol{b}_t$ contain the fields due to the other particles $j$ $(\neq i)$ and due to external sources. The electric and magnetic fields are needed up to order $c^{-2}$ and $c^{-1}$ respectively so as to describe all effects up to order $c^{-2}$.

Since we are interested in the motion of the composite particle as a whole, we shall define now a privileged point $\boldsymbol{X}$ which describes the position of the atom as a whole. In non-relativistic theory the centre of mass has been chosen as such a point. In relativity one should like to include beside the rest masses also internal kinetic and potential energies in the definition. This would lead to an energy centre, defined, up to order $c^{-2}$, as

$$\boldsymbol{X}^* = \frac{\sum_i (m_i + \frac{1}{2}c^{-2} m_i \dot{\boldsymbol{R}}_i^2 + c^{-2} \sum_{j(\neq i)} e_i e_j / 8\pi |\boldsymbol{R}_i - \boldsymbol{R}_j|)\boldsymbol{R}_i}{\sum_i (m_i + \frac{1}{2}c^{-2} m_i \dot{\boldsymbol{R}}_i^2 + c^{-2} \sum_{j(\neq i)} e_i e_j / 8\pi |\boldsymbol{R}_i - \boldsymbol{R}_j|)} . \tag{A88}$$

Such a definition, however, is still not convenient, since in this way the privileged point $\boldsymbol{X}^*$ would depend on the velocity of the observer with respect to the composite particle, and is hence not invariant. To overcome this drawback, let us consider the coordinate frame in which the composite particle is momentarily at rest, i.e. a frame which moves with the velocity $\dot{\boldsymbol{X}}^*$ with respect to the observer. If we now determine the energy centre in this momentary rest frame, we find a point $\boldsymbol{X}$ (different from $\boldsymbol{X}^*$). By repeating this procedure for each time $t$ one obtains a world line of energy centres $\boldsymbol{X}(t)$ [1]. From this construction we shall now derive the relation that determines the privileged point $\boldsymbol{X}$ in terms of the positions $\boldsymbol{R}_i$.

[1] One may ask whether this construction is the $c^{-2}$-limit of the relativistic definition of the centre of energy. In the relativistic case one takes the energy centre in space-like surfaces. As a weight function one uses the time-time component $t^{00}$ of the energy–momentum tensor (62). If one evaluates up to order $c^{-2}$ the expression $c^{-2} \int t^{00} \boldsymbol{R} \mathrm{d}\boldsymbol{R}$ one finds indeed the numerator of (A88) (v. problem 7). What remains to be checked is that the relativistic construction, in which one takes space-like surfaces normal to the total momentum $p^\alpha$, reduces to the present one, in which different space-like surfaces are employed, namely surfaces orthogonal to the four-velocity corresponding to $\boldsymbol{X}^*$. From (124) it follows that the space-components of $c$ times the normal unit vector i.e.: $cp^\alpha/\sqrt{-p^2}$ differs from the space-components of $u^\alpha \equiv \mathrm{d}X^\alpha/\mathrm{d}s$ by terms which are of order $c^{-2}$ and smaller. In the course of this appendix it will turn out that the velocity $\dot{\boldsymbol{X}}$ differs from the velocity $\dot{\boldsymbol{X}}^*$ by terms which are also of order $c^{-2}$ and smaller. Hence $c$ times the space components of the normal unit vector of the relativistic definition differs by terms which are of order $c^{-2}$ (and smaller) from $\dot{\boldsymbol{X}}^*$, which is used here. Since the construction given here will show that not the precise form (A88) of $\boldsymbol{X}^*$ is relevant, but only the fact that its $c^0$-terms give the non-relativistic centre of mass, it follows that the construction given here is indeed the $c^{-2}$-approximation to the relativistic definition.

Let us consider the points of the world lines $\boldsymbol{R}_i$, $\boldsymbol{X}$ and $\boldsymbol{X}^*$ which have the same time coordinate $t'$ in the coordinate frame (indicated by primes) which moves with the velocity $\dot{\boldsymbol{X}}^*(t^*)$ with respect to the observer (see fig. 2). In

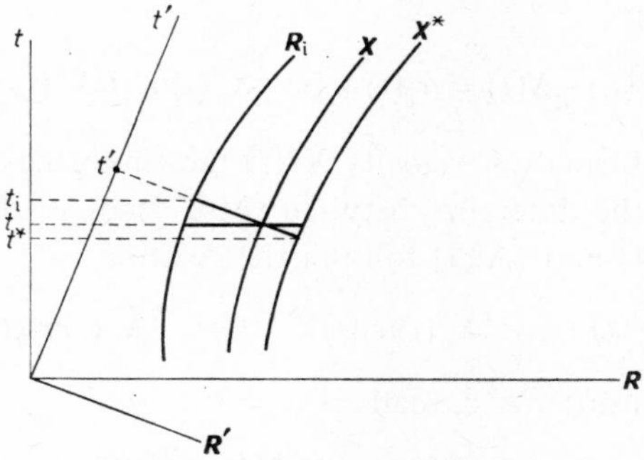


Fig. 2. The construction of the semi-relativistic energy centre.

the observer's frame these world points have the time components $t_i$, $t$ and $t^*$ respectively. The relative position vector in the primed frame

$$\boldsymbol{r}_i'(t) \equiv \boldsymbol{R}_i'(t_i) - \boldsymbol{X}'(t) \tag{A89}$$

of particle $i$ with respect to the energy centre $\boldsymbol{X}$ fulfils the relation

$$\sum_i \left\{ m_i + \tfrac{1}{2}c^{-2} m_i \dot{\boldsymbol{R}}_i'^2(t_i) + c^{-2} \sum_{j(\neq i)} \frac{e_i e_j}{8\pi |\boldsymbol{r}_i'(t) - \boldsymbol{r}_j'(t)|} \right\} \boldsymbol{r}_i'(t) = 0. \tag{A90}$$

In fact this is the defining formula of an energy centre up to order $c^{-2}$ (cf. (A88)). We now want to find what follows from this relation for the relative positions

$$\boldsymbol{r}_i(t) \equiv \boldsymbol{R}_i(t) - \boldsymbol{X}(t), \tag{A91}$$

in the observer's frame. From the fact that the world points $(\boldsymbol{R}_i(t_i), t_i)$ and $(\boldsymbol{X}(t), t)$ have the same time components in the primed frame, it follows, with the help of a Lorentz transformation, that, up to order $c^{-2}$, one has

$$t_i - t = c^{-2} \dot{\boldsymbol{X}}^*(t^*) \cdot \boldsymbol{r}_i'(t), \tag{A92}$$

where (A89) has been used. In the same way it follows that $t^* - t$ is of order $c^{-2}$, so that (A92) may be written as

$$t_i - t = c^{-2} \dot{\boldsymbol{X}}^*(t) \cdot \boldsymbol{r}_i'(t), \tag{A93}$$

up to order $c^{-2}$. The first terms of a Taylor expansion of $\boldsymbol{R}_i(t_i)$ around $\boldsymbol{R}_i(t)$

become with the help of (A93):

$$\boldsymbol{R}_i(t_i) = \boldsymbol{R}_i(t)+c^{-2}\dot{\boldsymbol{X}}^*(t)\cdot\boldsymbol{r}_i'(t)\dot{\boldsymbol{R}}_i(t). \tag{A94}$$

The same Lorentz transformation which has led to (A92) also yields with the notation (A89):

$$\boldsymbol{R}_i(t_i)-\boldsymbol{X}(t) = \boldsymbol{r}_i'(t)+\tfrac{1}{2}c^{-2}\dot{\boldsymbol{X}}^*(t)\cdot\boldsymbol{r}_i'(t)\dot{\boldsymbol{X}}^*(t), \tag{A95}$$

where in the transformation velocity $\dot{\boldsymbol{X}}^*(t)$ the time $t$ has been written instead of $t^*$, since the difference between these times is of order $c^{-2}$ only. From (A91), (A94) and (A95) follows the relation

$$\boldsymbol{r}_i(t) = \boldsymbol{r}_i'(t)+\tfrac{1}{2}c^{-2}\dot{\boldsymbol{X}}^*(t)\cdot\boldsymbol{r}_i'(t)\dot{\boldsymbol{X}}^*(t)-c^{-2}\dot{\boldsymbol{X}}^*(t)\cdot\boldsymbol{r}_i'(t)\dot{\boldsymbol{R}}_i(t). \tag{A96}$$

Its inversion, up to order $c^{-2}$, reads

$$\boldsymbol{r}_i'(t) = \boldsymbol{r}_i(t)-\tfrac{1}{2}c^{-2}\dot{\boldsymbol{X}}^*(t)\cdot\boldsymbol{r}_i(t)\dot{\boldsymbol{X}}^*(t)+c^{-2}\dot{\boldsymbol{X}}^*(t)\cdot\boldsymbol{r}_i(t)\dot{\boldsymbol{R}}_i(t). \tag{A97}$$

Apart from $\boldsymbol{r}_i'(t)$, which has now been found, the quantity $\dot{\boldsymbol{R}}_i'(t_i)$ also occurs in (A90). It may be written, up to order $c^0$, as

$$\dot{\boldsymbol{R}}_i'(t_i) = \dot{\boldsymbol{R}}_i(t_i)-\dot{\boldsymbol{X}}^*(t^*) = \dot{\boldsymbol{R}}_i(t)-\dot{\boldsymbol{X}}^*(t), \tag{A98}$$

where the fact has been used that both $t_i-t$ and $t^*-t$ are of order $c^{-2}$. Inserting (A97) and (A98) into (A90) we obtain now up to order $c^{-2}$:

$$\sum_i \left\{m_i+\tfrac{1}{2}c^{-2}m_i(\dot{\boldsymbol{R}}_i-\dot{\boldsymbol{X}}^*)^2+c^{-2}\sum_{j(\neq i)}\frac{e_i e_j}{8\pi|\boldsymbol{r}_i-\boldsymbol{r}_j|}\right\} (\boldsymbol{r}_i-\tfrac{1}{2}c^{-2}\dot{\boldsymbol{X}}^*\cdot\boldsymbol{r}_i\dot{\boldsymbol{X}}^*+c^{-2}\dot{\boldsymbol{X}}^*\cdot\boldsymbol{r}_i\dot{\boldsymbol{R}}_i) = 0, \tag{A99}$$

where the arguments $t$ of all quantities have been suppressed. From (A88) and (A99) it follows that the difference between $\dot{\boldsymbol{X}}^*$ and $\dot{\boldsymbol{X}}$ is of order $c^{-2}$. This permits us to write (A96) and (A99) with (A91), up to order $c^{-2}$, as

$$\boldsymbol{r}_i = \boldsymbol{r}_i'-\tfrac{1}{2}c^{-2}\dot{\boldsymbol{X}}\cdot\boldsymbol{r}_i'\dot{\boldsymbol{X}}-c^{-2}\dot{\boldsymbol{X}}\cdot\boldsymbol{r}_i'\dot{\boldsymbol{r}}_i \tag{A100}$$

and

$$\sum_i \left(m_i\boldsymbol{r}_i+\tfrac{1}{2}c^{-2}m_i\dot{\boldsymbol{r}}_i^2\boldsymbol{r}_i+c^{-2}\sum_{j(\neq i)}\frac{e_i e_j}{8\pi|\boldsymbol{r}_i-\boldsymbol{r}_j|}\boldsymbol{r}_i+c^{-2}m_i\dot{\boldsymbol{X}}\cdot\boldsymbol{r}_i\dot{\boldsymbol{r}}_i\right) = 0. \tag{A101}$$

The last relation defines a privileged point of the composite particle in a unique way. In fact if there would exist two different points $\boldsymbol{X}$ and $\boldsymbol{X}+\Delta\boldsymbol{X}$, one would have, apart from the relation (A101) as it stands, a relation like (A101) but with $\boldsymbol{X}+\Delta\boldsymbol{X}$ and $\boldsymbol{r}_i-\Delta\boldsymbol{X}$ instead of $\boldsymbol{X}$ and $\boldsymbol{r}_i$ respectively. Then from these two relations together it follows that $\Delta\boldsymbol{X}$ is at least of order $c^{-4}$.

Hence it is negligible in the framework of the present treatment, in which only effects of order $c^{-2}$ are considered. One may still ask whether the definition of the privileged point is biased by the original choice of the $(\boldsymbol{R}, t)$-frame as a starting point. Suppose in fact that one had started from an $(\boldsymbol{R}, t)$-frame which moves with a velocity $\boldsymbol{V}$ with respect to the observer to define the ancillary point $\hat{\boldsymbol{X}}^*$. Then one would have arrived at a privileged point $\hat{\boldsymbol{X}}(\hat{t})$ and relative positions

$$\hat{\boldsymbol{r}}_i(\hat{t}) = \hat{\boldsymbol{R}}_i(\hat{t}) - \hat{\boldsymbol{X}}(\hat{t}), \tag{A102}$$

which satisfy a relation like (A101) but 'circumflexed'. The Lorentz transform of the point $(\hat{\boldsymbol{X}}(\hat{t}), \hat{t})$ in the $(\hat{\boldsymbol{R}}, \hat{t})$-frame is the point $(\boldsymbol{X}(t), t)$ in the $(\boldsymbol{R}, t)$ frame. One may now ask which relation is satisfied by the relative positions $\boldsymbol{r}_i(t) = \boldsymbol{R}_i(t) - \boldsymbol{X}(t)$ with respect to the newly defined privileged point $\boldsymbol{X}(t)$. They are connected to the $\hat{\boldsymbol{r}}_i(\hat{t})$ by a relation which may be derived in the same way as (A97), and which reads

$$\hat{\boldsymbol{r}}_i(\hat{t}) = \boldsymbol{r}_i(t) - \tfrac{1}{2}c^{-2}\boldsymbol{V}\cdot\boldsymbol{r}_i(t)\boldsymbol{V} + c^{-2}\boldsymbol{V}\cdot\boldsymbol{r}_i(t)\dot{\boldsymbol{R}}_i(t). \tag{A103}$$

Substitution of this relation and the transformation formula $\dot{\hat{\boldsymbol{X}}} = \dot{\boldsymbol{X}} - \boldsymbol{V}$ into the circumflexed (A101) gives

$$\sum_i \left\{ m_i \boldsymbol{r}_i - \tfrac{1}{2}c^{-2} m_i \boldsymbol{V}\cdot\boldsymbol{r}_i \boldsymbol{V} + c^{-2} m_i \boldsymbol{V}\cdot\boldsymbol{r}_i \dot{\boldsymbol{R}}_i + \tfrac{1}{2}c^{-2} m_i \dot{\boldsymbol{r}}_i^2 \boldsymbol{r}_i \right.$$
$$\left. + c^{-2} \sum_{j(\neq i)} \frac{e_i e_j}{8\pi|\boldsymbol{r}_i - \boldsymbol{r}_j|} \boldsymbol{r}_i + c^{-2} m_i (\dot{\boldsymbol{X}} - \boldsymbol{V})\cdot\boldsymbol{r}_i \dot{\boldsymbol{r}}_i \right\} = 0. \tag{A104}$$

With $\dot{\boldsymbol{R}}_i = \dot{\boldsymbol{X}} + \dot{\boldsymbol{r}}_i$ the third term splits, such that the second part of it cancels together with the last term. Furthermore the first part of the third term and the second term are both of order $c^{-4}$, as follows because $\sum_i m_i \boldsymbol{r}_i$ is of order $c^{-2}$. As a result one is left with a relation of the same form as (A101). Since there is only one point which satisfies (A101), as was shown above, the choice of the $(\boldsymbol{R}, t)$ frame as a starting point does not cause a bias.

The relation (A101) will be used to derive an equation of motion for the atom as a whole. In fact by summation over $i$ of equation (A87) and the use of (A91) and (A101) one gets an equation of motion with at the left-hand side the time derivative of the quantity

$$(m + \tfrac{1}{2}c^{-2} m \dot{\boldsymbol{X}}^2 + \tfrac{1}{2}c^{-2} \sum_i m_i \dot{\boldsymbol{r}}_i^2)\dot{\boldsymbol{X}} + \sum_i c^{-2} m_i \dot{\boldsymbol{r}}_i\cdot\dot{\boldsymbol{X}}\dot{\boldsymbol{r}}_i + \sum_i \tfrac{1}{2}c^{-2} m_i \dot{\boldsymbol{r}}_i^2 \dot{\boldsymbol{r}}_i$$
$$- c^{-2} \frac{\mathrm{d}}{\mathrm{d}t} \sum_i \left( \tfrac{1}{2} m_i \dot{\boldsymbol{r}}_i^2 \boldsymbol{r}_i + \sum_{j(\neq i)} \frac{e_i e_j}{8\pi|\boldsymbol{r}_i - \boldsymbol{r}_j|} \boldsymbol{r}_i + m_i \dot{\boldsymbol{X}}\cdot\boldsymbol{r}_i \dot{\boldsymbol{r}}_i \right). \tag{A105}$$

This is the extension up to order $c^{-2}$ of the non-relativistic momentum $m\dot{\boldsymbol{X}}$

(the mass of the composite particle is $m = \sum_i m_i$). If the time derivation in (A105) is performed, one obtains – amongst other terms – second derivatives. These may be rewritten with the help of the zero order equations of motion

$$\begin{aligned} m_i \ddot{\boldsymbol{R}}_i &= e_i \boldsymbol{e}_t(\boldsymbol{R}_i, t), \\ m\ddot{\boldsymbol{X}} &= \sum_i e_i \boldsymbol{e}_t(\boldsymbol{R}_i, t), \end{aligned} \tag{A106}$$

which follow from (A87). In this way (A105) becomes

$$\begin{aligned} &(m+\tfrac{1}{2}c^{-2}m\dot{\boldsymbol{X}}^2+\tfrac{1}{2}c^{-2}\sum_i m_i \dot{\boldsymbol{r}}_i^2)\dot{\boldsymbol{X}}-c^{-2}\sum_i \Big[ \dot{\boldsymbol{r}}_i\cdot \Big\{ e_i \boldsymbol{e}_t(\boldsymbol{R}_i, t) \\ &-\frac{m_i}{m}\sum_j e_j \boldsymbol{e}_t(\boldsymbol{R}_j, t)\Big\}\boldsymbol{r}_i + \sum_{j(\neq i)} \frac{e_i e_j}{8\pi|\boldsymbol{r}_i-\boldsymbol{r}_j|}\Big\{\dot{\boldsymbol{r}}_i - \frac{(\boldsymbol{r}_i-\boldsymbol{r}_j)\cdot(\dot{\boldsymbol{r}}_i-\dot{\boldsymbol{r}}_j)\boldsymbol{r}_i}{|\boldsymbol{r}_i-\boldsymbol{r}_j|^2}\Big\} \\ &+\frac{m_i}{m}\sum_j e_j \boldsymbol{e}_t(\boldsymbol{R}_j, t)\cdot \boldsymbol{r}_i \dot{\boldsymbol{r}}_i + \dot{\boldsymbol{X}}\cdot\boldsymbol{r}_i \Big\{ e_i \boldsymbol{e}_t(\boldsymbol{R}_i, t) - \frac{m_i}{m}\sum_j e_j \boldsymbol{e}_t(\boldsymbol{R}_j, t)\Big\}\Big], \end{aligned} \tag{A107}$$

where the last term between the brackets may be omitted, because it is of order $c^{-4}$, since $\sum_i m_i \boldsymbol{r}_i$ is of order $c^{-2}$.

The total electric field $\boldsymbol{e}_t(\boldsymbol{R}_i, t)$, occurring in (A107), consists of the intra-atomic field, generated by the constituent particles $j$ ($\neq i$) of the atom itself, and the field $\boldsymbol{E}(\boldsymbol{R}_i, t)$ from outside the atom. Up to order $c^0$ we have thus

$$\boldsymbol{e}_t(\boldsymbol{R}_i, t) = \sum_{j(\neq i)} \frac{e_j(\boldsymbol{r}_i-\boldsymbol{r}_j)}{4\pi|\boldsymbol{r}_i-\boldsymbol{r}_j|^3} + \boldsymbol{E}(\boldsymbol{R}_i, t). \tag{A108}$$

Substitution of this expression into (A107) yields

$$\begin{aligned} &(m+\tfrac{1}{2}c^{-2}m\dot{\boldsymbol{X}}^2+\tfrac{1}{2}c^{-2}\sum_i m_i \dot{\boldsymbol{r}}_i^2)\dot{\boldsymbol{X}} \\ &-c^{-2}\sum_{i,j(i\neq j)} \frac{e_i e_j}{8\pi|\boldsymbol{r}_i-\boldsymbol{r}_j|}\Big\{\dot{\boldsymbol{r}}_i + \frac{(\boldsymbol{r}_i-\boldsymbol{r}_j)(\boldsymbol{r}_i-\boldsymbol{r}_j)\cdot\dot{\boldsymbol{R}}_i}{|\boldsymbol{r}_i-\boldsymbol{r}_j|^2}\Big\} \\ &-c^{-2}\sum_i e_i \Big\{\frac{\bar{\boldsymbol{s}}}{m}\wedge \boldsymbol{E}(\boldsymbol{R}_i, t) + \dot{\boldsymbol{r}}_i\cdot\boldsymbol{E}(\boldsymbol{R}_i, t)\boldsymbol{r}_i + \dot{\boldsymbol{X}}\cdot\boldsymbol{r}_i \boldsymbol{E}(\boldsymbol{R}_i, t)\Big\}, \end{aligned} \tag{A109}$$

where we introduced the non-relativistic inner angular momentum

$$\bar{\boldsymbol{s}} \equiv \sum_i m_i \boldsymbol{r}_i \wedge \dot{\boldsymbol{r}}_i. \tag{A110}$$

The time derivative of expression (A109) constitutes the left-hand side of the equation of motion for the composite particle. The right-hand side is the sum over all constituent particles $i$ of the Lorentz forces which appear in the right-hand side of (A87). In these forces the total fields are sums of intra-atomic fields and fields from outside the atom (cf. (A108)). Up to orders $c^{-2}$ and $c^{-1}$ the electric and magnetic fields are given by (v. (III.72))

$$\begin{aligned} \boldsymbol{e}_t(\boldsymbol{R}_i, t) = \sum_{j(\neq i)} e_j \Bigg[ \frac{\boldsymbol{r}_i - \boldsymbol{r}_j}{4\pi|\boldsymbol{r}_i - \boldsymbol{r}_j|^3} + c^{-2} \frac{(\boldsymbol{r}_i - \boldsymbol{r}_j)\dot{\boldsymbol{R}}_j^2}{8\pi|\boldsymbol{r}_i - \boldsymbol{r}_j|^3} - c^{-2} \frac{3(\boldsymbol{r}_i - \boldsymbol{r}_j)\{(\boldsymbol{r}_i - \boldsymbol{r}_j)\cdot\dot{\boldsymbol{R}}_j\}^2}{8\pi|\boldsymbol{r}_i - \boldsymbol{r}_j|^5} \\ - c^{-2} \frac{(\boldsymbol{r}_i - \boldsymbol{r}_j)(\boldsymbol{r}_i - \boldsymbol{r}_j)\cdot\ddot{\boldsymbol{R}}_j}{8\pi|\boldsymbol{r}_i - \boldsymbol{r}_j|^3} - c^{-2} \frac{\ddot{\boldsymbol{R}}_j}{8\pi|\boldsymbol{r}_i - \boldsymbol{r}_j|} \Bigg] + \boldsymbol{E}(\boldsymbol{R}_i, t), \qquad \text{(A111)} \end{aligned}$$

$$\boldsymbol{b}_t(\boldsymbol{R}_i, t) = c^{-1} \sum_{j(\neq i)} e_j \frac{\dot{\boldsymbol{R}}_j \wedge (\boldsymbol{r}_i - \boldsymbol{r}_j)}{4\pi|\boldsymbol{r}_i - \boldsymbol{r}_j|^3} + \boldsymbol{B}(\boldsymbol{R}_i, t),$$

where (A91) has been used. With those expressions for the total electromagnetic field the right-hand side of the equation of motion for the composite particle gets the form

$$-c^{-2} \frac{\mathrm{d}}{\mathrm{d}t} \left[ \sum_{i,j(i\neq j)} \frac{e_i e_j}{8\pi|\boldsymbol{r}_i - \boldsymbol{r}_j|} \left\{ \dot{\boldsymbol{R}}_i + \frac{(\boldsymbol{r}_i - \boldsymbol{r}_j)(\boldsymbol{r}_i - \boldsymbol{r}_j)\cdot\dot{\boldsymbol{R}}_i}{|\boldsymbol{r}_i - \boldsymbol{r}_j|^2} \right\} \right] + \sum_i e_i \{\boldsymbol{E}(\boldsymbol{R}_i, t) + c^{-1} \dot{\boldsymbol{R}}_i \wedge \boldsymbol{B}(\boldsymbol{R}_i, t)\}. \qquad \text{(A112)}$$

The equation of motion for the composite particle up to order $c^{-2}$ follows finally by equating the time derivative of (A109) and expression (A112):

$$\begin{aligned} \frac{\mathrm{d}}{\mathrm{d}t} \left\{ \left( m + \tfrac{1}{2}c^{-2} m \dot{\boldsymbol{X}}^2 + \tfrac{1}{2}c^{-2} \sum_i m_i \dot{\boldsymbol{r}}_i^2 + c^{-2} \sum_{i,j(i\neq j)} \frac{e_i e_j}{8\pi|\boldsymbol{r}_i - \boldsymbol{r}_j|} \right) \dot{\boldsymbol{X}} \right\} \\ = \sum_i e_i \{\boldsymbol{E}(\boldsymbol{R}_i, t) + c^{-1} \dot{\boldsymbol{R}}_i \wedge \boldsymbol{B}(\boldsymbol{R}_i, t)\} \\ + c^{-2} \frac{\mathrm{d}}{\mathrm{d}t} \left[ \sum_i e_i \left\{ \dot{\boldsymbol{r}}_i \cdot \boldsymbol{E}(\boldsymbol{R}_i, t) \boldsymbol{r}_i + \dot{\boldsymbol{X}} \cdot \boldsymbol{r}_i \boldsymbol{E}(\boldsymbol{R}_i, t) + \frac{\bar{\boldsymbol{s}}}{m} \wedge \boldsymbol{E}(\boldsymbol{R}_i, t) \right\} \right]. \qquad \text{(A113)} \end{aligned}$$

In the left-hand side one recognizes the time derivative of the velocity of the atom times its total energy divided by $c^2$ (its 'total mass'). At the right-hand side appears, apart from the sum of the Lorentz forces, a (total) time derivative of order $c^{-2}$. The last term of the quantity between brackets may be written in a compact form:

$$\sum_i e_i \frac{\bar{\boldsymbol{s}}}{m} \wedge \boldsymbol{E}(\boldsymbol{R}_i, t) = \bar{\boldsymbol{s}} \wedge \ddot{\boldsymbol{X}} \qquad \text{(A114)}$$

by using (A106) and (A108) (cf. the remark after eq. (126)).

In the case without fields $\boldsymbol{E}$ and $\boldsymbol{B}$ the right-hand side of (A113) vanishes, so that then the bracket in the left-hand side is conserved. According to equation (I.63) the time derivative of the factor between brackets (the 'total mass') is in this case at most of order $c^{-4}$ and hence negligible. This means that the velocity $\dot{X}$ of the atom as a whole is conserved, as one would expect in the field-free case.

Let us now consider the equation of motion (A113) for a composite particle in an external field $(\boldsymbol{E}, \boldsymbol{B})$ that changes slowly over the dimensions of the particle. Then the right-hand side of the equation of motion may be expanded in terms of the relative positions $\boldsymbol{r}_i$ (A91). In that way one may obtain a series expansion containing the non-relativistic multipole moments (I.15–16) of chapter I. These non-relativistic multipole moments however are not independent of the motion of the atoms in a theory in which all terms up to order $c^{-2}$ are taken into account. Therefore we want to introduce rest frame multipole moments, as in section 2*b* of the present chapter, which read

$$\begin{aligned} \boldsymbol{\mu}^{(n)} &= \frac{1}{n!}\sum_i e_i \boldsymbol{r}_i'^n, \qquad (n = 0, 1, 2, \ldots), \\ \boldsymbol{\nu}^{(n)} &= \frac{n}{(n+1)!}\sum_i e_i \boldsymbol{r}_i'^n \wedge \frac{\dot{\boldsymbol{r}}_i'}{c}, \qquad (n = 1, 2, \ldots). \end{aligned} \tag{A115}$$

The connexion (A100) between $\boldsymbol{r}_i$ and $\boldsymbol{r}_i'$ allows us to express the right-hand side of the equation of motion in terms of these rest frame multipole moments. If we confine ourselves to the contributions of the electric charge $\mu^{(0)} \equiv e$ and the electric and magnetic dipole moments $\boldsymbol{\mu}^{(1)} \equiv \boldsymbol{\mu}$ and $\boldsymbol{\nu}^{(1)} \equiv \boldsymbol{\nu}$, and if we make use of the homogeneous field equations for the external fields

$$\nabla \cdot \boldsymbol{B} = 0, \qquad c^{-1}\partial \boldsymbol{B}/\partial t + \nabla \wedge \boldsymbol{E} = 0, \tag{A116}$$

we obtain as the equation of motion up to order $c^{-2}$ for a composite charged dipole particle in an external field

$$\begin{aligned} &\frac{\mathrm{d}}{\mathrm{d}t}\left\{\left(m + \tfrac{1}{2}c^{-2}mv^2 + \tfrac{1}{2}c^{-2}\sum_i m_i \dot{\boldsymbol{r}}_i'^2 + c^{-2}\sum_{i,j(i\neq j)} \frac{e_i e_j}{8\pi|\boldsymbol{r}_i' - \boldsymbol{r}_j'|}\right)\boldsymbol{v}\right\} \\ &= e(\boldsymbol{E} + c^{-1}\boldsymbol{v}\wedge\boldsymbol{B}) + (\nabla\boldsymbol{E})\cdot(\boldsymbol{\mu} - \tfrac{1}{2}c^{-2}\boldsymbol{v}\boldsymbol{v}\cdot\boldsymbol{\mu} - c^{-1}\boldsymbol{\nu}\wedge\boldsymbol{v}) + (\nabla\boldsymbol{B})\cdot(\boldsymbol{\nu} + c^{-1}\boldsymbol{\mu}\wedge\boldsymbol{v}) \\ &+ c^{-1}\frac{\mathrm{d}}{\mathrm{d}t}(\boldsymbol{\mu}\wedge\boldsymbol{B}) - c^{-1}\frac{\mathrm{d}}{\mathrm{d}t}\left[\boldsymbol{\nu}\wedge\boldsymbol{E} - c^{-1}\boldsymbol{v}\cdot\boldsymbol{\mu}\boldsymbol{E} - c^{-1}\frac{\bar{\boldsymbol{s}}}{m}\wedge\{e\boldsymbol{E} + (\nabla\boldsymbol{E})\cdot\boldsymbol{\mu}\}\right], \end{aligned} \tag{A117}$$

where the fields depend on the position $\boldsymbol{X}$ and the time $t$ and where we have written $\boldsymbol{v}$ for the velocity $\dot{\boldsymbol{X}}$. This equation, which is indeed the $c^{-2}$ approximation of the relativistic equation (135) with (137), may be compared to the non-relativistic equation (55) of chapter I. At the left-hand side three additional terms appear, which describe $c^{-2}$ corrections to the inertial mass. At the right-hand side in the second term the Lorentz contracted electric dipole moment $\boldsymbol{\mu}_{\perp}+(1-\frac{1}{2}c^{-2}\boldsymbol{v}^2)\boldsymbol{\mu}_{//}$ up to order $c^{-2}$ appears (the dipole moment has been split into components orthogonal to and parallel with the velocity) and moreover a term due to the moving magnetic dipole moment. The last time derivative contains in the first place a magnetodynamic effect with $\boldsymbol{\nu}\wedge\boldsymbol{E}$ analogous to the electrodynamic effect with $\boldsymbol{\mu}\wedge\boldsymbol{B}$ of the penultimate time derivative. The third contribution to the last time derivative has the same form as the first contribution but with the 'normal magnetic dipole moment' $c^{-1}e\bar{s}/m$ instead of the (total) magnetic dipole moment $\boldsymbol{\nu}$, and with the opposite sign. Hence effectively only the 'anomalous magnetic moment' couples with the electric field (as has been found already in the relativistic treatment). For an ordinary atom the anomalous magnetic moment is nearly the same as the total magnetic moment $\boldsymbol{\nu}$, since the atomic mass $m$ is several thousand times greater than the masses of the particles which contribute to the inner angular momentum.

In the preceding all effects of order $c^{-2}$ were taken into account. The equation of motion is simplified if we consider the *semi-relativistic approximation*. The latter has been defined in section 1 of this appendix as the approximation which results if one retains terms up to order $c^{-1}$ only, considering the magnetic dipole moment as being of order $c^0$. In this way we get from (A117) the semi-relativistic equation of motion for a charged particle in a slowly varying external field:

$$m\dot{\boldsymbol{v}} = e(\boldsymbol{E}+c^{-1}\boldsymbol{v}\wedge\boldsymbol{B})+(\nabla\boldsymbol{E})\cdot(\boldsymbol{\mu}-c^{-1}\boldsymbol{\nu}\wedge\boldsymbol{v})+(\nabla\boldsymbol{B})\cdot(\boldsymbol{\nu}+c^{-1}\boldsymbol{\mu}\wedge\boldsymbol{v})$$
$$+c^{-1}\frac{\mathrm{d}}{\mathrm{d}t}(\boldsymbol{\mu}\wedge\boldsymbol{B}-\boldsymbol{\nu}\wedge\boldsymbol{E}). \quad \text{(A118)}$$

The dipole terms in this equation are symmetric with respect to electric and magnetic phenomena. This was not the case in the non-relativistic equation (I.55).

A semi-relativistic equation of motion has been given also by Coleman and Van Vleck[1], starting from the Darwin Hamiltonian (v. problem 6 of chapter III). They employ the point $\boldsymbol{X}^*$ (A88) as the centre of energy. This means that the relative coordinates with respect to the privileged point satisfy a relation like (A101) but without the fourth term. It follows however from the discussion leading to (A104) that such a definition is biased by the choice of the observer's frame: in other words such a definition of the centre of

[1] S. Coleman and J. H. Van Vleck, Phys. Rev. **171**(1968)1370.

energy is not even covariant up to order $c^{-2}$. Using this centre of energy they find an equation of motion up to order $c^{-2}$, which has a form similar to (A113) but with an extra term $c^{-2}(\mathrm{d}^2/\mathrm{d}t^2)(\sum_i m_i \dot{\boldsymbol{X}}\cdot\boldsymbol{r}_i\dot{\boldsymbol{r}}_i)$ at the left-hand side. Making a multipole expansion and taking the semi-relativistic limit, one arrives then at an equation like (A118) without the term $-c^{-1}(\nabla\boldsymbol{E})\cdot(\boldsymbol{v}\wedge\boldsymbol{v})$. Coleman and Van Vleck limit themselves to the case of a magnetic dipole in an electric field and thus find for the force only the term $-c^{-1}(\mathrm{d}/\mathrm{d}t)(\boldsymbol{v}\wedge\boldsymbol{E})$.

## b. *The energy equation*

The energy equation is obtained by multiplying the equation of motion (A87) with $\dot{\boldsymbol{R}}_i$ and summing over $i$:

$$\sum_i \frac{\mathrm{d}}{\mathrm{d}t}\{m_i(1+\tfrac{1}{2}c^{-2}\dot{\boldsymbol{R}}_i^2)\dot{\boldsymbol{R}}_i\}\cdot\dot{\boldsymbol{R}}_i = \sum_i e_i \dot{\boldsymbol{R}}_i\cdot\boldsymbol{e}_t(\boldsymbol{R}_i, t). \tag{A119}$$

If the relative positions $\boldsymbol{r}_i$ (A91) are introduced, and the relation (A101) is used, one gets as the left-hand side the time derivative of the quantity

$$\begin{aligned}&\tfrac{1}{2}m\dot{\boldsymbol{X}}^2+\tfrac{3}{8}c^{-2}m\dot{\boldsymbol{X}}^4+\sum_i \tfrac{1}{2}m_i\dot{\boldsymbol{r}}_i^2\\ &\quad -c^{-2}\sum_i\left\{\dot{\boldsymbol{X}}\cdot\frac{\mathrm{d}}{\mathrm{d}t}\left(\tfrac{1}{2}m_i\dot{\boldsymbol{r}}_i^2\boldsymbol{r}_i+\sum_{j(\neq i)}\frac{e_i e_j}{8\pi|\boldsymbol{r}_i-\boldsymbol{r}_j|}\boldsymbol{r}_i+m_i\dot{\boldsymbol{X}}\cdot\boldsymbol{r}_i\dot{\boldsymbol{r}}_i\right)\right.\\ &\qquad\left.-\tfrac{3}{4}m_i\dot{\boldsymbol{r}}_i^2\dot{\boldsymbol{X}}^2-\tfrac{3}{2}m_i(\dot{\boldsymbol{r}}_i\cdot\dot{\boldsymbol{X}})^2-\tfrac{3}{2}m_i\dot{\boldsymbol{r}}_i^2\dot{\boldsymbol{r}}_i\cdot\dot{\boldsymbol{X}}-\tfrac{3}{8}m_i\dot{\boldsymbol{r}}_i^4\right\}.\end{aligned} \tag{A120}$$

If the time derivative in the fourth term is performed, and the equations of motion (A106) up to order $c^0$ are employed, one obtains:

$$\begin{aligned}&\tfrac{1}{2}m\dot{\boldsymbol{X}}^2+\tfrac{3}{8}c^{-2}m\dot{\boldsymbol{X}}^4+\sum_i\tfrac{1}{2}m_i\dot{\boldsymbol{r}}_i^2+c^{-2}\sum_i\left[\tfrac{3}{4}m_i\dot{\boldsymbol{r}}_i^2\dot{\boldsymbol{X}}^2+\tfrac{1}{2}m_i(\dot{\boldsymbol{r}}_i\cdot\dot{\boldsymbol{X}})^2+m_i\dot{\boldsymbol{r}}_i^2\dot{\boldsymbol{r}}_i\cdot\dot{\boldsymbol{X}}\right.\\ &\quad+\tfrac{3}{8}m_i\dot{\boldsymbol{r}}_i^4-\sum_{j(\neq i)}\frac{e_i e_j}{8\pi|\boldsymbol{r}_i-\boldsymbol{r}_j|}\left\{\dot{\boldsymbol{r}}_i\cdot\dot{\boldsymbol{X}}-\frac{(\boldsymbol{r}_i-\boldsymbol{r}_j)\cdot(\dot{\boldsymbol{r}}_i-\dot{\boldsymbol{r}}_j)\boldsymbol{r}_i\cdot\dot{\boldsymbol{X}}}{|\boldsymbol{r}_i-\boldsymbol{r}_j|^2}\right\}\\ &\quad-\boldsymbol{r}_i\cdot\dot{\boldsymbol{X}}(\dot{\boldsymbol{X}}+\dot{\boldsymbol{r}}_i)\cdot\left\{e_i\boldsymbol{e}_t(\boldsymbol{R}_i,t)-\frac{m_i}{m}\sum_j e_j\boldsymbol{e}_t(\boldsymbol{R}_j,t)\right\}\\ &\qquad\left.-\frac{m_i}{m}\dot{\boldsymbol{r}}_i\cdot\dot{\boldsymbol{X}}\boldsymbol{r}_i\cdot\{\sum_j e_j\boldsymbol{e}_t(\boldsymbol{R}_j,t)\}\right].\end{aligned} \tag{A121}$$

The third term is the part of order $c^0$ of the internal kinetic energy in the observer's frame. We want to introduce the kinetic energy in the rest frame of the composite particle, since this is an invariant quantity. Instead of using the total time derivative of $\boldsymbol{r}'_i$, which would be obtained from (A100), it is more convenient to use the time derivative of $\boldsymbol{r}'_i$ at constant transformation velocity. It follows from (A96) or (A97) by taking $\dot{\boldsymbol{X}}^*(t)$ constant. Moreover we want to take into account the fact that time differentiations in the observer's frame and in the rest frame differ by a factor $1-\frac{1}{2}c^{-2}\{\dot{\boldsymbol{X}}^*(t)\}^2$. In this way we are led to the introduction[1] of the quantity $\dot{\boldsymbol{r}}'_i$ by means of

$$\dot{\boldsymbol{r}}_i = \dot{\boldsymbol{r}}'_i - \tfrac{1}{2}c^{-2}\dot{\boldsymbol{X}}^2\dot{\boldsymbol{r}}'_i + \tfrac{1}{2}c^{-2}\dot{\boldsymbol{X}}\cdot\dot{\boldsymbol{r}}_i\dot{\boldsymbol{X}} - c^{-2}\dot{\boldsymbol{X}}\cdot\dot{\boldsymbol{r}}_i\dot{\boldsymbol{R}}_i - c^{-2}\dot{\boldsymbol{X}}\cdot\boldsymbol{r}_i\ddot{\boldsymbol{R}}_i, \qquad \text{(A122)}$$

where we used the fact that $\dot{\boldsymbol{X}}^*$ and $\dot{\boldsymbol{X}}$ differ by terms of order $c^{-2}$ only. From (A122) and the equation of motion in zeroth order (A106) it follows that

$$\sum_i \tfrac{1}{2}m_i\dot{\boldsymbol{r}}_i^2 = \sum_i \tfrac{1}{2}m_i\dot{\boldsymbol{r}}_i'^2 - c^{-2}\sum_i \{\tfrac{1}{2}m_i\dot{\boldsymbol{r}}_i^2\dot{\boldsymbol{X}}^2 + \tfrac{1}{2}m_i(\dot{\boldsymbol{r}}_i\cdot\dot{\boldsymbol{X}})^2$$
$$+ m_i\dot{\boldsymbol{r}}_i^2\dot{\boldsymbol{r}}_i\cdot\dot{\boldsymbol{X}} + e_i\boldsymbol{r}_i\cdot\dot{\boldsymbol{X}}\dot{\boldsymbol{r}}_i\cdot\boldsymbol{e}_t(\boldsymbol{R}_i, t)\}. \qquad \text{(A123)}$$

Substituting this expression into (A121) and using (A101) we obtain as the left-hand side of the energy equation the time derivative of

$$\tfrac{1}{2}m\dot{\boldsymbol{X}}^2 + \tfrac{3}{8}c^{-2}m\dot{\boldsymbol{X}}^4 + \tfrac{1}{2}\sum_i m_i\dot{\boldsymbol{r}}_i'^2 + c^{-2}\sum_i \Bigg[\tfrac{1}{4}m_i\dot{\boldsymbol{r}}_i^2\dot{\boldsymbol{X}}^2 + \tfrac{3}{8}m_i\dot{\boldsymbol{r}}_i^4$$
$$-\sum_{j(\neq i)}\frac{e_ie_j}{8\pi|\boldsymbol{r}_i-\boldsymbol{r}_j|}\left\{\dot{\boldsymbol{r}}_i\cdot\dot{\boldsymbol{X}} - \frac{(\boldsymbol{r}_i-\boldsymbol{r}_j)\cdot(\dot{\boldsymbol{r}}_i-\dot{\boldsymbol{r}}_j)\boldsymbol{r}_i\cdot\dot{\boldsymbol{X}}}{|\boldsymbol{r}_i-\boldsymbol{r}_j|^2}\right\} - e_i\boldsymbol{r}_i\cdot\dot{\boldsymbol{X}}(\dot{\boldsymbol{X}}+2\dot{\boldsymbol{r}}_i)\cdot\boldsymbol{e}_t(\boldsymbol{R}_i, t)$$
$$+\frac{m_i}{m}(\boldsymbol{r}_i\cdot\dot{\boldsymbol{X}}\dot{\boldsymbol{r}}_i - \dot{\boldsymbol{r}}_i\cdot\dot{\boldsymbol{X}}\boldsymbol{r}_i)\cdot\sum_j e_j\boldsymbol{e}_t(\boldsymbol{R}_j, t)\Bigg]. \qquad \text{(A124)}$$

In the right-hand side of the energy equation (A119) we now substitute the expression (A111) for the total electric field. Then we obtain

$$-\frac{\mathrm{d}}{\mathrm{d}t}\sum_{i,j(i\neq j)}\frac{e_ie_j}{8\pi|\boldsymbol{r}_i-\boldsymbol{r}_j|}\Bigg(1+c^{-2}\Bigg[\tfrac{1}{2}\dot{\boldsymbol{r}}_i\cdot\dot{\boldsymbol{r}}_j + \frac{1}{2}\frac{(\boldsymbol{r}_i-\boldsymbol{r}_j)\cdot\dot{\boldsymbol{r}}_i(\boldsymbol{r}_i-\boldsymbol{r}_j)\cdot\dot{\boldsymbol{r}}_j}{|\boldsymbol{r}_i-\boldsymbol{r}_j|^2}$$
$$+\tfrac{1}{2}\dot{\boldsymbol{X}}^2 + \frac{1}{2}\frac{\{(\boldsymbol{r}_i-\boldsymbol{r}_j)\cdot\dot{\boldsymbol{X}}\}^2}{|\boldsymbol{r}_i-\boldsymbol{r}_j|^2} + \dot{\boldsymbol{r}}_i\cdot\dot{\boldsymbol{X}} + \frac{(\boldsymbol{r}_i-\boldsymbol{r}_j)\cdot\dot{\boldsymbol{r}}_i(\boldsymbol{r}_i-\boldsymbol{r}_j)\cdot\dot{\boldsymbol{X}}}{|\boldsymbol{r}_i-\boldsymbol{r}_j|^2}\Bigg]\Bigg)$$
$$+\sum_i e_i\dot{\boldsymbol{R}}_i\cdot\boldsymbol{E}(\boldsymbol{R}_i, t). \qquad \text{(A125)}$$

The equation of energy (A119) becomes finally with (A100), (A110), (A122),

[1] Earlier in this appendix we did not need to make a difference between $\dot{\boldsymbol{r}}_i$ and $\dot{\boldsymbol{r}}'_i$ since they only occurred with factors $c^{-1}$ or $c^{-2}$. The quantity $\dot{\boldsymbol{r}}'_i$ introduced here is the $c^{-2}$ approximation of (A34).

(A124) and (A125)

$$\frac{\mathrm{d}}{\mathrm{d}t}\Bigg(\tfrac{1}{2}m\dot{X}^2+\tfrac{3}{8}c^{-2}m\dot{X}^4+\sum_i\left(\tfrac{1}{2}m_i\dot{r}_i'^2+\tfrac{1}{4}c^{-2}m_i\dot{r}_i'^2\dot{X}^2+\tfrac{3}{8}c^{-2}m_i\dot{r}_i'^4\right)$$
$$+\sum_{i,j(i\neq j)}\frac{e_ie_j}{8\pi|r_i'-r_j'|}\left[1+\tfrac{1}{2}c^{-2}\left\{\dot{r}_i'\cdot\dot{r}_j'+\frac{(r_i'-r_j')\cdot\dot{r}_i'(r_i'-r_j')\cdot\dot{r}_j'}{|r_i'-r_j'|^2}+\dot{X}^2\right\}\right]\Bigg)$$
$$=\sum_i e_i\dot{R}_i\cdot E(R_i,t)+c^{-2}\frac{\mathrm{d}}{\mathrm{d}t}\left[\sum_i e_i\left\{\frac{\bar{s}}{m}\wedge E(R_i,t)\right.\right.$$
$$\left.\left.+2\dot{r}_i\cdot E(R_i,t)r_i+\dot{X}\cdot r_i\,E(R_i,t)\right\}\cdot\dot{X}\right]. \quad \text{(A126)}$$

This is the energy equation up to order $c^{-2}$ for a composite particle (cf. the equation of motion (A113)). It shows which corrections of order $c^{-2}$ arise as compared to the non-relativistic equation of chapter I. If the fields from outside the atom are slowly varying, we may perform a multipole expansion and introduce the multipole moments (A115). Just as for the equation of motion we shall confine ourselves to the contributions of the charge and the dipole moments. Moreover we introduce again the *semi-relativistic* approximation, as in the preceding subsection. We then obtain from (A126), using also the field equations (A116), and the notation $v$ for the atomic velocity $\dot{X}$:

$$\frac{\mathrm{d}}{\mathrm{d}t}\left(\tfrac{1}{2}mv^2+\sum_i\tfrac{1}{2}m_i\dot{r}_i'^2+\sum_{i,j(i\neq j)}\frac{e_ie_j}{8\pi|r_i'-r_j'|}\right)$$
$$=ev\cdot E+v\cdot(\nabla E)\cdot(\mu-c^{-1}\nu\wedge v)+v\cdot(\nabla B)\cdot(\nu+c^{-1}\mu\wedge v)$$
$$+\left\{\frac{\mathrm{d}}{\mathrm{d}t}(\mu-c^{-1}\nu\wedge v)\right\}\cdot E-(\nu+c^{-1}\mu\wedge v)\cdot\frac{\mathrm{d}B}{\mathrm{d}t}+2c^{-1}\frac{\mathrm{d}}{\mathrm{d}t}\{(\nu\wedge v)\cdot E\}, \quad \text{(A127)}$$

which is the semi-relativistic energy equation for a charged dipole particle in an external field. (It might have been obtained by taking the limit of the relativistic equation.) As compared to the non-relativistic equation (I.67) various new terms with magnetic dipoles in motion arise here.

### 3. *The angular momentum equation*

The angular momentum equation for a composite particle is obtained by multiplying the equation of motion (A87) with the position $R_i$ and summing over the index $i$ that labels the constituent particles:

$$\sum_i R_i\wedge\frac{\mathrm{d}}{\mathrm{d}t}\{(m_i(1+\tfrac{1}{2}c^{-2}\dot{R}_i^2)\dot{R}_i\}=\sum_i e_iR_i\wedge\{e_t(R_i,t)+c^{-1}\dot{R}_i\wedge b_t(R_i,t)\}. \quad \text{(A128)}$$

The left-hand side is the time derivative of the expression:

$$\sum_i m_i(1+\tfrac{1}{2}c^{-2}\dot{\boldsymbol{R}}_i^2)\boldsymbol{R}_i\wedge\dot{\boldsymbol{R}}_i. \tag{A129}$$

We introduce now the relative positions $\boldsymbol{r}_i$ (A91) and use the centre of energy condition (A101). This gives, up to order $c^{-2}$, for (A129):

$$\begin{aligned}
&m\boldsymbol{X}\wedge\dot{\boldsymbol{X}}+\bar{\boldsymbol{s}}+c^{-2}(\tfrac{1}{2}m\dot{\boldsymbol{X}}^2+\tfrac{1}{2}\sum_i m_i\dot{\boldsymbol{r}}_i^2)\boldsymbol{X}\wedge\dot{\boldsymbol{X}}\\
&\quad+c^{-2}\sum_i m_i(\tfrac{1}{2}\dot{\boldsymbol{X}}^2+\dot{\boldsymbol{X}}\cdot\dot{\boldsymbol{r}}_i+\tfrac{1}{2}\dot{\boldsymbol{r}}_i^2)\boldsymbol{r}_i\wedge\dot{\boldsymbol{r}}_i-c^{-2}(\bar{\boldsymbol{s}}\wedge\dot{\boldsymbol{X}})\wedge\dot{\boldsymbol{X}}\\
&\quad-c^{-2}\sum_{i,j(i\neq j)}\frac{e_i e_j}{8\pi|\boldsymbol{r}_i-\boldsymbol{r}_j|}\left\{\boldsymbol{r}_i\wedge\dot{\boldsymbol{X}}+\boldsymbol{X}\wedge\dot{\boldsymbol{r}}_i-\frac{(\boldsymbol{r}_i-\boldsymbol{r}_j)\cdot(\dot{\boldsymbol{r}}_i-\dot{\boldsymbol{r}}_j)}{|\boldsymbol{r}_i-\boldsymbol{r}_j|^2}\boldsymbol{X}\wedge\boldsymbol{r}_i\right\}\\
&\qquad\qquad-c^{-2}\sum_i m_i\boldsymbol{X}\wedge(\dot{\boldsymbol{r}}_i\cdot\ddot{\boldsymbol{r}}_i\boldsymbol{r}_i+\ddot{\boldsymbol{X}}\cdot\boldsymbol{r}_i\dot{\boldsymbol{r}}_i+\dot{\boldsymbol{X}}\cdot\boldsymbol{r}_i\ddot{\boldsymbol{r}}_i),
\end{aligned} \tag{A130}$$

where the non-relativistic inner angular momentum $\bar{\boldsymbol{s}}$ (A110) has been introduced. The accelerations in the last three terms may be eliminated by means of the equation of motion (A106) up to order $c^0$. For the electric field which then appears we write expression (A108). In this way we get as the left-hand side of the angular momentum equation the time derivative of the expression

$$\begin{aligned}
&m\boldsymbol{X}\wedge\dot{\boldsymbol{X}}+\bar{\boldsymbol{s}}+c^{-2}(\tfrac{1}{2}m\dot{\boldsymbol{X}}^2+\tfrac{1}{2}\sum_i m_i\dot{\boldsymbol{r}}_i^2)\boldsymbol{X}\wedge\dot{\boldsymbol{X}}\\
&\quad+c^{-2}\sum_i m_i(\tfrac{1}{2}\dot{\boldsymbol{X}}^2+\dot{\boldsymbol{X}}\cdot\dot{\boldsymbol{r}}_i+\tfrac{1}{2}\dot{\boldsymbol{r}}_i^2)\boldsymbol{r}_i\wedge\dot{\boldsymbol{r}}_i-c^{-2}(\bar{\boldsymbol{s}}\wedge\dot{\boldsymbol{X}})\wedge\dot{\boldsymbol{X}}\\
&\quad-c^{-2}\sum_i e_i\boldsymbol{X}\wedge\left\{\dot{\boldsymbol{r}}_i\cdot\boldsymbol{E}(\boldsymbol{R}_i,t)\boldsymbol{r}_i+\dot{\boldsymbol{X}}\cdot\boldsymbol{r}_i\boldsymbol{E}(\boldsymbol{R}_i,t)+\frac{\bar{\boldsymbol{s}}}{m}\wedge\boldsymbol{E}(\boldsymbol{R}_i,t)\right\}\\
&\quad-c^{-2}\sum_{i,j(i\neq j)}\frac{e_i e_j}{8\pi|\boldsymbol{r}_i-\boldsymbol{r}_j|}\left\{\boldsymbol{r}_i\wedge\dot{\boldsymbol{X}}+\boldsymbol{X}\wedge\dot{\boldsymbol{r}}_i+\frac{(\boldsymbol{r}_i-\boldsymbol{r}_j)\cdot\dot{\boldsymbol{R}}_i\boldsymbol{X}\wedge(\boldsymbol{r}_i-\boldsymbol{r}_j)}{|\boldsymbol{r}_i-\boldsymbol{r}_j|^2}\right\}.
\end{aligned} \tag{A131}$$

The right-hand side of the angular momentum equation (A128) contains the total electromagnetic field, for which we substitute the expressions (A111). In this way one finds for the right-hand side

$$\begin{aligned}
&-c^{-2}\sum_{i,j(i\neq j)}\frac{\mathrm{d}}{\mathrm{d}t}\left[\frac{e_i e_j}{8\pi|\boldsymbol{r}_i-\boldsymbol{r}_j|}\left\{\boldsymbol{R}_i\wedge\dot{\boldsymbol{R}}_j+\frac{(\boldsymbol{r}_i-\boldsymbol{r}_j)\cdot\dot{\boldsymbol{R}}_j\boldsymbol{R}_i\wedge(\boldsymbol{r}_i-\boldsymbol{r}_j)}{|\boldsymbol{r}_i-\boldsymbol{r}_j|^2}\right\}\right]\\
&\qquad\qquad+\sum_i e_i\boldsymbol{R}_i\wedge\{\boldsymbol{E}(\boldsymbol{R}_i,t)+c^{-1}\dot{\boldsymbol{R}}_i\wedge\boldsymbol{B}(\boldsymbol{R}_i,t)\}.
\end{aligned} \tag{A132}$$

The angular momentum equation (A128) becomes with (A131) and (A132):

$$\frac{\mathrm{d}}{\mathrm{d}t}\Bigg[m\boldsymbol{X}\wedge\dot{\boldsymbol{X}}+\bar{\boldsymbol{s}}+c^{-2}\left(\tfrac{1}{2}m\dot{\boldsymbol{X}}^2+\tfrac{1}{2}\sum_i m_i\dot{\boldsymbol{r}}_i^2+\sum_{i,j(i\neq j)}\frac{e_ie_j}{8\pi|\boldsymbol{r}_i-\boldsymbol{r}_j|}\right)\boldsymbol{X}\wedge\dot{\boldsymbol{X}}$$
$$+c^{-2}\sum_i m_i(\tfrac{1}{2}\dot{\boldsymbol{X}}^2+\dot{\boldsymbol{X}}\cdot\dot{\boldsymbol{r}}_i+\tfrac{1}{2}\dot{\boldsymbol{r}}_i^2)\boldsymbol{r}_i\wedge\dot{\boldsymbol{r}}_i-c^{-2}(\bar{\boldsymbol{s}}\wedge\dot{\boldsymbol{X}})\wedge\dot{\boldsymbol{X}}$$
$$+c^{-2}\sum_{i,j(i\neq j)}\frac{e_ie_j}{8\pi|\boldsymbol{r}_i-\boldsymbol{r}_j|}\left\{\boldsymbol{r}_i\wedge\dot{\boldsymbol{r}}_j-\frac{(\boldsymbol{r}_i-\boldsymbol{r}_j)\cdot\dot{\boldsymbol{R}}_j\,\boldsymbol{r}_i\wedge\boldsymbol{r}_j}{|\boldsymbol{r}_i-\boldsymbol{r}_j|^2}\right\}\Bigg]$$
$$=\sum_i e_i\boldsymbol{R}_i\wedge\{\boldsymbol{E}(\boldsymbol{R}_i,t)+c^{-1}\dot{\boldsymbol{R}}_i\wedge\boldsymbol{B}(\boldsymbol{R}_i,t)\}$$
$$+c^{-2}\frac{\mathrm{d}}{\mathrm{d}t}\left[\sum_i e_i\boldsymbol{X}\wedge\left\{\dot{\boldsymbol{r}}_i\cdot\boldsymbol{E}(\boldsymbol{R}_i,t)\boldsymbol{r}_i+\dot{\boldsymbol{X}}\cdot\boldsymbol{r}_i\,\boldsymbol{E}(\boldsymbol{R}_i,t)+\frac{\bar{\boldsymbol{s}}}{m}\wedge\boldsymbol{E}(\boldsymbol{R}_i,t)\right\}\right].\tag{A133}$$

This equation still contains the position $\boldsymbol{X}$ of the composite particle. It may be eliminated with the help of the equation which results if the equation of motion (A113) is multiplied (vectorially) by $\boldsymbol{X}$. Then we get

$$\frac{\mathrm{d}}{\mathrm{d}t}\Bigg[\bar{\boldsymbol{s}}+c^{-2}\sum_i m_i(\tfrac{1}{2}\dot{\boldsymbol{X}}^2+\dot{\boldsymbol{X}}\cdot\dot{\boldsymbol{r}}_i+\tfrac{1}{2}\dot{\boldsymbol{r}}_i^2)\boldsymbol{r}_i\wedge\dot{\boldsymbol{r}}_i-c^{-2}(\bar{\boldsymbol{s}}\wedge\dot{\boldsymbol{X}})\wedge\dot{\boldsymbol{X}}$$
$$+c^{-2}\sum_{i,j(i\neq j)}\frac{e_ie_j}{8\pi|\boldsymbol{r}_i-\boldsymbol{r}_j|}\left\{\boldsymbol{r}_i\wedge\dot{\boldsymbol{r}}_j-\frac{(\boldsymbol{r}_i-\boldsymbol{r}_j)\cdot\dot{\boldsymbol{R}}_j\,\boldsymbol{r}_i\wedge\boldsymbol{r}_j}{|\boldsymbol{r}_i-\boldsymbol{r}_j|^2}\right\}\Bigg]$$
$$=\sum_i e_i\boldsymbol{r}_i\wedge\{\boldsymbol{E}(\boldsymbol{R}_i,t)+c^{-1}\dot{\boldsymbol{R}}_i\wedge\boldsymbol{B}(\boldsymbol{R}_i,t)\}$$
$$+c^{-2}\sum_i e_i\boldsymbol{X}\wedge\left\{\dot{\boldsymbol{r}}_i\cdot\boldsymbol{E}(\boldsymbol{R}_i,t)\boldsymbol{r}_i+\dot{\boldsymbol{X}}\cdot\boldsymbol{r}_i\,\boldsymbol{E}(\boldsymbol{R}_i,t)+\frac{\bar{\boldsymbol{s}}}{m}\wedge\boldsymbol{E}(\boldsymbol{R}_i,t)\right\}.\tag{A134}$$

The non-relativistic inner angular momentum $\bar{\boldsymbol{s}}$ (A110), which occurs here, has been defined in terms of $\boldsymbol{r}_i$ and $\dot{\boldsymbol{r}}_i$. If we eliminate these quantities in the left-hand side in favour of $\boldsymbol{r}'_i$ and $\dot{\boldsymbol{r}}'_i$, with (A100) and (A122), we obtain as inner angular momentum equation

$$\frac{\mathrm{d}}{\mathrm{d}t}\Bigg[\sum_i m_i(1+\tfrac{1}{2}c^{-2}\dot{\boldsymbol{r}}_i'^2)\boldsymbol{r}'_i\wedge\dot{\boldsymbol{r}}'_i-\tfrac{1}{2}c^{-2}\{(\sum_i m_i\boldsymbol{r}'_i\wedge\dot{\boldsymbol{r}}'_i)\wedge\dot{\boldsymbol{X}}\}\wedge\dot{\boldsymbol{X}}$$
$$+c^{-2}\sum_{i,j(i\neq j)}\frac{e_ie_j}{8\pi|\boldsymbol{r}'_i-\boldsymbol{r}'_j|}\left\{\boldsymbol{r}'_i\wedge\dot{\boldsymbol{r}}'_j-\frac{(\boldsymbol{r}'_i-\boldsymbol{r}'_j)\cdot\dot{\boldsymbol{r}}'_j\,\boldsymbol{r}'_i\wedge\boldsymbol{r}'_j}{|\boldsymbol{r}'_i-\boldsymbol{r}'_j|^2}\right\}\Bigg]$$
$$=\sum_i e_i\boldsymbol{r}_i\wedge\{\boldsymbol{E}(\boldsymbol{R}_i,t)+c^{-1}\dot{\boldsymbol{R}}_i\wedge\boldsymbol{B}(\boldsymbol{R}_i,t)\}$$
$$+c^{-2}\sum_i e_i\dot{\boldsymbol{X}}\wedge\left\{\dot{\boldsymbol{r}}_i\cdot\boldsymbol{E}(\boldsymbol{R}_i,t)\boldsymbol{r}_i+\dot{\boldsymbol{X}}\cdot\boldsymbol{r}_i\,\boldsymbol{E}(\boldsymbol{R}_i,t)+\frac{\bar{\boldsymbol{s}}}{m}\wedge\boldsymbol{E}(\boldsymbol{R}_i,t)\right\}$$
$$+c^{-2}\frac{\mathrm{d}}{\mathrm{d}t}\{\sum_i e_i\boldsymbol{r}_i\cdot\dot{\boldsymbol{X}}\boldsymbol{r}_i\wedge\boldsymbol{E}(\boldsymbol{R}_i,t)\}.\tag{A135}$$

If the external fields are slowly varying we may perform a multipole expansion of the right-hand side and limit ourselves to dipole terms. Expressing the results (with the help of (A100)) in terms of the dipole moments (A115), and retaining only terms up to $c^{-1}$ (considering magnetic dipole moments as being of order $c^0$) we obtain the *semi-relativistic* inner angular momentum equation:

$$\frac{\mathrm{d}}{\mathrm{d}t}\Big(\sum_i m_i \boldsymbol{r}_i' \wedge \dot{\boldsymbol{r}}_i'\Big) = \boldsymbol{\mu} \wedge (\boldsymbol{E} + c^{-1}\boldsymbol{v} \wedge \boldsymbol{B}) + \boldsymbol{\nu} \wedge (\boldsymbol{B} - c^{-1}\boldsymbol{v} \wedge \boldsymbol{E}). \qquad \text{(A136)}$$

Here we have written $\boldsymbol{v}$ for the velocity $\dot{\boldsymbol{X}}$ of the composite particle. This equation (which may be found also from the relativistic theory namely from (136)) shows a symmetry between electric and magnetic phenomena, which was absent in the non-relativistic equation (I.79).

# PROBLEMS

**1.** It is possible to prove directly the covariance of the dipole contributions (48) and (49) to the polarization tensor (although the manifestly covariant derivation of these formulae guarantees their covariance already). To that end consider first the Lorentz transformation from the momentary atomic rest frame $(ct^{(0)}, \boldsymbol{R}^{(0)})$ to the observer's frame $(ct, \boldsymbol{R})$ which transforms the coordinates $\left(ct^{(0)}, \boldsymbol{R}_k^{(0)}(t^{(0)})\right)$ of a point of the world line of atom $k$ into $\left(ct_k, \boldsymbol{R}_k(t_k)\right)$, where in general $t_k$ differs from $t$. (The transformation velocity is $c\boldsymbol{\beta}_k(t_k) \equiv \mathrm{d}\boldsymbol{R}_k(t_k)/\mathrm{d}t_k$.) Show from the transformation formulae that

$$c(t_k - t) = \gamma_k^2(t)\boldsymbol{\beta}_k(t)\cdot\{\boldsymbol{R}_k(t) - \boldsymbol{R}\} + \ldots, \tag{P1}$$

where a Taylor expansion has been performed and where the dots stand for terms of higher order in $\boldsymbol{R}_k(t) - \boldsymbol{R}$. Prove then, again using the Lorentz transformation formulae, that

$$\boldsymbol{R}_k^{(0)}(t^{(0)}) - \boldsymbol{R}^{(0)} = \boldsymbol{\Omega}_k^{-1}(t)\cdot\{\boldsymbol{R}_k(t) - \boldsymbol{R}\} + \ldots, \tag{P2}$$

where $\boldsymbol{\Omega}^{-1}$ has been given in (A3). Show from the latter formula that

$$\delta\{\boldsymbol{R}_k^{(0)}(t^{(0)}) - \boldsymbol{R}^{(0)}\} = \gamma_k^{-1}(t)\delta\{\boldsymbol{R}_k(t) - \boldsymbol{R}\}, \tag{P3}$$

where (A16) has been used.

Show with (P1), (P3) and the transformation property (A5) of an antisymmetric tensor that one obtains the formulae (48) and (49) starting from the formulae

$$\begin{aligned} \boldsymbol{p}_k^{(0)}(\boldsymbol{R}^{(0)}, t^{(0)}) &= \boldsymbol{\mu}_k^{(1)}(t^{(0)})\delta\{\boldsymbol{R}_k^{(0)}(t^{(0)}) - \boldsymbol{R}^{(0)}\}, \\ \boldsymbol{m}_k^{(0)}(\boldsymbol{R}^{(0)}, t^{(0)}) &= \boldsymbol{\nu}_k^{(1)}(t^{(0)})\delta\{\boldsymbol{R}_k^{(0)}(t^{(0)}) - \boldsymbol{R}^{(0)}\}, \end{aligned} \tag{P4}$$

which are the atomic rest frame cases of (48) and (49).

**2.** Prove from (69) and (70) that, for sufficiently small forces, the world line determined by the centre of energy construction of section 3*b* is time-like. To that end consider the world line points $(R^0, \boldsymbol{X})$ and $(R^0 + \delta R^0, \boldsymbol{X} + \delta\boldsymbol{X})$ in the proper frame of $p^\alpha$ corresponding to the first point. The second point is a centre of energy in a plane surface $\Sigma'$ with normal parallel to $p^\alpha + \delta p^\alpha$. The proper frame of $p^\alpha + \delta p^\alpha$ is connected to the proper frame of $p^\alpha$ by relations like (71). The point $(R^0 + \delta R^0, \boldsymbol{X} + \delta\boldsymbol{X})$ has in the new frame the coordinates

$(\hat{R}^{0\prime}, \boldsymbol{X}'+\delta\boldsymbol{X}')$. Prove that in the present case one gets

$$\hat{R}^0 = R^0+\delta R^0+\boldsymbol{\varepsilon}\cdot(\boldsymbol{X}-\hat{\boldsymbol{R}}'),$$
$$\hat{\boldsymbol{R}} = \hat{\boldsymbol{R}}'-\boldsymbol{\varepsilon}R^0$$

in the same notation as in (76–77). Show then with the help of the equation of motion (64) that one has the relations

$$\int [\{\boldsymbol{\varepsilon}\cdot(\boldsymbol{X}-\boldsymbol{R})+\delta R^0\}\boldsymbol{f}(R^0, \boldsymbol{R})+\boldsymbol{\varepsilon}t^{00}(R^0, \boldsymbol{R})]\mathrm{d}\boldsymbol{R} = 0,$$

$$\int [\delta\boldsymbol{X}t^{00}(R^0, \boldsymbol{R})+(\boldsymbol{X}-\boldsymbol{R})\{\boldsymbol{\varepsilon}\cdot(\boldsymbol{X}-\boldsymbol{R})+\delta R^0\}f^0(R^0, \boldsymbol{R})]\mathrm{d}\boldsymbol{R}-c\boldsymbol{\varepsilon}\wedge\boldsymbol{s} = 0,$$

where (68) has been used. From the first of these relations it follows that $\boldsymbol{\varepsilon}$ vanishes if no forces are present, so that for small forces one has

$$\boldsymbol{\varepsilon} = -\delta R^0 \frac{\int \boldsymbol{f}\mathrm{d}\boldsymbol{R}}{\int t^{00}\mathrm{d}\boldsymbol{R}}$$

up to terms linear in the forces. Prove by substitution of this expression into the second relation that the infinitesimal change $\delta\boldsymbol{X}$ of the centre of energy and the change of time $c^{-1}\delta R^0$ are related as

$$\frac{\mathrm{d}\boldsymbol{X}}{\mathrm{d}R^0} = \frac{c^{-2}}{m}\int (\boldsymbol{R}-\boldsymbol{X})f^0\mathrm{d}\boldsymbol{R}-\frac{c^{-3}}{m^2}\left(\int \boldsymbol{f}\mathrm{d}\boldsymbol{R}\right)\wedge\boldsymbol{s},$$

where we introduced $m = c^{-2}\int t^{00}\mathrm{d}\boldsymbol{R}$ as an abbreviation (v. (120) and (123)). This formula shows that $\mathrm{d}\boldsymbol{X}/\mathrm{d}R^0$ tends to zero if the forces tend to zero, so that the world line is time-like if the forces are sufficiently small.

Discuss the connexion between the last formula and (124) (together with (120) and (123)).

**3.** Consider two plane surfaces $\Sigma(s)$ and $\Sigma(s+\mathrm{d}s)$, one through the position $X^\alpha(s)$ (in the same notation as used in § 3*c*) and with normal $n^\alpha(s)$, and the other through $X^\alpha(s+\mathrm{d}s)$ and with normal $n^\alpha(s+\mathrm{d}s)$. Their equations are therefore $n^\alpha(s)\{R_\alpha-X_\alpha(s)\} = 0$ and $n^\alpha(s+\mathrm{d}s)\{R_\alpha-X_\alpha(s+\mathrm{d}s)\} = 0$. Prove now that the volume $\mathrm{d}^4V$ of a parallelepiped with basis $\mathrm{d}^3\Sigma$ in $\Sigma(s)$ at the position $R^\alpha$, with edges parallel to $n^\alpha(s)$ and top in $\Sigma(s+\mathrm{d}s)$ is

$$\mathrm{d}^4V = -n_\alpha u^\alpha\left\{1-\frac{\dot{n}_\beta(R-X)^\beta}{n_\gamma u^\gamma}\right\}\mathrm{d}^3\Sigma\,\mathrm{d}s,$$

where $n^\alpha$, $X^\alpha$ and $u^\alpha \equiv \mathrm{d}X^\alpha/\mathrm{d}s$ all depend on $s$.

By choosing for $n^\alpha$ the vector $p^\alpha/\sqrt{-p^2}$ one recovers formula (94) with (95). (If one chooses for $n^\alpha$ the vector $c^{-1}u^\alpha$ one finds back the expression given in the footnote of chapter III after formula (129).)

**4.** Consider, as in section 2*a*, a composite particle consisting of point particles with charges $e_i$ (we omit the index $k$, which labelled the composite particle). Prove, by making a multipole expansion around a central world line $X^\alpha(s)$, the identity

$$\sum_i e_i u_i^\alpha f(R_i) = eu^\alpha f(X) + \sum_{n=1}^{\infty} (\mu^{\alpha_1\ldots\alpha_n}u^\alpha - \mu^{\alpha_1\ldots\alpha_{n-1}\alpha}u^{\alpha_n} + cv^{\alpha_1\ldots\alpha_n\alpha})\partial_{\alpha_1\ldots\alpha_n} f(X) + \sum_{n=0}^{\infty} \frac{\mathrm{d}}{\mathrm{d}s}\{\mu^{\alpha_1\ldots\alpha_n\alpha}\partial_{\alpha_1\ldots\alpha_n} f(X)\},$$

where $f$ is an arbitrary function and $\mu^{\alpha_1\ldots\alpha_n}$ and $v^{\alpha_1\ldots\alpha_n}$ are the multipole moments (9); $s$ is the proper time along the central world line $X^\alpha$ and $u^\alpha = \mathrm{d}X^\alpha/\mathrm{d}s$.

Apply this formula with $f(R_i) = \delta^{(4)}(R_i - R)$ and find then (7) with (5) and (10). Show furthermore that (15) and (16) are the multipole expansions of the integrand of (III.59).

Finally, by application to (106) with $f(R_i) = F^{\alpha\beta}(R_i)$ one may obtain (168) by making use of the homogeneous field equations (13).

**5.** Prove that, under the same circumstances as in the preceding problem, one has for an arbitrary function $f(R_i)$ the multipole expansion

$$\sum_i e_i r_i^\alpha u_i^\beta f(R_i) = \sum_{n=1}^{\infty} n\mu^{\alpha_1\ldots\alpha_{n-1}\alpha}\partial_{\alpha_1\ldots\alpha_{n-1}} u^\beta f(X) + \sum_{n=2}^{\infty} (n-1)\frac{\mathrm{d}}{\mathrm{d}s}(\mu^{\alpha_1\ldots\alpha_{n-2}\alpha\beta})\partial_{\alpha_1\ldots\alpha_{n-2}} f(X) + \sum_{n=1}^{\infty}\{-(n-1)v^{\alpha_1\ldots\alpha_{n-2}\alpha\beta\alpha_{n-1}} + v^{\alpha_1\ldots\alpha_{n-1}\alpha\beta}\}\partial_{\alpha_1\ldots\alpha_{n-1}} f(X).$$

Apply this formula to (107) with $f(R_i) = F^{\alpha\beta}(R_i)$ and find (173).

**6.** Prove from the definition (120) of $m$ by evaluating it in the rest frame of $p^\alpha$ (65) with (62) that one has up to order $c^{-2}$:

$$m = \sum_i m_i(1+\tfrac{1}{2}c^{-2}\dot{\boldsymbol{r}}_i^2) + c^{-2}\sum_{i,j(i\neq j)} \frac{e_i e_j}{8\pi|\boldsymbol{r}_i - \boldsymbol{r}_j|}\,.$$

Hint: The fields occurring in $t^{\alpha\beta}$ (62) follow from (III.72).

**7.** In the rest frame of $p^\alpha$ the definition (66) of the centre of energy reads $\boldsymbol{X} = \int \boldsymbol{R} t^{00} \mathrm{d}\boldsymbol{R} / \int t^{00} \mathrm{d}\boldsymbol{R}$. Prove that, up to order $c^{-2}$, this expression reduces to

$$\boldsymbol{X} = \frac{\sum_i m_i(1+\frac{1}{2}c^{-2}\dot{\boldsymbol{R}}_i^2)\boldsymbol{R}_i + c^{-2}\sum_{i,j(i\neq j)}(e_i e_j/8\pi|\boldsymbol{R}_i-\boldsymbol{R}_j|)\boldsymbol{R}_i}{\sum_i m_i(1+\frac{1}{2}c^{-2}\dot{\boldsymbol{R}}_i^2) + c^{-2}\sum_{i,j(i\neq j)}(e_i e_j/8\pi|\boldsymbol{R}_i-\boldsymbol{R}_j|)}.$$

Hint: the denominator follows along the same lines as in the preceding problem, and so does the material part of the numerator. For the field part of the numerator one finds first $\frac{1}{2}c^{-2}\int\sum_{i,j(i\neq j)}\boldsymbol{e}_i\cdot\boldsymbol{e}_j\boldsymbol{R}\mathrm{d}\boldsymbol{R}$ with $\boldsymbol{e}_i$ and $\boldsymbol{e}_j$ the Coulomb fields, due to particle $i$ and $j$. With the help of a partial integration, the application of Gauss's theorem and the integration prescription of section 3*b* (according to which one has to integrate over a spherical volume of unboundedly increasing dimension) one obtains the last term of the numerator.

**8.** Show from the definition (68) with (62) that up to order $c^{-2}$ the space-space components of the inner angular momentum tensor $s^{\alpha\beta}$ are given by

$$s^{kl} = \Bigg[\sum_i m_i \boldsymbol{r}_i' \wedge \dot{\boldsymbol{r}}_i'(1+\tfrac{1}{2}c^{-2}\dot{\boldsymbol{r}}_i'^2) - \tfrac{1}{2}c^{-2}\{(\sum_i m_i \boldsymbol{r}_i'\wedge\dot{\boldsymbol{r}}_i')\wedge\dot{\boldsymbol{X}}\}\wedge\dot{\boldsymbol{X}}$$
$$+c^{-2}\sum_{i,j(i\neq j)}\frac{e_i e_j}{8\pi|\boldsymbol{r}_i'-\boldsymbol{r}_j'|}\left\{\boldsymbol{r}_i'\wedge\dot{\boldsymbol{r}}_j' - \frac{(\boldsymbol{r}_i'-\boldsymbol{r}_j')\cdot\dot{\boldsymbol{r}}_j'(\boldsymbol{r}_i'\wedge\boldsymbol{r}_j')}{|\boldsymbol{r}_i'-\boldsymbol{r}_j'|^2}\right\}\Bigg]^m$$

(with $k, l, m = 1, 2, 3$ cycl.) in the same notation as in appendix V. The material part of this expression follows directly from the material part of $t^{\alpha\beta}$ (62), if one employs the defining relations(A100) and (A122) for $\boldsymbol{r}_i'$ and $\dot{\boldsymbol{r}}_i'$, and the transformation formulae (A5) for the antisymmetric tensor $s^{\alpha\beta}$. To find the field part, one should use the expressions (III.72) for the fields up to order $c^{-2}$ or rather the expressions (III.70) with (III.83) for the fields in terms of Coulomb gauge potentials. Furthermore one should employ here the integration prescription, as explained in § 3*b* (i.e. integrating over a sphere of increasing radius around the origin).

**9.** Prove from (A112) and the first line of (A106) with (A108) that one may write the sum of the total Lorentz forces on a set of particles $i$ with mass $m_i$, charge $e_i$, position $\boldsymbol{R}_i$, velocity $\dot{\boldsymbol{R}}_i$, acceleration $\ddot{\boldsymbol{R}}_i$ as

$$\sum_i e_i\{\boldsymbol{e}_t(\boldsymbol{R}_i, t) + c^{-1}\dot{\boldsymbol{R}}_i\wedge\boldsymbol{b}_t(\boldsymbol{R}_i, t)\}$$
$$= -c^{-2}\frac{\mathrm{d}^2}{\mathrm{d}t^2}\left(\sum_{i,j(i\neq j)}\frac{e_i e_j \boldsymbol{R}_i}{8\pi|\boldsymbol{R}_i-\boldsymbol{R}_j|}\right) - c^{-2}\frac{\mathrm{d}}{\mathrm{d}t}(\sum_i m_i\dot{\boldsymbol{R}}_i\cdot\ddot{\boldsymbol{R}}_i\boldsymbol{R}_i)$$
$$+\sum_i e_i\{\boldsymbol{E}(\boldsymbol{R}_i, t) + c^{-1}\dot{\boldsymbol{R}}_i\wedge\boldsymbol{B}(\boldsymbol{R}_i, t)\} + c^{-2}\frac{\mathrm{d}}{\mathrm{d}t}\{\sum_i e_i\dot{\boldsymbol{R}}_i\cdot\boldsymbol{E}(\boldsymbol{R}_i, t)\boldsymbol{R}_i\}. \qquad \text{(P5)}$$

Prove by insertion of this expression into (A87) (with a summation over $i$) that one obtains:

$$\frac{\mathrm{d}^2}{\mathrm{d}t^2}\left\{\sum_i \left(m_i+\tfrac{1}{2}c^{-2}m_i\dot{\boldsymbol{R}}_i^2+c^{-2}\sum_{j(\neq i)}\frac{e_i e_j}{8\pi|\boldsymbol{R}_i-\boldsymbol{R}_j|}\right)\boldsymbol{R}_i\right\}$$

$$=\sum_i e_i\{\boldsymbol{E}(\boldsymbol{R}_i,t)+c^{-1}\dot{\boldsymbol{R}}_i\wedge\boldsymbol{B}(\boldsymbol{R}_i,t)\}+c^{-2}\frac{\mathrm{d}}{\mathrm{d}t}\{\sum_i e_i\dot{\boldsymbol{R}}_i\cdot\boldsymbol{E}(\boldsymbol{R}_i,t)\boldsymbol{R}_i\}. \quad \text{(P6)}$$

Check that one recovers the equation of motion (A113) by employing (A91), (A101) and (A106).

*Remark.* At the right-hand side of (P6) two terms with the external fields $\boldsymbol{E}$ and $\boldsymbol{B}$ appear: the ordinary Lorentz force and an extra term. The latter has a form analogous to the term

$$c^{-2}\frac{\mathrm{d}}{\mathrm{d}t}\{\sum_i e_i\dot{\boldsymbol{r}}_i\cdot\boldsymbol{E}(\boldsymbol{R}_i,t)\boldsymbol{r}_i\},$$

which appears in the semi-relativistic equation of motion (A113). This term led, via the multipole expansion, to the term

$$-c^{-1}\frac{\mathrm{d}}{\mathrm{d}t}(\boldsymbol{v}\wedge\boldsymbol{E})$$

of the dipole equation of motion (A117), i.e. to the 'magnetodynamic effect' associated with the total magnetic moment. Some discussions (e.g. H. A. Haus and P. Penfield Jr., Physica **42**(1969)447) about the magnetodynamic effect limit themselves to the derivation of (part of) the terms given at the right-hand side of (P5), namely the terms with $\boldsymbol{E}$ and $\boldsymbol{B}$. The problem of deriving an equation of motion which involves the definition of a proper centre of energy is not solved then. The conclusion that also inner angular momentum terms appear in the equation, so that effectively only the anomalous magnetic moment contributes, is then missed.

**10.** Prove that the time-component of the equation (160) reads in three-dimensional notation:

$$m^*\boldsymbol{v}\cdot\frac{\mathrm{d}\boldsymbol{v}}{\mathrm{d}t}=\gamma^{-4}\left[e\boldsymbol{E}\cdot\gamma\boldsymbol{v}-\mathfrak{m}\cdot\left(\frac{\partial\boldsymbol{B}}{\partial t}-\boldsymbol{\beta}\wedge\frac{\partial\boldsymbol{E}}{\partial t}\right)\right.$$
$$\left.+\gamma^2\mathfrak{m}\cdot\left(\frac{\mathrm{d}\boldsymbol{B}}{\mathrm{d}t}-\boldsymbol{\beta}\wedge\frac{\mathrm{d}\boldsymbol{E}}{\mathrm{d}t}\right)-\gamma^2(\boldsymbol{\beta}\wedge\mathfrak{m}_{(\mathrm{a})})\cdot\left(\frac{\mathrm{d}\boldsymbol{E}}{\mathrm{d}t}+\boldsymbol{\beta}\wedge\frac{\mathrm{d}\boldsymbol{B}}{\mathrm{d}t}\right)\right].$$

Compare this result with that of problem 6 of chapter VIII.

CHAPTER V

# Covariant statistics: the laws for material media

## 1 *Introduction*

In order to find the covariant macroscopic laws from the corresponding microscopic equations one has to introduce a covariant averaging procedure. To that end covariant distribution functions will be employed that describe the statistical properties of collections of world lines in Minkowski space. In connexion with this, several covariant distribution functions of a particular type will ensue: 'synchronous', 'retarded' and 'advanced' distribution functions, which are useful to describe averages of certain microscopic quantities. The first of these is a direct generalization of the non-relativistic distribution function, while the latter have no non-relativistic counterparts.

With the help of the covariant averaging procedure we then derive the Maxwell equations, the energy–momentum and angular momentum balances and the thermodynamical laws. All macroscopic quantities occurring in these laws will be found as statistical expressions in terms of microscopic quantities. In particular we shall obtain in this way an expression for the macroscopic energy–momentum tensor of a polarized medium in the presence of electromagnetic fields. It will be shown that two ways of splitting the tensor in so-called field and material parts present themselves in a natural way. This result throws light on the much discussed controversy on the 'correct form' of the field part of the energy–momentum tensor. In fact this problem is not well posed if one does not bring into the discussion also the expression for the corresponding material energy–momentum tensor. Only the sum of the two parts of the tensor is physically significant. Nevertheless it is sometimes convenient to introduce a definite splitting to discuss certain physical properties of the system. The situation is analogous to the one encountered in the non-relativistic theory, where we found various expressions for the ponderomotive force density in a polarized medium, each with its corresponding material pressure tensor.

## 2 Covariant statistical mechanics

### a. *Covariant distribution functions*

A system of $N$ point particles $i = 1, 2, \ldots$ is completely specified by giving their world lines $R_i^\alpha(s_i)$ in Minkowski space. (The world lines are parametrized by means of their proper times $s_i$.) The number of world lines that intersect a three-surface element $\mathrm{d}^3\Sigma$ (with time-like normal $n^\alpha$, $n^0 > 0$) at the position $R_1^\alpha$, is given by

$$\sum_i \delta^{(3)}\{n; R_1 - R_i(s_i)\}|_{n\cdot\{R_1 - R_i(s_i)\} = 0}\, \mathrm{d}^3\Sigma. \tag{1}$$

The bar, with the equation $n\cdot\{R_1 - R_i(s_i)\} = 0$, indicates that one has to take in the delta function the proper times $s_{i0}$ which are the solutions of this equation. The delta function employed here is the generalization of the ordinary delta function $\delta\{\boldsymbol{R}_1 - \boldsymbol{R}_i(s_i)\}$; in fact it is equal to it if $n^\alpha$ is purely time-like: (1, 0, 0, 0). In the general case it is defined by writing

$$\delta^{(4)}(x) = \delta^{(3)}(n; x)\delta(n\cdot x), \tag{2}$$

where $x^\alpha$ is an arbitrary four-vector. If one wants to specify also the four-velocity and higher proper time derivatives as lying in the intervals ($R_1^{[1]\alpha}$, $R_1^{[1]\alpha} + \mathrm{d}^4 R_1^{[1]\alpha}$) etc.[1], one gets for the number of world lines crossing $\mathrm{d}^3\Sigma$:

$$\mathrm{d}N = \sum_i \delta^{(3)}\{n; R_1 - R_i(s_{i0})\}\delta^{(4)}\{R_1^{[1]} - R_i^{[1]}(s_{i0})\} \ldots \mathrm{d}^3\Sigma\, \mathrm{d}^4 R_1^{[1]} \ldots, \tag{3}$$

where $R_i^{[n]\alpha}(s_i) \equiv \mathrm{d}^n R_i^\alpha(s_i)/\mathrm{d}s_i^n$. By adding a factor $\delta[n\cdot\{R_1 - R_i(s_i)\}]$ and an integration over $n\cdot R_i(s_i)$ one obtains after introduction of the new integration variable $s_i$:

$$\mathrm{d}N = -\sum_i \int \delta^{(4)}\{R_1 - R_i(s_i)\} n\cdot R_i^{[1]}(s_i) \delta^{(4)}\{R_1^{[1]} - R_i^{[1]}(s_i)\} \ldots \\ \mathrm{d}s_i\, \mathrm{d}^3\Sigma\, \mathrm{d}^4 R_1^{[1]} \ldots. \tag{4}$$

We may write this as

$$\mathrm{d}N = -c^{-1} n\cdot R_1^{[1]} f_1(1) \mathrm{d}^3\Sigma\, \mathrm{d}^4 R_1^{[1]} \ldots \tag{5}$$

with the abbreviation

$$f_1(1) \equiv c \sum_i \int \delta^{(4)}\{R_1 - R_i(s_i)\}\delta^{(4)}\{R_1^{[1]} - R_i^{[1]}(s_i)\} \ldots \mathrm{d}s_i, \tag{6}$$

[1] The upper indices between square brackets indicate the number of differentiations with respect to proper time. The four-velocity $\mathrm{d}R_i^\alpha/\mathrm{d}s_i$ for instance is indicated as $R_i^{[1]\alpha}(s_i)$.

where the argument 1 at the left-hand side stands for the set of variables $R_1^\alpha$, $R_1^{[1]\alpha}$, .... This distribution function, which is a measure for the density of world lines, is an invariant with respect to Lorentz transformations and moreover independent of the normal vector $n^\alpha$. If the system contains many particles one may replace the discontinuous function (6) by a distribution function which is a smooth function of $R_1^\alpha$ and the independent components of $R_1^{[1]\alpha}$, etc. Formally such a coarse graining is achieved by employing for the one-point distribution function a weighted average of the right-hand side of (6)

$$f_1(1) = c\sum_\gamma w_\gamma \sum_i \int \delta^{(4)}\{R_1 - R_{i\gamma}(s_i)\}\delta^{(4)}\{R_1^{[1]} - R_{i\gamma}^{[1]}(s_i)\} \dots \mathrm{d}s_i, \qquad (7)$$

with $\sum_\gamma w_\gamma = 1$.

From (6) and (7) it is apparent that the distribution function vanishes if the $R_1^{[1]\alpha}$, $R_1^{[2]\alpha}$, ... do not satisfy simultaneously the set of relations

$$\begin{aligned} &R_1^{[1]}\cdot R_1^{[1]} + c^2 = 0, \\ &\sum_{i=0}^{j} \binom{j}{i} R_1^{[i+1]}\cdot R_1^{[j-i+1]} = 0, \qquad (j = 1, 2, \dots). \end{aligned} \qquad (8)$$

The components of $R_1^{[1]\alpha}$, $R_1^{[2]\alpha}$, ... are thus not all independent. In other words the distribution function is the product of a number of delta functions with as arguments the left-hand sides of the above relations and a function which is smooth after the coarse graining has been performed.

The distribution function (6) fulfils a continuity equation, which may be derived by writing the identity

$$c\sum_i \int \frac{\mathrm{d}}{\mathrm{d}s_i} [\delta^{(4)}\{R_1 - R_i(s_i)\}\delta^{(4)}\{R_1^{[1]} - R_i^{[1]}(s_i)\} \dots]\mathrm{d}s_i = 0. \qquad (9)$$

The differentiations with respect to $s_i$ follow by applying the chain rule. Then one gets for (9)

$$\left(R_1^{[1]}\cdot \frac{\partial}{\partial R_1} + R_1^{[2]}\cdot \frac{\partial}{\partial R_1^{[1]}} + \dots\right) f_1(1) = 0, \qquad (10)$$

which is the continuity equation. If the distribution function depends on a finite number of variables, i.e. $f_1(1) = f_1(R_1, \dots, R_1^{[n]})$, the continuity equation gets the form

$$\left(R_1^{[1]}\cdot \frac{\partial}{\partial R_1} + \dots + R_1^{[n]}\cdot \frac{\partial}{\partial R_1^{[n-1]}}\right) f_1(R_1, \dots, R_1^{[n]}) + \int \frac{\partial}{\partial R_1^{[n]}} \cdot R_1^{[n+1]} f_1(R_1, \dots, R_1^{[n+1]})\mathrm{d}^4 R_1^{[n+1]} = 0, \qquad (11)$$

where we supposed that $R_1^{[n+1]\alpha}$ was independent of the other variables[1].

The continuity equation was derived here starting from the representation (6) of the distribution function. The smoothed distribution function (7) will satisfy the same conservation law, since coarse graining, i.e. adding a summation with weights $w_\gamma$, will not change the proof.

In a similar way one derives that the joint probability (normalized to $N(N-1)$) to find a world line crossing a surface element $d^3\Sigma_1$ with normal $n_1^\alpha$ at the position $R_1^\alpha$ and four-velocity $R_1^{[1]\alpha}$ etc. and a different world line crossing $d^3\Sigma_2$ with normal $n_2^\alpha$ at $R_2^\alpha$ and $R_2^{[1]\alpha}$ etc. is:

$$c^{-2} n_1 \cdot R_1^{[1]} n_2 \cdot R_2^{[1]} f_2(1, 2) d^3\Sigma_1\, d^3\Sigma_2\, d^4R_1^{[1]} d^4R_2^{[1]} \ldots, \tag{12}$$

where $f_2(1, 2)$ is the two-point distribution function. Here $f_2(1, 2)$ is given by

$$f_2(1, 2) = c^2 \sum_\gamma w_\gamma \sum_{i,j(i \neq j)} \int \delta^{(4)}\{R_1 - R_{i\gamma}(s_i)\} \delta^{(4)}\{R_1^{[1]} - R_{i\gamma}^{[1]}(s_i)\} \ldots$$
$$\delta^{(4)}\{R_2 - R_{j\gamma}(s_j)\} \delta^{(4)}\{R_2^{[1]} - R_{j\gamma}^{[1]}(s_j)\} \ldots ds_i\, ds_j. \tag{13}$$

In the following we shall frequently use the two-point correlation function $c_2(1, 2)$, which is defined as

$$c_2(1, 2) = f_2(1, 2) - f_1(1) f_1(2). \tag{14}$$

The generalization to particles with structure and to mixtures of different particles is trivial. The distribution function will then depend also on the internal variables and will be labelled by an index numbering the species.

### b. *Definition of macroscopic quantities*

The microscopic quantities for which we want to define average values with the help of the covariant distribution function of the preceding subsection are sums of one-particle or two-particle quantities. The one-particle quantities have the form of integrals along world lines, so that their sums read

$$a(R) = \sum_i \int \alpha\{R_i(s_i), R_i^{[1]}(s_i), R_i^{[2]}(s_i), \ldots; R\} ds_i, \tag{15}$$

[1] If however the $R_1^{[n+1]\alpha}$ is dependent on $R_1, \ldots, R_1^{[n]\alpha}$, the integration in (11) may be performed with as a result the continuity equation

$$\left\{ R_1^{[1]} \cdot \frac{\partial}{\partial R_1} + \ldots + R_1^{[n]} \cdot \frac{\partial}{\partial R_1^{[n-1]}} + \frac{\partial}{\partial R_1^{[n]}} \cdot R_1^{[n+1]}(R_1, \ldots, R_1^{[n]}) \right\} f_1(R_1, \ldots, R_1^{[n]}) = 0.$$

This is the case which occurs in the kinetic theory of gases, where $n = 1$.

where $R_i^\mu(s_i)$ is the position four-vector of point particle $i$ with proper time $s_i$, $R_i^{[1]\mu}(s_i) \equiv dR_i^\mu/ds_i$ its four-velocity, $R_i^{[2]\mu}(s_i) \equiv d^2R_i^\mu/ds_i^2$ its four-acceleration, etc. The average value of such a quantity follows by taking the weighted average

$$A(R) \equiv \langle a(R) \rangle = \sum_\gamma w_\gamma \sum_i \int \alpha\{R_{i\gamma}(s_i), R_{i\gamma}^{[1]}(s_i), \ldots; R\} ds_i \tag{16}$$

with the weights $w_\gamma$ which have been introduced in the preceding subsection. This average may be written in terms of the distribution function (7):

$$A(R) = c^{-1} \int \alpha(1; R) f_1(1) d1, \tag{17}$$

where the argument 1 stands for the set of variables $R_1^\mu\ R_1^{[1]\mu}, \ldots$ and where d1 stands for $d^4R_1\, d^4R_1^{[1]} \ldots$.

Likewise for a sum of two-particle quantities

$$a(R) = \sum_{i,j(i\neq j)} \int \alpha\{R_i(s_i), R_i^{[1]}(s_i), \ldots, R_j(s_j), R_j^{[1]}(s_j), \ldots; R\} ds_i\, ds_j \tag{18}$$

one obtains in the same fashion the average

$$A(R) \equiv \langle a(R) \rangle = c^{-2} \int \alpha(1, 2; R) f_2(1, 2) d1\, d2, \tag{19}$$

where the two-point distribution function $f_2(1, 2)$ has been defined in (13).

Since the distribution functions $f_1(1)$ and $f_2(1, 2)$ are Lorentz invariant, as was shown in subsection *a*, the averages $A(R)$ (17) and (19) have the same tensorial character as the microscopic quantities $a(R)$ (15) and (18).

From the expressions (17) and (19) it is obvious that the average of a derivative of a quantity $a$ with respect to $R^\mu$ is equal to the derivative of the average quantity $A = \langle a \rangle$:

$$\langle \partial_\mu a \rangle = \partial_\mu \langle a \rangle \equiv \partial_\mu A. \tag{20}$$

This commutation property will be used frequently in the derivation of the macroscopic laws.

The covariant averages (17) and (19) may be cast in a particular form if the microscopic quantities have special properties, as will be shown in the following subsection.

### c. *Synchronous, retarded and advanced distribution functions*

Let us consider first a physical quantity of the form

$$a(n, \tau) = \sum_i \int \alpha\{R_i(s_i), \ldots\} \delta\{n \cdot R_i(s_i) + c\tau\} ds_i, \tag{21}$$

where $n^\mu$ is a time-like unit vector ($n^0 > 0$) and where $\tau$ is an arbitrary real number. (The charge–current density is an example of such a quantity.) In such a quantity only those points of the world lines of the particles $i$ contribute that lie in a plane three-surface $n\cdot R + c\tau = 0$. In particular if the normal $n^\mu$ has the form $\mathring{n}^\mu \equiv (1, 0, 0, 0)$ this equation for the plane reduces to $t = \tau$, i.e. only 'synchronous' points of the world lines contribute.

The integral in (21) may be performed by introducing the integration variable $n\cdot R_i(s_i)$ instead of $s_i$:

$$a(n, \tau) = -\sum_i \frac{\alpha\{R_i(s_i), \ldots\}}{n\cdot R_i^{[1]}(s_i)}\bigg|_{n\cdot R_i(s_i)+c\tau=0}, \tag{22}$$

where the suffix means that $s_i$ is the solution of the equation in question.

One often encounters physical quantities of the form (21) or (22) with $\alpha$ depending also explicitly on $R$ and the parameter $\tau$ equal to $-c^{-1}n\cdot R$. Moreover space-time derivatives of quantities of this type occur, for instance

$$\partial_\mu \sum_i \int \alpha\{R_i(s_i), \ldots; R\}\delta[n\cdot\{R_i(s_i)-R\}]\mathrm{d}s_i\,. \tag{23}$$

(The polarization tensor is a quantity of this type.) Such a quantity may be written in a form that shows that it is of the same 'synchronous' type as (21). To that purpose we use the identity for an arbitrary function $f(x)$ and an arbitrary four-vector $v^\mu$

$$\partial_\mu f[n\cdot\{R_i(s_i)-R\}] = -\frac{n_\mu}{n\cdot v}\, v\cdot\frac{\partial}{\partial R_i} f[n\cdot\{R_i(s_i)-R\}] \tag{24}$$

for the special choice $f(x) = \delta(x)$ and $v^\mu = R_i^{[1]\mu}(s_i)$. Then (23) becomes

$$\sum_i \int [\partial_\mu \alpha\{R_i(s_i), \ldots; R\}]\delta[n\cdot\{R_i(s_i)-R\}]\mathrm{d}s_i$$
$$-\sum_i \int \alpha\{R_i(s_i), \ldots; R\}\frac{n_\mu}{n\cdot R_i^{[1]}(s_i)}\frac{\mathrm{d}}{\mathrm{d}s_i}\delta[n\cdot\{R_i(s_i)-R\}]\mathrm{d}s_i\,. \tag{25}$$

After a partial integration one obtains

$$\sum_i \int \left\{\partial_\mu \alpha + n_\mu \frac{\mathrm{d}}{\mathrm{d}s_i}\left(\frac{\alpha}{n\cdot R_i^{[1]}}\right)\right\}\delta\{n\cdot(R_i - R)\}\mathrm{d}s_i\,, \tag{26}$$

so that it becomes apparent that the space-time derivatives (23) of 'synchronous quantities' are themselves synchronous quantities of the type (21). The latter fact could also have been seen from (22) with $\tau = -c^{-1}n\cdot R$ since differentiation of that function with implicit dependence yields

$$-\partial_\mu \left[\sum_i \frac{\alpha\{R_i(s_i), \ldots; R\}}{n\cdot R_i^{[1]}(s_i)}\bigg|_{n\cdot\{R_i(s_i)-R\}=0}\right]$$
$$= -\left[\sum_i \mathrm{d}_{i\mu}^{\mathrm{syn}} \frac{\alpha\{R_i(s_i), \ldots; R\}}{n\cdot R_i^{[1]}(s_i)}\right]\bigg|_{n\cdot\{R_i(s_i)-R\}=0}, \tag{27}$$

where the differentiation $\mathrm{d}_{i\mu}^{\mathrm{syn}}$ stands for

$$\mathrm{d}_{i\mu}^{\mathrm{syn}} \equiv \partial_\mu + \frac{n_\mu}{n\cdot R_i^{[1]}(s_i)} \frac{\mathrm{d}}{\mathrm{d}s_i}. \tag{28}$$

The right-hand side of (27) shows again that the quantity under consideration is of the synchronous type. (As it should be, the right-hand side of (27) may be shown to be equal to (26).)

The average of a quantity of the type (21), which according to (17) is

$$A(n, \tau) = c^{-1} \int \alpha(1)\delta(n\cdot R_1 + c\tau) f_1(1)\mathrm{d}1, \tag{29}$$

may be written as

$$A(n, \tau) = -\int \frac{\alpha(1)}{n\cdot R_1^{[1]}} f_1^{\mathrm{syn}}(1; n, \tau)\mathrm{d}1, \tag{30}$$

where we introduced the *synchronous distribution function*

$$f_1^{\mathrm{syn}}(1; n, \tau) \equiv -c^{-1} n\cdot R_1^{[1]} \delta(n\cdot R_1 + c\tau) f_1(1). \tag{31}$$

From (30) it is apparent that the average of a 'synchronous' quantity in the form (22) may be obtained by replacing $R_i^\mu(s_i)$, $R_i^{[1]\mu}(s_i)$, ... by the variables $R_1^\mu$, $R_1^{[1]\mu}$, ..., multiplying by the synchronous distribution function and integrating over all variables. From (31) and (5) the interpretation of $f_1^{\mathrm{syn}}(1; n, \tau)$ follows immediately: $f_1^{\mathrm{syn}}(1; n, \tau)\mathrm{d}1$ is the number of atoms with position four-vector satisfying the relation $n\cdot R_1 + c\tau = 0$, that lie in the volume element $\mathrm{d}1 \equiv \mathrm{d}^4 R_1 \mathrm{d}^4 R_1^{[1]} \ldots$ .

Let us next consider a 'retarded quantity' of the form:

$$a(R) = \sum_i \int \alpha\{R_i(s_i), \ldots\}\theta\{R - R_i(s_i)\}\delta[\{R - R_i(s_i)\}^2]\mathrm{d}s_i. \tag{32}$$

(The four-potential due to a point charge is a quantity of this type.) The combination of $\theta$- and $\delta$-function selects indeed world line points which lie on the negative light-cone with $R^\mu$ as top. If $R_i^\mu(s_i) \neq R^\mu$ for all $i$ and $s_i$, one may perform the integration in (32) by introducing the integration variable $\{R - R_i(s_i)\}^2$ instead of $s_i$. One obtains then:

$$a(R) = \sum_i \left. \frac{\alpha\{R_i(s_i), \ldots\}}{2R_i^{[1]}(s_i)\cdot\{R_i(s_i) - R\}} \right|_{\mathrm{ret}}, \tag{33}$$

where the suffix ret denotes the fact that one has to take $s_i$ as solution of the equations $\{R - R_i(s_i)\}^2 = 0$ and $R^0 - R_i^0(s_i) > 0$.

Apart from quantities of the type (32), there occur also quantities which are space-time derivatives:

$$\partial_\mu \sum_i \int \alpha\{R_i(s_i), \ldots; R\}\theta\{R-R_i(s_i)\}\delta[\{R-R_i(s_i)\}^2]\mathrm{d}s_i. \tag{34}$$

(The four-potentials due to electromagnetic multipoles and the electromagnetic fields are of this type). For $R^\mu \neq R_i^\mu(s_i)$ such a quantity may be written in a form which shows that it is a retarded quantity of the type (32). To this end one must use an identity valid for an arbitrary function $f$ and an arbitrary four-vector $v^\mu$:

$$\partial_\mu f[\{R-R(s_i)\}^2] = -\frac{\{R-R_i(s_i)\}_\mu}{v\cdot\{R-R_i(s_i)\}} v\cdot \frac{\partial}{\partial R_i} f[\{R-R_i(s_i)\}^2]. \tag{35}$$

If this relation for the choice $f(x) = \delta(x)$ and $v^\mu = R_i^{[1]\mu}(s_i)$ is used in (34) for $R^\mu \neq R_i^\mu(s_i)$ and a partial integration is performed one gets

$$\sum_i \int \left[\partial_\mu \alpha + \frac{\mathrm{d}}{\mathrm{d}s_i}\left\{\frac{(R-R_i)_\mu \alpha}{R_i^{[1]}\cdot(R-R_i)}\right\}\right] \theta(R-R_i)\delta\{(R-R_i)^2\}\mathrm{d}s_i, \tag{36}$$

so that for $R^\mu \neq R_i^\mu(s_i)$ the quantity (34) is indeed of the retarded type (32). This could also have been seen by differentiating, for $R^\mu \neq R_i^\mu(s_i)$, the implicit function (33) with respect to $R_\mu$:

$$\partial_\mu \left[\sum_i \frac{\alpha\{R_i(s_i), \ldots; R\}}{2R_i^{[1]}\cdot\{R_i(s_i)-R\}}\Bigg|_{\mathrm{ret}}\right] = \left[\sum_i \mathrm{d}_{i\mu}^{\mathrm{ret}} \frac{\alpha\{R_i(s_i), \ldots; R\}}{2R_i^{[1]}\cdot\{R_i(s_i)-R\}}\right]\Bigg|_{\mathrm{ret}}, \tag{37}$$

where the differentiation $\mathrm{d}_{i\mu}^{\mathrm{ret}}$ is (for $R^\mu \neq R_i^\mu(s_i)$)

$$\mathrm{d}_{i\mu}^{\mathrm{ret}} = \partial_\mu + \frac{\{R-R_i(s_i)\}_\mu}{R_i^{[1]}(s_i)\cdot\{R-R_i(s_i)\}} \frac{\mathrm{d}}{\mathrm{d}s_i}. \tag{38}$$

The right-hand side of (37) shows again that the quantity involved has retarded character for $R^\mu \neq R_i^\mu$. (One may prove that the right-hand side of (37) is indeed equal to (36).)

The average of the quantity (32), which reads according to (17)

$$A(R) = c^{-1} \int \alpha(1)\theta(R-R_1)\delta\{(R-R_1)^2\}f_1(1)\mathrm{d}1, \tag{39}$$

may be written as

$$A(R) = \int \frac{\alpha(1)}{2R_1^{[1]}\cdot(R_1-R)} f_1^{\mathrm{ret}}(1; R)\mathrm{d}1. \tag{40}$$

Here we introduced the *retarded distribution function*[1] defined as

$$f_1^{\mathrm{ret}}(1; R) \equiv 2c^{-1}R_1^{[1]}\cdot(R_1-R)\theta(R-R_1)\delta\{(R-R_1)^2\}f_1(1). \tag{41}$$

[1] Such a distribution function, but in its three-dimensional form (56), has been introduced by S. R. de Groot and J. Vlieger, Physica **31**(1965)254 and in four-dimensional form by L. G. Suttorp, thesis, Amsterdam (1968).

(It would seem from (40) that the integrand has a singularity for $R_1^\mu = R^\mu$, but the denominator appearing in front of $f_1^{\text{ret}}$ is compensated by a factor in $f_1^{\text{ret}}$ itself.) From (40) it appears that the average of a 'retarded' quantity which is given by (33) for $R_i^\mu(s_i) \neq R^\mu$ may be obtained by replacing $R_i^\mu(s_i)$, $R_i^{[1]\mu}(s_i)$, ... by the variables $R_1^\mu$, $R_1^{[1]\mu}$, ..., multiplying by the retarded distribution function and integrating over all variables. From (41) and (5) follows the interpretation of $f_1^{\text{ret}}(1; R)$, namely: $f_1^{\text{ret}}(1; R)\text{d}1$ is the number of atoms with position four-vector $R_1^\mu$ satisfying $(R-R_1)^2 = 0$ and $R^0 > R_1^0$, that lies in the volume element $\text{d}1 \equiv \text{d}^4R_1\,\text{d}^4R_1^{[1]} \ldots$.

We may treat in an analogous way the average of an 'advanced quantity'

$$a(R) = \sum_i \int \alpha\{R_i(s_i), \ldots\}\theta\{R_i(s_i)-R\}\delta[\{R-R_i(s_i)\}^2]\text{d}s_i, \tag{42}$$

which may be written for $R_i^\mu(s_i) \neq R^\mu$ as

$$a(R) = \sum_i \left.\frac{\alpha\{R_i(s_i), \ldots\}}{2R_i^{[1]}(s_i)\cdot\{R-R_i(s_i)\}}\right|_{\text{adv}}, \tag{43}$$

where the suffix adv indicates that $s_i$ is the solution of the positive light-cone equations $\{R-R_i(s_i)\}^2 = 0$ and $R_i^0(s_i)-R^0 > 0$. The average of such a quantity (42) reads according to (17)

$$A(R) = c^{-1}\int \alpha(1)\theta(R_1-R)\delta\{(R-R_1)^2\}f_1(1)\text{d}1. \tag{44}$$

It may be written as

$$A(R) = \int \frac{\alpha(1)}{2R_1^{[1]}\cdot(R-R_1)} f_1^{\text{adv}}(1; R)\text{d}1, \tag{45}$$

with the help of the *advanced distribution function*

$$f_1^{\text{adv}}(1; R) \equiv 2c^{-1}R_1^{[1]}\cdot(R-R_1)\theta(R_1-R)\delta\{(R-R_1)^2\}f_1(1). \tag{46}$$

The retarded, and likewise the advanced, distribution function may be written in terms of the synchronous distribution function. To that purpose we write first the identity valid for a four-vector $x^\mu$ ($\neq 0$):

$$\theta(x)\delta(x^2) = \frac{\delta(x^0-|\boldsymbol{x}|)}{2|\boldsymbol{x}|}, \tag{47}$$

where we used the property of the delta function

$$\delta\{f(x)\} = \sum_i \frac{1}{|\partial f/\partial x|}\,\delta(x-x_i) \tag{48}$$

with $x_i$ the (non-degenerate) roots of $f(x) = 0$. One may obtain the covariant form of (47) by noting that

$$x^2 = -(n\cdot x)^2 + |\Delta_n \cdot x|^2, \tag{49}$$

where $n^\mu$ is a time-like unit vector ($n^0 > 0$), $\Delta_n^{\mu\nu} \equiv g^{\mu\nu} + n^\mu n^\nu$ and $|y| \equiv \sqrt{(y\cdot y)}$ for a space-like vector $y^\mu$. If one substitutes (49) into the left-hand side of (47) and then uses the property (48) one obtains

$$\theta(x)\delta(x^2) = -\frac{\delta(n\cdot x + |\Delta_n \cdot x|)}{2n\cdot x}, \tag{50}$$

where the right-hand side is the covariant generalization of the right-hand side of (47). We now use relation (50) with $x^\mu = R^\mu - R_1^\mu$ in the expression (41) for the retarded distribution function. Then by comparison with the expression (31) for the synchronous distribution function one finds for $R^\mu \neq R_1^\mu$:

$$f_1^{\mathrm{ret}}(1; R) = -\frac{R_1^{[1]}\cdot(R-R_1)}{n\cdot(R-R_1)n\cdot R_1^{[1]}} f_1^{\mathrm{syn}}(1; n, -c^{-1}n\cdot R - c^{-1}|\Delta_n \cdot (R-R_1)|). \tag{51}$$

In particular if one chooses the unit vector $n^\mu$ as $R_1^{[1]\mu}/c$ one obtains unity for the factor in front of $f_1^{\mathrm{syn}}$. A different choice of $n^\mu$ which is often convenient, is $n^\mu = \mathring{n}^\mu \equiv (1, 0, 0, 0)$. Then (51) gets the form (for $R^\mu \neq R_1^\mu$)

$$f_1^{\mathrm{ret}}(1; R) = \kappa(1) f_1^{\mathrm{syn}}(1; \mathring{n}, t - c^{-1}\,|\boldsymbol{R} - \boldsymbol{R}_1|), \tag{52}$$

where we used the abbreviation

$$\kappa(1) \equiv 1 - \frac{\boldsymbol{\beta}_1 \cdot (\boldsymbol{R} - \boldsymbol{R}_1)}{|\boldsymbol{R} - \boldsymbol{R}_1|} \tag{53}$$

and the fact that $ct - ct_1 = |\boldsymbol{R} - \boldsymbol{R}_1|$, since $R_1^\mu - R^\mu$ lies on the negative light-cone. (The quantity $\boldsymbol{\beta}_1 \equiv \boldsymbol{R}_1^{[1]}/R_1^{[1]0}$ is the particle velocity divided by the speed of light.)

The synchronous, the retarded as well as the advanced distribution functions, given in (31), (41) and (46) respectively, depend on the four-vectors $R_1^\mu$, $R_1^{[1]\mu}, \ldots, R_1^{[n]\mu}$ of which not all components are independent. In fact the velocity and the higher derivatives satisfy the relations (8). Moreover the synchronous distribution function vanishes, unless $n\cdot R_1 + c\tau = 0$, whereas the retarded and advanced distribution functions vanish unless $(R-R_1)^2 = 0$ and $R^0 - R_1^0 > 0$ or $< 0$ respectively. Therefore it is often convenient to introduce distribution functions which result from the functions discussed above by integrating over the time-components of the four-vectors. Using then also the three-velocities, three-accelerations etc. instead of the space-components of the corresponding four-vectors, one writes the number of particles with position $\boldsymbol{R}_1$, velocity $c\boldsymbol{\beta}_1$, acceleration $c^2\partial_0\boldsymbol{\beta}_1$ etc. at time $\tau$ as

$$f_1^{\mathrm{syn}}(\boldsymbol{R}_1, \boldsymbol{\beta}_1, \partial_0\boldsymbol{\beta}_1, \ldots; \tau)\mathrm{d}\boldsymbol{R}_1\, \mathrm{d}\boldsymbol{\beta}_1\, \mathrm{d}\partial_0\boldsymbol{\beta}_1 \ldots, \tag{54}$$

where

$$f_1^{\mathrm{syn}}(\boldsymbol{R}_1, \boldsymbol{\beta}_1, \partial_0\boldsymbol{\beta}_1, \ldots; \tau) \equiv \frac{\partial(\boldsymbol{R}_1^{[1]}, \boldsymbol{R}_1^{[2]}, \ldots)}{\partial(\boldsymbol{\beta}_1, \partial_0\boldsymbol{\beta}_1, \ldots)} \int f_1^{\mathrm{syn}}(1; \mathring{n}, \tau)\mathrm{d}R_1^0\, \mathrm{d}R_1^{[1]0} \ldots, \tag{55}$$

where a Jacobian appears and where $\mathring{n}^\mu$ is the vector (1, 0, 0, 0). This shows that the synchronous distribution function is the generalization of the ordinary distribution function of non-relativistic theory.

Likewise one may write the number of particles with position $\boldsymbol{R}_1$, velocity $c\boldsymbol{\beta}_1$, acceleration $c^2\partial_0\boldsymbol{\beta}_1$, etc. at a time, related to the observer's time $t$ by the light-cone equation $t_1 = t - c^{-1}|\boldsymbol{R}-\boldsymbol{R}_1|$, as:

$$f_1^{\mathrm{ret}}(\boldsymbol{R}_1, \boldsymbol{\beta}_1, \partial_0\boldsymbol{\beta}_1, \ldots; \boldsymbol{R}, t)\mathrm{d}\boldsymbol{R}_1\mathrm{d}\boldsymbol{\beta}_1\mathrm{d}\partial_0\boldsymbol{\beta}_1 \ldots, \tag{56}$$

where

$$f_1^{\mathrm{ret}}(\boldsymbol{R}_1, \boldsymbol{\beta}_1, \partial_0\boldsymbol{\beta}_1, \ldots; \boldsymbol{R}, t) \equiv \frac{\partial(\boldsymbol{R}_1^{[1]}, \boldsymbol{R}_1^{[2]}, \ldots)}{\partial(\boldsymbol{\beta}_1, \partial_0\boldsymbol{\beta}_1, \ldots)} \int f_1^{\mathrm{ret}}(1; R)\mathrm{d}R_1^0\, \mathrm{d}R_1^{[1]0} \ldots. \tag{57}$$

The advanced distribution functions may be treated in a completely similar way.

The connexion between retarded and synchronous distribution functions in the form (52) may now be translated (for $R^\mu \neq R_1^\mu$) into:

$$f_1^{\mathrm{ret}}(\boldsymbol{R}_1, \boldsymbol{\beta}_1, \partial_0\boldsymbol{\beta}_1, \ldots; \boldsymbol{R}, t) = \kappa(1) f_1^{\mathrm{syn}}(\boldsymbol{R}_1, \boldsymbol{\beta}_1, \partial_0\boldsymbol{\beta}_1, \ldots; t - c^{-1}|\boldsymbol{R}-\boldsymbol{R}_1|), \tag{58}$$

as follows from (55) and (57).

## 3 *The Maxwell equations*

### a. *Derivation of the macroscopic field equations*

The atomic field tensor $f^{\alpha\beta}$ which has been given explicitly in (IV.14) is a quantity of the form (15), i.e. a sum of integrals along world lines of the atoms of the system. Therefore one may define the average in the way as defined in formula (17) of the preceding section:

$$F^{\alpha\beta} = \langle f^{\alpha\beta} \rangle \tag{59}$$

with components $(F^{01}, F^{02}, F^{03}) = \boldsymbol{E}$ and $(F^{23}, F^{31}, F^{12}) = \boldsymbol{B}$. Since the atomic fields $f^{\alpha\beta}$ are retarded quantities, one may alternatively write the macroscopic fields in terms of retarded distribution functions, as discussed in the preceding section[1].

The atomic charge–current density vector and the atomic polarization tensor which have been given in (IV.5) and (IV.8) are again quantities of the type (15). So again the average of these quantities may be defined as in formula (13):

$$J^{\alpha} = \langle j^{\alpha} \rangle, \tag{60}$$

$$M^{\alpha\beta} = \langle m^{\alpha\beta} \rangle, \tag{61}$$

with components $J^0 = c\varrho^{\mathrm{e}}$, $(J^1, J^2, J^3) = \boldsymbol{J}$, $(M^{01}, M^{02}, M^{03}) = -\boldsymbol{P}$ and $(M^{23}, M^{31}, M^{12}) = \boldsymbol{M}$.

The equations which govern these macroscopic quantities are obtained by averaging the atomic field equations (IV.7,13). This gives

$$\begin{aligned} \langle \partial_\beta f^{\alpha\beta} \rangle &= c^{-1} \langle j^{\alpha} \rangle + \langle \partial_\beta m^{\alpha\beta} \rangle, \\ \langle \partial^{\alpha} f^{\beta\gamma} \rangle + \langle \partial^{\beta} f^{\gamma\alpha} \rangle + \langle \partial^{\gamma} f^{\alpha\beta} \rangle &= 0. \end{aligned} \tag{62}$$

Since differentiation and averaging commute according to (20), one may write these equations as

$$\begin{aligned} \partial_\beta \langle f^{\alpha\beta} \rangle &= c^{-1} \langle j^{\alpha} \rangle + \partial_\beta \langle m^{\alpha\beta} \rangle, \\ \partial^{\alpha} \langle f^{\beta\gamma} \rangle + \partial^{\beta} \langle f^{\gamma\alpha} \rangle + \partial^{\gamma} \langle f^{\alpha\beta} \rangle &= 0. \end{aligned} \tag{63}$$

Now from these equations, with the notations (59), (60) and (61) for the macroscopic quantities, one obtains

$$\begin{aligned} \partial_\beta F^{\alpha\beta} &= c^{-1} J^{\alpha} + \partial_\beta M^{\alpha\beta}, \\ \partial^{\alpha} F^{\beta\gamma} + \partial^{\beta} F^{\gamma\alpha} + \partial^{\gamma} F^{\alpha\beta} &= 0. \end{aligned} \tag{64}$$

These are precisely Maxwell's equations

$$\begin{aligned} \nabla \cdot \boldsymbol{E} &= \varrho^{\mathrm{e}} - \nabla \cdot \boldsymbol{P}, \\ -\partial_0 \boldsymbol{E} + \nabla \wedge \boldsymbol{B} &= c^{-1} \boldsymbol{J} + \partial_0 \boldsymbol{P} + \nabla \wedge \boldsymbol{M}, \\ \nabla \cdot \boldsymbol{B} &= 0, \\ \partial_0 \boldsymbol{B} + \nabla \wedge \boldsymbol{E} &= 0. \end{aligned} \tag{65}$$

One can also introduce the macroscopic 'displacement tensor' (cf. (IV.24)):

$$H^{\alpha\beta} = F^{\alpha\beta} - M^{\alpha\beta} \tag{66}$$

[1] S. R. de Groot and J. Vlieger, op. cit.

with components $(H^{01}, H^{02}, H^{03}) = \boldsymbol{D}$ and $(H^{23}, H^{31}, H^{12}) = \boldsymbol{H}$. In other words (66) reads

$$\begin{aligned} \boldsymbol{D} &= \boldsymbol{E}+\boldsymbol{P}, \\ \boldsymbol{H} &= \boldsymbol{B}-\boldsymbol{M}. \end{aligned} \tag{67}$$

Then the equations (64) become

$$\begin{aligned} \partial_\beta H^{\alpha\beta} &= c^{-1} J^\alpha, \\ \partial^\alpha F^{\beta\gamma} + \partial^\beta F^{\gamma\alpha} + \partial^\gamma F^{\alpha\beta} &= 0, \end{aligned} \tag{68}$$

or in three-dimensional notation:

$$\begin{aligned} \nabla \cdot \boldsymbol{D} &= \varrho^{\mathrm{e}}, \\ -\partial_0 \boldsymbol{D} + \nabla \wedge \boldsymbol{H} &= c^{-1} \boldsymbol{J}, \\ \nabla \cdot \boldsymbol{B} &= 0, \\ \partial_0 \boldsymbol{B} + \nabla \wedge \boldsymbol{E} &= 0. \end{aligned} \tag{69}$$

The covariant nature of the Maxwell equations has now been obtained as a consequence of the covariant nature of the microscopic field equations. It needs no longer be postulated as in the traditional expositions of the Maxwell theory [1].

Finally one may derive the macroscopic law of conservation of charge by averaging the atomic conservation law (IV.26). One finds then, using the fact that differentiation and averaging commute and the notation (60):

$$\partial_\alpha J^\alpha = 0, \tag{70}$$

or

$$\frac{\partial \varrho^{\mathrm{e}}}{\partial t} + \nabla \cdot \boldsymbol{J} = 0 \tag{71}$$

in three-dimensional notation.

[1] Earlier, incomplete attempts to derive Maxwell's equations in a covariant way are due to Ph. Frank, Ann. Physik **27**(1908)1059; H. Minkowski and M. Born, Math. Ann. **68**(1910) 526; A. D. Fokker, Phil. Mag. **39**(1920)404; W. Dällenbach, Ann. Physik **58**(1919)523; J. Frenkel, Lehrbuch der Elektrodynamik II, (Springer, Berlin 1928); S. R. de Groot and J. Vlieger, Physica **31**(1965)254. For a discussion see: W. Pauli, Theory of relativity (Pergamon Press, London 1958); L. G. Suttorp, thesis, Amsterdam (1968); S. R. de Groot, The Maxwell equations (North-Holland Publ. Co., Amsterdam 1969).

b. *Explicit forms of the macroscopic current vector and polarization tensor*

The atomic four-current density given in (IV.5) is an example of a synchronous quantity of the type (21). Its form (22) with $n^\alpha = \mathring{n}^\alpha \equiv (1, 0, 0, 0)$ and $\tau = t$ was explicited in (IV.42) and (IV.43). Therefore according to (30) the average may be expressed in terms of the synchronous distribution function $f_1^{\text{syn}}(1; \mathring{n}, t)$ given in (31). Instead of this synchronous distribution function depending on the position and velocity four-vectors it is convenient to introduce the three-dimensional functions of the type (55). This gives for the macroscopic charge and current densities:

$$\begin{aligned} \varrho^{\mathrm{e}} &= \sum_a e_a \int \delta^{(3)}(\boldsymbol{R}_1 - \boldsymbol{R}) f_1^{\text{syn},a}(\boldsymbol{R}_1; t)\mathrm{d}\boldsymbol{R}_1 , \\ \boldsymbol{J} &= c \sum_a e_a \int \boldsymbol{\beta}_1 \delta^{(3)}(\boldsymbol{R}_1 - \boldsymbol{R}) f_1^{\text{syn},a}(\boldsymbol{R}_1 , \boldsymbol{\beta}_1 ; t)\mathrm{d}\boldsymbol{R}_1 \, \mathrm{d}\boldsymbol{\beta}_1 , \end{aligned} \tag{72}$$

(where we added a summation over different species $a$) or

$$\begin{aligned} \varrho^{\mathrm{e}} &= \sum_a e_a f_1^{\text{syn},a}(\boldsymbol{R}; t), \\ \boldsymbol{J} &= c \sum_a e_a \int \boldsymbol{\beta}_1 f_1^{\text{syn},a}(\boldsymbol{R}, \boldsymbol{\beta}_1 ; t)\mathrm{d}\boldsymbol{\beta}_1 . \end{aligned} \tag{73}$$

The polarization tensor, given in (IV.8) is likewise a synchronous quantity, this time of the type (23). Its form (27) with $n^\alpha = \mathring{n}^\alpha \equiv (1, 0, 0, 0)$ and $\tau = t$ has been evaluated in (IV.36–39). The average may thus be expressed, according to (30), with the help of the synchronous distribution function (31) or its three-dimensional form (55). In particular one finds with (IV.48) and (IV.49) for the dipole contribution to the polarization tensor

$$\begin{aligned} \boldsymbol{P}^{(1)} &= \int (\underline{\boldsymbol{\mu}}_1^{(1)} - \underline{\boldsymbol{\nu}}_1^{(1)} \wedge \boldsymbol{\beta}_1) f_1^{\text{syn}}(\boldsymbol{R}, \boldsymbol{\beta}_1 , \boldsymbol{\mu}_1^{(1)}, \boldsymbol{\nu}_1^{(1)}; t)\mathrm{d}\boldsymbol{\beta}_1 \, \mathrm{d}\boldsymbol{\mu}_1^{(1)}\mathrm{d}\boldsymbol{\nu}_1^{(1)}, \\ \boldsymbol{M}^{(1)} &= \int (\underline{\boldsymbol{\nu}}_1^{(1)} + \underline{\boldsymbol{\mu}}_1^{(1)} \wedge \boldsymbol{\beta}_1) f_1^{\text{syn}}(\boldsymbol{R}, \boldsymbol{\beta}_1 , \boldsymbol{\mu}_1^{(1)}, \boldsymbol{\nu}_1^{(1)}; t)\mathrm{d}\boldsymbol{\beta}_1 \, \mathrm{d}\boldsymbol{\mu}_1^{(1)}\mathrm{d}\boldsymbol{\nu}_1^{(1)}. \end{aligned} \tag{74}$$

The underlined quantities have been given in (IV.50, 51) as:

$$\begin{aligned} \underline{\boldsymbol{\mu}}_1^{(1)} &\equiv \boldsymbol{\Omega}_1 \cdot \boldsymbol{\mu}_1^{(1)} = \boldsymbol{\mu}_{1,\perp}^{(1)} + \sqrt{1 - \boldsymbol{\beta}_1^2}\, \boldsymbol{\mu}_{1,//}^{(1)} , \\ \underline{\boldsymbol{\nu}}_1^{(1)} &\equiv \boldsymbol{\Omega}_1 \cdot \boldsymbol{\nu}_1^{(1)} = \boldsymbol{\nu}_{1,\perp}^{(1)} + \sqrt{1 - \boldsymbol{\beta}_1^2}\, \boldsymbol{\nu}_{1,//}^{(1)} , \end{aligned} \tag{75}$$

where the electric and magnetic dipole moments $\boldsymbol{\mu}_1^{(1)}$ and $\boldsymbol{\nu}_1^{(1)}$ have been split into parts perpendicular and parallel to the atomic velocity $\boldsymbol{\beta}_1$.

The formulae (74–75) show the effects of the atomic velocity $\boldsymbol{\beta}_1 c$. Relativistic effects are $a$: the Lorentz contractions of the parallel components of both the electric and magnetic dipole moments, and furthermore $b$: the last term of $\boldsymbol{P}$, which describes the effect of moving magnetic dipoles on the electric polarization vector. Such an effect has been observed experimentally[1]. It forms the basis of the so-called unipolar induction machine in which a cylindrical permanent magnet with magnetization parallel to the axis is rotated around this axis. Then a potential difference arises between the mantle and the axis which can cause a current if the material is a conductor, as is the case for the iron used in such machines.

In fluids under ordinary circumstances $\beta_1$ is of the order of the sound velocity divided by the velocity of light, that is of the order of $10^{-6}$. The vibrations in solids may have circular frequencies $\omega$ of about $3\times 10^{13}\text{s}^{-1}$ (optical branch). The atomic velocity $\dot{R}_1$ is equal to about $R_1\omega$ with $R_1$ of the order of $10^{-8}$ cm. Thus $\beta_1 = \dot{R}_1/c$ is then about $10^{-5}$. The Lorentz contractions are then negligibly small, because $\sqrt{(1-\beta_1^2)}$ differs from unity only by an amount of the order of $10^{-12}$ for fluids, and $10^{-10}$ for solids. The effect of moving magnetic dipoles is not so small since it is proportional to $\beta_1$. Its magnitude compared to the main term is $\beta_1 \nu_1^{(1)}/\mu_1^{(1)}$. (Since we are concerned with upper limits on the orders of magnitude of the various effects, only absolute values are considered and no attention is paid to the vectorial character of the quantities.) The proportion $\nu_1^{(1)}/\mu_1^{(1)}$ is, according to (IV.27) and (IV.28), of the order of $10^{-2}$ (the fine structure constant). So the effect of moving magnetic dipoles has a relative magnitude as compared to the leading term of $10^{-8}$ for fluids and $10^{-7}$ for solids. (The corresponding effect of moving electric dipoles in the magnetization, which is a non-relativistic effect, is of order $10^{-4}$ for fluids and $10^{-3}$ for solids.)

If all atoms have the same velocity, say $\boldsymbol{\beta}c$, then the polarization may be expressed in this velocity and the macroscopic electric and magnetic dipole densities, defined as

$$\begin{aligned}\mathscr{P}^{(1)} &\equiv \int \boldsymbol{\mu}_1^{(1)} f_1^{\text{syn}}(\boldsymbol{R}, \boldsymbol{\mu}_1^{(1)}; t)\mathrm{d}\boldsymbol{\mu}_1^{(1)},\\ \mathscr{M}^{(1)} &\equiv \int \boldsymbol{\nu}_1^{(1)} f_1^{\text{syn}}(\boldsymbol{R}, \boldsymbol{\nu}_1^{(1)}; t)\mathrm{d}\boldsymbol{\nu}_1^{(1)}.\end{aligned} \tag{76}$$

Indeed with these definitions and $\boldsymbol{\beta}_1 = \boldsymbol{\beta}$, one gets for (74), using also the first equalities of (75):

[1] H. A. Wilson, Phil. Trans. A **204**(1904)121; H. A. Wilson and M. Wilson, Proc. Roy. Soc. A **89**(1913)99.

$$\begin{aligned} \boldsymbol{P}^{(1)} &= \boldsymbol{\Omega}\cdot\mathscr{P}^{(1)} - \mathscr{M}^{(1)} \wedge \boldsymbol{\beta}, \\ \boldsymbol{M}^{(1)} &= \boldsymbol{\Omega}\cdot\mathscr{M}^{(1)} + \mathscr{P}^{(1)} \wedge \boldsymbol{\beta}, \end{aligned} \tag{77}$$

where the tensor $\boldsymbol{\Omega} \equiv \mathbf{U} + (\gamma^{-1} - 1)\boldsymbol{\beta}\boldsymbol{\beta}/\beta^2$ with $\gamma^{-1} \equiv \sqrt{(1-\beta^2)}$ contains the common velocity $\boldsymbol{\beta}c$. These formulae may be applied to the case of rigid crystal lattices, where all carriers of electromagnetic moments have the same velocity if the vibrations are neglected.

The contributions from electric and magnetic quadrupole moments to the polarization tensor (61) become with (IV.52–53)

$$\begin{aligned} \boldsymbol{P}^{(2)} &= -\nabla\cdot\int (\boldsymbol{\mu}_1^{(2)} - \underline{\boldsymbol{\nu}}_1^{(2)} \wedge \boldsymbol{\beta}_1) f_1^{\mathrm{syn}}(\boldsymbol{R}, 1; t)\mathrm{d}1 \\ &\quad - \int [\gamma_1 \boldsymbol{\beta}_1\cdot\partial_0\{\gamma_1(\underline{\boldsymbol{\mu}}_1^{(2)} - \underline{\boldsymbol{\nu}}_1^{(2)} \wedge \boldsymbol{\beta}_1)\} + \tfrac{1}{2}\gamma_1\,\partial_0(\gamma_1 \boldsymbol{\beta}_1)\cdot\underline{\boldsymbol{\nu}}_1^{(2)} \wedge \boldsymbol{\beta}_1] f_1^{\mathrm{syn}}(\boldsymbol{R}, 1; t)\mathrm{d}1, \\ \boldsymbol{M}^{(2)} &= -\nabla\cdot\int (\underline{\boldsymbol{\nu}}_1^{(2)} + \underline{\boldsymbol{\mu}}_1^{(2)} \wedge \boldsymbol{\beta}_1) f_1^{\mathrm{syn}}(\boldsymbol{R}, 1; t)\mathrm{d}1 \\ &\quad - \int [\gamma_1 \boldsymbol{\beta}_1\cdot\partial_0\{\gamma_1(\underline{\boldsymbol{\nu}}_1^{(2)} + \underline{\boldsymbol{\mu}}_1^{(2)} \wedge \boldsymbol{\beta}_1)\} - \tfrac{1}{2}\gamma_1\,\partial_0(\gamma_1 \boldsymbol{\beta}_1)\cdot\underline{\boldsymbol{\nu}}_1^{(2)}] f_1^{\mathrm{syn}}(\boldsymbol{R}, 1; t)\mathrm{d}1, \end{aligned} \tag{78}$$

where 1 stands for the atomic quantities occurring in the integrands. The underlined quadrupole moments follow from (IV.54) as:

$$\begin{aligned} \underline{\boldsymbol{\mu}}_1^{(2)} &\equiv \boldsymbol{\Omega}_1\cdot\boldsymbol{\mu}_1^{(2)}\cdot\boldsymbol{\Omega}_1 = \boldsymbol{\mu}_{1,\perp\perp}^{(2)} + \sqrt{1-\beta_1^2}(\boldsymbol{\mu}_{1,\perp//}^{(2)} + \boldsymbol{\mu}_{1,//\perp}^{(2)}) + (1-\beta_1^2)\boldsymbol{\mu}_{1,////}^{(2)}, \\ \underline{\boldsymbol{\nu}}_1^{(2)} &\equiv \boldsymbol{\Omega}_1\cdot\boldsymbol{\nu}_1^{(2)}\cdot\boldsymbol{\Omega}_1 = \boldsymbol{\nu}_{1,\perp\perp}^{(2)} + \sqrt{1-\beta_1^2}(\boldsymbol{\nu}_{1,\perp//}^{(2)} + \boldsymbol{\nu}_{1,//\perp}^{(2)}) + (1-\beta_1^2)\boldsymbol{\nu}_{1,////}^{(2)}. \end{aligned} \tag{79}$$

This formula shows that the electric and magnetic quadrupole moments suffer a Lorentz contraction.

The first term of $\boldsymbol{P}$ and the first two terms of $\boldsymbol{M}$ occurred also in the non-relativistic theory (chapter II), but with the non-relativistic quadrupole moments $\overline{\boldsymbol{\mu}}_1^{(2)}$ and $\overline{\boldsymbol{\nu}}_1^{(2)}$. Thus relativistic effects of four different types appear in the expressions given here. In the first place two effects similar to those found for dipole substances occur:

1°: the Lorentz contractions of the longitudinal components of the quadrupole moments $\boldsymbol{\mu}_1^{(2)}$ and $\boldsymbol{\nu}_1^{(2)}$. Under common circumstances, the effects are quite small, for the same reasons as explained in the dipole case.

2°: the effect of moving magnetic quadrupoles, described by the second term of $\boldsymbol{P}$. Its relative magnitude as compared to the leading term of $\boldsymbol{P}$ is (again under normal circumstances) of the order of $10^{-8}$ for fluids and $10^{-7}$ for solids.

Two more relativistic effects, which were absent in the dipole case, are encountered in the expressions of $\boldsymbol{P}$ and $\boldsymbol{M}$:
3°: the multipole fluxion effect, which is connected with the occurrence of time derivatives of the quadrupole moments.
4°: the acceleration effect, which is due to the presence of terms with the atomic acceleration $\partial_0 \boldsymbol{\beta}_1$.

The effects depend on the magnitude of the atomic velocities and accelerations. Let us give some numerical estimations of the various effects in systems under normal circumstances. Let us first consider a solid with atomic vibrations in the optical branch. As mentioned above $\omega \simeq 3 \times 10^{13} \mathrm{s}^{-1}$ and $\beta_1 \simeq 10^{-5}$. The vibration acceleration is $\ddot{R}_1 \simeq R_1 \omega^2$ and therefore $\partial_0 \beta_1 = R_1 \omega^2 c^{-2} \simeq 10^{-2}\ \mathrm{cm}^{-1}$. The time derivative of the electric quadrupole moment $\partial_0 \mu_1^{(2)}$ is of the order $e_{ki} r_{ki} \dot{r}_{ki}/c$ or $10^6 e_{ki} r_{ki}^2$ (since $r_{ki} \simeq 10^{-8}$ cm and $\dot{r}_{ki}/c$ is of the order of the fine structure constant) or $10^6 \mu_1^{(2)}$. Similarly, since roughly $\ddot{r}_{ki}/\dot{r}_{ki} \simeq \dot{r}_{ki}/r_{ki}$, the time derivative of the magnetic quadrupole moment $\partial_0 \nu_1^{(2)}$ is of the order $10^6 \nu_1^{(2)}$. These numbers will allow us to estimate the relativistic effects in the polarization tensor. Just as in the dipole case the effects 1° and 2° are small compared to the leading terms which they accompany. Let us therefore consider here the other effects. The main terms in $\boldsymbol{P}$ and $\boldsymbol{M}$, including the non-relativistic effect, contain gradients. Their magnitude, if compared to the dipole terms, depends thus on the inhomogeneities of the material. A very rough guess is obtained in the following way. Let the electric quadrupole moment $\mu_1^{(2)}$ be of the order of $10^{-8}$ cm times $\mu_1^{(1)}$, because it contains one more factor $r_{ki}$ than the dipole moment $\mu_1^{(1)}$. Let the 'inhomogeneity length' be $10^{-2}$ to 1 cm. Then the magnitude of the (non-relativistic) $\nabla \cdot \boldsymbol{\mu}_1^{(2)}$ effect expressed in the $\mu_1^{(1)}$ effect is $10^{-6}$ to $10^{-8} \mu_1^{(1)}$. The relativistic effects are more interesting because they do not contain gradients and are thus independent of the inhomogeneities. The multipole fluxion effect in $\boldsymbol{P}$ is $\beta_1 \partial_0 \mu_1^{(2)} \simeq 10^{-7} \mu_1^{(1)}$. The acceleration effect in $\boldsymbol{P}$ is $(\partial_0 \beta_1) \nu_1^{(2)} \beta_1$. It contains the magnetic quadrupole moment which is about $10^{-2}$ (the fine structure constant $\simeq \dot{r}_{ki}/c$) of the electric quadrupole moment. The acceleration effect becomes of the order $10^{-17} \mu_1^{(1)}$.

Similarly in $\boldsymbol{M}$ the main (non-relativistic) term with $\nabla \cdot \boldsymbol{\nu}_1^{(2)}$ becomes (with the use of the same figures as above) $10^{-8}$ to $10^{-10} \mu_1^{(1)}$, the multipole fluxion effect $\beta_1 \partial_0 \nu_1^{(2)}$ becomes $10^{-9} \mu_1^{(1)}$ and the acceleration effect $(\partial_0 \beta_1) \nu_1^{(2)}$ (note that this effect contains the magnetic quadrupole moment both in $\boldsymbol{P}$ and $\boldsymbol{M}$) becomes $10^{-12} \mu_1^{(1)}$. The conclusion can be that all quadrupolar effects are very small, if compared to dipolar effects. But if a substance is studied which contains quadrupoles, but no dipoles, then the quadrupolar effects can only be compared amongst each other. The conclusion which one may draw from the figures given above is then that the relativistic multipole fluxion effect may exceed the non-relativistic effects under favourable physical circumstances.

The preceding concerned the case of a solid. In fluids $\beta_1$ is about ten times smaller and the collision frequency is perhaps a hundred times smaller. The relativistic effects are then exceedingly small.

If all atoms have the same velocity $\boldsymbol{\beta}$ and acceleration $\partial_0 \boldsymbol{\beta}$, the polarization can be expressed in terms of these quantities and of the macroscopic quadrupolar densities:

$$\mathscr{P}^{(2)} \equiv \int \boldsymbol{\mu}_1^{(2)} f_1^{\mathrm{syn}}(\boldsymbol{R}, \boldsymbol{\mu}_1^{(2)}; t) \mathrm{d}\boldsymbol{\mu}_1^{(2)},$$
$$\mathscr{M}^{(2)} \equiv \int \boldsymbol{\nu}_1^{(2)} f_1^{\mathrm{syn}}(\boldsymbol{R}, \boldsymbol{\nu}_1^{(2)}; t) \mathrm{d}\boldsymbol{\nu}_1^{(2)}. \tag{80}$$

Indeed with (79) and the abbreviations

$$\underline{\mathscr{P}}^{(2)} \equiv \boldsymbol{\Omega}\cdot\mathscr{P}^{(2)}\cdot\boldsymbol{\Omega} = \mathscr{P}^{(2)}_{\perp\perp} + \sqrt{1-\beta^2}(\mathscr{P}^{(2)}_{\perp//} + \mathscr{P}^{(2)}_{//\perp}) + (1-\beta^2)\mathscr{P}^{(2)}_{////},$$
$$\underline{\mathscr{M}}^{(2)} \equiv \boldsymbol{\Omega}\cdot\mathscr{M}^{(2)}\cdot\boldsymbol{\Omega} = \mathscr{M}^{(2)}_{\perp\perp} + \sqrt{1-\beta^2}(\mathscr{M}^{(2)}_{\perp//} + \mathscr{M}^{(2)}_{//\perp}) + (1-\beta^2)\mathscr{M}^{(2)}_{//}, \tag{81}$$

it follows that the quadrupole contributions to the polarization tensor (78) get the form

$$\boldsymbol{P}^{(2)} = -\boldsymbol{\nabla}\cdot(\underline{\mathscr{P}}^{(2)} - \underline{\mathscr{M}}^{(2)} \wedge \boldsymbol{\beta}) - \gamma\boldsymbol{\beta}\cdot\mathrm{d}_0\{\gamma(\underline{\mathscr{P}}^{(2)} - \underline{\mathscr{M}}^{(2)} \wedge \boldsymbol{\beta})\}$$
$$-\tfrac{1}{2}\gamma\partial_0(\gamma\boldsymbol{\beta})\cdot(\underline{\mathscr{M}}^{(2)} \wedge \boldsymbol{\beta}), \tag{82}$$
$$\boldsymbol{M}^{(2)} = -\boldsymbol{\nabla}\cdot(\underline{\mathscr{M}}^{(2)} + \underline{\mathscr{P}}^{(2)} \wedge \boldsymbol{\beta}) - \gamma\boldsymbol{\beta}\cdot\mathrm{d}_0\{\gamma(\underline{\mathscr{M}}^{(2)} + \underline{\mathscr{P}}^{(2)} \wedge \boldsymbol{\beta})\} + \tfrac{1}{2}\gamma\partial_0(\gamma\boldsymbol{\beta})\cdot\underline{\mathscr{M}}^{(2)},$$

where $\mathrm{d}_0 \equiv \partial_0 + \boldsymbol{\beta}\cdot\boldsymbol{\nabla}$ is the substantial time derivative. It should be remarked that in $\boldsymbol{M}$ the acceleration effect subsists even if the velocity $\boldsymbol{\beta} = 0$.

## 4 *The conservation of energy–momentum*

### a. *The conservation of rest mass*

The atomic conservation law (IV.176) of rest mass contains the mass flow four-vector, the average of which is, according to (17) with (15):

$$m_1^{\mathrm{rest}} \int u_1^\alpha \, \delta^{(4)}(X_1 - R) f_1(1) \mathrm{d}1. \tag{83}$$

Since this is a time-like four-vector it may be used to define the macroscopic four-velocity $U^\alpha$ (with $U_\alpha U^\alpha = -c^2$) and the macroscopic rest mass density

$\varrho'$ by writing

$$\varrho' U^\alpha = m_1^{\text{rest}} \int u_1^\alpha \delta^{(4)}(X_1 - R) f_1(1) \mathrm{d}1. \tag{84}$$

From the atomic mass conservation law (IV.176) it follows immediately, with (20), that

$$\partial_\alpha(\varrho' U^\alpha) = 0, \tag{85}$$

which is the macroscopic law of mass conservation.

b. *Energy–momentum conservation for a fluid system of neutral atoms*

The conservation law of energy–momentum will follow by taking the average of the atomic energy–momentum law (IV.177):

$$\begin{aligned} c\partial_\beta \left\langle \sum_k \int m_k u_k^\alpha u_k^\beta \delta^{(4)}(X_k - R) \mathrm{d}s_k \right\rangle \\ = c \left\langle \sum_k \int \mathfrak{f}_k^\alpha \delta^{(4)}(X_k - R) \mathrm{d}s_k \right\rangle \\ - c^{-1} \partial_\beta \left\langle \sum_k \int (s_k^{\alpha\gamma} \mathfrak{f}_{k\gamma}/m_k + \mathfrak{d}_k^{\alpha\gamma} u_{k\gamma}) u_k^\beta \delta^{(4)}(X_k - R) \mathrm{d}s_k \right\rangle . \end{aligned} \tag{86}$$

The total force and torque $\mathfrak{f}_k^\alpha$ and $\mathfrak{d}_k^{\alpha\beta}$ have been given in (IV.191–192) with (IV.193–194) and (IV.206–207). We shall confine ourselves in this subsection to systems of neutral atoms, since we are interested in the effects due to electromagnetic multipole moments. In the following we shall show that (86) may be cast into the form of a conservation law, i.e. as $\partial_\beta T^{\alpha\beta} = 0$, where $T^{\alpha\beta}$ is the energy–momentum tensor.

At the left-hand side of (86) appears the divergence $\partial_\beta T_{(\mathrm{m})\mathrm{I}}^{\alpha\beta}$ of the tensor

$$T_{(\mathrm{m})\mathrm{I}}^{\alpha\beta} \equiv \int m_1 u_1^\alpha u_1^\beta \delta^{(4)}(X_1 - R) f_1(1) \mathrm{d}1 \tag{87}$$

as follows from (17) with (15). This relativistic generalization of the kinetic energy and momentum densities and flows forms a first contribution (index I) to the macroscopic energy–momentum tensor. In particular, in view of its form, it is said to contribute to the *material* part of the energy–momentum tensor (denoted by the index (m))[1].

[1] Of course the characterization of (87) as being purely material does not exclude the fact that its value will depend in general on the macroscopic fields, since the distribution function will depend on these quantities. The classification by means of the index (m) for material, and later on (f) for field parts of the energy–momentum tensor is made only for convenience. The physical laws contain the *total* energy–momentum tensor, which will be specified in the following.

The right-hand side of (86) contains the total force and torque $\mathfrak{f}_k^\alpha$ and $\mathfrak{d}_k^{\alpha\beta}$ which have been split into long range and short range parts in (IV.191–192). We shall consider first the long range part which is obtained by introducing the total long range force and torque, specified in (IV.193–194). One gets thus the sum of external field and interatomic two-particle contributions. The latter, which contain the two-point distribution function $f_2(1, 2)$ (13), may be written as the sum of an uncorrelated term with $f_1(1)f_1(2)$ and a correlated term with $c_2(1, 2)$ (14). In this way one obtains in the first place the uncorrelated long range part of the right-hand side of (86). It contains the Maxwell fields (59) and reads

$$\int \tfrac{1}{2}(\partial^\alpha F^{\beta\gamma})\mathfrak{m}_{1\beta\gamma}\delta^{(4)}(X_1-R)f_1(1)\mathrm{d}1$$
$$-c^{-2}\partial_\beta \int \Big[(F^{\alpha\gamma}\mathfrak{m}_{1\gamma\varepsilon}u_1^\varepsilon - \Delta_1^{\alpha\gamma}\mathfrak{m}_{1\gamma\varepsilon}F^{\varepsilon\zeta}u_{1\zeta})u_1^\beta$$
$$+\frac{s_1^{\alpha\gamma}}{m_1}\Big\{\tfrac{1}{2}(\partial_\gamma F_{\varepsilon\zeta})\mathfrak{m}_1^{\varepsilon\zeta} - c^{-2}\frac{\mathrm{d}}{\mathrm{d}s_1}(F_{\gamma\varepsilon}\mathfrak{m}_1^{\varepsilon\zeta}u_{1\zeta})\Big\}u_1^\beta\Big]\delta^{(4)}(X_1-R)f_1(1)\mathrm{d}1, \quad (88)$$

where (IV.193–194) with (IV.182–183) have been used and where only dipole contributions have been taken into account. The latter limitation gives us the leading terms in the uncorrelated part of the macroscopic energy–momentum tensor. (In the correlated part such a limitation is not possible, since virtual multipoles of all orders cannot be excluded.) With the definitions (61) of the macroscopic polarization and (84) of the macroscopic velocity one may write (88) as

$$\tfrac{1}{2}(\partial^\alpha F^{\beta\gamma})M_{\beta\gamma} - c^{-2}\partial_\beta\{F^{\alpha\gamma}M_{\gamma\varepsilon}U^\varepsilon - \Delta^{\alpha\gamma}M_{\gamma\varepsilon}F^{\varepsilon\zeta}U_\zeta)U^\beta\} - \partial_\beta T^{\alpha\beta}_{(\mathrm{m})\mathrm{II}}, \quad (89)$$

with $\Delta^{\alpha\beta} \equiv g^{\alpha\beta} + c^{-2}U^\alpha U^\beta$. Here a second contribution to the material energy–momentum tensor appears:

$$T^{\alpha\beta}_{(\mathrm{m})\mathrm{II}} \equiv c^{-2}\int\Big[F^{\alpha\gamma}\mathfrak{m}_{1\gamma\varepsilon}(u_1^\varepsilon u_1^\beta - U^\varepsilon U^\beta) - \mathfrak{m}_{1\gamma\varepsilon}F^{\varepsilon\zeta}(\Delta_1^{\alpha\gamma}u_{1\zeta}u_1^\beta - \Delta^{\alpha\gamma}U_\zeta U^\beta)$$
$$+\frac{s_1^{\alpha\gamma}}{m_1}\Big\{\tfrac{1}{2}(\partial_\gamma F_{\varepsilon\zeta})\mathfrak{m}_1^{\varepsilon\zeta} - c^{-2}\frac{\mathrm{d}}{\mathrm{d}s_1}(F_{\gamma\varepsilon}\mathfrak{m}_1^{\varepsilon\zeta}u_{1\zeta})\Big\}u_1^\beta\Big]\delta^{(4)}(X_1-R)f_1(1)\mathrm{d}1. \quad (90)$$

This contribution contains in its first two terms velocity fluctuations $u_1^\alpha - U^\alpha$ while in the other two terms the inner angular momentum $s_1^{\alpha\beta}$ appears. On the other hand the first three terms of (89) contain exclusively macroscopic fields, polarizations and velocities. With the Maxwell equations (64) the first term of (89) may be written in the form of a divergence:

$$\tfrac{1}{2}(\partial^\alpha F^{\beta\gamma})M_{\beta\gamma} = F^{\alpha\gamma}\partial_\beta M_\gamma^{\cdot\beta} - \partial_\beta(F^{\alpha\gamma}M_\gamma^{\cdot\beta}) = \partial_\beta(F^{\alpha\gamma}F_\gamma^{\cdot\beta} - F^{\alpha\gamma}M_\gamma^{\cdot\beta} + \tfrac{1}{4}g^{\alpha\beta}F_{\gamma\varepsilon}F^{\gamma\varepsilon}), \tag{91}$$

where both the homogeneous and the inhomogeneous field equations have been employed. With (91) and (66) the first three terms of (89) become a divergence $-\partial_\beta T^{\alpha\beta}_{(\mathrm{f})}$, with the field energy–momentum tensor

$$T^{\alpha\beta}_{(\mathrm{f})} \equiv -F^{\alpha\gamma}H_\gamma^{\cdot\beta} - \tfrac{1}{4}g^{\alpha\beta}F_{\gamma\varepsilon}F^{\gamma\varepsilon} + c^{-2}(F^{\alpha\gamma}M_{\gamma\varepsilon}U^\varepsilon - \Delta^{\alpha\gamma}M_{\gamma\varepsilon}F^{\varepsilon\zeta}U_\zeta)U^\beta. \tag{92}$$

This is the only part of the total energy–momentum tensor which depends, in its explicit form, exclusively on the macroscopic fields, polarizations and velocities. Therefore we labelled it with the index (f), although, as remarked before, the division of the total energy–momentum tensor into a field part and a material part is not essential. We shall discuss the contents of (92) later on.

The uncorrelated long range part of the right-hand side of (86) has now been found, so that the correlated long range and the short range parts remain to be discussed. Let us start with the latter. The last terms of (86) are already in the form of a divergence; introducing a distribution function according to (19) and substituting (IV.206–207), one finds for them $-\partial_\beta T^{\alpha\beta}_{(\mathrm{m})\mathrm{III}}$, with a third contribution to the material energy–momentum tensor

$$T^{\alpha\beta}_{(\mathrm{m})\mathrm{III}} \equiv c^{-3}\int (s_1^{\alpha\gamma}\hat{\mathfrak{f}}^{\mathrm{S}}_{1;2\gamma}/m_1 + \hat{\mathfrak{d}}^{\mathrm{S}\alpha\gamma}_{1;2}u_{1\gamma})u_1^\beta\,\delta^{(4)}(X_1-R)f_2(1,2)\mathrm{d}1\,\mathrm{d}2. \tag{93}$$

In the short range part of the first term at the right-hand side of (86) we shall consider separately the contributions from the 'plus' and 'minus' fields. The plus part of the total force on atom $k$ is given explicitly in (IV.206) with (IV.204) and (IV.210). Introducing an extra variable $s^\alpha$ (with differential $\mathrm{d}^4 s \equiv \mathrm{d}s$) and a four-dimensional delta function we may write the short range plus field contribution as

$$c^{-1}\int \hat{\mathfrak{f}}^{\mathrm{S}\alpha}_{+1;2}(s)\delta^{(4)}(X_1-R)\delta^{(4)}(X_2-R+s)f_2(1,2)\mathrm{d}1\,\mathrm{d}2\,\mathrm{d}s, \tag{94}$$

where we employed the abbreviation

$$\begin{aligned}\hat{\mathfrak{f}}^{\mathrm{S}\alpha}_{+1;2}(s) \equiv{}& \frac{c^{-1}}{4\pi}\sum_{i,j} e_{1i}e_{2j}(u_{1i}\cdot u_{2j}\,\partial_s^\alpha - u_{2j}^\alpha\, u_{1i}\cdot\partial_s)\delta\{(s+r_{1i}-r_{2j})^2\}\\ &-\frac{c}{4\pi}\sum_{n,m=1}^{\infty}(-1)^m\left\{\mathfrak{m}_1^{\alpha_1\ldots\alpha_{n+1}}\partial_{s\alpha_n} + c^{-2}\left(\frac{\mathrm{d}}{\mathrm{d}s_1} + u_1\cdot\partial_s\right)\mathfrak{m}_1^{\alpha_1\ldots\alpha_{n+1}}u_{1\alpha_n}\right\}\\ &\left\{\mathfrak{m}_2^{\beta_1\ldots\beta_{m+1}}\partial_{s\beta_m} - c^{-2}\left(\frac{\mathrm{d}}{\mathrm{d}s_2} - u_2\cdot\partial_s\right)\mathfrak{m}_2^{\beta_1\ldots\beta_{m+1}}u_{2\beta_m}\right\}\partial_{s\alpha_1\ldots\alpha_{n-1}\beta_1\ldots\beta_{m-1}}\\ &(\partial_s^\alpha g_{\alpha_{n+1}\beta_{m+1}} - \partial_{s\alpha_{n+1}}g^\alpha_{\beta_{m+1}})\delta(s^2).\end{aligned} \tag{95}$$

In the non-relativistic theory the expression corresponding to (94) could be transformed into a divergence (a three-divergence and a time derivative) by making an appropriate Taylor expansion, which could be broken off after the first term since the integrand has short range as a function of $s^\alpha$. In fact it is the difference between the unexpanded and the multipole expanded atomic force so that it vanishes if the atoms are sufficiently far apart. We now have to generalize this procedure, due to Irving and Kirkwood, to the relativistic case. In (94) the two-point distribution function $f_2(1, 2)$ appears. As a result of the presence of the two delta functions, it contains the position four-vectors $R^\alpha$ and $R^\alpha - s^\alpha$ of the two atoms, so that its form is $f_2(R, 1, R-s, 2)$ (where now 1 and 2 denote the other dynamical properties of the atoms). As the relativistic generalization of the Irving–Kirkwood procedure we expand the two-point distribution function as a function of $R^\alpha$ in a Taylor series around $R^\alpha + \frac{1}{2}s^\alpha$:

$$f_2(R, 1, R-s, 2) = f_2(R+\tfrac{1}{2}s, 1, R-\tfrac{1}{2}s, 2) - \tfrac{1}{2}s^\alpha \partial_\alpha f_2(R+\tfrac{1}{2}s, 1, R-\tfrac{1}{2}s, 2) + \dots . \tag{96}$$

Only the first few terms contribute significantly since the integrand has short range as a function of $s^\alpha$.

If one introduces the expansion (96) into (94) one gets a sum of two terms: one term has the same form as (94), but with different delta functions:

$$c^{-1} \int \hat{\mathrm{f}}^{S\alpha}_{+1;2}(s)\delta^{(4)}(X_1 - R - \tfrac{1}{2}s)\delta^{(4)}(X_2 - R + \tfrac{1}{2}s) f_2(1, 2)\mathrm{d}1\,\mathrm{d}2\,\mathrm{d}s; \tag{97}$$

the other term may be written as $-\partial_\beta T^{\alpha\beta}_{(\mathrm{m})\mathrm{IV}}$, with a fourth contribution to the energy–momentum tensor

$$T^{\alpha\beta}_{(\mathrm{m})\mathrm{IV}} \equiv \tfrac{1}{2}c^{-1} \int s^\beta \hat{\mathrm{f}}^{S\alpha}_{+1;2}(s)\delta^{(4)}(X_1 - R - \tfrac{1}{2}s)\delta^{(4)}(X_2 - R + \tfrac{1}{2}s) f_2(1, 2)\mathrm{d}1\,\mathrm{d}2\,\mathrm{d}s. \tag{98}$$

If (95) is inserted into (97) one finds that the terms which contain $\partial^\alpha_s$ vanish for reasons of symmetry: by making the substitutions $s^\alpha \leftrightarrow -s^\alpha$ and $1 \leftrightarrow 2$ the integral changes into its opposite. As to the remaining terms of (97) with (95), one finds for the unexpanded part by integrating partially with respect to $s^\alpha$:

$$-\frac{c^{-2}}{4\pi}\int\sum_{i,j}e_{1i}e_{2j}u_{2j}^{\alpha}\left[\frac{\mathrm{d}r_{1i}}{\mathrm{d}s_1}\cdot\partial_s\delta\{(s+r_{1i}-r_{2j})^2\}-\delta\{(s+r_{1i}-r_{2j})^2\}u_1\cdot\partial_s\right]$$
$$\delta^{(4)}(X_1-R-\tfrac{1}{2}s)\delta^{(4)}(X_2-R+\tfrac{1}{2}s)f_2(1,2)\mathrm{d}1\,\mathrm{d}2\,\mathrm{d}s$$
$$=-\frac{c^{-2}}{4\pi}\int\sum_{i,j}e_{1i}e_{2j}u_{2j}^{\alpha}\frac{\mathrm{d}}{\mathrm{d}s_1}[\delta^{(4)}\{(s+r_{1i}-r_{2j})^2\}\delta^{(4)}(X_1-R-\tfrac{1}{2}s)]$$
$$\delta^{(4)}(X_2-R+\tfrac{1}{2}s)f_2(1,2)\mathrm{d}1\,\mathrm{d}2\,\mathrm{d}s$$
$$-\frac{c^{-2}}{8\pi}\partial_\beta\int u_1^{\beta}\sum_{i,j}e_{1i}e_{2j}u_{2j}^{\alpha}\delta\{(s+r_{1i}-r_{2j})^2\}\delta^{(4)}(X_1-R-\tfrac{1}{2}s)$$
$$\delta^{(4)}(X_2-R+\tfrac{1}{2}s)f_2(1,2)\mathrm{d}1\,\mathrm{d}2\,\mathrm{d}s. \quad (99)$$

The first term at the right-hand side vanishes as follows by integrating partially and employing the conservation of probability in the form (10), while the second term gives a contribution to $-\partial_\beta T_{(\mathrm{m})\mathrm{V}}^{\alpha\beta}$. The fifth part of the material energy–momentum tensor occurring here reads, if the multipole expanded part of the terms without $\partial_s^{\alpha}$ of (97) with (95) are treated in a similar fashion:

$$T_{(\mathrm{m})\mathrm{V}}^{\alpha\beta}\equiv\frac{c^{-2}}{8\pi}\int u_1^{\beta}\left[\sum_{i,j}e_{1i}e_{2j}u_{2j}^{\alpha}\delta\{(s+r_{1i}-r_{2j})^2\}+\sum_{n,m=1}^{\infty}(-1)^m\mathfrak{m}_1^{\alpha_1\ldots\alpha_{n+1}}\right.$$
$$\left.u_{1\alpha_{n+1}}\left\{\mathfrak{m}_2^{\beta_1\ldots\beta_m\alpha}\partial_{s\beta_m}-c^{-2}\left(\frac{\mathrm{d}}{\mathrm{d}s_2}-u_2\cdot\partial_s\right)\mathfrak{m}_2^{\beta_1\ldots\beta_m\alpha}u_{2\beta_m}\right\}\partial_{s\alpha_1\ldots\alpha_n\beta_1\ldots\beta_{m-1}}\delta(s^2)\right]$$
$$\delta^{(4)}(X_1-R-\tfrac{1}{2}s)\delta^{(4)}(X_2-R+\tfrac{1}{2}s)f_2(1,2)\mathrm{d}1\,\mathrm{d}2\,\mathrm{d}s. \quad (100)$$

The plus field short range contributions have thus been written in the form of a divergence. The essential step consisted in showing that owing to symmetry part of the terms of (97) with (95) vanished. This symmetry was intimately connected with the appearance of the delta function $\delta(s^2)$ which is invariant under the transformation $s^{\alpha}\leftrightarrow -s^{\alpha}$. If we had used from the beginning the complete retarded field, instead of only its plus part, an extra factor $\theta(s)$ would have appeared, which would have destroyed this invariance. Thus if we consider the minus field contribution a different procedure will have to be followed to transform it into a divergence[1].

The minus field short range part of the first term at the right-hand side of (86) is obtained by inserting the atomic expressions (IV.206) with (IV.208) and (IV.202). If one employs the inhomogeneous atomic field equations (IV.20–23) for the partial fields $f_{\pm}^{\alpha\beta}$ together with the homogeneous ones,

[1] In the non-relativistic theory this problem did not arise, because the non-relativistic terms were due exclusively to the plus field.

one finds for the unexpanded minus field part a divergence $-\partial_\beta T^{\alpha\beta}_{(m)VI}$, with the abbreviation

$$T^{\alpha\beta}_{(m)VI} \equiv -c^{-2}\sum_{i,j}\int \{\hat{f}^{\alpha\gamma}_{-2j}(R+r_{1i})\hat{f}^{\cdot\beta}_{+1i\gamma}(R+r_{1i})+\hat{f}^{\alpha\gamma}_{+1i}(R+r_{1i})\hat{f}^{\cdot\beta}_{-2j\gamma}(R+r_{1i})$$
$$+\tfrac{1}{2}g^{\alpha\beta}\hat{f}_{+1i\gamma\varepsilon}(R+r_{1i})\hat{f}^{\gamma\varepsilon}_{-2j}(R+r_{1i})\}f_2(1,2)\mathrm{d}1\,\mathrm{d}2$$
$$+\frac{c^{-2}}{4\pi}\sum_{i,j}e_{1i}\int \hat{f}^{\alpha\beta}_{-2j}(R+r_{1i})u_{1i}\cdot\partial\delta\{(R-X_1)^2\}f_2(1,2)\mathrm{d}1\,\mathrm{d}2, \quad (101)$$

Likewise one arrives by similar steps at the expression $-\partial_\beta T^{\alpha\beta}_{(m)VII}$ for the multipole expanded minus field short range part of the first term at the right-hand side of (86). The seventh contribution to the material energy–momentum tensor, which occurs here, has the form

$$T^{\alpha\beta}_{(m)VII} \equiv -\frac{c^{-1}}{4\pi}\sum_{n=1}^{\infty}\int\left[\left\{\mathfrak{m}_1^{\alpha_1\ldots\alpha_n\zeta}\partial_{\alpha_n}+c^{-2}\left(\frac{\mathrm{d}}{\mathrm{d}s_1}+u_1\cdot\partial\right)\mathfrak{m}_1^{\alpha_1\ldots\alpha_n\zeta}u_{1\alpha_n}\right\}\right.$$
$$\left.\partial_{\alpha_1\ldots\alpha_{n-1}}\hat{f}^{\gamma\varepsilon}_{-2(m)}\right]$$
$$\{g^\alpha_\gamma(\partial^\beta g_{\varepsilon\zeta}+g^\beta_\varepsilon\partial_\zeta-g^\beta_\zeta\partial_\varepsilon)+g^\beta_\gamma(\partial^\alpha g_{\varepsilon\zeta}-g^\alpha_\zeta\partial_\varepsilon)+g^{\alpha\beta}g_{\gamma\zeta}\partial^\varepsilon\}$$
$$\delta\{(R-X_1)^2\}f_2(1,2)\mathrm{d}1\,\mathrm{d}2. \quad (102)$$

In this way all short range parts of the right-hand side of (86) have been evaluated as divergences of various contributions to the material energy–momentum tensor.

The correlated long range parts of the right-hand side of (86) may be discussed along similar lines, the only difference being that one has to confine oneself to systems with short range correlations[1]. For such systems one may assume the validity of the generalized Irving–Kirkwood approximation, which states that the correlation function is slowly varying over distances that may be compared with the correlation length[2]. Then one may write a Taylor expansion for the correlation function (cf. (96)) and break it off after the second term. As a result one obtains for the correlated long range part of (86) a term $-\partial_\beta T^{\alpha\beta}_{(m)VIII}$. The material energy–momentum tensor appearing here consists of various contributions: first a term like (93) but with $\hat{\mathfrak{f}}^{S\alpha}_{1,2}$ and $\hat{\mathfrak{d}}^{S\alpha\beta}_{1;2}$ replaced by $\hat{\mathfrak{f}}^{L\alpha}_{1;2(mm)}$ and $\hat{\mathfrak{d}}^{L\alpha\beta}_{1;2(mm)}$ (v. (IV.202–203)), and $f_2(1,2)$ replaced

[1] The name long range referred to the atomic quantity. If such a quantity is multiplied by the correlation function, the long range character need no longer prevail: in fact if no long range correlations are present in the system, the correlation function and therefore also the product with the long range atomic quantity will have short range character.
[2] If long range correlations are present one may employ an artifice like that of chapter II, section 5*h* (cf. problem 9).

by $c_2(1, 2)$; secondly a term like (98) with (95) with similar alterations; thirdly a term like the multipole expanded part of (100) but with $f_2(1, 2)$ replaced by $-c_2(1, 2)$; and fourthly a term like (102) again with the same replacement.

In this way all contributions to the energy–momentum law (86) have been written in the form of divergences, so that we have obtained a conservation law of energy–momentum for a system of neutral atoms without long range correlations in an external electromagnetic field:

$$\partial_\beta T^{\alpha\beta} = 0 \tag{103}$$

with

$$T^{\alpha\beta} = T^{\alpha\beta}_{(\mathrm{f})} + T^{\alpha\beta}_{(\mathrm{m})}. \tag{104}$$

The energy–momentum tensor $T^{\alpha\beta}$ consists of nine contributions, which have been classified as one contribution $T^{\alpha\beta}_{(\mathrm{f})}$ (92), which depended solely on the macroscopic (Maxwell) fields, the polarizations and the velocities, and eight other terms. The latter form together what has been called here the material energy–momentum tensor $T^{\alpha\beta}_{(\mathrm{m})}$.

The case $\alpha = 0$ of (103) represents the energy conservation law which may be written (with $\partial_0 = \partial/\partial ct$ and $\partial_i = \nabla_i = \partial/\partial R^i$, $i = 1, 2, 3$) as

$$\frac{\partial}{\partial t} T^{00} + \nabla_i \cdot c T^{0i} = 0, \tag{105}$$

where $T^{00}$ is the energy density and where $cT^{0i}$ $(i = 1, 2, 3)$ are the components of the energy flow. The cases $\alpha = i = 1, 2, 3$ of (103) form the law of momentum conservation:

$$\frac{\partial}{\partial t} c^{-1} T^{i0} + \nabla_j T^{ij} = 0, \tag{106}$$

where $c^{-1}T^{i0}$ is the momentum density and $T^{ij}$ the momentum flow.

In subsection *d* we shall discuss the components of the energy–momentum tensor in more detail.

### c. *Energy–momentum conservation for a neutral plasma*

In the preceding we considered only one-component systems. The extension to a mixture is in particular necessary if one wants to study neutral plasmas, in which particles with different charges occur. The various species will be labelled by a special index. The starting point for the derivation of the conservation law of energy–momentum for plasmas is the atomic equation

(IV.177), where one has to take the inner angular momentum $s_k^{\alpha\beta}$ and the total torque $\mathfrak{d}_k^{\alpha\beta}$ as zero since the particles are considered to be point charges without structure (for the same reason we may denote the position of the particles simply by $R_k$ instead of $X_k$). The average of this atomic equation becomes then:

$$c\partial_\beta \left\langle \sum_k \int m_k u_k^\alpha u_k^\beta \delta^{(4)}(R_k - R) \mathrm{d}s_k \right\rangle = c \left\langle \sum_k \int \mathfrak{f}_k^\alpha \delta^{(4)}(R_k - R) \mathrm{d}s_k \right\rangle. \tag{107}$$

From the definition (15) with (17) of an average quantity it follows that the left-hand side of this equation can be written as $\partial_\beta T_{(\mathrm{m})\mathrm{I}}^{\alpha\beta}$ with

$$T_{(\mathrm{m})\mathrm{I}}^{\alpha\beta} \equiv \sum_a \int m_a u_1^\alpha u_1^\beta \delta^{(4)}(R_1 - R) f_1^a(1) \mathrm{d}1, \tag{108}$$

where $a$ labels the species. This quantity forms part of the material energy–momentum tensor. At the right-hand side of (107) the total force $\mathfrak{f}_k^\alpha$ on atom $k$ appears. It is given by (IV.191) with (IV.193), (IV.182) and (IV.195), but without the short range and the multipole terms:

$$\mathfrak{f}_k^\alpha = c^{-1} e_k F_\mathrm{e}^{\alpha\beta}(R_k) u_{k\beta} + c^{-1} \sum_{l(\neq k)} e_k f_l^{\alpha\beta}(R_k) u_{k\beta}, \tag{109}$$

where $F_\mathrm{e}^{\alpha\beta}$ is the external field and $f_l^{\alpha\beta}$ the retarded field generated by particle $l$. The latter was given in (IV.14, 15) as:

$$f_l^{\alpha\beta} = -\frac{e_l}{2\pi} \int (u_l^\alpha \partial^\beta - u_l^\beta \partial^\alpha) \delta\{(R - R_l)^2\} \theta(R - R_l) \mathrm{d}s_l. \tag{110}$$

We substitute (109) with (110) into the right-hand side of (107) and make use of the splitting of the two-point distribution function into the product of two one-point distribution functions and a correlation function. The uncorrelated part becomes

$$c^{-1} \sum_a e_a \int F^{\alpha\beta}(R) u_{1\beta} \delta^{(4)}(R_1 - R) f_1^a(1) \mathrm{d}1, \tag{111}$$

where $F^{\alpha\beta}(R)$ is the macroscopic (Maxwell) field. With the definition (60) with (IV.5) of the macroscopic charge–current density $J^\alpha$ this expression becomes

$$c^{-1} F^{\alpha\beta} J_\beta, \tag{112}$$

which is the macroscopic Lorentz force density. By using the inhomogeneous Maxwell equation one then finds $F^{\alpha\beta}\partial_\gamma F_\beta^{\cdot\gamma}$ or, with the homogeneous Maxwell equation, $-\partial_\beta T_{(\mathrm{f})}^{\alpha\beta}$ with the field energy–momentum tensor:

$$T_{(\mathrm{f})}^{\alpha\beta} \equiv -F^{\alpha\gamma} F_\gamma^{\cdot\beta} - \tfrac{1}{4} g^{\alpha\beta} F_{\gamma\varepsilon} F^{\gamma\varepsilon}. \tag{113}$$

These are the contributions which depend solely on the Maxwell fields. We now turn to the correlated part of the right-hand side of (107). Just as in the preceding subsection it will be convenient to split the interatomic field (110) into a plus and a minus part (cf. (IV.17) and (IV.18)). The plus field contribution to the correlated part at the right-hand side of (107) is:

$$-c^{-2}\sum_{a,b}\frac{e_a e_b}{4\pi}\int\{(u_2^\alpha\partial_s^\beta - u_2^\beta\partial_s^\alpha)u_{1\beta}\delta(s^2)\}$$
$$\delta^{(4)}(R_1-R)\delta^{(4)}(R_2-R+s)c_2^{ab}(1,2)\mathrm{d}1\,\mathrm{d}2\,\mathrm{d}s, \quad (114)$$

where we introduced an extra integration over a variable $s^\alpha$ and a four-dimensional delta function $\delta^{(4)}(R_2-R+s)$. We confine ourselves now to the case without long range correlations[1]. For a plasma which is neutral in its proper frame (i.e. in the frame in which $U^\alpha = (c, 0, 0, 0)$) and which is not too far from equilibrium this seems a reasonable assumption. Then we may make a Taylor expansion of the correlation function, just as in the preceding section, and retain only the first few terms. This procedure, which is the relativistic generalization of Irving and Kirkwood's method, brings (114) into the form

$$c^{-2}\sum_{a,b}\frac{e_a e_b}{4\pi}\int\{(u_1\cdot u_2\,\partial_s^\alpha - u_2^\alpha\,u_1\cdot\partial_s)\delta(s^2)\}\delta^{(4)}(R_1-R-\tfrac{1}{2}s)$$
$$\delta^{(4)}(R_2-R+\tfrac{1}{2}s)c_2^{ab}(1,2)\mathrm{d}1\,\mathrm{d}2\,\mathrm{d}s - \partial_\beta T^{\alpha\beta}_{(\mathrm{m})\mathrm{II}}, \quad (115)$$

where a second contribution to the material energy–momentum tensor arises:

$$T^{\alpha\beta}_{(\mathrm{m})\mathrm{II}} \equiv c^{-2}\sum_{a,b}\frac{e_a e_b}{8\pi}\int s^\beta\{(u_1\cdot u_2\,\partial_s^\alpha - u_2^\alpha\,u_1\cdot\partial_s)\delta(s^2)\}$$
$$\delta^{(4)}(R_1-R-\tfrac{1}{2}s)\delta^{(4)}(R_2-R+\tfrac{1}{2}s)c_2^{ab}(1,2)\mathrm{d}1\,\mathrm{d}2\,\mathrm{d}s. \quad (116)$$

The first part of the first term of (115) may be shown to vanish by using the transformation $s^\alpha \leftrightarrow -s^\alpha$, $1 \leftrightarrow 2$. The second part of the first term of (115) may be written after partial integrations, first with respect to $s^\alpha$, and subsequently with respect to $R_1^\alpha$, in the form

$$c^{-2}\sum_{a,b}\frac{e_a e_b}{4\pi}\int u_2^\alpha\,\delta(s^2)\delta^{(4)}(R_1-R-\tfrac{1}{2}s)\delta^{(4)}(R_2-R+\tfrac{1}{2}s)$$
$$u_1\cdot\frac{\partial}{\partial R_1}\,c_2^{ab}(1,2)\mathrm{d}1\,\mathrm{d}2\,\mathrm{d}s - \partial_\beta T^{\alpha\beta}_{(\mathrm{m})\mathrm{III}}. \quad (117)$$

[1] The case with long range correlations may be treated by making use of an artifice as employed in chapter II, section 5*h*.

With the help of the continuity equation for $c_2^{ab}(1, 2)$, which has the form (10), it appears that the first term of (117) vanishes. The second term contains a further contribution to the material energy–momentum tensor

$$T^{\alpha\beta}_{(\mathrm{m})\mathrm{III}} \equiv c^{-2} \sum_{a,b} \frac{e_a e_b}{8\pi} \int u_2^\alpha u_1^\beta \, \delta(s^2)\delta^{(4)}(R_1 - R - \tfrac{1}{2}s) \delta^{(4)}(R_2 - R + \tfrac{1}{2}s) c_2^{ab}(1, 2) \mathrm{d}1\,\mathrm{d}2\,\mathrm{d}s. \quad (118)$$

We finally have to treat the minus field contribution to the correlated part at the right-hand side of (107). It reads, according to (109) and the definition (19) with (18) of an average quantity,

$$c^{-2} \sum_a e_a \int \hat{f}^{\alpha\beta}_{-2}(R_1) u_{1\beta} \, \delta^{(4)}(R_1 - R) c_2^{ab}(1, 2) \mathrm{d}1\,\mathrm{d}2. \quad (119)$$

With the help of the atomic field equations (IV.13), (IV.20) and (IV.22) for the plus and minus fields $\hat{f}^{\alpha\beta}_+$ and $\hat{f}^{\alpha\beta}_-$, this expression may be transformed into a divergence $-\partial_\beta T^{\alpha\beta}_{(\mathrm{m})\mathrm{IV}}$, where a last contribution to the material energy–momentum tensor appears:

$$T^{\alpha\beta}_{(\mathrm{m})\mathrm{IV}} \equiv -c^{-2} \int \{ \hat{f}^{\alpha\gamma}_{-2} \hat{f}^{\cdot\beta}_{+1\gamma} + \hat{f}^{\alpha\gamma}_{+1} \hat{f}^{\cdot\beta}_{-2\gamma} + \tfrac{1}{2} g^{\alpha\beta} \hat{f}_{+1\gamma\varepsilon} \hat{f}^{\gamma\varepsilon}_{-2} \} c_2(1, 2) \mathrm{d}1\,\mathrm{d}2. \quad (120)$$

To summarize the results: the conservation law of energy–momentum for a neutral plasma

$$\partial_\beta T^{\alpha\beta} = 0, \quad (121)$$

with

$$T^{\alpha\beta} = T^{\alpha\beta}_{(\mathrm{f})} + T^{\alpha\beta}_{(\mathrm{m})} \quad (122)$$

has been found. The energy–momentum tensor $T^{\alpha\beta}$ consists of a 'field part' $T^{\alpha\beta}_{(\mathrm{f})}$ (113) and four contributions (108), (116), (118) and (120) to the 'material part' $T^{\alpha\beta}_{(\mathrm{m})}$. The law (121) contains the energy–momentum conservation law and the momentum conservation law as the $\alpha = 0$, and $\alpha = i = 1, 2, 3$ components respectively, as explained at the end of the preceding subsection.

### d. *The macroscopic energy–momentum tensor*

The macroscopic energy–momentum tensors (104) and (122) consist of field and material parts, which have been specified in the preceding.

The macroscopic field energy–momentum tensor for a fluid of dipole atoms in an external field is given in (92) as an expression involving the field and polarization tensors $F^{\alpha\beta}$, $H^{\alpha\beta}$ and $M^{\alpha\beta}$. Its components are the field energy density $T^{00}_{(f)}$, the field energy flow $cT^{0i}_{(f)}$, the field momentum density $c^{-1}T^{i0}_{(f)}$ and the field momentum flow $T^{ij}_{(f)}$. An alternative expression for the field energy–momentum tensor may be obtained if we define the four-vectors $E^\alpha$ and $B^\alpha$ in terms of the field tensor $F^{\alpha\beta}$ as:

$$E^\alpha = c^{-1}F^{\alpha\beta}U_\beta, \tag{123}$$

$$B^\alpha = -\tfrac{1}{2}c^{-1}\varepsilon^{\alpha\beta\gamma\zeta}F_{\beta\gamma}U_\zeta, \tag{124}$$

where $U^\alpha$ is the bulk velocity and $\varepsilon^{\alpha\beta\gamma\zeta}$ the Levi-Civita tensor with $\varepsilon^{0123} = -1$. From these definitions and the antisymmetry of $F^{\alpha\beta}$ and $\varepsilon^{\alpha\beta\gamma\zeta}$ follow the orthogonality relations

$$E_\alpha U^\alpha = 0, \qquad B_\alpha U^\alpha = 0. \tag{125}$$

Equations (123) and (124) may be inverted, with the result

$$F^{\alpha\beta} = c^{-1}(U^\alpha E^\beta - U^\beta E^\alpha + \varepsilon^{\alpha\beta\gamma\zeta}U_\gamma B_\zeta). \tag{126}$$

In the local momentary rest frame (denoted by (0)), where $U^\alpha = (c, 0, 0, 0)$ the definitions (123) and (124) reduce to

$$E^{\alpha(0)} = (0, \boldsymbol{E}^{(0)}); \qquad B^{\alpha(0)} = (0, \boldsymbol{B}^{(0)}), \tag{127}$$

so that $E^\alpha$ and $B^\alpha$ are four-vectors of which the space components in the local momentary rest frame are the electric and magnetic field respectively. In the observer's frame, where $U^\alpha = (\gamma c, \gamma c\boldsymbol{\beta})$, the four-vectors $E^\alpha$ and $B^\alpha$ read in three-dimensional notation:

$$E^\alpha = (\gamma\boldsymbol{\beta}\cdot\boldsymbol{E}, \gamma\boldsymbol{E} + \gamma\boldsymbol{\beta}\wedge\boldsymbol{B}), \tag{128}$$

$$B^\alpha = (\gamma\boldsymbol{\beta}\cdot\boldsymbol{B}, \gamma\boldsymbol{B} - \gamma\boldsymbol{\beta}\wedge\boldsymbol{E}). \tag{129}$$

In an analogous way we define $D^\alpha$ and $H^\alpha$ in terms of the excitation tensor $H^{\alpha\beta}$ as:

$$D^\alpha = c^{-1}H^{\alpha\beta}U_\beta, \tag{130}$$

$$H^\alpha = -\tfrac{1}{2}c^{-1}\varepsilon^{\alpha\beta\gamma\zeta}H_{\beta\gamma}U_\zeta, \tag{131}$$

with the properties

$$D_\alpha U^\alpha = 0, \qquad H_\alpha U^\alpha = 0 \tag{132}$$

and the inverse relation

$$H^{\alpha\beta} = c^{-1}(U^\alpha D^\beta - U^\beta D^\alpha + \varepsilon^{\alpha\beta\gamma\zeta}U_\gamma H_\zeta). \tag{133}$$

In the local momentary rest frame we have

$$D^{\alpha(0)} = (0, \boldsymbol{D}^{(0)}); \qquad H^{\alpha(0)} = (0, \boldsymbol{H}^{(0)}), \tag{134}$$

and in the observer's frame:

$$D^\alpha = (\gamma\boldsymbol{\beta}\cdot\boldsymbol{D},\ \gamma\boldsymbol{D}+\gamma\boldsymbol{\beta}\wedge\boldsymbol{H}), \tag{135}$$

$$H^\alpha = (\gamma\boldsymbol{\beta}\cdot\boldsymbol{H},\ \gamma\boldsymbol{H}-\gamma\boldsymbol{\beta}\wedge\boldsymbol{D}). \tag{136}$$

Finally we introduce $P^\alpha$ and $M^\alpha$ by the definitions involving the macroscopic polarization tensor $M^{\alpha\beta}$:

$$P^\alpha = -c^{-1}M^{\alpha\beta}U_\beta, \tag{137}$$

$$M^\alpha = -\tfrac{1}{2}c^{-1}\varepsilon^{\alpha\beta\gamma\zeta}M_{\beta\gamma}U_\zeta, \tag{138}$$

with the properties

$$P_\alpha U^\alpha = 0, \qquad M_\alpha U^\alpha = 0 \tag{139}$$

and the inverse relation

$$M^{\alpha\beta} = c^{-1}(-U^\alpha P^\beta + U^\beta P^\alpha + \varepsilon^{\alpha\beta\gamma\zeta}U_\gamma M_\zeta). \tag{140}$$

In the local momentary rest frame we have

$$P^{\alpha(0)} = (0, \boldsymbol{P}^{(0)}), \qquad M^{\alpha(0)} = (0, \boldsymbol{M}^{(0)}) \tag{141}$$

and in the observer's frame:

$$P^\alpha = (\gamma\boldsymbol{\beta}\cdot\boldsymbol{P},\ \gamma\boldsymbol{P}-\gamma\boldsymbol{\beta}\wedge\boldsymbol{M}), \tag{142}$$

$$M^\alpha = (\gamma\boldsymbol{\beta}\cdot\boldsymbol{M},\ \gamma\boldsymbol{M}+\gamma\boldsymbol{\beta}\wedge\boldsymbol{P}). \tag{143}$$

Since $H^{\alpha\beta} = F^{\alpha\beta} - M^{\alpha\beta}$ the four-vectors $E^\alpha$, $B^\alpha$, $D^\alpha$, $H^\alpha$, $P^\alpha$ and $M^\alpha$ are connected by the identities

$$E^\alpha + P^\alpha = D^\alpha, \qquad B^\alpha - M^\alpha = H^\alpha. \tag{144}$$

If we introduce the expressions (126), (133) and (140) into the macroscopic field energy–momentum tensor (92) we obtain an alternative expression for $T^{\alpha\beta}_{(\mathrm{f})}$:

$$\begin{aligned} T^{\alpha\beta}_{(\mathrm{f})} = {} & -E^\alpha D^\beta - H^\alpha B^\beta + \Delta^{\alpha\beta}(\tfrac{1}{2}E_\gamma E^\gamma + \tfrac{1}{2}B_\gamma B^\gamma - M_\gamma B^\gamma) \\ & + \tfrac{1}{2}c^{-2}U^\alpha U^\beta(E_\gamma E^\gamma + B_\gamma B^\gamma) - c^{-2}U^\alpha \varepsilon^{\beta\gamma\zeta\eta}E_\gamma H_\zeta U_\eta - c^{-2}U^\beta \varepsilon^{\alpha\gamma\zeta\eta}E_\gamma H_\zeta U_\eta . \end{aligned} \tag{145}$$

This expression shows that $T^{\alpha\beta}_{(\mathrm{f})}$ is in general asymmetric since $E^\alpha D^\beta + H^\alpha B^\beta$ is asymmetric. If however the medium is isotropic as far as polarization and magnetization are concerned, which means that $\boldsymbol{P}^{(0)} = \kappa\boldsymbol{E}^{(0)}$ and $\boldsymbol{M}^{(0)} = \chi\boldsymbol{B}^{(0)}$ (with susceptibilities $\kappa$ and $\chi$, which may depend on $\boldsymbol{E}^{(0)2}$ and $\boldsymbol{B}^{(0)2}$), it follows from (127) and (141) that

$$P^{\alpha} = \kappa E^{\alpha}, \qquad M^{\alpha} = \chi B^{\alpha}. \tag{146}$$

As a consequence of (144) these equalities imply that

$$D^{\alpha} = \varepsilon E^{\alpha}, \qquad H^{\alpha} = \mu^{-1} B^{\alpha}, \tag{147}$$

with the dielectric constant $\varepsilon = 1+\kappa$ and the (reciprocal) permeability $\mu^{-1} = 1-\chi$. With these relations it follows that $T^{\alpha\beta}_{(\mathrm{f})}$ is symmetric for substances that are isotropic as far as polarization and magnetization are concerned.

In the local momentary rest frame the components of the energy–momentum tensor (145) read in three-dimensional notation (with $i, j = 1, 2, 3$):

$$\begin{pmatrix} T^{00}_{(\mathrm{f})} & T^{0i}_{(\mathrm{f})} \\ T^{i0}_{(\mathrm{f})} & T^{ij}_{(\mathrm{f})} \end{pmatrix} = \begin{pmatrix} \frac{1}{2}(\boldsymbol{E}^2+\boldsymbol{B}^2) & (\boldsymbol{E}\wedge\boldsymbol{H})^i \\ (\boldsymbol{E}\wedge\boldsymbol{H})^i & -E^iD^j - H^iB^j + (\frac{1}{2}\boldsymbol{E}^2+\frac{1}{2}\boldsymbol{B}^2-\boldsymbol{M}\cdot\boldsymbol{B})g^{ij} \end{pmatrix}, \tag{148}$$

where we omitted the superscript (0) for brevity's sake.

For electric dipole substances ($\boldsymbol{M} = 0$ in the local momentary rest frame) the results (148) for $T^{i0}_{(\mathrm{f})}$ and $T^{ij}_{(\mathrm{f})}$ were already given by Lorentz[1] and by Einstein and Laub[1] on the basis of electron-theoretical arguments. Minkowski's[1] and Abraham's[1] tensors differ essentially from (148); both have for $T^{00}_{(\mathrm{f})}$ and in the bracket of $T^{ij}_{(\mathrm{f})}$ the expression $\frac{1}{2}\boldsymbol{E}\cdot\boldsymbol{D}+\frac{1}{2}\boldsymbol{B}\cdot\boldsymbol{H}$, Minkowski writes for $T^{i0}_{(\mathrm{f})}$ the vector $(\boldsymbol{D}\wedge\boldsymbol{B})^i$ and Abraham symmetrizes the pressure tensor $T^{ij}_{(\mathrm{f})}$ even for anisotropic substances. (For a discussion and for later literature v. section 7.)

The simple expression (148) is valid only in the local momentary rest frame. The general expression (145) contains the velocity $c\boldsymbol{\beta}$, taken at the observer's point $(ct, \boldsymbol{R})$. Its components read in three-dimensional notation (with $i, j = 1, 2, 3$):

$$T^{00}_{(\mathrm{f})} = \tfrac{1}{2}\boldsymbol{E}^2+\tfrac{1}{2}\boldsymbol{B}^2+\boldsymbol{P}\cdot\boldsymbol{E}-\gamma^2\boldsymbol{\beta}\cdot(\boldsymbol{P}\wedge\boldsymbol{B}-\boldsymbol{M}\wedge\boldsymbol{E})-\gamma^4(\boldsymbol{P}-\boldsymbol{\beta}\wedge\boldsymbol{M})\cdot\boldsymbol{\Omega}^2\cdot(\boldsymbol{E}+\boldsymbol{\beta}\wedge\boldsymbol{B}), \tag{149}$$

$$T^{0i}_{(\mathrm{f})} = (\boldsymbol{E}\wedge\boldsymbol{H})^i-\gamma^2\boldsymbol{\beta}\cdot(\boldsymbol{P}\wedge\boldsymbol{B}-\boldsymbol{M}\wedge\boldsymbol{E})\beta^i-\gamma^4(\boldsymbol{P}-\boldsymbol{\beta}\wedge\boldsymbol{M})\cdot\boldsymbol{\Omega}^2\cdot(\boldsymbol{E}+\boldsymbol{\beta}\wedge\boldsymbol{B})\beta^i, \tag{150}$$

$$T^{i0}_{(\mathrm{f})} = (\boldsymbol{E}\wedge\boldsymbol{H})^i-\gamma^2\boldsymbol{\beta}^2(\boldsymbol{P}\wedge\boldsymbol{B}-\boldsymbol{M}\wedge\boldsymbol{E})^i+\gamma^2\{\boldsymbol{\beta}\wedge(\boldsymbol{P}\wedge\boldsymbol{E}+\boldsymbol{M}\wedge\boldsymbol{B})\}^i -\gamma^4(\boldsymbol{P}-\boldsymbol{\beta}\wedge\boldsymbol{M})\cdot\boldsymbol{\Omega}^2\cdot(\boldsymbol{E}+\boldsymbol{\beta}\wedge\boldsymbol{B})\beta^i, \tag{151}$$

[1] H. A. Lorentz, Enc. Math. Wiss. V 2, fasc. 1 (Teubner, Leipzig 1904) 245; A. Einstein and J. Laub, Ann. Physik **26**(1908)541; H. Minkowski, Nachr. Ges. Wiss. Göttingen (1908) 53; Math. Ann. **68**(1910)472; M. Abraham, R. C. Circ. Mat. Palermo **28**(1909)1, **30**(1910) 33; Theorie der Elektrizität II (Teubner, Leipzig 1923) 300.

$$T_{(f)}^{ij} = -E^i D^j - H^i B^j + (\tfrac{1}{2}\boldsymbol{E}^2 + \tfrac{1}{2}\boldsymbol{B}^2 - \boldsymbol{M}\cdot\boldsymbol{B})g^{ij}$$
$$+\gamma^2\{\boldsymbol{\beta}\wedge(\boldsymbol{P}\wedge\boldsymbol{E}+\boldsymbol{M}\wedge\boldsymbol{B})-\boldsymbol{P}\wedge\boldsymbol{B}+\boldsymbol{M}\wedge\boldsymbol{E}\}^i\beta^j$$
$$-\gamma^4(\boldsymbol{P}-\boldsymbol{\beta}\wedge\boldsymbol{M})\cdot\boldsymbol{\Omega}^2\cdot(\boldsymbol{E}+\boldsymbol{\beta}\wedge\boldsymbol{B})\beta^i\beta^j, \quad (152)$$

where $\gamma = (1-\boldsymbol{\beta}^2)^{-\frac{1}{2}}$ and $\boldsymbol{\Omega}^2 = \mathsf{U} - \boldsymbol{\beta\beta}$ (v. (IV.A12)). In the non-relativistic limit one is interested in the quantities $T_{(f)}^{00}$, $cT_{(f)}^{0i}$, $c^{-1}T_{(f)}^{i0}$ and $T_{(f)}^{ij}$ up to order $c^{-1}$. Using the fact that the magnetization $\boldsymbol{M}$ is of order $c^{-1}$, one finds then from the above formulae the expressions that occur in the non-relativistic energy and momentum laws (II.109) and (II.118).

For a neutral plasma the macroscopic field energy–momentum tensor (113) has a simple form, as compared to (92), because now no polarization terms enter into the expression. The introduction of electric and magnetic field four-vectors (123) and (124) by means of (126) would be unpractical here, because then the four-velocity, which is absent from the original expression (113), would be artificially introduced. The fact that (113) depends only on the fields, not on the four-velocity, implies that its components

$$\begin{pmatrix} T_{(f)}^{00} & T_{(f)}^{0i} \\ T_{(f)}^{i0} & T_{(f)}^{ij} \end{pmatrix} = \begin{pmatrix} \frac{1}{2}(\boldsymbol{E}^2+\boldsymbol{B}^2) & (\boldsymbol{E}\wedge\boldsymbol{B})^i \\ (\boldsymbol{E}\wedge\boldsymbol{B})^i & -E^iE^j - B^iB^j + \frac{1}{2}(\boldsymbol{E}^2+\boldsymbol{B}^2)g^{ij} \end{pmatrix} \quad (153)$$

are form-invariant if a different Lorentz frame is chosen as coordinate system.

The remaining part of the total energy–momentum tensor has been called its material part. For a fluid system of dipole atoms it has been specified as a sum of eight contributions, while in the case of a plasma it consists of four terms. In order to get some insight into their structure it is instructive to study the way in which their non-relativistic limit is reached. It then turns out that for the dipole case the contributions (87), (90), (98–100) and the corresponding correlation terms lead (apart from rest energy terms) to the kinetic terms (labelled by K in chapter II), the field-dependent terms (F), the short range terms (S) and the correlation terms (C) of the non-relativistic approximation (cf. problem 8). The other terms of the material energy–momentum tensor give no contributions in the non-relativistic limit. Likewise one finds for neutral plasmas that the contributions (108) and (116–118) lead to the non-relativistic kinetic (K) and correlation terms (C) respectively (v. problem 7).

### e. *The ponderomotive force density*

The macroscopic conservation laws of energy and momentum (103) or (121) may be formulated in the form of a balance equation

$$\partial_\beta T^{\alpha\beta}_{(m)} = F^\alpha, \tag{154}$$

where $F^\alpha$ is defined as

$$F^\alpha \equiv -\partial_\beta T^{\alpha\beta}_{(f)}, \tag{155}$$

and is called the ponderomotive force density.

For a fluid system of neutral atoms the ponderomotive force that corresponds – according to (155) – to the field energy–momentum tensor (92) follows with (91):

$$F^\alpha = \tfrac{1}{2}(\partial^\alpha F^{\beta\gamma})M_{\beta\gamma} - c^{-2}\partial_\beta\{U^\beta(F^{\alpha\gamma}M_{\gamma\varepsilon}U^\varepsilon - \Delta^{\alpha\gamma}M_{\gamma\varepsilon}F^{\varepsilon\zeta}U_\zeta)\}, \tag{156}$$

where the projector $\Delta^{\alpha\beta}$ was defined as $g^{\alpha\beta}+c^{-2}U^\alpha U^\beta$. If we introduce the operator

$$D \equiv U^\alpha \partial_\alpha \tag{157}$$

and the specific volume

$$v' \equiv (\varrho')^{-1}, \tag{158}$$

we may write (156) in the form:

$$F^\alpha = \tfrac{1}{2}(\partial^\alpha F^{\beta\gamma})M_{\beta\gamma} - c^{-2}\varrho' D\{v'(F^{\alpha\beta}M_{\beta\gamma}U^\gamma - \Delta^{\alpha\beta}M_{\beta\gamma}F^{\gamma\varepsilon}U_\varepsilon)\}, \tag{159}$$

where (85) has been used.

If the four-vectors $E^\alpha$, $B^\alpha$, $P^\alpha_v \equiv v'P^\alpha$ and $M^\alpha_v \equiv v'M^\alpha$ are introduced with the help of (126) and (140) we obtain an alternative form for the ponderomotive force density:

$$\begin{aligned} F^\alpha = \varrho'[(\partial^\alpha E_\beta)P^\beta_v + (\partial^\alpha B_\beta)M^\beta_v - c^{-2}\varepsilon^{\alpha\beta\gamma\zeta} D\{(P_{v\beta}B_\gamma - M_{v\beta}E_\gamma)U_\zeta\} \\ - c^{-2}(\partial^\alpha U^\beta)\varepsilon_{\beta\gamma\zeta\eta}(P^\gamma_v B^\zeta - M^\gamma_v E^\zeta)U^\eta + c^{-2}D(U^\alpha E_\beta P^\beta_v), \end{aligned} \tag{160}$$

where we used (125), (139) and the identity $(\partial^\alpha U^\beta)U_\beta = 0$, which follows from $U_\alpha U^\alpha = -c^2$. Contraction with $U^\alpha$ yields the relation:

$$U_\alpha F^\alpha = -\varrho' E_\alpha D P^\alpha_v + (DB_\alpha)M^\alpha. \tag{161}$$

The components of the ponderomotive force density may be written in three-dimensional notation. From (160) with (128–129) and (142–143), or directly from (159), one finds with $U^\alpha = c\gamma(1, \boldsymbol{\beta})$:

$$\begin{aligned} F^0 = -(\partial_0 \boldsymbol{E})\cdot\boldsymbol{P} - (\partial_0 \boldsymbol{B})\cdot\boldsymbol{M} + \varrho'\gamma \mathrm{d}_0\{\gamma\boldsymbol{\beta}\cdot(\boldsymbol{P}_v \wedge \boldsymbol{B} - \boldsymbol{M}_v \wedge \boldsymbol{E})\} \\ + \varrho'\gamma \mathrm{d}_0\{\gamma^3(\boldsymbol{P}_v - \boldsymbol{\beta}\wedge\boldsymbol{M}_v)\cdot\boldsymbol{\Omega}^2\cdot(\boldsymbol{E}+\boldsymbol{\beta}\wedge\boldsymbol{B})\}, \end{aligned} \tag{162}$$

$$\begin{aligned}\boldsymbol{F} = (\nabla \boldsymbol{E})\cdot \boldsymbol{P} + (\nabla \boldsymbol{B})\cdot \boldsymbol{M} + \varrho' \gamma \mathrm{d}_0 \{\gamma(\boldsymbol{P}_v \wedge \boldsymbol{B} - \boldsymbol{M}_v \wedge \boldsymbol{E})\} \\ - \varrho' \gamma \mathrm{d}_0 \{\gamma \boldsymbol{\beta} \wedge (\boldsymbol{P}_v \wedge \boldsymbol{E} + \boldsymbol{M}_v \wedge \boldsymbol{B})\} \\ + \varrho' \gamma \mathrm{d}_0 \{\gamma^3 \boldsymbol{\beta}(\boldsymbol{P}_v - \boldsymbol{\beta} \wedge \boldsymbol{M}_v)\cdot \boldsymbol{\Omega}^2 \cdot (\boldsymbol{E} + \boldsymbol{\beta} \wedge \boldsymbol{B})\}, \quad (163)\end{aligned}$$

where $c\mathrm{d}_0$ is the substantial time derivative $c(\partial_0 + \boldsymbol{\beta}\cdot\nabla)$, $\boldsymbol{\Omega}^2 = \mathbf{U} - \boldsymbol{\beta\beta}$ and $\boldsymbol{P}_v$, $\boldsymbol{M}_v$ are the specific polarizations $v'\boldsymbol{P}$ and $v'\boldsymbol{M}$. These expressions get simple forms in the local momentary rest frame in which the local macroscopic velocity vanishes:

$$F^0 = \varrho' \boldsymbol{E}\cdot\partial_0 \boldsymbol{P}_v - (\partial_0 \boldsymbol{B})\cdot \boldsymbol{M} + 2(\partial_0 \boldsymbol{\beta})\cdot(\boldsymbol{E} \wedge \boldsymbol{M}), \quad (164)$$

$$\begin{aligned}\boldsymbol{F} = (\nabla \boldsymbol{E})\cdot \boldsymbol{P} + (\nabla \boldsymbol{B})\cdot \boldsymbol{M} + \varrho' \partial_0(\boldsymbol{P}_v \wedge \boldsymbol{B} - \boldsymbol{M}_v \wedge \boldsymbol{E}) \\ - (\partial_0 \boldsymbol{\beta}) \wedge (\boldsymbol{P} \wedge \boldsymbol{E} + \boldsymbol{M} \wedge \boldsymbol{B}) + (\partial_0 \boldsymbol{\beta})\boldsymbol{E}\cdot\boldsymbol{P}. \quad (165)\end{aligned}$$

In these expressions relativistic effects containing the acceleration occur. In the special case that $\boldsymbol{\beta}$ is constant in time and space one finds for the components of the ponderomotive force density in the rest frame:

$$F^0 = \boldsymbol{E}\cdot\partial_0 \boldsymbol{P} - (\partial_0 \boldsymbol{B})\cdot \boldsymbol{M}, \quad (166)$$

$$\boldsymbol{F} = (\nabla \boldsymbol{E})\cdot \boldsymbol{P} + (\nabla \boldsymbol{B})\cdot \boldsymbol{M} + \partial_0(\boldsymbol{P} \wedge \boldsymbol{B} - \boldsymbol{M} \wedge \boldsymbol{E}), \quad (167)$$

where we used the fact that $\partial_0 v'$ vanishes (since it is equal to $v'\nabla\cdot\boldsymbol{\beta}$ as follows from mass conservation in the rest frame).

From the general expressions (162) and (163) one may derive the non-relativistic and semi-relativistic expressions for the ponderomotive force. The latter follow by retaining terms of order $c^{-1}$ in $cF^0$ and $\boldsymbol{F}$ and considering the polarization and magnetization as being of order $c^0$. By using mass conservation one finds for $cF^0$ and $\boldsymbol{F}$ the expressions:

$$cF^0 = \frac{\partial \boldsymbol{P}}{\partial t}\cdot \boldsymbol{E} - \frac{\partial \boldsymbol{B}}{\partial t}\cdot \boldsymbol{M} + \nabla\cdot(\boldsymbol{vP}\cdot\boldsymbol{E}) + \frac{2\partial}{c\partial t}\{\boldsymbol{v}\cdot(\boldsymbol{E} \wedge \boldsymbol{M})\} + \frac{2}{c}\nabla\cdot\{\boldsymbol{vv}\cdot(\boldsymbol{E} \wedge \boldsymbol{M})\}, \quad (168)$$

$$\boldsymbol{F} = (\nabla \boldsymbol{E})\cdot \boldsymbol{P} + (\nabla \boldsymbol{B})\cdot \boldsymbol{M} + \frac{\partial}{c\partial t}(\boldsymbol{P} \wedge \boldsymbol{B} - \boldsymbol{M} \wedge \boldsymbol{E}) + \frac{1}{c}\nabla\cdot\{\boldsymbol{v}(\boldsymbol{P} \wedge \boldsymbol{B} - \boldsymbol{M} \wedge \boldsymbol{E})\}. \quad (169)$$

The non-relativistic expressions (II.114) and (II.106) follow from these by considering the magnetization as being of order $c^{-1}$ and again retaining terms of order $c^{-1}$. The difference between the non-relativistic and the semi-relativistic ponderomotive force densities is that the latter contains the magnetodynamic effect with the vector product of the magnetization $\boldsymbol{M}$ and the electric field $\boldsymbol{E}$.

For plasmas the ponderomotive force density that corresponds – according to (155) – to the field energy–momentum tensor (113) follows with (112) as:

$$F^{\alpha} = c^{-1} F^{\alpha\beta} J_{\beta}. \tag{170}$$

This is the Lorentz force density with components:

$$F^{0} = c^{-1} \boldsymbol{E} \cdot \boldsymbol{J}, \tag{171}$$

$$\boldsymbol{F} = \varrho^{e} \boldsymbol{E} + c^{-1} \boldsymbol{J} \wedge \boldsymbol{B}. \tag{172}$$

(The charge density in the proper frame of $U^{\alpha}$ was supposed to vanish; this need not be the case in other Lorentz frames.)

## 5 *The conservation of angular momentum*

### a. *The balance equation of inner angular momentum*

The inner angular momentum law for a system of neutral atoms will follow by taking the average of the atomic law (IV.178):

$$c\partial_{\gamma} \left\langle \sum_{k} \int s_{k}^{\alpha\beta} u_{k}^{\gamma} \delta^{(4)}(X_{k} - R) \mathrm{d}s_{k} \right\rangle$$
$$= c \left\langle \sum_{k} \int \{ \Delta_{k\gamma}^{\alpha} \Delta_{k\varepsilon}^{\beta} \mathfrak{d}_{k}^{\gamma\varepsilon} + c^{-2} (s_{k}^{\alpha\gamma} u_{k}^{\beta} - s_{k}^{\beta\gamma} u_{k}^{\alpha}) \mathfrak{f}_{k\gamma} / m_{k} \} \delta^{(4)}(X_{k} - R) \mathrm{d}s_{k} \right\rangle. \tag{173}$$

The averages in this equation may be written with the help of the covariant distribution functions that have been defined in section 2. The left-hand side becomes according to (15) with (17)

$$\partial_{\gamma} \int s_{1}^{\alpha\beta} u_{1}^{\gamma} \delta^{(4)}(X_{1} - R) f_{1}(1) \mathrm{d}1, \tag{174}$$

where $f_1(1)$ is the one-point distribution function (7). The macroscopic inner angular momentum density is defined as

$$S^{\alpha\beta} \equiv \int s_{1}^{\alpha\beta} \delta^{(4)}(X_{1} - R) f_{1}(1) \mathrm{d}1. \tag{175}$$

By splitting the atomic velocity $u_{1}^{\alpha}$ into the macroscopic velocity $U^{\alpha}$ defined in (84) and a velocity fluctuation $u_{1}^{\alpha} - U^{\alpha}$ we get for (174)

$$\partial_{\gamma} (S^{\alpha\beta} U^{\gamma} + J_{\mathrm{I}}^{\alpha\beta\gamma}), \tag{176}$$

with the abbreviation

$$J_{\mathrm{I}}^{\alpha\beta\gamma} \equiv \int s_1^{\alpha\beta}(u_1^\gamma - U^\gamma)\delta^{(4)}(X_1 - R)f_1(1)\mathrm{d}1. \tag{177}$$

The forces and torques $\mathfrak{f}_k^\alpha$ and $\mathfrak{d}_k^{\alpha\beta}$, which appear at the right-hand side of (173) have been specified in (IV.191–192) as sums of long range and short range contributions. The long range parts are given in (IV.193–194) as the sum of an external field and an interatomic field term. The latter part, being a two-point quantity, is multiplied by a two-point distribution function in (173) if the average is expressed with the help of distribution functions. The two-point distribution function is written in (14) as the sum of an uncorrelated and a correlated part. The sum of the external field part and the uncorrelated part of the long range contribution to the right-hand side follows from (IV.182–183) with (IV.200–203). Taking only dipole contributions and introducing the Maxwell fields we obtain then for this sum of terms

$$\int \Big[ \Delta_{1\gamma}^\alpha \Delta_{1\varepsilon}^\beta (F^{\gamma\zeta}\mathfrak{m}_{1\zeta}^{\cdot\varepsilon} - \mathfrak{m}_1^{\gamma\zeta}F_\zeta^{\cdot\varepsilon}) + \frac{1}{m_1 c^2}(s_1^{\alpha\gamma}u_1^\beta - s_1^{\beta\gamma}u_1^\alpha)\left\{\tfrac{1}{2}(\partial_\gamma F_{\varepsilon\zeta})\mathfrak{m}_1^{\varepsilon\zeta} - c^{-2}\frac{\mathrm{d}}{\mathrm{d}s_1}(F_{\gamma\varepsilon}\mathfrak{m}_1^{\varepsilon\zeta}u_{1\zeta})\right\}\Big] \delta^{(4)}(X_1 - R)f_1(1)\mathrm{d}1. \tag{178}$$

The terms with the macroscopic velocity in the first part of (178) may be written as (twice) the antisymmetric part

$$T_{(\mathrm{f})}^{\alpha\beta} - T_{(\mathrm{f})}^{\beta\alpha} \tag{179}$$

of the field energy–momentum tensor $T_{(\mathrm{f})}^{\alpha\beta}$ (92). The remaining terms of (178) may likewise be identified with (twice) the antisymmetric part of a term[1] of the total energy–momentum tensor, namely

$$T_{(\mathrm{m})\mathrm{II}}^{\alpha\beta} - T_{(\mathrm{m})\mathrm{II}}^{\beta\alpha}, \tag{180}$$

as follows from inspection of (90).

Now that the uncorrelated long range part of the right-hand side of (173) has been found, we consider its short range contribution. By employing two-point distribution functions according to (19) and inserting the atomic expressions (IV.206–207) one obtains for this contribution

$$c^{-1}\int \hat{\mathfrak{d}}_{1;2}^{\mathrm{S}\alpha\beta}\,\delta^{(4)}(X_1 - R)f_2(1,2)\mathrm{d}1\,\mathrm{d}2 + T_{(\mathrm{m})\mathrm{III}}^{\alpha\beta} - T_{(\mathrm{m})\mathrm{III}}^{\beta\alpha}, \tag{181}$$

[1] Note that the kinetic term $T_{(\mathrm{m})\mathrm{I}}^{\alpha\beta}$ of the material energy–momentum tensor does not contribute here, since it is symmetric.

where the tensor $T^{\alpha\beta}_{(m)III}$ appears, which has been given in (93). To discuss the first term of this expression, we split it into a plus field and a minus field part. The former follows by insertion of the atomic formulae (IV.211) with (IV. 205). If the generalized Irving–Kirkwood expansion (96) is employed one may write the plus field part of the first term of (181) as

$$c^{-1}\int \hat{\mathfrak{d}}^{S\alpha\beta}_{+1;2}(s)\delta^{(4)}(X_1-R-\tfrac{1}{2}s)\delta^{(4)}(X_2-R+\tfrac{1}{2}s)f_2(1,2)\mathrm{d}1\,\mathrm{d}2\,\mathrm{d}s-\partial_\gamma J^{\alpha\beta\gamma}_{\mathrm{II}} \quad (182)$$

with the abbreviations

$$\begin{aligned}\hat{\mathfrak{d}}^{S\alpha\beta}_{+1;2}(s) \equiv{}& \frac{c^{-1}}{4\pi}\sum_{i,j} e_{1i}e_{2j}r^{\alpha}_{1i}(u_{1i}\cdot u_{2j}\,\partial^{\beta}_{s}-u^{\beta}_{2j}u_{1i}\cdot\partial_s)\delta\{(s+r_{1i}-r_{2j})^2\}\\ &+\frac{c}{4\pi}\sum_{n,m=1}^{\infty}(-1)^m\Big\{\Delta^{\alpha}_{1\alpha_{n+1}}\mathfrak{m}_1^{\alpha_1\ldots\alpha_{n+1}}\partial_{s\alpha_{n-1}}-(n-1)\mathfrak{m}_1^{\alpha\alpha_1\ldots\alpha_n}\partial_{s\alpha_{n-1}}\\ &\qquad -c^{-2}(n-1)\left(\frac{\mathrm{d}}{\mathrm{d}s_1}+u_1\cdot\partial_s\right)\mathfrak{m}_1^{\alpha\alpha_1\ldots\alpha_n}u_{1\alpha_{n-1}}\Big\}\\ &\Big\{\mathfrak{m}_2^{\beta_1\ldots\beta_{m+1}}\partial_{s\beta_m}-c^{-2}\left(\frac{\mathrm{d}}{\mathrm{d}s_2}-u_2\cdot\partial_s\right)\mathfrak{m}_2^{\beta_1\ldots\beta_{m+1}}u_{2\beta_m}\Big\}\partial_{s\alpha_1\ldots\alpha_{n-2}\beta_1\ldots\beta_{m-1}}\\ &\qquad(\partial^{\beta}_{s}g_{\alpha_n\beta_{m+1}}-g^{\beta}_{\beta_{m+1}}\partial_{s\alpha_n})\delta(s^2)-(\alpha,\beta),\end{aligned} \quad (183)$$

and

$$J^{\alpha\beta\gamma}_{\mathrm{II}} \equiv \tfrac{1}{2}c^{-1}\int s^{\gamma}\hat{\mathfrak{d}}^{S\alpha\beta}_{+1;2}(s)\delta^{(4)}(X_1-R-\tfrac{1}{2}s)\delta^{(4)}(X_2-R+\tfrac{1}{2}s)f_2(1,2)\mathrm{d}1\,\mathrm{d}2\,\mathrm{d}s. \quad (184)$$

The last expression is a contribution to the inner angular momentum flow. If (183) is substituted into the first term of (182) one may distinguish various contributions. In the first place we consider the unexpanded term with $\partial^{\beta}_{s}$. Making use of the symmetry with respect to an interchange of 1 and 2, one may write it as

$$-\frac{c^{-2}}{8\pi}\int\sum_{i,j}e_{1i}e_{2j}s^{\alpha}u_{1i}\cdot u_{2j}[\partial^{\beta}_{s}\delta\{(s+r_{1i}-r_{2j})^2\}]\delta^{(4)}(X_1-R-\tfrac{1}{2}s)\\ \delta^{(4)}(X_2-R+\tfrac{1}{2}s)f_2(1,2)\mathrm{d}1\,\mathrm{d}2\,\mathrm{d}s-(\alpha,\beta), \quad (185)$$

where we used moreover the fact that $(s^{\alpha}+r^{\alpha}_{1i}-r^{\alpha}_{2j})\partial^{\beta}_{s}-(\alpha,\beta)$ acting on the delta function gives a vanishing result. Comparing (185) to (98) with (95) one obtains the result that it is equal to the unexpanded part with $\partial^{\alpha}_{s}$ or $\partial^{\beta}_{s}$ of $T^{\alpha\beta}_{(m)IV}-T^{\beta\alpha}_{(m)IV}$. Likewise one may derive that the corresponding multipole expanded part of (182) with (183) (i.e. again the part with $\partial^{\beta}_{s}$) is equal to the antisymmetric part of the multipole expanded terms with $\partial_s^\alpha$ or $\partial_s^\beta$ in $T^{\alpha\beta}_{(\mathrm{m})\mathrm{IV}} - T^{\beta\alpha}_{(\mathrm{m})\mathrm{IV}}$ given by (98) with (95). The latter result follows in the simplest way by making use of the identity

$$s^\beta \partial^\alpha_{s\alpha_1\ldots\alpha_n\beta_1\ldots\beta_m}\delta(s^2) - (\alpha, \beta) = -\sum_{i=1}^{n} \partial^\alpha_{s\alpha_1\ldots\alpha_{i-1}\alpha_{i+1}\ldots\alpha_n\beta_1\ldots\beta_m}\delta(s^2)\delta^\beta_{\alpha_i}$$

$$-\sum_{j=1}^{m} \partial^\alpha_{s\alpha_1\ldots\alpha_n\beta_1\ldots\beta_{j-1}\beta_{j+1}\ldots\beta_m}\delta(s^2)\delta^\beta_{\beta_j} - (\alpha, \beta), \tag{186}$$

(which follows from $\partial_\alpha\delta(s^2) = 2s_\alpha\delta'(s^2)$). The two parts of $T^{\alpha\beta}_{(\mathrm{m})\mathrm{IV}}$ which we have encountered up to now will be denoted by $T^{\alpha\beta}_{(\mathrm{m})\mathrm{IV}'}$ so that we have found for part of the first term of (182):

$$T^{\alpha\beta}_{(\mathrm{m})\mathrm{IV}'} - T^{\beta\alpha}_{(\mathrm{m})\mathrm{IV}'}. \tag{187}$$

We now consider the remaining parts of (183) (inserted into (182)). They may be transformed (by making use of (186) and of the conservation of probability) in such a way that they become

$$T^{\alpha\beta}_{(\mathrm{m})\mathrm{IV}''} + T^{\alpha\beta}_{(\mathrm{m})\mathrm{V}} - (\alpha, \beta) - \partial_\gamma J^{\alpha\beta\gamma}_{\mathrm{III}}. \tag{188}$$

Here the second part of the fourth together with the fifth part of the energy–momentum tensor appears (v. (98) and (100)). Furthermore we introduced a contribution to the inner angular momentum flow:

$$J^{\alpha\beta\gamma}_{\mathrm{III}} \equiv \frac{c^{-2}}{8\pi}\int u_1^\gamma \Bigg[\sum_{i,j} e_{1i}e_{2j}(r^\alpha_{1i} + \tfrac{1}{2}s^\alpha)u^\beta_{2j}\,\delta\{(s + r_{1i} - r_{2j})^2\}$$

$$+ \sum_{n,m=1}^{\infty} (-1)^n\{n\mathfrak{m}_1^{\alpha\alpha_1\ldots\alpha_n}u_{1\alpha_n} + \tfrac{1}{2}s^\alpha\mathfrak{m}_1^{\alpha_1\ldots\alpha_{n+1}}u_{1\alpha_{n+1}}\,\partial_{s\alpha_n}\}$$

$$\left\{\mathfrak{m}_2^{\beta_1\ldots\beta_m\beta}\partial_{s\beta_m} - c^{-2}\left(\frac{\mathrm{d}}{\mathrm{d}s_2} - u_2\cdot\partial_s\right)\mathfrak{m}_2^{\beta_1\ldots\beta_m\beta}u_{2\beta_m}\right\}\partial_{s\alpha_1\ldots\alpha_{n-1}\beta_1\ldots\beta_{m-1}}\delta(s^2)\Bigg]$$

$$\delta^{(4)}(X_1 - R - \tfrac{1}{2}s)\delta^{(4)}(X_2 - R + \tfrac{1}{2}s)f_2(1, 2)\,\mathrm{d}1\,\mathrm{d}2\,\mathrm{d}s - (\alpha, \beta). \tag{189}$$

As the plus field part of the first term of (181) we have found now from (182), (187) and (188):

$$T^{\alpha\beta}_{(\mathrm{m})\mathrm{IV}} + T^{\alpha\beta}_{(\mathrm{m})\mathrm{V}} - (\alpha, \beta) - \partial_\gamma(J^{\alpha\beta\gamma}_{\mathrm{II}} + J^{\alpha\beta\gamma}_{\mathrm{III}}). \tag{190}$$

Next we consider the minus field contribution to the first term of (181) in which one has to insert (IV.209) with (IV.203). For the unexpanded part (i.e. the part resulting from the minus field part of the first term of (IV.209)) one finds by making use of the atomic field equations (IV.20–23) and of the conservation of probability:

$$T^{\alpha\beta}_{(\mathrm{m})\mathrm{VI}} - (\alpha, \beta) - \partial_\gamma J^{\alpha\beta\gamma}_{\mathrm{IV}}, \tag{191}$$

where $T^{\alpha\beta}_{(\mathrm{m})\mathrm{VI}}$ (101) appears together with the divergence of a further contribution to the inner angular momentum flow:

$$\begin{aligned} J^{\alpha\beta\gamma}_{\mathrm{IV}} \equiv & -c^{-2} \sum_{i,j} \int r^{\alpha}_{1i} \{ \hat{f}^{\beta\varepsilon}_{-2j}(R+r_{1i}) \hat{f}^{\cdot\gamma}_{+1i\varepsilon}(R+r_{1i}) + \hat{f}^{\beta\varepsilon}_{+1i}(R+r_{1i}) \hat{f}^{\cdot\gamma}_{-2j\varepsilon}(R+r_{1i}) \\ & + \tfrac{1}{2} g^{\beta\gamma} \hat{f}_{-2j\varepsilon\zeta}(R+r_{1i}) \hat{f}^{\varepsilon\zeta}_{+1i}(R+r_{1i}) \} f_2(1, 2) \mathrm{d}1\, \mathrm{d}2 \\ & + c^{-2} \sum_{i,j} \frac{e_{1i}}{4\pi} \int r^{\alpha}_{1i} \hat{f}^{\beta\gamma}_{-2j}(R+r_{1i}) u_{1i} \cdot \partial \delta\{(R-X_1)^2\} f_2(1, 2) \mathrm{d}1\, \mathrm{d}2 - (\alpha, \beta) \\ & + c^{-2} \sum_{i,j} \frac{e_{1i}}{2\pi} \int \hat{f}^{\alpha\beta}_{-2j}(R+r_{1i}) \frac{\mathrm{d}r^{\gamma}_{1i}}{\mathrm{d}s_1} \delta\{(R-X_1)^2\} f_2(1, 2) \mathrm{d}1\, \mathrm{d}2. \end{aligned} \tag{192}$$

For the multipole expanded part of the minus field contribution to the first term of (181) one obtains along similar lines:

$$T^{\alpha\beta}_{(\mathrm{m})\mathrm{VII}} - (\alpha, \beta) - \partial_\gamma J^{\alpha\beta\gamma}_{\mathrm{V}} \tag{193}$$

with the material energy–momentum tensor (102) and the inner angular momentum flow

$$\begin{aligned} J^{\alpha\beta\gamma}_{\mathrm{V}} \equiv & -\frac{c^{-1}}{4\pi} \sum_{n=1}^{\infty} \int \Big[ \Big[ \Big\{ \Delta^{\alpha}_{1\alpha_n} \mathrm{m}_1^{\alpha_1 \ldots \alpha_{n-1}\varepsilon\alpha_n} \partial_{\alpha_{n-1}} - (n-1) \mathrm{m}_1^{\alpha\alpha_1 \ldots \alpha_{n-1}\varepsilon} \partial_{\alpha_{n-1}} \\ & + c^{-2}(n-1) \left( \frac{\mathrm{d}}{\mathrm{d}s_1} + u_1 \cdot \partial \right) \mathrm{m}_1^{\alpha\alpha_1 \ldots \alpha_{n-2}\varepsilon\alpha_n} u_{1\alpha_n} \Big\} \partial_{\alpha_1 \ldots \alpha_{n-2}} \hat{f}^{\zeta\eta}_{-2(\mathrm{m})} \Big] \\ & \{ \partial^\beta g_{\varepsilon\zeta} g^\gamma_\eta - \partial^\gamma g^\beta_\zeta g_{\varepsilon\eta} - \partial_\varepsilon g^\beta_\zeta g^\gamma_\eta + \partial_\zeta (g^{\beta\gamma} g_{\varepsilon\eta} - g^\beta_\eta g^\gamma_\varepsilon - g^\beta_\varepsilon g^\gamma_\eta) \} \\ & \delta\{(R-X_1)^2\} f_2(1, 2) \mathrm{d}1\, \mathrm{d}2 - (\alpha, \beta) \\ & - \frac{c^{-1}}{2\pi} \sum_{n=1}^{\infty} \int \Big[ \Big[ \Big\{ \mathrm{m}_1^{\alpha_1 \ldots \alpha_n \gamma} \partial_{\alpha_n} + c^{-2} \left( \frac{\mathrm{d}}{\mathrm{d}s_1} + u_1 \cdot \partial \right) \mathrm{m}_1^{\alpha_1 \ldots \alpha_n \gamma} u_{1\alpha_n} \\ & + c^{-2} u_1^\gamma \mathrm{m}_1^{\alpha_1 \ldots \alpha_{n+1}} u_{1\alpha_{n+1}} \partial_{\alpha_n} \Big\} \partial_{\alpha_1 \ldots \alpha_{n-1}} \hat{f}^{\alpha\beta}_{-2(\mathrm{m})} \Big] \delta\{(R-X_1)^2\} f_2(1, 2) \mathrm{d}1\, \mathrm{d}2. \end{aligned} \tag{194}$$

Now we have found the complete short range part of the right-hand side of (173):

$$T^{\alpha\beta}_{(\mathrm{m})\mathrm{III-VII}} - (\alpha, \beta) - \partial_\gamma J^{\alpha\beta\gamma}_{\mathrm{II-V}}, \tag{195}$$

where roman indices indicate sums of terms. The correlated part of the long range contribution to the right-hand side of (173) may now readily be found, since these terms have the same structure as the multipole expanded part of the short range contribution. By considering separately plus and minus field contributions one finds for systems of which the correlation function has short range character – so that the generalized Kirkwood approximation is valid[1] – as the correlated long range contribution to (173):

$$T^{\alpha\beta}_{(m)\mathrm{VIII}} - (\alpha, \beta) - \partial_\gamma J^{\alpha\beta\gamma}_{\mathrm{VI}}, \tag{196}$$

where the material energy–momentum contribution $T^{\alpha\beta}_{(m)\mathrm{VIII}}$ has been described in section 4*b*, while the inner angular momentum flow $J^{\alpha\beta\gamma}_{\mathrm{VI}}$ consists of three contributions: first a term like (184), but with $\hat{\mathfrak{d}}^{L\alpha\beta}_{+1;2}(s)$ instead of $\hat{\mathfrak{d}}^{S\alpha\beta}_{+1;2}(s)$ and the correlation function $c_2(1, 2)$ instead of the two-point distribution function $f_2(1, 2)$; secondly a term like (189), but only the multipole expanded part of it and $f_2(1, 2)$ replaced by $-c_2(1, 2)$; thirdly a term like (194), again with $-c_2(1, 2)$ instead of $f_2(1, 2)$.

Collecting the results we have reached the balance equation of inner angular momentum:

$$\partial_\gamma(S^{\alpha\beta}U^\gamma) = -\partial_\gamma J^{\alpha\beta\gamma} + T^{\alpha\beta} - T^{\beta\alpha}, \tag{197}$$

where the inner angular momentum flow consists of six contributions, given above, and where the source term is equal to twice the antisymmetric part of the total energy–momentum tensor.

The inner angular momentum law (197) has the form of a local balance equation, not of a conservation law, since in general the total energy–momentum tensor will not be symmetric. This is what one would expect since the total angular momentum contains an orbital part as well. The balance law for the orbital angular momentum density $R^\alpha T^{\beta\gamma} - R^\beta T^{\alpha\gamma}$ follows directly from the conservation of total energy–momentum $\partial_\beta T^{\alpha\beta} = 0$ (cf. (103)):

$$\partial_\gamma(R^\alpha T^{\beta\gamma} - R^\beta T^{\alpha\gamma}) = T^{\beta\alpha} - T^{\alpha\beta}. \tag{198}$$

By taking the sum of (197) and (198) one obtains:

$$\partial_\gamma(R^\alpha T^{\beta\gamma} - R^\beta T^{\alpha\gamma} + S^{\alpha\beta}U^\gamma + J^{\alpha\beta\gamma}) = 0, \tag{199}$$

which is the law of conservation of total angular momentum.

From the local laws (197–199) one may obtain global laws by integrating over three-space and using Gauss's theorem.

If one studies the non-relativistic limit of the inner angular momentum equation (197) one recovers indeed the equation (II.196) of the non-rela-

[1] We note again that the extension to systems with long range correlation presents no difficulties.

tivistic theory. In particular one finds for the non-relativistic limits of the space–space components of $S^{\alpha\beta}$ (175) the expression (II.166) of the non-relativistic treatment (v. problem 10). Furthermore the space–space–space components of $J_{\mathrm{I}}^{\alpha\beta\gamma}$ (177) reduce to $\mathbf{J}_s^{\mathrm{K}}$ given by (II.169) (in fact $J_{\mathrm{I}}^{ijk}$ reduces to $\varepsilon^{ijm} J_{s \cdot m}^{\mathrm{K}k}$ with $\varepsilon^{ijk}$ the Levi-Civita tensor) while the space–space–time component of $J_{\mathrm{I}}^{\alpha\beta\gamma}$ gives no contribution to the non-relativistic inner angular momentum law. Similarly one finds that $J_{\mathrm{II}}^{\alpha\beta\gamma}$ (184) and $J_{\mathrm{VI}}^{\alpha\beta\gamma}$ reduce to $\mathbf{J}_s^{\mathrm{S}}$ (II.180) and $\mathbf{J}_s^{\mathrm{C}}$ (II.183) respectively. The other parts of $J^{\alpha\beta\gamma}$ give no contribution in the non-relativistic limit (cf. problem 12).

For plasmas, where the internal structure of the particles is disregarded, the angular momentum laws reduce to simple forms since no inner angular momentum exists. Correspondingly (twice) the antisymmetric part $T^{\alpha\beta} - T^{\beta\alpha}$ of the total energy–momentum tensor may be written as $\partial_\gamma J^{\alpha\beta\gamma}$, as follows from inspection of its various terms (108), (113), (116), (118) and (120). In fact, only the part (116) is asymmetric; its antisymmetric part may be written as a divergence by making use of the conservation of probability. Therefore one finds analogously to (197) for a plasma the equation

$$T^{\alpha\beta} - T^{\beta\alpha} = \partial_\gamma J^{\alpha\beta\gamma}. \tag{200}$$

In spite of its resemblance to the inner angular momentum law (197) this equation has a different character; it is in fact only an identity valid for the antisymmetric part of the energy–momentum tensor[1]. Combining the identity (200) with the energy–momentum law (121) one finds for the angular momentum the conservation law

$$\partial_\gamma (R^\alpha T^{\beta\gamma} - R^\beta T^{\alpha\gamma} + J^{\alpha\beta\gamma}) = 0. \tag{201}$$

In the non-relativistic limit both the left-hand and the right-hand side of (200) tend separately to zero (v. problem 11).

### b. *The ponderomotive torque density*

The conservation law of total angular momentum (199) contains the total energy–momentum tensor $T^{\alpha\beta}$, which consists of two parts that we have called the ‘field’ and the ‘material’ energy–momentum tensors $T_{(\mathrm{f})}^{\alpha\beta}$ and $T_{(\mathrm{m})}^{\alpha\beta}$.

[1] It is possible to symmetrize the energy–momentum tensor by adding a divergenceless part (v. problem 13). To preserve the analogy one would then also have to change in a corresponding way the expressions for the dipole case. Such a change is feasible, but it leads to lengthy expressions. Moreover for the long range correlation case the non-relativistic limit of the symmetrized tensor would not have the same form as that of chapter II.

An alternative form of the conservation law is thus

$$\partial_\gamma(R^\alpha T^{\beta\gamma}_{(\mathrm{m})} - R^\beta T^{\alpha\gamma}_{(\mathrm{m})} + S^{\alpha\beta}U^\gamma + J^{\alpha\beta\gamma}) = R^\alpha F^\beta - R^\beta F^\alpha + T^{\alpha\beta}_{(\mathrm{f})} - T^{\beta\alpha}_{(\mathrm{f})}, \quad (202)$$

where we introduced the ponderomotive force density defined in (155) and given explicitly in (156) or (159). This formula shows at the right-hand side in the first place the torque exerted by the ponderomotive force density and in the second place a 'ponderomotive torque density'

$$D^{\alpha\beta} \equiv T^{\alpha\beta}_{(\mathrm{f})} - T^{\beta\alpha}_{(\mathrm{f})}. \quad (203)$$

Its explicit forms follow from the field energy–momentum tensor (92):

$$D^{\alpha\beta} = \Delta^\alpha_\varepsilon \Delta^\beta_\zeta (M^{\varepsilon\gamma}F^\zeta_{\cdot\gamma} - F^{\varepsilon\gamma}M^\zeta_{\cdot\gamma}) \quad (204)$$

or, written in terms of the field and polarization four-vectors (123–124) and (137–138)

$$D^{\alpha\beta} = P^\alpha E^\beta - P^\beta E^\alpha + M^\alpha B^\beta - M^\beta B^\alpha. \quad (205)$$

Its components read in three-dimensional notation (where $U^\alpha = c\gamma(1, \boldsymbol{\beta})$):

$$D^{0i} = [-\gamma^2\boldsymbol{\beta}\wedge\{\boldsymbol{\beta}\wedge(\boldsymbol{P}\wedge\boldsymbol{B}-\boldsymbol{M}\wedge\boldsymbol{E})\} - \gamma^2\boldsymbol{\beta}\wedge(\boldsymbol{P}\wedge\boldsymbol{E}+\boldsymbol{M}\wedge\boldsymbol{B})]^i, \quad (206)$$

$$D^{ij} = \{\gamma^2\boldsymbol{\Omega}^2\cdot(\boldsymbol{P}\wedge\boldsymbol{E}+\boldsymbol{M}\wedge\boldsymbol{B}) + \gamma^2\boldsymbol{\beta}\wedge(\boldsymbol{P}\wedge\boldsymbol{B}-\boldsymbol{M}\wedge\boldsymbol{E})\}^k, \quad (207)$$

where $i, j, k = 1, 2, 3$ (cycl.) and $\boldsymbol{\Omega}^2 = \mathbf{U} - \boldsymbol{\beta}\boldsymbol{\beta}$. In the local momentary rest frame (in which the local macroscopic velocity vanishes) these expressions reduce to

$$D^{0i} = 0, \quad (208)$$

$$D^{ij} = (\boldsymbol{P}\wedge\boldsymbol{E}+\boldsymbol{M}\wedge\boldsymbol{B})^k. \quad (209)$$

For substances which are isotropic as far as the electric and magnetic polarizations are concerned the torque density (209) vanishes.

From the general formula (207) one finds for the ponderomotive torque density in semi-relativistic approximation

$$D^{ij} = \{\boldsymbol{P}\wedge\boldsymbol{E}+\boldsymbol{M}\wedge\boldsymbol{B}+\boldsymbol{\beta}\wedge(\boldsymbol{P}\wedge\boldsymbol{B}-\boldsymbol{M}\wedge\boldsymbol{E})\}^k, \quad (210)$$

with $i, j, k = 1, 2, 3$ (cycl.). The non-relativistic limit follows by taking into account that $\boldsymbol{M}$ is of order $c^{-1}$. Then one recovers (II.189).

## 6 *Relativistic thermodynamics of polarized fluids and plasmas*

### a. *The first law*

In section 4 the macroscopic conservation laws of energy and momentum in a polarized medium have been derived from the atomic conservation laws. For systems with a correlation length which is small compared to macroscopic dimensions the conservation laws read:

$$\partial_\beta(T^{\alpha\beta}_{(\mathrm{f})}+T^{\alpha\beta}_{(\mathrm{m})}) = 0, \qquad (\alpha = 0, 1, 2, 3), \tag{211}$$

with a field energy–momentum tensor $T^{\alpha\beta}_{(\mathrm{f})}$ and a material energy–momentum tensor $T^{\alpha\beta}_{(\mathrm{m})}$. In terms of this material tensor $T^{\alpha\beta}_{(\mathrm{m})}$ we now define a scalar energy density $u'_v$, a heat flow four-vector $J^\alpha_q$, a momentum density four-vector $I^\alpha$ and a pressure four-tensor $P^{\alpha\beta}$ as:

$$u'_v = c^{-2}U_\alpha U_\beta T^{\alpha\beta}_{(\mathrm{m})}-\varrho'c^2, \tag{212}$$

$$J^\alpha_q = -U_\beta T^{\beta\gamma}_{(\mathrm{m})}\Delta^\alpha_\gamma, \tag{213}$$

$$I^\alpha = -c^{-2}\Delta^\alpha_\beta T^{\beta\gamma}_{(\mathrm{m})}U_\gamma, \tag{214}$$

$$P^{\alpha\beta} = \Delta^\alpha_\gamma \Delta^\beta_\varepsilon T^{\varepsilon\gamma}_{(\mathrm{m})}, \tag{215}$$

where $U^\alpha$ and $\varrho'$ are the bulk four-velocity and bulk rest mass density defined in (84), while $\Delta^\alpha_\beta$ stands for $\delta^\alpha_\beta+c^{-2}U^\alpha U_\beta$. From (213–215) the orthogonality relations

$$J^\alpha_q U_\alpha = 0, \qquad I^\alpha U_\alpha = 0, \qquad P^{\alpha\beta}U_\alpha = 0, \qquad P^{\alpha\beta}U_\beta = 0 \tag{216}$$

follow. In the local momentary rest frame, in which $U^\alpha$ has the components $(c, 0, 0, 0)$, the four-vectors $I^\alpha$, $J^\alpha_q$ and the four-tensor $P^{\alpha\beta}$ are hence purely space-like. In this frame the components of $T^{\alpha\beta}_{(\mathrm{m})}$ read:

$$T^{00(0)}_{(\mathrm{m})} = u'_v+\varrho'c^2, \tag{217}$$

$$T^{0i(0)}_{(\mathrm{m})} = c^{-1}J^{i(0)}_q, \tag{218}$$

$$T^{i0(0)}_{(\mathrm{m})} = cI^{i(0)}, \tag{219}$$

$$T^{ij(0)}_{(\mathrm{m})} = P^{ji(0)}, \tag{220}$$

(with $i, j = 1, 2, 3$), as follows from (212–215). In the $(ct, \boldsymbol{R})$-frame the expression for $T^{\alpha\beta}_{(\mathrm{m})}$ in terms of $u'_v$, $J^\alpha_q$, $I^\alpha$ and $P^{\alpha\beta}$ is:

$$T^{\alpha\beta}_{(\mathrm{m})} = c^{-2}(\varrho'c^2+u'_v)U^\alpha U^\beta+c^{-2}U^\alpha J^\beta_q+I^\alpha U^\beta+P^{\beta\alpha}. \tag{221}$$

If the conservation law (211) is multiplied by $U_\alpha$ and (221) is introduced we get:

$$U_\alpha \partial_\beta \{T_{(f)}^{\alpha\beta} + c^{-2}(\varrho' c^2 + u_v') U^\alpha U^\beta + c^{-2} U^\alpha J_q^\beta + I^\alpha U^\beta + P^{\beta\alpha}\} = 0. \tag{222}$$

The first term is equal to $-U_\alpha F^\alpha$ as follows from the definition (155). An explicit expression for this term has been given in (161):

$$-U_\alpha F^\alpha = \varrho' E_\alpha D(v' P^\alpha) - (D B_\alpha) M^\alpha, \tag{223}$$

where $E^\alpha$ and $B^\alpha$ are the field four-vectors defined in (123) and (124), while the polarization four-vectors $P^\alpha$ and $M^\alpha$ have been defined in (137) and (138). The symbol $D$ stands for the operator $U^\alpha \partial_\alpha$. The remaining terms of (222) may be put into the form:

$$-\varrho' D(v' u_v') - \partial_\alpha J_q^\alpha - I_\alpha D U^\alpha - P_{\alpha\beta} \partial^\alpha U^\beta, \tag{224}$$

where we used (85) and the orthogonality properties (216). Introducing the energy per unit rest mass

$$u' = v' u_v' \tag{225}$$

and inserting (223) and (224) we get from (222):

$$\varrho' D u' = -\partial_\alpha J_q^\alpha - I_\alpha D U^\alpha - P_{\alpha\beta} \partial^\alpha U^\beta + \varrho' E_\alpha D(v' P^\alpha) - (D B_\alpha) M^\alpha. \tag{226}$$

This is the first law of relativistic thermodynamics for polarized media; it gives an expression for the change in time of the energy $u'$. The right-hand side contains in the first place the divergence of the heat flow $J_q^\alpha$ together with Eckart's relativistic correction[1] and a term with the pressure tensor $P_{\alpha\beta}$. Furthermore terms with the electromagnetic fields $E^\alpha$, $B^\alpha$ and the polarizations $P^\alpha$, $M^\alpha$ occur.

Likewise one may derive the first law for a relativistic neutral plasma. One obtains

$$\varrho' D u' = -\partial_\alpha J_q^\alpha - I_\alpha D U^\alpha - P_{\alpha\beta} \partial^\alpha U^\beta + J_\alpha E^\alpha, \tag{227}$$

with $J^\alpha$ the electric four-current density.

### b. *The second law*

In chapter II the non-relativistic Gibbs relation has been derived from equilibrium statistical thermodynamics with the help of a canonical ensemble. Since no statistical derivation of a second law for relativistic systems (with

[1] C. Eckart, Phys. Rev. **58**(1940)919.

interactions) in equilibrium is available, we postulate in analogy with the non-relativistic law:

$$T'Ds' = Du' + p'Dv' - E_\alpha D(v'P^\alpha) + v'M_\alpha DB^\alpha \tag{228}$$

as the relativistic second law of thermodynamics for a dipole fluid of neutral atoms in local equilibrium. Here $T'$, $s'$, $u'$, $p'$, $v'$ are the temperature, specific entropy, specific energy, scalar equilibrium pressure and specific volume in the permanent local rest frame (denoted by a prime). (This frame in which matter is locally at rest all the time is a succession of Lorentz frames, not a Lorentz frame itself.) Furthermore $E^\alpha$, $B^\alpha$ and $P^\alpha$, $M^\alpha$ are the field and polarization four-vectors defined in (123), (124), (137) and (138). The derivative $D$ stands for $U^\alpha\partial_\alpha$, where $U^\alpha$ is the local bulk four-velocity. The quantities $u'$ and $p'$ are connected with the energy–momentum tensor $T^{\alpha\beta}_{(m)}$. The expression for the specific energy $u'$ follows from (225) with (212). The scalar pressure $p'$ will be connected to the pressure four-tensor $P^{\alpha\beta}$ (215). In fact the space–space part of the material energy–momentum tensor $T^{\alpha\beta}_{(m)}$ in the rest frame reduces in the non-relativistic limit to the pressure tensor $\mathbf{P}$ (v. problem 8), which is a scalar quantity $p$ for a fluid in equilibrium (II, section 7*b*). Taking over this property in the present theory one finds that for a fluid in equilibrium $T^{\alpha\beta}_{(m)}$ is a scalar $p'$ in the local momentary rest frame, as far as its space–space components are concerned. In the observer's frame we may express this, with (215), as:

$$p'\Delta^{\alpha\beta} = (\Delta^\alpha_\gamma \Delta^\beta_\varepsilon T^{\varepsilon\gamma}_{(m)})_{eq} \equiv P^{\alpha\beta}_{eq}. \tag{229}$$

From the combination of the first and second law the relativistic entropy balance may be obtained. In fact substitution of (226) in the right-hand side of (228) leads to the entropy balance equation for a polarized fluid of neutral atoms:

$$\varrho' Ds' = -\partial_\alpha S^\alpha + \sigma, \tag{230}$$

where we introduced the entropy flux:

$$S^\alpha = \frac{1}{T'} J^\alpha_q \tag{231}$$

and the entropy source strength $\sigma$ given by:

$$T'\sigma = -\frac{1}{T'} J_{q\alpha}\partial^\alpha T' - I_\alpha DU^\alpha - (P_{\alpha\beta} - p'\Delta_{\alpha\beta})\partial^\alpha U^\beta + \varrho'(E_\alpha - E_{eq,\alpha})D(v'P^\alpha) - (M_\alpha - M_{eq,\alpha})DB^\alpha. \tag{232}$$

Here we used $\partial_\alpha U^\alpha = \varrho' Dv'$, which is a consequence of the rest mass conservation (85). Furthermore we took into account that in the second law (228) the equilibrium values $E_{\text{eq},\alpha}$ and $M_{\text{eq},\alpha}$ are to be read for $E_\alpha$ and $M_\alpha$.

The entropy flux (231) is equal to the heat flow divided by the local temperature; the entropy source strength (232) contains contributions due to heat conduction, viscous phenomena and electric and magnetic relaxation.

For neutral plasmas we write in analogy with the non-relativistic theory the second law in the form

$$T'Ds' = Du' + p'Dv' \tag{233}$$

with the same connexions between $u'$ and $p'$ and the energy–momentum tensor $T^{\alpha\beta}_{(\text{m})}$ as given above. For the entropy balance one finds the same form as (230) with the entropy flux (231) and the entropy source strength

$$T'\sigma = -\frac{1}{T'} J_{q\alpha}\partial^\alpha T' - I_\alpha DU^\alpha - (P_{\alpha\beta} - p'\Delta_{\alpha\beta})\partial^\alpha U^\beta + J_\alpha E^\alpha, \tag{234}$$

where an extra term that represents the effect of Joule heat production appears.

c. *The free energy for systems with linear constitutive relations*

In this and the following sections some consequences of the relativistic first and second law of thermodynamics will be discussed, especially in connexion with the conservation laws of energy and momentum. The treatment will to some extent be similar to that of the non-relativistic theory of chapter II, § 8*a*. We shall confine ourselves to a polarized fluid of neutral atoms with linear constitutive relations of the form:

$$\begin{aligned} P^\alpha &= \kappa(v', T')E^\alpha, \\ M^\alpha &= \chi(v', T')B^\alpha. \end{aligned} \tag{235}$$

In three-dimensional notation and in the permanent local rest frame these relations read:

$$\begin{aligned} \boldsymbol{P}' &= \kappa(v', T')\boldsymbol{E}', \\ \boldsymbol{M}' &= \chi(v', T')\boldsymbol{B}'. \end{aligned} \tag{236}$$

From the Gibbs relation (228) we may derive an expression for the time derivative of the specific free energy

$$f' \equiv u' - T's'; \tag{237}$$

we obtain:

$$Df' = -p'Dv' - s'DT' + E_\alpha D(v'P^\alpha) - v'M_\alpha DB^\alpha. \tag{238}$$

This relation may be integrated at constant $v'$ and $T'$ with the result:

$$f' = f'_0 + v' \int_{\text{const. } v', T'} (E_\alpha \mathrm{d}P^\alpha - M_\alpha \mathrm{d}B^\alpha), \tag{239}$$

where $f'_0$ is the specific free energy for the same specific volume $v'$ and temperature $T'$ but at zero fields and polarizations. If the constitutive relations (235) are inserted, the integral in (239) may be carried out with the result

$$f' = f'_0 + \tfrac{1}{2}v'\kappa^{-1}P_\alpha P^\alpha - \tfrac{1}{2}v'\chi B_\alpha B^\alpha; \tag{240}$$

if $\kappa$ and $\chi$ are eliminated an alternative form is obtained:

$$f' = f'_0 + \tfrac{1}{2}v'E_\alpha P^\alpha - \tfrac{1}{2}v'B_\alpha M^\alpha. \tag{241}$$

The scalar equilibrium pressure follows from the specific free energy by differentiation with respect to the specific volume $v'$ at constant $T'$, specific polarization $v'P^\alpha$ and field $B^\alpha$, as (228) shows. Hence the pressure $p' = -\partial f'/\partial v'$ is connected with the pressure $p'_0 = -\partial f'_0/\partial v'$ for the same values of $v'$ and $T'$, but with switched-off fields, by a relation following from (240):

$$p' = p'_0 + \tfrac{1}{2}\kappa^{-1}P_\alpha P^\alpha + \tfrac{1}{2}\chi B_\alpha B^\alpha + \frac{v'}{2\kappa^2}\frac{\partial\kappa}{\partial v'} P_\alpha P^\alpha + \tfrac{1}{2}v'\frac{\partial\chi}{\partial v'} B_\alpha B^\alpha, \tag{242}$$

or with (235):

$$p' = p'_0 + \tfrac{1}{2}E_\alpha P^\alpha + \tfrac{1}{2}B_\alpha M^\alpha + \tfrac{1}{2}v'\frac{\partial\kappa}{\partial v'} E_\alpha E^\alpha + \tfrac{1}{2}v'\frac{\partial\chi}{\partial v'} B_\alpha B^\alpha. \tag{243}$$

The specific entropy follows from the specific free energy by differentiation with respect to temperature $T'$ at constant $v'$, $v'P^\alpha$ and $B^\alpha$. From (240) we obtain:

$$s' = s'_0 + \frac{v'}{2\kappa^2}\frac{\partial\kappa}{\partial T'} P_\alpha P^\alpha + \tfrac{1}{2}v'\frac{\partial\chi}{\partial T'} B_\alpha B^\alpha, \tag{244}$$

where $s' = -\partial f'/\partial T'$ and $s'_0 = -\partial f'_0/\partial T'$. Introducing $E^\alpha$ and $B^\alpha$ with (235) we may write this expression as:

$$s' = s'_0 + \tfrac{1}{2}v'\frac{\partial\kappa}{\partial T'} E_\alpha E^\alpha + \tfrac{1}{2}v'\frac{\partial\chi}{\partial T'} B_\alpha B^\alpha. \tag{245}$$

An expression for the specific energy follows from (237) with (241) and (245):

$$u' = u'_0 + \tfrac{1}{2}v'E_\alpha P^\alpha - \tfrac{1}{2}v'B_\alpha M^\alpha + \tfrac{1}{2}v'T'\frac{\partial\kappa}{\partial T'}E_\alpha E^\alpha + \tfrac{1}{2}v'T'\frac{\partial\chi}{\partial T'}B_\alpha B^\alpha, \tag{246}$$

where $u'_0 = f'_0 + T's'_0$ (cf. (237)). The energy density $u'_v = (v')^{-1}u'$ as compared with the energy density $u'_{v0} = (v')^{-1}u'_0$ at zero fields reads:

$$u'_v = u'_{v0} + \tfrac{1}{2}E_\alpha P^\alpha - \tfrac{1}{2}B_\alpha M^\alpha + \tfrac{1}{2}T'\frac{\partial\kappa}{\partial T'}E_\alpha E^\alpha + \tfrac{1}{2}T'\frac{\partial\chi}{\partial T'}B_\alpha B^\alpha. \tag{247}$$

We have obtained now the expressions (243) and (247) for the equilibrium pressure $p'$ and the energy density $u'_v$. The method which is employed here to derive these results is analogous to that given for the non-relativistic case. The formulae (243) and (247) will be used in the next subsection for a discussion of the material energy–momentum tensor.

d. *The energy–momentum tensor for a polarized fluid at local equilibrium*

The material energy–momentum tensor for a polarized fluid of neutral atoms has the general form (221). In view of the expression (232) for the entropy production, we shall suppose that in local equilibrium all thermodynamic flows $J^\alpha_q$, $I^\alpha$, $P^{\alpha\beta} - p'\Delta^{\alpha\beta}$, $E^\alpha - E^\alpha_{\text{eq}}$ and $M^\alpha - M^\alpha_{\text{eq}}$ vanish. Then (221) reduces to

$$T^{\alpha\beta}_{(\text{m})\text{eq}} = c^{-2}(\varrho'c^2 + u'_v)U^\alpha U^\beta + p'\Delta^{\alpha\beta}, \tag{248}$$

which is the energy–momentum tensor for a perfect fluid. The total energy–momentum tensor for a polarized fluid in local equilibrium is the sum of the field part (145) and the material part just given:

$$\begin{aligned} T^{\alpha\beta}_{\text{eq}} = {} & -E^\alpha D^\beta - H^\alpha B^\beta + \Delta^{\alpha\beta}(\tfrac{1}{2}E_\gamma E^\gamma + \tfrac{1}{2}B_\gamma B^\gamma - B_\gamma M^\gamma + p') \\ & + c^{-2}U^\alpha U^\beta(\tfrac{1}{2}E_\gamma E^\gamma + \tfrac{1}{2}B_\gamma B^\gamma + \varrho'c^2 + u'_v) \\ & - c^{-2}U^\alpha \varepsilon^{\beta\gamma\zeta\eta}E_\gamma H_\zeta U_\eta - c^{-2}U^\beta \varepsilon^{\alpha\gamma\zeta\eta}E_\gamma H_\zeta U_\eta. \end{aligned} \tag{249}$$

If the fields are switched off, while the temperature $T'$ and density $\varrho'$ are kept constant, the energy–momentum tensor becomes:

$$T^{\alpha\beta}_{\text{eq},0} = c^{-2}(\varrho'c^2 + u'_{v0})U^\alpha U^\beta + p'_0\Delta^{\alpha\beta}. \tag{250}$$

With the expressions (243) and (247) for $p'$ and $u'_v$, derived for a fluid with linear constitutive relations, the difference $T^{\alpha\beta}_{\text{eq}} - T^{\alpha\beta}_{\text{eq},0}$ between the energy–momentum tensor in the presence and in the absence of electromagnetic fields may be obtained. This tensor contains the complete effect of the switching-on of the fields. In view of this property it may be considered as the field part of the energy–momentum tensor; it will be called $T^{\alpha\beta}_{[f]}$ to distinguish it from $T^{\alpha\beta}_{(f)}$, that has been introduced earlier. The corresponding material energy–momentum tensor $T^{\alpha\beta}_{[m]}$ is then $T^{\alpha\beta}_{eq,0}$. Hence for a polarized fluid of neutral atoms in local equilibrium we have:

$$T^{\alpha\beta} = T^{\alpha\beta}_{[f]} + T^{\alpha\beta}_{[m]}, \tag{251}$$

$$T^{\alpha\beta}_{[m]} \equiv T^{\alpha\beta}_{eq,0}, \qquad T^{\alpha\beta}_{[f]} \equiv T^{\alpha\beta}_{eq} - T^{\alpha\beta}_{eq,0}. \tag{252}$$

We have introduced thus a second way of splitting the total energy–momentum tensor $T^{\alpha\beta}$ into a 'field' and a 'material' part. The difference between this splitting and that of (104) is that here we confine ourselves to (equilibrium) systems with linear constitutive relations. For such systems it is possible to specify the effect of the turning on of the electromagnetic fields. The splitting (104) was valid under more general circumstances, but did not permit the kind of disentangling, achieved with the present splitting of $T^{\alpha\beta}$. We may note here that a similar situation arose already in the non-relativistic theory in connexion with the material pressure tensor which has been defined there in two different ways (v. chapter II, section 8*a*).

From (251) and (252) with (243), (247), (249) and (250) we obtain explicit expressions for $T^{\alpha\beta}_{[f]}$ and $T^{\alpha\beta}_{[m]}$:

$$\begin{aligned} T^{\alpha\beta}_{[f]} = & -E^\alpha D^\beta - H^\alpha B^\beta + \tfrac{1}{2}\Delta^{\alpha\beta}\left(E_\gamma D^\gamma + B_\gamma H^\gamma + v'\frac{\partial\kappa}{\partial v'}E_\gamma E^\gamma + v'\frac{\partial\chi}{\partial v'}B_\gamma B^\gamma\right) \\ & + \tfrac{1}{2}c^{-2}U^\alpha U^\beta\left(E_\gamma D^\gamma + B_\gamma H^\gamma + T'\frac{\partial\kappa}{\partial T'}E_\gamma E^\gamma + T'\frac{\partial\chi}{\partial T'}B_\gamma B^\gamma\right) \\ & - c^{-2}U^\alpha \varepsilon^{\beta\gamma\zeta\eta}E_\gamma H_\zeta U_\eta - c^{-2}U^\beta\varepsilon^{\alpha\gamma\zeta\eta}E_\gamma H_\zeta U_\eta, \end{aligned} \tag{253}$$

$$T^{\alpha\beta}_{[m]} = c^{-2}(\varrho' c^2 + u'_{v0})U^\alpha U^\beta + p'_0\,\Delta^{\alpha\beta}, \tag{254}$$

where (144) has been used.

Often the magnetic susceptibility $\tilde{\chi}$ is defined by

$$M^\alpha = \tilde{\chi}(v', T')H^\alpha. \tag{255}$$

It is connected to $\chi$, defined in (235), by the relation:

$$\tilde{\chi} = \chi/(1-\chi), \tag{256}$$

as follows from (144). From (235), (255) and (256) one proves

$$B_\alpha B^\alpha \frac{\partial\chi}{\partial v'} = H_\alpha H^\alpha \frac{\partial\tilde{\chi}}{\partial v'}; \qquad B_\alpha B^\alpha \frac{\partial\chi}{\partial T'} = H_\alpha H^\alpha \frac{\partial\tilde{\chi}}{\partial T'}. \tag{257}$$

Hence the field energy–momentum tensor (253) may be written alternatively as:

$$T^{\alpha\beta}_{[f]} = -E^\alpha D^\beta - H^\alpha B^\beta + \tfrac{1}{2}\Delta^{\alpha\beta}\left(E_\gamma D^\gamma + B_\gamma H^\gamma + v' \frac{\partial \kappa}{\partial v'} E_\gamma E^\gamma + v' \frac{\partial \tilde{\chi}}{\partial v'} H_\gamma H^\gamma\right)$$

$$+ \tfrac{1}{2}c^{-2}U^\alpha U^\beta \left(E_\gamma D^\gamma + B_\gamma H^\gamma + T' \frac{\partial \kappa}{\partial T'} E_\gamma E^\gamma + T' \frac{\partial \tilde{\chi}}{\partial T'} H_\gamma H^\gamma\right)$$

$$- c^{-2}U^\alpha \varepsilon^{\beta\gamma\zeta\eta} E_\gamma H_\zeta U_\eta - c^{-2}U^\beta \varepsilon^{\alpha\gamma\zeta\eta} E_\gamma H_\zeta U_\eta . \tag{258}$$

This expression shows the symmetry of $T^{\alpha\beta}_{[f]}$ with respect to electric and magnetic phenomena: the equations (144), the first of (235), (255) and (258) are invariant under the transformations:

$$\begin{aligned} &E^\alpha \to H^\alpha; \quad &&P^\alpha \to M^\alpha; \quad &&D^\alpha \to B^\alpha; \quad &&\kappa \to \tilde{\chi};\\ &H^\alpha \to -E^\alpha; \quad &&M^\alpha \to -P^\alpha; \quad &&B^\alpha \to -D^\alpha; \quad &&\tilde{\chi} \to \kappa. \end{aligned} \tag{259}$$

(The Maxwell equations (68) without sources ($J^\alpha = 0$) are also invariant with respect to these transformations as may be proved with the help of (126) and (133).)

In the local permanent rest frame the tensors (253) and (254) take the form:

$$T'^{\alpha\beta}_{[f]} = \begin{bmatrix} \dfrac{1}{2}\left(\boldsymbol{E}'\cdot\boldsymbol{D}' + \boldsymbol{B}'\cdot\boldsymbol{H}' + T'\dfrac{\partial \kappa}{\partial T'}\boldsymbol{E}'^2 + T'\dfrac{\partial \chi}{\partial T'}\boldsymbol{B}'^2\right) & \boldsymbol{E}'\wedge\boldsymbol{H}' \\ \boldsymbol{E}'\wedge\boldsymbol{H}' & -\boldsymbol{E}'\boldsymbol{D}' - \boldsymbol{H}'\boldsymbol{B}' + \dfrac{1}{2}\left(\boldsymbol{E}'\cdot\boldsymbol{D}' + \boldsymbol{B}'\cdot\boldsymbol{H}' + v'\dfrac{\partial \kappa}{\partial v'}\boldsymbol{E}'^2 + v'\dfrac{\partial \chi}{\partial v'}\boldsymbol{B}'^2\right)\mathbf{U} \end{bmatrix}, \tag{260}$$

$$T'^{\alpha\beta}_{[m]} = \begin{bmatrix} \varrho' c^2 + u'_{v0} & 0 \\ 0 & p'_0\,\mathbf{U} \end{bmatrix}. \tag{261}$$

In terms of the susceptibility $\tilde{\chi}$ the field tensor reads:

$$T'^{\alpha\beta}_{[\mathrm{f}]} = \begin{bmatrix} \frac{1}{2}\Big(\boldsymbol{E}'\cdot\boldsymbol{D}'+\boldsymbol{B}'\cdot\boldsymbol{H}' & \boldsymbol{E}'\wedge\boldsymbol{H}' \\ \quad +T'\dfrac{\partial\kappa}{\partial T'}\boldsymbol{E}'^2+T'\dfrac{\partial\tilde{\chi}}{\partial T'}\boldsymbol{H}'^2\Big) & \\ & -\boldsymbol{E}'\boldsymbol{D}'-\boldsymbol{H}'\boldsymbol{B}' \\ & \quad +\frac{1}{2}\Big(\boldsymbol{E}'\cdot\boldsymbol{D}'+\boldsymbol{B}'\cdot\boldsymbol{H}' \\ \boldsymbol{E}'\wedge\boldsymbol{H}' & \quad +v'\dfrac{\partial\kappa}{\partial v'}\boldsymbol{E}'^2+v'\dfrac{\partial\tilde{\chi}}{\partial v'}\boldsymbol{H}'^2\Big)\,\mathsf{U} \end{bmatrix}. \tag{262}$$

The expression (253) for the field energy–momentum tensor gets a simpler form if the properties of the dipole fluid are further specified. This will be done in the next subsection.

### e. *Induced dipole and permanent dipole substances*

In the expressions (243), (247) and (253) for $p'$, $u'_v$ and $T^{\alpha\beta}_{[\mathrm{f}]}$ derivatives of the susceptibilities $\kappa$ and $\chi$ with respect to specific volume $v'$ and temperature $T'$ occur. These derivatives may be expressed in terms of $\kappa$ and $\chi$ themselves if more is known about the properties of the dipole fluid (cf. chapter II, section 8*a* of the non-relativistic treatment).

Let us consider first *induced dipole* substances that satisfy Clausius–Mossotti laws of the type:

$$\frac{\kappa}{\kappa+3} \sim \frac{1}{v'}, \qquad \frac{\chi}{3-2\chi} \sim \frac{1}{v'}, \tag{263}$$

while $\kappa$ and $\chi$ are independent of the temperature $T'$. With the help of these laws the partial derivatives of the susceptibilities may be evaluated; the expressions for $p'$ and $u'_v$ become:

$$p' = p'_0 - \tfrac{1}{6}P_\alpha P^\alpha + \tfrac{1}{3}M_\alpha M^\alpha, \tag{264}$$

$$u'_v = u'_{v0} + \tfrac{1}{2}E_\alpha P^\alpha - \tfrac{1}{2}B_\alpha M^\alpha. \tag{265}$$

The energy–momentum tensor $T^{\alpha\beta}_{[\mathrm{f}]}$ gets the form:

$$\begin{aligned} T^{\alpha\beta}_{[\mathrm{f}]} = & -E^\alpha D^\beta - H^\alpha B^\beta + \Delta^{\alpha\beta}(\tfrac{1}{2}E_\gamma E^\gamma + \tfrac{1}{2}B_\gamma B^\gamma - B_\gamma M^\gamma - \tfrac{1}{6}P_\gamma P^\gamma + \tfrac{1}{3}M_\gamma M^\gamma) \\ & + \tfrac{1}{2}c^{-2}U^\alpha U^\beta(E_\gamma D^\gamma + B_\gamma H^\gamma) - c^{-2}U^\alpha \varepsilon^{\beta\gamma\zeta\eta}E_\gamma H_\zeta U_\eta - c^{-2}U^\beta \varepsilon^{\alpha\gamma\zeta\eta}E_\gamma H_\zeta U_\eta. \end{aligned} \tag{266}$$

(The combination $\frac{1}{2}B_\gamma B^\gamma - B_\gamma M^\gamma + \frac{1}{3}M_\gamma M^\gamma$ may be written alternatively as $\frac{1}{2}H_\gamma H^\gamma - \frac{1}{6}M_\gamma M^\gamma$.) The energy–momentum tensor (266) reads in three-dimensional notation and in the local permanent rest frame:

$$T'^{\alpha\beta}_{[f]} = \begin{bmatrix} \frac{1}{2}(\boldsymbol{E}'\cdot\boldsymbol{D}' + \boldsymbol{B}'\cdot\boldsymbol{H}') & \boldsymbol{E}' \wedge \boldsymbol{H}' \\ & \\ & -\boldsymbol{E}'\boldsymbol{D}' - \boldsymbol{H}'\boldsymbol{B}' \\ \boldsymbol{E}' \wedge \boldsymbol{H}' & +(\frac{1}{2}\boldsymbol{E}'^2 + \frac{1}{2}\boldsymbol{B}'^2 - \boldsymbol{B}'\cdot\boldsymbol{M}' - \frac{1}{6}\boldsymbol{P}'^2 + \frac{1}{3}\boldsymbol{M}'^2)\mathsf{U} \end{bmatrix}. \quad (267)$$

The field energy density for induced dipole fluids is thus $\frac{1}{2}(\boldsymbol{E}'\cdot\boldsymbol{D}' + \boldsymbol{B}'\cdot\boldsymbol{H}')$, while the scalar part of the field pressure tensor contains $\frac{1}{2}\boldsymbol{E}'^2 + \frac{1}{2}\boldsymbol{B}'^2 - \boldsymbol{B}'\cdot\boldsymbol{M}' - \frac{1}{6}\boldsymbol{P}'^2 + \frac{1}{3}\boldsymbol{M}'^2$.

The relativistic first law for induced dipole fluids in equilibrium may be written in terms of $u_0' = v'u_{v0}'$ and $p_0'$. In fact, substituting (264) and (265) into equation (226), we obtain as the first law valid in equilibrium

$$\varrho' D u_0' = -p_0' \varrho' D v' + \tfrac{1}{2}\varrho' E_{\mathrm{L}\alpha} D(v'P^\alpha) \\ - \tfrac{1}{2}P_\alpha D E_{\mathrm{L}}^\alpha + \tfrac{1}{2}\varrho' B_{\mathrm{L}\alpha} D(v'M^\alpha) - \tfrac{1}{2}M_\alpha D B_{\mathrm{L}}^\alpha, \quad (268)$$

where we used the abbreviations

$$E_{\mathrm{L}}^\alpha = E^\alpha + \tfrac{1}{3}P^\alpha, \\ B_{\mathrm{L}}^\alpha = B^\alpha - \tfrac{2}{3}M^\alpha \; (= H^\alpha + \tfrac{1}{3}M^\alpha). \quad (269)$$

The relativistic second law (228) gets the form:

$$T' D s' = D u_0' + p_0' D v' - \tfrac{1}{2}E_{\mathrm{L}\alpha} D(v'P^\alpha) \\ + \tfrac{1}{2}v' P_\alpha D E_{\mathrm{L}}^\alpha - \tfrac{1}{2}B_{\mathrm{L}\alpha} D(v'M^\alpha) + \tfrac{1}{2}v' M_\alpha D B_{\mathrm{L}}^\alpha. \quad (270)$$

(From (268) and (270) one obtains $Ds' = 0$ in equilibrium.)

As a second case we consider fluids with *permanent dipoles* that satisfy Clausius–Mossotti laws of the type (263) and Langevin–Debye laws for the temperature dependence of the susceptibilities:

$$\frac{\kappa}{\kappa+3} \sim \frac{1}{T'}, \qquad \frac{\chi}{3-2\chi} \sim \frac{1}{T'}. \quad (271)$$

The expression for $p'$ (243) gets the form (264), while the expression (247) for $u_v'$ becomes:

$$u_v' = u_{v0}' - B_\alpha M^\alpha - \tfrac{1}{6}P_\alpha P^\alpha + \tfrac{1}{3}M_\alpha M^\alpha. \quad (272)$$

The expression (253) for $T^{\alpha\beta}_{[f]}$ reads now:

$$T^{\alpha\beta}_{[f]} = -E^\alpha D^\beta - H^\alpha B^\beta + \Delta^{\alpha\beta}(\tfrac{1}{2}E_\gamma E^\gamma + \tfrac{1}{2}B_\gamma B^\gamma - B_\gamma M^\gamma - \tfrac{1}{6}P_\gamma P^\gamma + \tfrac{1}{3}M_\gamma M^\gamma)$$
$$+c^{-2}U^\alpha U^\beta(\tfrac{1}{2}E_\gamma E^\gamma + \tfrac{1}{2}B_\gamma B^\gamma - B_\gamma M^\gamma - \tfrac{1}{6}P_\gamma P^\gamma + \tfrac{1}{3}M_\gamma M^\gamma)$$
$$-c^{-2}U^\alpha \varepsilon^{\beta\gamma\zeta\eta}E_\gamma H_\zeta U_\eta - c^{-2}U^\beta \varepsilon^{\alpha\gamma\zeta\eta}E_\gamma H_\zeta U_\eta, \quad (273)$$

or in three-dimensional notation and in the local permanent rest frame:

$$T'^{\alpha\beta}_{[f]} = \begin{bmatrix} \tfrac{1}{2}\boldsymbol{E}'^2 + \tfrac{1}{2}\boldsymbol{B}'^2 - \boldsymbol{B}'\cdot\boldsymbol{M}' - \tfrac{1}{6}\boldsymbol{P}'^2 + \tfrac{1}{3}\boldsymbol{M}'^2 & \boldsymbol{E}'\wedge\boldsymbol{H}' \\ & -\boldsymbol{E}'\boldsymbol{D}' - \boldsymbol{H}'\boldsymbol{B}' \\ \boldsymbol{E}'\wedge\boldsymbol{H}' & +(\tfrac{1}{2}\boldsymbol{E}'^2 + \tfrac{1}{2}\boldsymbol{B}'^2 - \boldsymbol{B}'\cdot\boldsymbol{M}' - \tfrac{1}{6}\boldsymbol{P}'^2 + \tfrac{1}{3}\boldsymbol{M}'^2)\mathsf{U} \end{bmatrix}. \quad (274)$$

The field energy density for permanent dipole fluids is hence $\frac{1}{2}\boldsymbol{E}'^2 + \frac{1}{2}\boldsymbol{B}'^2 - \boldsymbol{B}'\cdot\boldsymbol{M}' - \frac{1}{6}\boldsymbol{P}'^2 + \frac{1}{3}\boldsymbol{M}'^2$ in contrast with the expression $\frac{1}{2}(\boldsymbol{E}'\cdot\boldsymbol{D}' + \boldsymbol{B}'\cdot\boldsymbol{H}')$ for induced dipole fluids; no polarization energy of the form $\frac{1}{2}\boldsymbol{E}'_{\mathrm{L}}\cdot\boldsymbol{P}' + \frac{1}{2}\boldsymbol{B}'_{\mathrm{L}}\cdot\boldsymbol{M}'$ occurs here.

The relativistic first law for permanent dipole fluids in equilibrium reads in terms of $u'_0$ and $p'_0$:

$$\varrho' Du'_0 = -p'_0\,\varrho' Dv' + \varrho' E_{\mathrm{L}\alpha} D(v'P^\alpha) + \varrho' B_{\mathrm{L}\alpha} D(v'M^\alpha), \quad (275)$$

while the relativistic second law is:

$$T'Ds' = Du'_0 + p'_0\,Dv' - E_{\mathrm{L}\alpha} D(v'P^\alpha) - B_{\mathrm{L}\alpha} D(v'M^\alpha). \quad (276)$$

For *diluted* media terms quadratic in the susceptibilities may be neglected so that the expressions (266) and (273) may be further simplified. For a diluted fluid with induced dipoles we get:

$$T^{\alpha\beta}_{[f]} = -E^\alpha D^\beta - H^\alpha B^\beta + \Delta^{\alpha\beta}(\tfrac{1}{2}E_\gamma E^\gamma + \tfrac{1}{2}B_\gamma B^\gamma - B_\gamma M^\gamma)$$
$$+c^{-2}U^\alpha U^\beta(\tfrac{1}{2}E_\gamma D^\gamma + \tfrac{1}{2}B_\gamma H^\gamma) - c^{-2}U^\alpha \varepsilon^{\beta\gamma\zeta\eta}E_\gamma H_\zeta U_\eta - c^{-2}U^\beta \varepsilon^{\alpha\gamma\zeta\eta}E_\gamma H_\zeta U_\eta, \quad (277)$$

whereas for a diluted fluid with permanent dipoles the result is:

$$T^{\alpha\beta}_{[f]} = -E^\alpha D^\beta - H^\alpha B^\beta + \Delta^{\alpha\beta}(\tfrac{1}{2}E_\gamma E^\gamma + \tfrac{1}{2}B_\gamma B^\gamma - B_\gamma M^\gamma)$$
$$+c^{-2}U^\alpha U^\beta(\tfrac{1}{2}E_\gamma E^\gamma + \tfrac{1}{2}B_\gamma B^\gamma - B_\gamma M^\gamma) - c^{-2}U^\alpha \varepsilon^{\beta\gamma\zeta\eta}E_\gamma H_\zeta U_\eta$$
$$-c^{-2}U^\beta \varepsilon^{\alpha\gamma\zeta\eta}E_\gamma H_\zeta U_\eta. \quad (278)$$

(In the approximation used here the scalar $\frac{1}{2}B_\gamma B^\gamma - B_\gamma M^\gamma$ is equal to $\frac{1}{2}H_\gamma H^\gamma$.)

The tensor (277) for diluted induced dipole fluids shows some similarity to the tensors proposed by Minkowski and Abraham[1]. To make comparison easier we write in the local permanent rest frame:

$$T'^{\alpha\beta}_{[\mathrm{f}]} = \begin{bmatrix} \frac{1}{2}(\boldsymbol{E}'\cdot\boldsymbol{D}'+\boldsymbol{B}'\cdot\boldsymbol{H}') & \boldsymbol{E}'\wedge\boldsymbol{H}' \\ \boldsymbol{E}'\wedge\boldsymbol{H}' & -\boldsymbol{E}'\boldsymbol{D}'-\boldsymbol{H}'\boldsymbol{B}'+(\frac{1}{2}\boldsymbol{E}'^2+\frac{1}{2}\boldsymbol{B}'^2-\boldsymbol{B}'\cdot\boldsymbol{M}')\mathrm{U} \end{bmatrix}. \tag{279}$$

The rest frame expressions of Minkowski's and Abraham's tensor contain the same energy density and energy flow. However Minkowski's momentum density $c^{-1}\boldsymbol{D}'\wedge\boldsymbol{B}'$ is not found and neither is the field pressure tensor $-\boldsymbol{E}'\boldsymbol{D}'-\boldsymbol{H}'\boldsymbol{B}'+\frac{1}{2}(\boldsymbol{E}'\cdot\boldsymbol{D}'+\boldsymbol{B}'\cdot\boldsymbol{H}')\mathrm{U}$ proposed by both these authors[2] (v. also the discussion at the end of this chapter).

### f. *The generalized Helmholtz force density*

The energy–momentum conservation laws for a polarized fluid of neutral atoms in equilibrium may be written as:

$$\partial_\beta(T^{\alpha\beta}_{[\mathrm{f}]}+T^{\alpha\beta}_{[\mathrm{m}]}) = 0, \tag{280}$$

where $T^{\alpha\beta}_{[\mathrm{f}]}$ is the field energy–momentum tensor (253) and $T^{\alpha\beta}_{[\mathrm{m}]}$ the material energy–momentum tensor (254), which has been defined as the energy–momentum tensor in the absence of fields. The conservation laws can be put into the form:

$$\partial_\beta\, T^{\alpha\beta}_{[\mathrm{m}]} = \mathscr{F}^\alpha, \tag{281}$$

where a force density $\mathscr{F}^\alpha$ is introduced, which is given by

$$\mathscr{F}^\alpha \equiv -\partial_\beta\, T^{\alpha\beta}_{[\mathrm{f}]}. \tag{282}$$

It corresponds to a material pressure and internal energy density defined at zero fields and is therefore equal to the difference of the force densities in the presence and in the absence of electromagnetic fields.

With the help of (253) an explicit expression for the force density $\mathscr{F}^\alpha$ may be obtained. If use is made of the Maxwell equations (64), the definitions (126), (140) and the relations (235), one gets:

[1] H. Minkowski, op. cit.; M. Abraham, op. cit.

[2] As shown already in chapter II (v. equation (II.371)) a special case exists in which this (Maxwell–Heaviside) field pressure tensor shows up, namely if one considers a body immersed in an incompressible liquid.

$$\mathscr{F}^{\alpha} = -\tfrac{1}{2}E_{\gamma}E^{\gamma}\partial^{\alpha}\kappa - \tfrac{1}{2}B_{\gamma}B^{\gamma}\partial^{\alpha}\chi$$
$$-\tfrac{1}{2}\partial_{\beta}\left\{\Delta^{\alpha\beta}\left(v'\frac{\partial\kappa}{\partial v'}E_{\gamma}E^{\gamma}+v'\frac{\partial\chi}{\partial v'}B_{\gamma}B^{\gamma}\right)\right.$$
$$\left.+c^{-2}U^{\alpha}U^{\beta}\left(T'\frac{\partial\kappa}{\partial T'}E_{\gamma}E^{\gamma}+T'\frac{\partial\chi}{\partial T'}B_{\gamma}B^{\gamma}\right)\right\}$$
$$-c^{-2}\varepsilon^{\alpha\beta\gamma\zeta}\varrho' D\{v'(\kappa+\chi)E_{\beta}B_{\gamma}U_{\zeta}\} - c^{-2}(\partial^{\alpha}U^{\beta})\varepsilon_{\beta\gamma\zeta\eta}(\kappa+\chi)E^{\gamma}B^{\zeta}U^{\eta}. \tag{283}$$

In the local rest frame the components $\alpha = 0$ and $\alpha = 1, 2, 3$ of the force density $\mathscr{F}^{\alpha}$ read for a substance of constant and uniform velocity:

$$\mathscr{F}'^{0} = \tfrac{1}{2}\boldsymbol{E}'^{2}\partial_0'\kappa - \tfrac{1}{2}\partial_0'\left(T'\frac{\partial\kappa}{\partial T'}\boldsymbol{E}'^{2}\right) + \tfrac{1}{2}\boldsymbol{B}'^{2}\partial_0'\chi - \tfrac{1}{2}\partial_0'\left(T'\frac{\partial\chi}{\partial T'}\boldsymbol{B}'^{2}\right), \tag{284}$$

$$\mathscr{F}' = -\tfrac{1}{2}\boldsymbol{E}'^{2}\nabla'\kappa - \tfrac{1}{2}\nabla'\left(v'\frac{\partial\kappa}{\partial v'}\boldsymbol{E}'^{2}\right) - \tfrac{1}{2}\boldsymbol{B}'^{2}\nabla'\chi - \tfrac{1}{2}\nabla'\left(v'\frac{\partial\chi}{\partial v'}\boldsymbol{B}'^{2}\right)$$
$$+\partial_0'\{(\kappa+\chi)\boldsymbol{E}'\wedge\boldsymbol{B}'\}. \tag{285}$$

If we introduce the susceptibility $\tilde{\chi}$ as in (256) these expressions for $\mathscr{F}'^{0}$ and $\mathscr{F}'$ become:

$$\mathscr{F}'^{0} = \tfrac{1}{2}\boldsymbol{E}'^{2}\partial_0'\kappa - \tfrac{1}{2}\partial_0'\left(T'\frac{\partial\kappa}{\partial T'}\boldsymbol{E}'^{2}\right) + \tfrac{1}{2}\boldsymbol{H}'^{2}\partial_0'\tilde{\chi} - \tfrac{1}{2}\partial_0'\left(T'\frac{\partial\tilde{\chi}}{\partial T'}\boldsymbol{H}'^{2}\right), \tag{286}$$

$$\mathscr{F}' = -\tfrac{1}{2}\boldsymbol{E}'^{2}\nabla'\kappa - \tfrac{1}{2}\nabla'\left(v'\frac{\partial\kappa}{\partial v'}\boldsymbol{E}'^{2}\right) - \tfrac{1}{2}\boldsymbol{H}'^{2}\nabla'\tilde{\chi} - \tfrac{1}{2}\nabla'\left(v'\frac{\partial\tilde{\chi}}{\partial v'}\boldsymbol{H}'^{2}\right)$$
$$+\partial_0'\{(\kappa+\tilde{\chi}+\kappa\tilde{\chi})(\boldsymbol{E}'\wedge\boldsymbol{H}')\}. \tag{287}$$

The expression (285) or (287) may be called the relativistic Helmholtz ponderomotive force in view of its analogy with the non-relativistic expression (II.350). Comparison with this expression (for constant and uniform macroscopic velocity and equilibrium polarizations, i.e. the same physical situation as studied here) shows that the only difference consists in the appearance in (285) or (287) of a term $-\partial_0'(\boldsymbol{M}'\wedge\boldsymbol{E}')$ on a par with $\partial_0'(\boldsymbol{P}'\wedge\boldsymbol{B}')$ that figures already in the non-relativistic theory.

In this section the laws of thermodynamics for polarized systems have been obtained. From the second law expressions could be derived for the difference between the material pressure and the material energy density in the presence of fields and those in the absence of fields. The field and material part of the relativistic energy–momentum tensor could then be defined in such a way that the material tensor contains a pressure and an energy density without fields.

## 7 *On the uniqueness of the energy–momentum tensor*

The derivation of the energy–momentum laws in section 4 showed that they may be formulated in terms of a macroscopic energy–momentum tensor. Furthermore complete statistical expressions in terms of atomic quantities have been obtained for the components of this tensor.

Since only the four-divergence of the energy–momentum tensor occurs in the conservation laws

$$\partial_\beta T^{\alpha\beta} = 0, \tag{288}$$

the energy–momentum tensor is not uniquely determined: one may add a divergenceless tensor $\hat{T}^{\alpha\beta}$ without changing the contents of the laws. In the inner angular momentum law (section 5)

$$\partial_\gamma(S^{\alpha\beta}U^\gamma) = -\partial_\gamma J^{\alpha\beta\gamma} + T^{\alpha\beta} - T^{\beta\alpha} \tag{289}$$

such a change of $T^{\alpha\beta}$ into $T^{\alpha\beta}+\hat{T}^{\alpha\beta}$ may be compensated by a corresponding change of the inner angular momentum flow $J^{\alpha\beta\gamma}$ to $J^{\alpha\beta\gamma}+\hat{J}^{\alpha\beta\gamma}$ with a quantity $\hat{J}^{\alpha\beta\gamma}$ of the form $-R^\alpha\hat{T}^{\beta\gamma}+R^\beta\hat{T}^{\alpha\gamma}$. Then the right-hand side of (289) remains invariant. Thus such a change of the energy–momentum tensor and of the inner angular momentum flow does not alter the physical description by means of (288) and (289). The particular forms of $T^{\alpha\beta}$ and $J^{\alpha\beta\gamma}$ given in sections 4 and 5 have been adopted since the statistical expressions allow an interpretation which is analogous to that of the corresponding non-relativistic quantities, given in chapter II.

In the course of the treatment of section 4 it turned out to be convenient to call a certain part of the energy–momentum tensor its 'field part', and the remaining term its 'material part'. This nomenclature, which arose from the form of the various expressions, has of course no influence on the physical contents of the laws. Such a splitting of the energy–momentum tensor could be performed in different ways, which each have their particular advantages as shown in section 6. Exactly the same situation was encountered already in the non-relativistic theory, where the material pressure could be defined in different ways: Kelvin's and Helmholtz's, each with its own force density. It appeared there that both could be utilized to describe the physical phenomena.

Just as it is fruitless to discuss – in non-relativistic theory – the relative merits of the Kelvin and the Helmholtz forces without considering the ensuing difference in the corresponding material pressure tensors, a dispute on the correct form of the field energy–momentum tensor – in relativistic theory – is useless if one does not bring into the argument the form of the material energy–momentum tensor: the problem would then remain undetermined.

In the present statistical theory each choice of the field tensor determines the form of the corresponding material tensor explicitly, so that no ambiguity can arise. If the distribution functions are given, one may in principle calculate both the material and the field tensor. (In practice this requires a theory from which one may derive these distribution functions. A well-known example is the relativistic generalization of Boltzmann's kinetic theory.)

The history of the discussions on the energy–momentum tensor for polarized media goes back to the beginning of this century. Often only the field part of the total tensor was considered. As a consequence various authors could arrive at altogether different expressions: a manifestation of the inherent ambiguity which results, as explained, if one forgets about the material part.

After Lorentz's[1] original non-relativistic considerations on the electromagnetic forces in a polarized medium of electric dipoles at rest (v. chapter II) Einstein and Laub[2] were the first to try and give a relativistic expression for the force density in a polarized medium of electric and magnetic dipoles. By taking the same electric dipole terms as Lorentz and by postulating an analogy between electric and magnetic effects they arrived at an expression for the force density in a medium at rest which had a form as (167) apart from the second term, where they wrote $(\nabla \boldsymbol{H})\cdot\boldsymbol{M}$. The material part was not considered at all, so that their treatment suffers from the ambiguity mentioned above.

In the same year Minkowski[3] put forward an expression for the field energy–momentum tensor on the purely formal grounds that it should be form-invariant in all Lorentz frames. This implies that the field energy–momentum tensor should depend on the fields $F^{\alpha\beta}$ and $H^{\alpha\beta}$, but not on the four-velocity $U^{\alpha}$ of the polarized medium with respect to the observer. Furthermore the expressions for the field energy density, the field energy flow and the field pressure due to Maxwell, Poynting and Heaviside were

[1] H. A. Lorentz, Enc. Math. Wiss. V 2, fasc. 1 (Teubner, Leipzig 1904) 200.

[2] A. Einstein and J. Laub, Ann. Physik **26**(1908)541.

[3] H. Minkowski, Nachr. Ges. Wiss. Göttingen (1908)53; Math. Ann. **68**(1910)472.

taken over. In this way he arrived at the field energy–momentum tensor

$$T^{\alpha\beta}_{(f)M} = F^{\alpha\gamma}H^{\beta}_{.\gamma} - \tfrac{1}{4}F_{\gamma\varepsilon}H^{\gamma\varepsilon}g^{\alpha\beta}, \tag{290}$$

which in three-dimensional notation has the form

$$T^{\alpha\beta}_{(f)M} = \begin{pmatrix} \tfrac{1}{2}(\boldsymbol{E}\cdot\boldsymbol{D}+\boldsymbol{B}\cdot\boldsymbol{H}) & \boldsymbol{E}\wedge\boldsymbol{H} \\ \boldsymbol{D}\wedge\boldsymbol{B} & -\boldsymbol{E}\boldsymbol{D}-\boldsymbol{H}\boldsymbol{B}+\tfrac{1}{2}(\boldsymbol{E}\cdot\boldsymbol{D}+\boldsymbol{B}\cdot\boldsymbol{H})\mathrm{U} \end{pmatrix}. \tag{291}$$

The material tensor was again not considered. Moreover the principle of form invariance represents a mathematical requirement, which is foreign to the theory.

Abraham[1] abandoned the principle of form invariance but instead assumed that the field pressure tensor is symmetric in all Lorentz frames, even for anisotropic media. Maxwell's and Poynting's expressions were adopted as the field energy density and the field energy flow in the rest frame, whereas the field pressure tensor in the rest frame was taken to be represented by Hertz's symmetrized form. In this way a completely symmetric field energy–momentum tensor was obtained; it reads in covariant notation:

$$T^{\alpha\beta}_{(f)A} = \tfrac{1}{2}(F^{\alpha\gamma}H^{\beta}_{.\gamma}+F^{\beta\gamma}H^{\alpha}_{.\gamma}) - \tfrac{1}{4}F_{\gamma\varepsilon}H^{\gamma\varepsilon}g^{\alpha\beta} + \tfrac{1}{2}c^{-2}\{U^{\beta}(F^{\alpha\gamma}M_{\gamma\varepsilon}-M^{\alpha\gamma}F_{\gamma\varepsilon})+U^{\alpha}(F^{\beta\gamma}M_{\gamma\varepsilon}-M^{\beta\gamma}F_{\gamma\varepsilon})\}U^{\varepsilon}, \tag{292}$$

while the rest frame expression in three-dimensional notation is:

$$T'^{\alpha\beta}_{(f)A} = \begin{bmatrix} \tfrac{1}{2}(\boldsymbol{E}'\cdot\boldsymbol{D}'+\boldsymbol{B}'\cdot\boldsymbol{H}') & \boldsymbol{E}'\wedge\boldsymbol{H}' \\ \boldsymbol{E}'\wedge\boldsymbol{H}' & -\tfrac{1}{2}(\boldsymbol{E}'\boldsymbol{D}'+\boldsymbol{D}'\boldsymbol{E}'+\boldsymbol{H}'\boldsymbol{B}'+\boldsymbol{B}'\boldsymbol{H}') + \tfrac{1}{2}(\boldsymbol{E}'\cdot\boldsymbol{D}'+\boldsymbol{B}'\cdot\boldsymbol{H}')\mathrm{U} \end{bmatrix}. \tag{293}$$

In a subsequent paper Abraham[2] remarks that the expression for the field energy–momentum tensor should be derived from electron-theoretical considerations, but he limits himself to a discussion of several possible approaches to carry out this programme. As an argument in favour of the symmetry of the field energy–momentum tensor he mentions the fact that the microscopic energy–momentum tensor is symmetric. However this argument ensures only that the system of polarized matter and fields is closed, so that one expects macroscopic conservation of total angular momentum and hence the possibility to symmetrize[3] the total energy–momentum tensor.

[1] M. Abraham, R. C. Circ. Mat. Palermo **28**(1909)1, **30**(1910)33; Theorie der Elektrizität II (Teubner, Leipzig 1923) 300.

[2] M. Abraham, Ann. Physik **44**(1914)537.

[3] F. J. Belinfante, Physica **6**(1939)887; L. Rosenfeld, Mém. Acad. Roy. Belg. (Cl. Sc.) **18**(1940)6.

About the symmetry of the field part nothing can be found on these grounds.

Abraham's field energy–momentum tensor contains a field momentum density which (in the rest frame) is equal to the field energy flow (apart from a factor $c^{-2}$). Hence the field tensor has a property formulated as early as 1908 by Planck[1]; it is called sometimes the 'inertia law of energy'. This circumstance has often been considered as a strong argument in favour of Abraham's tensor[2]. However, Planck's law is valid only for a closed system[3], so that it may be applied to the total energy–momentum tensor only: the latter may be written in symmetrical form.

Soon after Minkowski's and Abraham's papers Dällenbach[4] tried to give a treatment of the energy–momentum laws on the basis of microscopic considerations valid for the electrostatic case only. These considerations are then generalized without further justification, with Minkowski's field tensor as a result. To explain its asymmetry he rightly remarks that only the sum of the field and the material energy–momentum tensor ought to be symmetrical. However the material tensor is not considered any further.

In Frenkel's[5] treatment microscopic concepts are employed together with macroscopic arguments. By a consideration of the forces exerted on surface charges and currents he obtains as the field energy–momentum tensor for a stationary medium an expression which is near to the result (148). However, as he is convinced that covariance should imply form invariance he rejects this tensor since it does not possess this property. Owing to this difficulty a definite conclusion on the field tensor is not reached. By postulating the symmetry of the field energy–momentum tensor with respect to time–space and space–time components in the rest frame, and employing a reasoning similar to Frenkel's, Rancoita[6] arrives at a field energy–momentum tensor of the form (148).

A much discussed argument in favour of the asymmetric Minkowski tensor was put forward in 1950 by Von Laue[7] following an old idea of

[1] M. Planck, Phys. Z. **9**(1908)828.

[2] M. von Laue, Die Relativitätstheorie I (Vieweg, Braunschweig 1919) 185; G. Marx and G. Györgyi, Acta Phys. Acad. Sci. Hung. **3**(1953)213; N. L. Balázs, Phys. Rev. **91**(1953) 408; G. Györgyi, Acta Phys. Acad. Sci. Hung. **4**(1954)121; G. Marx and G. Györgyi, Ann. Physik **16**(1955)241; J. Agudín, Phys. Letters **24A**(1967)761.

[3] C. Møller, Theory of relativity (Clarendon Press, Oxford 1952) 164, 189; cf. F. Beck, Naturwiss. **39**(1952)254; Z. Physik **134**(1953)136.

[4] W. Dällenbach, Ann. Physik **58**(1919)523.

[5] J. Frenkel, Lehrbuch der Elektrodynamik II (Springer-Verlag, Berlin 1928) 48–94.

[6] G. M. Rancoita, Suppl. N. Cim. **11**(1959)183.

[7] M. von Laue, Z. Physik **128**(1950)387; Die Relativitätstheorie I (Vieweg, Braunschweig 1952) 139.

Scheye[1]. According to this argument the energy transport velocity, which is the quotient of the energy flow and the energy density, should transform in such a way that the addition theorem for velocities is obeyed. Since the Minkowski field tensor satisfies this criterion, a number of authors[2] have advocated it. However Schöpf[3] remarked that Von Laue's criterion does not lead exclusively to Minkowski's tensor since it is satisfied by tensors of a different form as well. Furthermore Tang and Meixner[4] showed that even if Minkowski's tensor is adopted as the field part of the total energy–momentum tensor an explicit evaluation of the material contributions to the total energy density and flow leads to the conclusion that the total energy transport velocity does not satisfy Von Laue's criterion. Hence this criterion cannot be considered as a physical requirement to be imposed on an energy–momentum tensor.

Often reasonings which start from macroscopic variational principles are considered as derivations of the form of the energy–momentum tensor. In this way some authors try to derive the field energy–momentum tensor and arrive at Minkowski's[5] or Abraham's[6] tensor or still different tensors[7], whereas others[8] obtain expressions for the *total* energy–momentum tensor, which according to them is the only one that can be deduced from a variational principle. However against all such treatments the same objection may be raised: at the outset a macroscopic Lagrangian (or Hamiltonian) is postulated, not derived from first principles. Therefore arguments of this kind do not lead to a solution of the problem.

In the course of the discussions various *ad hoc* arguments of a macroscopic nature have been put forward in favour of one or the other of the field

[1] A. Scheye, Ann. Physik **30**(1909)805.

[2] H. Ott, Ann. Physik **11**(1952)33; F. Beck, loc. cit.; C. Møller, loc. cit. 206–211; E. Schmutzer, Ann. Physik **18**(1956)171; J. I. Horváth, Bull. Acad. Polon. Sci. **4**(1956)447; W. Pauli, Theory of relativity (Pergamon Press, London 1958) 216, note 11.

[3] H. G. Schöpf, Z. Physik **148**(1957)417.

[4] C. L. Tang and J. Meixner, Phys. Fluids **4**(1961)148.

[5] J. Ishiwara, Ann. Physik **42**(1913)986; W. Dällenbach, Ann. Physik **59**(1919)28; E. Schmutzer, Ann. Physik **20**(1957)349; U. E. Schröder, Z. Naturf. **24A**(1969)1356.

[6] E. Henschke, Ann. Physik **40**(1913)887; K. F. Novobátzky, Hung. Acta Phys. **1**(1949) fasc. 5; G. Marx, Acta Phys. Acad. Sci. Hung. **2**(1952)67; **3**(1953)75; G. Marx and G. Györgyi, Ann. Physik **16**(1955)241; H. G. Schöpf, Ann. Physik **13**(1964)41.

[7] K. Furutsu, Phys. Rev. **185**(1969)257.

[8] H. G. Schöpf, Ann. Physik **9**(1962)301; P. Penfield Jr. and H. A. Haus, Phys. Fluids **9**(1966)1195.

energy–momentum tensors[1]. An argument which was thought[2] to be in favour of Minkowski's tensor is the fact that the corresponding force density vanishes for a neutral, current-free and homogeneous medium with linear constitutive relations, if it moves with a uniform velocity. An argument which was claimed[3] to be in favour of Abraham's tensor is the fact that the field energy density of this tensor is positive for all values of the macroscopic velocity. As a consequence, if one performs a quantization of the macroscopic electromagnetic fields, one finds photons with positive energy[4]. However again the material energy–momentum was left out of consideration. Sometimes it was thought[5] that radiation pressure experiments can throw light on the correct form of the field momentum density. However, as shown in chapter II, the explanation of the experimental results is independent of the expression for the momentum density, since terms with time derivatives drop out from the equations[6]. (The results of chapter II are not altered in relativity theory since the extra term in the force density (167), as compared with (II.106), is also a time derivative.) Still other postulates[7] have been put forward in order to justify the choice of a particular form of the field

[1] A review of these and the other arguments in favour of one or the other field energy–momentum tensor was given by I. Brevik, Mat. Fys. Medd. Vid. Selsk. **37**(1970)no. 11, 13. In his first paper he seems to adopt Minkowski's tensor on the basis of *ad hoc* postulates, while in his second one he rightly remarks that only the total energy–momentum tensor has physical meaning. Yet he thinks that only Minkowski's and Abraham's field tensors do not run into conflict with experimental evidence. His argument to reject (122) with (92) as a useful splitting is that according to him only Helmholtz-type material pressures are in agreement with experiment. However, as explained in chapter II, the use of both the Kelvin and the Helmholtz pressures and forces is allowed, provided one employs them consistently.

[2] H. Ott, op. cit.; F. Beck, op. cit.; C. Møller, op. cit. p. 206–211; I. Brevik, op. cit.

[3] K. F. Novobátzky, op. cit.; M. von Laue, op. cit.; F. Beck, op. cit.; E. Schmutzer, Ann. Physik **18**(1956)171.

[4] K. Nagy, Acta Phys. Acad. Sci. Hung. **5**(1955)95. Minkowski's tensor leads to the possibility of negative energy densities and photons of negative energy: J. M. Jauch and K. M. Watson, Phys. Rev. **74**(1948)950, 1485; I. Brevik and B. Lautrup, Mat. Fys. Medd. Dan. Vid. Selsk. **38**(1970)no. 1.

[5] R. V. Jones and J. C. S. Richards, Proc. Roy. Soc. **221A**(1954)480; I. Brevik, op. cit., uses the experimental result of Jones and Richards as an argument pro Minkowski's momentum density (no. 11, p. 29), but elsewhere remarks (no. 11, p. 5) that it cannot exclude other forms of the field energy–momentum tensor.

[6] G. Marx and G. Györgyi, Acta Phys. Acad. Sci. Hung. **3**(1953)213.

[7] A. Rubinowicz, Acta Phys. Polon. **14**(1955)209, 225; G. Marx and K. Nagy, Bull. Acad. Polon. Sci. **4**(1956)79; J. I. Horváth, N. Cim. **7**(1958)628; O. Costa de Beauregard, C. R. Acad. Sci. Paris **260**(1965)6546; **263B**(1966)1007, 1279; N. Cim. **48B**(1967)293; W. Shockley, Proc. Nat. Acad. Sci. U.S.A. **60**(1968)807.

energy–momentum tensor, but again they are of an *ad hoc* and macroscopic character.

Arguments based on the propagation of light waves in connexion with the refraction of light, the Cherenkov effect, the Sagnac effect and Fizeau's experiments do not lead to a decision on the correct form of the field energy–momentum tensor[1], since these phenomena can be explained on the basis of the field equations alone.

A special class of theories is based on thermodynamical considerations. In the framework of the treatment of Kluitenberg and de Groot[2] a relativistic Gibbs relation and the symmetric character of the material energy–momentum tensor were postulated. As a result a field energy–momentum tensor was obtained which comes very near to that given in (92). The hypothesis about the symmetry of the material tensor is rather essential; if it is dropped different forms for the field energy–momentum tensor (for instance Minkowski's tensor) are justifiable from a thermodynamical point of view, as has been shown by Schmutzer[3]. De Sá[4] and Meixner[5] discuss various possibilities for the splitting of the total energy–momentum tensor into a material and a field part. They rightly conclude that thermodynamical considerations do not allow to specify the material part sufficiently well; the field part remains then undetermined. Chu, Haus and Penfield[6] postulate a form for the first law of thermodynamics together with the symmetrical character of the material tensor. Since this starting point is equivalent to Kluitenberg and de Groot's, their resulting field energy–momentum tensor is also the same apart from some diagonal terms.

In general it may be stated that the solution of the problem of deriving the energy–momentum and angular momentum laws for polarized media cannot be solved as long as macroscopic arguments are utilized. The problem is even undetermined if only the field energy–momentum tensor is considered without giving expressions for the material energy–momentum tensor. The complete programme can only be carried out if one starts from the microscopic laws. Then statistical expressions for the total energy–momentum tensor and

[1] G. Marx and G. Györgyi, Ann. Physik **16**(1955)241; G. Györgyi, Am. J. Phys. **28**(1960) 85; cf. however I. Brevik, op. cit.

[2] G. A. Kluitenberg and S. R. de Groot, Physica **20**(1954)199; **21**(1955)148, 169.

[3] E. Schmutzer, Ann. Physik **14**(1964)56; cf. G. Neugebauer, Wiss. Z. Friedrich Schiller Univ. Jena **13**(1964)209.

[4] B. de Sá, thesis, Aachen (1960).

[5] J. Meixner, Univ. Michigan Report, RL-184(1961); Z. Physik **229**(1969)352.

[6] L. J. Chu, H. A. Haus and P. Penfield jr., Proc. I.E.E.E. **54**(1966)920; P. Penfield jr. and H. A. Haus, Electrodynamics of moving media (M.I.T. Press, Cambridge, Mass. 1967).

the angular momentum density and flow can be derived, as shown in this chapter. These are then unambiguously defined (apart from terms which drop out from the conservation laws, as discussed in the beginning of this section); the splitting into a field and material part is a question of nomenclature only.

# PROBLEMS

**1.** Prove that the average $A(R)$ of the microscopic quantity

$$a(R) = \sum_i \int \alpha(i)\delta^{(4)}(R-R_i)\mathrm{d}s_i$$

may be written either as

$$A(R) = c^{-1}\int \alpha(1)\delta^{(4)}(R-R_1)f_1(1)\mathrm{d}1$$

or as

$$A(R) = -\int \frac{\alpha(1)}{n\cdot R_1^{[1]}}\,\delta^{(3)}(n; R-R_1)f_1^{\mathrm{syn}}(1; n, -c^{-1}n\cdot R)\mathrm{d}1,$$

where the three-dimensional delta function has been defined in (2).

Note that the first form shows that the second is independent of the unit four-vector $n^\alpha$. Show that the second expression may be written as

$$A(R) = c^{-1}\int \alpha(1)\gamma_1^{-1}f_1^{\mathrm{syn}}(\boldsymbol{R}, \boldsymbol{\beta}_1, \partial_0\boldsymbol{\beta}_1, \ldots; t)\mathrm{d}\boldsymbol{\beta}_1\,\mathrm{d}\partial_0\boldsymbol{\beta}_1 \ldots,$$

as follows by employing $\mathring{n}^\alpha = (1, 0, 0, 0)$ and (55). In spite of its appearance the derivation shows that this form of $A(R)$ has covariant character. (A particular case of physical importance is obtained by choosing $\alpha(i) = ce_i u_i^\alpha$ with $u_i^\alpha$ the four-velocity and $e_i$ the charge; then one gets the expressions (72).)

**2.** Calculate the Jacobian occurring in (55) for the velocities, i.e.,

$$\frac{\partial(\boldsymbol{R}_1^{[1]})}{\partial(\boldsymbol{\beta}_1)},$$

and for the velocities with accelerations, i.e.,

$$\frac{\partial(\boldsymbol{R}_1^{[1]}, \boldsymbol{R}_1^{[2]})}{\partial(\boldsymbol{\beta}_1, \partial_0\boldsymbol{\beta}_1)}.$$

**3.** Prove from (20) that one has for synchronous averages the identity

$$\partial_\mu\int \alpha(1; R)f_1^{\mathrm{syn}}(1; n, -c^{-1}n\cdot R)\mathrm{d}1 = \int \mathrm{d}_\mu^{\mathrm{syn}}\alpha(1; R)f_1^{\mathrm{syn}}(1; n, -c^{-1}n\cdot R)\mathrm{d}1,$$

where the notation (28) has been used. To prove this relation one should substitute a quantity of the type (21) (with $\tau = -c^{-1}n{\cdot}R$) into (20) and then employ the definition (30) of an average, using the transformation which leads from (23) to (26).

**4.** Prove (27) and (37) by differentiating the ancillary conditions, which procedure leads to

$$\partial_\mu s_i = \frac{n_\mu}{n{\cdot}R_i^{[1]}}$$

and

$$\partial_\mu s_i = \frac{(R-R_i)_\mu}{(R-R_i){\cdot}R_i^{[1]}},$$

respectively.

**5.** Prove from the definition (31) that synchronous distribution functions characterized by a different normal unit vector $n^\alpha$ and $n'^\alpha$ are related as

$$f_1^{\text{syn}}(1; n, \tau) = -\frac{n{\cdot}R_1^{[1]}}{n{\cdot}n'n'{\cdot}R_1^{[1]}} f_1^{\text{syn}}\left(1; n', -\frac{c^{-1}n{\cdot}\Delta'{\cdot}R_1+\tau}{n{\cdot}n'}\right),$$

where $\tau$ is a constant (independent of $R_1^\alpha$) and where $\Delta'^{\alpha\beta}$ is the tensor $g^{\alpha\beta}+n'^\alpha n'^\beta$. To prove this relation one should use the property (48).

Note that the relation is not symmetric in the distribution functions for $n^\alpha$ and $n'^\alpha$. This is due to the fact that the distribution function at the left-hand side contains as an argument a quantity $\tau$ independent of $R_1^\alpha$, while the distribution function at the right-hand side contains a third argument which does depend on $R_1^\alpha$. For that reason one may be interested in a relation which contains also at the left-hand side a distribution function of which the third argument is linear in $R_1^\alpha$. This relation, which may be proved along the same lines, reads

$$f_1^{\text{syn}}(1; n, \tau_0+\tau_1{\cdot}R_1)$$
$$= \frac{n{\cdot}R_1^{[1]}}{|n{\cdot}n'+cn'{\cdot}\tau_1|n'{\cdot}R_1^{[1]}} f_1^{\text{syn}}\left(1; n', -\frac{c^{-1}n{\cdot}\Delta'{\cdot}R_1+\tau_1{\cdot}\Delta'{\cdot}R_1+\tau_0}{n{\cdot}n'+cn'{\cdot}\tau_1}\right)$$

for arbitrary constants $\tau_0$ and $\tau_1^\alpha$. One may check that a repeated use of this formula leads to an identity. A symmetrical relation is obtained by choosing in particular

$$\tau_1^\alpha = c^{-1}\frac{n^\alpha n{\cdot}n'+n'^\alpha}{1-n{\cdot}n'}.$$

Then one gets:

$$f_1^{\text{syn}}\left(1;\, n,\, \tau_0+c^{-1}\frac{n\cdot R_1\, n\cdot n'+n'\cdot R_1}{1-n\cdot n'}\right)$$

$$=\frac{n\cdot R_1^{[1]}}{n'\cdot R_1^{[1]}}\, f_1^{\text{syn}}\left(1;\, n',\, \tau_0+c^{-1}\frac{n'\cdot R_1\, n\cdot n'+n\cdot R_1}{1-n\cdot n'}\right).$$

**6.** Prove from (159) that the inner product of the four-velocity $U^\alpha$ and the ponderomotive force $F^\alpha$ is given by

$$U_\alpha F^\alpha = \tfrac{1}{2}(DF^{\alpha\beta})M_{\alpha\beta}+c^{-2}(DU_\alpha)(F^{\alpha\beta}M_{\beta\gamma}-M^{\alpha\beta}F_{\beta\gamma})U^\gamma$$
$$-c^{-2}\varrho' D(v' U^\alpha F_{\alpha\beta} M^{\beta\gamma} U_\gamma).$$

Show, by introducing the splitting of $M^{\alpha\beta}$ into two parts, defined as

$$M^{(1)\alpha\beta} \equiv -c^{-2}(U^\alpha U_\gamma M^{\gamma\beta}+U^\beta U_\gamma M^{\alpha\gamma}), \qquad M^{(2)\alpha\beta} \equiv \Delta^\alpha_\gamma \Delta^\beta_\varepsilon M^{\gamma\varepsilon},$$

(so that in the rest frame $M^{(1)\alpha\beta}$ and $M^{(2)\alpha\beta}$ represent the electric and magnetic polarization respectively) that one may write this expression as

$$U_\alpha F^\alpha = -\tfrac{1}{2}F'_{\alpha\beta}\,\varrho' D(v' M^{(1)\prime\alpha\beta})+\tfrac{1}{2}M^{(2)\prime}_{\alpha\beta} DF^{\prime\alpha\beta},$$

where the primes indicate quantities in the permanent local rest frame. The proof follows from the mathematical identity

$$A'_{\alpha\beta} DB^{\prime\alpha\beta} = A_{\alpha\beta} DB^{\alpha\beta}+2c^{-2}U^\alpha(A_{\alpha\beta}B^{\beta\gamma}-B_{\alpha\beta}A^{\beta\gamma})DU_\gamma,$$

which itself is obtained by considering the Lorentz transformations from the permanent local rest frame to the observer's frame.

**7.** Show that the components of the kinetic contribution $T^{\alpha\beta}_{(\text{m})\text{I}}$ (108) to the material part of the energy–momentum tensor for a neutral plasma may be written as (cf. problem 1)

$$T^{00}_{(\text{m})\text{I}} = c^2\sum_a \int m_a \gamma_1 f_1^{\text{syn},a}(\boldsymbol{R}, \boldsymbol{\beta}_1; t)\mathrm{d}\boldsymbol{\beta}_1,$$

$$T^{0i}_{(\text{m})\text{I}} = T^{i0}_{(\text{m})\text{I}} = c^2\sum_a \int m_a \gamma_1 \beta_1^i f_1^{\text{syn},a}(\boldsymbol{R}, \boldsymbol{\beta}_1; t)\mathrm{d}\boldsymbol{\beta}_1,$$

$$T^{ij}_{(\text{m})\text{I}} = c^2\sum_a \int m_a \gamma_1 \beta_1^i \beta_1^j f_1^{\text{syn},a}(\boldsymbol{R}, \boldsymbol{\beta}_1; t)\mathrm{d}\boldsymbol{\beta}_1.$$

Prove that in the non-relativistic limit (i.e. up to order $c^{-1}$) the energy density, the energy flow, the momentum density and the momentum flow reduce to

$$
\begin{aligned}
T^{00}_{(m)I} &= \varrho c^2 + \tfrac{1}{2}\varrho v^2 + \varrho u^{K},\\
cT^{0i}_{(m)I} &= (\varrho c^2 \boldsymbol{v} + \boldsymbol{J}^{K}_{q} + \varrho u^{K}\boldsymbol{v} + \mathbf{P}^{K}\cdot\boldsymbol{v})^i,\\
c^{-1}T^{i0}_{(m)I} &= \varrho v^i,\\
T^{ij}_{(m)I} &= \varrho v^i v^j + P^{Kji}
\end{aligned}
$$

with non-relativistic quantities given by (II.119, 120, 122, 131, 132).

Prove that the non-relativistic limits of $T^{\alpha\beta}_{(m)II}$ (116) and $T^{\alpha\beta}_{(m)III}$ (118) have the forms

$$
\begin{aligned}
T^{00}_{(m)II+III} &= \varrho u^{C},\\
cT^{0i}_{(m)II+III} &= (\boldsymbol{J}^{C}_{q} + \varrho u^{C}\boldsymbol{v} + \mathbf{P}^{C}\cdot\boldsymbol{v})^i,\\
c^{-1}T^{i0}_{(m)II+III} &= 0,\\
T^{ij}_{(m)II+III} &= P^{Cji}
\end{aligned}
$$

with the non-relativistic quantities (II.127, 138, 139). Use in the proof that up to order $c^{-2}$ one has $\delta(s^2) = \delta(s^0)/|\boldsymbol{s}|$, where one should take into account that $s^0$ is of order $c$.

Show finally that the non-relativistic limits of the components of $T^{\alpha\beta}_{(m)IV}$ (120) vanish, as follows by considering the expressions for the minus fields, given in (III.75) with (III.72).

**8.** Show in a way analogous to that of the preceding problem, that the non-relativistic limits of the components of the material energy–momentum tensor for a dipole fluid are such that $T^{\alpha\beta}_{(m)I}$ (87) leads to the kinetic contributions K, given in (II.63, 78, 81) (together with mass terms), $T^{\alpha\beta}_{(m)II}$ (90) to the field dependent terms F (II.73, 89), $T^{\alpha\beta}_{(m)IV+V}$ (98, 100) to the short range terms S (II.94, 96, 97, 100) and $T^{\alpha\beta}_{(m)VIII}$ described below (102) to the correlation terms C (II.104, 111, 112). The terms $T^{\alpha\beta}_{(m)III}$, $T^{\alpha\beta}_{(m)VI}$ and $T^{\alpha\beta}_{(m)VII}$ give no contributions in the non-relativistic limit.

**9.** Show that in the case of long range correlations the generalized Irving–Kirkwood procedure may be written as

$$c_2(R, 1, R-s, 2) = c_2(R+\tfrac{1}{2}s, 1, R-\tfrac{1}{2}s, 2) - \tfrac{1}{2}s^\alpha \partial_\alpha \tilde{c}_2(R+\tfrac{1}{2}s, 1, R-\tfrac{1}{2}s, 2)$$

with the mean correlation function (cf. II.149)

$$\tilde{c}_2(R+\tfrac{1}{2}s, 1, R-\tfrac{1}{2}s, 2) \equiv \int_{-1}^{0} c_2\{R+\tfrac{1}{2}(\lambda+1)s, 1, R+\tfrac{1}{2}(\lambda-1)s, 2\}\mathrm{d}\lambda.$$

Apply this relation to the correlation terms of the energy–momentum law for a plasma and show that one obtains again an equation of the form (121) with a second contribution to the energy–momentum tensor, given by an expression like (116), but with $\tilde{c}_2^{ab}$ instead of $c_2^{ab}$. (A similar extension may be given for dipole substances.)

**10.** Show that the non-relativistic limits of the space–space components of the macroscopic inner angular momentum density $S^{\alpha\beta}$ (175) are given by (II.166) of the non-relativistic treatment. Write to that end $S^{\alpha\beta}$ with synchronous distribution functions of the type (55), namely as

$$S^{\alpha\beta} = \int \gamma_1^{-1} s_1^{\alpha\beta} f_1^{\text{syn}}(\boldsymbol{R}, s_1^{\gamma\varepsilon}; t) \mathrm{d}^6 s_1^{\gamma\varepsilon}$$

(cf. problem 1).

**11.** Show that the antisymmetric part of the energy–momentum tensor for a neutral plasma may be written in the form (200) with the quantity $J^{\alpha\beta\gamma}$ given as

$$J^{\alpha\beta\gamma} = c^{-2} \sum_{a,b} \frac{e_a e_b}{16\pi} \int (s^\alpha u_2^\beta - s^\beta u_2^\alpha) u_1^\gamma \, \delta(s^2) c_2^{ab}(R+\tfrac{1}{2}s, 1, R-\tfrac{1}{2}s, 2) \mathrm{d}1\, \mathrm{d}2\, \mathrm{d}s.$$

Prove that in the non-relativistic limit both $c^{-1}J^{ij0}$ and $J^{ijk}$ with $i, j, k = 1, 2, 3$ vanish.

**12.** Prove the statements on the non-relativistic limit of the inner angular momentum flow $J^{\alpha\beta\gamma}$ for a dipole fluid that are given below (199).

**13.** The energy–momentum tensor (122) for a neutral plasma is not symmetric since $T_{(\mathrm{m})\mathrm{II}}^{\alpha\beta}$ (116) contains a part that is asymmetric. By adding a divergenceless part to $T_{(\mathrm{m})\mathrm{II}}^{\alpha\beta}$ one may bring it into a symmetric form. Show that one may choose for this divergenceless term:

$$-c^{-2} \sum_{a,b} \frac{e_a e_b}{16\pi} \partial_\gamma \int s^\alpha u_2^\beta u_1^\gamma \, \delta(s^2) c_2^{ab}(R+\tfrac{1}{2}s, 1, R-\tfrac{1}{2}s, 2) \mathrm{d}1\, \mathrm{d}2\, \mathrm{d}s,$$

so that the second contribution to the energy–momentum tensor gets the symmetric form

$$\tfrac{1}{2}(T_{(\mathrm{m})\mathrm{II}}^{\alpha\beta} + T_{(\mathrm{m})\mathrm{II}}^{\beta\alpha}) - c^{-2} \sum_{a,b} \frac{e_a e_b}{32\pi} \partial_\gamma \int (s^\alpha u_2^\beta + s^\beta u_2^\alpha) \\ u_1^\gamma \, \delta(s^2) c_2^{ab}(R+\tfrac{1}{2}s, 1, R-\tfrac{1}{2}s, 2) \mathrm{d}1\, \mathrm{d}2\, \mathrm{d}s.$$

Show that in the non-relativistic limit the two tensors discussed coincide.

**14.** Derive a form of the first law for a polarized fluid of neutral atoms in local equilibrium by starting from the 'Helmholtz' splitting (251) of the total energy–momentum tensor and using the force density (282).

Hint: calculate first the inner product $U^{\alpha}\mathscr{F}_{\alpha}$.

# PART C

# *Non-relativistic quantum-mechanical electrodynamics*

CHAPTER VI

# The Weyl formulation of the microscopic laws

## 1 *Introduction*

The various laws of electrodynamics have been derived in the preceding on the basis of a classical model of matter. The question arises in how far these results retain their validity if a more realistic model based on quantum mechanics is adopted. This programme will be carried out in this chapter and the following, at first in the non-relativistic approximation.

Quantum mechanics is usually formulated in terms of operators and state vectors in Hilbert space. In the course of the treatment it will be convenient to introduce instead an equivalent formalism which employs functions in phase space. Then the physical quantities are represented by Weyl transforms, while the Wigner function takes over the role of the density operator.

## 2 *The field equations and the equation of motion of a set of charged point particles*

The starting point of the theory consists in propounding the Hamilton operator for a set of charged particles. For a system consisting of $N$ particles with masses $m_i$, charges $e_i$, coordinate operators $\boldsymbol{R}_{i,\mathrm{op}}$ and momentum operators $\boldsymbol{P}_{i,\mathrm{op}}$, in an external electromagnetic field with potentials $(\varphi_{\mathrm{e}}, \boldsymbol{A}_{\mathrm{e}})$ one writes the Hamilton operator up to terms of order $c^{-1}$ as:

$$H_{\mathrm{op}} = \sum_i \frac{\boldsymbol{P}_{i,\mathrm{op}}^2}{2m_i} + \sum_{i,j(i\neq j)} \frac{e_i e_j}{8\pi|\boldsymbol{R}_{i,\mathrm{op}}-\boldsymbol{R}_{j,\mathrm{op}}|} + \sum_i e_i \left[\varphi_{\mathrm{e}}(\boldsymbol{R}_{i,\mathrm{op}}, t) - \tfrac{1}{2}c^{-1}\left\{\frac{\boldsymbol{P}_{i,\mathrm{op}}}{m_i} \cdot, \boldsymbol{A}_{\mathrm{e}}(\boldsymbol{R}_{i,\mathrm{op}}, t)\right\}\right] \tag{1}$$

(cf. the classical expression (I.16)). In the last term the anticommutator of the momentum operator and the vector potential appears. As usual the dot indicates the scalar product of two vectors. This (hermitian) Hamilton operator governs the time development of the state vector $|\psi\rangle$ for the system via Schrödinger's equation.

The field equations, which in quantum mechanics replace the classical Lorentz equations (I.1) are the following differential equations for the hermitian operators $\boldsymbol{e}_{\text{op}}(\boldsymbol{R}, t)$ and $\boldsymbol{b}_{\text{op}}(\boldsymbol{R}, t)$ which represent the electric and magnetic fields:

$$
\begin{aligned}
\nabla \cdot \boldsymbol{e}_{\text{op}} &= \sum_i e_i\, \delta(\boldsymbol{R}_{i,\text{op}} - \boldsymbol{R}), \\
-\partial_0\, \boldsymbol{e}_{\text{op}} + \nabla \wedge \boldsymbol{b}_{\text{op}} &= \tfrac{1}{2} c^{-1} \sum_i e_i \{\dot{\boldsymbol{R}}_{i,\text{op}}, \delta(\boldsymbol{R}_{i,\text{op}} - \boldsymbol{R})\}, \\
\nabla \cdot \boldsymbol{b}_{\text{op}} &= 0, \\
\partial_0\, \boldsymbol{b}_{\text{op}} + \nabla \wedge \boldsymbol{e}_{\text{op}} &= 0,
\end{aligned}
\tag{2}
$$

where $c\partial_0$ and the fluxion dot should be interpreted as $(i/\hbar)$ times the commutator with the Hamiltonian plus the explicit time derivative:

$$
\dot{A}_{\text{op}} = c\partial_0\, A_{\text{op}} = \frac{i}{\hbar}\, [H_{\text{op}}, A_{\text{op}}] + \frac{\partial^{\text{e}} A_{\text{op}}}{\partial t}. \tag{3}
$$

(The time derivative of the expectation value of the operator $A_{\text{op}}$ is equal to the expectation value of $\dot{A}_{\text{op}}$, as follows from Schrödinger's equation.) In particular the fluxion $\dot{\boldsymbol{R}}_{i,\text{op}}$ is, up to order $c^0$, equal to $\boldsymbol{P}_{i,\text{op}}/m_i$, as follows from (1). The sources of the field equations (2) are hermitian operators, as is guaranteed by the fact that the anticommutator has been written in the second equation. From the field equations (2) one may find in principle the expressions for the electric and magnetic fields in terms of the coordinate and momentum operators of the particles.

The equation of motion for the charged particles follows by taking the second time derivative of the coordinate operator of particle $i$ in the sense defined by (3). With the Hamilton operator (1) one finds

$$
\begin{aligned}
m_i\, \dot{\boldsymbol{R}}_{i,\text{op}} &= \boldsymbol{P}_{i,\text{op}} - c^{-1} e_i\, \boldsymbol{A}_{\text{e}}(\boldsymbol{R}_{i,\text{op}}, t), \\
m_i\, \boldsymbol{R}_{i,\text{op}} &= -\nabla_{i,\text{op}} \sum_{j(\neq i)} \frac{e_i\, e_j}{4\pi |\boldsymbol{R}_{i,\text{op}} - \boldsymbol{R}_{j,\text{op}}|} \\
&\quad + e_i \left[ \boldsymbol{E}_{\text{e}}(\boldsymbol{R}_{i,\text{op}}, t) + \tfrac{1}{2} c^{-1} \left\{ \frac{\boldsymbol{P}_{i,\text{op}}}{m_i} \wedge, \boldsymbol{B}_{\text{e}}(\boldsymbol{R}_{i,\text{op}}, t) \right\} \right],
\end{aligned}
\tag{4}
$$

where we used the connexions of the fields and the potentials:

$$
\boldsymbol{E}_{\text{e}} = -\nabla \varphi_{\text{e}} - \partial_0\, \boldsymbol{A}_{\text{e}}, \qquad \boldsymbol{B}_{\text{e}} = \nabla \wedge \boldsymbol{A}_{\text{e}}. \tag{5}
$$

Furthermore the symbol $\nabla_{i,\mathrm{op}}$ denotes partial differentiation with respect to $\boldsymbol{R}_{i,\mathrm{op}}$ and the last anticommutator with the vector product stands for $(1/m_i)(\boldsymbol{P}_{i,\mathrm{op}} \wedge \boldsymbol{B}_\mathrm{e} - \boldsymbol{B}_\mathrm{e} \wedge \boldsymbol{P}_{i,\mathrm{op}})$.

The field equations (2) and the equations of motion (4) form the basis for the derivation of the macroscopic field equations and energy–momentum laws in this chapter and the following. The equations (2) and (4) will not be used in the form given so far in view of the fact that the handling of operators, in particular of anticommutators, leads to rather unwieldy expressions. Instead we shall use Weyl transforms.

## 3 *The Weyl transformation and the Wigner function*

### a. *The Weyl transformation*

Quantum mechanics in its usual form is concerned with the properties of vectors and operators in Hilbert space: each state of a system corresponds to a vector, each observable quantity to an operator.

However different formulations[1] are possible in which functions in a phase space are associated to both the states and the observable quantities. An example of such a formulation consists in employing for these functions the Wigner function and the Weyl transform respectively. This alternative formulation will be demonstrated for a one-particle system. The generalization to an $N$-particle system is trivial.

The eigenvectors $|\boldsymbol{p}\rangle$ and $|\boldsymbol{q}\rangle$ of the momentum and coordinate operators[2] $\boldsymbol{P}$ and $\boldsymbol{Q}$ satisfy the eigenvalue equations

$$\boldsymbol{P}|\boldsymbol{p}\rangle = \boldsymbol{p}|\boldsymbol{p}\rangle, \qquad \boldsymbol{Q}|\boldsymbol{q}\rangle = \boldsymbol{q}|\boldsymbol{q}\rangle, \tag{6}$$

where $\boldsymbol{p}$ and $\boldsymbol{q}$ are the eigenvalues. The complete set of eigenvectors is supposed to fulfil the closure relations

$$\begin{aligned} \int \mathrm{d}\boldsymbol{p}\, |\boldsymbol{p}\rangle\langle\boldsymbol{p}| &= I, \\ \int \mathrm{d}\boldsymbol{q}\, |\boldsymbol{q}\rangle\langle\boldsymbol{q}| &= I, \end{aligned} \tag{7}$$

[1] H. Weyl, Z. Physik **46**(1927)1; Gruppentheorie und Quantenmechanik (Hirzel, Leipzig 1931) p. 244; E. P. Wigner, Phys. Rev. **40**(1932)749; cf. also H. J. Groenewold, Physica **12**(1946)405; J. E. Moyal, Proc. Cambridge Phil. Soc. **45**(1949)99; K. Schram and B. R. A. Nijboer, Physica **25**(1959)733; B. Leaf, J. Math. Phys. **9**(1968)65, 769.

[2] In this section we use systematically capitals for operators, and lower case symbols for ordinary numbers.

(with $I$ the unit operator in Hilbert space) and the orthogonality relations

$$\langle \boldsymbol{p}|\boldsymbol{p}'\rangle = \delta(\boldsymbol{p}-\boldsymbol{p}'), \qquad \langle \boldsymbol{q}|\boldsymbol{q}'\rangle = \delta(\boldsymbol{q}-\boldsymbol{q}'). \tag{8}$$

The commutation rules of the operators $\boldsymbol{P}$ and $\boldsymbol{Q}$ read

$$[\boldsymbol{P}, \boldsymbol{P}] = 0, \qquad [\boldsymbol{Q}, \boldsymbol{Q}] = 0, \qquad [\boldsymbol{P}, \boldsymbol{Q}] = \frac{\hbar}{i}\,\mathsf{U}I, \tag{9}$$

where $\mathsf{U}$ is the unit Cartesian tensor with components $(i, j = 1, 2, 3)$. In the coordinate representation the momentum eigenvector has the wave function

$$\langle \boldsymbol{q}|\boldsymbol{p}\rangle = \frac{1}{h^{\frac{3}{2}}}\,\mathrm{e}^{(i/\hbar)\boldsymbol{p}\cdot\boldsymbol{q}}. \tag{10}$$

By using the closure relations (7) one may write the following identity[1] for an arbitrary operator $A$:

$$A = \int \mathrm{d}\boldsymbol{p}'\mathrm{d}\boldsymbol{p}''\mathrm{d}\boldsymbol{q}'\mathrm{d}\boldsymbol{q}''|\boldsymbol{q}''\rangle\langle \boldsymbol{q}''|\boldsymbol{p}''\rangle\langle \boldsymbol{p}''|A|\boldsymbol{p}'\rangle\langle \boldsymbol{p}'|\boldsymbol{q}'\rangle\langle \boldsymbol{q}'|. \tag{11}$$

Introducing the new integration variables

$$\begin{aligned} \boldsymbol{p}' &= \boldsymbol{p}-\tfrac{1}{2}\boldsymbol{u}, & \boldsymbol{q}' &= \boldsymbol{q}-\tfrac{1}{2}\boldsymbol{v}, \\ \boldsymbol{p}'' &= \boldsymbol{p}+\tfrac{1}{2}\boldsymbol{u}, & \boldsymbol{q}'' &= \boldsymbol{q}+\tfrac{1}{2}\boldsymbol{v}, \end{aligned} \tag{12}$$

with Jacobian equal to unity, one obtains for (11), using (10)

$$A = h^{-3}\int \mathrm{d}\boldsymbol{p}\,\mathrm{d}\boldsymbol{q}\; a(\boldsymbol{p}, \boldsymbol{q})\Delta(\boldsymbol{p}, \boldsymbol{q}) \tag{13}$$

with the function

$$a(\boldsymbol{p}, \boldsymbol{q}) \equiv \int \mathrm{d}\boldsymbol{u}\,\mathrm{e}^{(i/\hbar)\boldsymbol{q}\cdot\boldsymbol{u}}\langle \boldsymbol{p}+\tfrac{1}{2}\boldsymbol{u}|A|\boldsymbol{p}-\tfrac{1}{2}\boldsymbol{u}\rangle, \tag{14}$$

depending on $A$ and the hermitian operator

$$\Delta(\boldsymbol{p}, \boldsymbol{q}) \equiv \int \mathrm{d}\boldsymbol{v}\,\mathrm{e}^{(i/\hbar)\boldsymbol{p}\cdot\boldsymbol{v}}|\boldsymbol{q}+\tfrac{1}{2}\boldsymbol{v}\rangle\langle \boldsymbol{q}-\tfrac{1}{2}\boldsymbol{v}|, \tag{15}$$

which is independent of the operator $A$.

The function $a(\boldsymbol{p}, \boldsymbol{q})$ is called the *Weyl transform* of the operator $A$ with respect to the momentum and coordinate operators $\boldsymbol{P}$ and $\boldsymbol{Q}$. (The corre-

[1] B. Leaf, J. Math. Phys. 9(1968)65, on which paper most of the material in this section is based.

spondence between an operator and its Weyl transform will be denoted by the symbol $\rightleftarrows$.) In this way one has associated a $c$-number to each operator. If the operator $A$ is hermitian, the function $a(\boldsymbol{p}, \boldsymbol{q})$ is real, as follows from (14).

In (14) and (15) the variables $\boldsymbol{p}$ and $\boldsymbol{q}$ have not been treated on the same footing. We may obtain symmetrical forms for (15) by noting first that

$$|\boldsymbol{q}+\tfrac{1}{2}\boldsymbol{v}\rangle = \mathrm{e}^{-(i/\hbar)\boldsymbol{v}\cdot\boldsymbol{P}}|\boldsymbol{q}-\tfrac{1}{2}\boldsymbol{v}\rangle. \tag{16}$$

(The validity of this identity may be verified by multiplication with $\langle\boldsymbol{p}|$ and the use of (6) and (10).) If (16) is substituted into (15), one encounters the projection operator $|\boldsymbol{q}-\frac{1}{2}\boldsymbol{v}\rangle\langle\boldsymbol{q}-\frac{1}{2}\boldsymbol{v}|$ which may be transformed with the help of the identity

$$|\boldsymbol{q}\rangle\langle\boldsymbol{q}| = h^{-3}\int \mathrm{d}\boldsymbol{u}\,\mathrm{e}^{(i/\hbar)(\boldsymbol{q}-\boldsymbol{Q})\cdot\boldsymbol{u}}. \tag{17}$$

(This identity may be verified by letting it operate on $|\boldsymbol{q}'\rangle$ and by using (6).) In that way (15) becomes

$$\Delta(\boldsymbol{p}, \boldsymbol{q}) = h^{-3}\int \mathrm{d}\boldsymbol{u}\,\mathrm{d}\boldsymbol{v}\,\mathrm{e}^{(i/\hbar)(\boldsymbol{p}-\boldsymbol{P})\cdot\boldsymbol{v}}\mathrm{e}^{(i/\hbar)(\boldsymbol{q}-\frac{1}{2}\boldsymbol{v}-\boldsymbol{Q})\cdot\boldsymbol{u}}. \tag{18}$$

With the help of the operator identity, valid for operators $A$ and $B$ that commute with their commutator:

$$\mathrm{e}^{A}\mathrm{e}^{B} = \mathrm{e}^{A+B+\frac{1}{2}[A,B]} \tag{19}$$

(v. problem 1) and the commutation rule (9), one gets for (18) a symmetric form:

$$\Delta(\boldsymbol{p}, \boldsymbol{q}) = h^{-3}\int \mathrm{d}\boldsymbol{u}\,\mathrm{d}\boldsymbol{v}\,\mathrm{e}^{(i/\hbar)\{(\boldsymbol{q}-\boldsymbol{Q})\cdot\boldsymbol{u}+(\boldsymbol{p}-\boldsymbol{P})\cdot\boldsymbol{v}\}}. \tag{20}$$

From this expression one may find, interchanging the roles of $\boldsymbol{p}$ and $\boldsymbol{q}$, and retracing the argument given above, a form for $\Delta(\boldsymbol{p}, \boldsymbol{q})$ which is the counterpart of (15):

$$\Delta(\boldsymbol{p}, \boldsymbol{q}) = \int \mathrm{d}\boldsymbol{u}\,\mathrm{e}^{(i/\hbar)\boldsymbol{q}\cdot\boldsymbol{u}}|\boldsymbol{p}-\tfrac{1}{2}\boldsymbol{u}\rangle\langle\boldsymbol{p}+\tfrac{1}{2}\boldsymbol{u}|. \tag{21}$$

The Weyl transform may also be written in different forms. Since the trace of an operator may be written in terms of the complete set $|\boldsymbol{p}\rangle$ as

$$\mathrm{Tr}\,A = \int \mathrm{d}\boldsymbol{p}\langle\boldsymbol{p}|A|\boldsymbol{p}\rangle, \tag{22}$$

one has the identity

$$\langle \boldsymbol{p}'|A|\boldsymbol{p}''\rangle = \mathrm{Tr}\,(A|\boldsymbol{p}''\rangle\langle\boldsymbol{p}'|). \tag{23}$$

From the expression (14) for $a(\boldsymbol{p}, \boldsymbol{q})$ together with the expression (21) for $\Delta(\boldsymbol{p}, \boldsymbol{q})$ one then obtains the concise formula

$$a(\boldsymbol{p}, \boldsymbol{q}) = \mathrm{Tr}\,\{A\Delta(\boldsymbol{p}, \boldsymbol{q})\}. \tag{24}$$

With (15) for $\Delta(\boldsymbol{p}, \boldsymbol{q})$ and the alternative form of the trace

$$\mathrm{Tr}\,A = \int \mathrm{d}\boldsymbol{q}\langle\boldsymbol{q}|A|\boldsymbol{q}\rangle, \tag{25}$$

one finds the counterpart of (14):

$$a(\boldsymbol{p}, \boldsymbol{q}) = \int \mathrm{d}\boldsymbol{v}\, \mathrm{e}^{(i/\hbar)\boldsymbol{p}\cdot\boldsymbol{v}}\langle\boldsymbol{q}-\tfrac{1}{2}\boldsymbol{v}|A|\boldsymbol{q}+\tfrac{1}{2}\boldsymbol{v}\rangle. \tag{26}$$

The formulae (14) and (26) show that the set of operators $A$ in Hilbert space may be mapped upon a set of $c$-numbers: their Weyl transforms $a(\boldsymbol{p}, \boldsymbol{q})$. The reverse is also true: each function $a(\boldsymbol{p}, \boldsymbol{q})$ may generate an operator $A$ by means of formula (13) with (15) or (20) or (21). In particular if one uses (20) for the $\Delta$-operator, formula (13) becomes

$$A = h^{-6} \int \mathrm{d}\boldsymbol{p}\,\mathrm{d}\boldsymbol{q}\,\mathrm{d}\boldsymbol{u}\,\mathrm{d}\boldsymbol{v}\, a(\boldsymbol{p}, \boldsymbol{q})\mathrm{e}^{(i/\hbar)\{(\boldsymbol{q}-\boldsymbol{Q})\cdot\boldsymbol{u}+(\boldsymbol{p}-\boldsymbol{P})\cdot\boldsymbol{v}\}}. \tag{27}$$

One recognizes the Fourier transform $\tilde{a}(\boldsymbol{u}, \boldsymbol{v})$ of $a(\boldsymbol{p}, \boldsymbol{q})$:

$$\tilde{a}(\boldsymbol{u}, \boldsymbol{v}) = h^{-6} \int \mathrm{d}\boldsymbol{p}\,\mathrm{d}\boldsymbol{q}\, a(\boldsymbol{p}, \boldsymbol{q})\mathrm{e}^{(i/\hbar)(\boldsymbol{q}\cdot\boldsymbol{u}+\boldsymbol{p}\cdot\boldsymbol{v})}, \tag{28}$$

of which the inverse is

$$a(\boldsymbol{p}, \boldsymbol{q}) = \int \mathrm{d}\boldsymbol{u}\,\mathrm{d}\boldsymbol{v}\, \tilde{a}(\boldsymbol{u}, \boldsymbol{v})\mathrm{e}^{-(i/\hbar)(\boldsymbol{q}\cdot\boldsymbol{u}+\boldsymbol{p}\cdot\boldsymbol{v})}. \tag{29}$$

The expression (27) may hence be written as

$$A = \int \mathrm{d}\boldsymbol{u}\,\mathrm{d}\boldsymbol{v}\, \tilde{a}(\boldsymbol{u}, \boldsymbol{v})\mathrm{e}^{-(i/\hbar)(\boldsymbol{Q}\cdot\boldsymbol{u}+\boldsymbol{P}\cdot\boldsymbol{v})}. \tag{30}$$

The expressions (29) and (30) show the correspondence between an operator and its Weyl transform in an elegant way.

Weyl transforms of operators are especially simple, if the operator is a function of $\boldsymbol{P}$ or $\boldsymbol{Q}$ only. In fact the Weyl transform is then the same function of $\boldsymbol{p}$ and $\boldsymbol{q}$ respectively. An example of a Weyl correspondence which will be used frequently in the following is

$$\tfrac{1}{2}\{\boldsymbol{P}, f(\boldsymbol{Q})\} \rightleftarrows \boldsymbol{p} f(\boldsymbol{q}), \tag{31}$$

where the curly brackets indicate the anticommutator: $\{A, B\} \equiv AB+BA$ and where $f$ is an arbitrary function. It may be proved from (14) by insertion of a complete set $|\boldsymbol{q}'\rangle$ with the help of (7), and by application of (10).

One may ask whether taking the Weyl transform is an operation that is invariant under a change of canonical coordinates and momenta, i.e. whether one finds again the result (29) for the Weyl transform of an operator (30), if one performs first a transformation of coordinate and momentum operators in $A$, then takes the Weyl transform with respect to these new coordinates and momenta, and finally transforms back the Weyl transform to the old coordinates and momenta. For a linear transformation this invariance is guaranteed by the linear character of the exponentials in (29) and (30) (v. problem 2).

The operator $\Delta(\boldsymbol{p}, \boldsymbol{q})$ plays a special role in the Weyl correspondence, as is apparent from (13). By choosing in particular for $A$ the operator $\Delta(\boldsymbol{p}', \boldsymbol{q}')$ one finds that the Weyl transform of $\Delta(\boldsymbol{p}', \boldsymbol{q}')$ is essentially a delta function:

$$\Delta(\boldsymbol{p}', \boldsymbol{q}') \rightleftarrows h^3 \delta(\boldsymbol{p}-\boldsymbol{p}')\delta(\boldsymbol{q}-\boldsymbol{q}'). \tag{32}$$

The trace of the operator $\Delta(\boldsymbol{p}, \boldsymbol{q})$ and of products of $\Delta$-operators will be used frequently in the following. In particular one finds from (25) with (8) and (15) that

$$\operatorname{Tr} \Delta(\boldsymbol{p}, \boldsymbol{q}) = 1. \tag{33}$$

Furthermore from (24) with $A = \Delta(\boldsymbol{p}', \boldsymbol{q}')$ and (32) one has

$$\operatorname{Tr} \{\Delta(\boldsymbol{p}, \boldsymbol{q})\Delta(\boldsymbol{p}', \boldsymbol{q}')\} = h^3 \delta(\boldsymbol{p}-\boldsymbol{p}')\delta(\boldsymbol{q}-\boldsymbol{q}'). \tag{34}$$

The trace of the product of three $\Delta$-operators will also be useful. From (25) one finds with insertion of the completeness relation (7) for $|\boldsymbol{q}\rangle$, the expression (15) for $\Delta(\boldsymbol{p}, \boldsymbol{q})$ and with the use of (8):

$$\begin{aligned} &\operatorname{Tr} \{\Delta(\boldsymbol{p}, \boldsymbol{q})\Delta(\boldsymbol{p}', \boldsymbol{q}')\Delta(\boldsymbol{p}'', \boldsymbol{q}'')\} \\ &= \int \mathrm{d}\boldsymbol{q}_1\, \mathrm{d}\boldsymbol{q}_2\, \mathrm{d}\boldsymbol{q}_3\, \delta(\boldsymbol{q} - \tfrac{\boldsymbol{q}_1+\boldsymbol{q}_2}{2})\delta(\boldsymbol{q}' - \tfrac{\boldsymbol{q}_2+\boldsymbol{q}_3}{2})\delta(\boldsymbol{q}'' - \tfrac{\boldsymbol{q}_3+\boldsymbol{q}_1}{2}) \\ &\exp\left[\frac{i}{\hbar}\{\boldsymbol{p}\cdot(\boldsymbol{q}_1-\boldsymbol{q}_2)+\boldsymbol{p}'\cdot(\boldsymbol{q}_2-\boldsymbol{q}_3)+\boldsymbol{p}''\cdot(\boldsymbol{q}_3-\boldsymbol{q}_1)\}\right]. \end{aligned} \tag{35}$$

After introduction of new variables $\frac{1}{2}(\boldsymbol{q}_1+\boldsymbol{q}_2)$, $\frac{1}{2}(\boldsymbol{q}_2+\boldsymbol{q}_3)$ and $\frac{1}{2}(\boldsymbol{q}_3+\boldsymbol{q}_1)$ the integration over the delta functions may be performed with the result

$$\operatorname{Tr}\{\Delta(\boldsymbol{p},\boldsymbol{q})\Delta(\boldsymbol{p}',\boldsymbol{q}')\Delta(\boldsymbol{p}'',\boldsymbol{q}'')\}$$
$$= 2^6 \exp\left[\frac{2i}{\hbar}\{\boldsymbol{p}\cdot(\boldsymbol{q}''-\boldsymbol{q}')+\boldsymbol{p}'\cdot(\boldsymbol{q}-\boldsymbol{q}'')+\boldsymbol{p}''\cdot(\boldsymbol{q}'-\boldsymbol{q})\}\right]. \quad (36)$$

The trace of an arbitrary operator may be expressed in terms of its Weyl transform by making use of (13) and (33):

$$\operatorname{Tr} A = h^{-3}\int \mathrm{d}\boldsymbol{p}\,\mathrm{d}\boldsymbol{q}\; a(\boldsymbol{p},\boldsymbol{q}). \quad (37)$$

The trace of a product of two operators follows from (13) and (34):

$$\operatorname{Tr} AB = h^{-3}\int \mathrm{d}\boldsymbol{p}\,\mathrm{d}\boldsymbol{q}\; a(\boldsymbol{p},\boldsymbol{q})b(\boldsymbol{p},\boldsymbol{q}) = \operatorname{Tr} BA. \quad (38)$$

In quantum mechanics commutators and anticommutators play an important role. To find their Weyl transform we need to study the Weyl transform of a product of operators. The latter is not in general simply the product of the Weyl transforms of the operators. To find the Weyl transform of the product of operators $AB$ one may start from (24) and use (13) and (36):

$$AB \rightleftarrows \left(\frac{2}{h}\right)^6 \int \mathrm{d}\boldsymbol{p}'\mathrm{d}\boldsymbol{q}'\mathrm{d}\boldsymbol{p}''\mathrm{d}\boldsymbol{q}''a(\boldsymbol{p}',\boldsymbol{q}')b(\boldsymbol{p}'',\boldsymbol{q}'')$$
$$\exp\left[\frac{2i}{\hbar}\{(\boldsymbol{q}'-\boldsymbol{q})\cdot(\boldsymbol{p}''-\boldsymbol{p})-(\boldsymbol{p}'-\boldsymbol{p})\cdot(\boldsymbol{q}''-\boldsymbol{q})\}\right]. \quad (39)$$

Introducing new variables $\hat{\boldsymbol{p}} = \boldsymbol{p}''-\boldsymbol{p}$ and $\hat{\boldsymbol{q}} = \boldsymbol{q}''-\boldsymbol{q}$ and expanding $b(\boldsymbol{p}+\hat{\boldsymbol{p}}, \boldsymbol{q}+\hat{\boldsymbol{q}})$ into a Taylor series around $b(\boldsymbol{p},\boldsymbol{q})$, one gets

$$AB \rightleftarrows \left(\frac{2}{h}\right)^6 \int \mathrm{d}\boldsymbol{p}'\mathrm{d}\boldsymbol{q}'\mathrm{d}\hat{\boldsymbol{p}}\,\mathrm{d}\hat{\boldsymbol{q}}\; a(\boldsymbol{p}',\boldsymbol{q}')$$
$$\exp\left[\frac{2i}{\hbar}\{(\boldsymbol{q}'-\boldsymbol{q})\cdot\hat{\boldsymbol{p}}-(\boldsymbol{p}'-\boldsymbol{p})\cdot\hat{\boldsymbol{q}}\}\right]\left\{\exp\left(\hat{\boldsymbol{p}}\cdot\frac{\partial}{\partial\boldsymbol{p}}+\hat{\boldsymbol{q}}\cdot\frac{\partial}{\partial\boldsymbol{q}}\right)\right\} b(\boldsymbol{p},\boldsymbol{q}). \quad (40)$$

In the last exponential we may replace $\hat{\boldsymbol{p}}$ by the differential operator $-(\hbar/2i)\partial^{\leftarrow}/\partial\boldsymbol{q}$ and likewise $\hat{\boldsymbol{q}}$ by $(\hbar/2i)\partial^{\leftarrow}/\partial\boldsymbol{p}$, both acting to the left. Then the integration over $\hat{\boldsymbol{p}}$ and $\hat{\boldsymbol{q}}$ yields delta functions, so that the integrations over $\boldsymbol{p}'$ and $\boldsymbol{q}'$ may also be performed. In this way one gets[1]

$$AB \rightleftarrows a(\boldsymbol{p}, \boldsymbol{q}) \exp \left\{ \frac{\hbar}{2i} \left( \frac{\partial^{\leftarrow}}{\partial \boldsymbol{p}} \cdot \frac{\partial^{\rightarrow}}{\partial \boldsymbol{q}} - \frac{\partial^{\leftarrow}}{\partial \boldsymbol{q}} \cdot \frac{\partial^{\rightarrow}}{\partial \boldsymbol{p}} \right) \right\} b(\boldsymbol{p}, \boldsymbol{q}), \tag{41}$$

where for aesthetic reasons arrows pointing to the right have been added. An alternative notation of (41) is obtained by attaching indices $(a)$ and $(b)$ to the differential operators, indicating their objects:

$$AB \rightleftarrows \exp \left\{ \frac{\hbar}{2i} \left( \frac{\partial^{(a)}}{\partial \boldsymbol{p}} \cdot \frac{\partial^{(b)}}{\partial \boldsymbol{q}} - \frac{\partial^{(a)}}{\partial \boldsymbol{q}} \cdot \frac{\partial^{(b)}}{\partial \boldsymbol{p}} \right) \right\} a(\boldsymbol{p}, \boldsymbol{q}) b(\boldsymbol{p}, \boldsymbol{q}). \tag{42}$$

The right-hand side of (41) or (42) shows that the Weyl transform of a product of operators $AB$ is not in general equal to the product $a(\boldsymbol{p}, \boldsymbol{q}) b(\boldsymbol{p}, \boldsymbol{q})$ of their Weyl transforms.

Let us write as corollaries the Weyl transforms of the anticommutator and the commutator of $A$ and $B$:

$$\{A, B\} \rightleftarrows 2 \cos \left\{ \frac{\hbar}{2} \left( \frac{\partial^{(a)}}{\partial \boldsymbol{q}} \cdot \frac{\partial^{(b)}}{\partial \boldsymbol{p}} - \frac{\partial^{(a)}}{\partial \boldsymbol{p}} \cdot \frac{\partial^{(b)}}{\partial \boldsymbol{q}} \right) \right\} a(\boldsymbol{p}, \boldsymbol{q}) b(\boldsymbol{p}, \boldsymbol{q}), \tag{43}$$

$$[A, B] \rightleftarrows 2i \sin \left\{ \frac{\hbar}{2} \left( \frac{\partial^{(a)}}{\partial \boldsymbol{q}} \cdot \frac{\partial^{(b)}}{\partial \boldsymbol{p}} - \frac{\partial^{(a)}}{\partial \boldsymbol{p}} \cdot \frac{\partial^{(b)}}{\partial \boldsymbol{q}} \right) \right\} a(\boldsymbol{p}, \boldsymbol{q}) b(\boldsymbol{p}, \boldsymbol{q}). \tag{44}$$

These formulae show that the Weyl transforms of anticommutators and commutators are series in $\hbar^2$. The lowest order term of (43) is

$$2a(\boldsymbol{p}, \boldsymbol{q}) b(\boldsymbol{p}, \boldsymbol{q}), \tag{45}$$

while that of (44) is

$$i\hbar \left( \frac{\partial a}{\partial \boldsymbol{q}} \cdot \frac{\partial b}{\partial \boldsymbol{p}} - \frac{\partial a}{\partial \boldsymbol{p}} \cdot \frac{\partial b}{\partial \boldsymbol{q}} \right). \tag{46}$$

Hence the series for the Weyl transform of $\frac{1}{2}\{A, B\}$ starts off with the product of the Weyl transforms of $A$ and $B$, whereas the series for the Weyl transform of $-(i/\hbar)[A, B]$ starts off with the Poisson bracket of the Weyl transforms of $A$ and $B$.

An example, which will be frequently used in the following, is furnished by the time derivative $\dot{A}$, defined in (3) as $(i/\hbar)[H, A] + \partial^{e} A/\partial t$. It follows from (44) that:

$$\dot{A} \rightleftarrows \frac{2}{\hbar} \sin \left\{ \frac{\hbar}{2} \left( \frac{\partial^{(a)}}{\partial \boldsymbol{q}} \cdot \frac{\partial^{(h)}}{\partial \boldsymbol{p}} - \frac{\partial^{(a)}}{\partial \boldsymbol{p}} \cdot \frac{\partial^{(h)}}{\partial \boldsymbol{q}} \right) \right\} a(\boldsymbol{p}, \boldsymbol{q}; t) h(\boldsymbol{p}, \boldsymbol{q}) + \frac{\partial a(\boldsymbol{p}, \boldsymbol{q}; t)}{\partial t}. \tag{47}$$

[1] H. J. Groenewold, Physica **12**(1946)405.

Thus the Weyl transform of $\dot{A}$ is a series in $\hbar^2$, which starts off with the Poisson bracket of $a(\boldsymbol{p}, \boldsymbol{q}; t)$ and $h(\boldsymbol{p}, \boldsymbol{q})$, and the explicit time derivative of $a(\boldsymbol{p}, \boldsymbol{q}; t)$.

b. *The Wigner function*

Up to here we studied the one-to-one mapping of operators $A$ and functions in phase space $a(\boldsymbol{p}, \boldsymbol{q})$, given by the Weyl transformation. If one wants to find expectation values of an operator, one is interested in expressions which allow to find the expectation values directly from the Weyl transform of the operator.

A system may be described by its state vector $|\psi(t)\rangle$ in Hilbert space or alternatively by the density operator

$$P(t) = |\psi(t)\rangle\langle\psi(t)|. \tag{48}$$

The expectation value $\bar{A}$ of an operator $A$ is equal to $\langle\psi|A|\psi\rangle$ or

$$\bar{A}(t) = \mathrm{Tr}\,\{P(t)A\}. \tag{49}$$

Such a trace may be written in terms of Weyl transforms as is shown by (38). Let us denote the Weyl transform of the density operator $P$ by $h^3\rho(\boldsymbol{p}, \boldsymbol{q}; t)$:

$$P(t) \rightleftarrows h^3\rho(\boldsymbol{p}, \boldsymbol{q}; t). \tag{50}$$

Then the expectation value (49) becomes

$$\bar{A} = \bar{a}, \tag{51}$$

where the latter quantity is an integral over the product of the Weyl transform $a$ and the Wigner function:

$$\bar{a} \equiv \int \mathrm{d}\boldsymbol{p}\,\mathrm{d}\boldsymbol{q}\; a(\boldsymbol{p}, \boldsymbol{q})\rho(\boldsymbol{p}, \boldsymbol{q}; t). \tag{52}$$

From the expression (26) it follows that the Weyl transform $h^3\rho(\boldsymbol{p}, \boldsymbol{q}; t)$ may be written as

$$h^3\rho(\boldsymbol{p}, \boldsymbol{q}; t) = \int \mathrm{d}\boldsymbol{v}\, \mathrm{e}^{(i/\hbar)\boldsymbol{p}\cdot\boldsymbol{v}} \langle \boldsymbol{q} - \tfrac{1}{2}\boldsymbol{v}|\psi\rangle\langle\psi|\boldsymbol{q} + \tfrac{1}{2}\boldsymbol{v}\rangle \tag{53}$$

or, using the wave function notation $\psi(\boldsymbol{q}; t) \equiv \langle \boldsymbol{q}|\psi(t)\rangle$:

$$\rho(\boldsymbol{p}, \boldsymbol{q}; t) = h^{-3} \int \mathrm{d}\boldsymbol{v}\, \mathrm{e}^{(i/\hbar)\boldsymbol{p}\cdot\boldsymbol{v}} \psi(\boldsymbol{q} - \tfrac{1}{2}\boldsymbol{v}; t)\psi^*(\boldsymbol{q} + \tfrac{1}{2}\boldsymbol{v}; t). \tag{54}$$

At the right-hand side appears the function which Wigner introduced originally[1]. This formula together with (50) shows that the Wigner function is the Weyl transform of the density matrix (divided by $h^3$).

The Wigner function $\rho(\boldsymbol{p}, \boldsymbol{q}; t)$ is real as follows from the hermiticity of the density operator. Furthermore, since the trace of the density operator is unity:

$$\operatorname{Tr} P = 1, \tag{55}$$

(as follows from the normalization of the state vector $|\psi\rangle$) the Wigner function possesses the normalization property

$$\int \mathrm{d}\boldsymbol{p}\, \mathrm{d}\boldsymbol{q}\, \rho(\boldsymbol{p}, \boldsymbol{q}; t) = 1 \tag{56}$$

as a direct consequence of (37).

The relation (51) with (52) shows that one may calculate expectation values of an operator by evaluating an integral involving its Weyl transform and the Wigner function. In view of (51) with (52) and (56) one might be inclined to interpret the Wigner function as a probability density in phase space. However, such an interpretation is not possible since the Wigner function is not necessarily positive definite.

The time evolution of the state vector which describes the system is given by the Schrödinger equation

$$H|\psi(t)\rangle = -\frac{\hbar}{i}\frac{\partial}{\partial t}|\psi(t)\rangle, \tag{57}$$

or in terms of the density operator (48):

$$\frac{\partial P}{\partial t} = -\frac{i}{\hbar}[H, P] \tag{58}$$

(the left-hand side is the explicit time derivative denoted as $\partial^{\mathrm{e}}/\partial t$ in (3)). With the help of (44) this equation may be converted into an equation in terms of the Weyl transforms $h^3\rho(\boldsymbol{p}, \boldsymbol{q}; t)$ and $h(\boldsymbol{p}, \boldsymbol{q})$, namely

$$\frac{\partial \rho(\boldsymbol{p}, \boldsymbol{q}; t)}{\partial t} = \frac{2}{\hbar}\sin\left\{\frac{\hbar}{2}\left(\frac{\partial^{(h)}}{\partial \boldsymbol{q}}\cdot\frac{\partial^{(\rho)}}{\partial \boldsymbol{p}} - \frac{\partial^{(h)}}{\partial \boldsymbol{p}}\cdot\frac{\partial^{(\rho)}}{\partial \boldsymbol{q}}\right)\right\} h(\boldsymbol{p}, \boldsymbol{q})\rho(\boldsymbol{p}, \boldsymbol{q}; t). \tag{59}$$

This equation, which gives the evolution in time of the Wigner function, may be employed to find the time derivative of an expectation value. In fact

[1] E. P. Wigner, Phys. Rev. **40**(1932)749; cf. B. Leaf, op. cit.

one finds from (52) with (59) and a partial integration

$$\frac{\mathrm{d}\bar{a}(t)}{\mathrm{d}t} = \frac{2}{\hbar}\int \mathrm{d}\boldsymbol{p}\,\mathrm{d}\boldsymbol{q}\left[\sin\left\{\frac{\hbar}{2}\left(\frac{\partial^{(a)}}{\partial \boldsymbol{q}}\cdot\frac{\partial^{(h)}}{\partial \boldsymbol{p}} - \frac{\partial^{(a)}}{\partial \boldsymbol{p}}\cdot\frac{\partial^{(h)}}{\partial \boldsymbol{q}}\right)\right\} a(\boldsymbol{p}, \boldsymbol{q})h(\boldsymbol{p}, \boldsymbol{q})\right]\rho(\boldsymbol{p}, \boldsymbol{q}; t). \tag{60}$$

The left-hand side is equal to the time derivative of the expectation value $\bar{A}$ as follows from (51). The right-hand side is equal to the expectation value of the operator $\dot{A}$ (3) as follows from (47) and (51) with (52).

## 4 *The Weyl transforms of the field equations*

The field equations and the equation of motion, which have been written in operator form in section 2, will now be transformed to equations for phase space functions by using the Weyl transformation[1].

The Weyl transform of the Hamiltonian (1) follows by keeping in mind that functions of the coordinate or of the momentum operators transform into the same functions of the coordinates and momenta variables in phase space, and by using (31):

$$H_{\mathrm{op}} \rightleftarrows H(1, \dots, N; t) \equiv \sum_i \frac{\boldsymbol{P}_i^2}{2m_i} + \sum_{i,j(i\neq j)} \frac{e_i e_j}{8\pi|\boldsymbol{R}_i - \boldsymbol{R}_j|} + \sum_i e_i\left\{\varphi_{\mathrm{e}}(\boldsymbol{R}_i, t) - c^{-1}\frac{\boldsymbol{P}_i}{m_i}\cdot\boldsymbol{A}_{\mathrm{e}}(\boldsymbol{R}_i, t)\right\}, \tag{61}$$

where the arguments 1, ..., $N$ stand for $\boldsymbol{P}_1, \boldsymbol{R}_1 \dots \boldsymbol{P}_{\mathrm{N}}, \boldsymbol{R}_{\mathrm{N}}$, the variables of the $N$ particles.

The Weyl transforms of the field equations (2) become, according to (31) and (47)

$$\begin{aligned} \nabla\cdot\boldsymbol{e} &= \sum_i e_i\,\delta(\boldsymbol{R}_i - \boldsymbol{R}), \\ -\partial_0\,\boldsymbol{e} - c^{-1}\sin(\boldsymbol{e}; H) + \nabla\wedge\boldsymbol{b} &= c^{-1}\sum_i e_i\frac{\boldsymbol{P}_i}{m_i}\,\delta(\boldsymbol{R}_i - \boldsymbol{R}), \\ \nabla\cdot\boldsymbol{b} &= 0, \\ \partial_0\,\boldsymbol{b} + c^{-1}\sin(\boldsymbol{b}; H) + \nabla\wedge\boldsymbol{e} &= 0, \end{aligned} \tag{62}$$

[1] We resume the notation op of section 2 for operators, and use symbols without this index for Weyl transforms.

where as an abbreviation we wrote

$$\sin(a;b) \equiv \sum_{i=1}^{N} \frac{2}{\hbar} \sin\left\{\frac{\hbar}{2}\left(\frac{\partial^{(a)}}{\partial \boldsymbol{R}_i}\cdot\frac{\partial^{(b)}}{\partial \boldsymbol{P}_i} - \frac{\partial^{(a)}}{\partial \boldsymbol{P}_i}\cdot\frac{\partial^{(b)}}{\partial \boldsymbol{R}_i}\right)\right\} a(1, \ldots, N)b(1, \ldots, N) \quad (63)$$

and where $\boldsymbol{e}$ and $\boldsymbol{b}$ are functions of 1, …, $N$, $\boldsymbol{R}$ and $t$, while $H$ is given by (61).

To solve these equations, we note first that the last two equations have as solutions

$$\begin{aligned} \boldsymbol{e} &= -\nabla\varphi - \partial_0 \boldsymbol{a} - c^{-1}\sin(\boldsymbol{a}; H), \\ \boldsymbol{b} &= \nabla \wedge \boldsymbol{a} \end{aligned} \quad (64)$$

with functions $\varphi$ and $\boldsymbol{a}$ depending on 1, …, $N$, $\boldsymbol{R}$ and $t$. Insertion into the first two field equations leads, up to order $c^{-1}$, to:

$$\begin{aligned} \Delta\varphi + \partial_0 \nabla\cdot\boldsymbol{a} + c^{-1}\nabla\cdot\sin(\boldsymbol{a}; H) &= -\sum_i e_i\,\delta(\boldsymbol{R}_i - \boldsymbol{R}), \\ \Delta\boldsymbol{a} - \nabla\{\nabla\cdot\boldsymbol{a} + \partial_0\varphi + c^{-1}\sin(\varphi; H)\} &= -c^{-1}\sum_i e_i \frac{\boldsymbol{P}_i}{m_i}\,\delta(\boldsymbol{R}_i - \boldsymbol{R}). \end{aligned} \quad (65)$$

The formulae (64) show that a gauge transformation of the potentials

$$\begin{aligned} \varphi' &= \varphi - \partial_0\psi - c^{-1}\sin(\psi; H), \\ \boldsymbol{a}' &= \boldsymbol{a} + \nabla\psi, \end{aligned} \quad (66)$$

with $\psi(1, \ldots, N; \boldsymbol{R}, t)$ an arbitrary function, leads to the same fields, since the Weyl transform $H$ is independent of $\boldsymbol{R}$. Therefore it is allowed to require as a condition on the potentials

$$\partial_0\varphi + c^{-1}\sin(\varphi; H) + \nabla\cdot\boldsymbol{a} = 0, \quad (67)$$

because starting from potentials which do not satisfy this relation one may find $\psi$ such that the new potentials (66) do satisfy it.

With the use of (67) the equations (65) become separated in $\varphi$ and $\boldsymbol{a}$:

$$\begin{aligned} \Delta\varphi &= -\sum_i e_i\,\delta(\boldsymbol{R}_i - \boldsymbol{R}), \\ \Delta\boldsymbol{a} &= -c^{-1}\sum_i e_i \frac{\boldsymbol{P}_i}{m_i}\,\delta(\boldsymbol{R}_i - \boldsymbol{R}), \end{aligned} \quad (68)$$

where terms of order $c^{-2}$ have been neglected. Solving these equations and inserting the results into (64), we get for the Weyl transforms of the fields:

$$\begin{aligned} \boldsymbol{e} &= \sum_i \boldsymbol{e}_i, \qquad \boldsymbol{e}_i(\boldsymbol{R}_i; \boldsymbol{R}) = -\nabla \frac{e_i}{4\pi|\boldsymbol{R}_i - \boldsymbol{R}|}, \\ \boldsymbol{b} &= \sum_i \boldsymbol{b}_i, \qquad \boldsymbol{b}_i(\boldsymbol{P}_i, \boldsymbol{R}_i; \boldsymbol{R}) = c^{-1}\nabla \wedge \frac{e_i \boldsymbol{P}_i}{4\pi m_i|\boldsymbol{R}_i - \boldsymbol{R}|}. \end{aligned} \quad (69)$$

(Solutions of the sourceless equations may of course be added to these particular solutions.)

From the form of these solutions and that of the Weyl transform $H(1, \ldots, N; t)$ of the Hamiltonian it follows that the sine (63) appearing in the second equation of (62) reduces to

$$\sum_i \frac{\partial \boldsymbol{e}}{\partial \boldsymbol{R}_i} \cdot \frac{\partial H}{\partial \boldsymbol{P}_i}, \tag{70}$$

since the Weyl transform of the electric field is independent of the momenta and the Hamiltonian is quadratic in the momenta. The expression (70) is equal to the Poisson bracket of $\boldsymbol{e}$ and $H$, because Poisson brackets are defined as

$$\{a, b\}_{\mathrm{P}} \equiv \sum_{i=1}^{N} \left( \frac{\partial a}{\partial \boldsymbol{R}_i} \cdot \frac{\partial b}{\partial \boldsymbol{P}_i} - \frac{\partial a}{\partial \boldsymbol{P}_i} \cdot \frac{\partial b}{\partial \boldsymbol{R}_i} \right). \tag{71}$$

Let us introduce the abbreviation $\partial_{t\mathrm{P}} \equiv c\partial_{0\mathrm{P}}$ defined as

$$\partial_{t\mathrm{P}}\, a = c\partial_{0\mathrm{P}}\, a \equiv \{a, H\}_{\mathrm{P}} + \partial_t\, a, \tag{72}$$

where at the right-hand side the sum of the Poisson bracket of the Weyl transforms $a$ and $H$, and the explicit time derivative $\partial_t a \equiv \partial^e a/\partial t$ of $a$ appears. Then we may write the Weyl transforms of the field equations in the form

$$\begin{aligned} \nabla \cdot \boldsymbol{e} &= \sum_i e_i\, \delta(\boldsymbol{R}_i - \boldsymbol{R}), \\ -\partial_{0\mathrm{P}}\, \boldsymbol{e} + \nabla \wedge \boldsymbol{b} &= c^{-1} \sum_i e_i (\partial_{t\mathrm{P}}\, \boldsymbol{R}_i)\, \delta(\boldsymbol{R}_i - \boldsymbol{R}), \\ \nabla \cdot \boldsymbol{b} &= 0, \\ \partial_0\, \boldsymbol{b} + \nabla \wedge \boldsymbol{e} &= 0, \end{aligned} \tag{73}$$

where in the fourth equation only the explicit time derivative appears.

(From the solutions (69) of these equations (73) one may find, if one wishes, the operators for the fields. With (31) one obtains:

$$\begin{aligned} \boldsymbol{e}_{\mathrm{op}} &= \sum_i \boldsymbol{e}_{i,\mathrm{op}}, \qquad \boldsymbol{e}_{i,\mathrm{op}} = -\nabla \frac{e_i}{4\pi |\boldsymbol{R}_{i,\mathrm{op}} - \boldsymbol{R}|}, \\ \boldsymbol{b}_{\mathrm{op}} &= \sum_i \boldsymbol{b}_{i,\mathrm{op}}, \qquad \boldsymbol{b}_{i,\mathrm{op}} = \tfrac{1}{2} c^{-1} \left\{ \nabla \frac{e_i}{4\pi m_i |\boldsymbol{R}_{i,\mathrm{op}} - \boldsymbol{R}|} \wedge, \boldsymbol{P}_{i,\mathrm{op}} \right\}, \end{aligned} \tag{74}$$

where an anticommutator appears in the last expression.)

Equations for the expectation values may be found by multiplying (73) by a Wigner function and integrating over phase space (cf. (52)). Since (73) is equivalent to (62) one then finds with (60)

$$\begin{aligned}
\nabla\cdot\bar{e} &= \int \sum_i e_i\,\delta(\boldsymbol{R}_i-\boldsymbol{R})\rho(1, \ldots, N; t)\mathrm{d}1 \ldots \mathrm{d}N, \\
-\partial_0\bar{e}+\nabla\wedge\bar{b} &= c^{-1}\int \sum_i e_i\,\frac{\boldsymbol{P}_i}{m_i}\,\delta(\boldsymbol{R}_i-\boldsymbol{R})\rho(1, \ldots, N; t)\mathrm{d}1 \ldots \mathrm{d}N, \\
\nabla\cdot\bar{b} &= 0, \\
\partial_0\bar{b}+\nabla\wedge\bar{e} &= 0,
\end{aligned} \tag{75}$$

with the notations (52):

$$\begin{aligned}
\bar{e}(\boldsymbol{R}, t) &= \int e(1, \ldots, N; \boldsymbol{R})\rho(1, \ldots, N; t)\mathrm{d}1 \ldots \mathrm{d}N, \\
\bar{b}(\boldsymbol{R}, t) &= \int b(1, \ldots, N; \boldsymbol{R})\rho(1, \ldots, N; t)\mathrm{d}1 \ldots \mathrm{d}N,
\end{aligned} \tag{76}$$

where 1, ..., $N$ stands for all phase variables of the $N$-point system. (If external fields are present, the fields $\boldsymbol{e}$ and $\boldsymbol{b}$ depend also on $t$.)

It stands to reason that these results could also have been obtained directly from the operator equations (2) by taking the expectation values, as follows from (51–52) together with (31). The advantage of using the Weyl transform and the Wigner function will become apparent in the following when the atomic and macroscopic theories will be dealt with.

## 5 *The Weyl transform of the equation of motion*

The equation of motion in operator form has been given in (4). Its Weyl transform is obtained with the help of (31) and (47):

$$\begin{aligned}
&m_i \sin\{\sin(\boldsymbol{R}_i; H); H\}+m_i\frac{\partial}{\partial t}\{\sin(\boldsymbol{R}_i; H)\} \\
&= -\nabla_i \sum_{j(\neq i)} \frac{e_i e_j}{4\pi|\boldsymbol{R}_i-\boldsymbol{R}_j|}+e_i\left\{\boldsymbol{E}_{\mathrm{e}}(\boldsymbol{R}_i, t)+c^{-1}\frac{\boldsymbol{P}_i}{m_i}\wedge\boldsymbol{B}_{\mathrm{e}}(\boldsymbol{R}_i, t)\right\},
\end{aligned} \tag{77}$$

where the abbreviation (63) has been used. At the right-hand side one recognizes the total fields (cf. (69)) up to order $c^{-1}$ and $c^0$ respectively:

$$\begin{aligned}
\boldsymbol{e}_{\mathrm{t}}(\boldsymbol{R}_i, t) &= \sum_{j(\neq i)} \boldsymbol{e}_j(\boldsymbol{R}_j; \boldsymbol{R}_i)+\boldsymbol{E}_{\mathrm{e}}(\boldsymbol{R}_i, t), \\
\boldsymbol{b}_{\mathrm{t}}(\boldsymbol{R}_i, t) &= \boldsymbol{B}_{\mathrm{e}}(\boldsymbol{R}_i, t).
\end{aligned} \tag{78}$$

The equation (77) may thus also be written as

$$m_i \sin\{\sin(\boldsymbol{R}_i; H); H\} + m_i \frac{\partial}{\partial t}\{\sin(\boldsymbol{R}_i; H)\}$$
$$= e_i \left\{ \boldsymbol{e}_t(\boldsymbol{R}_i, t) + c^{-1} \frac{\boldsymbol{P}_i}{m_i} \wedge \boldsymbol{b}_t(\boldsymbol{R}_i, t) \right\}. \quad (79)$$

From the form of the Weyl transform $H$ (61) of the Hamiltonian and the definition (63) of the sine symbol it follows that the latter reduces here to Poisson brackets (71). This is most easily seen by writing the sine symbol (63) as

$$\sin(a; b) = \sum_{i=1}^{N} \frac{2}{\hbar} \left\{ \sin\left( \frac{\hbar}{2} \frac{\partial^{(a)}}{\partial \boldsymbol{R}_i} \cdot \frac{\partial^{(b)}}{\partial \boldsymbol{P}_i} \right) \cos\left( \frac{\hbar}{2} \frac{\partial^{(a)}}{\partial \boldsymbol{P}_i} \cdot \frac{\partial^{(b)}}{\partial \boldsymbol{R}_i} \right) \right.$$
$$\left. - \cos\left( \frac{\hbar}{2} \frac{\partial^{(a)}}{\partial \boldsymbol{R}_i} \cdot \frac{\partial^{(b)}}{\partial \boldsymbol{P}_i} \right) \sin\left( \frac{\hbar}{2} \frac{\partial^{(a)}}{\partial \boldsymbol{P}_i} \cdot \frac{\partial^{(b)}}{\partial \boldsymbol{R}_i} \right) \right\} a(1, \ldots, N) b(1, \ldots, N). \quad (80)$$

Therefore we find the Weyl transform (79) of the equation of motion as

$$m_i \partial_{t\mathrm{P}}^2 \boldsymbol{R}_i = e_i\{\boldsymbol{e}_t(\boldsymbol{R}_i, t) + c^{-1}(\partial_{t\mathrm{P}} \boldsymbol{R}_i) \wedge \boldsymbol{b}_t(\boldsymbol{R}_i, t)\}. \quad (81)$$

where the abbreviation (72) has been introduced. (An alternative way to derive this equation would have been to calculate directly the repeated Poisson bracket of $\boldsymbol{R}_i$ and $H$.)

An equation for the expectation value follows from (81) by multiplying with a Wigner function and integrating over phase space. Because of the equivalence of (81) and (79) one obtains in this way with (60)

$$m_i \frac{\mathrm{d}^2 \overline{\boldsymbol{R}}_i}{\mathrm{d}t^2} = e_i \int \left\{ \boldsymbol{e}_t(1, \ldots, N; \boldsymbol{R}_i, t) + c^{-1} \frac{\boldsymbol{P}_i}{m_i} \wedge \boldsymbol{b}_t(1, \ldots, N; \boldsymbol{R}_i, t) \right\}$$
$$\rho(1, \ldots, N; t) \mathrm{d}1 \ldots \mathrm{d}N \quad (82)$$

with the notation (52):

$$\overline{\boldsymbol{R}}_i = \int \boldsymbol{R}_i \rho(1, \ldots, N; t) \mathrm{d}1 \ldots \mathrm{d}N. \quad (83)$$

Again it may be remarked that the result (82) can be obtained in a more straightforward way by taking the expectation value of equation (4), as follows from (51–52), (31) and (78).

The Weyl transforms (73) of the field equations and (81) of the equation of motion have the same forms as the corresponding classical equations (I.1) and (I.12) respectively. This fact will be exploited in the following sections.

## 6 *The equations for the fields of composite particles*

In the preceding sections we derived the field equations and the equation of motion for a set of charged point particles in an external field. Their Weyl transforms turned out to have the same form as the classical equations. As a consequence of this feature we may now find the equations that govern the behaviour of stable groups of charged particles in a way completely analogous to the classical treatment. In this fashion the 'atomic level' of the quantum-mechanical theory will be reached.

The Weyl transforms of the field equations at the 'sub-atomic' level have been given in (73). They have the same form as the corresponding classical equations (I.1). Indeed, in the latter one may read, if one wishes so, the time derivations denoted by $c\partial_0$ or by a dot as the sum of a Poisson bracket with the Hamilton function (I.16) and an explicit time derivative. Since the Hamilton function (I.16) has the same form as the Weyl transform (61) of the quantum-mechanical Hamiltonian, the classical time derivation $c\partial_0$ (or the dot) acts on classical functions in exactly the same way as the operator $\partial_{tP} \equiv c\partial_{0P}$ (72) acts on Weyl transforms.

The right-hand sides of the Weyl transforms of the field equations may now be handled in the same way as the classical equations. In particular we expand the sources that are functions of $\boldsymbol{R}_i$ or, with a double indexing, of $\boldsymbol{R}_{ki}$ ($k$ numbering the atoms, $i$ their constituent particles) around a privileged coordinate $\boldsymbol{R}_k$ of atom $k$ (e.g. one may take for $\boldsymbol{R}_k$ the Weyl transform of the position operator of the nucleus or the centre of mass). Then with the same mathematical steps as in the classical treatment, we obtain for the Weyl transforms of the field equations (cf. (I.35)):

$$\begin{aligned} \nabla\cdot\boldsymbol{e} &= \rho^{\mathrm{e}} - \nabla\cdot\boldsymbol{p}, \\ -\partial_{0\mathrm{P}}\,\boldsymbol{e} + \nabla\wedge\boldsymbol{b} &= c^{-1}\boldsymbol{j} + \partial_{0\mathrm{P}}\,\boldsymbol{p} + \nabla\wedge\boldsymbol{m}, \\ \nabla\cdot\boldsymbol{b} &= 0, \\ \partial_0\,\boldsymbol{b} + \nabla\wedge\boldsymbol{e} &= 0, \end{aligned} \tag{84}$$

where $\rho^{\mathrm{e}}$ and $\boldsymbol{j}$ are given by (cf. (I.33))

$$\begin{aligned} \rho^{\mathrm{e}} &= \sum_k e_k\,\delta(\boldsymbol{R}_k - \boldsymbol{R}), \\ \boldsymbol{j} &= \sum_k e_k\,\boldsymbol{v}_k\,\delta(\boldsymbol{R}_k - \boldsymbol{R}) \end{aligned} \tag{85}$$

(with the abbreviation $\boldsymbol{v}_k \equiv \partial_{tP}\boldsymbol{R}_k$, v. (72)) and where $\boldsymbol{p}$ and $\boldsymbol{m}$ are the series (cf. (I.34))[1]:

$$\begin{aligned} \boldsymbol{p} &= \sum_{n=1}^{\infty} (-1)^{n-1} \nabla^{n-1} : \sum_k \overline{\boldsymbol{\mu}}_k^{(n)} \delta(\boldsymbol{R}_k - \boldsymbol{R}), \\ \boldsymbol{m} &= \sum_{n=1}^{\infty} (-1)^{n-1} \nabla^{n-1} : \sum_k (\overline{\boldsymbol{\nu}}_k^{(n)} + c^{-1} \overline{\boldsymbol{\mu}}_k^{(n)} \wedge \boldsymbol{v}_k) \delta(\boldsymbol{R}_k - \boldsymbol{R}). \end{aligned} \tag{86}$$

In these formulae we used the abbreviations (cf. I.31–32):

$$\begin{aligned} \overline{\boldsymbol{\mu}}_k^{(n)} &= \frac{1}{n!} \sum_i e_{ki} \boldsymbol{r}_{ki}^n, \qquad (n = 1, 2, \ldots), \\ \overline{\boldsymbol{\nu}}_k^{(n)} &= \frac{n}{(n+1)!} \sum_i e_{ki} \boldsymbol{r}_{ki}^n \wedge (\partial_{\mathrm{OP}} \boldsymbol{r}_{ki}), \qquad (n = 1, 2, \ldots), \end{aligned} \tag{87}$$

with $\boldsymbol{r}_{ki} \equiv \boldsymbol{R}_{ki} - \boldsymbol{R}_k$.

Thanks to the use of Weyl transforms we did not need to give the derivations since formally they are exactly the same as in classical theory. There is one difference with classical theory connected with the convergence of the series expansion. In the classical theory we considered only those points $\boldsymbol{R}$ that are outside the atoms (i.e. for which $|\boldsymbol{R}_{ki} - \boldsymbol{R}_k| < |\boldsymbol{R}_k - \boldsymbol{R}|$), since only then the series expansion of the sources converges. In the quantum-mechanical treatment convergence is guaranteed under (formally) the same condition that $|\boldsymbol{R}_{ki} - \boldsymbol{R}_k| < |\boldsymbol{R}_k - \boldsymbol{R}|$. Here $\boldsymbol{R}_{ki}$ is a running variable, so that the condition means that the equations are only valid in part of phase space. If expectation values are taken, i.e. if integrals (52) over the Weyl transforms multiplied by a Wigner function are calculated, the condition gets the meaning that only points $\boldsymbol{R}$ may be considered which have the property that the Wigner function is negligible for phase space points $\boldsymbol{R}_{ki}$ that do not satisfy the inequality given above. The condition on $\boldsymbol{R}$ is thus the quantum-mechanical version of the classical condition that the observation point should be outside the atoms.

From the Weyl transform (84) of the field equations one may derive equations for expectation values. In fact by multiplying with a Wigner function and integrating over phase space one obtains, with the notation (52)

$$\begin{aligned} \nabla \cdot \overline{\boldsymbol{e}} &= \overline{\rho^{\mathrm{e}}} - \nabla \cdot \overline{\boldsymbol{p}}, \\ -\overline{\partial_{\mathrm{OP}} \boldsymbol{e}} + \nabla \wedge \overline{\boldsymbol{b}} &= c^{-1} \overline{\boldsymbol{j}} + \overline{\partial_{\mathrm{OP}} \boldsymbol{p}} + \nabla \wedge \overline{\boldsymbol{m}}, \\ \nabla \cdot \overline{\boldsymbol{b}} &= 0, \\ \overline{\partial_0 \boldsymbol{b}} + \nabla \wedge \overline{\boldsymbol{e}} &= 0. \end{aligned} \tag{88}$$

[1] The bars over the symbols $\boldsymbol{\mu}$ and $\boldsymbol{\nu}$ indicate non-relativistic multipole moments, in the same fashion as in classical theory. They should not be confused with the symbol for expectation values.

Here the space derivations $\nabla$ could be written before the symbols with the bar, since the Wigner function does not depend upon $\boldsymbol{R}$. Furthermore the time derivative in the fourth equation can also be written in the form of $\partial_0 \overline{\boldsymbol{b}}$, because $\overline{\boldsymbol{b}} = \boldsymbol{b} = \boldsymbol{B}_e$, the external field in the present non-relativistic theory. It remains to discuss the quantities $\overline{\partial_{\text{OP}} \boldsymbol{e}}$ and $\overline{\partial_{\text{OP}} \boldsymbol{p}}$. Since both $\boldsymbol{e}$ and $\boldsymbol{p}$ are independent of the momenta (as follows from the first line of (86) for $\boldsymbol{p}$, and for $\boldsymbol{e}$ from the fact that it is the multipole expansion of the first line of (69)) and since moreover the Hamiltonian (61) is of second degree in the momenta, it follows from (60) that

$$\begin{aligned} \overline{\partial_{\text{OP}} \boldsymbol{e}} &= \partial_0 \overline{\boldsymbol{e}}, \\ \overline{\partial_{\text{OP}} \boldsymbol{p}} &= \partial_0 \overline{\boldsymbol{p}}. \end{aligned} \tag{89}$$

Therefore the atomic field equations (88) for the expectation values get the form

$$\begin{aligned} \nabla \cdot \overline{\boldsymbol{e}} &= \overline{\rho^{\mathrm{e}}} - \nabla \cdot \overline{\boldsymbol{p}}, \\ -\partial_0 \overline{\boldsymbol{e}} + \nabla \wedge \overline{\boldsymbol{b}} &= c^{-1} \overline{\boldsymbol{j}} + \partial_0 \overline{\boldsymbol{p}} + \nabla \wedge \overline{\boldsymbol{m}}, \\ \nabla \cdot \overline{\boldsymbol{b}} &= 0, \\ \partial_0 \overline{\boldsymbol{b}} + \nabla \wedge \overline{\boldsymbol{e}} &= 0. \end{aligned} \tag{90}$$

These atomic equations contain at the right-hand side expectation values of operators of which the Weyl transforms have been given in (85) and (86). From (85) and (31) it follows that the operators corresponding to $\rho^{\mathrm{e}}$ and $\boldsymbol{j}$ are

$$\begin{aligned} \rho^{\mathrm{e}}_{\text{op}} &= \sum_k e_k \, \delta(\boldsymbol{R}_{k,\text{op}} - \boldsymbol{R}), \\ \boldsymbol{j}_{\text{op}} &= \tfrac{1}{2} \sum_k e_k \{\boldsymbol{v}_{k,\text{op}}, \delta(\boldsymbol{R}_{k,\text{op}} - \boldsymbol{R})\}, \end{aligned} \tag{91}$$

where the velocity operator $\boldsymbol{v}_{k,\text{op}}$, which has the Weyl transform $\boldsymbol{v}_k \equiv \partial_{t\text{P}} \boldsymbol{R}_k$, is equal to

$$\boldsymbol{v}_{k,\text{op}} = \frac{i}{\hbar} [H_{\text{op}}, \boldsymbol{R}_{k,\text{op}}] = \frac{1}{m_k} \sum_i \{\boldsymbol{P}_{ki,\text{op}} - c^{-1} e_{ki} \boldsymbol{A}_{\mathrm{e}}(\boldsymbol{R}_{ki,\text{op}}, t)\}, \tag{92}$$

if one chooses the centre of mass as the central point, as follows from (44), (61) and (72). In view of the way in which the expectation values of these operators occur in the equations (90), they may be called the operators for the charge and current densities. Furthermore it follows from (86) and (31) that the operators corresponding to $\boldsymbol{p}$ and $\boldsymbol{m}$ are

$$\boldsymbol{p}_{\mathrm{op}} = \sum_{n=1}^{\infty} (-1)^{n-1} \nabla^{n-1} : \sum_{k} \overline{\boldsymbol{\mu}}_{k,\mathrm{op}}^{(n)} \delta(\boldsymbol{R}_{k,\mathrm{op}} - \boldsymbol{R}),$$

$$\boldsymbol{m}_{\mathrm{op}} = \tfrac{1}{2} \sum_{n=1}^{\infty} (-1)^{n-1} \nabla^{n-1} : \sum_{k} \{\overline{\boldsymbol{\nu}}_{k,\mathrm{op}}^{(n)} + \tfrac{1}{2} c^{-1} \{\overline{\boldsymbol{\mu}}_{k,\mathrm{op}}^{(n)} \wedge, \boldsymbol{v}_{k,\mathrm{op}}\}, \delta(\boldsymbol{R}_{k,\mathrm{op}} - \boldsymbol{R})\}, \tag{93}$$

which will be called the operators for the polarization densities. The atomic multipole moment operators that occur here are

$$\overline{\boldsymbol{\mu}}_{k,\mathrm{op}}^{(n)} = \frac{1}{n!} \sum_{i} e_{ki} \boldsymbol{r}_{ki,\mathrm{op}}^{n},$$

$$\overline{\boldsymbol{\nu}}_{k,\mathrm{op}}^{(n)} = \tfrac{1}{2} c^{-1} \frac{n}{(n+1)!} \sum_{i} e_{ki} \{\boldsymbol{r}_{ki,\mathrm{op}}^{n} \wedge, \dot{\boldsymbol{r}}_{ki,\mathrm{op}}\}, \tag{94}$$

with the notation (3):

$$\dot{\boldsymbol{r}}_{ki,\mathrm{op}} \equiv \frac{i}{\hbar} [H_{\mathrm{op}}, \boldsymbol{r}_{ki,\mathrm{op}}]. \tag{95}$$

## 7 *The momentum and energy equations for composite particles*

### a. *The equation of motion*

The starting point to derive the equation of motion for composite particles is the 'sub-atomic' equation (81). Again replacing the index $i$ by a double index $ki$, where $k$ numbers the atoms, and $i$ their constituent particles, one finds after a summation over $i$

$$m_k \partial_{t\mathrm{P}}^2 \boldsymbol{R}_k = \sum_{i} e_{ki} \{\boldsymbol{e}_{\mathrm{t}}(\boldsymbol{R}_{ki}, t) + (\partial_{\mathrm{OP}} \boldsymbol{R}_{ki}) \wedge \boldsymbol{b}_{\mathrm{t}}(\boldsymbol{R}_{ki}, t)\}, \tag{96}$$

where we introduced the total mass $m_k = \sum_i m_{ki}$ of the atom and the Weyl transform $\boldsymbol{R}_k$ of the centre of mass operator

$$\boldsymbol{R}_k = \sum_{i} m_{ki} \boldsymbol{R}_{ki} / m_k. \tag{97}$$

Following the same procedure as in the classical theory we obtain from (96) for the atomic equation of motion (cf. (I.50) with (I.54) and (I.52)):

$$m_k \partial_{t\mathrm{P}} \boldsymbol{v}_k = \boldsymbol{f}_k^{\mathrm{L}} + \boldsymbol{f}_k^{\mathrm{S}} \tag{98}$$

with the long range force

$$\begin{aligned}\boldsymbol{f}_k^{\mathrm{L}} = & -\sum_{l(\neq k)} \sum_{n,m=0}^{\infty} \overline{\boldsymbol{\mu}}_k^{(n)} : \boldsymbol{\nabla}_k^n \overline{\boldsymbol{\mu}}_l^{(m)} : \boldsymbol{\nabla}_l^m \boldsymbol{\nabla}_k \frac{1}{4\pi|\boldsymbol{R}_k - \boldsymbol{R}_l|} \\ & + e_k\{\boldsymbol{E}_{\mathrm{e}}(\boldsymbol{R}_k, t) + c^{-1}\boldsymbol{v}_k \wedge \boldsymbol{B}_{\mathrm{e}}(\boldsymbol{R}_k, t)\} + \{\boldsymbol{\nabla}_k \boldsymbol{E}_{\mathrm{e}}(\boldsymbol{R}_k, t)\}\cdot\overline{\boldsymbol{\mu}}_k^{(1)} \\ & + \{\boldsymbol{\nabla}_k \boldsymbol{B}_{\mathrm{e}}(\boldsymbol{R}_k, t)\}\cdot(\overline{\boldsymbol{v}}_k^{(1)} + c^{-1}\overline{\boldsymbol{\mu}}_k^{(1)} \wedge \boldsymbol{v}_k) + \partial_{\mathrm{OP}}\{\overline{\boldsymbol{\mu}}_k^{(1)} \wedge \boldsymbol{B}_{\mathrm{e}}(\boldsymbol{R}_k, t)\},\end{aligned} \tag{99}$$

and the short range force:

$$\begin{aligned}\boldsymbol{f}_k^{\mathrm{S}} = & -\sum_{l(\neq k)i,j} \boldsymbol{\nabla}_{ki} \frac{e_{ki}e_{lj}}{4\pi|\boldsymbol{R}_{ki} - \boldsymbol{R}_{lj}|} \\ & + \sum_{l(\neq k)} \sum_{n,m=0}^{\infty} \overline{\boldsymbol{\mu}}_k^{(n)} : \boldsymbol{\nabla}_k^n \overline{\boldsymbol{\mu}}_l^{(m)} : \boldsymbol{\nabla}_l^m \boldsymbol{\nabla}_k \frac{1}{4\pi|\boldsymbol{R}_k - \boldsymbol{R}_l|}.\end{aligned} \tag{100}$$

Here $\partial_{\mathrm{OP}}$ and $\partial_{t\mathrm{P}}$ have been defined in (72), while $\boldsymbol{v}_k$ stands for $\partial_{t\mathrm{P}}\boldsymbol{R}_k$. Furthermore the multipole moments $\overline{\boldsymbol{\mu}}_k$ and $\overline{\boldsymbol{\nu}}_k$, defined in (87), occur.

Thus it turns out that the Weyl transform (98) of the quantum-mechanical equation of motion has the same form as the classical equation with time derivatives replaced by the derivatives of the type (72). The latter contain the sum of an explicit derivative and the Poisson bracket with the Weyl transform of the Hamiltonian.

From (98) one finds an equation for the expectation values by multiplying with a Wigner function and integrating over phase space. Then, with the notation (52), one obtains

$$m_k \overline{\partial_{t\mathrm{P}} \boldsymbol{v}_k} = \overline{\boldsymbol{f}_k^{\mathrm{L}}} + \overline{\boldsymbol{f}_k^{\mathrm{S}}}. \tag{101}$$

From the form of the Weyl transform of the Hamiltonian (61), the definition $\boldsymbol{v}_k \equiv \partial_{t\mathrm{P}}\boldsymbol{R}_k$ (with $\partial_{t\mathrm{P}}$ defined in (72)), it follows with the expression (80) for the sine symbol that

$$\partial_{t\mathrm{P}} \boldsymbol{v}_k = \sin(\boldsymbol{v}_k; H) + \frac{\partial \boldsymbol{v}_k}{\partial t}. \tag{102}$$

Then with (60) we get for the equation (101)

$$m_k \frac{\mathrm{d}\overline{\boldsymbol{v}}_k}{\mathrm{d}t} = \overline{\boldsymbol{f}_k^{\mathrm{L}}} + \overline{\boldsymbol{f}_k^{\mathrm{S}}}. \tag{103}$$

At the left-hand side the time derivative of the expectation value of the velocity operator $\boldsymbol{v}_{k,\mathrm{op}}$ (92) appears, with $\boldsymbol{R}_{k,\mathrm{op}}$ the operator that has the Weyl transform $\boldsymbol{R}_k$, i.e. the right-hand side of (97) with $\boldsymbol{R}_{ki,\mathrm{op}}$ for $\boldsymbol{R}_{ki}$. Furthermore, at the right-hand side the expectation values of the operators that have the Weyl transforms (99) and (100) occur.

b. *The energy equation*

By scalar multiplication of the Weyl transformed equation of motion (81) (where $i$ is to be replaced by $ki$) by the quantity $\partial_{t\mathrm{P}}\boldsymbol{R}_{ki}$ one finds after summation over $i$:

$$\sum_i \tfrac{1}{2}m_{ki}\,\partial_{t\mathrm{P}}\{(\partial_{t\mathrm{P}}\boldsymbol{R}_{ki})^2\} = \sum_i e_{ki}\,\boldsymbol{e}_{\mathrm{t}}(\boldsymbol{R}_{ki},t)\cdot(\partial_{t\mathrm{P}}\boldsymbol{R}_{ki}). \tag{104}$$

With the same steps as have been taken to obtain the classical atomic energy equation we get then (cf. (I.63) with (I.65–66)):

$$\partial_{t\mathrm{P}}\left\{\tfrac{1}{2}m_k\,\boldsymbol{v}_k^2+\tfrac{1}{2}\sum_i m_{ki}(\partial_{t\mathrm{P}}\boldsymbol{r}_{ki})^2+\sum_{i,j(i\neq j)}\frac{e_{ki}\,e_{kj}}{8\pi|\boldsymbol{r}_{ki}-\boldsymbol{r}_{kj}|}\right\} = \psi_k^{\mathrm{L}}+\psi_k^{\mathrm{S}} \tag{105}$$

with two terms representing work exerted on the composite particle per unit of time: a long range term

$$\begin{aligned}\psi_k^{\mathrm{L}} = -\sum_{l(\neq k)}\sum_{m=0}^{\infty}\Big\{&\sum_{n=0}^{\infty}\overline{\boldsymbol{\mu}}_k^{(n)}:\nabla_k^n\,\boldsymbol{v}_k\cdot\nabla_k+\sum_{n=1}^{\infty}(\partial_{t\mathrm{P}}\overline{\boldsymbol{\mu}}_k^{(n)}):\nabla_k^n\Big\}\\ &\overline{\boldsymbol{\mu}}_l^{(m)}:\nabla_l^m\frac{1}{4\pi|\boldsymbol{R}_k-\boldsymbol{R}_l|}+e_k\,\boldsymbol{v}_k\cdot\boldsymbol{E}_{\mathrm{e}}(\boldsymbol{R}_k,t)+\boldsymbol{v}_k\cdot\{\nabla_k\boldsymbol{E}_{\mathrm{e}}(\boldsymbol{R}_k,t)\}\cdot\overline{\boldsymbol{\mu}}_k^{(1)}\\ &+(\partial_{t\mathrm{P}}\overline{\boldsymbol{\mu}}_k^{(1)})\cdot\boldsymbol{E}_{\mathrm{e}}(\boldsymbol{R}_k,t)-(\overline{\boldsymbol{\nu}}_k^{(1)}+c^{-1}\overline{\boldsymbol{\mu}}_k^{(1)}\wedge\boldsymbol{v}_k)\cdot\frac{\partial\boldsymbol{B}_{\mathrm{e}}(\boldsymbol{R}_k,t)}{\partial t},\end{aligned} \tag{106}$$

and a short range term

$$\begin{aligned}\psi_k^{\mathrm{S}} = &-\sum_{l(\neq k)i,j}(\partial_{t\mathrm{P}}\boldsymbol{R}_{ki})\cdot\nabla_{ki}\frac{e_{ki}\,e_{lj}}{4\pi|\boldsymbol{R}_{ki}-\boldsymbol{R}_{lj}|}\\ &+\sum_{l(\neq k)}\sum_{m=0}^{\infty}\Big\{\sum_{n=0}^{\infty}\overline{\boldsymbol{\mu}}_k^{(n)}:\nabla_k^n\,\boldsymbol{v}_k\cdot\nabla_k+\sum_{n=1}^{\infty}(\partial_{t\mathrm{P}}\overline{\boldsymbol{\mu}}_k^{(n)}):\nabla_k^n\Big\}\\ &\qquad\qquad\overline{\boldsymbol{\mu}}_l^{(m)}:\nabla_l^m\frac{1}{4\pi|\boldsymbol{R}_k-\boldsymbol{R}_l|},\end{aligned} \tag{107}$$

with $\boldsymbol{v}_k = \partial_{t\mathrm{P}}\boldsymbol{R}_k$ and $\partial_{0\mathrm{P}} = c^{-1}\partial_{t\mathrm{P}}$ defined by (72). Again formally this result is the same as that of the classical theory.

The corresponding equation for expectation values may be obtained from (105) by multiplying with a Wigner function and integrating over phase space. Then with the help of (60), (61), (72) and (80) one finds:

$$\frac{\mathrm{d}}{\mathrm{d}t}\left\{\tfrac{1}{2}m_k\,\overline{\boldsymbol{v}_k^2}+\tfrac{1}{2}\sum_i m_{ki}\overline{(\partial_{t\mathrm{P}}\boldsymbol{r}_{ki})^2}+\sum_{i,j(i\neq j)}\overline{\frac{e_{ki}\,e_{kj}}{8\pi|\boldsymbol{R}_{ki}-\boldsymbol{R}_{kj}|}}\right\} = \overline{\psi_k^{\mathrm{L}}}+\overline{\psi_k^{\mathrm{S}}}, \tag{108}$$

where the notation (52) has been employed. Since $\boldsymbol{v}_k$ is the Weyl transform of $\boldsymbol{v}_{k,\mathrm{op}}$ given by (92) and $\partial_{t\mathrm{P}}\boldsymbol{r}_{ki}$ is the Weyl transform of $\dot{\boldsymbol{r}}_{ki,\mathrm{op}}$ given by (95), one obtains with the help of (43) that the left-hand side contains the expectation values of the kinetic energy operator $\frac{1}{2}m_k\boldsymbol{v}_{k,\mathrm{op}}^2$, the internal kinetic energy operator $\frac{1}{2}\sum_i m_{ki}\dot{\boldsymbol{r}}_{ki,\mathrm{op}}^2$ and the internal Coulomb energy operator. At the right-hand side the expectation values of two operators appear of which the Weyl transforms are (106) and (107).

## 8 *The inner angular momentum equation for composite particles*

By vectorial multiplication of the Weyl transformed equation of motion (81) (with $i$ replaced by $ki$) with $\boldsymbol{r}_{ki} \equiv \boldsymbol{R}_{ki}-\boldsymbol{R}_k$, one finds with the help of (97) and a summation over $i$:

$$\partial_{t\mathrm{P}}\bar{\boldsymbol{s}}_k = \sum_i e_{ki}\boldsymbol{r}_{ki}\wedge\{\boldsymbol{e}_{\mathrm{t}}(\boldsymbol{R}_{ki},t)+(\partial_{0\mathrm{P}}\boldsymbol{R}_{ki})\wedge\boldsymbol{b}_{\mathrm{t}}(\boldsymbol{R}_{ki},t)\}, \tag{109}$$

where we introduced the quantity[1]

$$\boldsymbol{s}_k \equiv \sum_i m_{ki}\boldsymbol{r}_{ki}\wedge(\partial_{t\mathrm{P}}\boldsymbol{r}_{ki}). \tag{110}$$

Just as in the classical case one finds from this equation (cf. (I.76) with (I.77–78))

$$\partial_{t\mathrm{P}}\bar{\boldsymbol{s}}_k = \boldsymbol{d}_k^{\mathrm{L}}+\boldsymbol{d}_k^{\mathrm{S}} \tag{111}$$

with the long range moment

$$\boldsymbol{d}_k^{\mathrm{L}} = \sum_{l(\neq k)}\sum_{n,m=0}^{\infty} n\nabla_k\wedge\overline{\boldsymbol{\mu}}_k^{(n)} : \nabla_k^{n-1}\overline{\boldsymbol{\mu}}_l^{(m)} : \nabla_l^m\frac{1}{4\pi|\boldsymbol{R}_k-\boldsymbol{R}_l|}$$
$$+\overline{\boldsymbol{\mu}}_k^{(1)}\wedge\{\boldsymbol{E}_{\mathrm{e}}(\boldsymbol{R}_k,t)+c^{-1}\boldsymbol{v}_k\wedge\boldsymbol{B}_{\mathrm{e}}(\boldsymbol{R}_k,t)\}+\overline{\boldsymbol{\nu}}_k^{(1)}\wedge\boldsymbol{B}_{\mathrm{e}}(\boldsymbol{R}_k,t) \tag{112}$$

and the short range moment

$$\boldsymbol{d}_k^{\mathrm{S}} = -\sum_{l(\neq k)i,j}\boldsymbol{r}_{ki}\wedge\nabla_{ki}\frac{e_{ki}e_{lj}}{4\pi|\boldsymbol{R}_{ki}-\boldsymbol{R}_{lj}|}$$
$$-\sum_{l(\neq k)}\sum_{n,m=0}^{\infty} n\nabla_k\wedge\overline{\boldsymbol{\mu}}_k^{(n)} : \nabla_k^{n-1}\overline{\boldsymbol{\mu}}_l^{(m)} : \nabla_l^m\frac{1}{4\pi|\boldsymbol{R}_k-\boldsymbol{R}_l|}, \tag{113}$$

with $\boldsymbol{v}_k = \partial_{t\mathrm{P}}\boldsymbol{R}_k$ and $\partial_{0\mathrm{P}} = c^{-1}\partial_{t\mathrm{P}}$ defined in (72).

[1] The bar denotes a non-relativistic quantity, not an expectation value.

Multiplying (111) by a Wigner function and integrating over phase space one obtains

$$\overline{\partial_{tP}\,\bar{s}_k} = \overline{d_k^{\mathrm{L}}} + \overline{d_k^{\mathrm{S}}}, \tag{114}$$

where the notation (52) has been employed. With the help of (60), (61), (72), (80) and (110) we get for this equation

$$\frac{\mathrm{d}\overline{\bar{s}_k}}{\mathrm{d}t} = \overline{d_k^{\mathrm{L}}} + \overline{d_k^{\mathrm{S}}}. \tag{115}$$

At the left-hand side the time derivative of the expectation value of the operator $\bar{s}_{k,\mathrm{op}}$ for the internal angular momentum, with Weyl transform (110), occurs, while the right-hand side contains the expectation values of two operators for long and short range moments of which the Weyl transforms have been given in (112) and (113).

The field equations and the equation of motion for composite particles, as well as the ensuing energy and angular momentum equations, were obtained from the equations for point particles in formally the same fashion as in the classical treatment. This could be achieved because already at the sub-atomic level we translated the quantum-mechanical operator equations into their Weyl transforms. Therefore at this stage it becomes apparent that a transcription which leads away from the usual operator language gives rise to considerable formal simplification since *mutatis mutandis* the classical derivation may be taken over as such.

# Properties of the Weyl transformation and the Wigner function

## *1. A reformulation of quantum mechanics*

In the usual formulation of quantum mechanics one associates a vector in Hilbert space to each state of the system and to each physical quantity an operator acting in this Hilbert space. It is possible however to give an alternative description of quantum mechanics by using only ordinary functions in phase space for both the states and the physical quantities. An example of such an approach consists in introducing Weyl transforms instead of operators, and simultaneously the Wigner function instead of the state vector.

This programme will be carried out in the following appendix for a one-particle system. The reason for this limitation is merely to reduce slightly the length of the formulae. Indeed if $N$-particle systems are considered indices that label the particles and summation signs have to be added.

## *2. The Weyl transformation*

### a. *Preliminaries*

A few notions of ordinary quantum mechanics will be summarized here for use in the following. The momentum and coordinate operators[1] $\boldsymbol{P}$ and $\boldsymbol{Q}$ for a single point particle satisfy the commutation relations

$$[\boldsymbol{P}, \boldsymbol{P}] = 0, \qquad [\boldsymbol{Q}, \boldsymbol{Q}] = 0, \qquad [\boldsymbol{P}, \boldsymbol{Q}] = \frac{\hbar}{i}\, \mathbf{U} I, \tag{A1}$$

with $\mathbf{U}$ the unit three-tensor and $I$ the unit operator in Hilbert space. Their eigenvectors $|\boldsymbol{p}\rangle$ and $|\boldsymbol{q}\rangle$, defined by the eigenvalue equations

$$\boldsymbol{P}|\boldsymbol{p}\rangle = \boldsymbol{p}|\boldsymbol{p}\rangle, \qquad \boldsymbol{Q}|\boldsymbol{q}\rangle = \boldsymbol{q}|\boldsymbol{q}\rangle, \tag{A2}$$

[1] Throughout this appendix we use capitals to denote operators and lower case symbols for ordinary numbers.

satisfy completeness relations

$$\int \mathrm{d}\boldsymbol{p}|\boldsymbol{p}\rangle\langle\boldsymbol{p}| = I, \qquad \int \mathrm{d}\boldsymbol{q}|\boldsymbol{q}\rangle\langle\boldsymbol{q}| = I \tag{A3}$$

and inner product relations

$$\langle\boldsymbol{p}|\boldsymbol{p}'\rangle = \delta(\boldsymbol{p}-\boldsymbol{p}'), \qquad \langle\boldsymbol{q}|\boldsymbol{q}'\rangle = \delta(\boldsymbol{q}-\boldsymbol{q}'), \qquad \langle\boldsymbol{q}|\boldsymbol{p}\rangle = \frac{1}{h^{\frac{3}{2}}}\,\mathrm{e}^{(i/\hbar)\boldsymbol{p}\cdot\boldsymbol{q}}. \tag{A4}$$

The trace of an operator may be expressed in terms of the complete sets $|\boldsymbol{p}\rangle$ or $|\boldsymbol{q}\rangle$ as

$$\operatorname{Tr} A = \int \mathrm{d}\boldsymbol{p}\langle\boldsymbol{p}|A|\boldsymbol{p}\rangle = \int \mathrm{d}\boldsymbol{q}\langle\boldsymbol{q}|A|\boldsymbol{q}\rangle. \tag{A5}$$

b. *Definition*[1]

The Weyl transform of a quantum-mechanical operator $A$ for a single point particle is a scalar function $a(\boldsymbol{p}, \boldsymbol{q})$ defined as

$$a(\boldsymbol{p}, \boldsymbol{q}) = \int \mathrm{d}\boldsymbol{u}\,\mathrm{e}^{(i/\hbar)\boldsymbol{q}\cdot\boldsymbol{u}}\langle\boldsymbol{p}+\tfrac{1}{2}\boldsymbol{u}|A|\boldsymbol{p}-\tfrac{1}{2}\boldsymbol{u}\rangle, \tag{A6}$$

where $|\boldsymbol{p}\rangle$ is the eigenvector of the momentum operator with eigenvalue $\boldsymbol{p}$. Alternatively one may write

$$a(\boldsymbol{p}, \boldsymbol{q}) = \int \mathrm{d}\boldsymbol{v}\,\mathrm{e}^{(i/\hbar)\boldsymbol{p}\cdot\boldsymbol{v}}\langle\boldsymbol{q}-\tfrac{1}{2}\boldsymbol{v}|A|\boldsymbol{q}+\tfrac{1}{2}\boldsymbol{v}\rangle, \tag{A7}$$

where $|\boldsymbol{q}\rangle$ is the eigenvector of the coordinate operator with eigenvalue $\boldsymbol{q}$. From (A6) or (A7) it follows that the Weyl transform $a(\boldsymbol{p}, \boldsymbol{q})$ is real if the operator $A$ is hermitian.

The operator $A$ reads, in terms of the Weyl transform:

$$A = h^{-3}\int \mathrm{d}\boldsymbol{p}\,\mathrm{d}\boldsymbol{q}\,a(\boldsymbol{p}, \boldsymbol{q})\Delta(\boldsymbol{p}, \boldsymbol{q}), \tag{A8}$$

where the hermitian operator $\Delta(\boldsymbol{p}, \boldsymbol{q})$ is defined as

$$\Delta(\boldsymbol{p}, \boldsymbol{q}) = h^{-3}\int \mathrm{d}\boldsymbol{u}\,\mathrm{d}\boldsymbol{v}\,\mathrm{e}^{(i/\hbar)\{(\boldsymbol{q}-\boldsymbol{Q})\cdot\boldsymbol{u}+(\boldsymbol{p}-\boldsymbol{P})\cdot\boldsymbol{v}\}} \tag{A9}$$

[1] The derivations of (A6–A14) and of (A54) have been given in § 3 and will not be repeated here.

or alternatively as

$$\Delta(\boldsymbol{p}, \boldsymbol{q}) = \int \mathrm{d}\boldsymbol{u}\, \mathrm{e}^{(i/\hbar)\boldsymbol{q}\cdot\boldsymbol{u}} |\boldsymbol{p}-\tfrac{1}{2}\boldsymbol{u}\rangle\langle \boldsymbol{p}+\tfrac{1}{2}\boldsymbol{u}|, \tag{A10}$$

$$\Delta(\boldsymbol{p}, \boldsymbol{q}) = \int \mathrm{d}\boldsymbol{v}\, \mathrm{e}^{(i/\hbar)\boldsymbol{p}\cdot\boldsymbol{v}} |\boldsymbol{q}+\tfrac{1}{2}\boldsymbol{v}\rangle\langle \boldsymbol{q}-\tfrac{1}{2}\boldsymbol{v}|. \tag{A11}$$

The Weyl transform may also be written as a trace involving this operator $\Delta(\boldsymbol{p}, \boldsymbol{q})$

$$a(\boldsymbol{p}, \boldsymbol{q}) = \mathrm{Tr}\,\{A\Delta(\boldsymbol{p}, \boldsymbol{q})\}. \tag{A12}$$

A different formulation of the Weyl correspondence consists in giving both the operator and the Weyl transform as a Fourier integral

$$A = \int \mathrm{d}\boldsymbol{u}\, \mathrm{d}\boldsymbol{v}\, \tilde{a}(\boldsymbol{u}, \boldsymbol{v}) \mathrm{e}^{-(i/\hbar)(\boldsymbol{Q}\cdot\boldsymbol{u}+\boldsymbol{P}\cdot\boldsymbol{v})}, \tag{A13}$$

$$a(\boldsymbol{p}, \boldsymbol{q}) = \int \mathrm{d}\boldsymbol{u}\, \mathrm{d}\boldsymbol{v}\, \tilde{a}(\boldsymbol{u}, \boldsymbol{v}) \mathrm{e}^{-(i/\hbar)(\boldsymbol{q}\cdot\boldsymbol{u}+\boldsymbol{p}\cdot\boldsymbol{v})}, \tag{A14}$$

with the same function $\tilde{a}(\boldsymbol{u}, \boldsymbol{v})$ in both integrands.

A still different way to get the Weyl transform $a(\boldsymbol{p}, \boldsymbol{q})$ from the operator $A$ is obtained by starting from (A12) with (A9). Application of the identity

$$\mathrm{e}^{A+B} = \mathrm{e}^{A}\mathrm{e}^{B}\mathrm{e}^{-\frac{1}{2}[A,B]} \tag{A15}$$

for operators $A$ and $B$ that commute with their commutator (v. problem 1) yields with the help of (A5) and (A2)

$$a(\boldsymbol{p}, \boldsymbol{q}) = \frac{1}{h^3}\int \mathrm{d}\boldsymbol{u}\, \mathrm{d}\boldsymbol{v}\, \mathrm{d}\boldsymbol{p}'\, \mathrm{d}\boldsymbol{q}' \exp\left(-\frac{i}{2\hbar}\boldsymbol{u}\cdot\boldsymbol{v}\right) \exp\left[\frac{i}{\hbar}\{(\boldsymbol{p}-\boldsymbol{p}')\cdot\boldsymbol{v}+(\boldsymbol{q}-\boldsymbol{q}')\cdot\boldsymbol{u}\}\right] \langle \boldsymbol{q}'|A|\boldsymbol{p}'\rangle\langle \boldsymbol{p}'|\boldsymbol{q}'\rangle. \tag{A16}$$

The product $\boldsymbol{u}\cdot\boldsymbol{v}$ in the first exponent may be replaced by $-\hbar^2(\partial/\partial\boldsymbol{p}')\cdot(\partial/\partial\boldsymbol{q}')$ acting on the second exponential, or, by partial integration, by the same operator acting on the rest of the integrand. The integration over $\boldsymbol{u}$ and $\boldsymbol{v}$ then yields delta functions, so that the integration over $\boldsymbol{p}'$ and $\boldsymbol{q}'$ may also be performed, with the result

$$a(\boldsymbol{p}, \boldsymbol{q}) = h^3 \exp\left(-\frac{\hbar}{2i}\frac{\partial}{\partial\boldsymbol{p}}\cdot\frac{\partial}{\partial\boldsymbol{q}}\right)\langle \boldsymbol{q}|A|\boldsymbol{p}\rangle\langle \boldsymbol{p}|\boldsymbol{q}\rangle, \tag{A17}$$

or alternatively, if the roles of $\boldsymbol{p}$ and $\boldsymbol{q}$ are interchanged in the proof,

$$a(\boldsymbol{p}, \boldsymbol{q}) = h^3 \exp\left(\frac{\hbar}{2i}\frac{\partial}{\partial\boldsymbol{p}}\cdot\frac{\partial}{\partial\boldsymbol{q}}\right)\langle \boldsymbol{p}|A|\boldsymbol{q}\rangle\langle \boldsymbol{q}|\boldsymbol{p}\rangle. \tag{A18}$$

Finding the operator $A$ from the Weyl transform $a(\boldsymbol{p}, \boldsymbol{q})$ is in principle possible from the formulae (A8) with (A9), (A10) or (A11), or otherwise from (A13) with (A14). A method which is often more convenient will be indicated now. From (A8) with (A9) and the identity (A15) it follows that

$$A = \frac{1}{h^6}\int \mathrm{d}\boldsymbol{p}\,\mathrm{d}\boldsymbol{q}\,\mathrm{d}\boldsymbol{u}\,\mathrm{d}\boldsymbol{v}\, a(\boldsymbol{p}, \boldsymbol{q})\mathrm{e}^{(i/\hbar)(\boldsymbol{q}-\boldsymbol{Q})\cdot\boldsymbol{u}}\mathrm{e}^{(i/\hbar)(\boldsymbol{p}-\boldsymbol{P})\cdot\boldsymbol{v}}\mathrm{e}^{(i/2\hbar)\boldsymbol{u}\cdot\boldsymbol{v}}. \tag{A19}$$

Again the product $\boldsymbol{u}\cdot\boldsymbol{v}$ may be replaced by $-\hbar^2(\partial/\partial\boldsymbol{p})\cdot(\partial/\partial\boldsymbol{q})$ acting on the first two exponentials, or, after a partial integration, by the same operator acting on $a(\boldsymbol{p}, \boldsymbol{q})$, so that one has, after integration over $\boldsymbol{u}$ and $\boldsymbol{v}$,

$$A = \int \mathrm{d}\boldsymbol{p}\,\mathrm{d}\boldsymbol{q}\,\delta(\boldsymbol{q}-\boldsymbol{Q})\delta(\boldsymbol{p}-\boldsymbol{P})a_0(\boldsymbol{p}, \boldsymbol{q}) \tag{A20}$$

with

$$a_0(\boldsymbol{p}, \boldsymbol{q}) \equiv \exp\left(\frac{\hbar}{2i}\frac{\partial}{\partial\boldsymbol{p}}\cdot\frac{\partial}{\partial\boldsymbol{q}}\right) a(\boldsymbol{p}, \boldsymbol{q}). \tag{A21}$$

These formulae show that one may find the operator $A$ from its Weyl transform $a(\boldsymbol{p}, \boldsymbol{q})$ by calculating first $a_0(\boldsymbol{p}, \boldsymbol{q})$ from (A21) and then replacing the variables $\boldsymbol{q}$ and $\boldsymbol{p}$ by the operators $\boldsymbol{Q}$ and $\boldsymbol{P}$, always writing the coordinate operators at the left of the momentum operators. This shows that the procedure is only convenient if one has to do with binomials of $\boldsymbol{P}$ and $\boldsymbol{Q}$.

The Weyl transform $a(\boldsymbol{p}, \boldsymbol{q})$ of a quantum-mechanical operator $A$ may be employed to find the function in classical mechanics that corresponds to the operator $A$. This function is in general not simply the Weyl transform itself (since the latter depends in general on Planck's constant $\hbar$), but is obtained if one takes the limit for $\hbar \to 0$:

$$A \overset{\mathrm{cl}}{\to} a_{\mathrm{cl}}(\boldsymbol{p}, \boldsymbol{q}) = \lim_{\hbar\to 0} a(\boldsymbol{p}, \boldsymbol{q}). \tag{A22}$$

### c. *Examples*

One often encounters physical operators of which the Weyl transforms are of the form

$$a(p, q) = f(q)p^n \tag{A23}$$

with $f$ an arbitrary function and $n$ a natural number (for convenience we limit ourselves to the one-dimensional case in this subsection). Application of (A21) gives

$$a_0(p, q) = \sum_{k=0}^{n} \binom{n}{k}\left(\frac{\hbar}{2i}\right)^k \frac{\mathrm{d}^k f(q)}{\mathrm{d}q^k}\, p^{n-k}. \tag{A24}$$

Hence according to (A20) the corresponding operator is

$$A = \sum_{k=0}^{n} \binom{n}{k} \left(\frac{\hbar}{2i}\right)^k \frac{\mathrm{d}^k f(Q)}{\mathrm{d}Q^k} P^{n-k}. \tag{A25}$$

The result may be cast into the form:

$$A = \frac{1}{2^n} \sum_{k=0}^{n} \binom{n}{k} P^k f(Q) P^{n-k}, \tag{A26}$$

as may be seen in the following way. From the commutation relations (A1) one finds

$$P^k f(Q) = \sum_{j=0}^{k} \left(\frac{\hbar}{i}\right)^j \binom{k}{j} \frac{\mathrm{d}^j f(Q)}{\mathrm{d}Q^j} P^{k-j}. \tag{A27}$$

Inserting this expression into (A26) and using the identities

$$\binom{n}{k}\binom{k}{j} = \binom{n}{j}\binom{n-j}{k-j}, \qquad \frac{1}{2^m} \sum_{n=0}^{m} \binom{m}{n} = 1, \tag{A28}$$

one recovers indeed (A25), so that we have established the Weyl correspondence

$$f(q)p^n \rightleftarrows \frac{1}{2^n} \sum_{k=0}^{n} \binom{n}{k} P^k f(Q) P^{n-k}. \tag{A29}$$

It may also be formulated in terms of repeated anticommutators:

$$f(q)p^n \rightleftarrows \frac{1}{2^n} \{\{\dots \{f(Q), P\}, P\}, \dots, P\}. \tag{A30}$$

Special cases are for instance

$$p \rightleftarrows P, \tag{A31}$$

$$q \rightleftarrows Q, \tag{A32}$$

$$pq \rightleftarrows \tfrac{1}{2}(PQ+QP), \tag{A33}$$

$$p^2 q \rightleftarrows \tfrac{1}{4}(QP^2+2PQP+P^2Q), \tag{A34}$$

$$p^2 q^2 \rightleftarrows \tfrac{1}{4}(Q^2P^2+2PQ^2P+P^2Q^2). \tag{A35}$$

d. *The Δ-operator*

The Δ-operator (A9), (A10) or (A11) has as Weyl transform a product of delta functions as follows from (A8):

$$\Delta(\boldsymbol{p}', \boldsymbol{q}') \rightleftarrows h^3\delta(\boldsymbol{p}-\boldsymbol{p}')\delta(\boldsymbol{q}-\boldsymbol{q}'). \tag{A36}$$

Its matrix elements between eigenvectors of the coordinate and momentum operators follow with (A10), (A11) and (A4)

$$\langle \boldsymbol{q}'|\Delta(\boldsymbol{p}, \boldsymbol{q})|\boldsymbol{q}''\rangle = \delta\left(\boldsymbol{q} - \frac{\boldsymbol{q}'+\boldsymbol{q}''}{2}\right) \mathrm{e}^{(i/\hbar)\boldsymbol{p}\cdot(\boldsymbol{q}'-\boldsymbol{q}'')}, \tag{A37}$$

$$\langle \boldsymbol{p}'|\Delta(\boldsymbol{p}, \boldsymbol{q})|\boldsymbol{p}''\rangle = \delta\left(\boldsymbol{p} - \frac{\boldsymbol{p}'+\boldsymbol{p}''}{2}\right) \mathrm{e}^{-(i/\hbar)\boldsymbol{q}\cdot(\boldsymbol{p}'-\boldsymbol{p}'')}. \tag{A38}$$

In particular the diagonal elements are

$$\langle \boldsymbol{q}'|\Delta(\boldsymbol{p}, \boldsymbol{q})|\boldsymbol{q}'\rangle = \delta(\boldsymbol{q}-\boldsymbol{q}'), \tag{A39}$$

$$\langle \boldsymbol{p}'|\Delta(\boldsymbol{p}, \boldsymbol{q})|\boldsymbol{p}'\rangle = \delta(\boldsymbol{p}-\boldsymbol{p}'). \tag{A40}$$

Mixed matrix elements are found by using (A17) for $A = \Delta(\boldsymbol{p}', \boldsymbol{q}')$. Then with (A36) and (A4) one finds

$$\langle \boldsymbol{q}'|\Delta(\boldsymbol{p}, \boldsymbol{q})|\boldsymbol{p}'\rangle = \exp\left(\frac{i}{\hbar}\, \boldsymbol{p}'\cdot\boldsymbol{q}'\right) \exp\left(\frac{\hbar}{2i} \frac{\partial}{\partial \boldsymbol{p}} \cdot \frac{\partial}{\partial \boldsymbol{q}}\right) \delta(\boldsymbol{p}-\boldsymbol{p}')\delta(\boldsymbol{q}-\boldsymbol{q}'). \tag{A41}$$

Integration of the $\Delta$-operator (A11) or (A10) over $\boldsymbol{p}$ or $\boldsymbol{q}$ respectively yields

$$\int \mathrm{d}\boldsymbol{p}\, \Delta(\boldsymbol{p}, \boldsymbol{q}) = h^3|\boldsymbol{q}\rangle\langle \boldsymbol{q}|, \tag{A42}$$

$$\int \mathrm{d}\boldsymbol{q}\, \Delta(\boldsymbol{p}, \boldsymbol{q}) = h^3|\boldsymbol{p}\rangle\langle \boldsymbol{p}|, \tag{A43}$$

i.e. projection operators on a particular member of the complete set of eigenvectors of the coordinate and momentum operators. A second integration yields according to the closure relations (A3)

$$h^{-3}\int \mathrm{d}\boldsymbol{p}\,\mathrm{d}\boldsymbol{q}\, \Delta(\boldsymbol{p}, \boldsymbol{q}) = I. \tag{A44}$$

The trace of a product of $\Delta$-operators has different general forms for an even and an odd number of factors[1]:

$$\mathrm{Tr}\,\{\Delta(\boldsymbol{p}_1, \boldsymbol{q}_1) \dots \Delta(\boldsymbol{p}_n, \boldsymbol{q}_n)\}$$

$$= 2^{3(n-1)} \exp\left\{\frac{2i}{\hbar} \sum_{j,k=1(j<k)}^{n} (-1)^{j+k}(\boldsymbol{p}_j\cdot\boldsymbol{q}_k - \boldsymbol{p}_k\cdot\boldsymbol{q}_j)\right\}, \qquad (n \text{ odd}), \tag{A45}$$

[1] R. L. Stratonovich, Soviet Phys. JETP 4(1957)891.

$$\mathrm{Tr}\,\{\Delta(\boldsymbol{p}_1, \boldsymbol{q}_1) \dots \Delta(\boldsymbol{p}_n, \boldsymbol{q}_n)\}$$
$$= h^3\delta(\boldsymbol{p}_1 - \boldsymbol{p}_2 + \dots - \boldsymbol{p}_n)\delta(\boldsymbol{q}_1 - \boldsymbol{q}_2 + \dots - \boldsymbol{q}_n)$$
$$\mathrm{Tr}\,\{\Delta(\boldsymbol{p}_1, \boldsymbol{q}_1) \dots \Delta(\boldsymbol{p}_{n-1}, \boldsymbol{q}_{n-1})\}, \qquad (n \text{ even}), \quad \text{(A46)}$$

Proof: With (A5), insertion of complete sets $|\boldsymbol{q}_i'\rangle$ $(i = 2, \dots, n)$ by means of (A3) and the use of (A37) we may write the left-hand side of (A45) or (A46) as

$$\int \mathrm{d}\boldsymbol{q}_1' \dots \mathrm{d}\boldsymbol{q}_n' \delta \left(\boldsymbol{q}_1 - \frac{\boldsymbol{q}_1' + \boldsymbol{q}_2'}{2}\right) \dots \delta \left(\boldsymbol{q}_n - \frac{\boldsymbol{q}_n' + \boldsymbol{q}_1'}{2}\right)$$
$$\exp\left[\frac{i}{\hbar}\{\boldsymbol{p}_1\cdot(\boldsymbol{q}_1' - \boldsymbol{q}_2') + \dots \boldsymbol{p}_n\cdot(\boldsymbol{q}_n' - \boldsymbol{q}_1')\}\right]. \quad \text{(A47)}$$

It is convenient to introduce as new integration variables $\boldsymbol{q}_i'' = \frac{1}{2}(\boldsymbol{q}_i' + \boldsymbol{q}_{i+1}')$ with $i = 1, \dots, n$ $(\boldsymbol{q}_{n+1}' \equiv \boldsymbol{q}_1')$. The Jacobian of this transformation is equal to $2^{-3(n-1)}$ for $n$ = odd and 0 for $n$ = even. Therefore the proof continues for $n$ = odd only. In that case we obtain, writing also the exponential in terms of the new variables and performing the integrations, the right-hand side of (A45).

To find the result for $n$ even we use formula (A45) for $n+1$, and formula (A44) to write

$$\mathrm{Tr}\,\{\Delta(\boldsymbol{p}_1, \boldsymbol{q}_1) \dots \Delta(\boldsymbol{p}_n, \boldsymbol{q}_n)\}$$
$$= \left(\frac{2^n}{h}\right)^3 \int \mathrm{d}\boldsymbol{p}_{n+1}\, \mathrm{d}\boldsymbol{q}_{n+1} \exp\left[\frac{2i}{\hbar}\{\sum_{j,k=1(j<k)}^{n} (-1)^{j+k}(\boldsymbol{p}_j\cdot\boldsymbol{q}_k - \boldsymbol{p}_k\cdot\boldsymbol{q}_j)\right.$$
$$\left. + \sum_{j=1}^{n} (-1)^{j+1}(\boldsymbol{p}_j\cdot\boldsymbol{q}_{n+1} - \boldsymbol{p}_{n+1}\cdot\boldsymbol{q}_j)\}\right]. \quad \text{(A48)}$$

The integration over $\boldsymbol{p}_{n+1}$ and $\boldsymbol{q}_{n+1}$ may be performed to yield a product of two delta functions. In the remaining exponential the part with $k = n$ in the double sum vanishes because of the occurrence of the delta functions. Then one is left with the right-hand side of (A46) with the expression (A45) for $n-1$ inserted. Q.E.D.

The special cases $n$ = 1, 2 and 3 of (A45–46) have already been mentioned in the main text (33, 34, 36):

$$\mathrm{Tr}\,\Delta(\boldsymbol{p}_1, \boldsymbol{q}_1) = 1, \quad \text{(A49)}$$

$$\mathrm{Tr}\,\{\Delta(\boldsymbol{p}_1, \boldsymbol{q}_1)\Delta(\boldsymbol{p}_2, \boldsymbol{q}_2)\} = h^3\delta(\boldsymbol{p}_1 - \boldsymbol{p}_2)\delta(\boldsymbol{q}_1 - \boldsymbol{q}_2), \quad \text{(A50)}$$

$$\mathrm{Tr}\,\{\Delta(\boldsymbol{p}_1, \boldsymbol{q}_1)\Delta(\boldsymbol{p}_2, \boldsymbol{q}_2)\Delta(\boldsymbol{p}_3, \boldsymbol{q}_3)\}$$
$$= 2^6 \exp\left[\frac{2i}{\hbar}\{\boldsymbol{p}_1\cdot(\boldsymbol{q}_3-\boldsymbol{q}_2)+\boldsymbol{p}_2\cdot(\boldsymbol{q}_1-\boldsymbol{q}_3)+\boldsymbol{p}_3\cdot(\boldsymbol{q}_2-\boldsymbol{q}_1)\}\right]. \quad \text{(A51)}$$

The traces of products of operators follow by application of (A8), (A45) and (A46). In particular one finds thus from (A8), (A49) and (A50):

$$\mathrm{Tr}\, A = h^{-3}\int \mathrm{d}\boldsymbol{p}\,\mathrm{d}\boldsymbol{q}\; a(\boldsymbol{p}, \boldsymbol{q}), \quad \text{(A52)}$$

$$\mathrm{Tr}\, AB = h^{-3}\int \mathrm{d}\boldsymbol{p}\,\mathrm{d}\boldsymbol{q}\; a(\boldsymbol{p}, \boldsymbol{q})b(\boldsymbol{p}, \boldsymbol{q}) = \mathrm{Tr}\, BA. \quad \text{(A53)}$$

e. *Products of operators*

The Weyl transform of a product of the operators $A$ and $B$ follows from (A12), (A8) and (A51). One then obtains

$$AB \rightleftarrows \exp\left\{\frac{i\hbar}{2}\left(\frac{\partial^{(a)}}{\partial \boldsymbol{q}}\cdot\frac{\partial^{(b)}}{\partial \boldsymbol{p}} - \frac{\partial^{(a)}}{\partial \boldsymbol{p}}\cdot\frac{\partial^{(b)}}{\partial \boldsymbol{q}}\right)\right\} a(\boldsymbol{p}, \boldsymbol{q})b(\boldsymbol{p}, \boldsymbol{q}), \quad \text{(A54)}$$

where the indices $(a)$ and $(b)$ at the differential operators indicate which functions have to be differentiated. From this correspondence it follows that the Weyl transforms of the anticommutator and the commutator are given by

$$\tfrac{1}{2}\{A, B\} \rightleftarrows \cos\left\{\frac{\hbar}{2}\left(\frac{\partial^{(a)}}{\partial \boldsymbol{q}}\cdot\frac{\partial^{(b)}}{\partial \boldsymbol{p}} - \frac{\partial^{(a)}}{\partial \boldsymbol{p}}\cdot\frac{\partial^{(b)}}{\partial \boldsymbol{q}}\right)\right\} a(\boldsymbol{p}, \boldsymbol{q})b(\boldsymbol{p}, \boldsymbol{q}), \quad \text{(A55)}$$

$$-\frac{i}{\hbar}[A, B] \rightleftarrows \frac{2}{\hbar}\sin\left\{\frac{\hbar}{2}\left(\frac{\partial^{(a)}}{\partial \boldsymbol{q}}\cdot\frac{\partial^{(b)}}{\partial \boldsymbol{p}} - \frac{\partial^{(a)}}{\partial \boldsymbol{p}}\cdot\frac{\partial^{(b)}}{\partial \boldsymbol{q}}\right)\right\} a(\boldsymbol{p}, \boldsymbol{q})b(\boldsymbol{p}, \boldsymbol{q}). \quad \text{(A56)}$$

While the transform (A54) of a product $AB$ of operators is a series in $\hbar$, both (half) the anticommutator and $(-i/\hbar)$ times the commutator are series in $\hbar^2$.

The classical functions that correspond to the product $AB$ and the operators $\tfrac{1}{2}\{A, B\}$ and $-(i/\hbar)[A, B]$ are, according to (A22), the limits for $\hbar \to 0$ of the right-hand sides of (A54), (A55) and (A56)

$$AB \overset{\mathrm{cl}}{\rightarrow} a_{\mathrm{cl}}(\boldsymbol{p}, \boldsymbol{q})b_{\mathrm{cl}}(\boldsymbol{p}, \boldsymbol{q}), \quad \text{(A57)}$$

$$\tfrac{1}{2}\{A, B\} \overset{\mathrm{cl}}{\rightarrow} a_{\mathrm{cl}}(\boldsymbol{p}, \boldsymbol{q})b_{\mathrm{cl}}(\boldsymbol{p}, \boldsymbol{q}), \quad \text{(A58)}$$

$$-\frac{i}{\hbar}[A, B] \overset{\mathrm{cl}}{\rightarrow} \frac{\partial a_{\mathrm{cl}}(\boldsymbol{p}, \boldsymbol{q})}{\partial \boldsymbol{q}}\cdot\frac{\partial b_{\mathrm{cl}}(\boldsymbol{p}, \boldsymbol{q})}{\partial \boldsymbol{p}} - \frac{\partial a_{\mathrm{cl}}(\boldsymbol{p}, \boldsymbol{q})}{\partial \boldsymbol{p}}\cdot\frac{\partial b_{\mathrm{cl}}(\boldsymbol{p}, \boldsymbol{q})}{\partial \boldsymbol{q}}. \quad \text{(A59)}$$

The last expression obtained is the Poisson bracket of the classical functions $a_{\mathrm{cl}}(\boldsymbol{p}, \boldsymbol{q})$ and $b_{\mathrm{cl}}(\boldsymbol{p}, \boldsymbol{q})$.

### 3. *The Wigner function*

a. *Definition*

The state of a quantum-mechanical system may be described by means of a density operator

$$P(t) = |\psi(t)\rangle\langle\psi(t)|, \tag{A60}$$

where $|\psi(t)\rangle$ is the state vector in Hilbert space. The expectation value of an operator $A$ may be written in terms of this density operator as:

$$\bar{A}(t) = \mathrm{Tr}\,\{P(t)A\}. \tag{A61}$$

With the use of (A53) this expectation value becomes

$$\bar{A} = \bar{a} \tag{A62}$$

with the abbreviation

$$\bar{a}(t) \equiv \int \mathrm{d}\boldsymbol{p}\,\mathrm{d}\boldsymbol{q}\, a(\boldsymbol{p}, \boldsymbol{q})\rho(\boldsymbol{p}, \boldsymbol{q}; t). \tag{A63}$$

Here $h^3\rho(\boldsymbol{p}, \boldsymbol{q}; t)$ is the Weyl transform of the density operator

$$P(t) \rightleftarrows h^3\rho(\boldsymbol{p}, \boldsymbol{q}; t). \tag{A64}$$

Since the density operator is hermitian its Weyl transform $\rho(\boldsymbol{p}, \boldsymbol{q}; t)$ is real.

The function $\rho(\boldsymbol{p}, \boldsymbol{q}; t)$ is called the Wigner function. Its original form[1] follows from (A7) with (A60) inserted:

$$\rho(\boldsymbol{p}, \boldsymbol{q}; t) = h^{-3}\int \mathrm{d}\boldsymbol{v}\, \mathrm{e}^{(i/\hbar)\boldsymbol{p}\cdot\boldsymbol{v}}\psi(\boldsymbol{q}-\tfrac{1}{2}\boldsymbol{v}; t)\psi^*(\boldsymbol{q}+\tfrac{1}{2}\boldsymbol{v}; t) \tag{A65}$$

with the notation $\psi(\boldsymbol{q}; t) \equiv \langle\boldsymbol{q}|\psi(t)\rangle$, the wave function in the coordinate representation. From (A6), with (A60) inserted, one gets the Wigner function in terms of the wave functions $\varphi(\boldsymbol{p}; t) \equiv \langle\boldsymbol{p}|\psi(t)\rangle$ in the momentum representation

$$\rho(\boldsymbol{p}, \boldsymbol{q}; t) = h^{-3}\int \mathrm{d}\boldsymbol{u}\, \mathrm{e}^{(i/\hbar)\boldsymbol{q}\cdot\boldsymbol{u}}\varphi(\boldsymbol{p}+\tfrac{1}{2}\boldsymbol{u}; t)\varphi^*(\boldsymbol{p}-\tfrac{1}{2}\boldsymbol{u}; t). \tag{A66}$$

[1] E. P. Wigner, Phys. Rev. **40**(1932)749.

A third form is obtained by inserting (A60) into (A18) and using (A4). One then obtains (omitting the time dependence from now on):

$$\rho(\boldsymbol{p}, \boldsymbol{q}) = h^{-\frac{3}{2}} \exp\left(\frac{\hbar}{2i} \frac{\partial}{\partial \boldsymbol{p}} \cdot \frac{\partial}{\partial \boldsymbol{q}}\right) \{\varphi(\boldsymbol{p})\psi^*(\boldsymbol{q})\mathrm{e}^{(i/\hbar)\boldsymbol{p}\cdot\boldsymbol{q}}\}. \tag{A67}$$

Since $\rho$ is real one may also write the complex conjugate of the right-hand side:

$$\rho(\boldsymbol{p}, \boldsymbol{q}) = h^{-\frac{3}{2}} \exp\left(-\frac{\hbar}{2i} \frac{\partial}{\partial \boldsymbol{p}} \cdot \frac{\partial}{\partial \boldsymbol{q}}\right) \{\varphi^*(\boldsymbol{p})\psi(\boldsymbol{q})\mathrm{e}^{-(i/\hbar)\boldsymbol{p}\cdot\boldsymbol{q}}\}. \tag{A68}$$

The fact that the trace of the density operator is unity is reflected by the property that the Wigner function is normalized

$$\int \mathrm{d}\boldsymbol{p}\,\mathrm{d}\boldsymbol{q}\,\rho(\boldsymbol{p}, \boldsymbol{q}) = 1, \tag{A69}$$

as follows from (A52).

Expectation values of an operator $A$ are thus obtained as integrals (A62) with (A63) over its Weyl transform $a(\boldsymbol{p}, \boldsymbol{q})$ and the Wigner function $\rho(\boldsymbol{p}, \boldsymbol{q})$, which is normalized according to (A69).

### b. *Properties*

The function $\rho(\boldsymbol{p}, \boldsymbol{q})$ is real and normalized, but not necessarily positive definite. The integrals over the coordinates or momenta however are positive definite as follows from (A65) and (A66)

$$\int \rho(\boldsymbol{p}, \boldsymbol{q})\mathrm{d}\boldsymbol{p} = |\psi(\boldsymbol{q})|^2 \geqslant 0, \tag{A70}$$

$$\int \rho(\boldsymbol{p}, \boldsymbol{q})\mathrm{d}\boldsymbol{q} = |\varphi(\boldsymbol{p})|^2 \geqslant 0. \tag{A71}$$

Thus, in contrast to the function $\rho(\boldsymbol{p}, \boldsymbol{q})$ itself, these integrals may be interpreted as probability densities. (They are indeed normalized to unity.)

One may show that the Wigner function is limited: it satisfies the inequality

$$|\rho(\boldsymbol{p}, \boldsymbol{q})| \leqslant (2/h)^3. \tag{A72}$$

To prove this one starts from Schwarz's inequality, applied to the expression (A65) for the Wigner function

$$|\rho(\boldsymbol{p}, \boldsymbol{q})|^2 \leqslant h^{-6} \left\{\int \mathrm{d}\boldsymbol{v}|\mathrm{e}^{(i/\hbar)\boldsymbol{p}\cdot\boldsymbol{v}}\psi(\boldsymbol{q}-\tfrac{1}{2}\boldsymbol{v})|^2\right\}\left\{\int \mathrm{d}\boldsymbol{v}|\psi^*(\boldsymbol{q}+\tfrac{1}{2}\boldsymbol{v})|^2\right\}. \tag{A73}$$

From the normalization of the wave function it follows that the right-hand side of (A73) is indeed the square of the right-hand side of (A72).

Since $\rho$ is normalized, it follows from the inequality (A72) that $\rho$ is different from zero in a region of which the volume in phase space is at least equal to $(h/2)^3$, in other words its support has a volume larger than this volume. Hence the Wigner function can never be sharply localized both in $\boldsymbol{p}$ and $\boldsymbol{q}$: a delta function character is thus excluded. (This situation is a reflection of the uncertainty principle.)

The density operator has the form of a projection operator, so that

$$P^2 = P. \tag{A74}$$

As a consequence the trace of $P^2$ is equal to unity (since the trace of $P$ is unity). In terms of the Wigner function this property is

$$\int \mathrm{d}\boldsymbol{p}\, \mathrm{d}\boldsymbol{q} \{\rho(\boldsymbol{p}, \boldsymbol{q})\}^2 = h^{-3}, \tag{A75}$$

as follows from (A53).

The property (A74) has as a counterpart for the Wigner function

$$\rho(\boldsymbol{p}, \boldsymbol{q}) = h^3 \exp \left\{ \frac{i\hbar}{2} \left( \frac{\partial^{(a)}}{\partial \boldsymbol{q}} \cdot \frac{\partial^{(b)}}{\partial \boldsymbol{p}} - \frac{\partial^{(a)}}{\partial \boldsymbol{p}} \cdot \frac{\partial^{(b)}}{\partial \boldsymbol{q}} \right) \right\} \rho^{(a)}(\boldsymbol{p}, \boldsymbol{q}) \rho^{(b)}(\boldsymbol{p}, \boldsymbol{q}), \tag{A76}$$

as follows from (A54). The two factors $\rho$ at the right-hand side are in fact the same, but indices $(a)$ and $(b)$ have been employed to indicate the way in which the differential operators are acting.

To every wave function $\psi(\boldsymbol{q})$ corresponds one single Wigner function as is apparent from the expression (A65). The inverse is also true[1]: to every real function $f(\boldsymbol{p}, \boldsymbol{q})$ in phase space, which satisfies equations of the form (A69) and (A76) corresponds one single normalized complex function $g(\boldsymbol{q})$ of the coordinates (apart from a phase factor), in such a way that (cf. (A65)):

$$f(\boldsymbol{p}, \boldsymbol{q}) = h^{-3} \int \mathrm{d}\boldsymbol{v}\, \mathrm{e}^{(i/\hbar)\boldsymbol{p}\cdot\boldsymbol{v}} g(\boldsymbol{q} - \tfrac{1}{2}\boldsymbol{v}) g^*(\boldsymbol{q} + \tfrac{1}{2}\boldsymbol{v}). \tag{A77}$$

In other words every real function in phase space, satisfying (A69) and (A76), may play the role of a Wigner function. To prove (A77) we first define the function

$$\mathscr{K}(\boldsymbol{q} + \tfrac{1}{2}\boldsymbol{v}, \boldsymbol{q} - \tfrac{1}{2}\boldsymbol{v}) = \int \mathrm{d}\boldsymbol{p}\, \mathrm{e}^{-(i/\hbar)\boldsymbol{p}\cdot\boldsymbol{v}} f(\boldsymbol{p}, \boldsymbol{q}). \tag{A78}$$

[1] G. A. Baker jr., Phys. Rev. **109**(1958)2198.

Its inverse is

$$f(\boldsymbol{p}, \boldsymbol{q}) = h^{-3} \int \mathrm{d}\boldsymbol{v}\, \mathrm{e}^{(i/\hbar)\boldsymbol{p}\cdot\boldsymbol{v}} \mathscr{K}(\boldsymbol{q}+\tfrac{1}{2}\boldsymbol{v}, \boldsymbol{q}-\tfrac{1}{2}\boldsymbol{v}). \tag{A79}$$

We shall have to show that the kernel $\mathscr{K}(\boldsymbol{q}+\frac{1}{2}\boldsymbol{v}, \boldsymbol{q}-\frac{1}{2}\boldsymbol{v})$ factorizes, so that (A79) reduces to (A77). To that purpose we write the properties (A69) and (A76) for $f(\boldsymbol{p}, \boldsymbol{q})$ in terms of the kernel $\mathscr{K}$. Inserting (A78) and (A79) into (A69) and (A76) one finds:

$$\int \mathscr{K}(\boldsymbol{q}, \boldsymbol{q})\mathrm{d}\boldsymbol{q} = 1, \tag{A80}$$

$$\begin{aligned} &\mathscr{K}(\boldsymbol{q}+\tfrac{1}{2}\boldsymbol{v}, \boldsymbol{q}-\tfrac{1}{2}\boldsymbol{v}) \\ &\quad = h^{-3} \int \mathrm{d}\boldsymbol{v}'\mathrm{d}\boldsymbol{v}''\mathrm{d}\boldsymbol{p} \exp\left\{-\frac{i}{\hbar}\boldsymbol{p}\cdot\boldsymbol{v} + \tfrac{1}{2}i\hbar\left(\frac{\partial^{(a)}}{\partial\boldsymbol{q}}\cdot\frac{\partial^{(b)}}{\partial\boldsymbol{p}} - \frac{\partial^{(a)}}{\partial\boldsymbol{p}}\cdot\frac{\partial^{(b)}}{\partial\boldsymbol{q}}\right)\right\} \\ &\quad\quad \exp\left\{\frac{i}{\hbar}(\boldsymbol{p}^{(a)}\cdot\boldsymbol{v}' + \boldsymbol{p}^{(b)}\cdot\boldsymbol{v}'')\right\} \mathscr{K}^{(a)}(\boldsymbol{q}+\tfrac{1}{2}\boldsymbol{v}', \boldsymbol{q}-\tfrac{1}{2}\boldsymbol{v}')\mathscr{K}^{(b)}(\boldsymbol{q}+\tfrac{1}{2}\boldsymbol{v}'', \boldsymbol{q}-\tfrac{1}{2}\boldsymbol{v}''). \end{aligned} \tag{A81}$$

If the differentiations in (A81) with respect to the momentum variables are performed, the integration over these variables may be carried out. This gives

$$\begin{aligned} \mathscr{K}(\boldsymbol{q}+\tfrac{1}{2}\boldsymbol{v}, \boldsymbol{q}-\tfrac{1}{2}\boldsymbol{v}) = \int \mathrm{d}\boldsymbol{v}'\mathrm{d}\boldsymbol{v}''\delta(\boldsymbol{v}'+\boldsymbol{v}''-\boldsymbol{v}) \exp\left\{\frac{1}{2}\left(\boldsymbol{v}'\cdot\frac{\partial^{(b)}}{\partial\boldsymbol{q}} - \boldsymbol{v}''\cdot\frac{\partial^{(a)}}{\partial\boldsymbol{q}}\right)\right\} \\ \mathscr{K}^{(a)}(\boldsymbol{q}+\tfrac{1}{2}\boldsymbol{v}', \boldsymbol{q}-\tfrac{1}{2}\boldsymbol{v}')\mathscr{K}^{(b)}(\boldsymbol{q}+\tfrac{1}{2}\boldsymbol{v}'', \boldsymbol{q}-\tfrac{1}{2}\boldsymbol{v}''). \end{aligned} \tag{A82}$$

The exponential acting on the kernel may be seen as the operator which yields a Taylor expansion. Hence we may write now

$$\begin{aligned} \mathscr{K}(\boldsymbol{q}+\tfrac{1}{2}\boldsymbol{v}, \boldsymbol{q}-\tfrac{1}{2}\boldsymbol{v}) = \int \mathrm{d}\boldsymbol{v}'\mathrm{d}\boldsymbol{v}''\delta(\boldsymbol{v}'+\boldsymbol{v}''-\boldsymbol{v})\mathscr{K}\{\boldsymbol{q}+\tfrac{1}{2}(\boldsymbol{v}'-\boldsymbol{v}''), \boldsymbol{q}-\tfrac{1}{2}(\boldsymbol{v}'+\boldsymbol{v}'')\} \\ \mathscr{K}\{\boldsymbol{q}+\tfrac{1}{2}(\boldsymbol{v}'+\boldsymbol{v}''), \boldsymbol{q}+\tfrac{1}{2}(\boldsymbol{v}'-\boldsymbol{v}'')\}. \end{aligned} \tag{A83}$$

With new integration variables $\boldsymbol{v}''' = \frac{1}{2}(\boldsymbol{v}'-\boldsymbol{v}'')$, $\boldsymbol{v}'''' = \boldsymbol{v}'+\boldsymbol{v}''$ one finds, writing $\boldsymbol{v}'$ instead of $\boldsymbol{q}+\boldsymbol{v}'''$,

$$\mathscr{K}(\boldsymbol{q}+\tfrac{1}{2}\boldsymbol{v}, \boldsymbol{q}-\tfrac{1}{2}\boldsymbol{v}) = \int \mathrm{d}\boldsymbol{v}'\, \mathscr{K}(\boldsymbol{q}+\tfrac{1}{2}\boldsymbol{v}, \boldsymbol{v}')\mathscr{K}(\boldsymbol{v}', \boldsymbol{q}-\tfrac{1}{2}\boldsymbol{v}). \tag{A84}$$

A third property of the kernel $\mathscr{K}$ is its hermitian character:

$$\mathscr{K}^*(\boldsymbol{q}+\tfrac{1}{2}\boldsymbol{v}, \boldsymbol{q}-\tfrac{1}{2}\boldsymbol{v}) = \mathscr{K}(\boldsymbol{q}-\tfrac{1}{2}\boldsymbol{v}, \boldsymbol{q}+\tfrac{1}{2}\boldsymbol{v}), \tag{A85}$$

as follows from (A78). From (A84) it is seen that the eigenvalues $\lambda$ of the kernel $\mathscr{K}$ satisfy the equation $\lambda = \lambda^2$, so that $\lambda$ is 0 or 1. Furthermore it follows from (A80) that the sum of the eigenvalues is unity. Therefore only one eigenvalue is 1, and the rest 0. This means that the kernel may be expressed in terms of the normalized eigenfunction $g'$ that corresponds to the single eigenvalue 1:

$$\mathscr{K}(\boldsymbol{q}+\tfrac{1}{2}\boldsymbol{v}, \boldsymbol{q}-\tfrac{1}{2}\boldsymbol{v}) = g'(\boldsymbol{q}+\tfrac{1}{2}\boldsymbol{v})g'^*(\boldsymbol{q}-\tfrac{1}{2}\boldsymbol{v}), \tag{A86}$$

or, writing $g$ instead of $g'^*$ ($g$ is likewise normalized to unity),

$$\mathscr{K}(\boldsymbol{q}+\tfrac{1}{2}\boldsymbol{v}, \boldsymbol{q}-\tfrac{1}{2}\boldsymbol{v}) = g(\boldsymbol{q}-\tfrac{1}{2}\boldsymbol{v})g^*(\boldsymbol{q}+\tfrac{1}{2}\boldsymbol{v}), \tag{A87}$$

which completes the proof of (A77).

c. *Development in time*

The time evolution of the Wigner function is a direct consequence of the equation

$$\frac{\partial P}{\partial t} = -\frac{i}{\hbar}[H, P], \tag{A88}$$

which governs the time evolution of the density operator. In fact one finds from (A56) for the Weyl transform $h^3\rho(\boldsymbol{p}, \boldsymbol{q}; t)$ of the density operator:

$$\frac{\partial \rho(\boldsymbol{p}, \boldsymbol{q}; t)}{\partial t} = \frac{2}{\hbar}\sin\left\{\frac{\hbar}{2}\left(\frac{\partial^{(h)}}{\partial \boldsymbol{q}}\cdot\frac{\partial^{(\rho)}}{\partial \boldsymbol{p}} - \frac{\partial^{(h)}}{\partial \boldsymbol{p}}\cdot\frac{\partial^{(\rho)}}{\partial \boldsymbol{q}}\right)\right\} h(\boldsymbol{p}, \boldsymbol{q})\rho(\boldsymbol{p}, \boldsymbol{q}; t), \tag{A89}$$

where $h(\boldsymbol{p}, \boldsymbol{q})$ is the Weyl transform of the Hamilton operator. One may introduce a Liouville operator defined as

$$\mathscr{L}(\boldsymbol{p}, \boldsymbol{q}) \equiv -\frac{2}{\hbar}\sin\left\{\frac{\hbar}{2}\left(\frac{\partial^{(h)}}{\partial \boldsymbol{q}}\cdot\frac{\partial}{\partial \boldsymbol{p}} - \frac{\partial^{(h)}}{\partial \boldsymbol{p}}\cdot\frac{\partial}{\partial \boldsymbol{q}}\right)\right\} h(\boldsymbol{p}, \boldsymbol{q}). \tag{A90}$$

Then the time evolution of the Wigner function is described by

$$\frac{\partial \rho(\boldsymbol{p}, \boldsymbol{q}; t)}{\partial t} = -\mathscr{L}(\boldsymbol{p}, \boldsymbol{q})\rho(\boldsymbol{p}, \boldsymbol{q}; t). \tag{A91}$$

It has the formal solution, for a time-independent Hamiltonian,

$$\rho(\boldsymbol{p}, \boldsymbol{q}; t) = \exp\{-\mathscr{L}(\boldsymbol{p}, \boldsymbol{q})(t-t_0)\}\rho(\boldsymbol{p}, \boldsymbol{q}; t_0). \tag{A92}$$

The equation (A91) may be used to find an expression for the time derivative of the expectation value (A62) with (A63) of an operator $A$:

$$\frac{\mathrm{d}\bar{a}(t)}{\mathrm{d}t} = -\int \mathrm{d}\boldsymbol{p}\,\mathrm{d}\boldsymbol{q}\, a(\boldsymbol{p}, \boldsymbol{q})\mathscr{L}(\boldsymbol{p}, \boldsymbol{q})\rho(\boldsymbol{p}, \boldsymbol{q}; t). \tag{A93}$$

Partial integration gives

$$\frac{\mathrm{d}\bar{a}(t)}{\mathrm{d}t} = \int \mathrm{d}\boldsymbol{p}\,\mathrm{d}\boldsymbol{q}\{\mathscr{L}(\boldsymbol{p}, \boldsymbol{q})a(\boldsymbol{p}, \boldsymbol{q})\}\rho(\boldsymbol{p}, \boldsymbol{q}; t), \tag{A94}$$

or with the explicit Liouvillean (A90)

$$\frac{\mathrm{d}\bar{a}(t)}{\mathrm{d}t} = \frac{2}{\hbar}\int \mathrm{d}\boldsymbol{p}\,\mathrm{d}\boldsymbol{q}\left[\sin\left\{\frac{\hbar}{2}\left(\frac{\partial^{(a)}}{\partial \boldsymbol{q}}\cdot\frac{\partial^{(h)}}{\partial \boldsymbol{p}} - \frac{\partial^{(a)}}{\partial \boldsymbol{p}}\cdot\frac{\partial^{(h)}}{\partial \boldsymbol{q}}\right)\right\} a(\boldsymbol{p}, \boldsymbol{q})h(\boldsymbol{p}, \boldsymbol{q})\right] \rho(\boldsymbol{p}, \boldsymbol{q}; t). \tag{A95}$$

In particular if one chooses for $a(\boldsymbol{p}, \boldsymbol{q})$ the Weyl transforms $\boldsymbol{p}$ and $\boldsymbol{q}$ of the momentum and coordinate operators this relation gets the simple form

$$\frac{\mathrm{d}\bar{\boldsymbol{p}}(t)}{\mathrm{d}t} = -\int \mathrm{d}\boldsymbol{p}\,\mathrm{d}\boldsymbol{q}\,\frac{\partial h}{\partial \boldsymbol{q}}\,\rho(\boldsymbol{p}, \boldsymbol{q}; t), \tag{A96}$$

$$\frac{\mathrm{d}\bar{\boldsymbol{q}}(t)}{\mathrm{d}t} = \int \mathrm{d}\boldsymbol{p}\,\mathrm{d}\boldsymbol{q}\,\frac{\partial h}{\partial \boldsymbol{p}}\,\rho(\boldsymbol{p}, \boldsymbol{q}; t). \tag{A97}$$

These equations are the Weyl transform versions of the Ehrenfest equations of the ordinary formulation of quantum mechanics. Their connexion with the Hamilton equations will be discussed in the next subsection.

### d. *The classical limit*

Let us try to find those Wigner functions which are products of functions of the coordinates and momenta[1]

$$\rho(\boldsymbol{p}, \boldsymbol{q}) = \rho_1(\boldsymbol{p})\rho_2(\boldsymbol{q}). \tag{A98}$$

From the form (A65) of the Wigner function one finds for the Fourier transform of (A98), replacing $\frac{1}{2}\boldsymbol{v}$ by $\boldsymbol{v}$ for convenience,

$$\psi(\boldsymbol{q}-\boldsymbol{v})\psi^*(\boldsymbol{q}+\boldsymbol{v}) = \left\{\int \mathrm{d}\boldsymbol{p}\,\mathrm{e}^{-(2i/\hbar)\boldsymbol{p}\cdot\boldsymbol{v}}\rho_1(\boldsymbol{p})\right\}\rho_2(\boldsymbol{q}), \tag{A99}$$

where the expression at the right-hand side is a product of a function of $\boldsymbol{v}$ and a function of $\boldsymbol{q}$. Writing

$$\psi(\boldsymbol{q}) = \mathrm{e}^{\chi(\boldsymbol{q})}, \tag{A100}$$

$$\int \mathrm{d}\boldsymbol{p}\,\mathrm{e}^{-(2i/\hbar)\boldsymbol{p}\cdot\boldsymbol{v}}\rho_1(\boldsymbol{p}) = \mathrm{e}^{\alpha(\boldsymbol{v})}, \tag{A101}$$

$$\rho_2(\boldsymbol{q}) = \mathrm{e}^{\beta(\boldsymbol{q})}, \tag{A102}$$

[1] T. Takabayasi, Progr. Theor. Phys. **11**(1954)341.

one gets for the logarithm of (A99)

$$\chi(\boldsymbol{q}-\boldsymbol{v})+\chi^*(\boldsymbol{q}+\boldsymbol{v}) = \alpha(\boldsymbol{v})+\beta(\boldsymbol{q}). \tag{A103}$$

Developing the functions in the left-hand side in powers of $\boldsymbol{v}$, we get the identity in $\boldsymbol{q}$ and $\boldsymbol{v}$

$$\text{re}\,\chi(\boldsymbol{q})-i\boldsymbol{v}\cdot\frac{\partial}{\partial \boldsymbol{q}}\{\text{im}\,\chi(\boldsymbol{q})\}+\tfrac{1}{2}\boldsymbol{v}\boldsymbol{v}:\frac{\partial^2}{\partial \boldsymbol{q}\,\partial \boldsymbol{q}}\{\text{re}\,\chi(\boldsymbol{q})\}-\ldots = \tfrac{1}{2}\alpha(\boldsymbol{v})+\tfrac{1}{2}\beta(\boldsymbol{q}). \tag{A104}$$

From this identity it follows that

$$\begin{aligned}\frac{\partial^n}{\partial \boldsymbol{q}^n}\{\text{re}\,\chi(\boldsymbol{q})\} &= \mathbf{c}_n, \qquad (n = 2, 4, \ldots),\\ \frac{\partial^n}{\partial \boldsymbol{q}^n}\{\text{im}\,\chi(\boldsymbol{q})\} &= \mathbf{c}_n, \qquad (n = 1, 3, \ldots),\end{aligned} \tag{A105}$$

with (real) constants $\mathbf{c}_n$ independent of $\boldsymbol{q}$. These formulae yield for $n = 1$ and $n = 2$ respectively:

$$\begin{aligned}\text{im}\,\chi(\boldsymbol{q}) &= \boldsymbol{c}_1\cdot\boldsymbol{q}+d_0,\\ \text{re}\,\chi(\boldsymbol{q}) &= \tfrac{1}{2}\mathbf{c}_2:\boldsymbol{q}\boldsymbol{q}+\boldsymbol{d}_1\cdot\boldsymbol{q}+d_2,\end{aligned} \tag{A106}$$

with $\boldsymbol{c}_1$ and $\mathbf{c}_2$ constants which already occurred, and $d_0$, $\boldsymbol{d}_1$ and $d_2$ other real constant quantities. The expressions (A106) satisfy (A105) also for $n = 3, 4, \ldots$. From (A106) and (A100) one finds

$$\psi(\boldsymbol{q}) = \exp(\mathbf{a}_2:\boldsymbol{q}\boldsymbol{q}+\boldsymbol{a}_1\cdot\boldsymbol{q}+a_0) \tag{A107}$$

with $\mathbf{a}_2 = \frac{1}{2}\mathbf{c}_2$ real, and $\boldsymbol{a}_1 = \boldsymbol{d}_1+i\boldsymbol{c}_1$, $a_0 = d_2+id_0$ complex constants. By a rotation of the coordinate axes one may bring the quadratic expression in diagonal form. Subsequently the real part of the linear term is taken together with the quadratic term so as to form a square. In this way one finds

$$\psi(\boldsymbol{q}) = C\exp\left\{-\sum_{i=1}^{3}\frac{(\boldsymbol{q}-\boldsymbol{q}_0)_i^2}{4\Delta_i^2}+\frac{i\boldsymbol{p}_0\cdot\boldsymbol{q}}{\hbar}\right\} \tag{A108}$$

with real constants $\boldsymbol{p}_0$, $\boldsymbol{q}_0$ and $\Delta_i$ and a normalization constant $C$, which may be chosen to be real. If, for simplicity, we limit ourselves to the case $\Delta_1 = \Delta_2 = \Delta_3 = \Delta$, we have the wave function

$$\psi(\boldsymbol{q}) = \frac{1}{(2\pi)^{\frac{3}{4}}\Delta^{\frac{3}{2}}}\exp\left\{-\frac{(\boldsymbol{q}-\boldsymbol{q}_0)^2}{4\Delta^2}+\frac{i\boldsymbol{p}_0\cdot\boldsymbol{q}}{\hbar}\right\}, \tag{A109}$$

a 'minimum wave packet'. The Wigner function (A65) which corresponds to this wave function is

$$\rho(\boldsymbol{p}, \boldsymbol{q}) = (2/h)^3 \exp\left\{-\frac{(\boldsymbol{q}-\boldsymbol{q}_0)^2}{2\Delta^2} - \frac{2\Delta^2}{\hbar^2}(\boldsymbol{p}-\boldsymbol{p}_0)^2\right\}. \tag{A110}$$

This expression is the product of two Gaussians, one in the coordinate and one in the momentum space. The integrals of (A110) over $\boldsymbol{p}$ or $\boldsymbol{q}$ are the probability densities (A70) and (A71). They are Gaussians with widths $\Delta$ and $\hbar/2\Delta$ respectively. The product of their widths is $\frac{1}{2}\hbar$, which explains the name 'minimum wave packet' for (A109).

In the preceding section the classical limit of a physical quantity was defined as the limit, for $\hbar$ tending to zero, of the Weyl transform of the corresponding operator (v. (A22)). The situation is not exactly the same for the density operator, since physical states exist in quantum mechanics which have no classical counterpart. This means that taking the limit $\hbar \to 0$ of the Wigner function will not always lead to a classical function. Even if one utilizes minimum wave packets in their form (A110), the limit $\hbar \to 0$ has to be taken in a special way so as to obtain the classical limit. In fact one must let tend both $\hbar$ and $\Delta$ to zero, but the latter in such a way that $\hbar/\Delta$ tends to zero as well. In this way one finds from (A110) as the classical limit of the Wigner function

$$\rho(\boldsymbol{p}, \boldsymbol{q}) \stackrel{\text{cl}}{\to} \rho_{\text{cl}}(\boldsymbol{p}, \boldsymbol{q}) = \delta(\boldsymbol{p}-\boldsymbol{p}_0)\delta(\boldsymbol{q}-\boldsymbol{q}_0). \tag{A111}$$

With this classical function and (A22) we obtain as the classical limit of the expectation value from (A62) with (A63)

$$\bar{A} = \bar{a} \stackrel{\text{cl}}{\to} a_{\text{cl}}(\boldsymbol{p}_0, \boldsymbol{q}_0). \tag{A112}$$

The time derivative of an expectation value follows from (A94) with (A90). In the classical limit one finds then with (A111) and (A22)

$$\frac{\mathrm{d}\bar{a}(t)}{\mathrm{d}t} \stackrel{\text{cl}}{\to} \mathscr{L}_{\text{cl}}(\boldsymbol{p}_0, \boldsymbol{q}_0) a_{\text{cl}}(\boldsymbol{p}_0, \boldsymbol{q}_0). \tag{A113}$$

Here the classical Liouville operator is

$$\mathscr{L}_{\text{cl}}(\boldsymbol{p}, \boldsymbol{q}) \equiv \lim_{\hbar\to 0} \mathscr{L}(\boldsymbol{p}, \boldsymbol{q}) = \frac{\partial h_{\text{cl}}}{\partial \boldsymbol{p}} \cdot \frac{\partial}{\partial \boldsymbol{q}} - \frac{\partial h_{\text{cl}}}{\partial \boldsymbol{q}} \cdot \frac{\partial}{\partial \boldsymbol{p}}, \tag{A114}$$

where $h_{\text{cl}}(\boldsymbol{p}, \boldsymbol{q})$ is the classical limit (A22) of the Hamiltonian. Thus (A113) becomes

$$\frac{\mathrm{d}\bar{a}(t)}{\mathrm{d}t} \stackrel{\text{cl}}{\to} \frac{\partial h_{\text{cl}}(\boldsymbol{p}_0, \boldsymbol{q}_0)}{\partial \boldsymbol{p}_0} \cdot \frac{\partial a_{\text{cl}}(\boldsymbol{p}_0, \boldsymbol{q}_0)}{\partial \boldsymbol{q}_0} - \frac{\partial h_{\text{cl}}(\boldsymbol{p}_0, \boldsymbol{q}_0)}{\partial \boldsymbol{q}_0} \cdot \frac{\partial a_{\text{cl}}(\boldsymbol{p}_0, \boldsymbol{q}_0)}{\partial \boldsymbol{p}_0}, \tag{A115}$$

which is the Poisson bracket. With the help of (A112) we may write the classical limit of the left-hand side as:

$$\frac{\mathrm{d}\bar{a}(t)}{\mathrm{d}t} \overset{\text{cl}}{\to} \frac{\mathrm{d}a_{\text{cl}}(\boldsymbol{p}_0, \boldsymbol{q}_0)}{\mathrm{d}t}. \tag{A116}$$

Comparing (A115) and (A116) we obtain the classical equation:

$$\frac{\mathrm{d}a_{\text{cl}}(\boldsymbol{p}_0, \boldsymbol{q}_0)}{\mathrm{d}t} = \frac{\partial h_{\text{cl}}(\boldsymbol{p}_0, \boldsymbol{q}_0)}{\partial \boldsymbol{p}_0} \cdot \frac{\partial a_{\text{cl}}(\boldsymbol{p}_0, \boldsymbol{q}_0)}{\partial \boldsymbol{q}_0} - \frac{\partial h_{\text{cl}}(\boldsymbol{p}_0, \boldsymbol{q}_0)}{\partial \boldsymbol{q}_0} \cdot \frac{\partial a_{\text{cl}}(\boldsymbol{p}_0, \boldsymbol{q}_0)}{\partial \boldsymbol{p}_0}. \tag{A117}$$

In particular for the momenta and coordinates we get

$$\frac{\mathrm{d}\boldsymbol{p}_0}{\mathrm{d}t} = -\frac{\partial h_{\text{cl}}(\boldsymbol{p}_0, \boldsymbol{q}_0)}{\partial \boldsymbol{q}_0}, \tag{A118}$$

$$\frac{\mathrm{d}\boldsymbol{q}_0}{\mathrm{d}t} = \frac{\partial h_{\text{cl}}(\boldsymbol{p}_0, \boldsymbol{q}_0)}{\partial \boldsymbol{p}_0}, \tag{A119}$$

which are the classical Hamilton equations.

## e. *The propagator*

The density operator at time $t$ follows from the density operator at time $t_0$ with the help of the expression

$$P(t) = U(t, t_0)P(t_0)U^{\dagger}(t, t_0). \tag{A120}$$

The evolution operator $U(t, t_0)$ follows from the Schrödinger equation (for a time-independent Hamiltonian)

$$-\frac{\hbar}{i}\frac{\partial U(t, t_0)}{\partial t} = HU(t, t_0) \tag{A121}$$

with the initial condition $U(t_0, t_0) = 1$:

$$U(t, t_0) = \mathrm{e}^{-(i/\hbar)H(t-t_0)}. \tag{A122}$$

The Wigner function, which is the Weyl transform of the density operator (times $h^{-3}$) follows from (A12), with (A120) inserted,

$$\rho(\boldsymbol{p}, \boldsymbol{q}; t) = h^{-3}\,\mathrm{Tr}\,\{\Delta(\boldsymbol{p}, \boldsymbol{q})U(t, t_0)P(t_0)U^{\dagger}(t, t_0)\}. \tag{A123}$$

According to (A8) the third factor between the brackets may be written as an integral with a Wigner function at the time $t_0$, so that one has for (A123):

$$\rho(\boldsymbol{p}, \boldsymbol{q}; t) = \int \mathrm{d}\boldsymbol{p}_0 \, \mathrm{d}\boldsymbol{q}_0 \, \mathscr{P}(\boldsymbol{p}, \boldsymbol{q}, t|\boldsymbol{p}_0, \boldsymbol{q}_0, t_0)\rho(\boldsymbol{p}_0, \boldsymbol{q}_0; t_0) \tag{A124}$$

with the propagator of the Wigner function, defined as

$$\mathscr{P}(\boldsymbol{p}, \boldsymbol{q}, t|\boldsymbol{p}_0, \boldsymbol{q}_0, t_0) = h^{-3} \operatorname{Tr} \{\Delta(\boldsymbol{p}, \boldsymbol{q})U(t, t_0)\Delta(\boldsymbol{p}_0, \boldsymbol{q}_0)U^\dagger(t, t_0)\}. \tag{A125}$$

The propagator, which is a real function (as follows from its definition and the hermiticity of the $\Delta$-operator) is not necessarily positive. Hence an interpretation as a conditional probability is not justified. From the definition (A125) and (A12) it follows that it may be looked upon as the Weyl transform of the operator $h^{-3}U(t, t_0)\Delta(\boldsymbol{p}_0, \boldsymbol{q}_0)U^\dagger(t, t_0)$. Then as a consequence of (A8) one has

$$U(t, t_0)\Delta(\boldsymbol{p}_0, \boldsymbol{q}_0)U^\dagger(t, t_0) = \int \mathrm{d}\boldsymbol{p} \, \mathrm{d}\boldsymbol{q} \, \mathscr{P}(\boldsymbol{p}, \boldsymbol{q}, t|\boldsymbol{p}_0, \boldsymbol{q}_0, t_0)\Delta(\boldsymbol{p}, \boldsymbol{q}). \tag{A126}$$

In a similar way one finds:

$$U^\dagger(t, t_0)\Delta(\boldsymbol{p}, \boldsymbol{q})U(t, t_0) = \int \mathrm{d}\boldsymbol{p}_0 \, \mathrm{d}\boldsymbol{q}_0 \, \mathscr{P}(\boldsymbol{p}, \boldsymbol{q}, t|\boldsymbol{p}_0, \boldsymbol{q}_0, t_0)\Delta(\boldsymbol{p}_0, \boldsymbol{q}_0). \tag{A127}$$

The propagator is normalized, as follows from (A44) with (A49):

$$\begin{aligned} &\int \mathrm{d}\boldsymbol{p}_0 \, \mathrm{d}\boldsymbol{q}_0 \, \mathscr{P}(\boldsymbol{p}, \boldsymbol{q}, t|\boldsymbol{p}_0, \boldsymbol{q}_0, t_0) = 1, \\ &\int \mathrm{d}\boldsymbol{p} \, \mathrm{d}\boldsymbol{q} \, \mathscr{P}(\boldsymbol{p}, \boldsymbol{q}, t|\boldsymbol{p}_0, \boldsymbol{q}_0, t_0) = 1. \end{aligned} \tag{A128}$$

The symmetry of the propagator:

$$\mathscr{P}(\boldsymbol{p}_0, \boldsymbol{q}_0, t_0|\boldsymbol{p}, \boldsymbol{q}, t) = \mathscr{P}(\boldsymbol{p}, \boldsymbol{q}, t|\boldsymbol{p}_0, \boldsymbol{q}_0, t_0) \tag{A129}$$

follows from its definition with (A122).

For the initial value of the propagator one finds from (A122) and (A50):

$$\mathscr{P}(\boldsymbol{p}, \boldsymbol{q}, t_0|\boldsymbol{p}_0, \boldsymbol{q}_0, t_0) = \delta(\boldsymbol{p}-\boldsymbol{p}_0)\delta(\boldsymbol{q}-\boldsymbol{q}_0). \tag{A130}$$

Since the propagator $\mathscr{P}(\boldsymbol{p}, \boldsymbol{q}, t|\boldsymbol{p}_0, \boldsymbol{q}_0, t_0)$ is the Weyl transform of the operator $h^{-3}U(t, t_0)\Delta(\boldsymbol{p}_0, \boldsymbol{q}_0)U^\dagger(t, t_0)$ one may write, with (A53), (A122) and (A50), the orthogonality property:

$$\int \mathrm{d}\boldsymbol{p} \, \mathrm{d}\boldsymbol{q} \, \mathscr{P}(\boldsymbol{p}, \boldsymbol{q}, t|\boldsymbol{p}_0, \boldsymbol{q}_0, t_0)\mathscr{P}(\boldsymbol{p}, \boldsymbol{q}, t|\boldsymbol{p}'_0, \boldsymbol{q}'_0, t_0) = \delta(\boldsymbol{p}_0-\boldsymbol{p}'_0)\delta(\boldsymbol{q}_0-\boldsymbol{q}'_0), \tag{A131}$$

and similarly

$$\int \mathrm{d}\boldsymbol{p}_0 \, \mathrm{d}\boldsymbol{q}_0 \, \mathscr{P}(\boldsymbol{p}, \boldsymbol{q}, t | \boldsymbol{p}_0, \boldsymbol{q}_0, t_0) \mathscr{P}(\boldsymbol{p}', \boldsymbol{q}', t | \boldsymbol{p}_0, \boldsymbol{q}_0, t_0) = \delta(\boldsymbol{p}-\boldsymbol{p}')\delta(\boldsymbol{q}-\boldsymbol{q}'). \tag{A132}$$

From the definition (A125) and the explicit form (A122) for the evolution operator it follows that one may write for arbitrary $t_1$:

$$\mathscr{P}(\boldsymbol{p}, \boldsymbol{q}, t | \boldsymbol{p}_0, \boldsymbol{q}_0, t_0) = h^{-3} \operatorname{Tr} \{U^\dagger(t, t_1) \Delta(\boldsymbol{p}, \boldsymbol{q}) U(t, t_1) U(t_1, t_0) \Delta(\boldsymbol{p}_0, \boldsymbol{q}_0) U^\dagger(t_1, t_0)\}. \tag{A133}$$

With (A126), (A127) and (A50) this gives the convolution property

$$\mathscr{P}(\boldsymbol{p}, \boldsymbol{q}, t | \boldsymbol{p}_0, \boldsymbol{q}_0, t_0) = \int \mathrm{d}\boldsymbol{p}_1 \, \mathrm{d}\boldsymbol{q}_1 \, \mathscr{P}(\boldsymbol{p}, \boldsymbol{q}, t | \boldsymbol{p}_1, \boldsymbol{q}_1, t_1) \mathscr{P}(\boldsymbol{p}_1, \boldsymbol{q}_1, t_1 | \boldsymbol{p}_0, \boldsymbol{q}_0, t_0). \tag{A134}$$

Since $\mathscr{P}(\boldsymbol{p}, \boldsymbol{q}, t | \boldsymbol{p}_0, \boldsymbol{q}_0, t_0)$ is the propagator of the Wigner function, it satisfies the same equation for the time evolution as the Wigner function itself, i.e. (A91):

$$\frac{\partial \mathscr{P}(\boldsymbol{p}, \boldsymbol{q}, t | \boldsymbol{p}_0, \boldsymbol{q}_0, t_0)}{\partial t} = -\mathscr{L}(\boldsymbol{p}, \boldsymbol{q}) \mathscr{P}(\boldsymbol{p}, \boldsymbol{q}, t | \boldsymbol{p}_0, \boldsymbol{q}_0, t_0). \tag{A135}$$

Its formal solution follows with (A130):

$$\mathscr{P}(\boldsymbol{p}, \boldsymbol{q}, t | \boldsymbol{p}_0, \boldsymbol{q}_0, t_0) = \exp\{-\mathscr{L}(\boldsymbol{p}, \boldsymbol{q})(t-t_0)\} \delta(\boldsymbol{p}-\boldsymbol{p}_0)\delta(\boldsymbol{q}-\boldsymbol{q}_0). \tag{A136}$$

The symmetry (A129) permits to write this alternatively as

$$\mathscr{P}(\boldsymbol{p}, \boldsymbol{q}, t | \boldsymbol{p}_0, \boldsymbol{q}_0, t_0) = \exp\{\mathscr{L}(\boldsymbol{p}_0, \boldsymbol{q}_0)(t-t_0)\} \delta(\boldsymbol{p}-\boldsymbol{p}_0)\delta(\boldsymbol{q}-\boldsymbol{q}_0), \tag{A137}$$

which is the solution of

$$\frac{\partial \mathscr{P}(\boldsymbol{p}, \boldsymbol{q}, t | \boldsymbol{p}_0, \boldsymbol{q}_0, t_0)}{\partial t} = \mathscr{L}(\boldsymbol{p}_0, \boldsymbol{q}_0) \mathscr{P}(\boldsymbol{p}, \boldsymbol{q}, t | \boldsymbol{p}_0, \boldsymbol{q}_0, t_0). \tag{A138}$$

### 4. *Generalization to particles with spin*

#### a. *Introduction*

In the preceding we confined ourselves to the consideration of quantum mechanics for single point particles. In that case the three coordinate (or momentum) operators form a complete set of commuting operators. The eigenstates that could be characterized by the eigenvalues of the coordinate (or momentum) operators formed a basis of the complete Hilbert space of all possible states of the particle.

For particles with internal degrees of freedom this holds no longer true. Then several eigenstates correspond to each eigenvalue of the coordinate (or momentum) operator. They may be labelled by a new index $\kappa$. The eigenvalue equations for the momentum and coordinate operators $\boldsymbol{P}$ and $\boldsymbol{Q}$ read in this case

$$\boldsymbol{P}|\boldsymbol{p}, \kappa\rangle = \boldsymbol{p}|\boldsymbol{p}, \kappa\rangle, \qquad \boldsymbol{Q}|\boldsymbol{q}, \kappa\rangle = \boldsymbol{q}|\boldsymbol{q}, \kappa\rangle, \tag{A139}$$

with eigenvalues $\boldsymbol{p}$ and $\boldsymbol{q}$ and $\kappa = 1, 2, \ldots, n$. (For particles with spin for instance, which are described by the Schrödinger–Pauli theory one has $n = 2$, while in Dirac's relativistic theory of spin particles $n = 4$.) The vectors $|\boldsymbol{p}, \kappa\rangle$ or $|\boldsymbol{q}, \kappa\rangle$ with $\kappa$ one of the numbers $1, 2, \ldots, n$ form $n$ bases in $n$ Hilbert spaces of the same structure. The total Hilbert space is the direct sum of these $n$ spaces.

The closure relations for the bases $|\boldsymbol{p}, \kappa\rangle$ and $|\boldsymbol{q}, \kappa\rangle$ read now

$$\begin{aligned} \sum_\kappa \int \mathrm{d}\boldsymbol{p}|\boldsymbol{p}, \kappa\rangle\langle\boldsymbol{p}, \kappa| &= I, \\ \sum_\kappa \int \mathrm{d}\boldsymbol{q}|\boldsymbol{q}, \kappa\rangle\langle\boldsymbol{q}, \kappa| &= I, \end{aligned} \tag{A140}$$

while their inner products are

$$\begin{aligned} \langle\boldsymbol{p}, \kappa|\boldsymbol{p}', \kappa'\rangle &= \delta(\boldsymbol{p}-\boldsymbol{p}')\delta_{\kappa\kappa'}, \\ \langle\boldsymbol{q}, \kappa|\boldsymbol{q}', \kappa'\rangle &= \delta(\boldsymbol{q}-\boldsymbol{q}')\delta_{\kappa\kappa'}, \\ \langle\boldsymbol{q}, \kappa|\boldsymbol{p}, \kappa'\rangle &= \frac{1}{h^{\frac{3}{2}}}\, \mathrm{e}^{(i/\hbar)\boldsymbol{p}\cdot\boldsymbol{q}}\delta_{\kappa\kappa'}. \end{aligned} \tag{A141}$$

In the following we need the operator $\Omega_{\kappa\lambda}$ which is defined as

$$\Omega_{\kappa\lambda} = \int \mathrm{d}\boldsymbol{p}|\boldsymbol{p}, \kappa\rangle\langle\boldsymbol{p}, \lambda| = \int \mathrm{d}\boldsymbol{q}|\boldsymbol{q}, \kappa\rangle\langle\boldsymbol{q}, \lambda|. \tag{A142}$$

The last equality follows directly if one forms matrix elements of the operator $\Omega_{\kappa\lambda}$.

The trace of an operator may be expressed in terms of the bases $|\boldsymbol{p}, \kappa\rangle$ or $|\boldsymbol{q}, \kappa\rangle$:

$$\mathrm{Tr}\, A = \sum_\kappa \int \mathrm{d}\boldsymbol{p}\langle\boldsymbol{p}, \kappa|A|\boldsymbol{p}, \kappa\rangle = \sum_\kappa \int \mathrm{d}\boldsymbol{q}\langle\boldsymbol{q}, \kappa|A|\boldsymbol{q}, \kappa\rangle. \tag{A143}$$

b. *The Weyl transform*

The above generalization of the formulae for point particles permits to repeat the derivations of section 2 and 3 for the Weyl transforms and the Wigner functions. In the results the indices of the inner degrees of freedom will appear now and then.

Thus one finds for the Weyl transform of an operator $A$, instead of (A6), (A7) and (A12)

$$a_{\kappa\lambda}(\boldsymbol{p}, \boldsymbol{q}) = \int \mathrm{d}\boldsymbol{u}\, \mathrm{e}^{(i/\hbar)\boldsymbol{q}\cdot\boldsymbol{u}} \langle \boldsymbol{p}+\tfrac{1}{2}\boldsymbol{u}, \kappa | A | \boldsymbol{p}-\tfrac{1}{2}\boldsymbol{u}, \lambda \rangle, \tag{A144}$$

$$a_{\kappa\lambda}(\boldsymbol{p}, \boldsymbol{q}) = \int \mathrm{d}\boldsymbol{v}\, \mathrm{e}^{(i/\hbar)\boldsymbol{p}\cdot\boldsymbol{v}} \langle \boldsymbol{q}-\tfrac{1}{2}\boldsymbol{v}, \kappa | A | \boldsymbol{q}+\tfrac{1}{2}\boldsymbol{v}, \lambda \rangle, \tag{A145}$$

$$a_{\kappa\lambda}(\boldsymbol{p}, \boldsymbol{q}) = \mathrm{Tr}\, \{A\Delta_{\lambda\kappa}(\boldsymbol{p}, \boldsymbol{q})\}. \tag{A146}$$

The $\Delta$-operator which occurs here is given by

$$\Delta_{\kappa\lambda}(\boldsymbol{p}, \boldsymbol{q}) = h^{-3} \int \mathrm{d}\boldsymbol{u}\, \mathrm{d}\boldsymbol{v}\, \mathrm{e}^{(i/\hbar)\{(\boldsymbol{q}-\boldsymbol{Q})\cdot\boldsymbol{u}+(\boldsymbol{p}-\boldsymbol{P})\cdot\boldsymbol{v}\}} \Omega_{\kappa\lambda} \equiv \Delta(\boldsymbol{p}, \boldsymbol{q})\Omega_{\kappa\lambda}, \tag{A147}$$

with $\Delta(\boldsymbol{p}, \boldsymbol{q})$ as given in (A9). Alternative forms for $\Delta_{\kappa\lambda}(\boldsymbol{p}, \boldsymbol{q})$ are:

$$\Delta_{\kappa\lambda}(\boldsymbol{p}, \boldsymbol{q}) = \int \mathrm{d}\boldsymbol{u}\, \mathrm{e}^{(i/\hbar)\boldsymbol{q}\cdot\boldsymbol{u}} | \boldsymbol{p}-\tfrac{1}{2}\boldsymbol{u}, \kappa \rangle \langle \boldsymbol{p}+\tfrac{1}{2}\boldsymbol{u}, \lambda |, \tag{A148}$$

$$\Delta_{\kappa\lambda}(\boldsymbol{p}, \boldsymbol{q}) = \int \mathrm{d}\boldsymbol{v}\, \mathrm{e}^{(i/\hbar)\boldsymbol{p}\cdot\boldsymbol{v}} | \boldsymbol{q}+\tfrac{1}{2}\boldsymbol{v}, \kappa \rangle \langle \boldsymbol{q}-\tfrac{1}{2}\boldsymbol{v}, \lambda |, \tag{A149}$$

instead of (A10) and (A11). The operator $A$ is now a sum of integrals:

$$A = h^{-3} \sum_{\kappa,\lambda} \int \mathrm{d}\boldsymbol{p}\, \mathrm{d}\boldsymbol{q}\, a_{\kappa\lambda}(\boldsymbol{p}, \boldsymbol{q}) \Delta_{\kappa\lambda}(\boldsymbol{p}, \boldsymbol{q}) \tag{A150}$$

instead of (A8). The Weyl correspondence may also be expressed, as in (A13–14), by

$$A = \sum_{\kappa,\lambda} \int \mathrm{d}\boldsymbol{u}\, \mathrm{d}\boldsymbol{v}\, \tilde{a}_{\kappa\lambda}(\boldsymbol{u}, \boldsymbol{v}) \mathrm{e}^{-(i/\hbar)(\boldsymbol{Q}\cdot\boldsymbol{u}+\boldsymbol{P}\cdot\boldsymbol{v})} \Omega_{\kappa\lambda}, \tag{A151}$$

$$a_{\kappa\lambda}(\boldsymbol{p}, \boldsymbol{q}) = \int \mathrm{d}\boldsymbol{u}\, \mathrm{d}\boldsymbol{v}\, \tilde{a}_{\kappa\lambda}(\boldsymbol{u}, \boldsymbol{v}) \mathrm{e}^{-(i/\hbar)(\boldsymbol{q}\cdot\boldsymbol{u}+\boldsymbol{p}\cdot\boldsymbol{v})}. \tag{A152}$$

If an operator $A$ does not connect the different parts of Hilbert space labelled by $\kappa$, its Weyl transform is diagonal in the discrete indices, as follows from (A144) (or (A145)):

$$a_{\kappa\lambda}(\boldsymbol{p}, \boldsymbol{q}) = \delta_{\kappa\lambda} \int \mathrm{d}\boldsymbol{u}\, \mathrm{e}^{(i/\hbar)\boldsymbol{q}\cdot\boldsymbol{u}} \langle \boldsymbol{p}+\tfrac{1}{2}\boldsymbol{u}, \kappa | A | \boldsymbol{p}-\tfrac{1}{2}\boldsymbol{u}, \kappa \rangle. \tag{A153}$$

If the operator moreover acts in the same way in each subspace, the integral is independent of $\kappa$ and may be denoted as $a(\boldsymbol{p}, \boldsymbol{q})$:

$$a_{\kappa\lambda}(\boldsymbol{p}, \boldsymbol{q}) = \delta_{\kappa\lambda}\, a(\boldsymbol{p}, \boldsymbol{q}). \tag{A154}$$

Upon introduction into (A150) and the use of the relation $\sum_\kappa \Omega_{\kappa\kappa} = I$ that follows from (A140) and (A142), one recovers (A8).

As a different special case we consider operators $A$ which are independent of the coordinate and momentum operators, as for instance the spin operator. Then the Weyl transform (A144) or (A145) is independent of $\boldsymbol{p}$ and $\boldsymbol{q}$:

$$a_{\kappa\lambda}(\boldsymbol{p}, \boldsymbol{q}) = a_{\kappa\lambda}. \tag{A155}$$

Upon introduction into (A150) one finds with the help of (A44) that the operator may be written as

$$A = \sum_{\kappa,\lambda} a_{\kappa\lambda}\, \Omega_{\kappa\lambda}. \tag{A156}$$

As formula (A147) showed the basic operator $\Delta_{\kappa\lambda}(\boldsymbol{p}, \boldsymbol{q})$ has the simple form $\Delta_{\kappa\lambda}(\boldsymbol{p}, \boldsymbol{q}) = \Delta(\boldsymbol{p}, \boldsymbol{q})\Omega_{\kappa\lambda}$. Therefore one may derive properties for the $\Delta_{\kappa\lambda}(\boldsymbol{p}, \boldsymbol{q})$-operator from those of the $\Delta(\boldsymbol{p}, \boldsymbol{q})$-operator derived in section 2*d*. As corollaries one finds for the traces of an operator and a product of two operators (cf. (A52–53)):

$$\operatorname{Tr} A = h^{-3} \sum_{\kappa} \int \mathrm{d}\boldsymbol{p}\, \mathrm{d}\boldsymbol{q}\, a_{\kappa\kappa}(\boldsymbol{p}, \boldsymbol{q}), \tag{A157}$$

$$\operatorname{Tr} AB = h^{-3} \sum_{\kappa,\lambda} \int \mathrm{d}\boldsymbol{p}\, \mathrm{d}\boldsymbol{q}\, a_{\kappa\lambda}(\boldsymbol{p}, \boldsymbol{q}) b_{\lambda\kappa}(\boldsymbol{p}, \boldsymbol{q}) = \operatorname{Tr} BA. \tag{A158}$$

Furthermore one finds for the Weyl transform of a product of two operators (cf. (A54))

$$AB \rightleftarrows \exp\left\{\frac{i\hbar}{2}\left(\frac{\partial^{(a)}}{\partial \boldsymbol{q}} \cdot \frac{\partial^{(b)}}{\partial \boldsymbol{p}} - \frac{\partial^{(a)}}{\partial \boldsymbol{p}} \cdot \frac{\partial^{(b)}}{\partial \boldsymbol{q}}\right)\right\} \sum_{\mu} a_{\kappa\mu}(\boldsymbol{p}, \boldsymbol{q}) b_{\mu\lambda}(\boldsymbol{p}, \boldsymbol{q}). \tag{A159}$$

The commutator and the anticommutator of two operators are given by (cf. (A55) and (A56)):

$$\begin{aligned} \tfrac{1}{2}\{A, B\} \rightleftarrows \tfrac{1}{2} \cos\left\{\frac{\hbar}{2}\left(\frac{\partial^{(a)}}{\partial \boldsymbol{q}} \cdot \frac{\partial^{(b)}}{\partial \boldsymbol{p}} - \frac{\partial^{(a)}}{\partial \boldsymbol{p}} \cdot \frac{\partial^{(b)}}{\partial \boldsymbol{q}}\right)\right\} & \sum_{\mu} \{a_{\kappa\mu}(\boldsymbol{p}, \boldsymbol{q}) b_{\mu\lambda}(\boldsymbol{p}, \boldsymbol{q}) \\ & + b_{\kappa\mu}(\boldsymbol{p}, \boldsymbol{q}) a_{\mu\lambda}(\boldsymbol{p}, \boldsymbol{q})\} \\ + \frac{i}{2} \sin\left\{\frac{\hbar}{2}\left(\frac{\partial^{(a)}}{\partial \boldsymbol{q}} \cdot \frac{\partial^{(b)}}{\partial \boldsymbol{p}} - \frac{\partial^{(a)}}{\partial \boldsymbol{p}} \cdot \frac{\partial^{(b)}}{\partial \boldsymbol{q}}\right)\right\} & \sum_{\mu} \{a_{\kappa\mu}(\boldsymbol{p}, \boldsymbol{q}) b_{\mu\lambda}(\boldsymbol{p}, \boldsymbol{q}) \\ & - b_{\kappa\mu}(\boldsymbol{p}, \boldsymbol{q}) a_{\mu\lambda}(\boldsymbol{p}, \boldsymbol{q})\}, \end{aligned} \tag{A160}$$

$$-\frac{i}{\hbar}[A, B] \rightleftarrows \frac{1}{\hbar}\sin\left\{\frac{\hbar}{2}\left(\frac{\partial^{(a)}}{\partial \boldsymbol{q}}\cdot\frac{\partial^{(b)}}{\partial \boldsymbol{p}}-\frac{\partial^{(a)}}{\partial \boldsymbol{p}}\cdot\frac{\partial^{(b)}}{\partial \boldsymbol{q}}\right)\right\}\sum_{\mu}\{a_{\kappa\mu}(\boldsymbol{p}, \boldsymbol{q})b_{\mu\lambda}(\boldsymbol{p}, \boldsymbol{q}) + b_{\kappa\mu}(\boldsymbol{p}, \boldsymbol{q})a_{\mu\lambda}(\boldsymbol{p}, \boldsymbol{q})\}$$
$$-\frac{i}{\hbar}\cos\left\{\frac{\hbar}{2}\left(\frac{\partial^{(a)}}{\partial \boldsymbol{q}}\cdot\frac{\partial^{(b)}}{\partial \boldsymbol{p}}-\frac{\partial^{(a)}}{\partial \boldsymbol{p}}\cdot\frac{\partial^{(b)}}{\partial \boldsymbol{q}}\right)\right\}\sum_{\mu}\{a_{\kappa\mu}(\boldsymbol{p}, \boldsymbol{q})b_{\mu\lambda}(\boldsymbol{p}, \boldsymbol{q}) - b_{\kappa\mu}(\boldsymbol{p}, \boldsymbol{q})a_{\mu\lambda}(\boldsymbol{p}, \boldsymbol{q})\}. \quad \text{(A161)}$$

In the special case that the operators $A$ and $B$ do not act on the indices $\kappa$, i.e. if they have Weyl transforms of the type (A154), the expressions (A160) and (A161) reduce to

$$\tfrac{1}{2}\{A, B\} \rightleftarrows \cos\left\{\frac{\hbar}{2}\left(\frac{\partial^{(a)}}{\partial \boldsymbol{q}}\cdot\frac{\partial^{(b)}}{\partial \boldsymbol{p}}-\frac{\partial^{(a)}}{\partial \boldsymbol{p}}\cdot\frac{\partial^{(b)}}{\partial \boldsymbol{q}}\right)\right\} a(\boldsymbol{p}, \boldsymbol{q})b(\boldsymbol{p}, \boldsymbol{q})\delta_{\kappa\lambda}, \quad \text{(A162)}$$

$$-\frac{i}{\hbar}[A, B] \rightleftarrows \frac{2}{\hbar}\sin\left\{\frac{\hbar}{2}\left(\frac{\partial^{(a)}}{\partial \boldsymbol{q}}\cdot\frac{\partial^{(b)}}{\partial \boldsymbol{p}}-\frac{\partial^{(a)}}{\partial \boldsymbol{p}}\cdot\frac{\partial^{(b)}}{\partial \boldsymbol{q}}\right)\right\} a(\boldsymbol{p}, \boldsymbol{q})b(\boldsymbol{p}, \boldsymbol{q})\delta_{\kappa\lambda}, \quad \text{(A163)}$$

which have the same form as the right-hand sides of (A55) and (A56), apart from the trivial Kronecker deltas.

In the special case that the operators are independent of the coordinate and momentum operators, i.e. if their Weyl transforms are of the type (A155), one finds from (A160) and (A161)

$$\tfrac{1}{2}\{A, B\} \rightleftarrows \tfrac{1}{2}\sum_{\mu}(a_{\kappa\mu}b_{\mu\lambda}+b_{\kappa\mu}a_{\mu\lambda}), \quad \text{(A164)}$$

$$-\frac{i}{\hbar}[A, B] \rightleftarrows -\frac{i}{\hbar}\sum_{\mu}(a_{\kappa\mu}b_{\mu\lambda}-b_{\kappa\mu}a_{\mu\lambda}). \quad \text{(A165)}$$

While the special case (A162–163) has a classical limit of the form (A58–59), the special case (A164–165) has no classical limit.

## c. *The Wigner function*

As the Weyl transform of the density operator $P(t) = |\psi(t)\rangle\langle\psi(t)|$ the Wigner function will be equipped with indices if one considers a particle with internal degrees of freedom. Indeed from (A145) one finds (cf. (A65)):

$$\rho_{\kappa\lambda}(\boldsymbol{p}, \boldsymbol{q}; t) = h^{-3}\int \mathrm{d}\boldsymbol{v}\, \mathrm{e}^{(i/\hbar)\boldsymbol{p}\cdot\boldsymbol{v}}\psi_{\kappa}(\boldsymbol{q}-\tfrac{1}{2}\boldsymbol{v}; t)\psi_{\lambda}^{*}(\boldsymbol{q}+\tfrac{1}{2}\boldsymbol{v}; t), \quad \text{(A166)}$$

where the wave function $\langle \boldsymbol{q}, \kappa|\psi(t)\rangle$ in the coordinate representation has been written as $\psi_{\kappa}(\boldsymbol{q}; t)$. The expectation value of an operator $A$ follows now from (A158):

$$\bar{A}(t) = \bar{a}(t) \equiv \sum_{\kappa,\lambda}\int \mathrm{d}\boldsymbol{p}\,\mathrm{d}\boldsymbol{q}\,\rho_{\kappa\lambda}(\boldsymbol{p}, \boldsymbol{q}; t)a_{\lambda\kappa}(\boldsymbol{p}, \boldsymbol{q}) \quad \text{(A167)}$$

(cf. (A62) with (A63)). The Wigner function is normalized to unity (cf. (A69)) as follows from $\operatorname{Tr} P(t) = 1$ and (A157):

$$\sum_{\kappa} \int \mathrm{d}\boldsymbol{p}\, \mathrm{d}\boldsymbol{q}\, \rho_{\kappa\kappa}(\boldsymbol{p}, \boldsymbol{q}; t) = 1. \tag{A168}$$

The time evolution of the Wigner function follows by taking the Weyl transform of the equation for the time evolution of the density operator (A88):

$$\begin{aligned} \frac{\partial \rho_{\kappa\lambda}(\boldsymbol{p}, \boldsymbol{q}; t)}{\partial t} &= \frac{1}{\hbar} \sin \left\{ \frac{\hbar}{2} \left( \frac{\partial^{(h)}}{\partial \boldsymbol{q}} \cdot \frac{\partial^{(\rho)}}{\partial \boldsymbol{p}} - \frac{\partial^{(h)}}{\partial \boldsymbol{p}} \cdot \frac{\partial^{(\rho)}}{\partial \boldsymbol{q}} \right) \right\} \sum_{\mu} (h_{\kappa\mu} \rho_{\mu\lambda} + \rho_{\kappa\mu} h_{\mu\lambda}) \\ &- \frac{i}{\hbar} \cos \left\{ \frac{\hbar}{2} \left( \frac{\partial^{(h)}}{\partial \boldsymbol{q}} \cdot \frac{\partial^{(\rho)}}{\partial \boldsymbol{p}} - \frac{\partial^{(h)}}{\partial \boldsymbol{p}} \cdot \frac{\partial^{(\rho)}}{\partial \boldsymbol{q}} \right) \right\} \sum_{\mu} (h_{\kappa\mu} \rho_{\mu\lambda} - \rho_{\kappa\mu} h_{\mu\lambda}), \end{aligned} \tag{A169}$$

where $h_{\kappa\lambda}$ depends on $\boldsymbol{p}$ and $\boldsymbol{q}$ and $\rho_{\kappa\lambda}$ on $\boldsymbol{p}$, $\boldsymbol{q}$ and $t$. This equation may be used to find an expression for the time derivative of the expectation value of an operator (cf. (A95))

$$\begin{aligned} \frac{\mathrm{d}\bar{a}(t)}{\mathrm{d}t} = \frac{1}{\hbar} \sum_{\kappa,\lambda,\mu} \int \mathrm{d}\boldsymbol{p}\, \mathrm{d}\boldsymbol{q}\, \rho_{\lambda\kappa} \bigg[ &\sin \left\{ \frac{\hbar}{2} \left( \frac{\partial^{(a)}}{\partial \boldsymbol{q}} \cdot \frac{\partial^{(h)}}{\partial \boldsymbol{p}} - \frac{\partial^{(a)}}{\partial \boldsymbol{p}} \cdot \frac{\partial^{(h)}}{\partial \boldsymbol{q}} \right) \right\} (a_{\kappa\mu} h_{\mu\lambda} + h_{\kappa\mu} a_{\mu\lambda}) \\ &- i \cos \left\{ \frac{\hbar}{2} \left( \frac{\partial^{(a)}}{\partial \boldsymbol{q}} \cdot \frac{\partial^{(h)}}{\partial \boldsymbol{p}} - \frac{\partial^{(a)}}{\partial \boldsymbol{p}} \cdot \frac{\partial^{(h)}}{\partial \boldsymbol{q}} \right) \right\} (a_{\kappa\mu} h_{\mu\lambda} - h_{\kappa\mu} a_{\mu\lambda}) \bigg], \end{aligned} \tag{A170}$$

where $a_{\kappa\lambda}$ and $h_{\kappa\lambda}$ depend on $\boldsymbol{p}$ and $\boldsymbol{q}$, and $\rho_{\kappa\lambda}$ on $\boldsymbol{p}$, $\boldsymbol{q}$ and $t$.

If the operator $A$ does not act on the spin indices, i.e. if its Weyl transform is of the type (A154), expression (A170) becomes

$$\frac{\mathrm{d}\bar{a}(t)}{\mathrm{d}t} = \frac{2}{\hbar} \sum_{\kappa,\lambda} \int \mathrm{d}\boldsymbol{p}\, \mathrm{d}\boldsymbol{q} \left[ \sin \left\{ \frac{\hbar}{2} \left( \frac{\partial^{(a)}}{\partial \boldsymbol{q}} \cdot \frac{\partial^{(h)}}{\partial \boldsymbol{p}} - \frac{\partial^{(a)}}{\partial \boldsymbol{p}} \cdot \frac{\partial^{(h)}}{\partial \boldsymbol{q}} \right) \right\} a h_{\kappa\lambda} \right] \rho_{\lambda\kappa}, \tag{A171}$$

where $a$, $h_{\kappa\lambda}$ and $\rho_{\kappa\lambda}$ depend on $\boldsymbol{p}$ and $\boldsymbol{q}$, and $\rho_{\kappa\lambda}$ moreover on $t$. If on the other hand the operator $A$ is independent of the coordinate and momentum operators, so that its Weyl transform is of the type (A155), one has for (A170)

$$\frac{\mathrm{d}\bar{a}(t)}{\mathrm{d}t} = \frac{i}{\hbar} \sum_{\kappa,\lambda,\mu} \int \mathrm{d}\boldsymbol{p}\, \mathrm{d}\boldsymbol{q} (h_{\kappa\mu} a_{\mu\lambda} - a_{\kappa\lambda} h_{\mu\lambda}) \rho_{\lambda\kappa} \tag{A172}$$

with $h_{\kappa\lambda}$ depending on $\boldsymbol{p}$, $\boldsymbol{q}$ and $\rho_{\kappa\lambda}$ on $\boldsymbol{p}$, $\boldsymbol{q}$ and $t$. If the Weyl transform of the Hamiltonian is such that

$$\lim_{\hbar \to 0} h_{\kappa\lambda}(\boldsymbol{p}, \boldsymbol{q}) = \delta_{\kappa\lambda} h_{\mathrm{cl}}(\boldsymbol{p}, \boldsymbol{q}), \tag{A173}$$

the expression (A171) has a classical limit. The expression (A172) however has no classical limit.

# PROBLEMS

**1.** Prove the following theorems on exponentials of operators $A$ and $B$, which commute with their commutator $[A, B]$:

$$e^{A+B} = e^A e^B e^{-\frac{1}{2}[A,B]} = e^B e^A e^{\frac{1}{2}[A,B]}, \tag{P1}$$

$$e^{A+B} = e^{\frac{1}{2}A} e^B e^{\frac{1}{2}A} = e^{\frac{1}{2}B} e^A e^{\frac{1}{2}B}. \tag{P2}$$

Hint: Prove first the lemma ($\lambda$ is a number)

$$f(\lambda) \equiv e^{\lambda A} B e^{-\lambda A} = B + \lambda[A, B]. \tag{P3}$$

This follows by integration of $\partial f/\partial \lambda$. In the same way it follows, using also (P3), that

$$g(\lambda) \equiv e^{\lambda A} e^{\lambda B} = e^{\lambda(A+B)} e^{\frac{1}{2}\lambda^2[A,B]}, \tag{P4}$$

of which (P1) is a special case. Note the useful corollary of (P1):

$$e^A e^B = e^B e^A e^{[A,B]}. \tag{P5}$$

The relation (P2) follows if (P5) is used at the right-hand side, and then (P1) applied.

**2.** Show that taking the Weyl transform of an operator $A(\boldsymbol{P}, \boldsymbol{Q})$ and performing a linear transformation of the coordinate and momentum operators $\hat{\boldsymbol{Q}} = \mathbf{c}\cdot\boldsymbol{Q}$ and $\hat{\boldsymbol{P}} = \boldsymbol{P}\cdot\mathbf{c}^{-1}$ ($\mathbf{c}$ is a matrix of $c$-numbers) (or $\hat{\boldsymbol{q}} = \mathbf{c}\cdot\boldsymbol{q}$, $\hat{\boldsymbol{p}} = \boldsymbol{p}\cdot\mathbf{c}^{-1}$ of their Weyl transforms) are commuting operations. The proof consists in showing – on the basis of (29) and (30) – that one gets the same result if one takes the Weyl transform of the operator and then transforms $\boldsymbol{q}$ and $\boldsymbol{p}$ or if one transforms $\boldsymbol{Q}$ and $\boldsymbol{P}$ and then takes the Weyl transform.

**3.** Prove the relation (38) from (37) and (41).

**4.** A relation like (A29), but with the operator $P$ interchanged with $Q$ and the Weyl transform $p$ interchanged with $q$, is valid also. An example of this relation is

$$p^2 q \rightleftarrows \tfrac{1}{2}(P^2 Q + Q P^2).$$

Compare this result with (A34) and show that they are identical.

**5.** Prove from (A33) and (A35) that the square of the operator of which the Weyl transform is $pq$ is different from the operator of which the Weyl transform is $(pq)^2$ by explicitly evaluating both operators in terms of the operators $P$ and $Q$.

**6.** Prove the relation (A30) by making use of the Weyl transform (A55).

**7.** Prove that for the propagator

$$\mathscr{K} \equiv \langle \boldsymbol{q} | U(t, t_0) | \boldsymbol{q}_0 \rangle$$

of the Schrödinger equation $|\psi(t)\rangle = U(t, t_0)|\psi(t_0)\rangle$ with $U(t, t_0) = \exp\{-(i/\hbar)H(t-t_0)\}$ and the Hamiltonian operator $H = \boldsymbol{P}^2/2m + V(\boldsymbol{Q})$, with Weyl transform $h = \boldsymbol{p}^2/2m + V(\boldsymbol{q})$, one can derive the Feynman path integral

$$\mathscr{K} = \lim_{n\to\infty} \int \dots \int \mathrm{d}\boldsymbol{q}_1 \dots \mathrm{d}\boldsymbol{q}_{n-1} \prod_{j=0}^{n-1} \left\{ \frac{m}{ih(t_{j+1}-t_j)} \right\}^{\frac{3}{2}} \mathrm{e}^{(i/\hbar)(t_{j+1}-t_j)l_j}$$

with

$$l_j \equiv l\left(\frac{\boldsymbol{q}_{j+1}-\boldsymbol{q}_j}{t_{j+1}-t_j}, \frac{\boldsymbol{q}_j+\boldsymbol{q}_{j+1}}{2}\right) = \tfrac{1}{2}m\left(\frac{\boldsymbol{q}_{j+1}-\boldsymbol{q}_j}{t_{j+1}-t_j}\right)^2 - V\left(\frac{\boldsymbol{q}_j+\boldsymbol{q}_{j+1}}{2}\right),$$

which is sometimes[1] symbolically written as

$$\mathscr{K} = \sum_{\text{all paths}} \mathrm{e}^{(i/\hbar)\int l(\dot{\boldsymbol{q}}, \boldsymbol{q})\mathrm{d}t}.$$

Hint: The propagator may first be written as

$$\mathscr{K} = \lim_{n\to\infty} \int \dots \int \mathrm{d}\boldsymbol{q}_1 \dots \mathrm{d}\boldsymbol{q}_{n-1} \prod_{j=0}^{n-1} \langle \boldsymbol{q}_{j+1} | U(t_{j+1}, t_j) | \boldsymbol{q}_j \rangle$$

with $\boldsymbol{q}_n \equiv \boldsymbol{q}$ and $t \equiv t_n > t_{n-1} > \dots > t_1 > t_0$. Now use (13) for $U(t_{j+1}, t_j) = 1-(i/\hbar)(t_{j+1}-t_j)H$ up to terms linear in $t_{j+1}-t_j$ (its Weyl transform may be written again as an exponential) and apply the relation (A37). Integration over $\boldsymbol{p}$ will then yield the result.

*Note*: The result may be generalized to the case of a particle in a field, described by a Hamilton operator

$$H = \left(\boldsymbol{P} - \frac{e}{c}\boldsymbol{A}\right)^2 / 2m + V(\boldsymbol{Q}).$$

The theorem is not true for arbitrary Hamiltonians (v. Groenewold, loc. cit.).

[1] R. P. Feynman, Rev. Mod. Phys. **20**(1948)367; H. J. Groenewold, Mat. Fys. Medd. Dan. Vid. Selsk. **30**(1956)no. 19.

**8.** Show that for a free particle the wave packet, which at time $t_0$ is described by the Wigner function (A110) as a minimum wave packet, is at time $t \neq t_0$ described by the Wigner function

$$\rho(\boldsymbol{p}, \boldsymbol{q}; t) = (2/h)^3 \exp\left[-\frac{\{\boldsymbol{q}-\boldsymbol{q}_0-(\boldsymbol{p}/m)(t-t_0)\}^2}{2\Delta^2} - \frac{2\Delta^2(\boldsymbol{p}-\boldsymbol{p}_0)^2}{\hbar^2}\right],$$

by using the time evolution equation (A89) with $h = \boldsymbol{p}^2/2m$. Note that at times $t \neq t_0$ the wave packet shows correlations between the variables $\boldsymbol{p}$ and $\boldsymbol{q}$; in other words it is only a minimum wave packet at time $t = t_0$.

**9.** Prove that for Hamiltonians, which are at most quadratic in the coordinates and momenta the time evolution equation (A89) for the Wigner function reduces to the simple form

$$\frac{\partial \rho}{\partial t} = \{h, \rho\}_{\mathrm{P}}$$

with the Poisson bracket of the Weyl transform $h$ of the Hamiltonian and the Wigner function. Physical examples are: the free particle ($h = \boldsymbol{p}^2/2m$), the particle in a constant force field ($h = \boldsymbol{p}^2/2m - \boldsymbol{a}\cdot\boldsymbol{q}$) and the harmonic oscillator ($h = \boldsymbol{p}^2/2m + \frac{1}{2}m\omega^2\boldsymbol{q}^2$). It is well known that not all aspects of quantum mechanics show up clearly for these examples. The fact that the time evolution equation for the Wigner function, although quantum-mechanical, looks formally like the classical equation for the time evolution of a classical distribution function ($h$ is then to be replaced by the classical Hamiltonian) is another illustration of this feature.

Prove also that for these Hamiltonians the expression (A95) for the time derivative of an expectation value reduces to

$$\frac{\mathrm{d}\bar{a}(t)}{\mathrm{d}t} = \int \mathrm{d}\boldsymbol{p}\,\mathrm{d}\boldsymbol{q}\{a, h\}\rho(\boldsymbol{p}, \boldsymbol{q}; t),$$

which is again analogous to, but different from the classical case.

**10.** Prove the Weyl correspondence

$$ABC \rightleftarrows \exp\left\{\frac{i\hbar}{2}\left(\frac{\partial^{(a)}}{\partial \boldsymbol{q}}\cdot\frac{\partial^{(b)}}{\partial \boldsymbol{p}} - \frac{\partial^{(a)}}{\partial \boldsymbol{p}}\cdot\frac{\partial^{(b)}}{\partial \boldsymbol{q}} + \frac{\partial^{(a)}}{\partial \boldsymbol{q}}\cdot\frac{\partial^{(c)}}{\partial \boldsymbol{p}} - \frac{\partial^{(a)}}{\partial \boldsymbol{p}}\cdot\frac{\partial^{(c)}}{\partial \boldsymbol{q}}\right.\right.$$
$$\left.\left. + \frac{\partial^{(b)}}{\partial \boldsymbol{q}}\cdot\frac{\partial^{(c)}}{\partial \boldsymbol{p}} - \frac{\partial^{(b)}}{\partial \boldsymbol{p}}\cdot\frac{\partial^{(c)}}{\partial \boldsymbol{q}}\right)\right\} \sum_{\mu,\nu} a_{\kappa\mu}(\boldsymbol{p}, \boldsymbol{q}) b_{\mu\nu}(\boldsymbol{p}, \boldsymbol{q}) c_{\nu\lambda}(\boldsymbol{p}, \boldsymbol{q}),$$

valid for the product of three operators $A$, $B$ and $C$, acting on a spinor, with Weyl transforms $a_{\kappa\lambda}(\boldsymbol{p}, \boldsymbol{q})$, $b_{\kappa\lambda}(\boldsymbol{p}, \boldsymbol{q})$ and $c_{\kappa\lambda}(\boldsymbol{p}, \boldsymbol{q})$.

**11.** Verify that the expressions (74) for the field operators satisfy the equations (2).

CHAPTER VII

# Quantum statistical description of material media

## 1 *Introduction*

Macroscopic laws are obtained from laws at the atomic level by means of an appropriate averaging procedure. Since the atomic laws were formulated in terms of Weyl transforms, such an averaging will involve the use of Wigner functions for statistical, i.e. mixed states. In deriving the macroscopic laws in this way we take full advantage of the close analogy with classical theory. It will then turn out that the macroscopic quantum-mechanical laws obtained have the same form as the classical laws. The macroscopic quantities are also to a great extent analogous: most of them may be obtained from the classical quantities by replacing the distribution functions by Wigner functions.

## 2 *The Wigner function in statistical mechanics*

In quantum statistics a system is described by a density operator[1] $P$, which corresponds to a mixed state:

$$P(t) = \sum_{\gamma} w_{\gamma} |\psi_{\gamma}(t)\rangle\langle\psi_{\gamma}(t)| \tag{1}$$

with weights $w_{\gamma}$, normalized to unity,

$$\sum_{\gamma} w_{\gamma} = 1 \tag{2}$$

and pure state vectors $|\psi_{\gamma}(t)\rangle$ that form a complete orthonormal set. Average values of physical quantities, represented by operators $A$, are given by

$$\bar{A} = \sum_{\gamma} w_{\gamma} \langle\psi_{\gamma}|A|\psi_{\gamma}\rangle = \mathrm{Tr}\,(PA). \tag{3}$$

[1] In this section, as in section 3 of chapter VI, we use capitals for operators and lower case symbols for Weyl transforms.

Since in the preceding we used Weyl transforms rather than operators to represent physical quantities, we want to express also the average values (3) with the help of Weyl transforms. To that end we introduce the Wigner function as the Weyl transform (times $h^3$) of the density operator (1). Then one finds

$$\rho(\boldsymbol{p}, \boldsymbol{q}; t) = \sum_{\gamma} w_{\gamma} \rho_{\gamma}(\boldsymbol{p}, \boldsymbol{q}; t) \tag{4}$$

with the partial Wigner functions $\rho_{\gamma}(\boldsymbol{p}, \boldsymbol{q}; t)$ for pure states defined as (VI.54) in chapter VI, section 3*b*:

$$\rho_{\gamma}(\boldsymbol{p}, \boldsymbol{q}; t) = h^{-3} \int \mathrm{d}\boldsymbol{v}\, \mathrm{e}^{(i/\hbar)\boldsymbol{p}\cdot\boldsymbol{v}} \psi_{\gamma}(\boldsymbol{q} - \tfrac{1}{2}\boldsymbol{v}; t) \psi_{\gamma}^{*}(\boldsymbol{q} + \tfrac{1}{2}\boldsymbol{v}; t), \tag{5}$$

where $\psi_{\gamma}(\boldsymbol{q}; t)$ is the wave function of the pure state $|\psi_{\gamma}\rangle$.

With the help of this Wigner function (4) one may write the average (3) as

$$\bar{A} = \bar{a}, \tag{6}$$

where the right-hand side is the integral:

$$\bar{a} \equiv \int \mathrm{d}\boldsymbol{p}\, \mathrm{d}\boldsymbol{q}\, a(\boldsymbol{p}, \boldsymbol{q}) \rho(\boldsymbol{p}, \boldsymbol{q}; t), \tag{7}$$

as follows from (VI.51–52). Introducing (4) one may write (6) with (7) alternatively as

$$\bar{A} = \sum_{\gamma} w_{\gamma} \bar{a}^{\gamma} \tag{8}$$

with the pure state integral

$$\bar{a}^{\gamma} \equiv \int \mathrm{d}\boldsymbol{p}\, \mathrm{d}\boldsymbol{q}\, a(\boldsymbol{p}, \boldsymbol{q}) \rho_{\gamma}(\boldsymbol{p}, \boldsymbol{q}; t). \tag{9}$$

From the normalization of the partial Wigner functions (VI.56) and formula (2) it follows that also the total Wigner function (4) is normalized:

$$\int \mathrm{d}\boldsymbol{p}\, \mathrm{d}\boldsymbol{q}\, \rho(\boldsymbol{p}, \boldsymbol{q}; t) = 1. \tag{10}$$

Just as the partial Wigner functions for pure states, the total Wigner function introduced here cannot be interpreted as a probability density in phase space. The integrals over the momenta (or over the coordinates) however are positive definite and may be interpreted as probability densities:

$$\int \mathrm{d}\boldsymbol{p}\, \rho(\boldsymbol{p}, \boldsymbol{q}; t) = \sum_{\gamma} w_{\gamma} |\psi_{\gamma}(\boldsymbol{q})|^{2} \geqslant 0. \tag{11}$$

The time evolution of the Wigner function follows from the time evolution of the partial Wigner functions, given in (VI.59). As a result one gets an equation

$$\frac{\partial \rho(\boldsymbol{p}, \boldsymbol{q}; t)}{\partial t} = \frac{2}{\hbar} \sin \left\{ \frac{\hbar}{2} \left( \frac{\partial^{(h)}}{\partial \boldsymbol{q}} \cdot \frac{\partial^{(\rho)}}{\partial \boldsymbol{p}} - \frac{\partial^{(h)}}{\partial \boldsymbol{p}} \cdot \frac{\partial^{(\rho)}}{\partial \boldsymbol{q}} \right) \right\} h(\boldsymbol{p}, \boldsymbol{q}) \rho(\boldsymbol{p}, \boldsymbol{q}; t), \tag{12}$$

which has the same form as (VI.59). From this equation one finds an expression for the time derivative of an average quantity (6):

$$\frac{\mathrm{d}\bar{a}(t)}{\mathrm{d}t} = \frac{2}{\hbar} \int \mathrm{d}\boldsymbol{p}\, \mathrm{d}\boldsymbol{q} \left[ \sin \left\{ \frac{\hbar}{2} \left( \frac{\partial^{(a)}}{\partial \boldsymbol{q}} \cdot \frac{\partial^{(h)}}{\partial \boldsymbol{p}} - \frac{\partial^{(a)}}{\partial \boldsymbol{p}} \cdot \frac{\partial^{(h)}}{\partial \boldsymbol{q}} \right) \right\} a(\boldsymbol{p}, \boldsymbol{q}) h(\boldsymbol{p}, \boldsymbol{q}) \right] \rho(\boldsymbol{p}, \boldsymbol{q}; t), \tag{13}$$

formally the same as (VI.60).

For many-particle systems the averages of physical quantities can also be obtained with the help of the Weyl transforms of the operators corresponding to these quantities, and Wigner functions. Both the Weyl transform and the Wigner function depend then on the phase space variables of all particles.

## 3 *Reduced Wigner functions*

The formalism outlined in the preceding section will be applied to systems of $N$ atoms, labelled by $k = 1, 2, \ldots, N$, that consist of a number of point particles, labelled by $k, i$. Then the Weyl transform of the operator corresponding to a physical quantity will depend on all momentum and coordinate variables[1] $\boldsymbol{P}_{ki}$, $\boldsymbol{R}_{ki}$ ($k = 1, 2, \ldots, N$; $i = 1, 2, \ldots$) of the particles. The total set of variables will for short be denoted $1, 2, \ldots, N$.

Often the operators pertinent to physical quantities are sums of operators which depend on the variables of one single atom only, i.e. which are of the form

$$a_{\mathrm{op}} = \sum_k a_{k,\mathrm{op}} \,. \tag{14}$$

Here the quantities $a_{k,\mathrm{op}}$ depend on the coordinate and momentum operators $\boldsymbol{P}_{ki,\mathrm{op}}$ and $\boldsymbol{R}_{ki,\mathrm{op}}$ of the constituent particles $i = 1, 2, \ldots$ of atom $k$. The Weyl transform of such a quantity is equal to

$$a_{\mathrm{op}} \rightleftarrows a(1, \ldots, N) = \sum_k a_k(k), \tag{15}$$

where $k$ indicates all momentum and coordinate variables of atom $k$.

[1] We now return to the notation according to which operators are distinguished by a label op and quantities without such a label indicate ordinary numbers.

If the system consists of identical atoms the average (7) of the quantity (15) may be written as:

$$\bar{a} = N \int a_1(1)\rho(1, 2, \ldots, N; t)\mathrm{d}1 \ldots \mathrm{d}N, \tag{16}$$

or alternatively as:

$$\bar{a} = \int a_1(1) f_1(1; t)\mathrm{d}1, \tag{17}$$

where the reduced Wigner function defined as

$$f_1(1; t) \equiv N \int \rho(1, 2, \ldots, N; t)\mathrm{d}2 \ldots \mathrm{d}N \tag{18}$$

is a one-point function, normalized to $N$. In practice it is often convenient to introduce instead of the momentum and coordinate variables $\boldsymbol{P}_{1i}$ and $\boldsymbol{R}_{1i}$ $(i = 1, 2, \ldots)$ different variables in the integral (17). The Jacobian of such a transformation may then for convenience be absorbed in the new one-point function. For formal reasons one may then maintain the notation $f_1(1; t)$.

Furthermore one encounters two-point operators, which have Weyl transforms

$$a_{\mathrm{op}} \rightleftarrows a(1, \ldots, N) = \sum_{k,l(k \neq l)} a_{kl}(k, l). \tag{19}$$

Then (7) leads to the average value

$$\bar{a} = \int a_{12}(1, 2) f_2(1, 2; t)\mathrm{d}1\,\mathrm{d}2, \tag{20}$$

where the reduced Wigner function

$$f_2(1, 2; t) \equiv N(N-1) \int \rho(1, 2, \ldots, N; t)\mathrm{d}3 \ldots \mathrm{d}N \tag{21}$$

is a two-point function normalized to $N(N-1)$.

In contrast to the distribution functions of classical theory the one-point and two-point Wigner functions do not admit an interpretation in terms of probabilities. However such an interpretation is possible if one integrates away either all momentum or all coordinate variables.

In the following we want to have at our disposal also two-point 'correlation functions', which are defined as

$$c_2(1, 2; t) \equiv f_2(1, 2; t) - f_1(1; t) f_1(2; t). \tag{22}$$

For mixtures of several chemical components one needs an extra index to label the reduced one- and two-component distribution functions for the various species.

## 4 The Maxwell equations

The starting point for the derivation of the Maxwell equations is the set of equations (VI.90) for the expectation values of the operators representing the atomic fields.

In the set of equations (VI.90) all symbols denote in fact pure state integrals of the type (9). By making a weighted sum of these equations we get according to (3), or (8) with (9), the equations for average quantities

$$\begin{aligned} \nabla\cdot\boldsymbol{E} &= \varrho^{\mathrm{e}} - \nabla\cdot\boldsymbol{P}, \\ -\partial_0\boldsymbol{E}+\nabla\wedge\boldsymbol{B} &= c^{-1}\boldsymbol{J}+\partial_0\boldsymbol{P}+\nabla\wedge\boldsymbol{M}, \\ \nabla\cdot\boldsymbol{B} &= 0, \\ \partial_0\boldsymbol{B}+\nabla\wedge\boldsymbol{E} &= 0, \end{aligned} \tag{23}$$

where the macroscopic fields are defined as

$$\begin{aligned} \boldsymbol{E}(\boldsymbol{R}, t) &= \int \boldsymbol{e}(1, \ldots, N; \boldsymbol{R}, t)\rho(1, \ldots, N; t)\mathrm{d}1 \ldots \mathrm{d}N = \sum_\gamma w_\gamma \overline{\boldsymbol{e}}^\gamma, \\ \boldsymbol{B}(\boldsymbol{R}, t) &= \int \boldsymbol{b}(1, \ldots, N; \boldsymbol{R}, t)\rho(1, \ldots, N; t)\mathrm{d}1 \ldots \mathrm{d}N = \sum_\gamma w_\gamma \overline{\boldsymbol{b}}^\gamma. \end{aligned} \tag{24}$$

Furthermore the macroscopic charge and current densities are given by:

$$\begin{aligned} \varrho^{\mathrm{e}}(\boldsymbol{R}, t) &= \int \rho^{\mathrm{e}}(1, \ldots, N; \boldsymbol{R})\rho(1, \ldots, N; t)\mathrm{d}1 \ldots \mathrm{d}N = \sum_\gamma w_\gamma \overline{\rho^{\mathrm{e}\gamma}}, \\ \boldsymbol{J}(\boldsymbol{R}, t) &= \int \boldsymbol{j}(1, \ldots, N; \boldsymbol{R})\rho(1, \ldots, N; t)\mathrm{d}1 \ldots \mathrm{d}N = \sum_\gamma w_\gamma \overline{\boldsymbol{j}}^\gamma. \end{aligned} \tag{25}$$

Inserting (VI.85) and introducing the one-point reduced Wigner function (18) we may write these expressions as (cf. (II.19)):

$$\begin{aligned} \varrho^{\mathrm{e}}(\boldsymbol{R}, t) &= \sum_a e_a f_1^a(\boldsymbol{R}; t), \\ \boldsymbol{J}(\boldsymbol{R}, t) &= \sum_a \int e_a \boldsymbol{v}_1 f_1^a(\boldsymbol{R}, \boldsymbol{v}_1; t)\mathrm{d}\boldsymbol{v}_1, \end{aligned} \tag{26}$$

where $a$ labels the various species in the system.

The macroscopic polarization densities are defined as:

$$\begin{aligned} \boldsymbol{P}(\boldsymbol{R}, t) &= \int \boldsymbol{p}(1, \ldots, N; \boldsymbol{R})\rho(1, \ldots, N; t)\mathrm{d}1 \ldots \mathrm{d}N = \sum_{\gamma} w_{\gamma} \overline{\boldsymbol{p}}^{\gamma}, \\ \boldsymbol{M}(\boldsymbol{R}, t) &= \int \boldsymbol{m}(1, \ldots, N; \boldsymbol{R})\rho(1, \ldots, N; t)\mathrm{d}1 \ldots \mathrm{d}N = \sum_{\gamma} w_{\gamma} \overline{\boldsymbol{m}}^{\gamma}, \end{aligned} \tag{27}$$

or with (VI.86) and the one-point reduced Wigner function (18) (cf. (II.21)) as:

$$\begin{aligned} \boldsymbol{P}(\boldsymbol{R}, t) &= \sum_{a} \sum_{n=1}^{\infty} (-1)^{n-1} \nabla^{n-1} \vdots \int \overline{\boldsymbol{\mu}}_1^{(n)} f_1^a(\boldsymbol{R}, 1; t)\mathrm{d}1, \\ \boldsymbol{M}(\boldsymbol{R}, t) &= \sum_{a} \sum_{n=1}^{\infty} (-1)^{n-1} \nabla^{n-1} \vdots \int (\overline{\boldsymbol{\nu}}_1^{(n)} + \overline{\boldsymbol{\mu}}_1^{(n)} \wedge \boldsymbol{\beta}_1) f_1^a(\boldsymbol{R}, 1; t)\mathrm{d}1, \end{aligned} \tag{28}$$

where $\boldsymbol{\beta}_1 \equiv \boldsymbol{v}_1/c$.

In this way the Maxwell equations have been obtained in the framework of non-relativistic quantum mechanics. The macroscopic quantities are written as averages in terms of the Wigner function and Weyl transforms of operators on the atomic level. In particular the sources of the field equations contain the charge, current and polarization densities, given in (26) and (28) as expressions of the same form as the corresponding classical ones, but with one-point Wigner functions instead of one-point classical distribution functions. (The atomic operators of which these macroscopic quantities are the averages have been given in formulae (VI.91) and (VI.93) of the preceding chapter.)

As a consequence of the formal similarity of the classical and quantum-mechanical results of the non-relativistic treatments, one may also take over the applications to particular media which were given in chapter II, again replacing the classical distribution functions by Wigner functions.

The proof of the validity of the macroscopic Maxwell equations has been given here for the case of non-relativistic particles and fields, i.e. both described by expressions up to order $c^{-1}$. In chapter III it was shown that taking the non-relativistic limit of the fields meant that one confined oneself to situations in which two dimensionless parameters are small: the ratio $\beta$ of the source velocity to the velocity of light and the ratio of the retardation time to a characteristic time of the motion of the accelerated source. Sometimes however one is interested in an approximation in which only the source velocity is small, but not the retardation time. To derive the Maxwell equations in that case one must perform a second quantization of the fields[1].

[1] W. E. Brittin, Phys. Rev. **106**(1957)843; K. Schram, Physica **26**(1960)1080; J. M. Crowther and D. ter Haar, Proc. Kon. Ned. Akad. Wet. **B74**(1971)341, 351.

## 5 *The momentum and energy equations*

### a. *Introduction*

The macroscopic balance equations and conservation laws of momentum and energy will be derived from the atomic equation of motion and the atomic energy equation by using an averaging procedure involving a Wigner function. The use of the latter function simplifies again the calculations considerably in the sense that many derivations of classical theory can be taken over. The results of the present quantum-mechanical treatment will differ slightly more from classical theory than those of the preceding section where only the classical distribution functions had to be replaced by Wigner functions.

Just as in classical theory it will be convenient to derive the mass conservation law before turning to the momentum and energy equations. The treatment will be confined to one-component systems.

### b. *The mass conservation law*

The operator for the mass density on the atomic level has the form

$$\sum_k m_k \, \delta(\boldsymbol{R}_{k,\mathrm{op}} - \boldsymbol{R}), \tag{29}$$

where $m_k = m$ is the mass of the identical atoms and where $\boldsymbol{R}_{k,\mathrm{op}}$ is the mass centre operator $\sum_i m_{ki} \boldsymbol{R}_{ki,\mathrm{op}}/m_k$. Its Weyl transform is

$$\sum_k m_k \, \delta(\boldsymbol{R}_k - \boldsymbol{R}). \tag{30}$$

By taking the Poisson bracket with the Weyl transform of the Hamiltonian one finds with the notation (VI.72)

$$\partial_{t\mathrm{P}} \sum_k m_k \, \delta(\boldsymbol{R}_k - \boldsymbol{R}) = - \sum_k m_k (\partial_{t\mathrm{P}} \boldsymbol{R}_k) \cdot \nabla \delta(\boldsymbol{R}_k - \boldsymbol{R}). \tag{31}$$

By multiplication with a Wigner function and integration over phase space one finds

$$\frac{\partial}{\partial t} \int \sum_k m_k \, \delta(\boldsymbol{R}_k - \boldsymbol{R}) \rho(1, \ldots, N; t) \mathrm{d}1 \ldots \mathrm{d}N$$

$$= - \nabla \cdot \int \sum_k m_k \boldsymbol{v}_k \, \delta(\boldsymbol{R}_k - \boldsymbol{R}) \rho(1, \ldots, N; t) \mathrm{d}1 \ldots \mathrm{d}N, \tag{32}$$

where we used (VI.61), (VI.72), (13) and the abbreviation $\boldsymbol{v}_k \equiv \partial_{t\mathrm{P}} \boldsymbol{R}_k$.

With the introduction of reduced Wigner functions of the type (18), one may write (32) as

$$\frac{\partial \varrho}{\partial t} = -\boldsymbol{\nabla}\cdot(\varrho \boldsymbol{v}) \tag{33}$$

with the macroscopic mass density

$$\varrho = m f_1(\boldsymbol{R}; t) \tag{34}$$

and the macroscopic mass flow

$$\varrho \boldsymbol{v} = \int m \boldsymbol{v}_1 \, f_1(\boldsymbol{R}, \boldsymbol{v}_1; t) \mathrm{d}\boldsymbol{v}_1 \, . \tag{35}$$

The macroscopic mass density is the average of the atomic operator (29), while the macroscopic mass flow is the average of the operator

$$\tfrac{1}{2} \sum_k m\{\boldsymbol{v}_{k,\mathrm{op}}, \delta(\boldsymbol{R}_{k,\mathrm{op}} - \boldsymbol{R})\}, \tag{36}$$

where $\boldsymbol{v}_{k,\mathrm{op}}$ is defined in (VI.92) and where an anticommutator appears.

c. *The momentum balance*

The macroscopic momentum balance is obtained by multiplying the atomic equation of motion in its form (VI.98) by a delta function $\delta(\boldsymbol{R}_k - \boldsymbol{R})$, summing over $k$, multiplying by a Wigner function and integrating over phase space. Then one obtains

$$\int \sum_k m_k(\partial_{t\mathrm{P}} \boldsymbol{v}_k)\delta(\boldsymbol{R}_k - \boldsymbol{R})\rho(1, \ldots, N; t)\mathrm{d}1 \ldots \mathrm{d}N$$
$$= \int \sum_k (\boldsymbol{f}_k^{\mathrm{L}} + \boldsymbol{f}_k^{\mathrm{S}})\delta(\boldsymbol{R}_k - \boldsymbol{R})\rho(1, \ldots, N; t)\mathrm{d}1 \ldots \mathrm{d}N. \tag{37}$$

The left-hand side may be brought into relation with the time derivative of the mass flow, which is

$$\frac{\partial(\varrho \boldsymbol{v})}{\partial t} = \frac{\partial}{\partial t} \int \sum_k m_k \, \boldsymbol{v}_k \, \delta(\boldsymbol{R}_k - \boldsymbol{R})\rho(1, \ldots, N; t)\mathrm{d}1 \ldots \mathrm{d}N. \tag{38}$$

With the help of (13) and (VI.61) one finds with the notation (VI.72)

$$\frac{\partial(\varrho \boldsymbol{v})}{\partial t} = \int \sum_k m_k \, \partial_{t\mathrm{P}}\{\boldsymbol{v}_k \, \delta(\boldsymbol{R}_k - \boldsymbol{R})\}\rho(1, \ldots, N; t)\mathrm{d}1 \ldots \mathrm{d}N. \tag{39}$$

If one carries out the differentiations at the right-hand side this equation becomes

$$\frac{\partial(\varrho\boldsymbol{v})}{\partial t}=\int\sum_k m_k(\partial_{tP}\boldsymbol{v}_k)\delta(\boldsymbol{R}_k-\boldsymbol{R})\rho(1,\ldots,N;t)\mathrm{d}1\ldots\mathrm{d}N$$
$$-\nabla\cdot\int\sum_k m_k\boldsymbol{v}_k\boldsymbol{v}_k\delta(\boldsymbol{R}_k-\boldsymbol{R})\rho(1,\ldots,N;t)\mathrm{d}1\ldots\mathrm{d}N,\quad(40)$$

so that (37) may be written in the form

$$\frac{\partial(\varrho\boldsymbol{v})}{\partial t}=-\nabla\cdot\int\sum_k m_k\boldsymbol{v}_k\boldsymbol{v}_k\delta(\boldsymbol{R}_k-\boldsymbol{R})\rho(1,\ldots,N;t)\mathrm{d}1\ldots\mathrm{d}N$$
$$+\int\sum_k(\boldsymbol{f}_k^{\mathrm{L}}+\boldsymbol{f}_k^{\mathrm{S}})\delta(\boldsymbol{R}_k-\boldsymbol{R})\rho(1,\ldots,N;t)\mathrm{d}1\ldots\mathrm{d}N.\quad(41)$$

Splitting the Weyl transform $\boldsymbol{v}_k$ of the atomic velocity into the bulk velocity $\boldsymbol{v}(\boldsymbol{R},t)$, defined by (35) with (34), and a fluctuation term $\hat{\boldsymbol{v}}_k$:

$$\boldsymbol{v}_k=\boldsymbol{v}(\boldsymbol{R},t)+\hat{\boldsymbol{v}}_k(\boldsymbol{R},t)\quad(42)$$

and introducing reduced Wigner functions of the type (18), one finds as the momentum balance

$$\frac{\partial(\varrho\boldsymbol{v})}{\partial t}=-\nabla\cdot(\varrho\boldsymbol{v}\boldsymbol{v}+\mathsf{P}^{\mathrm{K}})+\boldsymbol{F}^{\mathrm{L}}+\boldsymbol{F}^{\mathrm{S}},\quad(43)$$

where we introduced a kinetic pressure tensor

$$\mathsf{P}^{\mathrm{K}}=\int m\hat{\boldsymbol{v}}_1\hat{\boldsymbol{v}}_1f_1(\boldsymbol{R},\boldsymbol{v}_1;t)\mathrm{d}\boldsymbol{v}_1\quad(44)$$

and the abbreviations for the long range and short range force densities

$$\boldsymbol{F}^{\mathrm{L,S}}=\int\sum_k\boldsymbol{f}_k^{\mathrm{L,S}}\delta(\boldsymbol{R}_k-\boldsymbol{R})\rho(1,\ldots,N;t)\mathrm{d}1\ldots\mathrm{d}N.\quad(45)$$

These expressions contain the atomic quantities (VI.99) and (VI.100), which have the same form as (I.54) and (I.52). For that reason we may take over the result of the classical evaluation of $\boldsymbol{F}^{\mathrm{L}}$ and $\boldsymbol{F}^{\mathrm{S}}$. The only step which we should investigate is the one which involves the expression

$$c^{-1}\int\sum_k\partial_{tP}\{\bar{\boldsymbol{\mu}}_k^{(1)}\wedge\boldsymbol{B}_{\mathrm{e}}(\boldsymbol{R}_k,t)\}\delta(\boldsymbol{R}_k-\boldsymbol{R})\rho(1,\ldots,N;t)\mathrm{d}1\ldots\mathrm{d}N.\quad(46)$$

(cf. the last term of (II.66)). Application of the relation (13) with the Weyl transform (VI.61) of the Hamiltonian leads to the identity

$$c^{-1}\int \sum_k \partial_{tP}\{\bar{\boldsymbol{\mu}}_k^{(1)}\wedge \boldsymbol{B}_e(\boldsymbol{R}_k, t)\}\delta(\boldsymbol{R}_k-\boldsymbol{R})\rho(1, \dots, N; t)\mathrm{d}1 \dots \mathrm{d}N$$

$$= c^{-1}\frac{\partial}{\partial t}\int \sum_k \bar{\boldsymbol{\mu}}_k^{(1)}\wedge \boldsymbol{B}_e(\boldsymbol{R}_k, t)\delta(\boldsymbol{R}_k-\boldsymbol{R})\rho(1, \dots, N; t)\mathrm{d}1 \dots \mathrm{d}N$$

$$+c^{-1}\boldsymbol{\nabla}\cdot\int \sum_k \boldsymbol{v}_k\bar{\boldsymbol{\mu}}_k^{(1)}\wedge \boldsymbol{B}_e(\boldsymbol{R}_k, t)\delta(\boldsymbol{R}_k-\boldsymbol{R})\rho(1, \dots, N; t)\mathrm{d}1 \dots \mathrm{d}N. \tag{47}$$

The right-hand side has the same form as the expression (II.67) of classical theory. This means that the long range force density, which occurs in (43), has the same form as (II.72):

$$\boldsymbol{F}^{\mathrm{L}} = \varrho^{\mathrm{e}}\boldsymbol{E}+c^{-1}\boldsymbol{J}\wedge\boldsymbol{B}+(\boldsymbol{\nabla}\boldsymbol{E})\cdot\boldsymbol{P}+(\boldsymbol{\nabla}\boldsymbol{B})\cdot\boldsymbol{M}$$

$$+c^{-1}\frac{\partial}{\partial t}(\boldsymbol{P}\wedge\boldsymbol{B})+c^{-1}\boldsymbol{\nabla}\cdot(\boldsymbol{v}\boldsymbol{P}\wedge\boldsymbol{B})-\boldsymbol{\nabla}\cdot\mathbf{P}^{\mathrm{F}}+\boldsymbol{F}^{\mathrm{C}}. \tag{48}$$

The macroscopic Maxwell fields, charge-current and polarization densities are the quantum-mechanical averages (24), (26) and (28). The pressure $\mathbf{P}^{\mathrm{F}}$ and the correlation contribution $\boldsymbol{F}^{\mathrm{C}}$ to the long range force density are given by (II.73) and (II.74) where now $f_1$ and $c_2$ stand for the reduced Wigner functions (18) and (22).

Likewise the short range term $\boldsymbol{F}^{\mathrm{S}}$ in (43) is given by (II.75), where $f_2$ now stands for the two-point Wigner function (21).

The quantities that occur in the momentum equation have been given here as integrals over the product of Weyl transforms of certain operators and Wigner functions. The advantage of this way of writing resides in the fact that their form is as simple as the classical one. If one wishes one may write down the operators of the Weyl transforms that occur here. For instance one may find the operator of which the kinetic pressure (44) is the average by making use of (VI.31) and (VI.92). Then this operator turns out to be of the form

$$\tfrac{1}{4}\sum_k m_k\{\boldsymbol{v}_{k,\mathrm{op}}-\boldsymbol{v}(\boldsymbol{R}, t), \{\boldsymbol{v}_{k,\mathrm{op}}-\boldsymbol{v}(\boldsymbol{R}, t), \delta(\boldsymbol{R}_{k,\mathrm{op}}-\boldsymbol{R})\}\}, \tag{49}$$

involving a double anticommutator.

### d. *The energy balance*

As in the derivation of the momentum law the macroscopic energy balance equation follows from the corresponding atomic equation (VI.105) by multiplication with the delta function $\delta(\boldsymbol{R}_k-\boldsymbol{R})$, summation over $k$, multiplication with a Wigner function and integration over the whole of phase space:

$$\int \sum_k \left[ \partial_{t\mathrm{P}} \left\{ \tfrac{1}{2} m_k \boldsymbol{v}_k^2 + \tfrac{1}{2} \sum_i m_{ki} (\partial_{t\mathrm{P}} \boldsymbol{r}_{ki})^2 + \sum_{i,j(i\neq j)} \frac{e_{ki} e_{kj}}{8\pi |\boldsymbol{r}_{ki}-\boldsymbol{r}_{kj}|} \right\} \right] \delta(\boldsymbol{R}_k-\boldsymbol{R}) \rho(1, \ldots, N; t) \mathrm{d}1 \ldots \mathrm{d}N$$

$$= \int \sum_k (\psi_k^{\mathrm{L}} + \psi_k^{\mathrm{S}}) \delta(\boldsymbol{R}_k-\boldsymbol{R}) \rho(1, \ldots, N; t) \mathrm{d}1 \ldots \mathrm{d}N. \tag{50}$$

In order to rewrite the left-hand side we consider the following time derivative

$$\frac{\partial}{\partial t} \int \sum_k \left\{ \tfrac{1}{2} m_k \boldsymbol{v}_k^2 + \tfrac{1}{2} \sum_i m_{ki} (\partial_{t\mathrm{P}} \boldsymbol{r}_{ki})^2 + \sum_{i,j(i\neq j)} \frac{e_{ki} e_{kj}}{8\pi |\boldsymbol{r}_{ki}-\boldsymbol{r}_{kj}|} \right\} \delta(\boldsymbol{R}_k-\boldsymbol{R}) \rho(1, \ldots, N; t) \mathrm{d}1 \ldots \mathrm{d}N. \tag{51}$$

This time derivative may be evaluated with the help of (13). Then one obtains, in view of the form (VI.61) of $H$, an integral over phase space with as integrand the product of the Wigner function $\rho(1, \ldots, N; t)$ and the expression

$$\left( \sum_{k,i} \left[ \frac{\partial}{\partial \boldsymbol{R}_{ki}} \cdot \frac{\partial^{(H)}}{\partial \boldsymbol{P}_{ki}} \left\{ 1 - \frac{\hbar^2}{8} \left( \sum_{l,j} \frac{\partial}{\partial \boldsymbol{P}_{lj}} \cdot \frac{\partial^{(H)}}{\partial \boldsymbol{R}_{lj}} \right)^2 \right\} - \frac{\partial}{\partial \boldsymbol{P}_{ki}} \cdot \frac{\partial^{(H)}}{\partial \boldsymbol{R}_{ki}} \right] H(1, \ldots, N; t) \right.$$

$$\left. + \frac{\partial}{\partial t} \right) \sum_m \left\{ \tfrac{1}{2} m_m \boldsymbol{v}_m^2 + \tfrac{1}{2} \sum_p m_{mp} (\partial_{t\mathrm{P}} \boldsymbol{r}_{mp})^2 + \sum_{p,s(p\neq s)} \frac{e_{mp} e_{ms}}{8\pi |\boldsymbol{r}_{mp}-\boldsymbol{r}_{ms}|} \right\} \delta(\boldsymbol{R}_m-\boldsymbol{R}), \tag{52}$$

where the differentiation symbols with label $(H)$ act only on the Weyl transform $H$ of the Hamiltonian and the other ones on all terms save $H$. With this result one then finds an expression for (51). If this expression is used in (50) we get the energy balance equation

$$\frac{\partial}{\partial t} (\tfrac{1}{2} \varrho \boldsymbol{v}^2 + \varrho u^{\mathrm{K}}) = -\nabla \cdot \{ \boldsymbol{v} (\tfrac{1}{2} \varrho \boldsymbol{v}^2 + \varrho u^{\mathrm{K}}) + \mathsf{P}^{\mathrm{K}} \cdot \boldsymbol{v} + \boldsymbol{J}_q^{\mathrm{K}} \} + \Psi^{\mathrm{L}} + \Psi^{\mathrm{S}}. \tag{53}$$

The energy density $\varrho u^{\mathrm{K}}$ is found as:

$$\varrho u^{\mathrm{K}} = \int \left\{ \tfrac{1}{2} m \hat{\boldsymbol{v}}_1^2 + \tfrac{1}{2} \sum_i m_i (\partial_{t\mathrm{P}} \boldsymbol{r}_{1i})^2 + \sum_{i,j(i\neq j)} \frac{e_i e_j}{8\pi |\boldsymbol{r}_{1i}-\boldsymbol{r}_{1j}|} \right\} f_1(\boldsymbol{R}, 1; t) \mathrm{d}1 \tag{54}$$

with the masses and charges $m_{ki} = m_i$ and $e_{ki} = e_i$ of the constituent particles of the identical atoms and the velocity fluctuation $\hat{\boldsymbol{v}}_1$ defined by (42). Furthermore the kinetic pressure $\mathbf{P}^{\mathrm{K}}$ has been given in (44) and the macroscopic power densities $\Psi^{\mathrm{L}}$ and $\Psi^{\mathrm{S}}$ stand for

$$\Psi^{\mathrm{L,S}} \equiv \int \sum_k \psi_k^{\mathrm{L,S}} \delta(\boldsymbol{R}_k - \boldsymbol{R}) \rho(1, \ldots, N; t) \mathrm{d}1 \ldots \mathrm{d}N \tag{55}$$

with atomic quantities given in (VI.106) and (VI.107). All quantities mentioned so far have the same form as those occurring in the classical theory with the distribution function replaced by a Wigner function. The heat flow $\boldsymbol{J}_q^{\mathrm{K}}$, however, has a form which differs from the corresponding classical expression (II.81):

$$\boldsymbol{J}_q^{\mathrm{K}} = \int \hat{\boldsymbol{v}}_1 \left\{ \tfrac{1}{2} m \hat{\boldsymbol{v}}_1^2 + \tfrac{1}{2} \sum_i m_i (\partial_{t\mathrm{P}} \boldsymbol{r}_{1i})^2 + \sum_{i,j(i \neq j)} \frac{e_i e_j}{8\pi |\boldsymbol{r}_{1i} - \boldsymbol{r}_{1j}|} \right\} f_1(\boldsymbol{R}, 1; t) \mathrm{d}1$$
$$+ \int \sum_i \frac{\hbar^2}{8} \frac{e_i}{m m_i c} \Delta_{1i} \boldsymbol{A}_{\mathrm{e}}(\boldsymbol{R}_{1i}, t) f_1(\boldsymbol{R}, 1; t) \mathrm{d}1. \tag{56}$$

The last term with the Laplacian of the vector potential $\boldsymbol{A}_{\mathrm{e}}$ has arisen from the second term of the expression (52). The occurrence of this rather exotic term does not destroy gauge invariance, as may be seen by writing down the operator of which (56) is the average (see problem 1):

$$\sum_k \sum_{i,j} \frac{1}{16 m_k m_{ki}} \{\boldsymbol{P}_{kj,\mathrm{op}} - c^{-1} e_{kj} \boldsymbol{A}_{\mathrm{e}}(\boldsymbol{R}_{kj,\mathrm{op}}, t) - m_{kj} \boldsymbol{v}(\boldsymbol{R}, t),$$
$$\{(\boldsymbol{P}_{ki,\mathrm{op}} - c^{-1} e_{ki} \boldsymbol{A}_{\mathrm{e}}(\boldsymbol{R}_{ki,\mathrm{op}}, t) - m_{ki} \boldsymbol{v}(\boldsymbol{R}, t))\cdot,$$
$$\{(\boldsymbol{P}_{ki,\mathrm{op}} - c^{-1} e_{ki} \boldsymbol{A}_{\mathrm{e}}(\boldsymbol{R}_{ki,\mathrm{op}}, t) - m_{ki} \boldsymbol{v}(\boldsymbol{R}, t)), \delta(\boldsymbol{R}_{k,\mathrm{op}} - \boldsymbol{R})\}\}\}$$
$$+ \sum_k \sum_i \frac{1}{2m_k} \Big\{ \boldsymbol{P}_{ki,\mathrm{op}} - c^{-1} e_{ki} \boldsymbol{A}_{\mathrm{e}}(\boldsymbol{R}_{ki,\mathrm{op}}, t) - m_{ki} \boldsymbol{v}(\boldsymbol{R}, t),$$
$$\sum_{p,s(p \neq s)} \frac{e_{kp} e_{ks}}{8\pi |\boldsymbol{r}_{kp,\mathrm{op}} - \boldsymbol{r}_{ks,\mathrm{op}}|} \delta(\boldsymbol{R}_{k,\mathrm{op}} - \boldsymbol{R}) \Big\}, \tag{57}$$

where the curly brackets indicate anticommutators. The fact that the vector potential appears in the traditional combination with the momentum operator guarantees the gauge invariance[1].

[1] As is well known the canonical commutation relations for coordinate and momentum operators are invariant under the von Neumann transformation $\boldsymbol{P}_{\mathrm{op}} \to \boldsymbol{P}'_{\mathrm{op}} \equiv \boldsymbol{P}_{\mathrm{op}} + \nabla\chi$ with $\chi$ an arbitrary function of the coordinate operators. As a consequence the combination $\boldsymbol{P}_{\mathrm{op}} - c^{-1} e \boldsymbol{A}_{\mathrm{op}}$ may be written as $\boldsymbol{P}'_{\mathrm{op}} - \nabla\chi - c^{-1} e \boldsymbol{A}_{\mathrm{op}}$. Since under a gauge transformation the vector potential $\boldsymbol{A}_{\mathrm{op}}$ becomes $\boldsymbol{A}'_{\mathrm{op}} = \boldsymbol{A}_{\mathrm{op}} + \nabla\psi$ (with arbitrary $\psi$), it follows that $\boldsymbol{P}_{\mathrm{op}} - c^{-1} e \boldsymbol{A}_{\mathrm{op}}$ can be written as $\boldsymbol{P}'_{\mathrm{op}} - c^{-1} e \boldsymbol{A}'_{\mathrm{op}}$, if $\chi$ is chosen as $c^{-1} e \psi$.

The energy balance equation (53) contains the long range and short range power densities (55), which may be evaluated along the same lines as in the classical treatment (section 5*d* of chapter II). The point which has to be checked in detail is the validity of the quantum-mechanical analogue of (II.83). This may be done in the same way as in (46–47). As a result one finds for the long range power density (55) (cf. (II.88))

$$\psi^{\mathrm{L}} = \boldsymbol{J}\cdot\boldsymbol{E} + \frac{\partial \boldsymbol{P}}{\partial t}\cdot\boldsymbol{E} + \nabla\cdot(\boldsymbol{v}\boldsymbol{P}\cdot\boldsymbol{E}) - \boldsymbol{M}\cdot\frac{\partial \boldsymbol{B}}{\partial t} - \nabla\cdot(\mathsf{P}^{\mathrm{F}}\cdot\boldsymbol{v} + \boldsymbol{J}_q^{\mathrm{F}}) + \Psi^{\mathrm{C}}, \tag{58}$$

where all quantities have the same form as in classical theory with the distribution functions replaced by Wigner functions and also fluxion dots by the operation $\partial_{t\mathrm{P}}$. The same applies to the short range power density $\Psi^{\mathrm{S}}$.

## e. *The short range terms*

Since the integrand in $\boldsymbol{F}^{\mathrm{S}}$, which is of the form (II.75), vanishes with increasing $|\boldsymbol{s}|$, one may expand the Wigner function in powers of $\boldsymbol{s}$ as the distribution function in the classical treatment. Then one finds that

$$\boldsymbol{F}^{\mathrm{S}} = -\nabla\cdot\mathsf{P}^{\mathrm{S}} \tag{59}$$

with a pressure contribution $\mathsf{P}^{\mathrm{S}}$ of the same form as (II.94).

As to the short range power density $\Psi^{\mathrm{S}}$, an expansion of the two-point Wigner function leads to a relation of the form (II.95). It may be written in an alternative way by considering the derivative of a quantity of the form (II.97) namely

$$\varrho u^{\mathrm{S}} = \int \sum_{k,l(k\neq l)} \left( \sum_{i,j} \frac{e_{ki}\,e_{lj}}{8\pi|\boldsymbol{s}+\boldsymbol{r}_{ki}-\boldsymbol{r}_{lj}|} - \sum_{n,m=0}^{\infty} (-1)^m \overline{\boldsymbol{\mu}}_k^{(n)} \vdots \nabla_s^n \overline{\boldsymbol{\mu}}_l^{(m)} \vdots \nabla_s^m \frac{1}{8\pi s} \right)$$
$$\delta(\boldsymbol{R}_k - \boldsymbol{R} - \tfrac{1}{2}\boldsymbol{s})\delta(\boldsymbol{R}_l - \boldsymbol{R} + \tfrac{1}{2}\boldsymbol{s})\rho(1, \ldots, N; t)\mathrm{d}1 \ldots \mathrm{d}N, \tag{60}$$

which contains the $N$-point Wigner function. Its time derivative follows by application of (13) with the Weyl transform (VI.61) of the Hamiltonian. One obtains an equation of the same form as (II.99) with distribution functions replaced by Wigner functions and fluxion dots by the symbol $\partial_{t\mathrm{P}}$. As a consequence the short range power density becomes (cf. (II.101)):

$$\Psi^{\mathrm{S}} = -\nabla\cdot(\boldsymbol{v}\varrho u^{\mathrm{S}} + \mathsf{P}^{\mathrm{S}}\cdot\boldsymbol{v} + \boldsymbol{J}_q^{\mathrm{S}}) - \frac{\partial \varrho u^{\mathrm{S}}}{\partial t} \tag{61}$$

with $\varrho u^{\mathrm{S}}$ given by (60) and $\boldsymbol{J}_q^{\mathrm{S}}$ by an expression of the same form as (II.96, 100).

f. *The correlation contributions*

The correlation contributions $\boldsymbol{F}^{\mathrm{C}}$ and $\Psi^{\mathrm{C}}$ will be studied here for a fluid system of neutral atoms. (The other cases, namely plasmas and systems with long range correlations, that have been studied in the classical treatment, may be generalized to quantum theory in an analogous fashion.)

In classical theory fluid systems of neutral atoms were characterized by the existence of a correlation length that is small compared to the distance over which the macroscopic quantities change appreciably. Therefore the correlation function could be expanded in a Taylor series, which could be broken off after a few terms (the Irving–Kirkwood approximation). In quantum mechanics two reasons for the occurrence of correlations exist: in the first place correlation effects that are the quantum-mechanical analogues of the classical correlation effects due to the interaction between the particles, and furthermore the correlation effects due to statistics, which take place even in a perfect gas. The latter effects are characterized by a 'correlation length' which is of the order of the thermal wave length $\lambda \equiv h/(2\pi mkT)^{\frac{1}{2}}$ as is shown in the appendix I. A necessary condition for the expansion of the correlation Wigner function is hence that the thermal wave length $\lambda$ is small compared to the distance $d$ over which the macroscopic quantities change appreciably, i.e. for sufficiently high temperature. (This does not mean that quantum effects due to statistics are now neglected altogether. That would require the smallness of the thermal wave length $\lambda$ with respect to the mean free path; the latter is always much smaller than the distance $d$ in the physical situations to which statistical mechanics applies.) The second condition for the possibility of expanding the correlation Wigner function is the smallness of the correlation length due to the atomic interactions with respect to the macroscopic distance $d$. The latter condition sets a limit to the applicability of the Irving–Kirkwood procedure of the same kind as in the classical case.

If indeed the system is such that the Wigner correlation function has short range in the sense described above, we may apply the Irving–Kirkwood approximation to the correlation force density $\boldsymbol{F}^{\mathrm{C}}$ and the correlation power density $\Psi^{\mathrm{C}}$. Then one may write

$$\boldsymbol{F}^{\mathrm{C}} = -\nabla\cdot\mathbf{P}^{\mathrm{C}}, \tag{62}$$

$$\Psi^{\mathrm{C}} = -\nabla\cdot(\boldsymbol{v}\varrho u^{\mathrm{C}} + \mathbf{P}^{\mathrm{C}}\cdot\boldsymbol{v} + \boldsymbol{J}_q^{\mathrm{C}}) - \frac{\partial \varrho u^{\mathrm{C}}}{\partial t}, \tag{63}$$

with $\mathbf{P}^{\mathrm{C}}$, $\boldsymbol{J}_q^{\mathrm{C}}$ and $\varrho u^{\mathrm{C}}$ given by formulae of the same form as (II.104), (II.111) and (II.112), but with a Wigner function instead of a classical distribution function and the Poisson bracket derivative $\partial_{t\mathrm{P}}$ instead of the fluxion dot.

(In taking the time derivative of the quantity $\varrho u^{\mathrm{C}}$, one has to apply the identity (13); owing to the special form of the integrand of $\varrho u^{\mathrm{C}}$ and the Weyl transform of the Hamiltonian, the sine operator reduces to the Poisson bracket operator.)

If the Wigner correlation function has long range character one may use an artifice of the same type as in chapter II, section 5*h* of the classical theory to derive again expressions of the form (62–63), but with a mean Wigner correlation function instead of the Wigner correlation function itself.

### g. *Substances with short range correlations*

Collecting the results of the preceding subsections we have found the momentum law for substances of neutral atoms of which the Wigner correlation functions have short range:

$$\frac{\partial \varrho \boldsymbol{v}}{\partial t} = -\nabla\cdot(\varrho \boldsymbol{v}\boldsymbol{v}+\mathbf{P})+(\nabla \boldsymbol{E})\cdot\boldsymbol{P}+(\nabla \boldsymbol{B})\cdot\boldsymbol{M}+c^{-1}\frac{\partial}{\partial t}(\boldsymbol{P}\wedge\boldsymbol{B})+c^{-1}\nabla\cdot(\boldsymbol{v}\boldsymbol{P}\wedge\boldsymbol{B}) \quad (64)$$

(cf. (II.105–106)). Hence the time derivative of the momentum density $\varrho \boldsymbol{v}$ is equal to the sum of a divergence of a material term, that contains the pressure tensor $\mathbf{P}$, and a force density due to the electromagnetic field.

The energy law for such substances is

$$\frac{\partial}{\partial t}(\tfrac{1}{2}\varrho v^2+\varrho u) = -\nabla\cdot\{\boldsymbol{v}(\tfrac{1}{2}\varrho v^2+\varrho u)+\mathbf{P}\cdot\boldsymbol{v}+\boldsymbol{J}_q\}$$
$$+\frac{\partial \boldsymbol{P}}{\partial t}\cdot\boldsymbol{E}+\nabla\cdot(\boldsymbol{v}\boldsymbol{P}\cdot\boldsymbol{E})-\boldsymbol{M}\cdot\frac{\partial \boldsymbol{B}}{\partial t}. \quad (65)$$

(cf. (II.113–114). The change of the total energy density with time is thus due to a material energy flow and a power density, which arises from the electromagnetic fields. The difference with the classical results consists in the replacement of the classical distribution functions by Wigner functions and fluxion dots by the operator $\partial_{t\mathrm{P}}$ in the expressions for the macroscopic quantities. Moreover an extra term with the Laplacian of the external vector potential appears in the contribution $\boldsymbol{J}_q^{\mathrm{K}}$ to the heat flow.

Just as in classical theory the balance equations (64) and (65) may be written in the form of conservation laws. One then obtains equations of the form (II.109) and (II.118).

The quantum-mechanical non-relativistic treatment of neutral plasma's and of substances with long range Wigner correlations presents no new aspects as compared to the classical non-relativistic theory.

Galilei invariance of the theory, i.e. invariance with respect to the coordinate transformation

$$\boldsymbol{R}' = \boldsymbol{R} + \boldsymbol{V}t, \qquad t' = t, \tag{66}$$

follows from the transformation of the wave function[1]

$$\psi'(\boldsymbol{R}', t') = \exp\left\{\frac{i}{\hbar}\left(\sum_{k,i} \boldsymbol{V}\cdot\boldsymbol{R}_{ki} + \tfrac{1}{2}\sum_{ki} m_{ki} V^2 t\right)\right\} \psi(\boldsymbol{R}, t). \tag{67}$$

In fact the Hamilton operator that governs the time behaviour of the wave function transforms as

$$H'_{\mathrm{op}}(\boldsymbol{P}_{ki,\mathrm{op}}, \boldsymbol{R}_{ki,\mathrm{op}}) = H_{\mathrm{op}}(\boldsymbol{P}_{ki,\mathrm{op}}, \boldsymbol{R}_{ki,\mathrm{op}} - \boldsymbol{V}t) + c^{-1} \sum_{k,i} e_{ki} \boldsymbol{V}\cdot\boldsymbol{A}_{\mathrm{e}}(\boldsymbol{R}_{ki,\mathrm{op}} - \boldsymbol{V}t, t). \tag{68}$$

The Weyl transform of this Hamiltonian occurs in the Poisson bracket $\partial_{t\mathrm{P}}$. The transformation of the Wigner function that follows from its definition (4–5) and (67) is

$$\rho'(\boldsymbol{P}_{ki}, \boldsymbol{R}_{ki}; t) = \rho(\boldsymbol{P}_{ki} - m_{ki}\boldsymbol{V}, \boldsymbol{R}_{ki} - \boldsymbol{V}t; t). \tag{69}$$

With (68) and (69) one may check the Galilei invariance of the equations.

## 6 *The angular momentum equations*

If one starts from the atomic angular momentum equation (VI.111) with (VI.112) and (VI.113) one finds the macroscopic angular momentum equation by multiplying with the delta function $\delta(\boldsymbol{R}_k - \boldsymbol{R})$, summing over $k$, multiplying with a Wigner function and integrating over phase space. Since the Weyl transform of the atomic inner angular momentum density $\sum_k \bar{\boldsymbol{s}}_k \delta(\boldsymbol{R}_k - \boldsymbol{R})$ is linear in the momenta, the sine occurring in the time derivative (13) reduces to the Poisson bracket. Therefore the angular momentum law has the same form as in classical theory, if one supposes again the Irving–Kirkwood approximation applicable. The result is then for a substance of neutral atoms with short range Wigner correlation functions:

$$\frac{\partial \boldsymbol{S}}{\partial t} = -\nabla\cdot(\boldsymbol{v}\boldsymbol{S} + \mathbf{J}_s) + \boldsymbol{D}_s + \boldsymbol{P}\wedge\boldsymbol{E} + \boldsymbol{M}\wedge\boldsymbol{B} + c^{-1}\boldsymbol{v}\wedge(\boldsymbol{P}\wedge\boldsymbol{B}). \tag{70}$$

This law has the same form as (II.185), but in the statistical expressions Wigner functions occur instead of classical distribution functions. It con-

[1] J. M. Lévy-Leblond, J. Math. Phys. 4(1963)776; see also problem 2.

tains the inner angular momentum density $\boldsymbol{S}$, the (conduction part of the) inner angular momentum flow $\boldsymbol{J}_s$, (which has the form (II.186)), the material part of the source term $\boldsymbol{D}_s$ (of the form (II.187)) and torque densities exerted by the electromagnetic fields $(\boldsymbol{E}, \boldsymbol{B})$ on the polarization densities $(\boldsymbol{P}, \boldsymbol{M})$.

The macroscopic quantities have been written throughout as integrals over Weyl transforms of atomic operators and Wigner functions. Therefore they are averages of operators that may be found from these Weyl transforms. In particular one finds that the macroscopic inner angular momentum density is the average of the atomic operator

$$\tfrac{1}{2}\sum_k \{\bar{\boldsymbol{s}}_{k,\mathrm{op}}, \delta(\boldsymbol{R}_{k,\mathrm{op}}-\boldsymbol{R})\}. \tag{71}$$

The inner angular momentum operator $\bar{\boldsymbol{s}}_{k,\mathrm{op}}$ has the Weyl transform (VI.110) and is hence:

$$\bar{\boldsymbol{s}}_{k,\mathrm{op}} = \tfrac{1}{2}\sum_i m_{ki}\{\boldsymbol{r}_{ki,\mathrm{op}}\wedge, \dot{\boldsymbol{r}}_{ki,\mathrm{op}}\}, \tag{72}$$

where $\boldsymbol{r}_{ki,\mathrm{op}}$ stands for $\boldsymbol{R}_{ki,\mathrm{op}}-\boldsymbol{R}_{k,\mathrm{op}}$ and where $\dot{\boldsymbol{r}}_{ki,\mathrm{op}}$ is given by (VI.95).

In the same way as in classical theory one may prove that the source term $\boldsymbol{D}_s$ is equal to minus the antisymmetric part of the pressure tensor

$$\boldsymbol{D}_s = -\boldsymbol{P}_\mathrm{A} \tag{73}$$

(cf. (II.195)). This fact has as a consequence that also the conservation law of total angular momentum in the form (II.197) is valid here.

## 7 *The laws of thermodynamics*

### a. *The first law*

The first law of thermodynamics is a direct consequence of the energy balance equation. Since the latter has the same form as in the classical treatment, one finds in the case of a system of neutral atoms with short range Wigner correlations an equation as (II.213), i.e.

$$\frac{\mathrm{d}q}{\mathrm{d}t} = \frac{\mathrm{d}u}{\mathrm{d}t} + v\tilde{\mathsf{P}} : \nabla \boldsymbol{v} - \frac{\mathrm{d}(v\boldsymbol{P}')}{\mathrm{d}t}\cdot\boldsymbol{E}' + v\boldsymbol{M}'\cdot\frac{\mathrm{d}\boldsymbol{B}'}{\mathrm{d}t}. \tag{74}$$

It shows that the heat supplied per unit mass and time is equal to the change of specific internal energy $u$ plus a viscous term with the pressure tensor $\mathsf{P}$ and the velocity gradients, plus two terms with the electromagnetic fields $\boldsymbol{E}'$, $\boldsymbol{B}'$ and the polarization densities $\boldsymbol{P}'$, $\boldsymbol{M}'$ (II.188) in the rest frame. The time derivatives at the right-hand side are material time derivatives $(\partial/\partial t + \boldsymbol{v}\cdot\nabla)$, and $v$ is the specific volume.

For neutral plasmas and for systems with long range correlations one finds laws of the form (II.216) and (II.217).

## b. *The second law*

In order to derive the second law for a fluid system of neutral atoms we consider a large polarized system at rest, divided into nearly uniform cells and describe these cells by canonical ensembles with the environments playing the role of heat baths. The cells are chosen to be ellipsoidal, so that the external fields due to the surroundings of the cells are uniform and given by (II.220) with (II.219).

The quantum-mechanical canonical ensemble is given by a density operator

$$P_{\mathrm{op}} = \exp\{(F^* - H^{\mathrm{W}}_{\mathrm{op}})/kT\}, \tag{75}$$

so that, since $\mathrm{Tr}\, P_{\mathrm{op}} = 1$, one has

$$\exp(-F^*/kT) = \mathrm{Tr}\exp(-H^{\mathrm{W}}_{\mathrm{op}}/kT). \tag{76}$$

Here $F^*$ is the free energy, $T$ the temperature and $H^{\mathrm{W}}_{\mathrm{op}}$ is the Hamilton operator of the system with the inclusion of wall potential operators

$$H^{\mathrm{W}}_{\mathrm{op}} = H_{\mathrm{op}} + \sum_k U^{\mathrm{W}}_{k,\mathrm{op}}(\boldsymbol{R}_{k,\mathrm{op}}). \tag{77}$$

The wall potential operator $U^{\mathrm{W}}_{k,\mathrm{op}}$, which depends on the centre of mass operator $\boldsymbol{R}_{k,\mathrm{op}}$ of atom $k$, is defined in such a way that it gives infinity if it operates on an eigenfunction of $\boldsymbol{R}_{k,\mathrm{op}}$ with eigenvalue $\boldsymbol{R}_k$ lying outside the volume $V$, while it gives zero if it operates on an eigenfunction of $\boldsymbol{R}_{k,\mathrm{op}}$ with eigenvalue $\boldsymbol{R}_k$ lying inside the volume $V$. The Hamilton operator $H_{\mathrm{op}}$ may be written as a function of the coordinate and momentum operators of all constituent particles of the system, or alternatively as a function of the centre of mass and relative coordinate operators and corresponding momentum operators. The latter form of the Hamilton operator has been derived in appendix II. Its Weyl transform is (A54) with $e_k = 0$:

$$\begin{aligned} H_{\mathrm{op}}(\boldsymbol{p}_{\mathrm{op}}, \boldsymbol{q}_{\mathrm{op}}, t) \rightleftarrows & \sum_k \left( \frac{\boldsymbol{P}_k^2}{2m_k} + \sum_{i=1}^{f-1} \frac{\boldsymbol{p}_{ki}^2}{2m_{ki}} - \sum_{i,j=1}^{f-1} \frac{\boldsymbol{p}_{ki}\cdot\boldsymbol{p}_{kj}}{2m_k} \right) \\ & + \sum_k \sum_{i,j=1(i\neq j)}^{f} \frac{e_{ki}e_{kj}}{8\pi|\boldsymbol{R}_{ki}-\boldsymbol{R}_{kj}|} + \sum_{k,l(k\neq l)} \sum_{i,j=1}^{f} \frac{e_{ki}e_{lj}}{8\pi|\boldsymbol{R}_{ki}-\boldsymbol{R}_{lj}|} \\ & - \sum_k \left\{ \bar{\boldsymbol{\mu}}_k^{(1)}\cdot \left( \boldsymbol{E}_{\mathrm{e}} + \tfrac{1}{2}c^{-1}\frac{\boldsymbol{P}_k}{m_k} \wedge \boldsymbol{B}_{\mathrm{e}} \right) + \tfrac{1}{2}c^{-1}(\partial_{t\mathrm{P}}\bar{\boldsymbol{\mu}}_k^{(1)})\cdot(\boldsymbol{B}_{\mathrm{e}} \wedge \boldsymbol{R}_k) + \bar{\boldsymbol{\nu}}_k^{(1)}\cdot\boldsymbol{B}_{\mathrm{e}} \right\} \end{aligned} \tag{78}$$

with the Weyl transforms of the electric and magnetic dipole moments:

$$\bar{\boldsymbol{\mu}}_k^{(1)}(\boldsymbol{q}) = \sum_{i=1}^{f} e_{ki}\{\boldsymbol{R}_{ki}(\boldsymbol{q}) - \boldsymbol{q}_{kf}\},$$
$$\bar{\boldsymbol{v}}_k^{(1)}(\boldsymbol{p}, \boldsymbol{q}) = \tfrac{1}{2}c^{-1}\sum_{i=1}^{f} e_{ki}\{\boldsymbol{R}_{ki}(\boldsymbol{q}) - \boldsymbol{q}_{kf}\} \wedge \left\{(1-\delta_{if})\frac{\boldsymbol{p}_{ki}}{m_{ki}} - \sum_{j=1}^{f-1}\frac{\boldsymbol{p}_{kj}}{m_k}\right\}. \tag{79}$$

The Hamilton operator (78) depends on the external electric and magnetic fields $(\boldsymbol{E}_\mathrm{e}, \boldsymbol{B}_\mathrm{e})$, and the wall potential operators on the boundary of the system. Hence the free energy depends on these quantities and on the temperature $T$. The partial derivative of the free energy with respect to the external electric field is:

$$\frac{\partial F^*}{\partial \boldsymbol{E}_\mathrm{e}} = -kT\mathrm{e}^{F^*/kT}\,\mathrm{Tr}\left\{\frac{\partial}{\partial \boldsymbol{E}_\mathrm{e}}\exp\left(-H_\mathrm{op}^\mathrm{W}/kT\right)\right\}. \tag{80}$$

We now apply an identity[1] for the derivative of an exponential operator:

$$\frac{\partial}{\partial\alpha}\mathrm{e}^{A_\mathrm{op}(\alpha)} = \sum_{n=0}^{\infty}\frac{1}{(n+1)!}A_\mathrm{op}^{(n)}(\alpha)\mathrm{e}^{A_\mathrm{op}(\alpha)} \tag{81}$$

with the operator $A_\mathrm{op}^{(n)}(\alpha)$ following from

$$A_\mathrm{op}^{(0)} \equiv \frac{\partial A_\mathrm{op}(\alpha)}{\partial\alpha}, \qquad A_\mathrm{op}^{(n)} = [A_\mathrm{op}, A_\mathrm{op}^{(n-1)}] \tag{82}$$

(cf. problem 3). Using this lemma in (80) one finds, since the traces of the commutators occurring vanish,

$$\frac{\partial F^*}{\partial \boldsymbol{E}_\mathrm{e}} = \mathrm{Tr}\left[\frac{\partial H_\mathrm{op}}{\partial \boldsymbol{E}_\mathrm{e}}\exp\{(F^* - H_\mathrm{op}^\mathrm{W})/kT\}\right], \tag{83}$$

where the fact that the wall potential operator is independent of the fields has been taken into account. This expression may be written in terms of Wigner functions and the Weyl transform of $\partial H_\mathrm{op}/\partial \boldsymbol{E}_\mathrm{e}$. The latter is equal to $\partial H/\partial \boldsymbol{E}_\mathrm{e}$:

$$\frac{\partial H_\mathrm{op}}{\partial \boldsymbol{E}_\mathrm{e}} \rightleftarrows \frac{\partial H}{\partial \boldsymbol{E}_\mathrm{e}} = -\sum_k \bar{\boldsymbol{\mu}}_k^{(1)}. \tag{84}$$

Thus one finds for (83)

$$\frac{\partial F^*}{\partial \boldsymbol{E}_\mathrm{e}} = -\int \bar{\boldsymbol{\mu}}_1^{(1)} f_1(1)\mathrm{d}1 = -V\boldsymbol{P}, \tag{85}$$

[1] R. J. Riddell jr. and G. E. Uhlenbeck, J. Chem. Phys. **18**(1950)1066.

where we employed the reduced Wigner function $f_1(1)$ and took the uniformity of the electric polarization density $\boldsymbol{P}$ (formula (28) for the dipole case) into account. Likewise we find for the derivative with respect to the magnetic field

$$\frac{\partial F^*}{\partial \boldsymbol{B}_e} = \mathrm{Tr}\left[\frac{\partial H_{op}}{\partial \boldsymbol{B}_e}\exp\{(F^* - H^{W}_{op})/kT\}\right]. \tag{86}$$

With the help of the Weyl correspondence

$$\frac{\partial H_{op}}{\partial \boldsymbol{B}_e} \rightleftarrows \frac{\partial H}{\partial \boldsymbol{B}_e} = -\sum_k \left\{\bar{\boldsymbol{v}}_k^{(1)} + \tfrac{1}{2}c^{-1}\bar{\boldsymbol{\mu}}_k^{(1)} \wedge \frac{\boldsymbol{P}_k}{m_k} - \tfrac{1}{2}c^{-1}(\partial_{tP}\bar{\boldsymbol{\mu}}_k^{(1)}) \wedge \boldsymbol{R}_k\right\} \tag{87}$$

(with $\partial_{tP}$ given by (VI.72)) that follows from (78), one obtains then

$$\frac{\partial F^*}{\partial \boldsymbol{B}_e} = -\int \left\{\bar{\boldsymbol{v}}_1^{(1)} + \tfrac{1}{2}c^{-1}\bar{\boldsymbol{\mu}}_1^{(1)} \wedge \frac{\boldsymbol{P}_1}{m_1} - \tfrac{1}{2}c^{-1}(\partial_{tP}\bar{\boldsymbol{\mu}}_1^{(1)}) \wedge \boldsymbol{R}_1\right\} f_1(1)\mathrm{d}1. \tag{88}$$

Since the canonical ensemble is stationary one derives by application of (13) that, up to order $c^0$,

$$\int (\partial_{tP}\bar{\boldsymbol{\mu}}_1^{(1)}) \wedge \boldsymbol{R}_1 \, f_1(1)\mathrm{d}1 = -\int \bar{\boldsymbol{\mu}}_1^{(1)} \wedge \frac{\boldsymbol{P}_1}{m_1} f_1(1)\mathrm{d}1, \tag{89}$$

where $\boldsymbol{P}_1/m_1 = \partial_{tP}\boldsymbol{R}_1 \equiv \boldsymbol{v}_1$ (up to order $c^0$). Therefore (88) may be written as

$$\frac{\partial F^*}{\partial \boldsymbol{B}_e} = -\int \left(\bar{\boldsymbol{v}}_1^{(1)} + c^{-1}\bar{\boldsymbol{\mu}}_1^{(1)} \wedge \frac{\boldsymbol{P}_1}{m_1}\right) f_1(1)\mathrm{d}1 = -V\boldsymbol{M}, \tag{90}$$

because of the definition (28) of the (uniform) magnetic polarization density $\boldsymbol{M}$ for the dipole case.

By definition the partial derivative of the free energy with respect to the temperature is equal to minus the entropy:

$$\frac{\partial F^*}{\partial T} = -S. \tag{91}$$

Just as in the classical case infinitesimal changes of the boundary will be described by a uniform deformation tensor $\delta\boldsymbol{\epsilon}$ in such a way that

$$\delta\boldsymbol{R} = \delta\boldsymbol{\epsilon}\cdot\boldsymbol{R}, \tag{92}$$

with the centre of the system as the origin of coordinates.

The total change of the free energy follows from (85), (90) and (91) as

$$\delta F^* = -S\delta T - V\boldsymbol{P}\cdot\delta\boldsymbol{E}_{\mathrm{e}} - V\boldsymbol{M}\cdot\delta\boldsymbol{B}_{\mathrm{e}} + \mathbf{A} : \delta\boldsymbol{\epsilon} \tag{93}$$

with an as yet unspecified tensor **A**.

The free energy is the difference of the canonical average of the Hamilton operator and the product of the temperature and the entropy. To find an expression for the average of the Hamilton operator, we shall employ its Weyl transform in the form (A57) with atomic charges $e_k = 0$:

$$H_{\mathrm{op}}(\boldsymbol{p}_{\mathrm{op}}, \boldsymbol{q}_{\mathrm{op}}) \rightleftarrows K + \sum_k \sum_{i,j=1(i\neq j)}^{f} \frac{e_{ki}e_{kj}}{8\pi|\boldsymbol{R}_{ki}-\boldsymbol{R}_{kj}|} + \sum_{k,l(k\neq l)} \sum_{i,j=1}^{f} \frac{e_{ki}e_{lj}}{8\pi|\boldsymbol{R}_{ki}-\boldsymbol{R}_{lj}|} - \sum_k \bar{\boldsymbol{\mu}}_k^{(1)}\cdot\boldsymbol{E}_{\mathrm{e}}, \tag{94}$$

with the quantity $K$ defined as

$$K = \sum_{k,i} \tfrac{1}{2} m_{ki} (\partial_{t\mathrm{P}} \boldsymbol{R}_{ki})^2. \tag{95}$$

The canonical average of the Hamilton operator follows by multiplication of the right-hand side of (94) with the Wigner function that belongs to the canonical ensemble and integration over phase space. Then one arrives at an expression which has the same form as that of the classical treatment. Along the same lines of reasoning as followed there one finds[1] for systems with a short range Wigner correlation function as the average Hamilton operator

$$V(\varrho u + \tfrac{1}{2}\boldsymbol{P}\boldsymbol{P} : \mathbf{L} - \boldsymbol{P}\cdot\boldsymbol{E}_{\mathrm{e}}), \tag{96}$$

where **L** is the depolarizing tensor and $u$ the specific internal energy.

The tensor **A**, which occurs in (93), is equal to

$$\mathbf{A} = -\int \sum_k \left( \frac{\partial H}{\partial \boldsymbol{P}_k} \boldsymbol{P}_k - \boldsymbol{R}_k \frac{\partial H}{\partial \boldsymbol{R}_k} \right) \rho(1, \ldots, N) \mathrm{d}1 \ldots \mathrm{d}N, \tag{97}$$

as shown in the appendix III. With the form (78) of the Weyl transform of the Hamilton operator one finds for **A**:

$$\mathbf{A} = -\int \sum_k \left[ m_k \boldsymbol{v}_k \left( \boldsymbol{v}_k - \tfrac{1}{2}c^{-1} \frac{\bar{\boldsymbol{\mu}}_k^{(1)}}{m_k} \wedge \boldsymbol{B}_{\mathrm{e}} \right) - \boldsymbol{R}_k \left\{ \nabla_k \sum_{l(\neq k)} \sum_{i,j=1}^{f} \frac{e_{ki}e_{lj}}{8\pi|\boldsymbol{R}_{ki}-\boldsymbol{R}_{lj}|} - \tfrac{1}{2}c^{-1}(\partial_{t\mathrm{P}} \bar{\boldsymbol{\mu}}_k^{(1)}) \wedge \boldsymbol{B}_{\mathrm{e}} \right\} \right] \rho(1, \ldots, N)\mathrm{d}1 \ldots \mathrm{d}N, \tag{98}$$

[1] The vanishing of the wave function outside the system leads to the vanishing of the Wigner function there, owing to the convexity of the ellipsoid (see the definition (5)).

with $\boldsymbol{v}_k \equiv \partial_{tP} \boldsymbol{R}_k$. From the stationary character of the canonical ensemble together with formula (13) it follows that

$$\int \sum_k \{\boldsymbol{v}_k \bar{\boldsymbol{\mu}}_k^{(1)} \wedge \boldsymbol{B}_e + \boldsymbol{R}_k(\partial_{tP} \bar{\boldsymbol{\mu}}_k^{(1)}) \wedge \boldsymbol{B}_e\} \rho(1, \ldots, N) \mathrm{d}1 \ldots \mathrm{d}N = 0, \tag{99}$$

so that the expression (98) now becomes:

$$\mathbf{A} = -\int \sum_k \left\{ m_k \boldsymbol{v}_k \left( \boldsymbol{v}_k - c^{-1} \frac{\bar{\boldsymbol{\mu}}_k^{(1)}}{m_k} \wedge \boldsymbol{B}_e \right) - \boldsymbol{R}_k \nabla_k \sum_{l(\neq k)} \sum_{i,j=1}^{f} \frac{e_{ki} e_{lj}}{8\pi |\boldsymbol{R}_{ki} - \boldsymbol{R}_{lj}|} \right\} \rho(1, \ldots, N) \mathrm{d}1 \ldots \mathrm{d}N. \tag{100}$$

This expression is again of the same form as that of classical theory (with the classical distribution function replaced by the Wigner function and the fluxion dot by the operator $\partial_{tP}$). Therefore we obtain, as in classical theory, for a system with short range Wigner correlations:

$$\mathbf{A} = -V(\mathbf{P} + \tfrac{1}{2}\mathbf{K} : \boldsymbol{P}\boldsymbol{P}), \tag{101}$$

with $\mathbf{P}$ the pressure tensor, $\mathbf{K}$ the tensor (II.236), $\boldsymbol{P}$ the electric polarization and $V$ the volume. Collecting the results (93), (96) and (101), and eliminating the external fields $(\boldsymbol{E}_e, \boldsymbol{B}_e)$ in favour of the Maxwell fields $(\boldsymbol{E}, \boldsymbol{B})$ (II.220), we get for the entropy change per unit of mass:

$$T\delta s = \delta u + v\mathbf{P} : \delta \boldsymbol{\epsilon} - \boldsymbol{E} \cdot \delta(v\boldsymbol{P}) + v\boldsymbol{M} \cdot \delta \boldsymbol{B}, \tag{102}$$

where $s$ is the specific entropy and $v$ the specific volume.

For fluids (isotropic in the absence of polarizations and fields) in which the polarization densities depend on the specific volume, the temperature and the Maxwell fields, one finds by the same reasoning as in the classical theory that the non-relativistic second law (or Gibbs relation) is

$$T\mathrm{d}s = \mathrm{d}u + p\mathrm{d}v - \boldsymbol{E} \cdot \mathrm{d}(v\boldsymbol{P}) + v\boldsymbol{M} \cdot \mathrm{d}\boldsymbol{B}. \tag{103}$$

The quantities occurring here are all defined for a system at rest. This formula shows that for such fluids at equilibrium the pressure tensor $\mathbf{P}$ is a multiple $p$ of the unit tensor.

From the combination of the first and second law (for local equilibrium) one may derive the entropy balance equation, just as in classical theory.

The second laws for amorphous or polycrystalline substances, for neutral plasmas and for systems with long range Wigner correlation functions have the same form as the corresponding classical laws, as may be shown by a reasoning analogous to that given above for a fluid with short range Wigner correlations.

# The Wigner function in statistical mechanics

## a. *Definition*

In statistical mechanics one does not consider pure states of $N$-particle systems, but an ensemble of pure states, i.e. a mixed state, described by a density operator[1]

$$P(t) = \sum_{\gamma} w_{\gamma} |\psi_{\gamma}(t)\rangle\langle\psi_{\gamma}(t)|, \tag{A1}$$

where the $|\psi_{\gamma}(t)\rangle$ form a complete orthonormal set and the $w_{\gamma}$ are statistical weights normalized to unity

$$\sum_{\gamma} w_{\gamma} = 1. \tag{A2}$$

Then the Wigner function, which is the Weyl transform of the density operator (times $h^3$) becomes a weighted sum of Wigner functions of pure states

$$\rho(\boldsymbol{p}, \boldsymbol{q}; t) = \sum_{\gamma} w_{\gamma} \rho_{\gamma}(\boldsymbol{p}, \boldsymbol{q}; t), \tag{A3}$$

where the partial Wigner function $\rho_{\gamma}$ is, according to (VI.A65)

$$\rho_{\gamma}(\boldsymbol{p}, \boldsymbol{q}; t) = h^{-3} \int \mathrm{d}\boldsymbol{v}\, \mathrm{e}^{(i/\hbar)\boldsymbol{p}\cdot\boldsymbol{v}} \psi_{\gamma}(\boldsymbol{q}-\tfrac{1}{2}\boldsymbol{v}; t)\psi_{\gamma}^{*}(\boldsymbol{q}+\tfrac{1}{2}\boldsymbol{v}; t). \tag{A4}$$

Alternatively one may use here (VI.A66–68).

The average value

$$\bar{A}(t) = \mathrm{Tr}\,\{P(t)A\} \tag{A5}$$

of a quantity, which is given by an operator $A$, may be written with the help of the Wigner function (A3) as:

$$\bar{A}(t) = \bar{a}(t) \equiv \int \mathrm{d}\boldsymbol{p}\,\mathrm{d}\boldsymbol{q}\, \rho(\boldsymbol{p}, \boldsymbol{q}; t) a(\boldsymbol{p}, \boldsymbol{q}), \tag{A6}$$

just as in (VI.A62–63), with $a(\boldsymbol{p}, \boldsymbol{q})$ the Weyl transform of the operator $A$. (For an $N$-particle system the integration is $6N$-fold.) From (A2) and the

[1] In this appendix we use capitals for operators and lower case symbols for $c$-numbers.

normalization (VI.A69) of the partial Wigner functions $\rho_\gamma(\boldsymbol{p}, \boldsymbol{q}; t)$ it follows that the total Wigner function is normalized

$$\int \mathrm{d}\boldsymbol{p}\,\mathrm{d}\boldsymbol{q}\,\rho(\boldsymbol{p}, \boldsymbol{q}; t) = 1. \tag{A7}$$

b. *Properties*

The Wigner function can not be interpreted as a probability density, since it is not necessarily positive definite, although it is real and normalized and permits to calculate average values according to (A6). The integrals over the coordinates or momenta however may be interpreted as probability densities, since they are positive definite:

$$\int \mathrm{d}\boldsymbol{p}\,\rho(\boldsymbol{p}, \boldsymbol{q}; t) = \sum_\gamma w_\gamma |\psi_\gamma(\boldsymbol{q}; t)|^2 \geqslant 0, \tag{A8}$$

$$\int \mathrm{d}\boldsymbol{q}\,\rho(\boldsymbol{p}, \boldsymbol{q}; t) = \sum_\gamma w_\gamma |\varphi_\gamma(\boldsymbol{p}; t)|^2 \geqslant 0, \tag{A9}$$

as follows from (A3) and (A4) and the analogous formula with the wave function in the momentum representation.

The partial Wigner functions $\rho_\gamma(\boldsymbol{p}, \boldsymbol{q}; t)$ for pure states fulfil an inequality of the form (VI.A72). Since the total Wigner function is the weighted sum of pure state Wigner functions, it fulfils the same inequality

$$|\rho(\boldsymbol{p}, \boldsymbol{q}; t)| \leqslant (2/h)^3. \tag{A10}$$

For mixed states the density operator is not idempotent, since now

$$P^2(t) = \sum_\gamma w_\gamma^2 |\psi_\gamma(t)\rangle\langle\psi_\gamma(t)|, \tag{A11}$$

as follows from (A1) and the orthonormality of $|\psi_\gamma(t)\rangle$. Clearly this is only equal to $P(t)$ (A1) if $w_\gamma$ is 0 or 1 and hence according to (A2) if only one single $w_\gamma$ is 1 and the rest zero (a pure state). It follows from (A11) and (A1–2) with the orthogonality of the state vectors $|\psi_\gamma(t)\rangle$ and the inequality $\sum_\gamma w_\gamma^2 \leqslant \sum_\gamma w_\gamma$ that

$$\mathrm{Tr}\,\{P(t)^2\} \leqslant 1. \tag{A12}$$

Here the equality sign refers to the pure state. From (A12), (VI.A53) and the fact that the Wigner function is the Weyl transform of the density operator one finds that

$$\int \mathrm{d}\boldsymbol{p}\,\mathrm{d}\boldsymbol{q}\{\rho(\boldsymbol{p}, \boldsymbol{q}; t)\}^2 \leqslant h^{-3}. \tag{A13}$$

Again the equality sign holds for the pure state (cf. (VI.A75)).

The development in time of the Wigner function for a mixed state is governed by the same laws as those for a pure state (v. (VI.A89).

c. *Reduced Wigner functions*

For the description of systems containing $N$ identical particles it is useful to introduce reduced Wigner functions. These functions are obtained from the Wigner function $\rho(1, 2, \ldots, N; t)$, which depends on all momentum and coordinate variables of the particles, by integrating over the momentum and coordinate variables of a number of particles. They are convenient if one considers the averages of physical quantities which are sum functions. For instance if a physical quantity has the operator form

$$A = \sum_{i=1}^{N} A_i, \tag{A14}$$

where $A_i$ depends on the coordinate and momentum operators of particle $i$, its Weyl transform is

$$A \rightleftarrows a(1, \ldots, N) = \sum_{i=1}^{N} a_i(i) \tag{A15}$$

and its average is according to (A6)

$$\bar{A}(t) = N \int a_1(1)\rho(1, 2, \ldots, N; t)\mathrm{d}1 \ldots \mathrm{d}N. \tag{A16}$$

This may be written as

$$\bar{A}(t) = \int a_1(1) f_1(1; t)\mathrm{d}1 \tag{A17}$$

with the one-point reduced Wigner function defined as

$$f_1(1; t) = N \int \rho(1, 2, \ldots, N; t)\mathrm{d}2 \ldots \mathrm{d}N, \tag{A18}$$

normalized to $N$.

Likewise if a physical quantity is a two-point function, i.e. if

$$A = \sum_{i,j=1(i \neq j)}^{N} A_{ij}, \tag{A19}$$

so that its Weyl transform is of the form:

$$a(1, \ldots, N) = \sum_{i,j=1(i \neq j)}^{N} a_{ij}(i, j), \tag{A20}$$

its average value may be written as

$$\bar{A}(t) = \int a_{12}(1, 2) f_2(1, 2; t) \mathrm{d}1\, \mathrm{d}2 \tag{A21}$$

with the two-point reduced Wigner function

$$f_2(1, 2; t) = N(N-1) \int \rho(1, 2, \ldots, N; t) \mathrm{d}3 \ldots \mathrm{d}N, \tag{A22}$$

normalized to $N(N-1)$.

The two-point correlation function is defined as

$$c_2(1, 2; t) = f_2(1, 2; t) - f_1(1; t) f_1(2; t), \tag{A23}$$

normalized to $-N$.

### d. *The reduced Wigner function for a perfect gas*[1]

The Wigner function for a mixed state is a weighted sum (A3) of Wigner functions (A4) for pure states. Often the density operator and hence also the Wigner function is a superposition of energy states, i.e. the index $\gamma$ in (A1–4) labels the energy states. The wave functions corresponding to the eigenstates of the energy for an $N$-particle system have simple forms in the case of a perfect gas:

$$\psi(\boldsymbol{q}_1, \ldots, \boldsymbol{q}_N; t) = \sqrt{\frac{n_1!\, n_2!\, n_3! \ldots n_m!}{N!}}$$

$$\sum_{\mathrm{P}} (\pm 1)^{\mathrm{P}} \mathrm{P} \{ u_1(\boldsymbol{q}_1) u_1(\boldsymbol{q}_2) \ldots u_1(\boldsymbol{q}_{n_1}) u_2(\boldsymbol{q}_{n_1+1}) \ldots u_m(\boldsymbol{q}_N) \} \mathrm{e}^{-(i/\hbar)Et}, \tag{A24}$$

where the upper and lower sign refer to boson and fermion systems respectively. The orthonormal functions $u_k(\boldsymbol{q}_i)$ are one-particle eigenfunctions of the one-particle Hamiltonian for particle $i$ with energy value $E_k$. In the product the eigenfunction $u_k$ occurs $n_k$ times (with $n_k$ the occupation number of energy level $k$; $\sum_{k=1}^{m} n_k = N$). In the fermion case $n_k$ can only assume the values 0 or 1. The sum is extended over those permutations of the arguments $\boldsymbol{q}_i$ of the functions $u_k$ which yield different terms. The number of these permutations is therefore $N!/n_1!n_2! \ldots n_m!$. The factor $(-1)^{\mathrm{P}}$ is plus or minus 1 for an even or odd permutation respectively. As a consequence of the orthonormality of the one-particle eigenfunctions the normalization of $\psi$ is

[1] Cf. J. E. Moyal, Proc. Cambr. Phil. Soc. **45**(1949)99.

guaranteed. The wave function $\psi$ is an eigenfunction of the total Hamiltonian with eigenvalue $E = \sum_{k=1}^{m} n_k E_k$.

Let us now consider the Wigner function (A4) of the state (A24). We shall study only the reduced Wigner functions of the type (A18) and (A22). The one-point reduced Wigner function gets the form:

$$f_1(\boldsymbol{p}_1, \boldsymbol{q}_1) = h^{-3N} N \frac{n_1! n_2! \ldots n_m!}{N!} \sum_{\mathrm{P}} \sum_{\mathrm{P}'} (\pm 1)^{\mathrm{P}} (\pm 1)^{\mathrm{P}'}$$

$$\int \mathrm{d}\boldsymbol{v}_1 \ldots \mathrm{d}\boldsymbol{v}_N \,\mathrm{d}\boldsymbol{p}_2 \ldots \mathrm{d}\boldsymbol{p}_N \,\mathrm{d}\boldsymbol{q}_2 \ldots \mathrm{d}\boldsymbol{q}_N \exp\left(\frac{i}{\hbar} \sum_{i=1}^{N} \boldsymbol{p}_i \cdot \boldsymbol{v}_i\right)$$

$$\mathrm{P}\{u_1(\boldsymbol{q}_1 - \tfrac{1}{2}\boldsymbol{v}_1) u_1(\boldsymbol{q}_2 - \tfrac{1}{2}\boldsymbol{v}_2) \ldots u_m(\boldsymbol{q}_N - \tfrac{1}{2}\boldsymbol{v}_N)\}$$

$$\mathrm{P}'\{u_1^*(\boldsymbol{q}_1 + \tfrac{1}{2}\boldsymbol{v}_1) u_1^*(\boldsymbol{q}_2 + \tfrac{1}{2}\boldsymbol{v}_2) \ldots u_m^*(\boldsymbol{q}_N + \tfrac{1}{2}\boldsymbol{v}_N)\}. \tag{A25}$$

In order to evaluate this expression it is convenient to introduce ancillary functions defined as

$$f_{kl}(\boldsymbol{p}, \boldsymbol{q}) = h^{-3} \int \mathrm{d}\boldsymbol{v}\, \mathrm{e}^{(i/\hbar)\boldsymbol{p}\cdot\boldsymbol{v}} u_k(\boldsymbol{q} - \tfrac{1}{2}\boldsymbol{v}) u_l^*(\boldsymbol{q} + \tfrac{1}{2}\boldsymbol{v}). \tag{A26}$$

As a consequence of the orthonormality of the $u_k$, this function has the property

$$\int \mathrm{d}\boldsymbol{p}\, \mathrm{d}\boldsymbol{q}\, f_{kl}(\boldsymbol{p}, \boldsymbol{q}) = \delta_{kl}. \tag{A27}$$

With (A26) and (A27) one finds from (A25)

$$f_1(1) = \sum_{k=1}^{m} n_k f_{kk}(1), \tag{A28}$$

where 1 stands for $\boldsymbol{p}_1$ and $\boldsymbol{q}_1$.

In an analogous way one may derive the two-point reduced Wigner function (A22) that corresponds to the wave function (A24):

$$f_2(1, 2) = \sum_{k,l=1 (k \neq l)}^{m} n_k n_l f_{kk}(1) f_{ll}(2) + \sum_{k=1}^{m} n_k (n_k - 1) f_{kk}(1) f_{kk}(2)$$
$$\pm \sum_{k,l=1 (k \neq l)}^{m} n_k n_l f_{kl}(1) f_{lk}(2). \tag{A29}$$

If the system is described by a mixture of energy states of the type (A24), one finds from (A3) that the one-point and two-point reduced Wigner functions are:

$$f_1(1) = \sum_{k} \bar{n}_k f_{kk}(1), \tag{A30}$$

$$f_2(1,2) = \sum_{k,l(k\neq l)} \overline{n_k n_l}\, f_{kk}(1) f_{ll}(2) + \sum_k \overline{n_k(n_k-1)}\, f_{kk}(1) f_{kk}(2) \pm \sum_{k,l(k\neq l)} \overline{n_k n_l}\, f_{kl}(1) f_{lk}(2), \tag{A31}$$

where the bars denote weighted averages over the pure states $\gamma$ with weights $w_\gamma$.

If one uses as the mixed state the grand-canonical ensemble[1] of energy states with temperature $T$ one finds for the average occupation numbers

$$\bar{n}_k = \frac{1}{\xi e^{E_k/kT} \mp 1}, \tag{A32}$$

where the constant $\xi$ follows from $\sum_k \bar{n}_k = \bar{N}$. Furthermore one finds then for the other averages of occupation numbers which occur in (A31)[2]

$$\overline{n_k n_l} = \bar{n}_k \bar{n}_l, \qquad (k \neq l), \tag{A33}$$

$$\overline{n_k(n_k-1)} = (1\pm 1)(\bar{n}_k)^2. \tag{A34}$$

With the use of (A33) and (A34) the expression (A31) may be written as

$$f_2(1,2) = \sum_{k,l} \bar{n}_k \bar{n}_l\, f_{kk}(1) f_{ll}(2) \pm \sum_{k,l} \bar{n}_k \bar{n}_l\, f_{kl}(1) f_{lk}(2), \tag{A35}$$

where in the double sums the case $k = l$ is included. In this way both the one-point Wigner function (A30) and the two-point function (A35) have been expressed in terms of mean occupation numbers (A32) and the ancillary functions (A26).

By substituting (A30) into (A35) we obtain

$$f_2(1,2) = f_1(1) f_1(2) \pm \sum_{k,l} \bar{n}_k \bar{n}_l\, f_{kl}(1) f_{lk}(2). \tag{A36}$$

For the correlation function $c_2(1, 2)$, defined in (A23), we have thus found

$$c_2(1,2) = \pm \sum_{k,l} \bar{n}_k \bar{n}_l\, f_{kl}(1) f_{lk}(2). \tag{A37}$$

In order to obtain explicit expressions we must find values for the ancillary

[1] We prefer to use here the grand-canonical ensemble rather than the canonical ensemble in view of the fact that then the averages, occurring in (A30) and (A31), may be found in a simple way. Since in this ensemble the total number of particles is not fixed, one finds for the normalization of the reduced Wigner functions $f_1(1), f_2(1,2)$ and $c_2(1,2)$ the values $\bar{N}$, $\overline{N(N-1)}$ and $\overline{N(N-1)}-(\bar{N})^2$ instead of $N$, $N(N-1)$ and $-N$ respectively. This follows directly from (A30) and (A31) with (A27).

[2] Excluding the ground level $k = 0$ for the Bose–Einstein case.

functions $f_{kl}(1)$ (A26). We choose for the functions $u_k(\boldsymbol{q})$ the plane waves

$$u_{\boldsymbol{n}}(\boldsymbol{q}) = V^{-\frac{1}{2}} e^{(2\pi i/a)\boldsymbol{n}\cdot\boldsymbol{q}}, \tag{A38}$$

where the components $n_x$, $n_y$ and $n_z$ of the vector $\boldsymbol{n}$ may assume the values $0, \pm 1, \pm 2, \ldots$. These plane waves are eigenfunctions of the free particle Hamiltonian with eigenvalues

$$E_{\boldsymbol{n}} = \frac{h^2 \boldsymbol{n}^2}{2ma^2}. \tag{A39}$$

The plane waves (A38) are chosen such that periodic boundary conditions involving a cube with edge $a$ and volume $V \equiv a^3$ are satisfied. Then according to (A26) we get

$$f_{\boldsymbol{n}\boldsymbol{n}'} = V^{-1} e^{(2\pi i/a)(\boldsymbol{n}-\boldsymbol{n}')\cdot\boldsymbol{q}} \delta\left\{\boldsymbol{p} - \frac{h(\boldsymbol{n}+\boldsymbol{n}')}{2a}\right\}. \tag{A40}$$

This is to be inserted into (A30) with (A32). If the summation over the vector $\boldsymbol{n}$ is replaced by an integral by introducing the integration variable $\boldsymbol{p}' = h\boldsymbol{n}/a$, one obtains

$$f_1(\boldsymbol{p}, \boldsymbol{q}) = h^{-3} \int \mathrm{d}\boldsymbol{p}' \, \bar{n}(\boldsymbol{p}') \delta(\boldsymbol{p} - \boldsymbol{p}') = h^{-3} \bar{n}(\boldsymbol{p}). \tag{A41}$$

Furthermore one finds for $f_2(1, 2)$ given by (A36) with (A32):

$$\begin{aligned} f_2(\boldsymbol{p}_1, \boldsymbol{q}_1, \boldsymbol{p}_2, \boldsymbol{q}_2) &= h^{-6} \bar{n}(\boldsymbol{p}_1) \bar{n}(\boldsymbol{p}_2) \\ &\pm h^{-6} \int \mathrm{d}\boldsymbol{p}' \, \mathrm{d}\boldsymbol{p}'' \, \bar{n}(\boldsymbol{p}') \bar{n}(\boldsymbol{p}'') e^{(i/\hbar)(\boldsymbol{p}'-\boldsymbol{p}'')\cdot(\boldsymbol{q}_1-\boldsymbol{q}_2)} \\ &\qquad \delta\left(\boldsymbol{p}_1 - \frac{\boldsymbol{p}'+\boldsymbol{p}''}{2}\right) \delta\left(\boldsymbol{p}_2 - \frac{\boldsymbol{p}'+\boldsymbol{p}''}{2}\right) \\ &= h^{-6} \bar{n}(\boldsymbol{p}_1) \bar{n}(\boldsymbol{p}_2) \pm h^{-6} \delta(\boldsymbol{p}_1 - \boldsymbol{p}_2) \int \mathrm{d}\boldsymbol{p}' \, \bar{n}(\boldsymbol{p}_1 + \tfrac{1}{2}\boldsymbol{p}') \bar{n}(\boldsymbol{p}_1 - \tfrac{1}{2}\boldsymbol{p}') e^{(i/\hbar)\boldsymbol{p}'\cdot(\boldsymbol{q}_1-\boldsymbol{q}_2)}, \end{aligned} \tag{A42}$$

so that the correlation function (A37) becomes

$$\begin{aligned} c_2(\boldsymbol{p}_1, \boldsymbol{q}_1, \boldsymbol{p}_2, \boldsymbol{q}_2) \\ = \pm h^{-6} \delta(\boldsymbol{p}_1 - \boldsymbol{p}_2) \int \mathrm{d}\boldsymbol{p}' \, \bar{n}(\boldsymbol{p}_1 + \tfrac{1}{2}\boldsymbol{p}') \bar{n}(\boldsymbol{p}_1 - \tfrac{1}{2}\boldsymbol{p}') e^{(i/\hbar)\boldsymbol{p}'\cdot(\boldsymbol{q}_1-\boldsymbol{q}_2)}. \end{aligned} \tag{A43}$$

The mean occupation numbers occurring in this expression are almost constants if the integration variable changes by an amount which is smaller than $(2\pi mkT)^{\frac{1}{2}}$. Furthermore the exponential oscillates rapidly in this interval if $|\boldsymbol{q}_1 - \boldsymbol{q}_2|$ is large compared to the thermal wave length $h(2\pi mkT)^{-\frac{1}{2}}$. Hence one may conclude that the range of the Wigner correlation function (A43) as a function of $|\boldsymbol{q}_1 - \boldsymbol{q}_2|$ is of the order of the thermal wave length.

# The Hamilton operator for a system of composite particles in an external field

The non-relativistic Hamilton operator for a set of charged particles $ki$ grouped into stable entities $k$ is given by the expression (VI.1) (compare also the classical Hamiltonian (II.A26)):

$$H_{\text{op}}(\boldsymbol{P}_{ki,\text{op}}, \boldsymbol{R}_{ki,\text{op}}, t) = \sum_{k,i} \frac{\boldsymbol{P}^2_{ki,\text{op}}}{2m_{ki}}$$

$$+ \sum_{k} \sum_{i,j(i \neq j)} \frac{e_{ki}\, e_{kj}}{8\pi|\boldsymbol{R}_{ki,\text{op}} - \boldsymbol{R}_{kj,\text{op}}|} + \sum_{k,l(k \neq l)} \sum_{i,j} \frac{e_{ki}\, e_{lj}}{8\pi|\boldsymbol{R}_{ki,\text{op}} - \boldsymbol{R}_{lj,\text{op}}|}$$

$$+ \sum_{k,i} e_{ki} \left[ \varphi_{\text{e}}(\boldsymbol{R}_{ki,\text{op}}, t) - \tfrac{1}{2} c^{-1} \left\{ \frac{\boldsymbol{P}_{ki,\text{op}}}{m_{ki}} \cdot, \boldsymbol{A}_{\text{e}}(\boldsymbol{R}_{ki,\text{op}}, t) \right\} \right], \tag{A44}$$

with $m_{ki}$ the mass of particle $ki$, $e_{ki}$ its charge, $\boldsymbol{R}_{ki,\text{op}}$ its coordinate operator, $\boldsymbol{P}_{ki,\text{op}}$ its momentum operator. Furthermore $\varphi_{\text{e}}$ and $\boldsymbol{A}_{\text{e}}$ are scalar and vector potentials of the external electromagnetic field. The last term contains a scalar product denoted by a dot, and an anticommutator, denoted by curly brackets and a comma. In analogy with the classical treatment one may introduce new coordinate and momentum operators for the particles belonging to group $k$, such that the new coordinate operators are the centre of mass operator and the relative coordinate operators of the constituent particles $ki$ with respect to this centre of mass operator:

$$\begin{aligned}
\boldsymbol{q}_{ki,\text{op}} &= \boldsymbol{R}_{ki,\text{op}} - \boldsymbol{R}_{k,\text{op}} \equiv \boldsymbol{R}_{ki,\text{op}} - \sum_{j=1}^{f} (m_{kj}/m_k) \boldsymbol{R}_{kj,\text{op}}, \quad (i = 1, \ldots, f-1), \\
\boldsymbol{q}_{kf,\text{op}} &= \boldsymbol{R}_{k,\text{op}} \equiv \sum_{j=1}^{f} (m_{kj}/m_k) \boldsymbol{R}_{kj,\text{op}}, \\
\boldsymbol{p}_{ki,\text{op}} &= \boldsymbol{P}_{ki,\text{op}} - (m_{kj}/m_{kf}) \boldsymbol{P}_{kf,\text{op}}, \quad (i = 1, \ldots, f-1), \\
\boldsymbol{p}_{kf,\text{op}} &= \boldsymbol{P}_{k,\text{op}} \equiv \sum_{i=1}^{f} \boldsymbol{P}_{ki,\text{op}}.
\end{aligned} \tag{A45}$$

If one inverts these relations one gets

$$\begin{aligned} \boldsymbol{R}_{ki,\mathrm{op}} &= \boldsymbol{R}_{k,\mathrm{op}} + (1-\delta_{if})\boldsymbol{q}_{ki,\mathrm{op}} - \delta_{if} \sum_{j=1}^{f-1} (m_{kj}/m_{kf})\boldsymbol{q}_{kj,\mathrm{op}}, \\ \boldsymbol{P}_{ki,\mathrm{op}} &= (m_{ki}/m_k)\boldsymbol{P}_{k,\mathrm{op}} + (1-\delta_{if})\boldsymbol{p}_{ki,\mathrm{op}} - (m_{ki}/m_k) \sum_{j=1}^{f-1} \boldsymbol{p}_{kj,\mathrm{op}}. \end{aligned} \tag{A46}$$

The transformed Hamilton operator is obtained by insertion of (A46) into (A44). Then one obtains for the Weyl transform of the new Hamilton operator (cf. the classical expression (II.A29)):

$$\begin{aligned} H_{\mathrm{op}}(\boldsymbol{p}_{\mathrm{op}}, \boldsymbol{q}_{\mathrm{op}}, t) &\rightleftarrows \sum_k \left( \frac{\boldsymbol{P}_k^2}{2m_k} + \sum_{i=1}^{f-1} \frac{\boldsymbol{p}_{ki}^2}{2m_{ki}} - \sum_{i,j=1}^{f-1} \frac{\boldsymbol{p}_{ki} \cdot \boldsymbol{p}_{kj}}{2m_k} \right) \\ &+ \sum_k \sum_{i,j=1(i\neq j)}^{f} \frac{e_{ki}\, e_{kj}}{8\pi |\boldsymbol{R}_{ki}(\boldsymbol{q}) - \boldsymbol{R}_{kj}(\boldsymbol{q})|} + \sum_{k,l(k\neq l)} \sum_{i,j=1}^{f} \frac{e_{ki}\, e_{lj}}{8\pi |\boldsymbol{R}_{ki}(\boldsymbol{q}) - \boldsymbol{R}_{lj}(\boldsymbol{q})|} \\ &+ \sum_k e_k \left\{ \varphi_{\mathrm{e}}(\boldsymbol{R}_k, t) - c^{-1} \frac{\boldsymbol{P}_k}{m_k} \cdot \boldsymbol{A}_{\mathrm{e}}(\boldsymbol{R}_k, t) \right\} \\ &+ \sum_k \sum_{i=1}^{f} e_{ki} \left[ \{\boldsymbol{R}_{ki}(\boldsymbol{q}) - \boldsymbol{R}_k\} \cdot \nabla_k \left\{ \varphi_{\mathrm{e}}(\boldsymbol{R}_k, t) - c^{-1} \frac{\boldsymbol{P}_k}{m_k} \cdot \boldsymbol{A}_{\mathrm{e}}(\boldsymbol{R}_k, t) \right\} \right. \\ &- c^{-1} \left\{ \frac{\boldsymbol{P}_{ki}(\boldsymbol{p})}{m_{ki}} - \frac{\boldsymbol{P}_k}{m_k} \right\} \cdot \boldsymbol{A}_{\mathrm{e}}(\boldsymbol{R}_k, t) \\ &\left. - c^{-1} \{\boldsymbol{R}_{ki}(\boldsymbol{q}) - \boldsymbol{R}_k\} \cdot \nabla_k \boldsymbol{A}_{\mathrm{e}}(\boldsymbol{R}_k, t) \cdot \left\{ \frac{\boldsymbol{P}_{ki}(\boldsymbol{p})}{m_{ki}} - \frac{\boldsymbol{P}_k}{m_k} \right\} \right], \end{aligned} \tag{A47}$$

up to terms with derivatives of the potentials. At the right-hand side the symbols $\boldsymbol{R}_{ki}(\boldsymbol{q})$ and $\boldsymbol{P}_{ki}(\boldsymbol{p})$ stand for the right-hand sides of (A46) but without the index op. Since the Weyl correspondence is invariant under a linear transformation of coordinates and momenta (v. problem 2 of chapter VI), the correspondence sign may be understood either as a Weyl correspondence with respect to the old coordinates and momenta $\boldsymbol{R}_{ki}$, $\boldsymbol{P}_{ki}$ or with respect to the new coordinates (A45). A second, non-linear transformation of coordinates and momenta as employed in the classical treatment of appendix II of chapter II, will not be performed here, because the Weyl correspondence is not invariant under that non-linear transformation.

Let us consider the Weyl transform $K$ defined as

$$K \equiv \sum_k \sum_{i=1}^{f} \tfrac{1}{2} m_{ki} (\partial_{t\mathrm{P}} \boldsymbol{R}_{ki})^2, \tag{A48}$$

where the symbol $\partial_{t\mathrm{P}}$ stands for the Poisson bracket (cf. (VI.72))

$$\partial_{t\mathrm{P}} a = \{a, H\}_{\mathrm{P}} \tag{A49}$$

for $a$ independent of $t$. The function $H$ is the Weyl transform of the Hamilton operator (A44) or alternatively the right-hand side of (A47). According to (A45) we may write (A48) as

$$K = \sum_k \tfrac{1}{2} m_k (\partial_{t\mathrm{P}} \boldsymbol{R}_k)^2 + \sum_k \left\{ \sum_{i=1}^{f-1} \tfrac{1}{2} m_{ki} (\partial_{t\mathrm{P}} \boldsymbol{q}_{ki})^2 + \sum_{i,j=1}^{f-1} \frac{1}{2} \frac{m_{ki} m_{kj}}{m_{kf}} (\partial_{t\mathrm{P}} \boldsymbol{q}_i) \cdot (\partial_{t\mathrm{P}} \boldsymbol{q}_{kj}) \right\}. \tag{A50}$$

The Poisson brackets may be evaluated in terms of either the old coordinates and momenta $(\boldsymbol{R}_{ki}, \boldsymbol{P}_{ki})$ or the new variables $(\boldsymbol{q}_{ki}, \boldsymbol{p}_{ki})$. If one chooses the latter set of variables and for the Weyl transform of the Hamilton operator the right-hand side of (A47) one finds for (A50), with $e_k = \sum_i e_{ki}$:

$$\begin{aligned} K = \sum_k \Bigg[ & \frac{\boldsymbol{P}_k^2}{2m_k} - \frac{c^{-1} e_k}{m_k} \boldsymbol{P}_k \cdot \boldsymbol{A}_{\mathrm{e}}(\boldsymbol{R}_k, t) \\ & - c^{-1} \sum_{i=1}^{f} e_{ki} \{\boldsymbol{R}_{ki}(\boldsymbol{q}) - \boldsymbol{R}_k\} \cdot \nabla_k \frac{\boldsymbol{P}_k}{m_k} \cdot \boldsymbol{A}_{\mathrm{e}}(\boldsymbol{R}_k, t) \Bigg] \\ & + \sum_k \Bigg[ \sum_{i=1}^{f-1} \frac{\boldsymbol{p}_{ki}^2}{2m_{ki}} - \sum_{i,j=1}^{f-1} \frac{\boldsymbol{p}_{ki} \cdot \boldsymbol{p}_{kj}}{2m_k} - c^{-1} \sum_{i=1}^{f} e_{ki} \left\{ \frac{\boldsymbol{P}_{ki}(\boldsymbol{p})}{m_{ki}} - \frac{\boldsymbol{P}_k}{m_k} \right\} \cdot \boldsymbol{A}_{\mathrm{e}}(\boldsymbol{R}_k, t) \\ & - c^{-1} \sum_{i=1}^{f} e_{ki} \{\boldsymbol{R}_{ki}(\boldsymbol{q}) - \boldsymbol{R}_k\} \cdot \nabla_k \boldsymbol{A}_{\mathrm{e}}(\boldsymbol{R}_k, t) \cdot \left\{ \frac{\boldsymbol{P}_{ki}(\boldsymbol{p})}{m_{ki}} - \frac{\boldsymbol{P}_k}{m_k} \right\} \Bigg]. \end{aligned} \tag{A51}$$

If this is used in the right-hand side of (A47), one finds

$$\begin{aligned} H_{\mathrm{op}}(\boldsymbol{p}_{\mathrm{op}}, \boldsymbol{q}_{\mathrm{op}}, t) \rightleftarrows K & + \sum_k \sum_{i,j=1 (i \neq j)}^{f} \frac{e_{ki} e_{kj}}{8\pi |\boldsymbol{R}_{ki}(\boldsymbol{q}) - \boldsymbol{R}_{kj}(\boldsymbol{q})|} \\ & + \sum_{k,l (k \neq l)} \sum_{i,j=1}^{f} \frac{e_{ki} e_{lj}}{8\pi |\boldsymbol{R}_{ki}(\boldsymbol{q}) - \boldsymbol{R}_{lj}(\boldsymbol{q})|} \\ & + \sum_k e_k \varphi_{\mathrm{e}}(\boldsymbol{R}_k, t) + \sum_k \sum_{i=1}^{f} e_{ki} \{\boldsymbol{R}_{ki}(\boldsymbol{q}) - \boldsymbol{R}_k\} \cdot \nabla_k \varphi_{\mathrm{e}}(\boldsymbol{R}_k, t). \end{aligned} \tag{A52}$$

If the external electromagnetic fields $\boldsymbol{E}_{\mathrm{e}}$ and $\boldsymbol{B}_{\mathrm{e}}$ are uniform and time-independent, one may choose as potentials $\varphi_{\mathrm{e}}(\boldsymbol{R}, t)$ and $\boldsymbol{A}_{\mathrm{e}}(\boldsymbol{R}, t)$:

$$\begin{aligned} \varphi_{\mathrm{e}}(\boldsymbol{R}) &= -\boldsymbol{R} \cdot \boldsymbol{E}_{\mathrm{e}}, \\ \boldsymbol{A}_{\mathrm{e}}(\boldsymbol{R}) &= \tfrac{1}{2} \boldsymbol{B}_{\mathrm{e}} \wedge \boldsymbol{R}, \end{aligned} \tag{A53}$$

which are time-independent[1]. If these expressions are substituted into the right-hand side of (A47) one finds

$$H_{\text{op}}(\boldsymbol{p}_{\text{op}}, \boldsymbol{q}_{\text{op}}, t) \rightleftarrows \sum_k \left( \frac{\boldsymbol{P}_k^2}{2m_k} + \sum_{i=1}^{f-1} \frac{\boldsymbol{p}_{ki}^2}{2m_{ki}} - \sum_{i,j=1}^{f-1} \frac{\boldsymbol{p}_{ki} \cdot \boldsymbol{p}_{kj}}{2m_k} \right)$$
$$+ \sum_k \sum_{i,j=1(i \neq j)}^{f} \frac{e_{ki} e_{kj}}{8\pi |\boldsymbol{R}_{ki}(\boldsymbol{q}) - \boldsymbol{R}_{kj}(\boldsymbol{q})|}$$
$$+ \sum_{k,l(k\neq l)} \sum_{i,j=1}^{f} \frac{e_{ki} e_{lj}}{8\pi |\boldsymbol{R}_{ki}(\boldsymbol{q}) - \boldsymbol{R}_{lj}(\boldsymbol{q})|} - \sum_k \left\{ e_k \boldsymbol{R}_k \cdot \left( \boldsymbol{E}_{\text{e}} + \tfrac{1}{2} c^{-1} \frac{\boldsymbol{P}_k}{m_k} \wedge \boldsymbol{B}_{\text{e}} \right) \right.$$
$$\left. + \bar{\boldsymbol{\mu}}_k^{(1)} \cdot \left( \boldsymbol{E}_{\text{e}} + \tfrac{1}{2} c^{-1} \frac{\boldsymbol{P}_k}{m_k} \wedge \boldsymbol{B}_{\text{e}} \right) + \tfrac{1}{2} c^{-1} \partial_{t\text{P}} \bar{\boldsymbol{\mu}}_k^{(1)} \cdot (\boldsymbol{B}_{\text{e}} \wedge \boldsymbol{R}_k) + \bar{\boldsymbol{v}}_k^{(1)} \cdot \boldsymbol{B}_{\text{e}} \right\} \tag{A54}$$

with the abbreviations (cf. (II.A33))

$$\bar{\boldsymbol{\mu}}_k^{(1)}(\boldsymbol{q}) = \sum_{i=1}^{f} e_{ki} \{ \boldsymbol{R}_{ki}(\boldsymbol{q}) - \boldsymbol{q}_{kf} \},$$
$$\bar{\boldsymbol{v}}_k^{(1)}(\boldsymbol{p}, \boldsymbol{q}) = \tfrac{1}{2} c^{-1} \sum_{i=1}^{f} e_{ki} \{ \boldsymbol{R}_{ki}(\boldsymbol{q}) - \boldsymbol{q}_{kf} \} \wedge \left\{ (1 - \delta_{if}) \frac{\boldsymbol{p}_{ki}}{m_{ki}} - \sum_{j=1}^{f-1} \frac{\boldsymbol{p}_{kj}}{m_k} \right\}. \tag{A55}$$

These quantities are the Weyl transforms of the electric and magnetic dipole moment operators. They are indeed equal to the case $n = 1$ of (VI.94), since $\boldsymbol{q}_{kf} \equiv \boldsymbol{R}_k$ and since, up to order $c^0$, one has

$$(1 - \delta_{if}) \frac{\boldsymbol{p}_{ki}}{m_{ki}} - \sum_{j=1}^{f-1} \frac{\boldsymbol{p}_{kj}}{m_k} = \partial_{t\text{P}} \{ \boldsymbol{R}_{ki}(\boldsymbol{q}) - \boldsymbol{q}_{kf} \}. \tag{A56}$$

Finally one obtains for (A52):

$$H_{\text{op}}(\boldsymbol{p}_{\text{op}}, \boldsymbol{q}_{\text{op}}, t) \rightleftarrows K + \sum_k \sum_{i,j=1(i\neq j)}^{f} \frac{e_{ki} e_{kj}}{8\pi |\boldsymbol{R}_{ki}(\boldsymbol{q}) - \boldsymbol{R}_{kj}(\boldsymbol{q})|}$$
$$+ \sum_{k,l(k\neq l)} \sum_{i,j=1}^{f} \frac{e_{ki} e_{lj}}{8\pi |\boldsymbol{R}_{ki}(\boldsymbol{q}) - \boldsymbol{R}_{lj}(\boldsymbol{q})|} - \sum_k e_k \boldsymbol{R}_k \cdot \boldsymbol{E}_{\text{e}} - \sum_k \bar{\boldsymbol{\mu}}_k^{(1)} \cdot \boldsymbol{E}_{\text{e}}. \tag{A57}$$

Whereas the form (A55) shows explicitly the dependence on the external magnetic field, this dependence is now hidden in the quantity $K$.

[1] Other time-independent potentials, which might be used as well in the Hamiltonian, are of the form

$$\varphi_{\text{e}}' = -\boldsymbol{R} \cdot \boldsymbol{E}_{\text{e}} + \chi$$
$$\boldsymbol{A}_{\text{e}}' = \tfrac{1}{2} \boldsymbol{B}_{\text{e}} \wedge \boldsymbol{R} + \nabla \psi(\boldsymbol{R}),$$

where $\chi$ and $\psi(\boldsymbol{R})$ may depend also on the external fields. Although the Hamiltonian becomes more complicated then, the final results, derived in the main text, remain the same.

# Deformations and free energy in quantum theory

In quantum statistical mechanics the free energy $F^*$ of a system of atoms described by a canonical ensemble with temperature $T$ is given by the expression

$$e^{-F^*/kT} = \mathrm{Tr}\,\{\exp(-H^{\mathrm{W}}_{\mathrm{op}}/kT)\}. \tag{A58}$$

Here the total Hamilton operator $H^{\mathrm{W}}_{\mathrm{op}}$ is the sum of the Hamilton operator $H_{\mathrm{op}}$ and the wall potential operator $U^{\mathrm{W}}_{\mathrm{op}}$:

$$U^{\mathrm{W}}_{\mathrm{op}} = \sum_k U^{\mathrm{W}}_{k,\mathrm{op}}(\boldsymbol{R}_{k,\mathrm{op}}), \tag{A59}$$

where $\boldsymbol{R}_{k,\mathrm{op}}$ is the centre of mass operator of atom $k$ and the sum is extended over all atoms of the system. The wall potential operator $U^{\mathrm{W}}_{k,\mathrm{op}}$ has the same eigenfunctions as $\boldsymbol{R}_{k,\mathrm{op}}$ and eigenvalues $\infty$ or 0 if the eigenvalues of $\boldsymbol{R}_{k,\mathrm{op}}$ denote positions outside or inside the boundary of the system respectively.

If the position of the boundary is deformed according to the formula:

$$\boldsymbol{R}^{\mathrm{W}'} = \{\mathbf{U}+\delta\boldsymbol{\epsilon}(\boldsymbol{R}^{\mathrm{W}})\}\cdot\boldsymbol{R}^{\mathrm{W}}, \tag{A60}$$

with $\mathbf{U}$ the unit tensor and $\delta\boldsymbol{\epsilon}(\boldsymbol{R}^{\mathrm{W}})$ the deformation tensor, the wall potential becomes:

$$U^{\mathrm{W}'}_{\mathrm{op}} = \sum_k U^{\mathrm{W}}_{k,\mathrm{op}}[\{\mathbf{U}-\delta\boldsymbol{\epsilon}(\boldsymbol{R}_{k,\mathrm{op}})\}\cdot\boldsymbol{R}_{k,\mathrm{op}}]. \tag{A61}$$

Deformation of the boundary leads to a change of the free energy:

$$\delta_\varepsilon F^* = -kT e^{F^*/kT}\delta\,\mathrm{Tr}\,\{\exp(-H^{\mathrm{W}}_{\mathrm{op}}/kT)\}, \tag{A62}$$

according to (A58). We now use the identity (81) with (82) for the derivative of an exponential operator, so as to derive

$$\delta_\varepsilon F^* = \mathrm{Tr}\,\{e^{(F^*-H_{\mathrm{op}})/kT}\delta U^{\mathrm{W}}_{\mathrm{op}}\}, \tag{A63}$$

where we used the fact that the traces of the commutators occurring here vanish and where $\delta U^{\mathrm{W}}_{\mathrm{op}}$ stands for $U^{\mathrm{W}'}_{\mathrm{op}} - U^{\mathrm{W}}_{\mathrm{op}}$. With the insertion of (A61) we obtain for (A63)

$$\delta_\varepsilon F^* = -\mathrm{Tr}\,\{e^{(F^*-H_{\mathrm{op}})/kT}\sum_k \boldsymbol{R}_{k,\mathrm{op}}\cdot\delta\tilde{\boldsymbol{\epsilon}}(\boldsymbol{R}_{k,\mathrm{op}})\cdot\boldsymbol{\nabla}_k U^{\mathrm{W}}_{k,\mathrm{op}}(\boldsymbol{R}_{k,\mathrm{op}})\}. \tag{A64}$$

Writing this average in terms of the Wigner function of the canonical ensemble we get

$$\delta_\varepsilon F^* = -\int \sum_k \boldsymbol{R}_k \cdot \delta\tilde{\boldsymbol{\epsilon}}(\boldsymbol{R}_k) \cdot \boldsymbol{\nabla}_k\, U_k^{\mathrm{W}}(\boldsymbol{R}_k) \rho(1, \ldots, N) \mathrm{d}1 \ldots \mathrm{d}N. \qquad \text{(A65)}$$

Now according to (13) and the stationarity of the canonical ensemble

$$0 = \frac{\partial}{\partial t} \int \sum_k \boldsymbol{P}_k \cdot \delta\boldsymbol{\epsilon}(\boldsymbol{R}_k) \cdot \boldsymbol{R}_k\, \rho(1, \ldots, N) \mathrm{d}1 \ldots \mathrm{d}N$$

$$= \int \sum_k \{H^{\mathrm{W}}, \boldsymbol{P}_k \cdot \delta\boldsymbol{\epsilon}(\boldsymbol{R}_k) \cdot \boldsymbol{R}_k\}_{\mathrm{P}}\, \rho(1, \ldots, N) \mathrm{d}1 \ldots \mathrm{d}N, \qquad \text{(A66)}$$

where $H^{\mathrm{W}}$ is the Weyl transform of the total Hamiltonian $H_{\mathrm{op}}^{\mathrm{W}}$ and where Poisson brackets appear in the last member. Splitting $H^{\mathrm{W}}$ into the Weyl transform $H$ of the Hamilton operator and $U^{\mathrm{W}}$ of the wall potential operator, and evaluating the Poisson brackets we get for (A66)

$$\int \sum_k \boldsymbol{R}_k \cdot \delta\tilde{\boldsymbol{\epsilon}}(\boldsymbol{R}_k) \cdot \boldsymbol{\nabla}_k\, U_k^{\mathrm{W}}(\boldsymbol{R}_k) \rho(1, \ldots, N) \mathrm{d}1 \ldots \mathrm{d}N$$

$$= \int \sum_k \left\{ \boldsymbol{P}_k \cdot \delta\mathbf{e}(\boldsymbol{R}_k) \cdot \frac{\partial H}{\partial \boldsymbol{P}_k} - \frac{\partial H}{\partial \boldsymbol{R}_k} \cdot \delta\boldsymbol{\epsilon}(\boldsymbol{R}_k) \cdot \boldsymbol{R}_k \right\} \rho(1, \ldots, N) \mathrm{d}1 \ldots \mathrm{d}N, \qquad \text{(A67)}$$

where the deformation gradient tensor $\delta\mathbf{e}$, defined in (II.A51) has been used. Substituting this result into (A65) we get finally

$$\delta_\varepsilon F^* = -\int \sum_k \left\{ \boldsymbol{P}_k \cdot \delta\mathbf{e}(\boldsymbol{R}_k) \cdot \frac{\partial H}{\partial \boldsymbol{P}_k} - \frac{\partial H}{\partial \boldsymbol{R}_k} \cdot \delta\boldsymbol{\epsilon}(\boldsymbol{R}_k) \cdot \boldsymbol{R}_k \right\} \rho(1, \ldots, N) \mathrm{d}1 \ldots \mathrm{d}N. \qquad \text{(A68)}$$

For uniform deformations $\delta\boldsymbol{\epsilon} = \delta\mathbf{e}$, so that one has for the change of the free energy

$$\delta_\varepsilon F^* = \mathbf{A} : \delta\boldsymbol{\epsilon} \qquad \text{(A69)}$$

with the tensor $\mathbf{A}$ given by

$$\mathbf{A} = -\int \sum_k \left( \frac{\partial H}{\partial \boldsymbol{P}_k} \boldsymbol{P}_k - \boldsymbol{R}_k \frac{\partial H}{\partial \boldsymbol{R}_k} \right) \rho(1, \ldots, N) \mathrm{d}1 \ldots \mathrm{d}N, \qquad \text{(A70)}$$

which proves formula (97) of the main text.

# PROBLEMS

**1.** Prove that (56) is the average of the operator given in (57). Show first, with the help of the expression (VI.61) for the Weyl transform of the Hamiltonian, that one has

$$m_{ki}\partial_{t\mathrm{P}}\boldsymbol{R}_{ki} = \boldsymbol{P}_{ki}-c^{-1}e_{ki}\boldsymbol{A}_{\mathrm{e}}(\boldsymbol{R}_{ki}, t),$$

so that one finds for $\boldsymbol{v}_k \equiv \partial_{t\mathrm{P}}\boldsymbol{R}_k$:

$$\boldsymbol{v}_k = \frac{1}{m_k}\sum_i \{\boldsymbol{P}_{ki}-c^{-1}e_{ki}\boldsymbol{A}_{\mathrm{e}}(\boldsymbol{R}_{ki}, t)\}.$$

**2.** Consider a particle in an external electromagnetic field. Its wave function satisfies the Schrödinger equation $H_{\mathrm{op}}\psi(\boldsymbol{R}, t) = -(\hbar/i)\partial\psi(\boldsymbol{R}, t)/\partial t$ with the Hamilton operator up to order $c^{-1}$

$$H_{\mathrm{op}}(\boldsymbol{P}_{\mathrm{op}}, \boldsymbol{R}_{\mathrm{op}}) = \frac{\boldsymbol{P}_{\mathrm{op}}^2}{2m} - \frac{e}{2mc}\{\boldsymbol{P}_{\mathrm{op}}\cdot, \boldsymbol{A}_{\mathrm{e}}(\boldsymbol{R}_{\mathrm{op}}, t)\}+e\varphi_{\mathrm{e}}(\boldsymbol{R}_{\mathrm{op}}, t).$$

Show that the transformation

$$\psi'(\boldsymbol{R}', t') = \exp\left\{\frac{i}{\hbar}(m\boldsymbol{V}\cdot\boldsymbol{R}+\tfrac{1}{2}mV^2t)\right\}\psi(\boldsymbol{R}, t)$$

of the wave function with respect to a Galilei transformation (66) leaves the Schrödinger equation invariant, at least up to order $c^{-1}$. Use to that end the transformation rules (II.23) (which imply $\boldsymbol{P}'_{\mathrm{op}} = \boldsymbol{P}_{\mathrm{op}}$) and the transformation properties of the potentials (cf. (II.27))

$$\boldsymbol{A}'_{\mathrm{e}}(\boldsymbol{R}', t') = \boldsymbol{A}_{\mathrm{e}}(\boldsymbol{R}, t)+c^{-1}\boldsymbol{V}\varphi_{\mathrm{e}}(\boldsymbol{R}, t),$$

$$\varphi'_{\mathrm{e}}(\boldsymbol{R}', t') = \varphi_{\mathrm{e}}(\boldsymbol{R}, t)+c^{-1}\boldsymbol{V}\cdot\boldsymbol{A}_{\mathrm{e}}(\boldsymbol{R}, t).$$

The transformed Hamiltonian is thus

$$\boldsymbol{H}'_{\mathrm{op}}(\boldsymbol{P}'_{\mathrm{op}}, \boldsymbol{R}'_{\mathrm{op}}) = \boldsymbol{H}_{\mathrm{op}}(\boldsymbol{P}_{\mathrm{op}}, \boldsymbol{R}_{\mathrm{op}})+c^{-1}e\boldsymbol{V}\cdot\boldsymbol{A}_{\mathrm{e}}(\boldsymbol{R}_{\mathrm{op}}, t).$$

Since $\boldsymbol{P}'_{\mathrm{op}} = \boldsymbol{P}_{\mathrm{op}}$ and $\boldsymbol{R}'_{\mathrm{op}} = \boldsymbol{R}_{\mathrm{op}}+\boldsymbol{V}t$ one finds (68).

Prove furthermore (69) from the transformation of the wave function.

**3.** Prove the identity (81) with (82). Prove first by induction with respect to $n$:

$$\sum_{k=0}^{n-1} \frac{1}{n!} A_{\text{op}}^{k} \frac{\partial A_{\text{op}}}{\partial \alpha} A_{\text{op}}^{n-k-1} = \sum_{k=0}^{n-1} \frac{1}{(k+1)!(n-k-1)!} A_{\text{op}}^{(k)} A_{\text{op}}^{n-k-1}.$$

# PART D

# *Covariant and semi-relativistic quantum-mechanical electrodynamics*

CHAPTER VIII

# Dirac and Klein-Gordon particles in external fields

## 1 *Introduction*

In relativistic quantum mechanics free particles are described by wave equations: the Klein–Gordon equation for particles without spin and the Dirac equation for particles with spin $\frac{1}{2}$. If the particles move in external electromagnetic fields, generated by classical sources, the interaction of these fields and the particles is described by adding appropriate terms to the wave equations. If one wants to confine oneself to single particle theories the electromagnetic fields should change relatively slowly in space and time so as to avoid effects due to particle production.

The purpose of this chapter is to find equations of motion and of spin for Klein–Gordon and Dirac particles of what one may call the Ehrenfest type, i.e. expressions for the time derivatives of the expectation values of the position and spin operators. The latter will be uniquely determined by imposing their transformation character with respect to the Poincaré group. The Hamiltonian, which governs the time behaviour of the expectation values, will be brought into a form which allows to distinguish between positive- and negative-energy solutions; it will be given up to terms with the first derivatives of the potentials (the fields).

In sections 2 and 3 the equations of motion and of spin for a particle with spin $\frac{1}{2}$ are derived[1]. They will turn out to have forms analogous to those found in chapter IV for a classical composite particle with inner angular momentum. For composite particles without inner angular momentum the equations of that chapter simplify considerably. In sections 4 and 5 it will appear that equations of that simple type may indeed be derived for a quantum particle without spin, i.e. a particle described by the equation of Klein and Gordon.

[1] L. G. Suttorp and S. R. de Groot, N. Cim. **65A**(1970)245, on which paper the discussion in these sections will be based.

## 2 *The free Dirac particle*

### a. *Invariances of the Dirac equation*

It is useful to study first the free particle case, since it contains a number of aspects which it has in common with the problem of a particle in a field.

In Dirac's theory for a single particle the states are described by four-component wave functions $\psi(\boldsymbol{R}, t)$ that depend on space and time in the coordinate representation. The time evolution of these wave functions is governed by the Dirac equation. This equation may be written as

$$H_{\text{op}}\psi(\boldsymbol{R}, t) = -\frac{\hbar}{i}\frac{\partial\psi(\boldsymbol{R}, t)}{\partial t}, \tag{1}$$

where the Hamilton operator for a free particle with mass $m$ ($\neq 0$) is given by

$$H_{\text{op}} = c\boldsymbol{\alpha}\cdot\boldsymbol{P}_{\text{op}} + \beta mc^2. \tag{2}$$

The symbols $\boldsymbol{\alpha}$ and $\beta$ stand for hermitian $4\times4$ matrices which obey the anti-commutation rules

$$\{\boldsymbol{\alpha}, \boldsymbol{\alpha}\} = 2\mathbf{U}, \qquad \{\boldsymbol{\alpha}, \beta\} = 0, \qquad \{\beta, \beta\} = 2. \tag{3}$$

(where $\mathbf{U}$ is the unit tensor and 2 stands for twice the unit $4\times4$ matrix). Furthermore $\boldsymbol{P}_{\text{op}}$ is the momentum operator, which reads

$$\boldsymbol{P}_{\text{op}} = \frac{\hbar}{i}\frac{\partial}{\partial\boldsymbol{R}} \tag{4}$$

in the coordinate representation.

Physical quantities are represented by operators acting on wave functions. The expectation value of an operator $\Omega_{\text{op}}$ (which is a function of the coordinate $\boldsymbol{R}$ and the momentum operator $(\hbar/i)\partial/\partial\boldsymbol{R}$) in a state characterized by a wave function $\psi(\boldsymbol{R}, t)$ is defined as

$$\bar{\Omega}_{\text{op}} = \int \psi^{\dagger}(\boldsymbol{R}, t)\Omega_{\text{op}}\left(\boldsymbol{R}, \frac{\hbar}{i}\frac{\partial}{\partial\boldsymbol{R}}\right)\psi(\boldsymbol{R}, t)\mathrm{d}\boldsymbol{R}. \tag{5}$$

The Dirac equation (1) with (2) is covariant under the transformations of the Poincaré group, of which we shall study in particular spatial translations, spatial rotations, spatial inversions, time reversal and pure Lorentz transformations.

In a coordinate frame which is connected to the original frame by an infinitesimal *translation*

$$\begin{aligned} \boldsymbol{R}' &= \boldsymbol{R}+\boldsymbol{\varepsilon}, \\ t' &= t, \end{aligned} \tag{6}$$

(where $\boldsymbol{\varepsilon}$ is an infinitesimal vector), the wave function transforms as

$$\psi'(\boldsymbol{R}', t') = \psi(\boldsymbol{R}, t), \tag{7}$$

since then it follows that (1) with (2) is invariant, i.e. valid for quantities with primes throughout. Moreover the inner product $\int \psi_1^\dagger \psi_2 \mathrm{d}\boldsymbol{R}$ of two wave functions is invariant under this transformation.

The expectation value in the new coordinate frame

$$\int \psi'^\dagger(\boldsymbol{R}', t')\Omega_{\mathrm{op}}\left(\boldsymbol{R}', \frac{\hbar}{i}\frac{\partial}{\partial \boldsymbol{R}'}\right)\psi'(\boldsymbol{R}', t')\mathrm{d}\boldsymbol{R}' \tag{8}$$

may be written (up to first order in $\boldsymbol{\varepsilon}$) as

$$\int \psi^\dagger(\boldsymbol{R}', t)\Omega_{\mathrm{op}}\left(\boldsymbol{R}', \frac{\hbar}{i}\frac{\partial}{\partial \boldsymbol{R}'}\right)\psi(\boldsymbol{R}', t)\mathrm{d}\boldsymbol{R}'$$

$$+\boldsymbol{\varepsilon}\cdot\int \psi^\dagger(\boldsymbol{R}', t)\left[\frac{\partial}{\partial \boldsymbol{R}'}, \Omega_{\mathrm{op}}\left(\boldsymbol{R}', \frac{\hbar}{i}\frac{\partial}{\partial \boldsymbol{R}'}\right)\right]\psi(\boldsymbol{R}', t)\mathrm{d}\boldsymbol{R}', \tag{9}$$

where a partial integration has led to a commutator. We shall write this as

$$\overline{\Omega}_{\mathrm{op}} + \overline{\delta\Omega}_{\mathrm{op}}, \tag{10}$$

where the first term is equal to (5). The second is the expectation value of the operator

$$\delta\Omega_{\mathrm{op}} = \frac{i}{\hbar}\boldsymbol{\varepsilon}\cdot[\boldsymbol{P}_{\mathrm{op}}, \Omega_{\mathrm{op}}]. \tag{11}$$

This shows that $\boldsymbol{P}_{\mathrm{op}}$ is the generator of spatial translations.

A coordinate frame which is related to the original frame by an infinitesimal *rotation* has coordinates

$$\begin{aligned} \boldsymbol{R}' &= \boldsymbol{R}+\boldsymbol{\varepsilon}\wedge\boldsymbol{R}, \\ t' &= t, \end{aligned} \tag{12}$$

with $\boldsymbol{\varepsilon}$ an infinitesimal vector. The wave function in the new frame is

$$\psi'(\boldsymbol{R}', t') = (1-\tfrac{1}{2}i\boldsymbol{\varepsilon}\cdot\boldsymbol{\sigma})\psi(\boldsymbol{R}, t), \tag{13}$$

where the matrix $\boldsymbol{\sigma}$ is defined as

$$\boldsymbol{\sigma} = -\tfrac{1}{2}i\boldsymbol{\alpha} \wedge \boldsymbol{\alpha}. \tag{14}$$

Indeed one may check that the transformations (12) and (13) leave the Dirac equation (1) with (2) and the inner product of two wave functions invariant.

The expectation value of an operator $\Omega_{\text{op}}$ in the new coordinate frame becomes, upon introduction of (12) and (13) into (8), of the form (10) with the operator

$$\delta\Omega_{\text{op}} = \frac{i}{\hbar}\,\boldsymbol{\varepsilon}\cdot[\boldsymbol{R} \wedge \boldsymbol{P}_{\text{op}} + \tfrac{1}{2}\hbar\boldsymbol{\sigma}, \Omega_{\text{op}}]. \tag{15}$$

This shows that $\boldsymbol{R} \wedge \boldsymbol{P}_{\text{op}} + \frac{1}{2}\hbar\boldsymbol{\sigma}$ is the generator of spatial rotations.

A vector operator is characterized by

$$\delta\boldsymbol{\Omega}_{\text{op}} = \boldsymbol{\varepsilon} \wedge \boldsymbol{\Omega}_{\text{op}}, \tag{16}$$

(cf. (12)). For such operators (15) becomes

$$[(\boldsymbol{R} \wedge \boldsymbol{P}_{\text{op}} + \tfrac{1}{2}\hbar\boldsymbol{\sigma})^i, \Omega^j_{\text{op}}] = i\hbar\varepsilon^{ijk}\Omega_{k,\text{op}}, \tag{17}$$

with $\varepsilon^{ijk}$ the Levi-Civita symbol.

For *spatial inversion*

$$\begin{aligned} \boldsymbol{R}' &= -\boldsymbol{R}, \\ t' &= t \end{aligned} \tag{18}$$

the wave function transforms as[1]

$$\psi'(\boldsymbol{R}', t') = \beta\psi(\boldsymbol{R}, t). \tag{19}$$

Indeed (18) and (19) leave the Dirac equation (1) with (2) and the inner product invariant.

The expectation value of an operator $\Omega_{\text{op}}$ in the new frame reads:

$$\int \psi'^{\dagger}(\boldsymbol{R}', t')\Omega_{\text{op}}\left(\boldsymbol{R}', \frac{\hbar}{i}\frac{\partial}{\partial\boldsymbol{R}'}\right)\psi'(\boldsymbol{R}', t')\mathrm{d}\boldsymbol{R}' = \int \psi^{\dagger}(\boldsymbol{R}, t)\beta\Omega_{\text{op}}\left(-\boldsymbol{R}, -\frac{\hbar}{i}\frac{\partial}{\partial\boldsymbol{R}}\right)\beta\psi(\boldsymbol{R}, t)\mathrm{d}\boldsymbol{R}. \tag{20}$$

A polar or axial vector operator is characterized by the property that the expectation value in the new frame is equal to minus or plus the expectation

[1] Phase factors will be left out since they do not affect the expectation values considered here.

value in the old frame. Hence it follows from (20) that polar or axial vector operators satisfy the relation

$$\boldsymbol{\Omega}_{\text{op}}(-\boldsymbol{R}, -\boldsymbol{P}_{\text{op}}) = \mp\beta\boldsymbol{\Omega}_{\text{op}}(\boldsymbol{R}, \boldsymbol{P}_{\text{op}})\beta. \tag{21}$$

For *time inversion*

$$\begin{aligned} \boldsymbol{R}' &= \boldsymbol{R}, \\ t' &= -t \end{aligned} \tag{22}$$

the wave function transforms as

$$\psi'(\boldsymbol{R}', t') = T\psi^*(\boldsymbol{R}, t), \tag{23}$$

where the asterisk indicates the complex conjugate. The matrix $T$ is such that it transforms the Dirac matrices in the following way:

$$T^{-1}\boldsymbol{\alpha}T = -\boldsymbol{\alpha}^*, \qquad T^{-1}\beta T = \beta^*. \tag{24}$$

From these relations it follows that $T^*T$ is a multiple $\lambda$ of the $4\times 4$ unit matrix. The anti-linear transformation (23) with (22) and (24) leaves the Dirac equation (1) with (2) invariant. The inner product of two wave functions changes into its complex conjugate, so that its absolute value remains the same.

The expectation value of an operator $\Omega_{\text{op}}$ in the new frame is

$$\int \psi'^{\dagger}(\boldsymbol{R}', t')\Omega_{\text{op}}\left(\boldsymbol{R}', \frac{\hbar}{i}\frac{\partial}{\partial \boldsymbol{R}'}\right)\psi'(\boldsymbol{R}', t')\mathrm{d}\boldsymbol{R}' = \int \psi^{\dagger}(\boldsymbol{R}, t)\tilde{T}\tilde{\Omega}_{\text{op}}\left(\boldsymbol{R}, -\frac{\hbar}{i}\frac{\partial}{\partial \boldsymbol{R}}\right)T^*\psi(\boldsymbol{R}, t)\mathrm{d}\boldsymbol{R}, \tag{25}$$

as follows by inserting (23) in the first member and taking the transpose, which is denoted by a tilde. Choosing for $\Omega_{\text{op}}$ the unit operator and requiring the normalization of the wave function to be invariant one finds that the matrix $T$ is unitary ($T^{\dagger}T = 1$). From this property together with $T^*T = \lambda$ it follows that $\tilde{T} = \pm T$ and also $T^* = \pm T^{-1}$. Thus instead of (25) one may write

$$\int \psi'^{\dagger}(\boldsymbol{R}', t')\Omega_{\text{op}}\left(\boldsymbol{R}', \frac{\hbar}{i}\frac{\partial}{\partial \boldsymbol{R}'}\right)\psi'(\boldsymbol{R}', t')\mathrm{d}\boldsymbol{R}' = \int \psi^{\dagger}(\boldsymbol{R}, t)T\tilde{\Omega}_{\text{op}}\left(\boldsymbol{R}, -\frac{\hbar}{i}\frac{\partial}{\partial \boldsymbol{R}}\right)T^{-1}\psi(\boldsymbol{R}, t)\mathrm{d}\boldsymbol{R}. \tag{26}$$

We now turn to the discussion of *pure Lorentz transformations*. Under an infinitesimal pure Lorentz transformation the space–time coordinates transform according to

$$\begin{aligned} \boldsymbol{R}' &= \boldsymbol{R} - \boldsymbol{\varepsilon} ct, \\ ct' &= ct - \boldsymbol{\varepsilon} \cdot \boldsymbol{R}, \end{aligned} \tag{27}$$

with $\boldsymbol{\varepsilon}$ an infinitesimal vector. The wave function then transforms as

$$\psi'(\boldsymbol{R}', t') = (1 - \tfrac{1}{2}\boldsymbol{\varepsilon} \cdot \boldsymbol{\alpha})\psi(\boldsymbol{R}, t), \tag{28}$$

since the Dirac equation (1) with (2) and the inner product is invariant for the transformation (27–28).

We now want to compare the expectation value (5) of an operator $\Omega_{\text{op}}$ at the time $t$ with an expectation value (8) at the time $\hat{t}'$, which is numerically equal to $t$:

$$\int \psi'^{\dagger}(\hat{\boldsymbol{R}}', \hat{t}')\Omega_{\text{op}}\left(\hat{\boldsymbol{R}}', \frac{\hbar}{i}\frac{\partial}{\partial \hat{\boldsymbol{R}}'}\right)\psi'(\hat{\boldsymbol{R}}', \hat{t}')\mathrm{d}\boldsymbol{R}'. \tag{29}$$

Here the variables $\hat{\boldsymbol{R}}'$ and $\hat{t}'$ occur, which correspond to the variables $\hat{\boldsymbol{R}}$ and $\hat{t}$ in the old frame. The latter follow from the inverse of the Lorentz transformation (27). One has, with $\hat{t}' = t$:

$$\begin{aligned} \hat{\boldsymbol{R}} &= \hat{\boldsymbol{R}}' + \boldsymbol{\varepsilon} ct, \\ c\hat{t} &= ct + \boldsymbol{\varepsilon} \cdot \hat{\boldsymbol{R}}', \end{aligned} \tag{30}$$

up to first order in $\boldsymbol{\varepsilon}$. From (28) with circumflexes, i.e. from

$$\psi'(\hat{\boldsymbol{R}}', \hat{t}') = (1 - \tfrac{1}{2}\boldsymbol{\varepsilon} \cdot \boldsymbol{\alpha})\psi(\hat{\boldsymbol{R}}, \hat{t}) \tag{31}$$

and (30) it follows with the Dirac equation (1) that one has:

$$\begin{aligned} \psi'(\hat{\boldsymbol{R}}', \hat{t}') = \psi(\hat{\boldsymbol{R}}', t) - \tfrac{1}{2}\boldsymbol{\varepsilon} \cdot \boldsymbol{\alpha}\psi(\hat{\boldsymbol{R}}', t) \\ + ct\boldsymbol{\varepsilon} \cdot \frac{\partial \psi(\hat{\boldsymbol{R}}', t)}{\partial \hat{\boldsymbol{R}}'} - \frac{i}{\hbar c}\boldsymbol{\varepsilon} \cdot \hat{\boldsymbol{R}}' H_{\text{op}}\left(\hat{\boldsymbol{R}}', \frac{\hbar}{i}\frac{\partial}{\partial \hat{\boldsymbol{R}}'}\right)\psi(\boldsymbol{R}', t). \end{aligned} \tag{32}$$

Inserting this expression and its hermitian conjugate into (29) we obtain as the expectation value in the new coordinate frame an expression of the form (10) with

$$\delta\Omega_{\text{op}} = \frac{i}{\hbar}\boldsymbol{\varepsilon} \cdot [\boldsymbol{N}_{\text{op}} - ct\boldsymbol{P}_{\text{op}}, \Omega_{\text{op}}], \tag{33}$$

where we introduced the abbreviation

$$\boldsymbol{N}_{\text{op}} \equiv c^{-1}\boldsymbol{R}H_{\text{op}} - \tfrac{1}{2}i\hbar\boldsymbol{\alpha} \tag{34}$$

or, with the use of (2),

$$\boldsymbol{N}_{\text{op}} = \tfrac{1}{2}c^{-1}\{\boldsymbol{R}, H_{\text{op}}\}, \tag{35}$$

where the curly brackets indicate the anticommutator. In this way we found that $\boldsymbol{N}_{\text{op}} - ct\boldsymbol{P}_{\text{op}}$ is the generator of pure Lorentz transformations.

b. *Covariance requirements on position and spin*

For the description of the behaviour of the particle we need operators for the position and for the spin. A number of constraints upon these operators will follow from requirements about their transformation properties with respect to the Poincaré group, in particular with respect to spatial translations, spatial rotations, spatial inversions, time reversal and pure Lorentz transformations.

As transformation properties with respect to infinitesimal translations (6) we require that the expectation value of the position operator $\boldsymbol{X}_{\text{op}}$ change by an amount $\boldsymbol{\varepsilon}$ and that that of the spin operator $\boldsymbol{s}_{\text{op}}$ be invariant. Then from (11) it follows that

$$[\boldsymbol{P}_{\text{op}}, \boldsymbol{X}_{\text{op}}] = \frac{\hbar}{i}\,\mathbf{U}, \tag{36}$$

$$[\boldsymbol{P}_{\text{op}}, \boldsymbol{s}_{\text{op}}] = 0. \tag{37}$$

As to the rotation properties we require that both the position and spin operator be vectors, so that one has from (17)

$$[(\boldsymbol{R}\wedge\boldsymbol{P}_{\text{op}} + \tfrac{1}{2}\hbar\boldsymbol{\sigma})^i, X^j_{\text{op}}] = i\hbar\varepsilon^{ijk}X_{k,\text{op}}, \tag{38}$$

$$[(\boldsymbol{R}\wedge\boldsymbol{P}_{\text{op}} + \tfrac{1}{2}\hbar\boldsymbol{\sigma})^i, s^j_{\text{op}}] = i\hbar\varepsilon^{ijk}s_{k,\text{op}}. \tag{39}$$

As regards the transformation properties with respect to spatial inversion we postulate that the position operator be a polar vector and the spin operator an axial vector. In view of (21) this means that we require

$$\boldsymbol{X}_{\text{op}}(-\boldsymbol{R}, -\boldsymbol{P}_{\text{op}}) = -\beta\boldsymbol{X}_{\text{op}}(\boldsymbol{R}, \boldsymbol{P}_{\text{op}})\beta, \tag{40}$$

$$\boldsymbol{s}_{\text{op}}(-\boldsymbol{R}, -\boldsymbol{P}_{\text{op}}) = \beta\boldsymbol{s}_{\text{op}}(\boldsymbol{R}, \boldsymbol{P}_{\text{op}})\beta. \tag{41}$$

For the transformation property with respect to time reversal we require that the expectation value of the position operator be invariant, while that of the spin operator should change sign. In view of (26) this means that we postulate

$$\boldsymbol{X}_{\text{op}}(\boldsymbol{R}, -\boldsymbol{P}_{\text{op}}) = T\tilde{\boldsymbol{X}}_{\text{op}}(\boldsymbol{R}, \boldsymbol{P}_{\text{op}})T^{-1}, \tag{42}$$

$$\boldsymbol{s}_{\text{op}}(\boldsymbol{R}, -\boldsymbol{P}_{\text{op}}) = -T\tilde{\boldsymbol{s}}_{\text{op}}(\boldsymbol{R}, \boldsymbol{P}_{\text{op}})T^{-1}, \tag{43}$$

where the tilde indicates the transposed of a $4\times 4$ matrix.

We now turn to a discussion of the transformation properties of the position and spin operators under pure Lorentz transformations. The expectation value of the position operator $\boldsymbol{X}_{\text{op}}$ will be required to change under the infinitesimal pure Lorentz transformation (27) by an amount[1] which is equal to the expectation value of the operator

$$\delta \boldsymbol{X}_{\text{op}} = -\boldsymbol{\varepsilon} ct + \frac{i}{2\hbar c} \{\boldsymbol{\varepsilon}\cdot\boldsymbol{X}_{\text{op}}, [H_{\text{op}}, \boldsymbol{X}_{\text{op}}]\}. \tag{44}$$

Therefore it follows from (10) with (33) that the covariance condition[2] for the position operator is:

$$[N^i_{\text{op}}, X^j_{\text{op}}] = \tfrac{1}{2}c^{-1}\{X^i_{\text{op}}, [H_{\text{op}}, X^j_{\text{op}}]\}, \tag{45}$$

where (36) has been employed. The use of the latter formula had as a consequence that the terms with the time $t$ cancelled, so that the requirement (45) contains only the three-vector $\boldsymbol{X}_{\text{op}}$ for the position operator.

For the spin operator $\boldsymbol{s}_{\text{op}}$ we require that its expectation value change under the pure Lorentz transformation (27) by a term[3] which is the expectation

[1] In the classical theory of a composite particle the set of centres of energy at successive times determines a world line independent of the Lorentz frame. As a result the positions observed in different Lorentz frames are connected in a particular way (cf. M. H. L. Pryce, Proc. Roy. Soc. **A 195**(1949)62). In fact, let us consider the two points $t$, $\boldsymbol{X}(t)$ and $\bar{t}$, $\boldsymbol{X}(\bar{t})$ on the world line of which the time coordinates $t$ in the reference frame and $\bar{t}'$ in an infinitesimally different frame have the same numerical value. Thus from (27), $t \equiv \bar{t}' = \bar{t} - c^{-1}\boldsymbol{\varepsilon}\cdot\boldsymbol{X}(\bar{t})$ and $\boldsymbol{X}'(t) \equiv \boldsymbol{X}'(\bar{t}') = \boldsymbol{X}(\bar{t}) - \boldsymbol{\varepsilon} c\bar{t}$. From the first of these equations one has up to first order in $\boldsymbol{\varepsilon}$ that $c\bar{t} = ct + \boldsymbol{\varepsilon}\cdot\boldsymbol{X}(t)$. With the help of this relation the second equation becomes upon Taylor expansion up to first order

$$\boldsymbol{X}'(t) - \boldsymbol{X}(t) = -\boldsymbol{\varepsilon} ct + c^{-1}\boldsymbol{\varepsilon}\cdot\boldsymbol{X}(t)\frac{\mathrm{d}\boldsymbol{X}(t)}{\mathrm{d}t}.$$

This expression is equal to the expectation value of (44) for a narrow wave packet in the limit $\hbar \to 0$, since then the expectation value of a (symmetrized) product of operators is equal to the product of expectation values.

[2] Cf. T. F. Jordan and N. Mukunda, Phys. Rev. **132**(1963)1842; G. Lugarini and M. Pauri, N. Cim. **47A**(1967)299.

[3] In the classical theory of a composite particle the inner angular momentum $\boldsymbol{s}$ is the space-space part $(s^{23}, s^{31}, s^{12})$ of an antisymmetric tensor $s^{\alpha\beta}$ of which the space–time components $(s^{10}, s^{20}, s^{30})$ are denoted as $\boldsymbol{t}$. With the same notation as used above we find that in a Lorentz frame which is connected to the observer's frame by an infinitesimal Lorentz transformation, the inner angular momentum $\boldsymbol{s}'$ at the time $\bar{t}'$ which is numerically equal to $t$ is $\boldsymbol{s}'(t) = \boldsymbol{s}(\bar{t}) + \boldsymbol{\varepsilon}\wedge\boldsymbol{t}(\bar{t})$. With a Taylor expansion this relation becomes

$$\boldsymbol{s}'(t) - \boldsymbol{s}(t) = \boldsymbol{\varepsilon}\wedge\boldsymbol{t}(t) + c^{-1}\boldsymbol{\varepsilon}\cdot\boldsymbol{X}(t)\frac{\mathrm{d}\boldsymbol{s}(t)}{\mathrm{d}t}.$$

value of the operator

$$\delta \boldsymbol{s}_{\mathrm{op}} = \boldsymbol{\varepsilon} \wedge \boldsymbol{t}_{\mathrm{op}} + \frac{i}{2\hbar c} \{\boldsymbol{\varepsilon} \cdot \boldsymbol{X}_{\mathrm{op}}, [H_{\mathrm{op}}, \boldsymbol{s}_{\mathrm{op}}]\} \tag{46}$$

with the three-vector operator $\boldsymbol{t}_{\mathrm{op}}$ such that

$$\delta \boldsymbol{t}_{\mathrm{op}} = -\boldsymbol{\varepsilon} \wedge \boldsymbol{s}_{\mathrm{op}} + \frac{i}{2\hbar c} \{\boldsymbol{\varepsilon} \cdot \boldsymbol{X}_{\mathrm{op}}, [H_{\mathrm{op}}, \boldsymbol{t}_{\mathrm{op}}]\}. \tag{47}$$

With (10) and (33) it now follows that the covariance condition for the spin operator is

$$[N^i_{\mathrm{op}}, s^j_{\mathrm{op}}] = \tfrac{1}{2} c^{-1} \{X^i_{\mathrm{op}}, [H_{\mathrm{op}}, s^j_{\mathrm{op}}]\} + i\hbar \varepsilon^{ijk} t_{k,\mathrm{op}} \tag{48}$$

together with the relation

$$[N^i_{\mathrm{op}}, t^j_{\mathrm{op}}] = \tfrac{1}{2} c^{-1} \{X^i_{\mathrm{op}}, [H_{\mathrm{op}}, t^j_{\mathrm{op}}]\} - i\hbar \varepsilon^{ijk} s_{k,\mathrm{op}}, \tag{49}$$

where (37), which is valid both for $\boldsymbol{s}_{\mathrm{op}}$ and $\boldsymbol{t}_{\mathrm{op}}$, has been used.

The conditions (36–43), (45) and (48–49) will be employed for the determination of the form of the position and spin operators.

### c. *Transformation of the Hamiltonian to even form; the position and spin operators*

In the Pauli representation the Dirac matrices $\boldsymbol{\alpha}$ and $\beta$ of the Hamilton operator (2) are written as

$$\boldsymbol{\alpha} = \rho_1 \boldsymbol{\sigma}, \qquad \beta = \rho_3, \tag{50}$$

where the matrices $\boldsymbol{\sigma}$ are the $4 \times 4$ matrices

$$\boldsymbol{\sigma} = \begin{pmatrix} \boldsymbol{\tau} & 0 \\ 0 & \boldsymbol{\tau} \end{pmatrix} \tag{51}$$

with $\boldsymbol{\tau}$ the $2 \times 2$ Pauli matrices

$$\tau_1, \tau_2, \tau_3 = \begin{pmatrix} 0 & 1 \\ 1 & 0 \end{pmatrix}, \quad \begin{pmatrix} 0 & -i \\ i & 0 \end{pmatrix}, \quad \begin{pmatrix} 1 & 0 \\ 0 & -1 \end{pmatrix}. \tag{52}$$

---

Likewise one finds

$$\boldsymbol{t}'(t) - \boldsymbol{t}(t) = -\boldsymbol{\varepsilon} \wedge \boldsymbol{s}(t) + c^{-1} \boldsymbol{\varepsilon} \cdot \boldsymbol{X}(t) \frac{\mathrm{d}\boldsymbol{t}(t)}{\mathrm{d}t}.$$

For narrow wave packets in the limit $\hbar \to 0$ one finds that these expressions are the expectation values of (46) and (47).

The matrices $\rho$ are the $4\times 4$ matrices:

$$\rho_1, \rho_2, \rho_3 = \begin{pmatrix} 0 & 1 \\ 1 & 0 \end{pmatrix}, \quad \begin{pmatrix} 0 & -i \\ i & 0 \end{pmatrix}, \quad \begin{pmatrix} 1 & 0 \\ 0 & -1 \end{pmatrix}, \tag{53}$$

where 1 stands for the $2\times 2$ unit matrix. (The advantage of the use of $\boldsymbol{\rho}$- and $\boldsymbol{\sigma}$-matrices is that the three $\boldsymbol{\rho}$-matrices commute with the three $\boldsymbol{\sigma}$-matrices, while the product rules for the $\boldsymbol{\rho}$-matrices and for the $\boldsymbol{\sigma}$-matrices amongst each other are the same as those for the Pauli matrices $\boldsymbol{\tau}$.) The Dirac matrices (50) are thus of the form:

$$\boldsymbol{\alpha} = \begin{pmatrix} 0 & \boldsymbol{\tau} \\ \boldsymbol{\tau} & 0 \end{pmatrix}, \quad \beta = \begin{pmatrix} 1 & 0 \\ 0 & -1 \end{pmatrix}. \tag{54}$$

The Dirac equation (1) with (2) may now be written as a set of two equations for the upper two and lower two components $\psi_1$ and $\psi_2$ of the four-component wave function $\psi$:

$$\begin{aligned} c\boldsymbol{\tau}\cdot\boldsymbol{P}_{\text{op}}\,\psi_2 + mc^2\psi_1 &= -\frac{\hbar}{i}\frac{\partial\psi_1}{\partial t}, \\ c\boldsymbol{\tau}\cdot\boldsymbol{P}_{\text{op}}\,\psi_1 - mc^2\psi_2 &= -\frac{\hbar}{i}\frac{\partial\psi_2}{\partial t}. \end{aligned} \tag{55}$$

These equations are coupled because the matrices $\boldsymbol{\alpha}$, which occur in the Hamilton operator, have 'odd' character in the representation (54), i.e. they couple the upper and lower components of the wave function $\psi$. The equations for upper and lower components may be uncoupled by performing a unitary transformation due to Pryce[1] and Foldy–Wouthuysen[2]:

$$U_{\text{op}} \equiv \frac{E_{\text{op}} + mc^2 + c\beta\boldsymbol{\alpha}\cdot\boldsymbol{P}_{\text{op}}}{\{2E_{\text{op}}(E_{\text{op}} + mc^2)\}^{\frac{1}{2}}} \tag{56}$$

with the abbreviation

$$E_{\text{op}} \equiv (c^2\boldsymbol{P}_{\text{op}}^2 + m^2c^4)^{\frac{1}{2}}. \tag{57}$$

(One may find the expression (56) by solving the eigenvalue problem of the Hamiltonian (2) and using the complex conjugates of the eigenvectors as the rows of the matrix $U_{\text{op}}$.)

Indeed with the transformation (56) the Hamilton operator becomes

$$\hat{H}_{\text{op}} \equiv U_{\text{op}} H_{\text{op}} U_{\text{op}}^{\dagger} = \beta E_{\text{op}}, \tag{58}$$

[1] M. H. L. Pryce, Proc. Roy. Soc. **A195**(1949)62.

[2] L. L. Foldy and S. A. Wouthuysen, Phys. Rev. **78**(1950)29.

which has 'even' form, since $\beta$ has 'even' character. The circumflex will be employed to indicate operators in the Pryce–Foldy–Wouthuysen (P–FW) picture in order to distinguish them from the original operators in the Dirac picture.

Since now the Hamilton operator has the simple form (58) its eigenvalues are immediately seen to be $\pm(c^2\boldsymbol{p}^2+m^2c^4)^{\frac{1}{2}}$ with $\boldsymbol{p}$ the eigenvalue of the momentum operator $\boldsymbol{P}_{\text{op}}$. The positive- and negative-energy eigenfunctions have now the property that the lower two or upper two components vanish.

If one calculates the expectation value of a physical quantity for a positive- or a negative-energy solution only the part of the corresponding operator that is even in the P–FW picture plays a role. In particular if one wants to define the *position operator* only its 'even' part is of importance. This even part, which we shall simply denote by the symbol $\boldsymbol{X}_{\text{op}}$ from now on, is completely determined if we impose a number of conditions. In the first place, from the transformation properties of translation (36), rotation (38), spatial inversion (40) and time reversal (42) with $T = \sigma_2$ in the Pauli representation[1] it follows that in the P–FW picture $\boldsymbol{X}_{\text{op}}$ has the form

$$\hat{\boldsymbol{X}}_{\text{op}} \equiv U_{\text{op}}\boldsymbol{X}_{\text{op}}U^{\dagger}_{\text{op}} = \boldsymbol{R}+\{f_1(E_{\text{op}})+\beta f_2(E_{\text{op}})\}\boldsymbol{\sigma}\wedge\boldsymbol{P}_{\text{op}}, \tag{59}$$

where $f_1(E_{\text{op}})$ and $f_2(E_{\text{op}})$ are arbitrary real functions of $E_{\text{op}}$. Indeed $\boldsymbol{P}_{\text{op}}$ and $\boldsymbol{\sigma}$ are the only vectors available and hence $\boldsymbol{P}_{\text{op}}$ and $\boldsymbol{\sigma}\wedge\boldsymbol{P}_{\text{op}}$ the only polar vectors. If one limits oneself to vectors that have the right transformation character under time reversal one is left with $\boldsymbol{\sigma}\wedge\boldsymbol{P}_{\text{op}}$ only[2].

The transformation character under pure Lorentz transformations is determined by the commutation rule (45), which reads in the P–FW picture

$$[\hat{N}^i_{\text{op}}, \hat{X}^j_{\text{op}}] = \tfrac{1}{2}c^{-1}[\hat{X}^i_{\text{op}}, [\hat{H}_{\text{op}}, \hat{X}^j_{\text{op}}]\}. \tag{60}$$

The left-hand side contains the generator $\hat{\boldsymbol{N}}_{\text{op}}$ in the P–FW picture. It has been given in (35) in the Dirac picture. In the P–FW picture the Hamiltonian $\hat{H}_{\text{op}}$ is given by (58) while the Dirac coordinate gets the form

$$\hat{\boldsymbol{R}}_{\text{op}} \equiv U_{\text{op}}\boldsymbol{R}U^{\dagger}_{\text{op}} = \boldsymbol{R}+\boldsymbol{\xi}_{\text{op}}, \tag{61}$$

where

$$\boldsymbol{\xi}_{\text{op}} \equiv \frac{\hbar}{i}\frac{\partial U_{\text{op}}}{\partial \boldsymbol{P}_{\text{op}}}U^{\dagger}_{\text{op}} = -\frac{\hbar}{i}U_{\text{op}}\frac{\partial U^{\dagger}_{\text{op}}}{\partial \boldsymbol{P}_{\text{op}}} \equiv \boldsymbol{\xi}_{\text{e,op}}+\boldsymbol{\xi}_{\text{o,op}}. \tag{62}$$

[1] The $T$-matrix may be chosen as $\sigma_2$ in the Pauli representation (50), as follows from (24). Indeed $\sigma_2$ fulfils the relations $T^*T = \lambda$ and $T^{\dagger}T = 1$, with $\lambda = -1$.

[2] If one does not impose time reversal invariance from the beginning one should write additional terms $\{f_3(E_{\text{op}})+\beta f_4(E_{\text{op}})\}\boldsymbol{P}_{\text{op}}$ in (59). If the requirement (60) is imposed on (59) with these terms added one finds that $f_3(E_{\text{op}})$ and $f_4(E_{\text{op}})$ vanish. Hence, strictly spoken, time reversal invariance need not be invoked to obtain a unique position operator. The latter is also true for the obtention of the spin operator.

The explicit forms for the even and odd parts $\xi_{e,op}$ and $\xi_{o,op}$ follow with (56)

$$\begin{aligned}\xi_{e,op} &= \frac{\hbar c^2 \boldsymbol{P}_{op} \wedge \boldsymbol{\sigma}}{2E_{op}(E_{op}+mc^2)},\\ \xi_{o,op} &= -\frac{\hbar ic\beta\boldsymbol{\alpha}}{2E_{op}} + \frac{\hbar ic^3 \beta\boldsymbol{\alpha}\cdot\boldsymbol{P}_{op}\boldsymbol{P}_{op}}{2E_{op}^2(E_{op}+mc^2)}.\end{aligned} \tag{63}$$

The generator $\hat{\boldsymbol{N}}_{op}$ in the P–FW picture becomes now:

$$\hat{\boldsymbol{N}}_{op} = \tfrac{1}{2}c^{-1}\beta\{\boldsymbol{R}, E_{op}\} + \frac{\beta\hbar c\boldsymbol{P}_{op}\wedge\boldsymbol{\sigma}}{2(E_{op}+mc^2)}. \tag{64}$$

If (59) and (64) are inserted into (60) one obtains the result that a certain linear combination of the independent tensors

$$P^i_{op}(\boldsymbol{\sigma}\wedge\boldsymbol{P}_{op})^j,\quad P^j_{op}(\boldsymbol{\sigma}\wedge\boldsymbol{P}_{op})^i,\quad \varepsilon^{ijk}\sigma_k \tag{65}$$

vanishes. The tensor $\varepsilon^{ijk}P_{k,op}\boldsymbol{P}_{op}\cdot\boldsymbol{\sigma}$ depends upon these, as a consequence of the relation:

$$P^i_{op}(\boldsymbol{\sigma}\wedge\boldsymbol{P}_{op})^j - P^j_{op}(\boldsymbol{\sigma}\wedge\boldsymbol{P}_{op})^i - \boldsymbol{P}^2_{op}\,\varepsilon^{ijk}\sigma_{k,op} + \varepsilon^{ijk}P_{k,op}\boldsymbol{P}_{op}\cdot\boldsymbol{\sigma} = 0. \tag{66}$$

All coefficients of the independent tensors (65) have to be zero. This leads to the solution:

$$f_1(E_{op}) = \frac{\hbar}{2m(E_{op}+mc^2)},\qquad f_2(E_{op}) = 0. \tag{67}$$

If this is substituted into (59) we obtain as the position operator in the P–FW picture:

$$\hat{\boldsymbol{X}}_{op} = \boldsymbol{R} + \frac{\hbar\boldsymbol{\sigma}\wedge\boldsymbol{P}_{op}}{2m(E_{op}+mc^2)}. \tag{68}$$

The expression $\boldsymbol{X}_{op}$ in the Dirac picture may be found with (56), which implies

$$U^\dagger_{op} = \beta U_{op}\beta \tag{69}$$

and

$$\hat{\boldsymbol{\sigma}}_{op} \equiv U_{op}\boldsymbol{\sigma}U^\dagger_{op} = \boldsymbol{\sigma} + \frac{ic\beta\boldsymbol{\alpha}\wedge\boldsymbol{P}_{op}}{E_{op}} - \frac{c^2\boldsymbol{P}_{op}\wedge(\boldsymbol{\sigma}\wedge\boldsymbol{P}_{op})}{E_{op}(E_{op}+mc^2)}. \tag{70}$$

Using also (61–63) we get then from (68) the Dirac picture position operator

$$\boldsymbol{X}_{op} = \boldsymbol{R} + \frac{i\hbar}{2mc}\beta\left(\boldsymbol{\alpha} - c^2\frac{\boldsymbol{\alpha}\cdot\boldsymbol{P}_{op}\boldsymbol{P}_{op}}{E^2_{op}}\right). \tag{71}$$

In this way the even part of the position operator has been obtained in a unique way by imposing its transformation character. A position operator of this form has been put forward by Pryce[1].

The part of the spin operator that is even in the P–FW picture may likewise be determined by means of the covariance conditions of the preceding subsection. Indeed from the requirements (37), (39), (41) and (43) for the translation, rotation, spatial inversion and time reversal properties, it follows that the even part of the spin operator, which we shall denote by the symbol $\boldsymbol{s}_{op}$ in the Dirac picture and by $\hat{\boldsymbol{s}}_{op}$ in the P–FW picture, has the form[2]

$$\hat{\boldsymbol{s}}_{op} = \{f_1(E_{op})+\beta f_2(E_{op})\}\boldsymbol{\sigma}+\{f_3(E_{op})+\beta f_4(E_{op})\}\boldsymbol{P}_{op}\boldsymbol{P}_{op}\cdot\boldsymbol{\sigma} \tag{72}$$

with arbitrary real functions $f_i(E_{op})$ $(i = 1, \ldots, 4)$. (Indeed $\boldsymbol{\sigma}$ and $\boldsymbol{P}_{op}\boldsymbol{P}_{op}\cdot\boldsymbol{\sigma}$ are the only axial vectors available. Moreover they have the right time reversal behaviour.) We now substitute this expression into the covariance condition (48), which in the P–FW picture reads

$$[\hat{N}^i_{op}, \hat{s}^j_{op}] = \tfrac{1}{2}c^{-1}\{\hat{X}^i_{op}, [H_{op}, \hat{s}^j_{op}]\} - \frac{\hbar}{i}\varepsilon^{ijk}\hat{t}_{k,op}. \tag{73}$$

Then one finds, by noting that the coefficients of the independent (symmetrical) tensors $P^i_{op}P^j_{op}\boldsymbol{P}_{op}\cdot\boldsymbol{\sigma}$ and $\delta^{ij}\boldsymbol{P}_{op}\cdot\boldsymbol{\sigma}$ must vanish, the form of the functions $f_i(E_{op})$. In this way the expression (72) becomes

$$\hat{\boldsymbol{s}}_{op} = (\lambda+\beta\mu)E_{op}\boldsymbol{\sigma} - \frac{\lambda+\beta\mu}{E_{op}+mc^2}c^2\boldsymbol{P}_{op}\boldsymbol{P}_{op}\cdot\boldsymbol{\sigma}, \tag{74}$$

with arbitrary real constants $\lambda$ and $\mu$. Furthermore one obtains from (73) for $\hat{\boldsymbol{t}}_{op}$:

$$\hat{\boldsymbol{t}}_{op} = c\beta(\lambda+\beta\mu)\boldsymbol{P}_{op}\wedge\boldsymbol{\sigma}. \tag{75}$$

The expressions (74) and (75) fulfil the relation (49) (with circumflexes).

From the transformation properties we have found an expression for the spin operator $\hat{\boldsymbol{s}}_{op}$, which still contains two arbitrary constants. (The reason for the occurrence of such multiplicative constants is the fact that the covariance requirements (37), (39), (41), (43) and (48) are all linear and homogeneous in $\boldsymbol{s}_{op}$ and $\boldsymbol{t}_{op}$.) To fix the scale we impose a final condition. We require that the sum of the orbital angular momentum $\boldsymbol{X}_{op}\wedge\boldsymbol{P}_{op}$ and the spin angular momentum $\boldsymbol{s}_{op}$ be equal to the total angular momentum which is the

[1] M. H. L. Pryce, op. cit.

[2] If time reversal is not imposed nothing changes in the expression (72), in contrast with the situation for the position operator.

generator of rotations given by (15):

$$\boldsymbol{X}_{op} \wedge \boldsymbol{P}_{op} + \boldsymbol{s}_{op} = \boldsymbol{R} \wedge \boldsymbol{P}_{op} + \tfrac{1}{2}\hbar\boldsymbol{\sigma}. \tag{76}$$

Transforming from the Dirac picture to the P–FW picture we find for this condition

$$\hat{\boldsymbol{X}}_{op} \wedge \boldsymbol{P}_{op} + \hat{\boldsymbol{s}}_{op} = \boldsymbol{R} \wedge \boldsymbol{P}_{op} + \tfrac{1}{2}\hbar\boldsymbol{\sigma}. \tag{77}$$

We could omit some circumflexes, because one has $\hat{\boldsymbol{P}}_{op} = \boldsymbol{P}_{op}$ and

$$\hat{\boldsymbol{R}} \wedge \boldsymbol{P}_{op} + \tfrac{1}{2}\hbar\hat{\boldsymbol{\sigma}}_{op} = \boldsymbol{R} \wedge \boldsymbol{P}_{op} + \tfrac{1}{2}\hbar\boldsymbol{\sigma}, \tag{78}$$

as follows from (61–63) and (70). Substituting (68) and (74) into (77) one finds that the constants are $\lambda = \hbar/2mc^2$ and $\mu = 0$, so that finally the even part of the spin operator (74) becomes

$$\hat{\boldsymbol{s}}_{op} = \frac{\hbar E_{op}}{2mc^2}\boldsymbol{\sigma} - \frac{\hbar \boldsymbol{P}_{op}\boldsymbol{P}_{op}\cdot\boldsymbol{\sigma}}{2m(E_{op}+mc^2)}, \tag{79}$$

while (75) gets the form:

$$\hat{\boldsymbol{t}}_{op} = \frac{\hbar}{2mc}\beta\boldsymbol{P}_{op} \wedge \boldsymbol{\sigma}. \tag{80}$$

The spin operator (79) is conserved since it commutes with the Hamiltonian (58).

In the preceding we showed that the covariance properties alone sufficed to fix the position operator $\hat{\boldsymbol{X}}_{op}$ and to find the spin operator $\hat{\boldsymbol{s}}_{op}$ apart from multiplicative constants. It turned out to be possible to choose these constants in such a way that also the total angular momentum condition (77) could be satisfied. (Of course, since in the present case $\hat{\boldsymbol{X}}_{op}$ is completely fixed by the covariance requirements the condition (77) alone would have been sufficient to determine $\hat{\boldsymbol{s}}_{op}$. However, in view of the fact that such a procedure is not possible in the case with fields – to be considered later – we have not followed this line of reasoning to determine $\hat{\boldsymbol{s}}_{op}$.)

The operators (79) and (80) are connected by the relation

$$\hat{\boldsymbol{t}}_{op} = \frac{\beta c\boldsymbol{P}_{op}}{E_{op}} \wedge \hat{\boldsymbol{s}}_{op}. \tag{81}$$

By introducing the velocity operator in the P–FW picture

$$\hat{\boldsymbol{v}}_{op} \equiv \frac{i}{\hbar}[\hat{H}_{op}, \hat{\boldsymbol{X}}_{op}] = \frac{\beta c^2\boldsymbol{P}_{op}}{E_{op}} \tag{82}$$

(where (58) and (68) have been used), one may write the relation (81) also as

$$\hat{\boldsymbol{t}}_{\text{op}} = c^{-1}\hat{\boldsymbol{v}}_{\text{op}} \wedge \hat{\boldsymbol{s}}_{\text{op}}. \tag{83}$$

It is the quantum-mechanical counterpart of the classical relation $p_\alpha s^{\alpha\beta} = 0$ (IV.67) with $p^\alpha = mu^\alpha$ (IV.119) for the field-free case.

In the Dirac picture the operators $\boldsymbol{s}_{\text{op}}$ and $\boldsymbol{t}_{\text{op}}$ follow from (79) and (80):

$$\boldsymbol{s}_{\text{op}} = \tfrac{1}{2}\hbar\boldsymbol{\sigma} - \frac{i\hbar\beta\boldsymbol{\alpha} \wedge \boldsymbol{P}_{\text{op}}}{2mc}, \tag{84}$$

$$\boldsymbol{t}_{\text{op}} = \frac{\hbar}{2mc}\beta\boldsymbol{P}_{\text{op}} \wedge \boldsymbol{\sigma}, \tag{85}$$

where we used (70) and:

$$U_{\text{op}}\beta U^\dagger_{\text{op}} = \frac{-c\boldsymbol{\alpha}\cdot\boldsymbol{P}_{\text{op}} + \beta mc^2}{E_{\text{op}}}, \tag{86}$$

which follows from (2), (58) and (69). The spin operator (84) has been given already by Pryce[1].

The operators (84) and (85) are the space–space and space–time parts of an antisymmetric tensor

$$s^{\mu\nu}_{\text{op}} = \tfrac{1}{2}\hbar\sigma^{\mu\nu} + \frac{\hbar}{2mc}(\gamma^\mu P^\nu_{\text{op}} - \gamma^\nu P^\mu_{\text{op}}), \tag{87}$$

where we introduced Dirac matrices $\gamma^\mu$ ($\mu = 0, 1, 2, 3$) defined as:

$$\gamma^0 \equiv -i\beta, \qquad \boldsymbol{\gamma} \equiv -i\beta\boldsymbol{\alpha} \tag{88}$$

and the abbreviation

$$\sigma^{\mu\nu} \equiv -\tfrac{1}{2}i[\gamma^\mu, \gamma^\nu]. \tag{89}$$

The zero-component $P^0_{\text{op}}$ of $P^\mu_{\text{op}}$ is defined to be equal to $H_{\text{op}}/c$ ($= \boldsymbol{\alpha}\cdot\boldsymbol{P}_{\text{op}} + \beta mc$). The spin tensor (87) has been found by Fradkin and Good[2] and by Hilgevoord and Wouthuysen[3] starting from a different basis.

The components of the position operator (68) do not commute; in fact one finds the commutation rule

$$[\hat{X}^i_{\text{op}}, \hat{X}^j_{\text{op}}] = i\hbar\varepsilon^{ijk}\left(\frac{\hat{s}_{k,\text{op}}c^2}{E^2_{\text{op}}} + \frac{P_{k,\text{op}}\boldsymbol{P}_{\text{op}}\cdot\hat{\boldsymbol{s}}_{\text{op}}}{m^2E^2_{\text{op}}}\right) \tag{90}$$

[1] M. H. L. Pryce, op. cit.

[2] D. M. Fradkin and R. H. Good jr., N. Cim. **22**(1961)643.

[3] J. Hilgevoord and S. A. Wouthuysen, Nucl. Phys. **40**(1963)1.

(it has the same form in the Dirac picture, i.e. without circumflexes). For the components of the spin operator one gets a commutation rule of the form

$$[\hat{s}^i_{\text{op}}, \hat{s}^j_{\text{op}}] = i\hbar\varepsilon^{ijk}\left(\hat{s}_{k,\text{op}} + \frac{P_{k,\text{op}}\boldsymbol{P}_{\text{op}}\cdot\hat{\boldsymbol{s}}_{\text{op}}}{m^2c^2}\right), \tag{91}$$

(again it has the same form in the Dirac picture).

The requirement of covariant behaviour imposed on the part of the position operator that is even in the P–FW picture has led to non-commuting components. This state of affairs is different from the situation in non-relativistic theory, where one is acquainted with position operators that possess commuting components. (Indeed the right-hand side of (90) is of order $c^{-2}$: namely of the order of the square of the Compton wavelength.) Correspondingly the commutation relations for the components of the spin operator do not have the same form as those for the components of the generator $\boldsymbol{R}\wedge\boldsymbol{P}_{\text{op}}+\frac{1}{2}\hbar\boldsymbol{\sigma}$ of spatial rotations[1].

If in Dirac theory one would impose as a condition the commutation of the Cartesian components of the even part of the position operator one finds, following a similar line of reasoning as above, a position operator which is that of Newton and Wigner[2] (v. appendix):

$$\hat{\boldsymbol{X}}_{\text{op,NW}} = \boldsymbol{R}. \tag{92}$$

This operator however does not possess covariant properties as does (68).

It has been tried to reconcile the requirement of covariance and commutation of the components of the position operator. This can only be achieved through an interplay of even and odd parts of the position operator. One obtains in this way the Dirac position and spin operators (v. appendix and [3]). However the even parts alone of these operators violate the covariance condition, and since only these parts occur in the expectation values for positive (or negative) energy solutions the latter will not possess covariant properties. (Still a different position operator may be proposed[4] if apart from the requirement of evenness also the commutivity condition is abandoned.)

[1] M. H. L. Pryce, op. cit., showed that the commutation relations (90) and (91) have classical counterparts in Poisson bracket relations for the components of the centre of energy and the inner angular momentum of a composite particle in classical theory, as discussed in chapter IV.

[2] T. D. Newton and E. P. Wigner, Rev. Mod. Phys. **21**(1949)400.

[3] T. F. Jordan and N. Mukunda, op. cit.; G. Lugarini and M. Pauri, op. cit.

[4] M. Bunge, N. Cim. **1**(1955)977; H. Yamasaki, Progr. Theor. Phys. **31**(1964)322, 324; M. Kolsrud, Phys. Norv. **2**(1967)141, 149.

## 3 *The Dirac particle in a field*

### a. *Invariance properties*

The Dirac Hamiltonian for a particle in an electromagnetic field $\boldsymbol{E}(\boldsymbol{R}, t)$, $\boldsymbol{B}(\boldsymbol{R}, t)$, with potentials $\varphi(\boldsymbol{R}, t)$, $\boldsymbol{A}(\boldsymbol{R}, t)$, reads

$$H_{\mathrm{op}} = c\boldsymbol{\alpha}\cdot\boldsymbol{\pi}_{\mathrm{op}} + \beta mc^2 + e\varphi + H_{\mathrm{a,op}}, \tag{93}$$

$$H_{\mathrm{a,op}} \equiv \tfrac{1}{2}(g-2)\mu_{\mathrm{B}}(i\beta\boldsymbol{\alpha}\cdot\boldsymbol{E} - \beta\boldsymbol{\sigma}\cdot\boldsymbol{B}), \tag{94}$$

where $\boldsymbol{\pi}_{\mathrm{op}}$ stands for $\boldsymbol{P}_{\mathrm{op}} - (e/c)\boldsymbol{A}$ and $\mu_{\mathrm{B}}$ is the Bohr magneton $e\hbar/2mc$. The Pauli term $H_{\mathrm{a,op}}$ represents the coupling of the anomalous magnetic moment with the field.

The Dirac equation (1) with this Hamilton operator is covariant under the transformations of the Poincaré group. One may find (just as in section 2) expressions for the change of the expectation value of an operator under these transformations. In particular we are interested in the change of the expectation value of an operator $\Omega_{\mathrm{op}}$ that depends on the coordinates $\boldsymbol{R}$, the momentum operator $\boldsymbol{P}_{\mathrm{op}} \equiv (\hbar/i)\partial/\partial\boldsymbol{R}$ and the potentials $\varphi(\boldsymbol{R}, t)$ and $\boldsymbol{A}(\boldsymbol{R}, t)$.

Under an infinitesimal *translation* (6) the expectation value of $\Omega_{\mathrm{op}}$ changes by an amount which is the expectation value of the operator

$$\delta\Omega_{\mathrm{op}} = \frac{i}{\hbar}\,\boldsymbol{\varepsilon}\cdot[\boldsymbol{P}_{\mathrm{op}}, \Omega_{\mathrm{op}}] + \Omega_{\mathrm{op}}(\boldsymbol{R}, \boldsymbol{P}_{\mathrm{op}}, \boldsymbol{A} - \boldsymbol{\varepsilon}\cdot\nabla\boldsymbol{A}, \varphi - \boldsymbol{\varepsilon}\cdot\nabla\varphi) - \Omega_{\mathrm{op}}(\boldsymbol{R}, \boldsymbol{P}_{\mathrm{op}}, \boldsymbol{A}, \varphi). \tag{95}$$

This may be derived in the same way as (11) is derived in section 2, if one uses the transformation property of $\varphi$ and $\boldsymbol{A}$ under translations:

$$\begin{aligned} \varphi'(\boldsymbol{R}', t') &= \varphi(\boldsymbol{R}, t), \\ \boldsymbol{A}'(\boldsymbol{R}', t') &= \boldsymbol{A}(\boldsymbol{R}, t). \end{aligned} \tag{96}$$

Up to terms with the potentials, but without derivatives of the potentials, (95) simplifies to

$$\delta\Omega_{\mathrm{op}} = \frac{i}{\hbar}\,\boldsymbol{\varepsilon}\cdot[\boldsymbol{P}_{\mathrm{op}}, \Omega_{\mathrm{op}}], \tag{97}$$

which is the same expression as (11) for the free particle.

Under an infinitesimal *rotation* (12) the expectation value changes by a term which is the expectation value of the operator

$$\begin{aligned} \delta\Omega_{\mathrm{op}} = {} & \frac{i}{\hbar}\,\boldsymbol{\varepsilon}\cdot[\boldsymbol{R}\wedge\boldsymbol{P}_{\mathrm{op}} + \tfrac{1}{2}\hbar\boldsymbol{\sigma}, \Omega_{\mathrm{op}}] \\ & + \Omega_{\mathrm{op}}(\boldsymbol{R}, \boldsymbol{P}_{\mathrm{op}}, \boldsymbol{A} + \boldsymbol{\varepsilon}\wedge\boldsymbol{A} - (\boldsymbol{\varepsilon}\wedge\boldsymbol{R})\cdot\nabla\boldsymbol{A}, \varphi - (\boldsymbol{\varepsilon}\wedge\boldsymbol{R})\cdot\nabla\varphi) - \Omega_{\mathrm{op}}(\boldsymbol{R}, \boldsymbol{P}_{\mathrm{op}}, \boldsymbol{A}, \varphi), \end{aligned} \tag{98}$$

where we used the transformation property of $\varphi$ and $\boldsymbol{A}$ under rotations

$$\begin{aligned}\varphi'(\boldsymbol{R}', t') &= \varphi(\boldsymbol{R}, t),\\ \boldsymbol{A}'(\boldsymbol{R}', t') &= \boldsymbol{A}(\boldsymbol{R}, t)+\boldsymbol{\varepsilon}\wedge\boldsymbol{A}(\boldsymbol{R}, t).\end{aligned} \tag{99}$$

Up to terms without derivatives of the potentials the expression (98) is

$$\delta\Omega_{\mathrm{op}} = \frac{i}{\hbar}\,\boldsymbol{\varepsilon}\cdot[\boldsymbol{R}\wedge\boldsymbol{P}_{\mathrm{op}}+\tfrac{1}{2}\hbar\boldsymbol{\sigma}, \Omega_{\mathrm{op}}]+(\boldsymbol{\varepsilon}\wedge\boldsymbol{A})\cdot\frac{\partial\Omega_{\mathrm{op}}}{\partial\boldsymbol{A}}. \tag{100}$$

Owing to the presence of the last term, this expression differs from (15) for the field-free case. For a vector operator, which is characterized by (16), the relation (100) becomes:

$$[(\boldsymbol{R}\wedge\boldsymbol{P}_{\mathrm{op}}+\tfrac{1}{2}\hbar\boldsymbol{\sigma})^i, \Omega^j_{\mathrm{op}}]-i\hbar\varepsilon^{imn}A_m\frac{\partial\Omega^j_{\mathrm{op}}}{\partial A^n} = i\hbar\varepsilon^{ijk}\Omega_{k,\mathrm{op}}, \tag{101}$$

which is the generalization of (17) to the case with fields.

In the special case that $\Omega_{\mathrm{op}}$ is independent of $\varphi$ and depends on $\boldsymbol{A}$ only via $\boldsymbol{\pi}_{\mathrm{op}} \equiv \boldsymbol{P}_{\mathrm{op}}-(e/c)\boldsymbol{A}$ (i.e. $\Omega_{\mathrm{op}} = \Omega_{\mathrm{op}}(\boldsymbol{R}, \boldsymbol{\pi}_{\mathrm{op}})$) we have for the last term of (100)

$$-\frac{e}{c}(\boldsymbol{\varepsilon}\wedge\boldsymbol{A})\cdot\frac{\partial\Omega_{\mathrm{op}}}{\partial\boldsymbol{\pi}_{\mathrm{op}}} = \frac{ie}{\hbar c}[(\boldsymbol{\varepsilon}\wedge\boldsymbol{A})\cdot\boldsymbol{R}, \Omega_{\mathrm{op}}]. \tag{102}$$

(The differential quotient stands for the limit $\lambda\to 0$ of $\lambda^{-1}\{\Omega_{\mathrm{op}}(\boldsymbol{R}, \boldsymbol{\pi}_{\mathrm{op}}+\boldsymbol{\lambda}) - \Omega_{\mathrm{op}}(\boldsymbol{R}, \boldsymbol{\pi}_{\mathrm{op}})\}$, where $\boldsymbol{\lambda} = (\lambda, 0, 0)$ and cycl.) Therefore one may write the expression (100) for this special class of operators $\Omega_{\mathrm{op}}$ as

$$\delta\Omega_{\mathrm{op}} = \frac{i}{\hbar}\,\boldsymbol{\varepsilon}\cdot[\boldsymbol{R}\wedge\boldsymbol{\pi}_{\mathrm{op}}+\tfrac{1}{2}\hbar\boldsymbol{\sigma}, \Omega_{\mathrm{op}}], \tag{103}$$

which has a form analogous to (15), with $\boldsymbol{P}_{\mathrm{op}}$ replaced by $\boldsymbol{\pi}_{\mathrm{op}}$.

For *spatial inversion* (18) one finds for the expectation value of $\Omega_{\mathrm{op}}$ in the new frame the expectation value of the operator (cf. (20) for the field-free case):

$$\beta\Omega_{\mathrm{op}}(-\boldsymbol{R}, -\boldsymbol{P}_{\mathrm{op}}, -\boldsymbol{A}, \varphi)\beta, \tag{104}$$

where we used the transformation of the potentials:

$$\begin{aligned}\varphi'(\boldsymbol{R}', t') &= \varphi(\boldsymbol{R}, t),\\ \boldsymbol{A}'(\boldsymbol{R}', t') &= -\boldsymbol{A}(\boldsymbol{R}, t).\end{aligned} \tag{105}$$

Therefore for a polar or axial vector operator one has the relation (cf. (21)):

$$\boldsymbol{\Omega}_{\mathrm{op}}(-\boldsymbol{R}, -\boldsymbol{P}_{\mathrm{op}}, -\boldsymbol{A}, \varphi) = \mp\beta\boldsymbol{\Omega}_{\mathrm{op}}(\boldsymbol{R}, \boldsymbol{P}_{\mathrm{op}}, \boldsymbol{A}, \varphi)\beta. \tag{106}$$

Under *time reversal* (22) one obtains for the expectation value of $\Omega_{op}$ in the new frame the expectation value of the operator (cf. (26) for the field-free case):

$$T\tilde{\Omega}_{op}(\boldsymbol{R}, -\boldsymbol{P}_{op}, -\boldsymbol{A}, \varphi)T^{-1}, \tag{107}$$

where we used the transformation property of the potentials

$$\begin{aligned} \varphi'(\boldsymbol{R}', t') &= \varphi(\boldsymbol{R}, t), \\ \boldsymbol{A}'(\boldsymbol{R}', t') &= -\boldsymbol{A}(\boldsymbol{R}, t). \end{aligned} \tag{108}$$

The behaviour of an expectation value under *pure Lorentz transformations* follows from the transformation of the wave function, which is given by an expression as (32) but now with a Hamilton operator $H_{op}$ which depends on the potentials $\varphi(\hat{\boldsymbol{R}}', t)$ and $\boldsymbol{A}(\hat{\boldsymbol{R}}', t)$. This expression has to be substituted into the transformed expectation value, which reads as (29) but with an operator $\Omega_{op}$ which depends also upon the transformed potentials $\varphi'(\hat{\boldsymbol{R}}', \hat{t}')$ and $\boldsymbol{A}'(\hat{\boldsymbol{R}}', \hat{t}')$. If one uses the transformation formulae for the potentials

$$\begin{aligned} \boldsymbol{A}'(\hat{\boldsymbol{R}}', \hat{t}') &= \boldsymbol{A}(\hat{\boldsymbol{R}}', t) - \boldsymbol{\varepsilon}\varphi(\hat{\boldsymbol{R}}', t) + ct\boldsymbol{\varepsilon}\cdot \frac{\partial \boldsymbol{A}(\hat{\boldsymbol{R}}', t)}{\partial \boldsymbol{R}'} + c^{-1}\boldsymbol{\varepsilon}\cdot\hat{\boldsymbol{R}}' \frac{\partial \boldsymbol{A}(\hat{\boldsymbol{R}}', t)}{\partial t}, \\ \varphi'(\hat{\boldsymbol{R}}', \hat{t}') &= \varphi(\hat{\boldsymbol{R}}', t) - \boldsymbol{\varepsilon}\cdot\boldsymbol{A}(\hat{\boldsymbol{R}}', t) + ct\boldsymbol{\varepsilon}\cdot \frac{\partial \varphi(\hat{\boldsymbol{R}}', t)}{\partial \hat{\boldsymbol{R}}'} + c^{-1}\boldsymbol{\varepsilon}\cdot\hat{\boldsymbol{R}}' \frac{\partial \varphi(\hat{\boldsymbol{R}}', t)}{\partial t}, \end{aligned} \tag{109}$$

one finds that the expectation value changes by a quantity which is the expectation value of the operator (cf. (33) for the field-free case):

$$\begin{aligned} \delta\Omega_{op} &= \frac{i}{\hbar}\boldsymbol{\varepsilon}\cdot[\boldsymbol{N}_{op} - ct\boldsymbol{P}_{op}, \Omega_{op}] \\ &\quad + \Omega_{op}\left(\boldsymbol{R}, \boldsymbol{P}_{op}, \boldsymbol{A} - \boldsymbol{\varepsilon}\varphi + ct\boldsymbol{\varepsilon}\cdot\nabla\boldsymbol{A} + c^{-1}\boldsymbol{\varepsilon}\cdot\boldsymbol{R}\frac{\partial \boldsymbol{A}}{\partial t},\right. \\ &\qquad \left.\varphi - \boldsymbol{\varepsilon}\cdot\boldsymbol{A} + ct\boldsymbol{\varepsilon}\cdot\nabla\varphi + c^{-1}\boldsymbol{\varepsilon}\cdot\boldsymbol{R}\frac{\partial\varphi}{\partial t}\right) - \Omega_{op}(\boldsymbol{R}, \boldsymbol{P}_{op}, \boldsymbol{A}, \varphi), \end{aligned} \tag{110}$$

where $\boldsymbol{N}_{op}$ is given by (34) or (35) with (93).

If one confines oneself to terms without the derivatives of the potentials, this expression reduces to

$$\delta\Omega_{op} = \frac{i}{\hbar}\boldsymbol{\varepsilon}\cdot[\boldsymbol{N}_{op} - ct\boldsymbol{P}_{op}, \Omega_{op}] - \boldsymbol{\varepsilon}\cdot\frac{\partial\Omega_{op}}{\partial \boldsymbol{A}}\varphi - \frac{\partial\Omega_{op}}{\partial\varphi}\boldsymbol{\varepsilon}\cdot\boldsymbol{A}. \tag{111}$$

In particular if the operator $\Omega_{op}$ is independent of $\varphi$ and depends on $\boldsymbol{A}$ only through $\boldsymbol{\pi}_{\text{op}} \equiv \boldsymbol{P}_{\text{op}} - (e/c)\boldsymbol{A}$, the last two terms become

$$\frac{e}{c}\,\boldsymbol{\varepsilon}\cdot\frac{\partial \Omega_{\text{op}}}{\partial \boldsymbol{\pi}_{\text{op}}}\,\varphi = -\frac{ie}{\hbar c}\,[\boldsymbol{\varepsilon}\cdot\boldsymbol{R}\varphi,\, \Omega_{\text{op}}], \tag{112}$$

so that (111) gets the form:

$$\delta\Omega_{\text{op}} = \frac{i}{\hbar}\,\boldsymbol{\varepsilon}\cdot[\boldsymbol{N}_{\text{op}}^{(\varphi)} - ct\boldsymbol{P}_{\text{op}},\, \Omega_{\text{op}}], \tag{113}$$

where we used the abbreviation

$$\boldsymbol{N}_{\text{op}}^{(\varphi)} \equiv \boldsymbol{N}_{\text{op}} - (e/c)\boldsymbol{R}\varphi = \tfrac{1}{2}c^{-1}\{\boldsymbol{R},\, H_{\text{op}} - e\varphi\}. \tag{114}$$

In the last member the definition (34) or (35) has been used.

b. *Covariance requirements on the position and spin operators for a particle in a field*

The position and spin operators for a free Dirac particle have been found in section 2. If the particle moves in an electromagnetic field the problem of the derivation of the position and spin operators should be reconsidered from the beginning. The expression for these operators should reduce to those of the field-free case if the fields are switched off. In the presence of fields the position and spin operators will contain additional terms with the potentials and their derivatives whith respect to time and space coordinates. In the following we shall be interested only in those additional terms which contain the potentials, not their derivatives. These additional terms will be determined by imposing a number of conditions just as in the field-free case. In this section we shall be concerned with the requirements of covariance with respect to the Poincaré group.

As *translation* properties we impose again (36–37) on the position and spin operator. As *rotation* properties we require that both the position and spin operator be vector operators, so that we have, in view of (101),

$$[(\boldsymbol{R}\wedge\boldsymbol{P}_{\text{op}} + \tfrac{1}{2}\hbar\boldsymbol{\sigma})^i,\, X_{\text{op}}^j] - i\hbar\varepsilon^{imn}A_m\frac{\partial X_{\text{op}}^j}{\partial A^n} = i\hbar\varepsilon^{ijk}X_{k,\text{op}}, \tag{115}$$

$$[(\boldsymbol{R}\wedge\boldsymbol{P}_{\text{op}} + \tfrac{1}{2}\hbar\boldsymbol{\sigma})^i,\, s_{\text{op}}^j] - i\hbar\varepsilon^{imn}A_m\frac{\partial s_{\text{op}}^j}{\partial A^n} = i\hbar\varepsilon^{ijk}s_{k,\text{op}}. \tag{116}$$

As to the properties under *spatial inversion*, we require that the position operator be a polar vector and the spin operator an axial vector. We have thus from (106)

$$\boldsymbol{X}_{\text{op}}(-\boldsymbol{R}, -\boldsymbol{P}_{\text{op}}, -\boldsymbol{A}, \varphi) = -\beta \boldsymbol{X}_{\text{op}}(\boldsymbol{R}, \boldsymbol{P}_{\text{op}}, \boldsymbol{A}, \varphi)\beta, \tag{117}$$

$$\boldsymbol{s}_{\text{op}}(-\boldsymbol{R}, -\boldsymbol{P}_{\text{op}}, -\boldsymbol{A}, \varphi) = \beta \boldsymbol{s}_{\text{op}}(\boldsymbol{R}, \boldsymbol{P}_{\text{op}}, \boldsymbol{A}, \varphi)\beta. \tag{118}$$

Under *time reversal* the expectation value of the position operator should remain invariant, while the spin operator must change sign, so that one must have in view of (107)

$$\boldsymbol{X}_{\text{op}}(\boldsymbol{R}, -\boldsymbol{P}_{\text{op}}, -\boldsymbol{A}, \varphi) = T\tilde{\boldsymbol{X}}_{\text{op}}(\boldsymbol{R}, \boldsymbol{P}_{\text{op}}, \boldsymbol{A}, \varphi)T^{-1}, \tag{119}$$

$$\boldsymbol{s}_{\text{op}}(\boldsymbol{R}, -\boldsymbol{P}_{\text{op}}, -\boldsymbol{A}, \varphi) = -T\tilde{\boldsymbol{s}}_{\text{op}}(\boldsymbol{R}, \boldsymbol{P}_{\text{op}}, \boldsymbol{A}, \varphi)T^{-1}. \tag{120}$$

For the transformation property of the position operator under a *pure Lorentz transformation* we postulate, just as in the field-free case (cf. (44)), that

$$\delta \boldsymbol{X}_{\text{op}} = -\boldsymbol{\varepsilon} ct + \frac{i}{2\hbar c}\{\boldsymbol{\varepsilon}\cdot\boldsymbol{X}_{\text{op}}, [H_{\text{op}}, \boldsymbol{X}_{\text{op}}]\}, \tag{121}$$

but where now the Hamilton operator $H_{\text{op}}$ stands for the expression (93). With (111) for $\boldsymbol{\Omega}_{\text{op}} = \boldsymbol{X}_{\text{op}}$ and (36) we find from (121)

$$[N^i_{\text{op}}, X^j_{\text{op}}] - \frac{\hbar}{i}\left(\frac{\partial X^j_{\text{op}}}{\partial A_i}\varphi + \frac{\partial X^j_{\text{op}}}{\partial \varphi}A^i\right) = \tfrac{1}{2}c^{-1}\{X^i_{\text{op}}, [H_{\text{op}}, X^j_{\text{op}}]\}. \tag{122}$$

For the transformation property of the spin operator under a *pure Lorentz transformation* we also postulate an equation of the same form as in the field-free case i.e. (46–47), but with $H_{\text{op}}$ (93) inserted. With (37) and (111) for $\boldsymbol{\Omega}_{\text{op}} = \boldsymbol{s}_{\text{op}}$ we find from (46–47):

$$[N^i_{\text{op}}, s^j_{\text{op}}] - \frac{\hbar}{i}\left(\frac{\partial s^j_{\text{op}}}{\partial A_i}\varphi + \frac{\partial s^j_{\text{op}}}{\partial \varphi}A^i\right) = \tfrac{1}{2}c^{-1}\{X^i_{\text{op}}, [H_{\text{op}}, s^j_{\text{op}}]\} - \frac{\hbar}{i}\varepsilon^{ijk}t_{k,\text{op}}, \tag{123}$$

with the three-vector operator $\boldsymbol{t}_{\text{op}}$ such that

$$[N^i_{\text{op}}, t^j_{\text{op}}] - \frac{\hbar}{i}\left(\frac{\partial t^j_{\text{op}}}{\partial A_i}\varphi + \frac{\partial t^j_{\text{op}}}{\partial \varphi}A^i\right) = \tfrac{1}{2}c^{-1}\{X^i_{\text{op}}, [H_{\text{op}}, t^j_{\text{op}}]\} + \frac{\hbar}{i}\varepsilon^{ijk}s_{k,\text{op}}. \tag{124}$$

In the following we shall find the position and spin operators up to terms in the potentials by using the covariance requirements given above. We first have to transform the Hamilton operator for a Dirac particle in a field. It will be convenient to use Weyl transforms in the course of the reasoning. For that reason a short digression on Weyl transforms, in particular their generalization to operators pertaining to particles with internal degrees of freedom, will now be given.

c. *Weyl transforms for particles with spin*

In chapter VI the theory of Weyl transforms of operators for point particles was considered. If the particles have structure, this method has to be generalized somewhat. (See also the appendix of chapter VI for details.) Indeed to every eigenvalue $\boldsymbol{p}$ and $\boldsymbol{q}$ of the momentum and coordinate operators $\boldsymbol{P}$ and $\boldsymbol{Q}$ [1] now correspond several eigenstates, which will be labelled by an extra index:

$$\boldsymbol{P}|\boldsymbol{p}, \kappa\rangle = \boldsymbol{p}|\boldsymbol{p}, \kappa\rangle, \qquad \boldsymbol{Q}|\boldsymbol{q}, \kappa\rangle = \boldsymbol{q}|\boldsymbol{q}, \kappa\rangle. \tag{125}$$

In Dirac theory $\kappa$ assumes the values 1, 2, 3 or 4. From this basis we construct the operator

$$\Omega_{\kappa\lambda} = \int \mathrm{d}\boldsymbol{p}|\boldsymbol{p}, \kappa\rangle\langle\boldsymbol{p}, \lambda| = \int \mathrm{d}\boldsymbol{q}|\boldsymbol{q}, \kappa\rangle\langle\boldsymbol{q}, \lambda|, \tag{126}$$

which transforms the subspace of Hilbert space labelled by $\lambda$ into that labelled by $\kappa$.

The Weyl transform of an operator $A$ may be defined in a way which is analogous to that of the theory of point particles. One gets (cf. (VI.14) and (VI.26)):

$$a_{\kappa\lambda}(\boldsymbol{p}, \boldsymbol{q}) = \int \mathrm{d}\boldsymbol{u}\, \mathrm{e}^{(i/\hbar)\boldsymbol{q}\cdot\boldsymbol{u}}\langle\boldsymbol{p}+\tfrac{1}{2}\boldsymbol{u}, \kappa|A|\boldsymbol{p}-\tfrac{1}{2}\boldsymbol{u}, \lambda\rangle$$
$$= \int \mathrm{d}\boldsymbol{v}\, \mathrm{e}^{(i/\hbar)\boldsymbol{p}\cdot\boldsymbol{v}}\langle\boldsymbol{q}-\tfrac{1}{2}\boldsymbol{v}, \kappa|A|\boldsymbol{q}+\tfrac{1}{2}\boldsymbol{v}, \lambda\rangle. \tag{127}$$

The Weyl transform thus depends on a pair of labels $\kappa\lambda$. From the Weyl transform one may recover the operator (cf. (VI.13)):

$$A = h^{-3}\sum_{\kappa,\lambda}\int \mathrm{d}\boldsymbol{p}\,\mathrm{d}\boldsymbol{q}\; a_{\kappa\lambda}(\boldsymbol{p}, \boldsymbol{q})\Delta_{\kappa\lambda}(\boldsymbol{p}, \boldsymbol{q}). \tag{128}$$

Here the operator $\Delta_{\kappa\lambda}(\boldsymbol{p}, \boldsymbol{q})$ is given by

$$\Delta_{\kappa\lambda}(\boldsymbol{p}, \boldsymbol{q}) = \Delta(\boldsymbol{p}, \boldsymbol{q})\Omega_{\kappa\lambda} \tag{129}$$

with the two operators $\Delta(\boldsymbol{p}, \boldsymbol{q})$ (VI.15) and $\Omega_{\kappa\lambda}$ (126).

In the special case that the operator $A$ does not connect the different parts of Hilbert space labelled by $\kappa$ and acts moreover in each subspace in the same way, one finds from (127) that its Weyl transform has the form

$$a_{\kappa\lambda}(\boldsymbol{p}, \boldsymbol{q}) = \delta_{\kappa\lambda}\, a(\boldsymbol{p}, \boldsymbol{q}), \tag{130}$$

where $a(\boldsymbol{p}, \boldsymbol{q})$ is independent of $\kappa$ and $\lambda$.

[1] In this subsection we use capitals for operators and lower case symbols for *c*-numbers.

If the operator $A$ is independent of the coordinate and momentum operators (as for instance the Dirac matrices) its Weyl transform (127) is independent of $\boldsymbol{p}$ and $\boldsymbol{q}$:

$$a_{\kappa\lambda}(\boldsymbol{p}, \boldsymbol{q}) = a_{\kappa\lambda}. \tag{131}$$

For the Weyl transform of a product of operators one obtains (cf. (VI.42))

$$AB \rightleftarrows \exp\left\{\frac{i\hbar}{2}\left(\frac{\partial^{(a)}}{\partial \boldsymbol{q}}\cdot\frac{\partial^{(b)}}{\partial \boldsymbol{p}} - \frac{\partial^{(a)}}{\partial \boldsymbol{p}}\cdot\frac{\partial^{(b)}}{\partial \boldsymbol{q}}\right)\right\} \sum_{\mu} a_{\kappa\mu}(\boldsymbol{p}, \boldsymbol{q}) b_{\mu\lambda}(\boldsymbol{p}, \boldsymbol{q}). \tag{132}$$

This expression permits to find the Weyl transforms of the commutator and anticommutator of two operators as well (see chapter VI, formulae (A160–161).

## d. *Transformation of the Hamilton operator*

The Hamiltonian (93) will be put to even form by three successive transformations. First a transformation will be performed[1] with the operator

$$S_{1,\mathrm{op}} \rightleftarrows \frac{E_\pi + mc^2 + c\beta\boldsymbol{\alpha}\cdot\boldsymbol{\pi}}{\{2E_\pi(E_\pi + mc^2)\}^{\frac{1}{2}}}, \tag{133}$$

where we used the abbreviation

$$E_\pi \equiv (c^2\boldsymbol{\pi}^2 + m^2c^4)^{\frac{1}{2}}, \tag{134}$$

with $\boldsymbol{\pi} \equiv \boldsymbol{P} - (e/c)\boldsymbol{A}$. At the right-hand side of (133) the Weyl transform has been written. If $e = 0$, the operator $S_{1,\mathrm{op}}$ reduces to $U_{\mathrm{op}}$ (56) of which the Weyl transform will be denoted by $U$. If only terms linear in $e$ and without second and higher derivatives of the potentials are taken into account we find

$$\begin{aligned} S_{1,\mathrm{op}} H_{\mathrm{op}} S^\dagger_{1,\mathrm{op}} \rightleftarrows S_1 H S^\dagger_1 &+ \frac{ie\hbar}{2c}\varepsilon_{ijk}\frac{\partial U}{\partial P_i}\frac{\partial H}{\partial P_j} U^\dagger B^k + \frac{ie\hbar}{2c}\varepsilon_{ijk} U \frac{\partial H}{\partial P_i}\frac{\partial U^\dagger}{\partial P_j} B^k \\ &+ \frac{ie\hbar}{2c}\varepsilon_{ijk}\frac{\partial U}{\partial P_i} H \frac{\partial U^\dagger}{\partial P_j} B^k - \frac{ie\hbar}{2}\frac{\partial U}{\partial P_i}\frac{\partial \varphi}{\partial R^i} U^\dagger + \frac{ie\hbar}{2} U \frac{\partial \varphi}{\partial R_i}\frac{\partial U^\dagger}{\partial P^i}, \end{aligned} \tag{135}$$

since the Weyl transform of $S_{1,\mathrm{op}}$ depends on $\boldsymbol{P}$ and $\boldsymbol{R}$ only through $\boldsymbol{\pi}$. Here the same symbols are used for operators (l.h.s.) and their Weyl transforms (r.h.s.). The operator $S_{1,\mathrm{op}}$ is not unitary since

$$S_{1,\mathrm{op}} S^\dagger_{1,\mathrm{op}} \rightleftarrows 1 + \frac{ie\hbar}{2c}\varepsilon_{ijk}\frac{\partial U}{\partial P_i}\frac{\partial U^\dagger}{\partial P_j} B^k. \tag{136}$$

[1] E. I. Blount, Phys. Rev. **126**(1962)1636, **128**(1962)2454.

However the product $U_{1,\mathrm{op}} = S_{2,\mathrm{op}} S_{1,\mathrm{op}}$ is unitary (up to terms linear in $e$ and without second and higher derivatives of the potentials), if $S_{2,\mathrm{op}}$ is chosen such that

$$S_{2,\mathrm{op}} \rightleftarrows 1 - \frac{ie\hbar}{4c}\,\varepsilon_{ijk}\,\frac{\partial U}{\partial P_i}\,\frac{\partial U^\dagger}{\partial P_j}\,B^k. \tag{137}$$

The transformed Hamiltonian becomes

$$\begin{aligned} U_{1,\mathrm{op}} H_{\mathrm{op}} U^\dagger_{1,\mathrm{op}} \rightleftarrows\; & S_1 H S_1^\dagger + \frac{ie\hbar}{2c}\,\varepsilon_{ijk}\,\frac{\partial U}{\partial P_i}\,\frac{\partial H}{\partial P_j}\,U^\dagger B^k + \frac{ie\hbar}{2c}\,\varepsilon_{ijk}\,U\,\frac{\partial H}{\partial P_i}\,\frac{\partial U^\dagger}{\partial P_j}\,B^k \\ & + \frac{ie\hbar}{2c}\,\varepsilon_{ijk}\,\frac{\partial U}{\partial P_i}\,H\,\frac{\partial U^\dagger}{\partial P_j}\,B^k - \frac{ie\hbar}{2}\,\frac{\partial U}{\partial P_i}\,\frac{\partial \varphi}{\partial R^i}\,U^\dagger + \frac{ie\hbar}{2}\,U\,\frac{\partial \varphi}{\partial R_i}\,\frac{\partial U^\dagger}{\partial P^i} \\ & - \frac{ie\hbar}{4c}\,\varepsilon_{ijk}\,\frac{\partial U}{\partial P_i}\,\frac{\partial U^\dagger}{\partial P_j}\,B^k \beta E - \frac{ie\hbar}{4c}\,\beta E \varepsilon_{ijk}\,\frac{\partial U}{\partial P_i}\,\frac{\partial U^\dagger}{\partial P_j}\,B^k. \end{aligned} \tag{138}$$

We now introduce the abbreviation $\xi$ (62) and employ the identity

$$U\,\frac{\partial H}{\partial \boldsymbol{P}}\,U^\dagger = \frac{\partial}{\partial \boldsymbol{P}}(UHU^\dagger) - \frac{\partial U}{\partial \boldsymbol{P}}\,HU^\dagger - UH\,\frac{\partial U^\dagger}{\partial \boldsymbol{P}} = \beta\,\frac{c^2\boldsymbol{P}}{E} + \frac{2i}{\hbar}\,E\beta\xi_{\mathrm{o}}, \tag{139}$$

where o (and e) denote odd (and even) parts. In the last member we used the Weyl transform of (58) and (62). Then (138) becomes

$$\begin{aligned} U_{1,\mathrm{op}} H_{\mathrm{op}} U^\dagger_{1,\mathrm{op}} \rightleftarrows\; & \beta E_\pi + e\varphi - \frac{ieE}{2\hbar c}\,\varepsilon_{ijk}\{\xi^i, \beta\xi^j_{\mathrm{o}}\}B^k \\ & - \frac{ec}{E}\,\beta\xi_{\mathrm{e}}\cdot(\boldsymbol{P}\wedge\boldsymbol{B}) + e\xi\cdot\frac{\partial\varphi}{\partial\boldsymbol{R}} + \tfrac{1}{2}(g-2)\mu_{\mathrm{B}}\,U(i\beta\boldsymbol{\alpha}\cdot\boldsymbol{E} - \beta\boldsymbol{\sigma}\cdot\boldsymbol{B})U^\dagger. \end{aligned} \tag{140}$$

Since the time derivative of the transformed wave function is determined by $\{U_1 H U_1^\dagger - (\hbar/i)(\partial U_1/\partial t)U_1^\dagger\}_{\mathrm{op}}$ we also need $\partial U_{1,\mathrm{op}}/\partial t$ of which the Weyl transform is $-(e/c)(\partial U/\partial \boldsymbol{P})\cdot(\partial \boldsymbol{A}/\partial t)$ (up to terms linear in $e$ and without second derivatives of the potentials). We obtain thus

$$\begin{aligned} U_{1,\mathrm{op}} H_{\mathrm{op}} U^\dagger_{1,\mathrm{op}} - (\hbar/i)(\partial U_{1,\mathrm{op}}/\partial t)U^\dagger_{1,\mathrm{op}} \rightleftarrows\; & \beta E_\pi + e\varphi - \frac{ieE}{2\hbar c}\,\varepsilon_{ijk}\{\xi^i, \beta\xi^j_{\mathrm{o}}\}B^k \\ & - \frac{ec}{E}\,\beta\xi_{\mathrm{e}}\cdot(\boldsymbol{P}\wedge\boldsymbol{B}) - e\xi\cdot\boldsymbol{E} + \tfrac{1}{2}(g-2)\mu_{\mathrm{B}}\,U(i\beta\boldsymbol{\alpha}\cdot\boldsymbol{E} - \beta\boldsymbol{\sigma}\cdot\boldsymbol{B})U^\dagger. \end{aligned} \tag{141}$$

Here the odd terms which depend on the fields may be transformed away by means of a final unitary (up to terms linear in $e$ and without second and higher derivatives of the potentials) transformation

$$U_{2,\mathrm{op}} \rightleftarrows 1 - \frac{ie}{4\hbar c}\,\varepsilon_{ijk}\{\xi_{\mathrm{e}}^{i},\,\xi_{\mathrm{o}}^{j}\}B^{k}$$

$$-\frac{e}{2E}\,\beta\xi_{\mathrm{o}}\cdot\boldsymbol{E} + \frac{(g-2)\mu_{\mathrm{B}}}{4E}\,\beta\{U(i\beta\boldsymbol{\alpha}\cdot\boldsymbol{E}-\beta\boldsymbol{\sigma}\cdot\boldsymbol{B})U^{\dagger}\}_{\mathrm{o}}. \quad (142)$$

The explicit transformed Hamiltonian is obtained if we substitute the expressions (63). The result is – up to terms linear in $e$ and without second and higher derivatives of the potentials –

$$\hat{H}_{\mathrm{op}} \equiv U_{2,\mathrm{op}}U_{1,\mathrm{op}}H_{\mathrm{op}}U_{1,\mathrm{op}}^{\dagger}U_{2,\mathrm{op}}^{\dagger} - \frac{\hbar}{i}\,\frac{\partial(U_{2,\mathrm{op}}U_{1,\mathrm{op}})}{\partial t}\,U_{1,\mathrm{op}}^{\dagger}U_{2,\mathrm{op}}^{\dagger}$$

$$\rightleftarrows \beta E_{\pi} + e\varphi - \mu_{\mathrm{B}}\frac{mc^{2}}{E}\,\beta\boldsymbol{\sigma}\cdot\boldsymbol{B} - \mu_{\mathrm{B}}\frac{mc^{3}}{E(E+mc^{2})}\,(\boldsymbol{P}\wedge\boldsymbol{\sigma})\cdot\boldsymbol{E}$$

$$-\tfrac{1}{2}(g-2)\mu_{\mathrm{B}}\left\{\beta\boldsymbol{\sigma}\cdot\boldsymbol{B} - \frac{\beta c^{2}\boldsymbol{P}\cdot\boldsymbol{\sigma}\boldsymbol{P}\cdot\boldsymbol{B}}{E(E+mc^{2})} + \frac{c(\boldsymbol{P}\wedge\boldsymbol{\sigma})\cdot\boldsymbol{E}}{E}\right\}, \quad (143)$$

which is the relativistic generalization of the expression derived by Foldy–Wouthuysen[1]. Apart from the anomalous terms it has been found by Blount[2]. We shall call it the Hamiltonian in the Blount picture.

## e. *Covariant position and spin operators*

In order to obtain the equation of motion up to second order derivatives of the potentials an expression for the position operator including terms with the potentials will be needed. Since the Hamiltonian (143) in the Blount picture is even only the even part $\hat{X}_{\mathrm{op}}$ of the position operator in the Blount picture is relevant. This part is fixed by a set of conditions. In the first place the expression $\hat{\boldsymbol{X}}_{\mathrm{op}}$ in the Blount picture should reduce to the form (68) for the field-free case. If furthermore the transformation properties of $\boldsymbol{X}_{\mathrm{op}}$ under translations (36), rotations (115), spatial inversion (117), time reversal (119) with $T = \sigma_2$ in the Pauli representation and pure Lorentz transformations (122) are taken into account it follows after a straightforward but rather long calculation, that in the Blount picture $\hat{\boldsymbol{X}}_{\mathrm{op}}$ has the form

$$\hat{\boldsymbol{X}}_{\mathrm{op}} \rightleftarrows \boldsymbol{R} + \frac{\hbar\boldsymbol{\sigma}\wedge\boldsymbol{\pi}}{2m(E_{\pi}+mc^{2})}$$

$$+(a_1+\beta a_2)\left\{\frac{c}{E}\,\boldsymbol{P}\wedge\boldsymbol{\sigma}\boldsymbol{P}\cdot\boldsymbol{A} + \frac{m^{2}c^{3}}{E}\,\boldsymbol{A}\wedge\boldsymbol{\sigma} - \frac{mc^{3}\boldsymbol{P}\wedge\boldsymbol{A}\boldsymbol{P}\cdot\boldsymbol{\sigma}}{E(E+mc^{2})} - \beta\boldsymbol{P}\wedge\boldsymbol{\sigma}\varphi\right\}, \quad (144)$$

[1] L. L. Foldy and S. A. Wouthuysen, op. cit.

[2] E. I. Blount, op. cit.

where $a_1$ and $a_2$ are real arbitrary constants. The velocity operator (up to terms with the potentials) corresponding to this position operator follows with (143):

$$\hat{\boldsymbol{v}}_{\text{op}} \equiv \frac{i}{\hbar}[\hat{H}_{\text{op}}, \hat{\boldsymbol{X}}_{\text{op}}] \rightleftarrows \beta c^2 \boldsymbol{\pi}/E_\pi \equiv \hat{\boldsymbol{v}}. \tag{145}$$

In an analogous way the even part $\hat{\boldsymbol{s}}_{\text{op}}$ of the spin operator in the Blount picture up to terms with the potentials (which should reduce to (79) in the field-free case) is found by fixing its transformation properties under translations (37), rotations (116), spatial inversion (118), time reversal (120) with $T = \sigma_2$ and pure Lorentz transformations (123–124). We obtain

$$\hat{\boldsymbol{s}}_{\text{op}} \rightleftarrows \frac{\hbar E_\pi}{2mc^2}\boldsymbol{\sigma} - \frac{\hbar \boldsymbol{\pi}\boldsymbol{\pi}\cdot\boldsymbol{\sigma}}{2m(E_\pi + mc^2)} + (b_1 + \beta b_2)\left(\boldsymbol{A}\boldsymbol{P}\cdot\boldsymbol{\sigma} - mc\beta\boldsymbol{\sigma}\varphi - \frac{\beta c \boldsymbol{P}\boldsymbol{P}\cdot\boldsymbol{\sigma}\varphi}{E+mc^2}\right)$$
$$+(b_3 + \beta b_4)\left(\frac{c\boldsymbol{P}\boldsymbol{P}\cdot\boldsymbol{A}\boldsymbol{P}\cdot\boldsymbol{\sigma}}{E+mc^2} - c^{-1}E\boldsymbol{\sigma}\boldsymbol{P}\cdot\boldsymbol{A} - \frac{\beta E \boldsymbol{P}\boldsymbol{P}\cdot\boldsymbol{\sigma}\varphi}{E+mc^2} + c^{-2}\beta E^2\boldsymbol{\sigma}\varphi\right). \tag{146}$$

The operator $\boldsymbol{t}_{\text{op}}$, which is connected with $\boldsymbol{s}_{\text{op}}$ according to (123–124), is in the Blount picture:

$$\hat{\boldsymbol{t}}_{\text{op}} \rightleftarrows \frac{\hbar\beta\boldsymbol{\pi}\wedge\boldsymbol{\sigma}}{2mc} + (b_1 + \beta b_2)\left(mc\beta\boldsymbol{\sigma}\wedge\boldsymbol{A} + \frac{c\beta\boldsymbol{P}\wedge\boldsymbol{A}\boldsymbol{P}\cdot\boldsymbol{\sigma}}{E+mc^2}\right)$$
$$+(b_3 + \beta b_4)(\beta\boldsymbol{\sigma}\wedge\boldsymbol{P}\boldsymbol{P}\cdot\boldsymbol{A} + c^{-1}E\varphi\boldsymbol{P}\wedge\boldsymbol{\sigma}). \tag{147}$$

A further constraint on the spin operator follows from the orthogonality condition

$$c^{-1}\hat{\boldsymbol{v}}_{\text{op}} \wedge \hat{\boldsymbol{s}}_{\text{op}} = \hat{\boldsymbol{t}}_{\text{op}}, \tag{148}$$

which is the quantum-mechanical counterpart (up to terms with the potentials) of the classical condition $p_\alpha s^{\alpha\beta} = 0$ (IV.67) with (IV.152). It is satisfied if $b_1$ and $b_2$ vanish.

The position operator and the spin operator are not independent of each other. As a generalization of the field-free case we shall require that the sum of the orbital angular momentum $\boldsymbol{X}_{\text{op}} \wedge \boldsymbol{\pi}_{\text{op}}$ and the spin $\boldsymbol{s}_{\text{op}}$ be equal to the operator $\boldsymbol{R} \wedge \boldsymbol{\pi}_{\text{op}} + \frac{1}{2}\hbar\boldsymbol{\sigma}$ in the Dirac picture. As shown in (103) this quantity is the generator of rotations for a special class of operators. The requirement reads written in the Blount picture

$$\hat{\boldsymbol{X}}_{\text{op}} \wedge \boldsymbol{\pi}_{\text{op}} + \hat{\boldsymbol{s}}_{\text{op}} = \boldsymbol{R} \wedge \boldsymbol{\pi}_{\text{op}} + \tfrac{1}{2}\hbar\boldsymbol{\sigma}. \tag{149}$$

If (144) and (146) are inserted we get the result that the remaining constants

$a_1$, $a_2$, $b_3$ and $b_4$ vanish as well, so that finally we obtain in the Blount picture

$$\hat{\boldsymbol{X}}_{\mathrm{op}} \rightleftarrows \boldsymbol{R} + \frac{\hbar\boldsymbol{\sigma}\wedge\boldsymbol{\pi}}{2m(E_\pi+mc^2)}, \tag{150}$$

$$\hat{\boldsymbol{s}}_{\mathrm{op}} \rightleftarrows \frac{\hbar E_\pi}{2mc^2}\boldsymbol{\sigma} - \frac{\hbar\boldsymbol{\pi}\boldsymbol{\pi}\cdot\boldsymbol{\sigma}}{2m(E_\pi+mc^2)} \tag{151}$$

and in the Dirac picture

$$\boldsymbol{X}_{\mathrm{op}} \rightleftarrows \boldsymbol{R} + \frac{i\hbar}{2mc}\beta\left(\boldsymbol{\alpha} - c^2\frac{\boldsymbol{\alpha}\cdot\boldsymbol{\pi}\boldsymbol{\pi}}{E_\pi^2}\right), \tag{152}$$

$$\boldsymbol{s}_{\mathrm{op}} \rightleftarrows \tfrac{1}{2}\hbar\boldsymbol{\sigma} - \frac{i\hbar\beta\boldsymbol{\alpha}\wedge\boldsymbol{\pi}}{2mc} \tag{153}$$

as position and spin operators.

With the help of these final results for the position and spin operators up to terms with the potentials we shall derive equations of motion and spin.

### f. *Equations of motion and of spin*

The *equation of motion* for the Dirac particle is obtained by taking twice the total time derivative (with the use of the Hamiltonian (143)) of the position operator in the Blount picture. In the first place we have to evaluate the velocity operator. The Weyl transform of the commutator $[\hat{H}_{\mathrm{op}}, \hat{\boldsymbol{X}}_{\mathrm{op}}]$ can be expressed in terms of the Weyl transforms of $\hat{H}_{\mathrm{op}}$ and $\hat{\boldsymbol{X}}_{\mathrm{op}}$ with the use of (132). Up to terms linear in $e$ and without field derivatives one obtains for the velocity operator (cf. (145)):

$$\begin{aligned}\hat{\boldsymbol{v}}_{\mathrm{op}} &\equiv \frac{\mathrm{d}\hat{\boldsymbol{X}}_{\mathrm{op}}}{\mathrm{d}t} \equiv \frac{i}{\hbar}[\hat{H}_{\mathrm{op}}, \hat{\boldsymbol{X}}_{\mathrm{op}}] + \frac{\partial\hat{\boldsymbol{X}}_{\mathrm{op}}}{\partial t} \\ &\rightleftarrows \frac{\beta\boldsymbol{\pi}c^2}{E_\pi} - \frac{e\hbar c\beta\boldsymbol{\sigma}\boldsymbol{P}\cdot\boldsymbol{B}}{2mE(E+mc^2)} + \frac{e\hbar c\beta\boldsymbol{P}\boldsymbol{\sigma}\cdot\boldsymbol{B}(E^2+Emc^2+m^2c^4)}{2mE^3(E+mc^2)} \\ &\quad + \frac{e\hbar c^2\boldsymbol{P}(\boldsymbol{P}\wedge\boldsymbol{\sigma})\cdot\boldsymbol{E}}{2mE^3} + \tfrac{1}{2}(g-2)\mu_{\mathrm{B}}\left\{\frac{\beta\boldsymbol{P}\cdot\boldsymbol{\sigma}\boldsymbol{B}}{mE} - \frac{\beta c^2\boldsymbol{P}\boldsymbol{P}\cdot\boldsymbol{\sigma}\boldsymbol{P}\cdot\boldsymbol{B}}{mE^3}\right. \\ &\quad \left. + \frac{c\boldsymbol{E}\wedge\boldsymbol{\sigma}}{E} + \frac{c^3\boldsymbol{P}(\boldsymbol{P}\wedge\boldsymbol{\sigma})\cdot\boldsymbol{E}}{E^3} - \frac{c\boldsymbol{P}\cdot\boldsymbol{\sigma}\boldsymbol{P}\wedge\boldsymbol{E}}{mE(E+mc^2)}\right\}.\end{aligned} \tag{154}$$

Likewise the acceleration operator may be calculated up to terms linear in $e$ and with first derivatives of the field

$$\frac{\mathrm{d}^2\hat{\boldsymbol{X}}_{\text{op}}}{\mathrm{d}t^2} \equiv \frac{\mathrm{d}\hat{\boldsymbol{v}}_{\text{op}}}{\mathrm{d}t} \equiv \frac{i}{\hbar}[\hat{H}_{\text{op}}, \hat{\boldsymbol{v}}_{\text{op}}] + \frac{\partial \hat{\boldsymbol{v}}_{\text{op}}}{\partial t} \rightleftarrows \frac{c^2}{E}\left(\mathsf{U} - \frac{c^2\boldsymbol{P}\boldsymbol{P}}{E^2}\right)$$

$$\cdot \Bigg[ e\beta\boldsymbol{E} + \frac{ec}{E}\boldsymbol{P}\wedge\boldsymbol{B} + \mu_{\text{B}}\left\{\frac{mc^2}{E}(\boldsymbol{\nabla}\boldsymbol{B})\cdot\boldsymbol{\sigma} + \frac{\beta mc^3}{E(E+mc^2)}(\boldsymbol{\nabla}\boldsymbol{E})\cdot(\boldsymbol{P}\wedge\boldsymbol{\sigma})\right\}$$

$$+\mu_{\text{B}}\left(\frac{\partial}{\partial t} + \frac{\beta c^2\boldsymbol{P}\cdot\boldsymbol{\nabla}}{E}\right)\left\{\frac{\boldsymbol{P}(\boldsymbol{P}\wedge\boldsymbol{\sigma})\cdot\boldsymbol{E}}{m^2c^3} - \frac{\beta\boldsymbol{\sigma}\boldsymbol{P}\cdot\boldsymbol{B}}{E+mc^2} + \beta\frac{\boldsymbol{P}\boldsymbol{\sigma}\cdot\boldsymbol{B}(E^2+Emc^2+m^2c^4)}{m^2c^4(E+mc^2)}\right.$$

$$\left. - \frac{\beta\boldsymbol{P}\boldsymbol{P}\cdot\boldsymbol{\sigma}\boldsymbol{P}\cdot\boldsymbol{B}}{m^2c^2(E+mc^2)}\right\} + \tfrac{1}{2}(g-2)\mu_{\text{B}}\left\{(\boldsymbol{\nabla}\boldsymbol{B})\cdot\boldsymbol{\sigma} - \frac{c^2(\boldsymbol{\nabla}\boldsymbol{B})\cdot\boldsymbol{P}\boldsymbol{P}\cdot\boldsymbol{\sigma}}{E(E+mc^2)} - \beta\frac{c(\boldsymbol{\nabla}\boldsymbol{E})\cdot(\boldsymbol{\sigma}\wedge\boldsymbol{P})}{E}\right\}$$

$$+\tfrac{1}{2}(g-2)\mu_{\text{B}}\left(\frac{\partial}{\partial t} + \frac{\beta c^2\boldsymbol{P}\cdot\boldsymbol{\nabla}}{E}\right)\left\{c^{-1}\boldsymbol{E}\wedge\boldsymbol{\sigma} + \frac{\beta\boldsymbol{\sigma}\cdot\boldsymbol{P}\boldsymbol{B}}{mc^2} - \frac{\boldsymbol{P}\cdot\boldsymbol{\sigma}\boldsymbol{P}\wedge\boldsymbol{E}}{mc(E+mc^2)}\right\}\Bigg]. \tag{155}$$

(The time and space derivations act only on the fields.) The first terms at the right-hand side contain the fields $\boldsymbol{E}(\boldsymbol{R}, t)$ and $\boldsymbol{B}(\boldsymbol{R}, t)$ as functions of the space coordinate in the Blount picture. Since the position of the particle is given by $\hat{\boldsymbol{X}}$ (150), we now wish to introduce the fields as functions of $\boldsymbol{X}$. Then we obtain for the first two terms on the right-hand side of (155) up to terms linear in $e$ and with first derivatives of the fields

$$e\beta\boldsymbol{E}(\hat{\boldsymbol{X}}, t) + \frac{ec}{E}\boldsymbol{P}\wedge\boldsymbol{B}(\hat{\boldsymbol{X}}, t) - \mu_{\text{B}}\left\{\frac{\beta c(\boldsymbol{\sigma}\wedge\boldsymbol{P})\cdot\boldsymbol{\nabla}\boldsymbol{E}}{E+mc^2} + \frac{c^2(\boldsymbol{\sigma}\wedge\boldsymbol{P})\cdot\boldsymbol{\nabla}(\boldsymbol{P}\wedge\boldsymbol{B})}{E(E+mc^2)}\right\}. \tag{156}$$

(The non-commutative character of the components of $\hat{\boldsymbol{X}}$ does not cause trouble here because of the limitation to first derivatives of the fields.) Furthermore we introduce the spin operator $\hat{\boldsymbol{s}}$ (151) instead of $\boldsymbol{\sigma}$; since in (155) $\boldsymbol{\sigma}$ is only needed up to order $e^0$ we write

$$\tfrac{1}{2}\hbar\boldsymbol{\sigma} = \frac{mc^2}{E}\hat{\boldsymbol{s}} + \frac{c^2\boldsymbol{P}\boldsymbol{P}\cdot\hat{\boldsymbol{s}}}{E(E+mc^2)}. \tag{157}$$

Substituting (156) and (157) into (155), using the Maxwell equation $\boldsymbol{\nabla}\wedge\boldsymbol{E} = -\partial_0\boldsymbol{B}$ and introducing the abbreviations $\boldsymbol{\beta} \equiv c\boldsymbol{P}/E$, $\gamma \equiv (1-\boldsymbol{\beta}^2)^{-\frac{1}{2}}$ and $\partial_0 \equiv c^{-1}\partial/\partial t$ we obtain as the equation of motion:

$$m\frac{\mathrm{d}\hat{\boldsymbol{v}}_{\text{op}}}{\mathrm{d}t} \rightleftarrows \gamma^{-2}(\mathsf{U}-\boldsymbol{\beta}\boldsymbol{\beta})\cdot\Bigg[\gamma e\boldsymbol{E}(\hat{\boldsymbol{X}}, t) + \gamma e\boldsymbol{\beta}\wedge\boldsymbol{B}(\hat{\boldsymbol{X}}, t)$$

$$+\frac{ge}{2mc}\{(\boldsymbol{\nabla}\boldsymbol{B})\cdot\hat{\boldsymbol{s}} + (\boldsymbol{\nabla}\boldsymbol{E})\cdot(\boldsymbol{\beta}\wedge\hat{\boldsymbol{s}}) + \gamma^2(\partial_0+\boldsymbol{\beta}\cdot\boldsymbol{\nabla})\boldsymbol{\beta}\hat{\boldsymbol{s}}\cdot(\boldsymbol{B}-\boldsymbol{\beta}\wedge\boldsymbol{E})\}$$

$$-\frac{(g-2)e}{2mc}\gamma^2(\partial_0+\boldsymbol{\beta}\cdot\boldsymbol{\nabla})\{\hat{\boldsymbol{s}}\wedge(\boldsymbol{E}+\boldsymbol{\beta}\wedge\boldsymbol{B}) - \hat{\boldsymbol{s}}\wedge\boldsymbol{\beta}\boldsymbol{\beta}\cdot\boldsymbol{E}\}\Bigg]. \tag{158}$$

Here we have limited ourselves to the 'upper left' part of the matrix expression (i.e. $\beta$ replaced by 1) which is the relevant part if expectation values for positive-energy solutions are evaluated. (Again the time and space derivations act only on the fields.)

At the right-hand side various terms which represent forces appear. (The factor $\mathbf{U}-\boldsymbol{\beta}\boldsymbol{\beta}$ arose, because we considered the time derivative of the velocity operator. In classical theory one also encounters a similar factor in that case.) In the first place one recognizes the Lorentz force on the particle with charge $e$. Its velocity independent part is equal to $e\boldsymbol{E}$. The quantity $\boldsymbol{\beta}$ is, up to order $e^0$ and for positive-energy solutions, the Weyl transform of the velocity operator times $c^{-1}$, as (154) shows. Next, two terms with space derivatives of the fields $\boldsymbol{E}$ and $\boldsymbol{B}$ appear. The velocity independent part is the 'Kelvin force' $(\nabla\boldsymbol{B})\cdot\hat{\mathfrak{m}}$, where $\hat{\mathfrak{m}}$ stands for the total magnetic moment $(ge/2mc)\hat{\mathfrak{s}}$ in the Blount picture. Finally two terms with the total time derivation $\partial_0+\boldsymbol{\beta}\cdot\nabla$ of the fields appear. The velocity independent part is $-\partial_0(\hat{\mathfrak{m}}_a\wedge\boldsymbol{E})$ with $\hat{\mathfrak{m}}_a$ the anomalous magnetic moment $\{(g-2)e/2mc\}\hat{\mathfrak{s}}$ in the Blount picture. This magnetodynamic effect is seen to contain the vector product of the electric field $\boldsymbol{E}$ and the anomalous part of the magnetic moment[1].

The magnetodynamic effect was discussed extensively in recent years. Some authors[2] found, in contrast with the result obtained, that also the normal magnetic moment (or half it) contributes to this effect. This is a consequence of the fact that their treatment was based on non-covariant position operators, such as Newton–Wigner's or the even part of Dirac's operator (v. problems 4 and 5)[3].

The *spin* equation follows by taking the total time derivative of the spin operator (151) in the Blount picture. Using the Hamiltonian (143) we obtain for its Weyl transform

$$\frac{\mathrm{d}\hat{\mathfrak{s}}_{\mathrm{op}}}{\mathrm{d}t}\equiv\frac{i}{\hbar}[\hat{H}_{\mathrm{op}},\hat{\mathfrak{s}}_{\mathrm{op}}]+\frac{\partial\hat{\mathfrak{s}}_{\mathrm{op}}}{\partial t}\rightleftarrows\mu_{\mathrm{B}}\left\{\beta\boldsymbol{\sigma}\wedge\boldsymbol{B}-\frac{\beta c^2\boldsymbol{P}\wedge\boldsymbol{B}\boldsymbol{P}\cdot\boldsymbol{\sigma}}{E(E+mc^2)}+\left(\frac{c\boldsymbol{P}}{E}\wedge\boldsymbol{\sigma}\right)\wedge\boldsymbol{E}\right\}$$
$$+\tfrac{1}{2}(g-2)\mu_{\mathrm{B}}\left\{\beta\boldsymbol{\sigma}\wedge\boldsymbol{B}+\frac{\beta\boldsymbol{P}\wedge\boldsymbol{B}\boldsymbol{P}\cdot\boldsymbol{\sigma}}{m(E+mc^2)}-\frac{c\boldsymbol{P}\boldsymbol{\sigma}\cdot\boldsymbol{E}}{E}+\frac{E\boldsymbol{P}\cdot\boldsymbol{\sigma}}{mc}-\frac{c\boldsymbol{P}\boldsymbol{P}\cdot\boldsymbol{\sigma}\boldsymbol{P}\cdot\boldsymbol{E}}{mE(E+mc^2)}\right\},\qquad(159)$$

[1] L. G. Suttorp and S. R. de Groot, op. cit.

[2] A. Conort, Compt. Rend. **266** B(1968)1184; H. Bacry, Compt. Rend. **267** B(1968)89; W. Shockley, Phys. Rev. Lett. **20**(1968)343; W. Shockley and K. K. Thornber, Phys. Lett. **27** A(1968)534; J. H. Van Vleck and N. L. Huang, Phys. Lett. **28** A(1969)768.

[3] P. Hraskó, N. Cim. **3B**(1971)213, avoids this problem by studying particles with an anomalous magnetic moment only, using the Newton–Wigner position operator.

where only terms linear in $e$ and without derivatives of the fields have been included. Upon introduction of $\hat{\mathbf{s}}$ with the help of (157) the equation becomes

$$\frac{\mathrm{d}\hat{\mathbf{s}}_{\mathrm{op}}}{\mathrm{d}t} \rightleftarrows \frac{ge}{2mc}\frac{mc^2}{E}\left\{\beta\hat{\mathbf{s}}\wedge\boldsymbol{B}+\left(\frac{c\boldsymbol{P}}{E}\wedge\hat{\mathbf{s}}\right)\wedge\boldsymbol{E}\right\}$$
$$+\frac{(g-2)e}{2mc}\left\{\beta\frac{\boldsymbol{P}\cdot\hat{\mathbf{s}}}{mE}\boldsymbol{P}\wedge\boldsymbol{B}-\frac{mc^3}{E^2}\boldsymbol{P}\wedge(\hat{\mathbf{s}}\wedge\boldsymbol{E})-\frac{c}{mE^2}\boldsymbol{P}\wedge(\boldsymbol{P}\wedge\boldsymbol{E})\boldsymbol{P}\cdot\hat{\mathbf{s}}\right\}. \quad (160)$$

Using the same abbreviations as in (158) and replacing again the matrix $\beta$ by 1 (i.e. limiting ourselves to the part occurring in the expectation value for the positive-energy solutions) we get finally

$$\frac{\mathrm{d}\hat{\mathbf{s}}_{\mathrm{op}}}{\mathrm{d}t} \rightleftarrows \frac{ge}{2mc}\gamma^{-1}\{\hat{\mathbf{s}}\wedge\boldsymbol{B}+(\boldsymbol{\beta}\wedge\hat{\mathbf{s}})\wedge\boldsymbol{E}\}$$
$$+\frac{(g-2)e}{2mc}\gamma^{-1}\{\gamma^2\boldsymbol{\beta}\cdot\hat{\mathbf{s}}(\boldsymbol{E}+\boldsymbol{\beta}\wedge\boldsymbol{B})-\hat{\mathbf{s}}\boldsymbol{\beta}\cdot\boldsymbol{E}-\gamma^2\boldsymbol{\beta}\boldsymbol{\beta}\cdot\hat{\mathbf{s}}\boldsymbol{\beta}\cdot\boldsymbol{E}\}. \quad (161)$$

At the right-hand side various terms appear which express the torques exerted by the fields on the particle with magnetic moment. The first two terms contain the total magnetic moment $\hat{\mathfrak{m}}$, the remaining ones the anomalous part $\hat{\mathfrak{m}}_{\mathrm{a}}$ of it only. The velocity independent term is simply $\hat{\mathfrak{m}}\wedge\boldsymbol{B}$.

Equations (158) and (161) are the quantum-mechanical equations of motion and of spin for a Dirac particle with both a normal and an anomalous magnetic moment moving in an external electromagnetic field. They are essentially operator equations: at the right-hand side Weyl transforms appear from which the corresponding operators might be retraced.

From the operator equations one obtains directly equations for expectation values. They contain expectation values of products of operators. In general such expectation values are not equal to the product of expectation values of the individual operators. In the *classical* limit the expectation value of a product of operators does become equal – for narrow wave packets – to the product of expectation values. Then one obtains equations for expectation values, which have precisely the same form as the corresponding equations derived in classical theory (v. (IV.162–163))[1].

[1] Earlier discussions on the derivation of equations of classical form from quantum theory include a paper by D. M. Fradkin and R. H. Good jr. (Rev. Mod. Phys. **33**(1961) 343) on the motion of wave packets in homogeneous fields. Furthermore WKB methods have been used, for homogeneous fields, by S. I. Rubinow and J. B. Keller (Phys. Rev. **131**(1963)2789) and by K. Rafanelli and R. Schiller (Phys. Rev. **135B**(1964)279). W. G. Dixon (N. Cim. **38**(1965) 1616) employed position and spin operators without specifying the contributions due to electromagnetic potentials; he limited himself to a particle with a

## 4 *The free Klein–Gordon particle*

### a. *The Klein–Gordon equation and its transformation properties*

In this section and the following we shall derive the equation of motion for a relativistic particle without spin, i.e. a particle which is described by the Klein–Gordon equation. It will turn out that a treatment that is to a large extent analogous to the one given above may be followed. Again we shall study first the free particle before discussing the particle under the influence of an electromagnetic field.

The Klein–Gordon equation of a free particle without spin and with mass $m$ ($\neq 0$) reads in the coordinate representation

$$\left(\Box - \frac{m^2c^2}{\hbar^2}\right)\psi(\boldsymbol{R}, t) = 0, \tag{162}$$

with $\Box = \Delta - c^{-2}\partial^2/\partial t^2$ the d'Alembertian and $\psi(\boldsymbol{R}, t)$ the wave function. The inner product of two wave functions $\psi_1$ and $\psi_2$

$$\langle\psi_1|\psi_2\rangle \equiv \frac{i\hbar}{mc}\int \psi_1^* \overleftrightarrow{\partial_0} \psi_2 \, \mathrm{d}\boldsymbol{R} \equiv \frac{i\hbar}{mc^2}\int \left(\psi_1^* \frac{\partial \psi_2}{\partial t} - \frac{\partial \psi_1^*}{\partial t}\psi_2\right) \mathrm{d}\boldsymbol{R} \tag{163}$$

is defined in such a way, that it is conserved if $\psi_1$ and $\psi_2$ fulfil the Klein–Gordon equation. (In the following we shall consider only normalized wave functions $\psi$, i.e. with $\langle\psi|\psi\rangle = 1$.) The expectation value of an operator $\Omega_{\mathrm{op}}$ is defined as

$$\bar{\Omega}_{\mathrm{op}} = \frac{i\hbar}{mc^2}\int \left(\psi^* \Omega_{\mathrm{op}} \frac{\partial \psi}{\partial t} - \frac{\partial \psi^*}{\partial t}\Omega_{\mathrm{op}}\psi\right) \mathrm{d}\boldsymbol{R}. \tag{164}$$

The transformation properties of the wave function which leave the Klein–Gordon equation and the absolute value of the inner product (163) invariant are the following. Under *spatial translations* (6), $\boldsymbol{R}' = \boldsymbol{R} + \varepsilon$, $t' = t$, the wave function is invariant:

$$\psi'(\boldsymbol{R}', t') = \psi(\boldsymbol{R}, t). \tag{165}$$

---

normal magnetic moment in a homogeneous field. H. C. Corben (Phys. Rev. **121**(1961) 1833), M. Kolsrud (N. Cim. **39**(1965)504), E. Plahte (Suppl. N. Cim. **4**(1966)246, 291; **5**(1967)944), H. Yamasaki (Progr. Theor. Phys. **39**(1968)372) and K. Rafanelli (N. Cim. **67A**(1970)48) introduce proper time into Dirac theory without solving the difficulties of interpretation pertinent to this notion.

Under *rotations* (12), $\boldsymbol{R}' = \boldsymbol{R}+\boldsymbol{\varepsilon}\wedge\boldsymbol{R}$, $t' = t$, the wave function is also invariant

$$\psi'(\boldsymbol{R}', t') = \psi(\boldsymbol{R}, t). \tag{166}$$

Under *spatial inversion* (18), $\boldsymbol{R}' = -\boldsymbol{R}$, $t' = t$, we have once again[1]

$$\psi'(\boldsymbol{R}', t') = \psi(\boldsymbol{R}, t). \tag{167}$$

Under *time reversal* (22), $\boldsymbol{R}' = \boldsymbol{R}$, $t' = -t$, the wave function undergoes an anti-linear transformation

$$\psi'(\boldsymbol{R}', t') = \psi^*(\boldsymbol{R}, t). \tag{168}$$

Finally, under *pure Lorentz transformations* (27), $\boldsymbol{R}' = \boldsymbol{R}-\boldsymbol{\varepsilon}ct$, $ct' = ct-\boldsymbol{\varepsilon}\cdot\boldsymbol{R}$, we have

$$\psi'(\boldsymbol{R}', t') = \psi(\boldsymbol{R}, t). \tag{169}$$

b. *Feshbach and Villars's formulation*

The Klein–Gordon equation is a differential equation of second order in the time. It may be written in the form of a first order equation for a two-component wave function by introducing[2] the functions:

$$\begin{aligned} u &\equiv \frac{1}{\sqrt{2}}\left(\psi - \frac{\hbar}{imc^2}\frac{\partial\psi}{\partial t}\right), \\ v &\equiv \frac{1}{\sqrt{2}}\left(\psi + \frac{\hbar}{imc^2}\frac{\partial\psi}{\partial t}\right). \end{aligned} \tag{170}$$

With the two-component wave function

$$\Psi = \begin{pmatrix} u \\ v \end{pmatrix} \tag{171}$$

one may write the Klein–Gordon equation (162) for a free particle as

$$H_{\text{op}}\Psi = -\frac{\hbar}{i}\frac{\partial\Psi}{\partial t}, \tag{172}$$

with the Hamilton operator

$$H_{\text{op}} \equiv (\tau_3 + i\tau_2)\frac{\boldsymbol{P}_{\text{op}}^2}{2m} + mc^2\tau_3, \tag{173}$$

[1] Phase factors are ignored, just as before, since they have no influence on the expectation values considered.

[2] H. Feshbach and F. Villars, Rev. Mod. Phys. **30**(1958)24; cf. M. Taketani and S. Sakata, Proc. Phys. Math. Soc. Japan **22**(1940)757.

where $\tau_i$ are the Pauli matrices (52), and $\boldsymbol{P}_{\text{op}}$ is the momentum operator $(\hbar/i)\partial/\partial\boldsymbol{R}$.

The inner product (163) may now be written as

$$\langle\Psi_1|\Psi_2\rangle \equiv \int \Psi_1^\dagger \tau_3 \Psi_2 \,\mathrm{d}\boldsymbol{R}, \tag{174}$$

where the obelisk denotes the hermitian conjugate. This follows by insertion of the components (170) of the wave functions $\Psi_1$ and $\Psi_2$. The expectation value of an operator will be defined as

$$\bar{\Omega}_{\text{op}} = \int \Psi^\dagger \tau_3 \Omega_{\text{op}} \Psi \,\mathrm{d}\boldsymbol{R}. \tag{175}$$

This definition reduces to (164) for operators which are a multiple of the $2\times 2$ unit matrix. An operator has real expectation values (175) if one has

$$\Omega_{\text{op}}^\dagger = \tau_3 \Omega_{\text{op}} \tau_3 . \tag{176}$$

In particular one may notice that $H_{\text{op}}$ (173) satisfies this relation.

The transformation properties (165–169) may be expressed in terms of the new wave function $\Psi$. For translations (165), rotations (166) and spatial inversion (167) we get each time

$$\Psi'(\boldsymbol{R}', t') = \Psi(\boldsymbol{R}, t). \tag{177}$$

Under *translations* the expectation value of an operator changes therefore by an amount which is the expectation value of

$$\delta\Omega_{\text{op}} = \frac{i}{\hbar}\, \boldsymbol{\varepsilon}\cdot[\boldsymbol{P}_{\text{op}}, \Omega_{\text{op}}], \tag{178}$$

while under *rotations* we have

$$\delta\Omega_{\text{op}} = \frac{i}{\hbar}\, \boldsymbol{\varepsilon}\cdot[\boldsymbol{R}\wedge\boldsymbol{P}_{\text{op}}, \Omega_{\text{op}}]. \tag{179}$$

A vector operator is characterized by its property

$$[(\boldsymbol{R}\wedge\boldsymbol{P}_{\text{op}})^i, \Omega_{\text{op}}^j] = i\hbar\varepsilon^{ijk}\Omega_{k,\text{op}} . \tag{180}$$

Under *spatial inversion* the expectation value changes according to

$$\int \Psi'^\dagger(\boldsymbol{R}', t')\tau_3 \Omega_{\text{op}}\left(\boldsymbol{R}', \frac{\hbar}{i}\frac{\partial}{\partial\boldsymbol{R}'}\right)\Psi'(\boldsymbol{R}', t')\mathrm{d}\boldsymbol{R}'$$
$$= \int \Psi^\dagger(\boldsymbol{R}, t)\tau_3 \Omega_{\text{op}}\left(-\boldsymbol{R}, -\frac{\hbar}{i}\frac{\partial}{\partial\boldsymbol{R}}\right)\Psi(\boldsymbol{R}, t)\mathrm{d}\boldsymbol{R}. \tag{181}$$

In particular a polar or axial vector operator satisfies the relation

$$\boldsymbol{\Omega}_{op}(-\boldsymbol{R}, -\boldsymbol{P}_{op}) = \mp\boldsymbol{\Omega}_{op}(\boldsymbol{R}, \boldsymbol{P}_{op}). \tag{182}$$

The transformation property of the wave function under *time reversal* follows from (168):

$$\Psi'(\boldsymbol{R}', t') = \Psi^*(\boldsymbol{R}, t), \tag{183}$$

so that the expectation value of an operator $\Omega_{op}$ transforms according to

$$\int \Psi'^\dagger(\boldsymbol{R}', t')\tau_3\, \Omega_{op}\left(\boldsymbol{R}', \frac{\hbar}{i}\frac{\partial}{\partial \boldsymbol{R}'}\right)\Psi'(\boldsymbol{R}', t')\mathrm{d}\boldsymbol{R}'$$
$$= \int \Psi^\dagger(\boldsymbol{R}, t)\tilde{\Omega}_{op}\left(\boldsymbol{R}, -\frac{\hbar}{i}\frac{\partial}{\partial \boldsymbol{R}}\right)\tau_3\, \Psi(\boldsymbol{R}, t)\mathrm{d}\boldsymbol{R}, \tag{184}$$

where the tilde indicates the transposed matrix.

Finally under *pure Lorentz transformations* the two-component wave function transforms as

$$\Psi'(\boldsymbol{R}', t') = \left\{1 - \tfrac{1}{2}(\tau_3 + i\tau_2)\frac{\boldsymbol{\varepsilon}\cdot\boldsymbol{P}_{op}}{mc}\right\}\Psi(\boldsymbol{R}, t), \tag{185}$$

as follows from (169) with (170) and (171). The change of the expectation value of an operator $\Omega_{op}$ under pure Lorentz transformations may be found now by proceeding along similar lines as followed in (29–33). One finds that the change of the expectation value is given by the expectation value of the operator

$$\delta\Omega_{op} = \frac{i}{\hbar}\,\boldsymbol{\varepsilon}\cdot[\boldsymbol{N}_{op} - ct\boldsymbol{P}_{op}, \Omega_{op}], \tag{186}$$

where the operator $\boldsymbol{N}_{op}$ stands for

$$\boldsymbol{N}_{op} \equiv \tfrac{1}{2}c^{-1}\{\boldsymbol{R}, H_{op}\}. \tag{187}$$

The curly brackets indicate an anticommutator and $H_{op}$ is given by (173).

### c. *Covariance requirements on the position operator*

The position operator $\boldsymbol{X}_{op}$ for the Klein–Gordon particle will be obtained by imposing a number of conditions. We require in the first place that it be a polar vector operator with the usual property under translations:

$$[\boldsymbol{P}_{op}, \boldsymbol{X}_{op}] = \frac{\hbar}{i}\,\mathrm{U}, \tag{188}$$

$$[(\boldsymbol{R} \wedge \boldsymbol{P}_{\text{op}})^i, X^j_{\text{op}}] = i\hbar\varepsilon^{ijk} X_{k,\text{op}}, \tag{189}$$

$$\boldsymbol{X}_{\text{op}}(-\boldsymbol{R}, -\boldsymbol{P}_{\text{op}}) = -\boldsymbol{X}_{\text{op}}(\boldsymbol{R}, \boldsymbol{P}_{\text{op}}), \tag{190}$$

where (178), (180) and (182) have been used. As to its time reversal property we require that the expectation value of the position operator be invariant, i.e., according to (184),

$$\boldsymbol{X}_{\text{op}}(\boldsymbol{R}, -\boldsymbol{P}_{\text{op}}) = \tau_3 \tilde{\boldsymbol{X}}_{\text{op}}(\boldsymbol{R}, \boldsymbol{P}_{\text{op}})\tau_3 . \tag{191}$$

For the pure Lorentz transformation character we postulate in view of (44) (which is valid for any spin) and (186)

$$[N^i_{\text{op}}, X^j_{\text{op}}] = \tfrac{1}{2}c^{-1}\{X^i_{\text{op}}, [H_{\text{op}}, X^j_{\text{op}}]\}, \tag{192}$$

where we applied the translation property (188).

d. *Transformation to even form of the Hamilton operator; the position operator*

The Feshbach–Villars Hamiltonian (173) contains the odd matrix $\tau_2$. As a consequence the wave equation (172) consists of two coupled differential equations. They may be uncoupled by performing a transformation[1] which is the analogue of the Pryce–Foldy–Wouthuysen transformation for the Dirac particle. If one transforms the wave function according to

$$\hat{\Psi} = U_{\text{op}} \Psi, \tag{193}$$

the operators $\Omega_{\text{op}}$ should transform in such a way that the expectation values are invariant, i.e. as a consequence of (175),

$$\int \hat{\Psi}_1^\dagger \tau_3 \hat{\Omega}_{\text{op}} \hat{\Psi}_2 \,\mathrm{d}\boldsymbol{R} = \int \Psi_1^\dagger \tau_3 \Omega_{\text{op}} \Psi_2 \,\mathrm{d}\boldsymbol{R}, \tag{194}$$

so that, with (193), we have

$$\hat{\Omega}_{\text{op}} = \tau_3 (U^\dagger_{\text{op}})^{-1} \tau_3 \Omega_{\text{op}} U^{-1}_{\text{op}}. \tag{195}$$

In particular choosing $\Omega_{\text{op}}$ as the unit operator and requiring that $\hat{\Omega}_{\text{op}}$ be the unit operator as well (so that the inner product (174) is invariant) one has

$$1 = \tau_3 (U^\dagger_{\text{op}})^{-1} \tau_3 U^{-1}_{\text{op}}, \tag{196}$$

or equivalently

$$U_{\text{op}} = \tau_3 (U^\dagger_{\text{op}})^{-1} \tau_3 , \tag{197}$$

[1] H. Feshbach and F. Villars, op. cit.

so that (195) may be written as

$$\hat{\Omega}_{op} = U_{op}\,\Omega_{op}\,U_{op}^{-1}. \tag{198}$$

The Hamiltonian, which governs the time behaviour of $\hat{\Psi}$, follows from (172) and (193):

$$\hat{H}_{op} = U_{op}\,H_{op}\,U_{op}^{-1} - \frac{\hbar}{i}\,\frac{\partial U_{op}}{\partial t}\,U_{op}^{-1}. \tag{199}$$

(One should note that this transformation of the Hamiltonian is of course not of the type (198); only if $U_{op}$ is independent of the time – as it will turn out to be in the present field-free case – this expression reduces to its first term.)

The Hamiltonian $H_{op}$ (173) may be diagonalized with the help of an operator $U_{op}$ which is such that its inverse $U_{op}^{-1}$ contains in its columns the eigenvectors of $H_{op}$. Then one finds that a possible choice for $U_{op}$ is:

$$U_{op} = \tfrac{1}{2}(1+\tau_1)\left(\frac{E_{op}}{mc^2}\right)^{\frac{1}{2}} + \tfrac{1}{2}(1-\tau_1)\left(\frac{mc^2}{E_{op}}\right)^{\frac{1}{2}}, \tag{200}$$

where the energy operator is

$$E_{op} \equiv (\boldsymbol{P}_{op}^2 c^2 + m^2 c^4)^{\frac{1}{2}}. \tag{201}$$

The form (200) satisfies (197). Indeed the transformed Hamiltonian (199), that follows from (173) by applying the time-independent transformation operator (200), is now

$$\hat{H}_{op} = \tau_3\,E_{op} \tag{202}$$

and has thus diagonal form. Hence the positive- and negative-energy solutions in this new picture (indicated by circumflexes) are no longer mixed. For that reason only that part of the operators for physical quantities that is even in the new picture comes into play if one considers its expectation value for positive- (or negative-) energy solutions.

The even part of the position operator, which will be denoted as $\hat{\boldsymbol{X}}_{op}$ in the new picture, follows from the requirements (188–192). From the requirements (188) and (189) alone – translation and rotation covariance – one has the general form

$$\hat{\boldsymbol{X}}_{op} = \boldsymbol{R} + f_1(E_{op})\boldsymbol{P}_{op} + f_2(E_{op})\tau_3\,\boldsymbol{P}_{op}, \tag{203}$$

with arbitrary functions $f_1$ and $f_2$. The requirement (190) about spatial inversion does not restrict (203) any further. The requirement of time reversal (191) makes both $f_1$ and $f_2$ vanish. Thus the even part of the position operator in the new picture is found to be

$$\hat{\boldsymbol{X}}_{\text{op}} = \boldsymbol{R}. \tag{204}$$

One should still check whether this result satisfies the Lorentz covariance condition (192) (with circumflexes). To that end we write first the transformed coordinate

$$\hat{\boldsymbol{R}}_{\text{op}} \equiv U_{\text{op}} \boldsymbol{R} U_{\text{op}}^{-1} = \boldsymbol{R} + \xi_{\text{op}}, \tag{205}$$

with the latter quantity given by

$$\xi_{\text{op}} = -\frac{\hbar}{i} U_{\text{op}} \frac{\partial U_{\text{op}}^{-1}}{\partial \boldsymbol{P}_{\text{op}}} = \frac{\hbar}{i} \frac{\partial U_{\text{op}}}{\partial \boldsymbol{P}_{\text{op}}} U_{\text{op}}^{-1} \tag{206}$$

or explicitly, with (200) inserted,

$$\xi_{\text{op}} = \frac{\hbar c^2 \boldsymbol{P}_{\text{op}}}{2iE_{\text{op}}^2} \tau_1 . \tag{207}$$

Then the operator $\hat{N}_{\text{op}}$ (187) becomes with (202), (205) and (207):

$$\hat{\boldsymbol{N}}_{\text{op}} = \tfrac{1}{2} c^{-1} \tau_3 \{\boldsymbol{R}, E_{\text{op}}\}. \tag{208}$$

Substituting this expression, (202) and (204) into (192) with circumflexes, one finds an identity, so that indeed the Lorentz covariance condition is satisfied.

We note that the position operator (204) reads in the original, Feshbach–Villars picture

$$\boldsymbol{X}_{\text{op}} = \boldsymbol{R} - \frac{\hbar c^2 \boldsymbol{P}_{\text{op}}}{2iE_{\text{op}}^2} \tau_1 . \tag{209}$$

A few remarks may be made about the position operator obtained. In the first place we note that its components commute (in contrast with the components of the position operator for the Dirac particle). Furthermore the orbital angular momentum $\boldsymbol{R} \wedge \boldsymbol{P}_{\text{op}}$, which according to (179) is the generator of rotations, may be written as the vector product $\boldsymbol{X}_{\text{op}} \wedge \boldsymbol{P}_{\text{op}}$ of the position operator $\boldsymbol{X}_{\text{op}}$ and the momentum operator $\boldsymbol{P}_{\text{op}}$, as follows from (209).

The position operator found here is the same as the operator obtained by Newton and Wigner[1], as may be checked by translating it into the momentum representation, which they employ.

[1] T. D. Newton and E. P. Wigner, op. cit.

## 5 *The Klein–Gordon particle in a field*

### a. *Invariance properties*

The wave function for the Klein–Gordon particle with charge $e$ in an external electromagnetic field, described by the four-potential $A^\mu = (\varphi, \boldsymbol{A})$ satisfies the equation

$$\left\{\left(\partial_\mu - \frac{ie}{\hbar c} A_\mu\right)\left(\partial^\mu - \frac{ie}{\hbar c} A^\mu\right) - \frac{m^2c^2}{\hbar^2}\right\} \psi(\boldsymbol{R}, t) = 0. \qquad (210)$$

The transformation properties (165–169) of the wave function remain valid in this case if one transforms the four-potential in the right way, i.e. as a four-vector under Lorentz transformation and as $(\varphi'(\boldsymbol{R}', t'), \boldsymbol{A}'(\boldsymbol{R}', t')) = (\varphi(\boldsymbol{R}, t), -\boldsymbol{A}(\boldsymbol{R}, t))$ both under spatial inversion and time reversal.

The Klein–Gordon equation (210) for a particle in an external field may be transformed by employing an artifice, similar to that for the free particle. In fact if one defines

$$u = \frac{1}{\sqrt{2}}\left\{\psi - \frac{\hbar}{imc^2}\left(\frac{\partial}{\partial t} + \frac{ie}{\hbar}\varphi\right)\psi\right\},$$
$$v = \frac{1}{\sqrt{2}}\left\{\psi + \frac{\hbar}{imc^2}\left(\frac{\partial}{\partial t} + \frac{ie}{\hbar}\varphi\right)\psi\right\} \qquad (211)$$

and uses (171), one obtains for (210) an equation of the form (172), but with the Hamiltonian[1]

$$H_{\text{op}} = (\tau_3 + i\tau_2)\frac{\pi_{\text{op}}^2}{2m} + mc^2\tau_3 + e\varphi, \qquad (212)$$

where we introduced the abbreviation

$$\pi_{\text{op}} \equiv \boldsymbol{P}_{\text{op}} - \frac{e}{c}\boldsymbol{A}. \qquad (213)$$

Let us study the invariance properties of expectation values for operators that depend not only on the coordinate and momentum operators but also on the potentials.

Under *translations* (6) one finds that the expectation value of an operator changes by an amount which is the expectation value of the operator

$$\delta\Omega_{\text{op}} = \frac{i}{\hbar}\,\boldsymbol{\varepsilon}\cdot[\boldsymbol{P}_{\text{op}}, \Omega_{\text{op}}] + \Omega_{\text{op}}(\boldsymbol{R}, \boldsymbol{P}_{\text{op}}, \boldsymbol{A} - \boldsymbol{\varepsilon}\cdot\nabla\boldsymbol{A}, \varphi - \boldsymbol{\varepsilon}\cdot\nabla\varphi) - \Omega_{\text{op}}(\boldsymbol{R}, \boldsymbol{P}_{\text{op}}, \boldsymbol{A}, \varphi), \qquad (214)$$

[1] H. Feshbach and F. Villars, op. cit.

as follows from (96) and (177). If one limits oneself to terms with the potentials but without their derivatives this expression reduces to

$$\delta\Omega_{\text{op}} = \frac{i}{\hbar}\,\boldsymbol{\varepsilon}\cdot[\boldsymbol{P}_{\text{op}}, \Omega_{\text{op}}]. \tag{215}$$

Under *rotations* (12), the change of expectation values is governed by the operator

$$\delta\Omega_{\text{op}} = \frac{i}{\hbar}\,\boldsymbol{\varepsilon}\cdot[\boldsymbol{R}\wedge\boldsymbol{P}_{\text{op}}, \Omega_{\text{op}}] + \Omega_{\text{op}}(\boldsymbol{R}, \boldsymbol{P}_{\text{op}}, \boldsymbol{A}+\boldsymbol{\varepsilon}\wedge\boldsymbol{A}-(\boldsymbol{\varepsilon}\wedge\boldsymbol{R})\cdot\nabla\boldsymbol{A},$$
$$\varphi-(\boldsymbol{\varepsilon}\wedge\boldsymbol{R})\cdot\nabla\varphi) - \Omega_{\text{op}}(\boldsymbol{R}, \boldsymbol{P}_{\text{op}}, \boldsymbol{A}, \varphi), \tag{216}$$

as follows from (99) and (177). Up to potentials only it becomes

$$\delta\Omega_{\text{op}} = \frac{i}{\hbar}\,\boldsymbol{\varepsilon}\cdot[\boldsymbol{R}\wedge\boldsymbol{P}_{\text{op}}, \Omega_{\text{op}}] + (\boldsymbol{\varepsilon}\wedge\boldsymbol{A})\cdot\frac{\partial\Omega_{\text{op}}}{\partial\boldsymbol{A}}. \tag{217}$$

If the operator $\boldsymbol{\Omega}_{\text{op}}$ is a vector operator, it satisfies the relation

$$[(\boldsymbol{R}\wedge\boldsymbol{P}_{\text{op}})^i, \Omega^j_{\text{op}}] - i\hbar\varepsilon^{imn}A_m\frac{\partial\Omega^j_{\text{op}}}{\partial A^n} = i\hbar\varepsilon^{ijk}\Omega_{k,\text{op}}. \tag{218}$$

A particular case of (217) arises if the operator $\Omega_{\text{op}}$ is independent of $\varphi$ and depends on $\boldsymbol{A}$ only in the combination $\boldsymbol{\pi}_{\text{op}}$ (213). Then (217) reads

$$\delta\Omega_{\text{op}} = \frac{i}{\hbar}\,\boldsymbol{\varepsilon}\cdot[\boldsymbol{R}\wedge\boldsymbol{\pi}_{\text{op}}, \Omega_{\text{op}}]. \tag{219}$$

Under *spatial inversion* (18) the expectation value in the new frame is the expectation value of

$$\Omega_{\text{op}}(-\boldsymbol{R}, -\boldsymbol{P}_{\text{op}}, -\boldsymbol{A}, \varphi), \tag{220}$$

as follows from (105) and (177). In particular a polar or an axial vector operator is characterized by

$$\boldsymbol{\Omega}_{\text{op}}(-\boldsymbol{R}, -\boldsymbol{P}_{\text{op}}, -\boldsymbol{A}, \varphi) = \mp\boldsymbol{\Omega}_{\text{op}}(\boldsymbol{R}, \boldsymbol{P}_{\text{op}}, \boldsymbol{A}, \varphi). \tag{221}$$

Under *time reversal* (22) the expectation value in the new frame is the expectation value of

$$\tau_3\tilde{\Omega}_{\text{op}}(\boldsymbol{R}, -\boldsymbol{P}_{\text{op}}, -\boldsymbol{A}, \varphi)\tau_3, \tag{222}$$

where we have used (108) and (183).

Under *pure Lorentz transformations* (27) the two-component wave function transforms in a way that is slightly different from the transformation

(185) for the free particle case, namely

$$\Psi'(\boldsymbol{R}', t') = \left\{1 - \tfrac{1}{2}(\tau_3 + i\tau_2)\frac{\boldsymbol{\varepsilon}\cdot\boldsymbol{\pi}_{\text{op}}}{mc}\right\} \Psi(\boldsymbol{R}, t), \tag{223}$$

as follows from (169) and the definitions (211) (note that the latter contain the scalar potential which transforms also, and thus yield a term with the vector potential). For the expectation value of an operator in the new frame one finds, by a reasoning which is analogous to that of (29–33), an expression which differs from that in the old frame by a quantity which is the expectation value of the operator

$$\delta\Omega_{\text{op}} = \frac{i}{\hbar}\,\boldsymbol{\varepsilon}\cdot[\boldsymbol{N}_{\text{op}} - ct\boldsymbol{P}_{\text{op}}, \Omega_{\text{op}}] + \Omega_{\text{op}}\left(\boldsymbol{R}, \boldsymbol{P}_{\text{op}}, \boldsymbol{A} - \boldsymbol{\varepsilon}\varphi + ct\boldsymbol{\varepsilon}\cdot\nabla\boldsymbol{A} + c^{-1}\boldsymbol{\varepsilon}\cdot\boldsymbol{R}\frac{\partial \boldsymbol{A}}{\partial t},\right.$$
$$\left.\varphi - \boldsymbol{\varepsilon}\cdot\boldsymbol{A} + ct\boldsymbol{\varepsilon}\cdot\nabla\varphi + c^{-1}\boldsymbol{\varepsilon}\cdot\boldsymbol{R}\frac{\partial\varphi}{\partial t}\right) - \Omega_{\text{op}}(\boldsymbol{R}, \boldsymbol{P}_{\text{op}}, \boldsymbol{A}, \varphi), \tag{224}$$

where (109) has been used. Here $\boldsymbol{N}_{\text{op}}$ is given by (187), but with the Hamiltonian (212). Up to terms with the potentials this expression reduces to

$$\delta\Omega_{\text{op}} = \frac{i}{\hbar}\,\boldsymbol{\varepsilon}\cdot[\boldsymbol{N}_{\text{op}} - ct\boldsymbol{P}_{\text{op}}, \Omega_{\text{op}}] - \boldsymbol{\varepsilon}\cdot\frac{\partial\Omega_{\text{op}}}{\partial\boldsymbol{A}}\varphi - \frac{\partial\Omega_{\text{op}}}{\partial\varphi}\boldsymbol{\varepsilon}\cdot\boldsymbol{A}. \tag{225}$$

If $\Omega_{\text{op}}$ depends only on $\boldsymbol{R}$ and $\boldsymbol{\pi}_{\text{op}}$, this expression becomes

$$\delta\Omega_{\text{op}} = \frac{i}{\hbar}\,\boldsymbol{\varepsilon}\cdot[\boldsymbol{N}^{(\varphi)}_{\text{op}} - ct\boldsymbol{P}_{\text{op}}, \Omega_{\text{op}}] \tag{226}$$

with the definition

$$\boldsymbol{N}^{(\varphi)}_{\text{op}} \equiv \tfrac{1}{2}c^{-1}\{\boldsymbol{R}, H_{\text{op}} - e\varphi\}, \tag{227}$$

where we used (187).

b. *Covariance requirements on the position operator*

Just as in the field-free case (section 4*c*) we shall list the covariance requirements for the position operator, in which we now include terms with the potentials.

The *translation* property remains of the form (188), while the *rotation* and *spatial inversion* properties follow by stipulating that the position operator be a polar vector:

$$[(\boldsymbol{R}\wedge\boldsymbol{P}_{\text{op}})^i, X^j_{\text{op}}] - i\hbar\varepsilon^{imn}A_m\frac{\partial X^j_{\text{op}}}{\partial A_n} = i\hbar\varepsilon^{ijk}X_{k,\text{op}}, \tag{228}$$

$$\boldsymbol{X}_{\text{op}}(-\boldsymbol{R}, -\boldsymbol{P}_{\text{op}}, -\boldsymbol{A}, \varphi) = -\boldsymbol{X}_{\text{op}}(\boldsymbol{R}, \boldsymbol{P}_{\text{op}}, \boldsymbol{A}, \varphi) \tag{229}$$

(cf. (218) and (221)). *Time reversal* invariance of the position operator implies (cf. (222))

$$X_{op}(R, -P_{op}, -A, \varphi) = \tau_3 \tilde{X}_{op}(R, P_{op}, A, \varphi)\tau_3 . \tag{230}$$

As to the *Lorentz covariance* we require the relation

$$[N^i_{op}, X^j_{op}] - \frac{\hbar}{i}\left(\frac{\partial X^j_{op}}{\partial A_i}\varphi + \frac{\partial X^j_{op}}{\partial \varphi}A^i\right) = \tfrac{1}{2}c^{-1}\{X^i_{op}, [H_{op}, X^j_{op}]\}, \tag{231}$$

which follows from (44) and (225).

c. *The transformed Hamilton operator*

The Hamilton operator (212) which contains the odd matrix $\tau_2$ may be brought into even form by means of two successive transformations. In the first place we employ a transformation operator $U_{1,op}$ of which the Weyl transform is analogous to the Weyl transform of (200):

$$U_{1,op} \rightleftarrows \tfrac{1}{2}(1+\tau_1)\left(\frac{E_\pi}{mc^2}\right)^{\frac{1}{2}} + \tfrac{1}{2}(1-\tau_1)\left(\frac{mc^2}{E_\pi}\right)^{\frac{1}{2}} \equiv U_1 . \tag{232}$$

Here we used the abbreviation $E_\pi$ (134) with $\pi = p - (e/c)A$ the Weyl transform of $\pi_{op}$ (213). Then, since the Weyl transform of $H_{op}$ (212) is

$$H_{op} \rightleftarrows (\tau_3 + i\tau_2)\frac{\pi^2}{2m} + mc^2\tau_3 + e\varphi \equiv H, \tag{233}$$

one finds up to terms with the derivatives of the potentials

$$\begin{aligned} U_{1,op} H_{op}\tau_3 U^\dagger_{1,op}\tau_3 &\rightleftarrows U_1 H\tau_3 U^\dagger_1\tau_3 \\ &+ \frac{ie\hbar}{2c}\varepsilon_{ijk}\frac{\partial U}{\partial P_i}\frac{\partial H}{\partial P_j}\tau_3 U^\dagger\tau_3 B^k + \frac{ie\hbar}{2c}\varepsilon_{ijk} U\frac{\partial H}{\partial P_i}\tau_3\frac{\partial U^\dagger}{\partial P_j}\tau_3 B^k \\ &+ \frac{ie\hbar}{2c}\varepsilon_{ijk}\frac{\partial U}{\partial P_i}H\tau_3\frac{\partial U^\dagger}{\partial P_j}\tau_3 B^k \\ &- \frac{ie\hbar}{2}\frac{\partial U}{\partial P_i}\frac{\partial\varphi}{\partial R^i}\tau_3 U^\dagger\tau_3 + \frac{ie\hbar}{2}U\frac{\partial\varphi}{\partial R_i}\tau_3\frac{\partial U^\dagger}{\partial P^i}\tau_3 \end{aligned} \tag{234}$$

with $U$ the Weyl transform of (200). Apart from the matrices $\tau_3$ this expression is formally identical with (135). The operator $U_{1,op}$ fulfils the relation (196) since, up to terms with the derivatives of the potentials:

$$U_{1,op}\tau_3 U^\dagger_{1,op}\tau_3 \rightleftarrows U_1\tau_3 U^\dagger_1\tau_3 + \frac{ie\hbar}{2c}\varepsilon_{ijk}\frac{\partial U}{\partial P_i}\tau_3\frac{\partial U^\dagger}{\partial P_j}\tau_3 B^k = 1, \tag{235}$$

where we used the explicit expression (232) to show that the first term in the middle member is equal to 1, while the vanishing of the last term in the middle member follows from (206) with (207). This result shows that $\tau_3 U_{1,\mathrm{op}}^{\dagger} \tau_3$ is equal to $U_{1,\mathrm{op}}^{-1}$. Substituting (232), (233), (202), (206) and (207) into the right-hand side of (234) we get

$$U_{1,\mathrm{op}} H_{\mathrm{op}} U_{1,\mathrm{op}}^{-1} \rightleftarrows \tau_3 E_\pi + e\varphi + e \frac{\partial \varphi}{\partial \boldsymbol{R}} \cdot \xi. \tag{236}$$

Since the time derivative of the transformed wave function is determined by (199), we also need $\partial U_{1,\mathrm{op}}/\partial t$ of which the Weyl transform is $-(e/c)(\partial U/\partial \boldsymbol{p})\cdot(\partial \boldsymbol{A}/\partial t)$ (up to terms linear in $e$ and without second derivatives of the potentials), as follows from (232). Therefore we find

$$U_{1,\mathrm{op}} H U_{1,\mathrm{op}}^{-1} - \frac{\hbar}{i} \frac{\partial U_{1,\mathrm{op}}}{\partial t} U_{1,\mathrm{op}}^{-1} \rightleftarrows \tau_3 E_\pi + e\varphi - e\boldsymbol{E}\cdot\xi. \tag{237}$$

A second transformation with the operator

$$U_{2,\mathrm{op}} \rightleftarrows 1 - \frac{e}{2E} \tau_3 \boldsymbol{E}\cdot\xi, \tag{238}$$

which fulfils (196) (up to terms linear in $e$), brings the Hamiltonian to the even form

$$\hat{H}_{\mathrm{op}} \equiv U_{2,\mathrm{op}} U_{1,\mathrm{op}} H_{\mathrm{op}} U_{1,\mathrm{op}}^{-1} U_{2,\mathrm{op}}^{-1} - \frac{\hbar}{i} \frac{\partial (U_{2,\mathrm{op}} U_{1,\mathrm{op}})}{\partial t} U_{1,\mathrm{op}}^{-1} U_{2,\mathrm{op}}^{-1}$$
$$\rightleftarrows \tau_3 E_\pi + e\varphi, \tag{239}$$

up to terms linear in $e$ and without second derivatives of the potentials. This result shows that the transformed Hamiltonian contains only terms with the potentials and not with the fields (although in the derivation terms with the derivatives of the potentials have been taken into account). This situation is different from that of the Dirac particle, as (143) shows.

### d. *The position operator and the equation of motion*

From the translation, rotation, spatial inversion, time reversal and Lorentz covariance properties (188) and (228–231), it follows that the part of the position operator that is even in the new picture is, up to terms with the potentials,

$$\hat{X}_{\mathrm{op}} \rightleftarrows \boldsymbol{R}, \tag{240}$$

or, if one transforms back to the original picture,

$$\boldsymbol{X}_{\text{op}} \rightleftarrows \boldsymbol{R} - \frac{\hbar c^2 \boldsymbol{\pi}}{2iE_\pi^2}\tau_1 \tag{241}$$

(cf. (209) for the field-free case).

The time derivative of the expectation value of the position operator (240) is the expectation value of the velocity operator

$$\hat{\boldsymbol{v}}_{\text{op}} \equiv \frac{\mathrm{d}\hat{\boldsymbol{X}}_{\text{op}}}{\mathrm{d}t} \equiv \frac{i}{\hbar}[\hat{H}_{\text{op}}, \hat{\boldsymbol{X}}_{\text{op}}]. \tag{242}$$

This may be seen by writing for an arbitrary operator $\hat{\Omega}_{\text{op}}$

$$\frac{\partial}{\partial t}\int \hat{\Psi}^\dagger \tau_3 \hat{\Omega}_{\text{op}} \hat{\Psi}\, \mathrm{d}\boldsymbol{R} = \int \hat{\Psi}^\dagger \tau_3 \left\{\frac{i}{\hbar}[\hat{H}_{\text{op}}, \hat{\Omega}_{\text{op}}] + \frac{\partial \hat{\Omega}_{\text{op}}}{\partial t}\right\} \hat{\Psi}\, \mathrm{d}\boldsymbol{R}, \tag{243}$$

as follows from (172) with circumflexes and the fact that the Hamiltonian (239) commutes with $\tau_3$. From (239) and (240) one finds for (242):

$$\hat{\boldsymbol{v}}_{\text{op}} \rightleftarrows \tau_3 \frac{c^2 \boldsymbol{\pi}}{E_\pi}. \tag{244}$$

For the second time derivative of $\hat{\boldsymbol{X}}_{\text{op}}$ one finds with (239):

$$\frac{\mathrm{d}^2 \hat{\boldsymbol{X}}_{\text{op}}}{\mathrm{d}t^2} \equiv \frac{\mathrm{d}\hat{\boldsymbol{v}}_{\text{op}}}{\mathrm{d}t} \equiv \frac{i}{\hbar}[\hat{H}_{\text{op}}, \hat{\boldsymbol{v}}_{\text{op}}] + \frac{\partial \hat{\boldsymbol{v}}_{\text{op}}}{\partial t} \rightleftarrows \frac{c^2}{E}\left(\mathsf{U} - \frac{c^2 \boldsymbol{PP}}{E^2}\right)\cdot\left(\tau_3 e\boldsymbol{E} + e\frac{c\boldsymbol{P}}{E}\wedge \boldsymbol{B}\right). \tag{245}$$

If only the positive energy solutions are considered, one may replace $\tau_3$ by 1. Then (245) becomes

$$m\frac{\mathrm{d}\hat{\boldsymbol{v}}_{\text{op}}}{\mathrm{d}t} \rightleftarrows \gamma^{-1}(\mathsf{U} - \boldsymbol{\beta\beta})\cdot(e\boldsymbol{E} + e\boldsymbol{\beta}\wedge\boldsymbol{B}), \tag{246}$$

where we introduced the abbreviations $\boldsymbol{\beta} \equiv c\boldsymbol{P}/E$ and $\gamma \equiv (1-\boldsymbol{\beta}^2)^{-\frac{1}{2}}$. Up to order $e^0$ and for positive energy solutions $\boldsymbol{\beta}$ is the Weyl transform of $c^{-1}$ times the velocity operator.

The right-hand side of (246) contains the Lorentz force, but no terms with derivatives of the fields, although such terms with first derivatives have been taken into account in the derivation. (For the Dirac particle they did occur, as (158) shows.) The factor $\mathsf{U} - \boldsymbol{\beta\beta}$ is a consequence of the fact that we studied the time derivative of the velocity operator: in classical theory one encounters a factor of the same type.

The same general remarks on the connexion with classical theory as made at the end of section 3 for the Dirac particle apply also here.

# On covariance properties of physical quantities for the Dirac and Klein-Gordon particles

## a. *The Dirac equation in covariant notation*

The Dirac equation (1) with (2) for a free particle may be written in covariant form, by introducing the matrices $\gamma^0 = -i\beta$ and $\boldsymbol{\gamma} = -i\beta\boldsymbol{\alpha}$, which fulfil the anticommutation relations

$$\{\gamma^\mu, \gamma^\nu\} = 2g^{\mu\nu}, \tag{A1}$$

where the metric tensor $g^{\mu\nu}$ has components $g^{00} = -1$, $g^{ii} = 1$ $(i = 1, 2, 3)$ and the others zero. Since $\boldsymbol{\alpha}$ and $\beta$ are hermitian, $\gamma^0$ is anti-hermitian and $\boldsymbol{\gamma}$ hermitian. By multiplication of the Dirac equation (1) with (2) by $\beta/\hbar c$ one obtains the form

$$\left(\gamma^\mu\partial_\mu + \frac{mc}{\hbar}\right)\psi = 0, \tag{A2}$$

with $\partial_\mu = \partial/\partial R^\mu$. The covariance properties of this equation follow by considering an infinitesimal Poincaré transformation

$$R'^\mu = (\delta^\mu_\nu + \varepsilon^\mu_{\ ,\nu})R^\nu + \eta^\mu \tag{A3}$$

with $\eta^\mu$ an infinitesimal four-vector and $\varepsilon^{\mu\nu}$ an infinitesimal antisymmetric four-tensor. The Dirac equation (A2) is covariant with respect to the Poincaré group if one transforms the wave function as

$$\psi'(\boldsymbol{R}', t') = (1 + \tfrac{1}{4}i\varepsilon^{\mu\nu}\sigma_{\mu\nu})\psi(\boldsymbol{R}, t) \tag{A4}$$

with $\sigma^{\mu\nu} \equiv -\frac{1}{2}i[\gamma^\mu, \gamma^\nu]$. Indeed the covariance of the Dirac equation follows, if one substitutes this expression and the transformed four-derivative (which follows from (A3)) into the equation (A2) written with primes and if use is made of the commutation rule

$$[\gamma^\mu, \tfrac{1}{4}i\varepsilon^{\lambda\rho}\sigma_{\lambda\rho}] = \varepsilon^{\mu\lambda}\gamma_\lambda. \tag{A5}$$

In particular if $\varepsilon^{ij} = -\varepsilon^{ijk}\varepsilon_k$ (with $\varepsilon^{ijk}$ the antisymmetric unit tensor and $\boldsymbol{\varepsilon}$ an infinitesimal three-vector), $\varepsilon^{i0} = 0$ and $\eta^\mu = 0$ one finds for (A3) the expression (12) and for (A4) the expression (13). The latter fact follows because $\varepsilon^{ij}\sigma_{ij} = -\varepsilon^{ijk}\varepsilon_k\sigma_{ij} = -2\boldsymbol{\varepsilon}\cdot\boldsymbol{\sigma}$ with $\boldsymbol{\sigma} = -\frac{1}{2}i\boldsymbol{\gamma}\wedge\boldsymbol{\gamma} = -\frac{1}{2}i\boldsymbol{\alpha}\wedge\boldsymbol{\alpha}$.

In the special case $\varepsilon^{i}_{.0} = \varepsilon^{0}_{.i} = -\varepsilon^{i}$ (with an infinitesimal three-vector $\boldsymbol{\varepsilon}$), $\varepsilon^{ij} = 0$ and $\eta^{\mu} = 0$ one recovers (27) from (A3), and (28) from (A4).

One may also prove the invariance of the Dirac equation under spatial inversion

$$t' = t, \qquad \boldsymbol{R}' = -\boldsymbol{R}. \tag{A6}$$

Indeed with the transformation of the wave function

$$\psi'(\boldsymbol{R}', t') = i\gamma^0\psi(\boldsymbol{R}, t) \tag{A7}$$

(instead of $i\gamma^0 = \beta$ one might as well use $\gamma^0$ times a different phase factor) one finds that the Dirac equation is valid with primes throughout.

Finally the Dirac equation is invariant under time reversal

$$t' = -t, \qquad \boldsymbol{R}' = \boldsymbol{R}. \tag{A8}$$

With the anti-linear transformation

$$\psi'(\boldsymbol{R}', t') = T\psi^*(\boldsymbol{R}, t) \tag{A9}$$

(the asterisk indicates the complex conjugate), where $T$ is a matrix such that

$$T^{-1}\gamma^0 T = -\gamma^{0*}, \qquad T^{-1}\gamma^i T = \gamma^{i*}, \tag{A10}$$

one finds that the Dirac equation is valid for primed quantities. (In the Pauli representation one has $\gamma^0 = -i\rho_3$ and $\boldsymbol{\gamma} = \rho_2\boldsymbol{\sigma}$, so that $\boldsymbol{\alpha} = \rho_1\boldsymbol{\sigma}$ and $\beta = \rho_3$; one finds from (A10) that a possible choice for $T$ is $\sigma_2$.)

The Dirac equation for a particle with an anomalous magnetic moment (1) with (93) becomes in covariant notation

$$\left\{\gamma^{\mu}\left(\partial_{\mu} - \frac{ie}{\hbar c}A_{\mu}\right) + \frac{mc}{\hbar} - \frac{(g-2)e}{8mc^2}\sigma^{\mu\nu}F_{\mu\nu}\right\}\psi = 0. \tag{A11}$$

The covariance of this equation follows by using the same arguments as above.

## b. *Local covariance and Klein's theorem*

From the transformation character of the four-component wave function $\psi$ under pure Lorentz transformations we shall prove the following lemma:

Let $\Omega_{\text{op}}^{\nu_1 \ldots \nu_n}$ (with $\nu_1, \ldots, \nu_n$ assuming the values 0, 1, 2, 3) be a set of operators depending on Dirac matrices and the momentum operator $\boldsymbol{P}_{\text{op}} = (\hbar/i)\partial/\partial\boldsymbol{R}$. Then the quantity

$$\bar{\psi}(R)\gamma^{\mu}\Omega_{\text{op}}^{\nu_1 \ldots \nu_n}(\boldsymbol{P}_{\text{op}})\psi(R) \tag{A12}$$

(with $\bar{\psi} \equiv \psi^\dagger \gamma^0$) is – for all solutions $\psi$ of the Dirac equation for a free particle – a local tensor density under pure Lorentz transformations if and only if the operator relation:

$$[N^i_{\mathrm{op}}, \Omega^{\nu_1\ldots\nu_n}_{\mathrm{op}}] = c^{-1}R^i[H_{\mathrm{op}}, \Omega^{\nu_1\ldots\nu_n}_{\mathrm{op}}] + i\hbar \sum_{j=1}^{n} (g^{i\nu_j}\Omega^{\nu_1\ldots\nu_{j-1}0\nu_{j+1}\ldots\nu_n}_{\mathrm{op}} - g^{0\nu_j}\Omega^{\nu_1\ldots\nu_{j-1}i\nu_{j+1}\ldots\nu_n}_{\mathrm{op}}), \quad \text{(A13)}$$

(with $\boldsymbol{N}_{\mathrm{op}}$ given in (34) as $\frac{1}{2}c^{-1}\{\boldsymbol{R}, H_{\mathrm{op}}\}$) holds true.

*Proof*: Under the pure Lorentz transformation (27) the quantity (A12) is, according to (28), transformed to

$$\bar{\psi}'(R')\gamma^\mu \Omega^{\nu_1\ldots\nu_n}_{\mathrm{op}}(\boldsymbol{P}'_{\mathrm{op}})\psi'(R')$$
$$= \bar{\psi}(R)(1+\tfrac{1}{2}\boldsymbol{\varepsilon}\cdot\boldsymbol{\alpha})\gamma^\mu(1-\tfrac{1}{2}\boldsymbol{\varepsilon}\cdot\boldsymbol{\alpha})\Omega^{\nu_1\ldots\nu_n}_{\mathrm{op}}(\boldsymbol{P}'_{\mathrm{op}})\psi(R) - \bar{\psi}(R)\gamma^\mu[\Omega^{\nu_1\ldots\nu_n}_{\mathrm{op}}(\boldsymbol{P}_{\mathrm{op}}), \tfrac{1}{2}\boldsymbol{\varepsilon}\cdot\boldsymbol{\alpha}]\psi(R), \quad \text{(A14)}$$

(up to first order in $\boldsymbol{\varepsilon}$). Now, since

$$\tfrac{1}{2}[\boldsymbol{\varepsilon}\cdot\boldsymbol{\alpha}, \gamma^\mu] = \varepsilon^{\mu}_{\cdot\nu}\gamma^\nu, \quad \text{(A15)}$$

with $\varepsilon^{i}_{\cdot 0} = \varepsilon^{0}_{\cdot i} = -\varepsilon^i$ and $\varepsilon^{ij} = 0$, one may write this as:

$$\bar{\psi}'(R')\gamma^\mu \Omega^{\nu_1\ldots\nu_n}_{\mathrm{op}}(\boldsymbol{P}'_{\mathrm{op}})\psi'(R')$$
$$= \bar{\psi}(R)(\delta^\mu_\nu + \varepsilon^{\mu}_{\cdot\nu})\gamma^\nu \Omega^{\nu_1\ldots\nu_n}_{\mathrm{op}}(\boldsymbol{P}_{\mathrm{op}})\psi(R) + \bar{\psi}(R)\gamma^\mu\{\Omega^{\nu_1\ldots\nu_n}_{\mathrm{op}}(\boldsymbol{P}'_{\mathrm{op}}) - \Omega^{\nu_1\ldots\nu_n}_{\mathrm{op}}(\boldsymbol{P}_{\mathrm{op}})\}\psi(R) - \bar{\psi}(R)\gamma^\mu[\Omega^{\nu_1\ldots\nu_n}_{\mathrm{op}}(\boldsymbol{P}_{\mathrm{op}}), \tfrac{1}{2}\boldsymbol{\varepsilon}\cdot\boldsymbol{\alpha}]\psi(R). \quad \text{(A16)}$$

The second term at the right-hand side may be written in a different form by using the relation that follows from (27) and the Dirac equation (1):

$$\Omega_{\mathrm{op}}(\boldsymbol{P}'_{\mathrm{op}}) - \Omega_{\mathrm{op}}(\boldsymbol{P}_{\mathrm{op}}) = -c^{-1}\boldsymbol{\varepsilon}\cdot\frac{\partial \Omega_{\mathrm{op}}}{\partial \boldsymbol{P}_{\mathrm{op}}} H_{\mathrm{op}}, \quad \text{(A17)}$$

or, with the commutation rule $[\boldsymbol{P}_{\mathrm{op}}, \boldsymbol{R}] = -i\hbar \mathsf{U}$,

$$\Omega_{\mathrm{op}}(\boldsymbol{P}'_{\mathrm{op}}) - \Omega_{\mathrm{op}}(\boldsymbol{P}_{\mathrm{op}}) = \frac{i}{\hbar c}[\boldsymbol{\varepsilon}\cdot\boldsymbol{R}H_{\mathrm{op}}, \Omega_{\mathrm{op}}] - \frac{i}{\hbar c}\boldsymbol{\varepsilon}\cdot\boldsymbol{R}[H_{\mathrm{op}}, \Omega_{\mathrm{op}}]. \quad \text{(A18)}$$

Then the second and third term at the right-hand side of (A16) become together

$$\frac{i}{\hbar}\bar{\psi}(R)\gamma^\mu([\boldsymbol{\varepsilon}\cdot\boldsymbol{N}_{\mathrm{op}}, \Omega^{\nu_1\ldots\nu_n}_{\mathrm{op}}] - c^{-1}\boldsymbol{\varepsilon}\cdot\boldsymbol{R}[H_{\mathrm{op}}, \Omega^{\nu_1\ldots\nu_n}_{\mathrm{op}}])\psi(R), \quad \text{(A19)}$$

where (34) has been used. From (A16) with this result and the fact that $\psi$ is arbitrary (in its dependence on space coordinates) it follows by considering the coefficients of the components of $\varepsilon$ that the lemma as stated above is proved.

In the main text it was shown that the expectation value of an operator $\Omega_{op}$ that depends on the momentum operator and on Dirac matrices changes under pure Lorentz transformations (27) by an amount which is the expectation value of the operator (33):

$$\delta\Omega_{op} = \frac{i}{\hbar}[\boldsymbol{\varepsilon}\cdot\boldsymbol{N}_{op}, \Omega_{op}]. \tag{A20}$$

On the other hand the expectation value of $\Omega_{op}^{\nu_1\ldots\nu_n}$ transforms as a tensor if one has

$$\delta\Omega_{op}^{\nu_1\ldots\nu_n} = -\varepsilon_i \sum_{j=1}^{n} (g^{i\nu_j}\Omega_{op}^{\nu_1\ldots\nu_{j-1}0\nu_{j+1}\ldots\nu_n} - g^{0\nu_j}\Omega_{op}^{\nu_1\ldots\nu_{j-1}i\nu_{j+1}\ldots\nu_n}). \tag{A21}$$

Hence the expectation value of $\Omega_{op}$ transforms as a tensor if and only if the right-hand sides of (A20) and (A21) are equal or if

$$[N_{op}^i, \Omega_{op}^{\nu_1\ldots\nu_n}] = i\hbar \sum_{j=1}^{n} (g^{i\nu_j}\Omega_{op}^{\nu_1\ldots\nu_{j-1}0\nu_{j+1}\ldots\nu_n} - g^{0\nu_j}\Omega_{op}^{\nu_1\ldots\nu_{j-1}i\nu_{j+1}\ldots\nu_n}). \tag{A22}$$

Comparison of this condition with the lemma condition (A13) shows that the statement that (A12) has tensor character is then and only then equivalent with the statement that the expectation value of $\Omega_{op}$ transforms as a four-tensor, if the operator $\Omega_{op}$ commutes with the Hamiltonian $H_{op}$, i.e. if $\Omega_{op}$ is a conserved quantity. This is the theorem of Felix Klein. (This theorem may alternatively be proved by considering the local conservation law that follows from the commutation of $\Omega_{op}$ with the Hamiltonian; see problems 1 and 2.) The treatment given above shows explicitly how local covariance and covariance of expectation values are connected in the general case in which the quantity $\Omega_{op}$ is not conserved.

In the following we shall need an extension of the lemma given in (A12–A13) to the case of an operator $\breve{\Omega}_{op}^{\nu}$ which depends also on coordinates and on time in such a way that

$$\begin{aligned} [P_{op}^i, \breve{\Omega}_{op}^{\nu}] &= \frac{\hbar}{i} g^{i\nu}, \\ \left[-\frac{\hbar}{ic}\frac{\partial}{\partial t}, \breve{\Omega}_{op}^{\nu}\right] &= \frac{\hbar}{i} g^{0\nu}. \end{aligned} \tag{A23}$$

This means that $\breve{\Omega}^{\nu}_{\text{op}}$ is of the form

$$\breve{\Omega}^{\nu}_{\text{op}}(\boldsymbol{P}_{\text{op}}, R) = R^{\nu} + \Omega^{\nu}_{\text{op}}(\boldsymbol{P}_{\text{op}}), \tag{A24}$$

where $\Omega^{\nu}_{\text{op}}(\boldsymbol{P}_{\text{op}})$ is an operator of the type discussed earlier.

Since the quantity $\bar{\psi}\gamma^{\mu}R^{\nu}\psi$ transforms as a local tensor density, the statement that $\bar{\psi}\gamma^{\mu}\breve{\Omega}^{\nu}_{\text{op}}\psi$ is a local tensor density under pure Lorentz transformations is equivalent to the statement that $\bar{\psi}\gamma^{\mu}\Omega^{\nu}_{\text{op}}\psi$ transforms as a tensor. Hence according to the lemma (A12–13) one finds that $\Omega^{\nu}_{\text{op}}$ fulfils (A13). Now one may check by using (34) that $R^{\nu}$ satisfies the identity

$$[N^{i}_{\text{op}} - ctP^{i}_{\text{op}}, R^{\nu}] = c^{-1}R^{i}[H_{\text{op}} - i\hbar\, \partial/\partial t, R^{\nu}] + i\hbar(g^{i\nu}R^{0} - g^{0\nu}R^{i}). \tag{A25}$$

From this relation and (A24) it follows that (A13) is for the present case equivalent with

$$[N^{i}_{\text{op}} - ctP^{i}_{\text{op}}, \breve{\Omega}^{\nu}_{\text{op}}] = c^{-1}R^{i}[H_{\text{op}} - i\hbar\, \partial/\partial t, \breve{\Omega}^{\nu}_{\text{op}}] + i\hbar(g^{i\nu}\breve{\Omega}^{0}_{\text{op}} - g^{0\nu}\breve{\Omega}^{i}_{\text{op}}), \tag{A26}$$

where we used the relations $[P^{i}_{\text{op}}, \Omega^{\nu}_{\text{op}}] = 0$ and $[\partial/\partial t, \Omega^{\nu}_{\text{op}}] = 0$. In other words we have derived the generalized lemma:

Let $\breve{\Omega}^{\nu}_{\text{op}}$ be a set of operators depending on Dirac matrices, the momentum operator, the coordinates and time in such a way that (A23) is fulfilled. Then the quantity

$$\bar{\psi}(R)\gamma^{\mu}\breve{\Omega}^{\nu}_{\text{op}}(\boldsymbol{P}_{\text{op}}, R)\psi(R) \tag{A27}$$

is – for all solutions $\psi$ of the Dirac equation for a free particle – a local tensor density under pure Lorentz transformation if and only if the operator relation

$$[N^{i}_{\text{op}} - ctP^{i}_{\text{op}}, \breve{\Omega}^{\nu}_{\text{op}}] = c^{-1}R^{i}[H_{\text{op}} - i\hbar\, \partial/\partial t, \breve{\Omega}^{\nu}_{\text{op}}] + i\hbar(g^{i\nu}\breve{\Omega}^{0}_{\text{op}} - g^{0\nu}\breve{\Omega}^{i}_{\text{op}}) \tag{A28}$$

is valid.

c. *Covariance requirements on the position and spin operator of the free Dirac particle*

In the main text we derived the position and spin operator from a number of requirements which it should fulfil. Among them were the conditions (45) and (48–49) which we called the covariance conditions. They were inspired by the analogy with classical reasonings. The imposing of these conditions led to position and spin operators, which (when generalized to the case of a particle in a field) obeyed equations of motion that have the same form as the classical equations of motion for a composite particle.

One may ask oneself how these covariance conditions are related to local covariance and covariance of expectation values, discussed in the preceding subsection in connexion with Klein's theorem. In particular one may wonder whether it is possible to find a position operator $\boldsymbol{X}_{op}$ which is the space part of a set of four operators $\breve{X}^{\nu}_{op}$ ($\nu = 0, 1, 2, 3$) such that $\bar{\psi}\gamma^{\mu}X^{\nu}_{op}\psi$ is a local tensor density. One knows that the space part $\boldsymbol{X}_{op}$ should satisfy the property of translation covariance (36). Let us assume moreover that $\boldsymbol{X}_{op}$ does not explicitly depend on the time $t$ and that the time component $X^{0}_{op}$ does not depend explicitly on the coordinates $\boldsymbol{R}$ and on the time $t$ in a way specified by

$$\left[\frac{\partial}{c\partial t}, X^{0}_{op}\right] = 1. \tag{A29}$$

Then it follows from the lemma (A27–28) that local covariance of $\bar{\psi}\gamma^{\mu}X^{\nu}_{op}\psi$ leads to four conditions ($\nu = 0, 1, 2, 3$)

$$[N^{i}_{op}-ctP^{i}_{op}, X^{\nu}_{op}] = c^{-1}R^{i}[H_{op}-i\hbar\, \partial/\partial t, X^{\nu}_{op}]+i\hbar(g^{i\nu}X^{0}_{op}-g^{0\nu}X^{i}_{op}). \tag{A30}$$

Since we are interested in even operators, i.e. operators which contain only even Dirac matrices in the P–FW picture, and since $\hat{\boldsymbol{N}}_{op}$ (64), $\hat{H}_{op}$ (48) are even but $\hat{\boldsymbol{R}}_{op}$ (61) contains an odd part given in (63), it follows that (A30) cannot be satisfied. Hence four operators $X^{\nu}_{op}$ ($\nu = 0, 1, 2, 3$) which would lead to a local four-tensor density $\bar{\psi}\gamma^{\mu}X^{\nu}_{op}\psi$ cannot be found.

Instead one might look for a set of four operators of which the zero component is simply $ct$ on the argument that in the usual formulation of quantum mechanics the time plays a role which is essentially different from that of the space coordinates. One may try then to impose (A30) for $\nu = 1, 2, 3$ only, but putting $X^{0}$ equal to $ct$. If again one assumes that $\boldsymbol{X}_{op}$ does not depend explicitly on time, one finds

$$[N^{i}_{op}-ctP^{i}_{op}, X^{j}_{op}] = c^{-1}R^{i}[H_{op}, X^{j}_{op}]+i\hbar ctg^{ij}. \tag{A31}$$

For the same reasons as given above an even operator $X^{i}_{op}$ satisfying this condition does not exist either.

The condition (A31) is equivalent to the following two conditions, which are half the sum and half the difference of (A31) and its hermitian conjugate

$$0 = [R^{i}, [H_{op}, X^{j}_{op}]], \tag{A32}$$

$$[N^{i}_{op}-ctP^{i}_{op}, X^{j}_{op}] = \tfrac{1}{2}c^{-1}\{R^{i}, [H_{op}, X^{j}_{op}]\}+i\hbar ctg^{ij}. \tag{A33}$$

As seen above no even operator $X^{j}_{op}$ exists satisfying both of these relations. The condition (A32) alone is sufficient already to exclude the existence of a solution, as follows from the general form (59) for the position operator with (58) and (61). Therefore one may try to impose only (A33), while forgetting about (A32). Using the translation property (36) one may write (A33) as

$$[N^i_{\text{op}}, X^j_{\text{op}}] = \tfrac{1}{2}c^{-1}\{R^i, [H_{\text{op}}, X^j_{\text{op}}]\}. \tag{A34}$$

However by insertion of the general form (59) with (58), (61) and (64) one gets contradictory equations for the form factors $f_1(E_{\text{op}})$ and $f_2(E_{\text{op}})$ so that no even position operator that satisfies (A34) exists.

A different line of approach would consist in requiring the covariance of the expectation value of the position operator instead of local covariance. Then, as follows from (33), we must have

$$[N^i_{\text{op}} - ctP^i_{\text{op}}, X^j_{\text{op}}] = i\hbar ctg^{ij}. \tag{A35}$$

Once more with (59) and (64) inserted one finds a negative result. So this road is blocked as well.

Returning to the condition (A34) one notices that the coordinate $R^i$ looks like a foreign element since elsewhere in the condition the position $X^i_{\text{op}}$ occurs. So one is led to replace $R^i$ in (A34) by $X^i_{\text{op}}$. In that case one gets

$$[N^i_{\text{op}}, X^j_{\text{op}}] = \tfrac{1}{2}c^{-1}\{X^i_{\text{op}}, [H_{\text{op}}, X^j_{\text{op}}]\}. \tag{A36}$$

This is precisely the condition of the main text, which has a solution explicitly given there. The considerations given here were only intended to show that various other conceivable requirements do not lead to results.

We now turn to the discussion of the covariance requirements on the spin operator. Since the translation property (37) implies that the spin operator is independent of the coordinates (in the Dirac picture and hence in the P–FW picture), the evenness of the operator in the P–FW picture implies that it is conserved, as follows from (58). This means that the lemma (A12–13) for the spin operator $s^{\nu_1\nu_2}_{\text{op}}$ (which will be assumed to be a quantity with two indices in which it is antisymmetric) says that the quantity $\bar{\psi}\gamma^\mu s^{\nu_1\nu_2}_{\text{op}}\psi$ is a local covariant tensor if and only if

$$[N^i_{\text{op}}, s^{\nu_1\nu_2}_{\text{op}}] = i\hbar(g^{i\nu_1}s^{0\nu_2}_{\text{op}} - g^{i\nu_2}s^{0\nu_1}_{\text{op}} - g^{0\nu_1}s^{i\nu_2}_{\text{op}} + g^{0\nu_2}s^{i\nu_1}_{\text{op}}). \tag{A37}$$

From (A22) it then follows that this condition is equivalent with the requirement that the expectation value of the spin operator transforms as a tensor. (This is the application of Klein's theorem to the case of the spin operator.) Written in terms of the operators $s^i_{\text{op}} = \tfrac{1}{2}\varepsilon^{ijk}s_{jk,\text{op}}$ and $t^i_{\text{op}} = s^{i0}_{\text{op}}$ the condition (A37) reads

$$[N^i_{\text{op}}, s^j_{\text{op}}] = i\hbar\varepsilon^{ijk}t_{k,\text{op}}, \tag{A38}$$

$$[N^i_{\text{op}}, t^j_{\text{op}}] = -i\hbar\varepsilon^{ijk}s_{k,\text{op}}. \tag{A39}$$

These are the same as (48) and (49), since $\boldsymbol{s}_{op}$ and $\boldsymbol{t}_{op}$ are both conserved so that the conditions (48) and (49) are equivalent with both local covariance and covariance of the expectation value of the spin operator. The coinciding of the various possible requirements on the spin operator made the problem to find it essentially simpler than that of the position operator.

d. *Three mutually excluding requirements on the position operator for the free Dirac particle*

In the main text we found that the position operator had non-commuting components. Hence the requirement of commutation of these components is not consistent with the requirements of Lorentz covariance, at least not for an even position operator. One may ask what happens if one would require from the beginning the commutation and forget about covariance. To find the position operator which has this property we insert the general expression (59) for the position operator in the P–FW picture into the commutation rule

$$[\hat{X}^i_{op}, \hat{X}^j_{op}] = 0. \tag{A40}$$

This yields a number of differential equations for the functions $f_1$ and $f_2$. There are three independent solutions. They give rise to the following forms for the position operator in the P–FW picture:

$$\hat{\boldsymbol{X}}_{op} = \boldsymbol{R}, \tag{A41}$$

$$\hat{\boldsymbol{X}}_{op} = \boldsymbol{R} - \frac{\hbar c^2}{E_{op}^2 - m^2 c^4} \boldsymbol{\sigma} \wedge \boldsymbol{P}_{op}, \tag{A42}$$

$$\hat{\boldsymbol{X}}_{op} = \boldsymbol{R} - \frac{\hbar c^2}{2(E_{op}^2 - m^2 c^4)} (1 \pm \beta) \boldsymbol{\sigma} \wedge \boldsymbol{P}_{op}. \tag{A43}$$

The second and third solutions have the unwanted property to be singular for zero momentum. If one discards them for that reason, one is left with (A41), which is the position operator of Newton and Wigner[1]. Hence we conclude that if commutation of Cartesian components – as well as a regularity con-

[1] T. D. Newton and E. P. Wigner, Rev. Mod. Phys. **21**(1949)400; K. Bardakci and R. Acharya, N. Cim. **21**(1961)802; P. M. Mathews and A. Sankaranarayanan, Progr. Theor. Phys. **26**(1961)499; **27**(1962)1063; W. Weidlich and A. K. Mitra, N. Cim. **30**(1963)385; U. Schröder, Ann. Physik **14**(1964)91; T. O. Philips, Phys. Rev. **136B**(1964)893; A. Galindo, N. Cim. **37**(1965)413; R. A. Berg, J. Math. Phys. **6**(1965)34; A. Sankaranarayanan and R. H. Good Jr., Phys. Rev. **140B**(1965)509; P. M. Mathews, Phys. Rev. **143**(1966)985; M. Lunn, J. Phys. **A2**(1969)17.

dition – is imposed on the position operator (apart from its translation, rotation, spatial inversion and time reversal properties), one finds the operator of Newton and Wigner. However covariance is then lost.

If one wants to impose both covariance and commutation of the components one is obliged to leave the domain of operators which are even in the P–FW picture. Then one gets afflicted with the interplay of positive and negative energy solutions. If in spite of this one remains interested in the possible forms of operators which fulfil the requirements of covariance and commutation, one may start from the general expression which satisfies the requirements of translation, rotation, spatial inversion and time reversal. In the P–FW picture one has then

$$\hat{\boldsymbol{X}}_{\mathrm{op}} = \boldsymbol{R} + \{f_1(E_{\mathrm{op}}) + \beta f_2(E_{\mathrm{op}})\}\boldsymbol{\sigma}\wedge\boldsymbol{P}_{\mathrm{op}} + f_3(E_{\mathrm{op}})\rho_2\,\boldsymbol{\sigma} + f_4(E_{\mathrm{op}})\rho_2\,\boldsymbol{P}_{\mathrm{op}}\,\boldsymbol{P}_{\mathrm{op}}\cdot\boldsymbol{\sigma}. \tag{A44}$$

By imposing the Lorentz covariance condition (60) one finds equations for the $f_i(E_{\mathrm{op}})$ $(i = 1, 2, 3, 4)$. Upon substitution of the solutions into the above expression one obtains two types of possible position operators

$$\hat{\boldsymbol{X}}_{\mathrm{op}} = \boldsymbol{R} + \frac{\hbar\boldsymbol{\sigma}\wedge\boldsymbol{P}_{\mathrm{op}}}{2m(E_{\mathrm{op}}+mc^2)} + f_4(E_{\mathrm{op}})\rho_2\,\boldsymbol{P}_{\mathrm{op}}\,\boldsymbol{P}_{\mathrm{op}}\cdot\boldsymbol{\sigma}, \tag{A45}$$

$$\hat{\boldsymbol{X}}_{\mathrm{op}} = \boldsymbol{R} - \frac{\hbar c^2\boldsymbol{\sigma}\wedge\boldsymbol{P}_{\mathrm{op}}}{2E_{\mathrm{op}}(E_{\mathrm{op}}+mc^2)} \pm \frac{\hbar c}{2E_{\mathrm{op}}}\,\rho_2\left\{\boldsymbol{\sigma} - \frac{c^2\boldsymbol{P}_{\mathrm{op}}\,\boldsymbol{P}_{\mathrm{op}}\cdot\boldsymbol{\sigma}}{E_{\mathrm{op}}(E_{\mathrm{op}}+mc^2)}\right\}. \tag{A46}$$

If one imposes moreover the condition of commutativity of the components of $\hat{\boldsymbol{X}}_{\mathrm{op}}$, one finds that the solution (A45) does not fulfil this requirement (for any $f_4$), while (A46) fulfils it as it stands for *both* possible signs. The latter, with the upper sign, is the Dirac position operator in the P–FW picture, as follows from (62) with (63) (keeping in mind that one has $i\beta\boldsymbol{\alpha} \equiv -\rho_2\boldsymbol{\sigma}$ in the Pauli representation). It is simply $\boldsymbol{R}$ in the Dirac picture. The expression (A46) with the minus sign is as good a solution as the Dirac position operator. It reads, in the Dirac picture

$$\boldsymbol{X}_{\mathrm{op}} = \boldsymbol{R} - \frac{\hbar c^2\boldsymbol{\sigma}\wedge\boldsymbol{P}_{\mathrm{op}}}{E_{\mathrm{op}}^2} + \frac{i\hbar mc^3\beta\boldsymbol{\alpha}}{E_{\mathrm{op}}^2}. \tag{A47}$$

Incidentally it may be remarked that if we impose the evenness instead of the commutativity, one finds as only possibility (A45) with $f_4 = 0$, i.e., as it should be, the position operator employed in the main text.

### e. *On the uniqueness of the position operator of the free Klein–Gordon particle*

In the main text we found the even part of the position operator from the requirements of translation, rotation, spatial inversion and time reversal alone. It turned out to satisfy the Lorentz covariance requirement and to possess commuting components. Hence these two latter properties are compatible for even operators in the case of the Klein–Gordon particle.

For purely academic reasons one may still ask which position operators are possible if odd operators are allowed to play a role. Then, from the requirements of translation, rotation, space inversion and time reversal alone, one has for the position operator the form

$$\hat{\boldsymbol{X}}_{\mathrm{op}} = \boldsymbol{R} + f_1(E_{\mathrm{op}})\tau_1 \boldsymbol{P}_{\mathrm{op}}. \tag{A48}$$

This position operator satisfies the Lorentz covariance condition and has also commuting components, for arbitrary function $f_1$. Its even part is $\boldsymbol{R}$, which also separately fulfils the requirements of Lorentz covariance and has commuting components.

The possibility of fulfilling simultaneously the requirements of covariance and commutativity for the even part of the position operator makes the treatment of a particle without spin essentially simpler than that of a particle with spin.

# PROBLEMS

**1.** Show from the Dirac equation (1) that $j^\alpha \equiv \bar{\psi}\gamma^\alpha \Omega_{\mathrm{op}}\psi$ (with $\Omega_{\mathrm{op}}$ independent of the time) is conserved ($\partial_\alpha j^\alpha = 0$) if and only if the operator $\Omega_{\mathrm{op}}$ commutes with the Hamiltonian $H_{\mathrm{op}}$.

**2.** Prove F. Klein's theorem in its standard form: if a tensor $t^{\alpha\beta\ldots\mu\nu}$ satisfies a local conservation law $\partial_\nu t^{\alpha\beta\ldots\mu\nu} = 0$ then the quantity $\int t^{\alpha\beta\ldots\mu 0}\mathrm{d}\boldsymbol{R}$ is conserved and is a tensor of the type $u^{\alpha\beta\ldots\mu}$ if the integrand tends to zero at infinity in such a way that surface integrals vanish there.

Hint: since the indices $\alpha\beta\ldots\mu$ appear everywhere in the same way, the theorem is proved if one shows for a vector $j^\alpha$ with $\partial_\alpha j^\alpha = 0$ that $\int j^0 \mathrm{d}\boldsymbol{R}$ is conserved and is a scalar invariant, provided that $j^0$ vanishes sufficiently quickly at infinity. The proof follows by considering two space-like surfaces $\sigma$ and $\hat{\sigma}$, quantities $\int j^0 \mathrm{d}\boldsymbol{R} = \int j^\alpha \mathrm{d}\sigma_\alpha$ and $\int j^{0\prime} \mathrm{d}\boldsymbol{R}' = \int j^{\alpha\prime} \mathrm{d}\hat{\sigma}'_\alpha = \int j^\alpha \mathrm{d}\hat{\sigma}_\alpha$ and by application of Gauss's theorem.

**3.** Find the unitary transformation (56) by considering the plane wave solutions of the Dirac equation (1) with Hamiltonian (2).

**4.** Derive the equation of motion that would result if one takes for the position operator of the Dirac particle in an electromagnetic field the operator $\boldsymbol{R}$ in the Blount picture. Choose for convenience $g = 2$ and derive first the Weyl transform of the velocity operator

$$\hat{\boldsymbol{v}}_{\mathrm{op}} \equiv \frac{i}{\hbar}[\hat{H}_{\mathrm{op}}, \boldsymbol{R}] \rightleftarrows \frac{\beta\boldsymbol{\pi}c^2}{E_\pi} + \frac{e\hbar c^3 \beta \boldsymbol{P}\boldsymbol{\sigma}\cdot\boldsymbol{B}}{2E^3} - \frac{e\hbar c^2 \boldsymbol{\sigma}\wedge\boldsymbol{E}}{2E(E+mc^2)} + \frac{e\hbar c^4(2E+mc^2)\boldsymbol{P}(\boldsymbol{P}\wedge\boldsymbol{\sigma})\cdot\boldsymbol{E}}{2E^3(E+mc^2)^2}.$$

Derive then the equation of motion

$$m\frac{\mathrm{d}\hat{\boldsymbol{v}}_{\mathrm{op}}}{\mathrm{d}t} \equiv \frac{i}{\hbar}[\hat{H}_{\mathrm{op}}, m\hat{\boldsymbol{v}}_{\mathrm{op}}] + m\frac{\partial\hat{\boldsymbol{v}}_{\mathrm{op}}}{\partial t} \rightleftarrows \gamma^{-2}(\mathsf{U}-\boldsymbol{\beta}\boldsymbol{\beta})\cdot\Bigg(\gamma e\boldsymbol{E} + \gamma e\boldsymbol{\beta}\wedge\boldsymbol{B} + \frac{e\hbar}{2mc}\Bigg[(\nabla\boldsymbol{B})\cdot\boldsymbol{\sigma} + \frac{\gamma}{\gamma+1}(\nabla\boldsymbol{E})\cdot(\boldsymbol{\beta}\wedge\boldsymbol{\sigma}) + \gamma^2(\partial_0+\boldsymbol{\beta}\cdot\nabla)\left\{\boldsymbol{\beta}\boldsymbol{\sigma}\cdot\boldsymbol{B} - \frac{\boldsymbol{\sigma}\wedge\boldsymbol{E}}{\gamma(\gamma+1)} + \frac{\gamma^2\boldsymbol{\beta}\boldsymbol{\beta}\cdot(\boldsymbol{\sigma}\wedge\boldsymbol{E})}{(\gamma+1)^2}\right\}\Bigg]\Bigg),$$

where we limited ourselves to the upper left part of the matrix expression, i.e. $\beta$ replaced by 1 (necessary for expectation values of the positive energy solutions) and where we used the same abbreviations as given above formula (158). The fields depend on the position $\boldsymbol{R}$ and the time $t$.

Corresponding to this choice of the position operator one takes now for the spin operator (v. (149)): $\frac{1}{2}\hbar\boldsymbol{\sigma}$. Show that the equation for this spin operator becomes (again with $g = 2$)

$$\frac{\mathrm{d}\boldsymbol{\sigma}}{\mathrm{d}t} \equiv \frac{i}{\hbar}[\hat{H}_{\mathrm{op}}, \boldsymbol{\sigma}] \rightleftarrows \frac{e}{mc}\left\{\gamma^{-1}\boldsymbol{\sigma}\wedge\boldsymbol{B} - \frac{1}{\gamma+1}\,\boldsymbol{\sigma}\wedge(\boldsymbol{\beta}\wedge\boldsymbol{E})\right\},$$

where again $\beta$ has been replaced by 1.

The right-hand sides of the equations of motion and spin given above are not covariant. To prove this fact one should show in the first place that the right-hand side of the equation of motion without the factor $\gamma^{-2}(\mathbf{U}-\boldsymbol{\beta}\boldsymbol{\beta})$ is not the space part of a four-vector. Show this for the terms with field derivatives by writing first the rest frame expression:

$$(\boldsymbol{\nabla}'\boldsymbol{B}')\cdot\boldsymbol{\sigma}' - \tfrac{1}{2}\boldsymbol{\sigma}'\wedge\partial_0'\boldsymbol{E}'$$

and transforming this with the help of the Lorentz transformation formulae for $(\partial_0', \boldsymbol{\nabla}')$ and $(\boldsymbol{E}', \boldsymbol{B}')$, assuming the transformation character of $\boldsymbol{\sigma}'$ to have the general form

$$\boldsymbol{\sigma}' = F\boldsymbol{\sigma} + G\boldsymbol{\beta}\boldsymbol{\beta}\cdot\boldsymbol{\sigma}$$

with $F$ and $G$ functions of $|\boldsymbol{\beta}|$ or $\gamma$. It turns out then that the difference between this transformed expression and the field derivative terms in the equation of motion is not parallel to the velocity $\boldsymbol{\beta}$, so that the non-covariance is then proved.

The non-covariance of the spin equation may be proved along similar lines.

(The covariance of the equations of motion and spin as derived in the main text may be proved with the same technique. Strictly spoken this is not necessary since we started from covariant position and spin operators; moreover the resulting equations have the same form as the manifestly covariant classical equations.)

**5.** The same questions as those of the preceding problem arise if a still different position operator is adopted, namely one that is connected with the Dirac position operator. Prove first that the operator which is simply $\boldsymbol{R}$ in the Dirac picture gets the form (for $g = 2$)

$$\hat{\boldsymbol{R}} \rightleftarrows \boldsymbol{R} + \frac{\hbar c^2 \boldsymbol{\pi} \wedge \boldsymbol{\sigma}}{2E_\pi(E_\pi + mc^2)} + \frac{e\hbar^2 c^3(2E^2 - m^2c^4)\boldsymbol{P} \wedge \boldsymbol{B}}{8E^4(E+mc^2)^2} - \frac{\beta e\hbar^2 c^2}{4E^3}\left(\boldsymbol{E} - \frac{\boldsymbol{P}\boldsymbol{P}\cdot\boldsymbol{E}}{E^2}\right) + \hat{\boldsymbol{R}}_\mathrm{o},$$

in the Blount picture. Here $\hat{\boldsymbol{R}}_\mathrm{o}$ is the odd part of the transformed Dirac position. Its explicit form is irrelevant if one is interested in the expectation value of positive (or negative) energy solutions: then only the even part of the Dirac position operator comes into play.

Show that the velocity operator corresponding to the even part of the Dirac position is:

$$\hat{\boldsymbol{v}}_\mathrm{op} \rightleftarrows \frac{\beta \boldsymbol{\pi} c^2}{E_\pi} + \frac{e\hbar mc^5 \beta \boldsymbol{P}\boldsymbol{\sigma}\cdot\boldsymbol{B}}{2E^3(E+mc^2)} + \frac{e\hbar c^3 \beta \boldsymbol{\sigma}\boldsymbol{P}\cdot\boldsymbol{B}}{2E^2(E+mc^2)} - \frac{e\hbar mc^4}{2E^3}\boldsymbol{\sigma} \wedge \boldsymbol{E} - \frac{e\hbar c^4 \boldsymbol{P} \wedge \boldsymbol{E}\boldsymbol{P}\cdot\boldsymbol{\sigma}}{2E^3(E+mc^2)}.$$

Prove then the equation of motion for this choice of the position operator

$$m\frac{\mathrm{d}\hat{\boldsymbol{v}}_\mathrm{op}}{\mathrm{d}t} \rightleftarrows \gamma^{-2}(\mathsf{U} - \boldsymbol{\beta}\boldsymbol{\beta})\cdot\Bigg(\gamma e\boldsymbol{E} + \gamma e\boldsymbol{\beta} \wedge \boldsymbol{B} + \frac{e\hbar}{2mc}\Bigg[(\boldsymbol{\nabla}\boldsymbol{B})\cdot\boldsymbol{\sigma} + \frac{\gamma}{\gamma+1}(\boldsymbol{\nabla}\boldsymbol{E})\cdot(\boldsymbol{\beta} \wedge \boldsymbol{\sigma})$$

$$+ (\partial_0 + \boldsymbol{\beta}\cdot\boldsymbol{\nabla})\Bigg\{\frac{\gamma}{\gamma+1}\boldsymbol{\sigma}\boldsymbol{\beta}\cdot\boldsymbol{B} + \frac{\gamma^2}{\gamma+1}\boldsymbol{\beta}\boldsymbol{\sigma}\cdot\boldsymbol{B} + \frac{\gamma^3}{\gamma+1}\boldsymbol{\beta}\boldsymbol{\beta}\cdot\boldsymbol{\sigma}\boldsymbol{\beta}\cdot\boldsymbol{B}$$

$$- \gamma^{-1}\boldsymbol{\sigma} \wedge \boldsymbol{E} - \frac{\gamma}{\gamma+1}\boldsymbol{\beta} \wedge \boldsymbol{E}\boldsymbol{\beta}\cdot\boldsymbol{\sigma} - \gamma\boldsymbol{\beta}\boldsymbol{\beta}\cdot(\boldsymbol{\sigma} \wedge \boldsymbol{E})\Bigg\}\Bigg]\Bigg).$$

Again $\beta$ was replaced by 1. The fields in this equation depend on $\boldsymbol{R}$ and $t$. Introduce now the fields at the position $\hat{\boldsymbol{R}}$ and time $t$. Introduce moreover the even part $\frac{1}{2}\hbar\hat{\boldsymbol{\sigma}}_\mathrm{e,op}$ of the spin operator that corresponds to the Dirac position operator, i.e. the operator which is $\frac{1}{2}\hbar\boldsymbol{\sigma}$ in the Dirac picture. Prove first that the Blount picture expression of this spin operator is, up to terms without potentials and fields (needed only for the equation of motion):

$$\hat{\boldsymbol{\sigma}}_\mathrm{op} \rightleftarrows \hat{\boldsymbol{\sigma}}_\mathrm{e} + \hat{\boldsymbol{\sigma}}_\mathrm{o}$$

with the even part

$$\hat{\boldsymbol{\sigma}}_\mathrm{e} = \boldsymbol{\sigma} - \frac{c^2\boldsymbol{P} \wedge (\boldsymbol{\sigma} \wedge \boldsymbol{P})}{E(E+mc^2)} = \boldsymbol{\sigma} - \frac{\gamma\boldsymbol{\beta} \wedge (\boldsymbol{\sigma} \wedge \boldsymbol{\beta})}{\gamma+1}$$

(which might be written as $\hat{\boldsymbol{\sigma}}_\mathrm{e} = \gamma^{-1}\boldsymbol{\Omega}^{-1}\cdot\boldsymbol{\sigma}$) and $\boldsymbol{\sigma}_\mathrm{o}$ the odd part which need not be specified. Show that the equation of motion then gets the form

$$m\frac{\mathrm{d}\hat{\mathfrak{v}}_{\mathrm{op}}}{\mathrm{d}t} \rightleftarrows \gamma^{-2}(\mathsf{U}-\boldsymbol{\beta}\boldsymbol{\beta})\cdot\Big(\gamma e\boldsymbol{E}+\gamma e\boldsymbol{\beta}\wedge\boldsymbol{B}+\frac{e\hbar}{2mc}[(\nabla\boldsymbol{B})\cdot\hat{\boldsymbol{\sigma}}_{\mathrm{e}}$$
$$+\gamma^2(\partial_0+\boldsymbol{\beta}\cdot\nabla)\{\boldsymbol{\beta}\hat{\boldsymbol{\sigma}}_{\mathrm{e}}\cdot\boldsymbol{B}-\gamma^{-2}\hat{\boldsymbol{\sigma}}_{\mathrm{e}}\wedge\boldsymbol{E}-\boldsymbol{\beta}\boldsymbol{\beta}\cdot(\hat{\boldsymbol{\sigma}}_{\mathrm{e}}\wedge\boldsymbol{E})\}]\Big).$$

To derive the spin equation show first that the even part of the spin operator which is simply $\frac{1}{2}\hbar\boldsymbol{\sigma}$ in the Dirac picture, reads (up to terms with potentials) in the Blount picture (again $g=2$):

$$\hat{\boldsymbol{\sigma}}_{\mathrm{e}}=\boldsymbol{\sigma}-\frac{c^2\boldsymbol{\pi}\wedge(\boldsymbol{\sigma}\wedge\boldsymbol{\pi})}{E_\pi(E_\pi+mc^2)}.$$

Derive then the equation of motion for this spin operator:

$$\frac{\mathrm{d}\hat{\boldsymbol{\sigma}}_{\mathrm{e,op}}}{\mathrm{d}t} \rightleftarrows \frac{e}{mc}\gamma^{-1}\{\hat{\boldsymbol{\sigma}}_{\mathrm{e}}\wedge\boldsymbol{B}+(\hat{\boldsymbol{\sigma}}_{\mathrm{e}}\wedge\boldsymbol{E})\wedge\boldsymbol{\beta}\}.$$

Just as in the preceding problem one may prove the non-covariance of the right-hand sides of the equations of motion and spin.

**6.** Prove from (158) that one has for the inner product of $\hat{\mathfrak{v}}_{\mathrm{op}}$ and $m\,\mathrm{d}\hat{\mathfrak{v}}_{\mathrm{op}}/\mathrm{d}t$ the equation:

$$\tfrac{1}{2}m\left\{\hat{\mathfrak{v}}_{\mathrm{op}}\cdot,\frac{\mathrm{d}\hat{\mathfrak{v}}_{\mathrm{op}}}{\mathrm{d}t}\right\} \rightleftarrows \gamma^{-4}\Big[c\gamma e\boldsymbol{\beta}\cdot\boldsymbol{E}(\hat{\boldsymbol{X}},t)+\frac{ge}{2m}\{\boldsymbol{\beta}\cdot(\nabla\boldsymbol{B})\cdot\hat{\mathfrak{s}}+\boldsymbol{\beta}\cdot(\nabla\boldsymbol{E})\cdot(\boldsymbol{\beta}\wedge\hat{\mathfrak{s}})$$
$$+\gamma^2\beta^2(\partial_0+\boldsymbol{\beta}\cdot\nabla)\hat{\mathfrak{s}}\cdot(\boldsymbol{B}-\boldsymbol{\beta}\wedge\boldsymbol{E})\}-\frac{(g-2)e}{2m}\gamma^2(\partial_0+\boldsymbol{\beta}\cdot\nabla)(\boldsymbol{\beta}\wedge\hat{\mathfrak{s}})\cdot(\boldsymbol{E}+\boldsymbol{\beta}\wedge\boldsymbol{B})\Big].$$

Compare this result with that of problem 10 of chapter IV.

CHAPTER IX

# Semi-relativistic description of particles with spin

## *1 Introduction*

For single particles that move in external electromagnetic fields equations of motion and of spin have been derived in a completely covariant way in the preceding chapter. If a set of particles moves under the combined influence of external fields and mutual interactions, a covariant description can be obtained only in the framework of quantum electrodynamics. This would take us outside the scope of the present treatise. However magnetic effects are found already if the non-relativistic theory (in which only electrostatic terms are effectively taken into account) is extended with terms up to and including those of order $c^{-2}$. Such a description will be the subject of this and the following chapter. In point of fact we shall not even have to consider all terms of order $c^{-2}$, but only those which contain at least one magnetic multipole term. Such an approximation can alternatively be described by declaring magnetic multipole moments as being of order $c^0$, and subsequently retaining only terms up to order $c^{-1}$: the so-called semi-relativistic approximation.

The present chapter will be devoted to the study of the semi-relativistic theory for a set of point particles with spin grouped into stable entities, while the next chapter will contain the corresponding theory of continuous media.

## *2 The Hamilton operator up to order $c^{-2}$ for a system of Dirac and Klein–Gordon particles in an external field*

The wave equation for a single Dirac particle with mass $m$ and charge $e$ in an external electromagnetic field with potentials $\varphi_e$ and $\boldsymbol{A}_e$ is

$$H_{op}\psi(\boldsymbol{R}, t) = -\frac{\hbar}{i}\frac{\partial\psi(\boldsymbol{R}, t)}{\partial t}, \tag{1}$$

where $\psi$ is a four-component wave function and where the Hamilton operator has the form

$$H_{\text{op}} = c\boldsymbol{\alpha}\cdot\boldsymbol{\pi}_{\text{op}} + \beta mc^2 + e\varphi_{\text{e}}(\boldsymbol{R}, t). \tag{2}$$

Here we used the $4 \times 4$ Dirac matrices $\boldsymbol{\alpha}$ and $\beta$, and the abbreviation

$$\boldsymbol{\pi}_{\text{op}} \equiv \boldsymbol{P}_{\text{op}} - \frac{e}{c}\boldsymbol{A}_{\text{e}}(\boldsymbol{R}, t). \tag{3}$$

(No anomalous magnetic moment term has been included in the Hamiltonian.)

If one wants to describe a system of two Dirac particles one needs a wave function which is an element of the direct product space of the wave functions for particles 1 and 2, and hence a 16-component wave function, labelled by two indices each running from 1 to 4. For such a system the Hamiltonian describing the time behaviour of the wave function will be the sum of two Hamiltonians of the type (2) and an interaction term. An approximate form for the latter (tantamount to taking only terms up to order $c^{-2}$ into account) has been written down by Breit[1] in close analogy to the classical Darwin Hamiltonian (which is valid up to order $c^{-2}$, see problem 6 of chapter III). It reads

$$H_{\text{int,op}} \equiv \frac{e_1 e_2}{4\pi|\boldsymbol{R}_1 - \boldsymbol{R}_2|}\{1 - \tfrac{1}{2}\boldsymbol{\alpha}_1\cdot\mathbf{T}(\boldsymbol{R}_1 - \boldsymbol{R}_2)\cdot\boldsymbol{\alpha}_2\}, \tag{4}$$

where $e_1$ and $e_2$ are the charges of the two particles, $\boldsymbol{R}_1$ and $\boldsymbol{R}_2$ their coordinates and where the three-tensor $\mathbf{T}$ is given by

$$\mathbf{T}(\boldsymbol{s}) \equiv \mathbf{U} + \frac{\boldsymbol{s}\boldsymbol{s}}{s^2}. \tag{5}$$

Furthermore $\boldsymbol{\alpha}_1$ and $\boldsymbol{\alpha}_2$ are Dirac matrices operating on the first and second index of the wave function respectively. (In the direct product notation, they may be written as $\boldsymbol{\alpha}\otimes 1$ and $1\otimes\boldsymbol{\alpha}$ respectively.) The total Hamiltonian of the two-particle system has thus the form

$$H_{\text{op}} = \sum_{i=1}^{2}\{c\boldsymbol{\alpha}_i\cdot\boldsymbol{\pi}_{i,\text{op}} + \beta_i m_i c^2 + e_i\varphi_{\text{e}}(\boldsymbol{R}_i, t)\} + \frac{e_1 e_2}{4\pi|\boldsymbol{R}_1 - \boldsymbol{R}_2|}\{1 - \tfrac{1}{2}\boldsymbol{\alpha}_1\cdot\mathbf{T}(\boldsymbol{R}_1 - \boldsymbol{R}_2)\cdot\boldsymbol{\alpha}_2\} \tag{6}$$

[1] G. Breit, Phys. Rev. **34**(1929)553, **36**(1930)383, **39**(1932)616; for a discussion from the point of view of quantum electrodynamics, see for instance H. A. Bethe and E. E. Salpeter, Quantum mechanics of one- and two-electron atoms (Springer-Verlag, Berlin 1957) p. 170ff, or A. I. Achieser and W. B. Berestezki, Quantenelektrodynamik (Teubner, Leipzig 1962) p. 428ff.

with

$$\pi_{i,\mathrm{op}} = \boldsymbol{P}_{i,\mathrm{op}} - \frac{e}{c}\boldsymbol{A}_{\mathrm{e}}(\boldsymbol{R}_i, t). \tag{7}$$

The Dirac matrices $\beta_1$ and $\beta_2$ should be understood in the same way as $\boldsymbol{\alpha}_1$ and $\boldsymbol{\alpha}_2$.

The Hamiltonian (6) contains the matrices $\boldsymbol{\alpha}_1$ and $\boldsymbol{\alpha}_2$, which are odd in the Pauli representation of the Dirac matrices that has been used in the preceding chapter. As a consequence the wave equation is in fact a set of 16 coupled wave equations. A similar situation of coupled equations arose for the one-particle system with its four-component wave function. There it turned out to be possible to write the set of four coupled equations as two uncoupled pairs of coupled equations, one for the upper two components and one for the lower two of the wave function. The advantage of this procedure was that then positive and negative energy solutions could be considered separately in a convenient way. In the present case we want to execute a similar programme, again uncoupling the upper-upper part and the lower-lower part from the other components of the wave function[1]. To that end we first transform the wave function by means of a product of two Blount transformations, as for the one-particle case. Then the first two terms of the Hamiltonian get even form, as in (VIII.143), if one considers only terms linear in the external fields (or in the charges) and without derivatives of the fields. If one is not interested in terms of higher order than bilinear in the charges, the last term of the Hamiltonian (6) transforms into an expression which is obtained by utilizing the product of the two Blount transformations up to order $e^0$, i.e. the product of the P–FW transformation operators (VIII.56) both for particle 1 and particle 2. Then we obtain

$$\hat{H}_{\mathrm{op}} \rightleftarrows \sum_{i=1}^{2} \left\{ \beta_i E_{\pi i} + e_i \varphi_{\mathrm{e}}(\boldsymbol{R}_i, t) - \frac{e_i \hbar c}{2E_i} \beta_i \boldsymbol{\sigma}_i \cdot \boldsymbol{B}_{\mathrm{e}}(\boldsymbol{R}_i, t) \right.$$
$$\left. - \frac{e_i \hbar c^2}{2E_i(E_i + m_i c^2)} (\boldsymbol{P}_i \wedge \boldsymbol{\sigma}_i) \cdot \boldsymbol{E}_{\mathrm{e}}(\boldsymbol{R}_i, t) \right\} + \hat{H}_{\mathrm{int}} \equiv \hat{H}, \tag{8}$$

where at the right-hand side the Weyl transform $\hat{H}$ of $\hat{H}_{\mathrm{op}}$ has been written. The first terms represent the two one-particle Hamiltonians of (VIII.143). The last term is the Weyl transform of the transformed interaction Hamiltonian. In view of (4) this interaction Hamiltonian consists of two terms

$$\hat{H}_{\mathrm{int}} = \hat{H}_{\mathrm{I,int}} + \hat{H}_{\mathrm{II,int}}, \tag{9}$$

[1] It is unnecessary to decouple also the upper-lower part and the lower-upper part from each other, v. Z. V. Chraplyvy, Phys. Rev. **91**(1953)388, **92**(1953)1310.

with

$$\hat{H}_{\mathrm{I,int,op}} \equiv U_{\mathrm{op}}(1)U_{\mathrm{op}}(2)\,\frac{e_1 e_2}{4\pi|\boldsymbol{R}_1-\boldsymbol{R}_2|}\,U^{\dagger}_{\mathrm{op}}(1)U^{\dagger}_{\mathrm{op}}(2) \rightleftarrows H_{\mathrm{I,int}}, \tag{10}$$

$$\hat{H}_{\mathrm{II,int,op}} \equiv -U_{\mathrm{op}}(1)U_{\mathrm{op}}(2)\,\frac{e_1 e_2}{8\pi|\boldsymbol{R}_1-\boldsymbol{R}_2|}\,\boldsymbol{\alpha}_1\cdot\mathbf{T}(\boldsymbol{R}_1-\boldsymbol{R}_2)\cdot\boldsymbol{\alpha}_2\,U^{\dagger}_{\mathrm{op}}(1)U^{\dagger}_{\mathrm{op}}(2) \rightleftarrows H_{\mathrm{II,int}}. \tag{11}$$

From the rules for Weyl transforms (VIII.132) and the abbreviations (VIII.62)

$$\xi_i \equiv \frac{\hbar}{i}\,\frac{\partial U(i)}{\partial \boldsymbol{P}_i}\,U^{\dagger}(i) = -\frac{\hbar}{i}\,U(i)\,\frac{\partial U^{\dagger}(i)}{\partial \boldsymbol{P}_i}, \tag{12}$$

one finds for (10)

$$\hat{H}_{\mathrm{I,int}} = \left(1+\xi_1\cdot\nabla_1+\xi_2\cdot\nabla_2+\xi_1\cdot\nabla_1\,\xi_2\cdot\nabla_2+\tfrac{1}{2}\xi_1\,\xi_1 : \nabla_1\nabla_1 + \tfrac{1}{2}\xi_2\,\xi_2 : \nabla_2\nabla_2+\dots\right)\frac{e_1 e_2}{4\pi|\boldsymbol{R}_1-\boldsymbol{R}_2|}. \tag{13}$$

Furthermore we find in the same fashion for the second interaction term (11)

$$\begin{aligned}\hat{H}_{\mathrm{II,int}} = {} & F+\tfrac{1}{2}\xi_1\cdot\nabla_1 F+\tfrac{1}{2}(\nabla_1 F)\cdot\xi_1+\tfrac{1}{2}\xi_2\cdot\nabla_2 F+\tfrac{1}{2}(\nabla_2 F)\cdot\xi_2 \\ & +\tfrac{1}{4}\xi_1\cdot\nabla_1\,\xi_2\cdot\nabla_2 F+\tfrac{1}{4}\xi_1\cdot\nabla_1(\nabla_2 F)\cdot\xi_2+\tfrac{1}{4}\xi_2\cdot\nabla_2(\nabla_1 F)\cdot\xi_1 \\ & +\tfrac{1}{4}(\nabla_1\nabla_2 F) : \xi_1\,\xi_2-\tfrac{1}{8}\left(\hbar i\,\frac{\partial\xi_1}{\partial\boldsymbol{P}_1}-\xi_1\,\xi_1\right) : \nabla_1\nabla_1 F \\ & -\tfrac{1}{8}(\nabla_1\nabla_1 F) : \left(-\hbar i\,\frac{\partial\xi_1}{\partial\boldsymbol{P}_1}-\xi_1\,\xi_1\right)+\tfrac{1}{4}\xi_1\cdot\nabla_1(\nabla_1 F)\cdot\xi_1 \\ & -\tfrac{1}{8}\left(\hbar i\,\frac{\partial\xi_2}{\partial\boldsymbol{P}_2}-\xi_2\,\xi_2\right) : \nabla_2\nabla_2 F-\tfrac{1}{8}(\nabla_2\nabla_2 F) : \left(-\hbar i\,\frac{\partial\xi_2}{\partial\boldsymbol{P}_2}-\xi_2\,\xi_2\right) \\ & +\tfrac{1}{4}\xi_2\cdot\nabla_2(\nabla_2 F)\cdot\xi_2+\dots\end{aligned} \tag{14}$$

with the abbreviation

$$F \equiv -\frac{e_1 e_2\,\hat{\boldsymbol{\alpha}}_1\cdot\mathbf{T}(\boldsymbol{R}_1-\boldsymbol{R}_2)\cdot\hat{\boldsymbol{\alpha}}_2}{8\pi|\boldsymbol{R}_1-\boldsymbol{R}_2|}, \tag{15}$$

where

$$\hat{\boldsymbol{\alpha}}_i \equiv U(i)\boldsymbol{\alpha}_i\,U^{\dagger}(i). \tag{16}$$

In deriving (13) and (14) we employed the identity, which follows from (12),

$$\frac{\partial^2 U(i)}{\partial\boldsymbol{P}_i\,\partial\boldsymbol{P}_i}\,U^{\dagger}(i) = \frac{\partial}{\partial\boldsymbol{P}_i}\left(\frac{i}{\hbar}\,\xi_i\right)-\frac{1}{\hbar^2}\,\xi_i\,\xi_i. \tag{17}$$

The transformed Hamiltonian, given by (8) with (9), (13) and (14) is not in closed form. It contains series in the derivatives of the interparticle potential, occurring in (13) and of the function $F$ (15), occurring in (14). In the case of one single particle in an external field, such series (but there with the derivatives of the external field) also appeared, at least in principle. But there (as well as here for the external field terms) these series could be broken off on the assumption that the external field did not change rapidly (on the scale of the Compton wave length). Here, for the interaction terms (13–14), we are not in such a simple situation, since the interparticle fields may change rapidly. The series may still be broken off if a stronger assumption is adopted namely that only terms up to a certain order in $c^{-1}$ should be taken into account. In fact both (13) and (14) may be seen as series in powers of $c^{-1}$, if one realizes that $\xi_i$ is of order $c^{-1}$ and $\partial \xi_i / \partial \boldsymbol{P}_i$ of order $c^{-2}$, as follows from the expressions (VIII.63). Therefore we shall limit ourselves from now on to terms up to and including those of order $c^{-2}$ (terms of higher order in $c^{-1}$ can only be obtained consistently if one starts from an expression of more general validity than (4)). In that case the expression (8) becomes

$$\hat{H}_{\text{op}} \rightleftarrows \sum_{i=1}^{2} \Big[ \beta_i m_i c^2 + \beta_i \frac{\boldsymbol{P}_i^2}{2m_i} - \beta_i \frac{\boldsymbol{P}_i^4}{8m_i^3 c^2} - \beta_i e_i \frac{\boldsymbol{P}_i}{m_i c} \cdot \boldsymbol{A}_{\text{e}}(\boldsymbol{R}_i, t)$$

$$+ e_i \varphi_{\text{e}}(\boldsymbol{R}_i, t) - \frac{e_i \hbar}{2m_i c} \boldsymbol{\sigma}_i \cdot \Big\{ \boldsymbol{B}_{\text{e}}(\boldsymbol{R}_i, t)\beta_i - \frac{\boldsymbol{P}_i}{2m_i c} \wedge \boldsymbol{E}_{\text{e}}(\boldsymbol{R}_i, t) \Big\} \Big] + \hat{H}_{\text{int}}, \quad (18)$$

where only terms up to first order in the external fields have been retained and where now $\hat{H}_{\text{int}}$ is given by (9) with (13–14), the latter without the terms indicated by dots.

The Hamiltonian in (18) contains a number of terms which are already of even-even form, i.e. they do not couple the upper-upper and the lower-lower parts of the wave function with the upper-lower and lower-upper parts. The interaction Hamiltonian $\hat{H}_{\text{int}}$ however contains also parts which are of even-odd, odd-even and odd-odd character and hence do couple the upper-upper and lower-lower components with each other and with the mixed components. The general form of (18) is thus

$$\hat{H}_{\text{op}} \rightleftarrows \mathscr{E}_1 \mathscr{E}_2 + \mathscr{E}_1 \mathscr{O}_2 + \mathscr{O}_1 \mathscr{E}_2 + \mathscr{O}_1 \mathscr{O}_2, \quad (19)$$

where $\mathscr{E}_i$ and $\mathscr{O}_i$ indicate even and odd parts with respect to the matrix indices pertinent to particle $i$. The odd parts in (19) may be brought into a more convenient form in the following way. Let us transform the Hamiltonian by means of an operator

$$V_{\text{op}} \rightleftarrows 1 + \lambda_1 \beta_2 \mathscr{E}_1 \mathscr{O}_2 + \lambda_2 \beta_1 \mathscr{O}_1 \mathscr{E}_2 + (\lambda_3 \beta_1 + \lambda_4 \beta_2) \mathscr{O}_1 \mathscr{O}_2. \quad (20)$$

Up to order $e_1 e_2$ this operator is unitary, if the $\lambda_i$ are real coefficients. This is seen by taking into account that the last three parts of the Hamiltonian (19) are of order $e_1 e_2$.

If the Weyl transform of $V_{\text{op}}\hat{H}_{\text{op}} V_{\text{op}}^{-1}$ is calculated, one finds in the first place a term which is simply the product $V\hat{H}V^{-1}$ of the Weyl transforms given by (19) and (20). Up to order $e_1 e_2$ this product is:

$$V\hat{H}V^{-1} = \mathscr{E}_1\mathscr{E}_2+\mathscr{E}_1\mathscr{O}_2+\mathscr{O}_1\mathscr{E}_2+\mathscr{O}_1\mathscr{O}_2-2\lambda_1 E_2\mathscr{E}_1\mathscr{O}_2-2\lambda_2 E_1\mathscr{O}_1\mathscr{E}_2$$
$$-2\{\lambda_3(E_1+\beta_1\beta_2 E_2)+\lambda_4(E_2+\beta_1\beta_2 E_1)\}\mathscr{O}_1\mathscr{O}_2, \quad (21)$$

where we used the (anti)commutation property of $\beta_i$ with even (odd) matrices $\mathscr{E}_i\,(\mathscr{O}_i)$ and the fact that the terms independent of $e_1$ and $e_2$ in $\hat{H}$ are $\beta_1 E_1+\beta_2 E_2$. We require that the transformed Hamiltonian, part of which is written in (21), contains no $\mathscr{E}_1\mathscr{O}_2$ and $\mathscr{O}_1\mathscr{E}_2$ terms, so we choose $\lambda_1 = 1/2E_2$ and $\lambda_2 = 1/2E_1$. Then the upper-upper terms (and the lower-lower terms) are no longer coupled to the mixed components through terms of the type $\mathscr{E}_1\mathscr{O}_2$ and $\mathscr{O}_1\mathscr{E}_2$. In order to achieve that the odd-odd terms do not give rise to unwanted couplings, one may try to make them vanish as well. This gives rise to a unitary operator which is singular and has therefore to be rejected. However the odd-odd terms are harmless[1] already if they are multiplied by $(1-\beta_1\beta_2)$, because then this term gives zero if it operates on a wave function which has only upper-upper or only lower-lower components. This is accomplished by choosing $\lambda_3$ and $\lambda_4$ as $1/\{4(E_1+E_2)\}$. Hence the unitary operator (20) which we employ is

$$V_{\text{op}} \rightleftarrows 1+\frac{1}{2E_2}\beta_2\mathscr{E}_1\mathscr{O}_2+\frac{1}{2E_1}\beta_1\mathscr{O}_1\mathscr{E}_2+\frac{\beta_1+\beta_2}{4(E_1+E_2)}\mathscr{O}_1\mathscr{O}_2. \quad (22)$$

Then, up to order $c^{-2}$ and $e_1 e_2$, the transformed Hamiltonian becomes, employing the rule (VIII.132) for the Weyl transform of a product,

$$V_{\text{op}}\hat{H}_{\text{op}}V_{\text{op}}^{-1} \rightleftarrows \mathscr{E}_1\mathscr{E}_2+\tfrac{1}{2}(1-\beta_1\beta_2)\mathscr{O}_1\mathscr{O}_2+\frac{\hbar^2c^2}{16E_1E_2}(1+\beta_1\beta_2)\Delta_1\mathscr{O}_1\mathscr{O}_2, \quad (23)$$

where we used the fact that the terms with $\mathscr{E}_1\mathscr{O}_2$ and $\mathscr{O}_1\mathscr{E}_2$ in (19) are both of order $c^{-1}$. (The Laplacian operating on the coordinates of particle 1 has been denoted by $\Delta_1$.) Hence in the transformed Hamiltonian again a term (namely the last) appears that is of the unwanted kind discussed above.

[1] Z. V. Chraplyvy, op. cit.

However, in contrast with the $\mathcal{O}_1\mathcal{O}_2$ term of (19) (which is of order $c^0$) it is only of order $c^{-2}$. It may be made to disappear by a final unitary transformation with an operator

$$W_{\text{op}} \rightleftarrows 1+\frac{\beta_1+\beta_2}{4(E_1+E_2)}\frac{\hbar^2c^2}{16E_1E_2}(1+\beta_1\beta_2)\Delta_1\mathcal{O}_1\mathcal{O}_2. \qquad (24)$$

Then the Hamiltonian gets the form (up to order $c^{-2}$ and $e_1e_2$):

$$\hat{\hat{H}}_{\text{op}} \equiv W_{\text{op}}V_{\text{op}}\hat{H}_{\text{op}}V_{\text{op}}^{-1}W_{\text{op}}^{-1} \rightleftarrows \mathscr{E}_1\mathscr{E}_2+\tfrac{1}{2}(1-\beta_1\beta_2)\mathcal{O}_1\mathcal{O}_2. \qquad (25)$$

If we confine ourselves to positive energy solutions, i.e. to wave functions with only an upper-upper part, we may replace $\beta_1$ and $\beta_2$ by 1 so that the last term drops out. Then we have for the Weyl transform of the Hamiltonian effectively only the even-even part $\mathscr{E}_1\mathscr{E}_2$ of the right-hand side of (18). The even-even part of (13), up to order $c^{-2}$, follows by making use of the approximate expressions for the even and odd parts of $\xi_i$ (v. (VIII.63)):

$$\xi_{i,\text{e}} = \frac{\hbar \boldsymbol{P}_i\wedge\boldsymbol{\sigma}_i}{4m_i^2c^2}, \qquad \xi_{i,\text{o}} = \frac{-\hbar i\beta_i\boldsymbol{\alpha}_i}{2m_ic}. \qquad (26)$$

Substituting these expressions into (13) we find

$$\hat{H}_{\text{I,int}} = \left[1+\sum_{i=1}^{2}\left\{\frac{\hbar(\boldsymbol{P}_i\wedge\boldsymbol{\sigma}_i)\cdot\boldsymbol{\nabla}_i}{4m_i^2c^2}+\frac{\hbar^2}{8m_i^2c^2}\Delta_i\right\}\right]\frac{e_1e_2}{4\pi|\boldsymbol{R}_1-\boldsymbol{R}_2|}. \qquad (27)$$

If we use moreover the approximate expressions for $\hat{\boldsymbol{\alpha}}_i$ (16):

$$\hat{\boldsymbol{\alpha}}_{i,\text{e}} = \beta\frac{\boldsymbol{P}_i}{mc}, \qquad \hat{\boldsymbol{\alpha}}_{i,\text{o}} = \boldsymbol{\alpha}, \qquad (28)$$

we find for (14) with (15), up to order $c^{-2}$,

$$\hat{H}_{\text{II,int}} = -\beta_1\beta_2\{\boldsymbol{P}_1\boldsymbol{P}_2+\tfrac{1}{2}\hbar(\boldsymbol{\sigma}_1\wedge\boldsymbol{\nabla}_1)\boldsymbol{P}_2+\tfrac{1}{2}\hbar(\boldsymbol{\sigma}_2\wedge\boldsymbol{\nabla}_2)\boldsymbol{P}_1$$
$$+\tfrac{1}{4}\hbar^2(\boldsymbol{\sigma}_1\wedge\boldsymbol{\nabla}_1)(\boldsymbol{\sigma}_2\wedge\boldsymbol{\nabla}_2)\} : \frac{e_1e_2\mathbf{T}(\boldsymbol{R}_1-\boldsymbol{R}_2)}{8\pi m_1m_2c^2|\boldsymbol{R}_1-\boldsymbol{R}_2|}. \qquad (29)$$

If we replace $\beta_1$ and $\beta_2$ by 1, we have found now for the complete Hamiltonian for positive-energy solutions, up to terms bilinear in the charges and up to order $c^{-2}$ (omitting from now on the double circumflexes over $H_{\text{op}}$):

$$H_{\mathrm{op}} \rightleftarrows \sum_{i=1}^{2} \left[ m_i c^2 + \frac{\boldsymbol{P}_i^2}{2m_i} - \frac{\boldsymbol{P}_i^4}{8m_i^3 c^2} - e_i \frac{\boldsymbol{P}_i}{m_i c} \cdot \boldsymbol{A}_{\mathrm{e}}(\boldsymbol{R}_i, t) + e_i \varphi_{\mathrm{e}}(\boldsymbol{R}_i, t) \right.$$

$$\left. - \frac{e_i \hbar}{2m_i c} \boldsymbol{\sigma}_i \cdot \left\{ \boldsymbol{B}_{\mathrm{e}}(\boldsymbol{R}_i, t) - \frac{\boldsymbol{P}_i}{2m_i c} \wedge \boldsymbol{E}_{\mathrm{e}}(\boldsymbol{R}_i, t) \right\} \right]$$

$$+ \left[ 1 + \sum_{i=1}^{2} \left\{ \frac{\hbar (\boldsymbol{P}_i \wedge \boldsymbol{\sigma}_i) \cdot \nabla_i}{4m_i^2 c^2} + \frac{\hbar^2}{8m_i^2 c^2} \Delta_i \right\} \right] \frac{e_1 e_2}{4\pi |\boldsymbol{R}_1 - \boldsymbol{R}_2|}$$

$$- \frac{e_1 e_2 \boldsymbol{P}_1 \cdot \mathrm{T}(\boldsymbol{R}_1 - \boldsymbol{R}_2) \cdot \boldsymbol{P}_2}{8\pi m_1 m_2 c^2 |\boldsymbol{R}_1 - \boldsymbol{R}_2|} - \{ \hbar (\boldsymbol{\sigma}_1 \wedge \nabla_1) \cdot \boldsymbol{P}_2 + \hbar (\boldsymbol{\sigma}_2 \wedge \nabla_2) \cdot \boldsymbol{P}_1$$

$$+ \tfrac{1}{2} \hbar^2 (\boldsymbol{\sigma}_1 \wedge \nabla_1) \cdot (\boldsymbol{\sigma}_2 \wedge \nabla_2) \} \frac{e_1 e_2}{8\pi m_1 m_2 c^2 |\boldsymbol{R}_1 - \boldsymbol{R}_2|}, \tag{30}$$

where in the last term we used the property of the tensor $\mathbf{T}(\boldsymbol{s})$ (5):

$$(\boldsymbol{a} \wedge \nabla_s) \cdot \frac{\mathrm{T}(\boldsymbol{s})}{4\pi s} = 2(\boldsymbol{a} \wedge \nabla_s) \frac{1}{4\pi s} \tag{31}$$

for an arbitrary vector $\boldsymbol{a}$.

The Hamiltonian[1] obtained contains in the first place the approximation up to order $c^{-2}$ of the one-particle Hamiltonians for particles 1 and 2. It includes, apart from the non-relativistic terms, a kinetic term with the fourth power of the momentum and two terms which couple the spin with the electromagnetic field. Furthermore interaction terms appear, which apart from the Coulomb term, are all of order $c^{-2}$. One recognizes spin–orbit coupling terms with the vector product of momentum and spin. Next terms with the Laplacians acting on the Coulomb expression appear. They may be written alternatively as[2]

$$- \sum_{i=1}^{2} \frac{\hbar^2}{8m_i^2 c^2} e_1 e_2 \, \delta(\boldsymbol{R}_1 - \boldsymbol{R}_2). \tag{32}$$

Furthermore one encounters the quantum-mechanical analogue of a term in the Darwin Hamiltonian of classical theory (see problem 6 of chapter III). Finally two terms that couple the spin of one particle to the momentum of the other, and a spin–spin interaction term occur. The last term may be

[1] V. e.g. H. A. Bethe and E. E. Salpeter, op. cit. p. 181; A. I. Achieser and W. B. Berestezki, op. cit. p. 431.

[2] In the past confusion about this term existed. Instead of the operator corresponding to (32) one found a non-hermitian operator by employing an elimination procedure for the lower components of the wave function. Then the normalization of the wave function is lost, so that non-hermitian terms appear (cf. A. I. Achieser and W. B. Berestezki, op. cit.).

written in an alternative form by using the ancillary formula

$$\boldsymbol{\nabla}_s \boldsymbol{\nabla}_s \frac{1}{4\pi s} = \mathscr{P}_{\text{sph}} \boldsymbol{\nabla}_s \boldsymbol{\nabla}_s \frac{1}{4\pi s} - \tfrac{1}{3}\mathbf{U}\delta(\boldsymbol{s}), \tag{33}$$

which expresses the double nabla operator acting on $(4\pi s)^{-1}$ in its principal value and a term with the three-dimensional delta function multiplied by the unit tensor $\mathbf{U}$ (v. problem 2 of chapter II). Then the last term of (30) becomes

$$\mathscr{P}_{\text{sph}} \boldsymbol{\sigma}_1 \cdot \boldsymbol{\nabla}_1 \boldsymbol{\sigma}_2 \cdot \boldsymbol{\nabla}_2 \frac{e_1 e_2 \hbar^2}{16\pi m_1 m_2 c^2 |\boldsymbol{R}_1 - \boldsymbol{R}_2|} - \boldsymbol{\sigma}_1 \cdot \boldsymbol{\sigma}_2 \frac{e_1 e_2}{6 m_1 m_2 c^2} \delta(\boldsymbol{R}_1 - \boldsymbol{R}_2). \tag{34}$$

The Hamilton operator, given in (30) describes a system consisting of two Dirac particles. The generalization to $N$ Dirac particles is obvious. The same procedure may be followed to bring the Hamiltonian into a form which does not couple the upper-...-upper part of the wave function with the other parts. The result is an expression like (30) but now with $N$ one-particle contributions and $\frac{1}{2}N(N-1)$ pair contributions of the type given there.

The physical systems consist usually of electrons and nuclei. The couplings of the electrons with the external field and with each other are described by terms of the form (30). Since the nuclei are much heavier than the electrons, their spin effects can often be neglected: hyperfine splittings are small corrections only. Therefore we shall describe the nuclei from now on as particles without spin, i.e., as Klein–Gordon particles. To find the Hamilton operators for the interaction of a Dirac and a Klein–Gordon particle and of two Klein–Gordon particles in a form comparable to (30) (for the interaction of two Dirac particles) one has to start from an expression that comes instead of (4). Just as the latter formula for two Dirac particles can be derived from quantum electrodynamics, one may obtain for the interaction of two Klein–Gordon particles 1 and 2:

$$\frac{e_1 e_2}{4\pi |\boldsymbol{R}_1 - \boldsymbol{R}_2|} - \frac{e_1 e_2 (\tau_3 + i\tau_2)_1 (\tau_3 + i\tau_2)_2}{8c^2} \left\{ \frac{\boldsymbol{P}_{1,\text{op}}}{m_1}, \left\{ \frac{\boldsymbol{P}_{2,\text{op}}}{m_2}, \frac{:\mathbf{T}(\boldsymbol{R}_1 - \boldsymbol{R}_2)}{4\pi |\boldsymbol{R}_1 - \boldsymbol{R}_2|} \right\} \right\}, \tag{35}$$

where the Feshbach–Villars representation (chapter VIII, section 4) has been employed. The interaction between a Klein–Gordon particle 1 and a Dirac particle 2 is given by

$$\frac{e_1 e_2}{4\pi |\boldsymbol{R}_1 - \boldsymbol{R}_2|} - \frac{e_1 e_2 (\tau_3 + i\tau_2)_1}{4c} \left\{ \frac{\boldsymbol{P}_{1,\text{op}}}{m_1}, \frac{\cdot \mathbf{T}(\boldsymbol{R}_1 - \boldsymbol{R}_2) \cdot \boldsymbol{\alpha}_2}{4\pi |\boldsymbol{R}_1 - \boldsymbol{R}_2|} \right\}. \tag{36}$$

(As compared to (4) one finds here that the Dirac matrix $\boldsymbol{\alpha}$ for a Dirac particle is to be replaced by $(\tau_3 + i\tau_2)\boldsymbol{P}_{\text{op}}/mc$ for a Klein–Gordon particle. Moreover anticommutators have to be added for the latter case.) If one takes the same steps as those which led from (4) to (30), one finds from (35) and (36) an expression like (30) for the Klein–Gordon particles but without $\boldsymbol{\sigma}$-terms and the term with the Laplacian. In this way one finds for a collection of electrons and nuclei (described as Dirac and Klein–Gordon particles in the present model) that the Weyl transform of the total Hamiltonian is given by

$$\begin{aligned} H_{\text{op}} \rightleftarrows & \sum_i \frac{\boldsymbol{P}_i^2}{2m_i} + \sum_{i,j(i\neq j)} \frac{e_i e_j}{8\pi|\boldsymbol{R}_i - \boldsymbol{R}_j|} + \sum_i e_i \left\{\varphi_{\text{e}}(\boldsymbol{R}_i, t) - \frac{\boldsymbol{P}_i}{m_i c}\cdot \boldsymbol{A}_{\text{e}}(\boldsymbol{R}_i, t)\right\} \\ & - \sum_i \frac{\boldsymbol{P}_i^4}{8m_i^3c^2} - \sum_{i,j(i\neq j)} \frac{e_i e_j}{16\pi c^2|\boldsymbol{R}_i - \boldsymbol{R}_j|} \frac{\boldsymbol{P}_i}{m_i}\cdot \mathsf{T}(\boldsymbol{R}_i - \boldsymbol{R}_j)\cdot \frac{\boldsymbol{P}_j}{m_j} \\ & + \sum_{i,j(i\neq j)}' \frac{e_i e_j \hbar}{4m_i c^2} \left\{\left(\frac{\boldsymbol{P}_i}{m_i} \wedge \boldsymbol{\sigma}_i\right)\cdot \nabla_i - 2\left(\frac{\boldsymbol{P}_j}{m_j} \wedge \boldsymbol{\sigma}_i\right)\cdot \nabla_i\right\} \frac{1}{4\pi|\boldsymbol{R}_i - \boldsymbol{R}_j|} \\ & - \sum_{i,j(i\neq j)}'' \frac{e_i e_j \hbar^2}{8m_i m_j c^2} (\boldsymbol{\sigma}_i \wedge \nabla_i)\cdot(\boldsymbol{\sigma}_j \wedge \nabla_j) \frac{1}{4\pi|\boldsymbol{R}_i - \boldsymbol{R}_j|} \\ & - \sum_{i,j(i\neq j)}' \frac{e_i e_j \hbar^2}{8m_i^2c^2} \delta(\boldsymbol{R}_i - \boldsymbol{R}_j) - \sum_i' \frac{e_i \hbar}{2m_i c} \boldsymbol{\sigma}_i\cdot \left\{\boldsymbol{B}_{\text{e}}(\boldsymbol{R}_i, t) - \frac{\boldsymbol{P}_i}{2m_i c} \wedge \boldsymbol{E}_{\text{e}}(\boldsymbol{R}_i, t)\right\} \\ & = H(1, \ldots, N; t), \end{aligned} \tag{37}$$

where the rest energy terms have been suppressed. The primes indicate that the summations concerned are extended only over the electrons. (One should note that a single prime at a double summation sign means that only the summation over the first index is to be limited to the electrons.)

## 3 *The field equations and the equations of motion for a set of spin particles*

In this section we shall first study the equations of motion and in connexion with them the field equations. The Hamilton operator that specifies the system is given by expression (37).

The equations of motion for the electrons and nuclei will follow by evaluating the commutators of their position operator with the Hamiltonian. The position operator for electron $i$ reads up to order $c^{-2}$:

$$\boldsymbol{X}_{i,\text{op}} = \boldsymbol{R}_i + \frac{\hbar\boldsymbol{\sigma}_i \wedge \boldsymbol{P}_{i,\text{op}}}{4m_i^2c^2}, \tag{38}$$

as follows from (VIII.68) of the preceding chapter. The position operator found there is valid only for a free particle. However by comparison with (VIII.150) it may be seen that up to order $c^{-2}$ the position operator does not change (at least not with terms in the potentials) if an external field is switched on. For the nuclei the position operator is simply

$$\boldsymbol{X}_{i,\mathrm{op}} = \boldsymbol{R}_i , \tag{39}$$

as follows from (VIII.204).

The equation of motion for an electron $i$ will follow by taking first the commutator of (38) with the Hamiltonian (37). With the rules for Weyl transforms (in particular (VI.A161)) one finds up to order $c^{-2}$:

$$\boldsymbol{v}_{i,\mathrm{op}} \equiv \frac{\mathrm{d}\boldsymbol{X}_{i,\mathrm{op}}}{\mathrm{d}t} \equiv \frac{i}{\hbar}[H_{\mathrm{op}}, \boldsymbol{X}_{i,\mathrm{op}}] \rightleftarrows \frac{\partial H}{\partial \boldsymbol{P}_i} - \frac{\hbar}{4m_i^2c^2}\boldsymbol{\sigma}_i \wedge \frac{\partial H}{\partial \boldsymbol{R}_i}, \tag{40}$$

or explicitly

$$\begin{aligned}\boldsymbol{v}_{i,\mathrm{op}} \rightleftarrows{} & \frac{\boldsymbol{P}_i}{m_i} - \frac{e_i}{m_i c}\boldsymbol{A}_{\mathrm{e}}(\boldsymbol{R}_i, t) - \frac{P_i^2\boldsymbol{P}_i}{2m_i^3c^2} \\ & - \sum_{j(\neq i)} \frac{e_i e_j \boldsymbol{P}_j\cdot\mathsf{T}(\boldsymbol{R}_i-\boldsymbol{R}_j)}{8\pi m_i m_j c^2|\boldsymbol{R}_i-\boldsymbol{R}_j|} - \sum_{j(\neq i)}{}' \frac{e_i e_j \hbar}{2m_i m_j c^2}\boldsymbol{\sigma}_j \wedge \nabla_j \frac{1}{4\pi|\boldsymbol{R}_i-\boldsymbol{R}_j|} .\end{aligned} \tag{41}$$

The second time derivative is found by taking once more the commutator with the Hamiltonian and adding an explicit time derivative. Then one finds an equation of the form

$$m_i \frac{\mathrm{d}\boldsymbol{v}_{i,\mathrm{op}}}{\mathrm{d}t} \equiv m_i \frac{i}{\hbar}[H_{\mathrm{op}}, \boldsymbol{v}_{i,\mathrm{op}}] + m_i \frac{\partial \boldsymbol{v}_{i,\mathrm{op}}}{\partial t} \rightleftarrows \boldsymbol{f}_i , \tag{42}$$

where $\boldsymbol{f}_i$ is the Weyl transform of the force on the electron. It is convenient to study first its part $\boldsymbol{f}_{i\mathrm{e}}$, which depends on the external fields. Retaining only terms linear in these fields (and linear in the charge $e_i$) we obtain

$$\begin{aligned}\boldsymbol{f}_{i\mathrm{e}} ={} & e_i \boldsymbol{E}_{\mathrm{e}}(\boldsymbol{R}_i, t) + \frac{e_i}{m_i c}\boldsymbol{P}_i \wedge \boldsymbol{B}_{\mathrm{e}}(\boldsymbol{R}_i, t) \\ & - \frac{e_i}{m_i^2c^2}\boldsymbol{P}_i\boldsymbol{P}_i\cdot\boldsymbol{E}_{\mathrm{e}}(\boldsymbol{R}_i, t) - \frac{e_i}{2m_i^2c^2}P_i^2\boldsymbol{E}_{\mathrm{e}}(\boldsymbol{R}_i, t) \\ & + \frac{e_i\hbar}{2m_i c}\left[\{\nabla_i \boldsymbol{B}_{\mathrm{e}}(\boldsymbol{R}_i, t)\}\cdot\boldsymbol{\sigma}_i + \tfrac{1}{2}\{\nabla_i \boldsymbol{E}_{\mathrm{e}}(\boldsymbol{R}_i, t)\}\cdot\left(\frac{\boldsymbol{P}_i}{m_i c}\wedge\boldsymbol{\sigma}_i\right)\right] .\end{aligned} \tag{43}$$

The total force has a similar structure. It becomes:

$$f_i = e_i \boldsymbol{e}_t(\boldsymbol{R}_i, t) + \frac{e_i}{m_i c} \boldsymbol{P}_i \wedge \boldsymbol{b}_t(\boldsymbol{R}_i, t)$$

$$- \frac{e_i}{m_i^2 c^2} \boldsymbol{P}_i \boldsymbol{P}_i \cdot \boldsymbol{e}_t(\boldsymbol{R}_i, t) - \frac{e_i}{2m_i^2 c^2} \boldsymbol{P}_i^2 \boldsymbol{e}_t(\boldsymbol{R}_i, t) + \frac{e_i \hbar}{2m_i c} \Big[ \{\nabla_i \boldsymbol{b}_t(\boldsymbol{R}_i, t)\} \cdot \boldsymbol{\sigma}_i$$

$$+ \tfrac{1}{2} \{\nabla_i \boldsymbol{e}_t(\boldsymbol{R}_i, t)\} \cdot \left( \frac{\boldsymbol{P}_i}{m_i c} \wedge \boldsymbol{\sigma}_i \right) \Big] + \frac{e_i \hbar^2}{4m_i^2 c^2} \Delta_i \boldsymbol{e}_t(\boldsymbol{R}_i, t), \tag{44}$$

if only terms bilinear in the charges are included. Only the last term of (44) is of a type that did not occur in (43). The other terms contain, instead of the external fields $\boldsymbol{E}_e$ and $\boldsymbol{B}_e$, the Weyl transforms of quantities $\boldsymbol{e}_t$ and $\boldsymbol{b}_t$, which will be called the total fields acting on the particle. They are of the form

$$\begin{aligned} \boldsymbol{e}_t(\boldsymbol{R}, t) &= \boldsymbol{E}_e(\boldsymbol{R}, t) + \sum_{j(\neq i)} \boldsymbol{e}_j(\boldsymbol{R}, t), \\ \boldsymbol{b}_t(\boldsymbol{R}, t) &= \boldsymbol{B}_e(\boldsymbol{R}, t) + \sum_{j(\neq i)} \boldsymbol{b}_j(\boldsymbol{R}, t), \end{aligned} \tag{45}$$

with partial fields

$$\begin{aligned} \boldsymbol{e}_j(\boldsymbol{R}, t) &= -e_j \nabla \frac{1}{4\pi|\boldsymbol{R}_j - \boldsymbol{R}|} + \frac{e_j}{2m_j^2 c^2} \boldsymbol{P}_j \cdot \nabla \frac{\boldsymbol{P}_j \cdot \mathsf{T}(\boldsymbol{R}_j - \boldsymbol{R})}{4\pi|\boldsymbol{R}_j - \boldsymbol{R}|} \\ &\quad + \frac{e_j \hbar}{4m_j^2 c^2} \nabla(\boldsymbol{P}_j \wedge \boldsymbol{\sigma}_j) \cdot \nabla \frac{1}{4\pi|\boldsymbol{R}_j - \boldsymbol{R}|} \\ &\quad - \frac{e_j \hbar}{2m_j^2 c^2} \boldsymbol{P}_j \cdot \nabla \boldsymbol{\sigma}_j \wedge \nabla \frac{1}{4\pi|\boldsymbol{R}_j - \boldsymbol{R}|} + \frac{e_j \hbar^2}{8m_j^2 c^2} \nabla \delta(\boldsymbol{R}_j - \boldsymbol{R}), \\ \boldsymbol{b}_j(\boldsymbol{R}, t) &= \frac{e_j}{m_j c} \nabla \wedge \left( \frac{\boldsymbol{P}_j}{4\pi|\boldsymbol{R}_j - \boldsymbol{R}|} \right) - \frac{e_j \hbar}{2m_j c} \nabla \wedge (\boldsymbol{\sigma}_j \wedge \nabla) \frac{1}{4\pi|\boldsymbol{R}_j - \boldsymbol{R}|} . \end{aligned} \tag{46}$$

The partial fields generated by particle $j$ have been written here for the case that $j$ is an electron. In the sums of (45) all particles occur. For nuclei one has to retain only the first two terms of $\boldsymbol{e}_j$ and the first term of $\boldsymbol{b}_j$.

The total force (44) on electron $i$ gets a simpler interpretation if one introduces the fields at the position $\boldsymbol{X}_i$ rather than at $\boldsymbol{R}_i$. Since the difference between these quantities is of order $c^{-2}$, such a change of arguments of the fields affects only the first term in (44). One has, with (38),

$$\boldsymbol{e}(\boldsymbol{X}_i, t) = \boldsymbol{e}(\boldsymbol{R}_i, t) + \frac{\hbar \boldsymbol{\sigma}_i \wedge \boldsymbol{P}_i}{4m_i^2 c^2} \cdot \nabla_i \boldsymbol{e}(\boldsymbol{R}_i, t). \tag{47}$$

Then the equation of motion (42) with (44) may be written as

$$m_i \frac{\mathrm{d}\boldsymbol{v}_{i,\mathrm{op}}}{\mathrm{d}t} \rightleftarrows \boldsymbol{f}_i = e_i \boldsymbol{e}_\mathrm{t}(\boldsymbol{X}_i, t) + \frac{e_i}{m_i c} \boldsymbol{P}_i \wedge \boldsymbol{b}_\mathrm{t}(\boldsymbol{X}_i, t) - \frac{e_i}{m_i^2 c^2} \boldsymbol{P}_i \boldsymbol{P}_i \cdot \boldsymbol{e}_\mathrm{t}(\boldsymbol{X}_i, t)$$
$$- \frac{e_i}{2m_i^2 c^2} \boldsymbol{P}_i^2 \boldsymbol{e}_\mathrm{t}(\boldsymbol{X}_i, t) + \frac{e_i \hbar}{2m_i c} \Big[ \{\boldsymbol{\nabla}_i \boldsymbol{b}_\mathrm{t}(\boldsymbol{X}_i, t)\} \cdot \boldsymbol{\sigma}_i$$
$$+ \{\boldsymbol{\nabla}_i \boldsymbol{e}_\mathrm{t}(\boldsymbol{X}_i, t)\} \cdot \left( \frac{\boldsymbol{P}_i}{m_i c} \wedge \boldsymbol{\sigma}_i \right) \Big] + \frac{e_i \hbar^2}{4m_i^2 c^2} \Delta_i \boldsymbol{e}_\mathrm{t}(\boldsymbol{X}_i, t). \tag{48}$$

The Weyl transform of the force contains in the first place the Lorentz force, present already in non-relativistic theory, supplemented here by two relativistic corrections, connected with the motion of the particle. Then two spin terms appear, which couple the space derivative of the magnetic field with the spin of the particle and the space derivative of the electric field with the spin in motion. One should note that here, in analogy with the classical theory for the orbital magnetic moment, the vector product of the momentum (divided by $m_i c$) and the spin magnetic dipole moment $e_i \hbar \boldsymbol{\sigma}_i / 2m_i c$ occurs and not just half of it as in (44).

The equation of motion for the nuclei may likewise be derived from the Hamiltonian (37). For the velocity operator one finds then the same result as (41). For the second derivative one obtains, after multiplication with $m_i$, an equation like (48) but without spin terms and with a factor 8 instead of 4 in the denominator of the last term.

In deriving the equations of motion we encountered certain expressions (45) with (46) which have been called the Weyl transforms of the total fields (acting on particle $i$) and which occurred in the equations of motion (48). The sums $\boldsymbol{e}$ and $\boldsymbol{b}$ of the external fields ($\boldsymbol{E}_\mathrm{e}, \boldsymbol{B}_\mathrm{e}$) and all partial fields ($\boldsymbol{e}_j, \boldsymbol{b}_j$):

$$\begin{aligned} \boldsymbol{e}(\boldsymbol{R}, t) &= \boldsymbol{E}_\mathrm{e}(\boldsymbol{R}, t) + \sum_j \boldsymbol{e}_j(\boldsymbol{R}, t), \\ \boldsymbol{b}(\boldsymbol{R}, t) &= \boldsymbol{B}_\mathrm{e}(\boldsymbol{R}, t) + \sum_j \boldsymbol{b}_j(\boldsymbol{R}, t) \end{aligned} \tag{49}$$

satisfy equations that follow from the explicit expressions (46), namely

$$\begin{aligned} \boldsymbol{\nabla} \cdot \boldsymbol{e} &= \rho^\mathrm{e} - \boldsymbol{\nabla} \cdot \boldsymbol{p}, \\ -\partial_{0\mathrm{P}} \boldsymbol{e} + \boldsymbol{\nabla} \wedge \boldsymbol{b} &= c^{-1} \boldsymbol{j} + \boldsymbol{\nabla} \wedge \boldsymbol{m}, \\ \boldsymbol{\nabla} \cdot \boldsymbol{b} &= 0, \\ \partial_{0\mathrm{P}} \boldsymbol{b} + \boldsymbol{\nabla} \wedge \boldsymbol{e} &= 0, \end{aligned} \tag{50}$$

where $\partial_{0\mathrm{P}} a \equiv c^{-1}\partial_{t\mathrm{P}} a = c^{-1}\{a, H\}_{\mathrm{P}} + \partial_0 a$ (cf. (VI.72)) with the Hamiltonian $H$ (37) and where the sources have the form

$$\begin{aligned}
\rho^{\mathrm{e}} &= \sum_j e_j \delta(\boldsymbol{X}_j - \boldsymbol{R}), \\
c^{-1}\boldsymbol{j} &= \sum_j \frac{e_j}{m_j c} \boldsymbol{P}_j \delta(\boldsymbol{X}_j - \boldsymbol{R}), \\
\boldsymbol{p} &= \sum_j{}' \frac{e_j \hbar}{2m_j^2 c^2} (\boldsymbol{P}_j \wedge \boldsymbol{\sigma}_j - \tfrac{1}{4}\hbar \nabla)\delta(\boldsymbol{X}_j - \boldsymbol{R}), \\
\boldsymbol{m} &= \sum_j{}' \frac{e_j \hbar}{2m_j c} \boldsymbol{\sigma}_j \delta(\boldsymbol{X}_j - \boldsymbol{R}).
\end{aligned} \tag{51}$$

(The primes at the summation signs indicate that the sum has to be extended over the electrons only.) The delta functions occurring here contain the position $\boldsymbol{X}_j$ (38) of the electrons in their arguments. One should understand them as an abbreviation of the expression (cf. (47))

$$\delta(\boldsymbol{X}_j - \boldsymbol{R}) = \delta(\boldsymbol{R}_j - \boldsymbol{R}) + \frac{\hbar \boldsymbol{\sigma}_j \wedge \boldsymbol{P}_j}{4m_j^2 c^2} \cdot \nabla_j \delta(\boldsymbol{R}_j - \boldsymbol{R}). \tag{52}$$

(The use of $\boldsymbol{X}_j$ instead of $\boldsymbol{R}_j$ is only significant in the first expression of (51), since in the others it gives rise to terms of order $c^{-3}$.) In the derivation of (50) with (51) we retained only terms linear in the charges, as is consistent with the fact that in the Hamiltonian (37) only terms bilinear in the charges have been included.

The equations (50) with (51) contain in their source terms charge and current densities of the same form as the non-relativistic ones (v. (VI.73)), except for the occurrence of $\boldsymbol{X}_j$, and moreover polarization and magnetization densities $\boldsymbol{p}$ and $\boldsymbol{m}$ due to the presence of spin. Terms of this type occur here already at the sub-atomic level in contrast with what was the case for point particles, as treated in chapter VI.

Owing to the use of the position operator $\boldsymbol{X}_j$ (which has covariant character; v. the preceding chapter), the polarization $\boldsymbol{p}$ contains a term due to the spins in motion which has a form similar to that of the relativistic classical expression for a composite particle, namely with the vector product of the velocity $\boldsymbol{P}_j/m_j$ and the magnetic moment $e_j \hbar \boldsymbol{\sigma}/2m_j c$ [1].

The equations of motion (48) and the field equations (50–51) show – as compared to their non-relativistic counterparts (VI.73) and (VI.81) –

[1] If the non-covariant position $\boldsymbol{R}_j$ had been used one would find only half of this term: cf. J. M. Crowther and D. ter Haar, Proc. Kon. Ned. Akad. Wet. **B74**(1971)341, 351.

which terms have to be added if spin effects are included (up to order $c^{-2}$).

In addition we have to discuss the equation which describes the change in time of the spin for the electrons. It may be obtained by calculating the commutator of the spin operator with the Hamiltonian. From (VIII.79) of the preceding chapter it follows that up to order $c^{-2}$ the spin operator for electron $i$ is

$$\boldsymbol{s}_{i,\mathrm{op}} = \tfrac{1}{2}\hbar \left\{ \boldsymbol{\sigma}_i + \frac{(\boldsymbol{P}_{i,\mathrm{op}} \wedge \boldsymbol{\sigma}_i) \wedge \boldsymbol{P}_{i,\mathrm{op}}}{2m_i^2 c^2} \right\}. \tag{53}$$

Although the spin operator, given there, pertains to a free particle, it follows by comparison with (VIII.151) that up to order $c^{-2}$ the spin operator does not change if external fields are present.

The spin equation follows by taking the commutator of (53) with the Hamilton operator (37). One finds, with the rule for Weyl transforms, up to order $c^{-2}$,

$$\frac{\mathrm{d}\boldsymbol{s}_{i,\mathrm{op}}}{\mathrm{d}t} \equiv \frac{i}{\hbar}[H_{\mathrm{op}}, \boldsymbol{s}_{i,\mathrm{op}}] \rightleftarrows \frac{e_i \hbar}{2m_i c} \left\{ \boldsymbol{\sigma}_i \wedge \boldsymbol{b}_{\mathrm{t}}(\boldsymbol{R}_i, t) + c^{-1} \left( \frac{\boldsymbol{P}_i}{m_i} \wedge \boldsymbol{\sigma}_i \right) \wedge \boldsymbol{e}_{\mathrm{t}}(\boldsymbol{R}_i, t) \right\}. \tag{54}$$

The fields $\boldsymbol{e}_{\mathrm{t}}$ and $\boldsymbol{b}_{\mathrm{t}}$ at the right-hand side are given by (45) with (46). They are taken at the position $\boldsymbol{R}_i$. Application of formulae like (47) shows that one may write $\boldsymbol{X}_i$ instead of $\boldsymbol{R}_i$ if one wishes: the difference leads to terms of order $c^{-3}$. For the same reason one may replace $\frac{1}{2}\hbar\boldsymbol{\sigma}_i$ by $\boldsymbol{s}_i$.

The equation (54) shows that a moment is exerted on the spin if it is not parallel to the magnetic field and if its vector product with the velocity $\boldsymbol{P}_i/m_i$ is not parallel to the electric field.

## 4 *The semi-relativistic approximation*

To derive equations for stable groups of particles we start from the equations for the Weyl transforms of quantities pertaining to point particles with spin, that have been given in the preceding sections. By making Taylor expansions of quantities occurring in the latter equations, we obtain expressions which contain multipole moments of orbital and spin character that characterize the stable groups as a whole.

In the following not all terms of order $c^{-2}$ will be retained in the multipole expanded quantities, but only those for which at least one of the factors $c^{-1}$ is contained in a magnetic orbital or spin multipole moment. An alternative way to express this procedure consists in considering both the electric and magnetic multipole moments as quantities of order $c^0$, and subsequently retaining only terms of order $c^{-1}$. In this way we shall obtain a set of approximate equations, which we shall call the 'semi-relativistic limit' of the theory. (In classical theory (v. chapter IV) we employed a similar approximation.) The reason for considering such a truncated form of the $c^{-2}$-equations is that in this way all magnetic interaction terms, especially those due to magnetic multipole moments in motion, are taken into account, while effects as the Lorentz contraction of the electric multipole moments are left out. The latter effects are indeed much smaller than the former, since they contain the velocity of the atom as a whole instead of an intra-atomic velocity. As a result one finds then expressions which show an analogy between electric and magnetic contributions (v. sections 5 and 6).

## 5 *The equations for the fields due to composite particles*

The first pair of sub-atomic field equations (50) with (51) reads, if instead of the summation index $j$ we introduce a double index $ki$ where $k$ labels the stable groups (atoms) and $i$ their constituent particles:

$$\nabla\cdot\boldsymbol{e} = \sum_{k,i} e_{ki}\,\delta(\boldsymbol{X}_{ki}-\boldsymbol{R}) - \sum_{k,i}{}' \frac{e_{ki}\hbar}{2m_{ki}^2c^2}(\boldsymbol{P}_{ki}\wedge\boldsymbol{\sigma}_{ki})\cdot\nabla\delta(\boldsymbol{X}_{ki}-\boldsymbol{R}) + \sum_{k,i}{}' \frac{e_{ki}\hbar^2}{8m_{ki}^2c^2}\Delta\delta(\boldsymbol{X}_{ki}-\boldsymbol{R}), \tag{55}$$

$$-\partial_{0\mathrm{P}}\boldsymbol{e}+\nabla\wedge\boldsymbol{b} = \sum_{k,i}\frac{e_{ki}\boldsymbol{P}_{ki}}{m_{ki}c}\delta(\boldsymbol{X}_{ki}-\boldsymbol{R}) + \sum_{k,i}{}'\frac{e_{ki}\hbar}{2m_{ki}c}\nabla\wedge\boldsymbol{\sigma}_{ki}\delta(\boldsymbol{X}_{ki}-\boldsymbol{R}).$$

The quantity $\boldsymbol{P}_{ki}/m_{ki}$ occurring at the right-hand sides of the equations may be replaced by $\partial_{t\mathrm{P}}\boldsymbol{X}_{ki}$, where the symbol $\partial_{t\mathrm{P}}$ stands for a Poisson bracket with the Hamiltonian $H$ (37). This is justified since we limited ourselves in the right-hand sides of (55) to terms linear in the charges and up to order $c^{-2}$.

We now introduce a privileged point $\boldsymbol{R}_k$ for each atom $k$. The relative coordinates of the particles with respect to this point will be denoted by

$$\boldsymbol{r}_{ki} = \boldsymbol{X}_{ki}-\boldsymbol{R}_k. \tag{56}$$

Then, by making a Taylor expansion of the right-hand sides of (55) one gets the expressions

$$\rho^{\mathrm{e}}-\sum_{n=1}^{\infty}(-1)^{n-1}\nabla^{n}:\sum_{k}(\hat{\mu}_{k}^{(n)}-c^{-1}\hat{\nu}_{k,\mathrm{spin}}^{(n)}\wedge \boldsymbol{v}_{k})\delta(\boldsymbol{R}_{k}-\boldsymbol{R}),$$

$$c^{-1}\boldsymbol{j}+\partial_{\mathrm{OP}}\{\sum_{n=1}^{\infty}(-1)^{n-1}\nabla^{n-1}:\sum_{k}\hat{\mu}_{k}^{(n)}\delta(\boldsymbol{R}_{k}-\boldsymbol{R})\}$$
$$+\nabla\wedge\{\sum_{n=1}^{\infty}(-1)^{n-1}\nabla^{n-1}:\sum_{k}(\hat{\nu}_{k}^{(n)}+c^{-1}\hat{\mu}_{k}^{(n)}\wedge \boldsymbol{v}_{k})\delta(\boldsymbol{R}_{k}-\boldsymbol{R})\}, \tag{57}$$

where the definition of the semi-relativistic approximation has been employed to suppress a number of terms. The atomic charge and current densities[1], that occur here, are given by

$$\rho^{\mathrm{e}}=\sum_{k}e_{k}\,\delta(\boldsymbol{R}_{k}-\boldsymbol{R}),$$
$$\boldsymbol{j}=\sum_{k}e_{k}\,\boldsymbol{v}_{k}\,\delta(\boldsymbol{R}_{k}-\boldsymbol{R}) \tag{58}$$

with $\boldsymbol{v}_k$ defined as $\partial_{t\mathrm{P}}\boldsymbol{R}_k$, i.e. by a Poisson bracket of $\boldsymbol{R}_k$ with the Weyl transform of the Hamiltonian plus an explicit time derivative. Furthermore the expressions (57) contain the electric and magnetic multipole moments, defined as

$$\hat{\mu}_{k}^{(n)}=\hat{\mu}_{k,\mathrm{orb}}^{(n)}+\hat{\mu}_{k,\mathrm{spin}}^{(n)},\qquad \hat{\nu}_{k}^{(n)}=\hat{\nu}_{k,\mathrm{orb}}^{(n)}+\hat{\nu}_{k,\mathrm{spin}}^{(n)} \tag{59}$$

with their orbital and spin parts

$$\hat{\mu}_{k,\mathrm{orb}}^{(n)}\equiv\frac{1}{n!}\sum_{i}e_{ki}\,\boldsymbol{r}_{ki}^{n},$$
$$\hat{\mu}_{k,\mathrm{spin}}^{(n)}\equiv\frac{1}{(n-1)!}\,\mathscr{S}\sum_{i}{}'\,\frac{e_{ki}\,\hbar}{2m_{ki}\,c}\,\boldsymbol{r}_{ki}^{n-1}\{(\partial_{\mathrm{OP}}\boldsymbol{r}_{ki})\wedge\boldsymbol{\sigma}_{ki}\},$$
$$\hat{\nu}_{k,\mathrm{orb}}^{(n)}\equiv\frac{n}{(n+1)!}\sum_{i}e_{ki}\,\boldsymbol{r}_{ki}^{n-1}\boldsymbol{r}_{ki}\wedge(\partial_{\mathrm{OP}}\boldsymbol{r}_{ki}),$$
$$\hat{\nu}_{k,\mathrm{spin}}^{(n)}\equiv\frac{1}{(n-1)!}\sum_{i}{}'\,\frac{e_{ki}\,\hbar}{2m_{ki}\,c}\,\boldsymbol{r}_{ki}^{n-1}\boldsymbol{\sigma}_{ki}. \tag{60}$$

The symbol $\mathscr{S}$ indicates that a symmetrization has to be performed on the asymmetric tensor in front of which it appears. The moments are all defined with the help of the internal coordinates $\boldsymbol{r}_{ki}$ (56). We note that here such purely space-like quantities have been employed as internal coordinates, in contrast with what was done in classical semi-relativistic theory. For that reason

[1] These atomic quantities should not be confused with the sub-atomic quantities (51), denoted by the same symbols.

we now want to introduce multipole moments expressed in terms of quantities $\boldsymbol{r}'_{ki}$. The latter will be defined in terms of $\boldsymbol{r}_{ki}$ and its derivative in a way completely analogous to the classical treatment (v. (IV.A100)):

$$\boldsymbol{r}'_{ki} = \boldsymbol{r}_{ki}+\tfrac{1}{2}c^{-2}(\partial_{t\mathrm{P}}\boldsymbol{R}_k)\cdot\boldsymbol{r}_{ki}(\partial_{t\mathrm{P}}\boldsymbol{R}_k)+c^{-2}(\partial_{t\mathrm{P}}\boldsymbol{R}_k)\cdot\boldsymbol{r}_{ki}\,\partial_{t\mathrm{P}}\boldsymbol{r}_{ki}\,. \tag{61}$$

Substituting the inverse of this expression (up to order $c^{-2}$) into (57) with (59) and (60) one finds

$$\begin{aligned} &\rho^{\mathrm{e}}-\nabla\cdot\boldsymbol{p},\\ &c^{-1}\boldsymbol{j}+\partial_{0\mathrm{P}}\boldsymbol{p}+\nabla\wedge\boldsymbol{m}, \end{aligned} \tag{62}$$

with the electric and magnetic polarization densities

$$\begin{aligned} \boldsymbol{p} &= \sum_{n=1}^{\infty}(-1)^{n-1}\nabla^{n-1}:\sum_k(\boldsymbol{\mu}_k^{(n)}-c^{-1}\boldsymbol{\nu}_k^{(n)}\wedge\boldsymbol{v}_k)\delta(\boldsymbol{R}_k-\boldsymbol{R}),\\ \boldsymbol{m} &= \sum_{n=1}^{\infty}(-1)^{n-1}\nabla^{n-1}:\sum_k(\boldsymbol{\nu}_k^{(n)}+c^{-1}\boldsymbol{\mu}_k^{(n)}\wedge\boldsymbol{v}_k)\delta(\boldsymbol{R}_k-\boldsymbol{R}). \end{aligned} \tag{63}$$

The semi-relativistic multipole moments $\boldsymbol{\mu}_k^{(n)}$ and $\boldsymbol{\nu}_k^{(n)}$ that occur here are defined by expressions of the same form as (59–60), but with $\boldsymbol{r}'_{ki}$ instead of $\boldsymbol{r}_{ki}$:

$$\boldsymbol{\mu}_k^{(n)} = \boldsymbol{\mu}_{k,\mathrm{orb}}^{(n)}+\boldsymbol{\mu}_{k,\mathrm{spin}}^{(n)}, \qquad \boldsymbol{\nu}_k^{(n)} = \boldsymbol{\nu}_{k,\mathrm{orb}}^{(n)}+\boldsymbol{\nu}_{k,\mathrm{spin}}^{(n)}, \tag{64}$$

with orbital and spin parts:

$$\begin{aligned} \boldsymbol{\mu}_{k,\mathrm{orb}}^{(n)} &\equiv \frac{1}{n!}\sum_i e_{ki}\boldsymbol{r}'^{n}_{ki},\\ \boldsymbol{\mu}_{k,\mathrm{spin}}^{(n)} &\equiv \frac{1}{(n-1)!}\,\mathscr{S}\sum_i{}'\,\frac{e_{ki}\hbar}{2m_{ki}c}\,\boldsymbol{r}'^{n-1}_{ki}\{(\partial_{0\mathrm{P}}\boldsymbol{r}'_{ki})\wedge\boldsymbol{\sigma}_{ki}\} = \hat{\boldsymbol{\mu}}_{k,\mathrm{spin}}^{(n)},\\ \boldsymbol{\nu}_{k,\mathrm{orb}}^{(n)} &\equiv \frac{n}{(n+1)!}\sum_i e_{ki}\boldsymbol{r}'^{n-1}_{ki}\boldsymbol{r}'_{ki}\wedge(\partial_{0\mathrm{P}}\boldsymbol{r}'_{ki}) = \hat{\boldsymbol{\nu}}_{k,\mathrm{orb}}^{(n)},\\ \boldsymbol{\nu}_{k,\mathrm{spin}}^{(n)} &\equiv \frac{1}{(n-1)!}\sum_i{}'\,\frac{e_{ki}\hbar}{2m_{ki}c}\,\boldsymbol{r}'^{n-1}_{ki}\boldsymbol{\sigma}_{ki} = \hat{\boldsymbol{\nu}}_{k,\mathrm{spin}}^{(n)}. \end{aligned} \tag{65}$$

Since all terms of order $c^{-3}$ are to be neglected, in fact only the orbital electric multipole moments $\hat{\boldsymbol{\mu}}_{k,\mathrm{orb}}^{(n)}$ and $\boldsymbol{\mu}_{k,\mathrm{orb}}^{(n)}$ are different.

The expressions (63) show a complete symmetry between the electric and magnetic multipole moments, just as the corresponding classical expressions (IV.57) and (IV.58). In particular one finds now a contribution to the electric polarization, due to magnetic multipole moments in motion. The multipole moments that occur here contain contributions due to the occurrence of spin:

the definitions (65) show how multipole moments for spin particles have to be defined such that they add to the orbital multipole moments that occur already in non-relativistic theory (v. (VI.87)).

## 6 *The laws of motion for composite particles*

### a. *The equation of motion*

The equation of motion for a composite particle in a field will be obtained from the equation (48) for its constituent particles, which may carry spin. From the derivation given in (38–48) it follows by inspection that the left-hand side of (48) has as Weyl transform $m_i \partial_{tP}^2 \boldsymbol{X}_i + (e_i \hbar^2 / 8 m_i^2 c^2) \Delta_i \boldsymbol{e}_t(\boldsymbol{X}_i, t)$ so that (replacing $i$ by $ki$) one may write (48) in the form:

$$\begin{aligned}
\partial_{tP}[m_{ki}\{1+\tfrac{1}{2}c^{-2}(\partial_{tP}\boldsymbol{X}_{ki})^2\}\partial_{tP}\boldsymbol{X}_{ki}] \\
&= e_{ki}\boldsymbol{e}_t(\boldsymbol{X}_{ki}, t) + c^{-1}e_{ki}(\partial_{tP}\boldsymbol{X}_{ki}) \wedge \boldsymbol{b}_t(\boldsymbol{X}_{ki}, t) \\
&\quad + \frac{e_{ki}\hbar}{2m_{ki}c}[\{\nabla_{ki}\boldsymbol{b}_t(\boldsymbol{X}_{ki}, t)\}\cdot\boldsymbol{\sigma}_{ki} + c^{-1}\{\nabla_{ki}\boldsymbol{e}_t(\boldsymbol{X}_{ki}, t)\}\cdot\{(\partial_{tP}\boldsymbol{X}_{ki}) \wedge \boldsymbol{\sigma}_{ki}\}] \\
&\quad + \frac{e_{ki}\hbar^2}{8m_{ki}^2c^2}\Delta_{ki}\boldsymbol{e}_t(\boldsymbol{X}_{ki}, t), \qquad (66)
\end{aligned}$$

where the fact has been used that in the right-hand side of (48) $\boldsymbol{P}_{ki}/m_{ki}$ could be replaced by $\partial_{tP}\boldsymbol{X}_{ki}$ (up to terms of order $c^{-2}$ and bilinear in the charges). The fields which occur here are given by the expressions (45) with (46). The equation (66) is valid for the electrons. For the nuclei an equation like (66) but without spin terms and without the last term follows immediately from (48).

The equation (66) bears a close resemblance to the classical equation (IV.A87), the difference being that extra spin terms occur here and that all quantities are Weyl transforms of operators. Just as in chapter IV we want to define a central point that characterizes the composite particle as a whole. To that end we introduce now an operator $\boldsymbol{X}_{k,\mathrm{op}}$ with Weyl transform $\boldsymbol{X}_k$ in such a way that the Weyl transform of the relative position $\boldsymbol{r}_{ki} = \boldsymbol{X}_{ki} - \boldsymbol{X}_k$ satisfies the equation (cf. (IV.A101)):

$$\begin{aligned}
\sum_i \Big\{ & m_{ki}\boldsymbol{r}_{ki} + \tfrac{1}{2}c^{-2}m_{ki}(\partial_{tP}\boldsymbol{r}_{ki})^2\boldsymbol{r}_{ki} \\
& + c^{-2}\sum_{j(\neq i)}\frac{e_{ki}e_{kj}}{8\pi|\boldsymbol{X}_{ki}-\boldsymbol{X}_{kj}|}\boldsymbol{r}_{ki} + c^{-2}m_{ki}(\partial_{tP}\boldsymbol{X}_k)\cdot\boldsymbol{r}_{ki}(\partial_{tP}\boldsymbol{r}_{ki})\Big\} = 0. \qquad (67)
\end{aligned}$$

(The order of the matrices occurring here does not matter, since only terms up to order $c^{-2}$ are to be retained.) The factor $|\boldsymbol{X}_{ki}-\boldsymbol{X}_{kj}|^{-1}$ is defined in a similar way as in (47). From (67) follows the expression for the central point $\boldsymbol{X}_k$ in terms of $\boldsymbol{X}_{ki}$ (38–39)

$$\boldsymbol{X}_k = \frac{1}{m_k}\sum_i m_{ki}\boldsymbol{X}_{ki} + \frac{1}{m_k c^2}\sum_i \left\{\tfrac{1}{2}m_{ki}(\partial_{t\mathrm{P}}\boldsymbol{r}_{ki})^2\boldsymbol{r}_{ki} + \sum_{j(\neq i)} \frac{e_{ki}e_{kj}}{8\pi|\boldsymbol{r}_{ki}-\boldsymbol{r}_{kj}|}\boldsymbol{r}_{ki} + m_{ki}(\partial_{t\mathrm{P}}\boldsymbol{X}_k)\cdot\boldsymbol{r}_{ki}(\partial_{t\mathrm{P}}\boldsymbol{r}_{ki})\right\}. \quad (68)$$

At the right-hand side the relative positions $\boldsymbol{r}_{ki}$ are to be understood as $\boldsymbol{X}_{ki}-m_k^{-1}\sum_i m_{ki}\boldsymbol{X}_{ki}$ (or $\boldsymbol{R}_{ki}-m_k^{-1}\sum_i m_{ki}\boldsymbol{R}_{ki}$), since only terms up to order $c^{-2}$ are to be included.

From (66) one may derive an equation for the atoms as a whole by taking a sum over the electrons and nuclei. If one introduces now quantities $\boldsymbol{r}'_{ki}$, defined in (61) (with $\boldsymbol{X}_k$ instead of $\boldsymbol{R}_k$), one finds as the left-hand side of the equation of motion the Poisson bracket derivation $\partial_{t\mathrm{P}}$ of a quantity that has the same form as (IV.A109). The right-hand side of the equation of motion may be split again into three parts: an intra-atomic, an external and an interatomic field contribution. The Weyl transforms of the fields are given by (45) with (46) instead of (IV.A111). In the former spin terms and a derivative of a delta function occur for the electron contributions, which are absent in the latter. Another difference is that the latter contains terms with accelerations, which are missing in the former, because they are effectively quadratic in the charges. As a consequence the intra-atomic contributions to the right-hand side of the equation of motion are the sum of terms that are the counterparts of those of (IV.A112) and an extra spin contribution $-\partial_{t\mathrm{P}}\boldsymbol{g}_k$ with $\boldsymbol{g}_k$ given by

$$\boldsymbol{g}_k \equiv \sum_{i,j(i\neq j)}{}' \frac{e_{ki}e_{kj}\hbar}{2m_{ki}c^2}\boldsymbol{\sigma}_{ki}\wedge\nabla_{ki}\frac{1}{4\pi|\boldsymbol{r}'_{ki}-\boldsymbol{r}'_{kj}|}. \quad (69)$$

For the Weyl transform of the equation of motion for the composite particle as a whole we obtain thus on a par with (IV.A113):

$$\begin{aligned}\partial_{t\mathrm{P}}&\left[\left\{m_k+\tfrac{1}{2}c^{-2}m_k\boldsymbol{v}_k^2+\tfrac{1}{2}c^{-2}\sum_i m_{ki}(\partial_{t\mathrm{P}}\boldsymbol{r}_{ki})^2+c^{-2}\sum_{i,j(i\neq j)}\frac{e_{ki}e_{kj}}{8\pi|\boldsymbol{r}'_{ki}-\boldsymbol{r}'_{kj}|}\right\}\boldsymbol{v}_k+\boldsymbol{g}_k\right]\\
&=\sum_i e_{ki}\{\boldsymbol{e}(\boldsymbol{X}_{ki},t)+c^{-1}(\partial_{t\mathrm{P}}\boldsymbol{X}_{ki})\wedge\boldsymbol{b}(\boldsymbol{X}_{ki},t)\}\\
&+c^{-2}\partial_{t\mathrm{P}}\left[\sum_i e_{ki}\left\{(\partial_{t\mathrm{P}}\boldsymbol{r}_{ki})\cdot\boldsymbol{e}(\boldsymbol{X}_{ki},t)\boldsymbol{r}_{ki}+(\partial_{t\mathrm{P}}\boldsymbol{X}_k)\cdot\boldsymbol{r}_{ki}\boldsymbol{e}(\boldsymbol{X}_{ki},t)\right.\right.\\
&\left.\left.+\frac{\bar{\boldsymbol{s}}_k}{m_k}\wedge\boldsymbol{e}(\boldsymbol{X}_{ki},t)\right\}\right]+\sum_i{}'\frac{e_{ki}\hbar}{2m_{ki}c}\left[\{\nabla_{ki}\boldsymbol{b}(\boldsymbol{X}_{ki},t)\}\cdot\boldsymbol{\sigma}_{ki}\right.\\
&\left.+c^{-1}\{\nabla_{ki}\boldsymbol{e}(\boldsymbol{X}_{ki},t)\}\cdot\{(\partial_{t\mathrm{P}}\boldsymbol{X}_{ki})\wedge\boldsymbol{\sigma}_{ki}\}+\frac{\hbar}{4m_{ki}c}\Delta_{ki}\boldsymbol{e}(\boldsymbol{X}_{ki},t)\right],\end{aligned} \quad (70)$$

with $\boldsymbol{v}_k \equiv \partial_{t\mathrm{P}} \boldsymbol{X}_k$ the Weyl transform of the velocity and $\bar{\boldsymbol{s}}_k \equiv \sum_i m_{ki} \boldsymbol{r}_{ki} \wedge \partial_{t\mathrm{P}} \boldsymbol{r}_{ki}$ the (non-relativistic) orbital inner angular momentum. The quantities $\boldsymbol{e}$ and $\boldsymbol{b}$ are the Weyl transforms of the interatomic and external fields.

If the external fields change slowly one may make a multipole expansion of the external field terms at the right-hand side of (70) and retain only the charge and dipole terms. Then one finds, in semi-relativistic approximation, with the definitions (64) and (65) for the semi-relativistic multipole moments defined with respect to $\boldsymbol{X}_k$, for the external field contribution $\boldsymbol{f}^{\mathrm{L}}_{k\mathrm{e}}$ of the right-hand side of (70) (cf. (IV.A118)):

$$\begin{aligned}\boldsymbol{f}^{\mathrm{L}}_{k\mathrm{e}} = {} & e_k\{\boldsymbol{E}_\mathrm{e}(\boldsymbol{X}_k, t) + c^{-1}\boldsymbol{v}_k \wedge \boldsymbol{B}_\mathrm{e}(\boldsymbol{X}_k, t)\} \\ & + \{\boldsymbol{\nabla}_k \boldsymbol{E}_\mathrm{e}(\boldsymbol{X}_k, t)\}\cdot(\boldsymbol{\mu}^{(1)}_k - c^{-1}\boldsymbol{\nu}^{(1)}_k \wedge \boldsymbol{v}_k) + \{\boldsymbol{\nabla}_k \boldsymbol{B}_\mathrm{e}(\boldsymbol{X}_k, t)\}\cdot(\boldsymbol{\nu}^{(1)}_k + c^{-1}\boldsymbol{\mu}^{(1)}_k \wedge \boldsymbol{v}_k) \\ & + c^{-1}\partial_{t\mathrm{P}}\{\boldsymbol{\mu}^{(1)}_k \wedge \boldsymbol{B}_\mathrm{e}(\boldsymbol{X}_k, t) - \boldsymbol{\nu}^{(1)}_{k,\mathrm{orb}} \wedge \boldsymbol{E}_\mathrm{e}(\boldsymbol{X}_k, t)\}, \qquad (71)\end{aligned}$$

where the Maxwell equation $\boldsymbol{\nabla} \wedge \boldsymbol{E}_\mathrm{e} = -\partial_0 \boldsymbol{B}_\mathrm{e}$ for the external fields has been employed. If a single composite particle moves in an external field the semi-relativistic (Weyl-transformed) equation of motion becomes thus:

$$\partial_{t\mathrm{P}}(m_k \boldsymbol{v}_k + \boldsymbol{g}_k) = \boldsymbol{f}^{\mathrm{L}}_{k\mathrm{e}}, \qquad (72)$$

where at the left-hand side only those $c^{-2}$-terms that contain spin vectors have been retained. As compared to the non-relativistic equation (VI.98–100) one finds here, apart from an extra term $\boldsymbol{g}_k$ at the left-hand side due to the presence of spin, a term that couples the magnetic dipoles in motion with the gradient of the electric field at the right-hand side and moreover a term with the vector product of the orbital magnetic moment and the electric field. The latter is the magnetodynamic effect, from which the spin part is absent here, since the electrons are supposed to carry only normal magnetic moments (v. (VIII.158)). Furthermore both the electric and magnetic dipole moments contain spin contributions. The fields $\boldsymbol{E}_\mathrm{e}$ and $\boldsymbol{B}_\mathrm{e}$ are taken at the centre $\boldsymbol{X}_k$ of the composite particle.

For a set of composite particles which move in each other's fields, there also exist interatomic contributions to the right-hand side of (70). If the atoms are outside each other one may make a multipole expansion both for the sources of the interatomic fields and for the particles on which the fields act. One finds then in the semi-relativistic approximation a double multipole expansion in terms of electric–electric, magnetic–magnetic and electric–magnetic multipole moments. Only the former two terms are written down here for brevity's sake:

$$\begin{aligned}\boldsymbol{f}^{\mathrm{L}}_k - \boldsymbol{f}^{\mathrm{L}}_{k\mathrm{e}} = {} & -\sum_{l(\neq k)} \boldsymbol{\nabla}_k \sum_{n,m=0}^{\infty} \boldsymbol{\nabla}^n_k \vdots \boldsymbol{\mu}^{(n)}_k \boldsymbol{\nabla}^m_l \vdots \boldsymbol{\mu}^{(m)}_l \frac{1}{4\pi|\boldsymbol{X}_k - \boldsymbol{X}_l|} \\ & + \sum_{l(\neq k)} \boldsymbol{\nabla}_k \sum_{n,m=1}^{\infty} (\boldsymbol{\nabla}^{n-1}_k \vdots \boldsymbol{\nu}^{(n)}_k \wedge \boldsymbol{\nabla}_k)\cdot(\boldsymbol{\nabla}^{m-1}_l \vdots \boldsymbol{\nu}^{(m)}_l \wedge \boldsymbol{\nabla}_l) \frac{1}{4\pi|\boldsymbol{X}_k - \boldsymbol{X}_l|}. \qquad (73)\end{aligned}$$

In the general case that the atoms are at arbitrary distances of each other one may write the sum of the forces due to the external and interatomic fields as the sum of a long range and a short range part. The long range part $\boldsymbol{f}_k^{\mathrm{L}}$ is given by the sum of (71) and (73). The short range part $\boldsymbol{f}_k^{\mathrm{S}}$ equals the difference of the unexpanded interatomic field contribution and the expanded one (73). We write down only those terms of $\boldsymbol{f}_k^{\mathrm{S}}$ that are the unexpanded counterparts of $\boldsymbol{f}_k^{\mathrm{L}}-\boldsymbol{f}_{k\mathrm{e}}^{\mathrm{L}}$. They read

$$\boldsymbol{f}_k^{\mathrm{S}} = -\sum_{l(\neq k)}\sum_{i,j}\Bigg[1-c^{-2}(\partial_{t\mathrm{P}}\boldsymbol{r}_{ki})\cdot(\partial_{t\mathrm{P}}\boldsymbol{r}_{lj})$$
$$+c^{-2}\{\partial_{t\mathrm{P}}(\boldsymbol{r}_{ki}-\boldsymbol{r}_{lj})\}\cdot\left\{\left(\frac{\hbar}{2m_{ki}}\boldsymbol{\sigma}_{ki}+\frac{\hbar}{2m_{lj}}\boldsymbol{\sigma}_{lj}\right)\wedge\nabla_{ki}\right\}$$
$$+c^{-2}\frac{\hbar^2}{4m_{ki}m_{lj}}(\boldsymbol{\sigma}_{ki}\wedge\nabla_{ki})\cdot(\boldsymbol{\sigma}_{lj}\wedge\nabla_{ki})\Bigg]\nabla_{ki}\frac{e_{ki}e_{lj}}{4\pi|\boldsymbol{X}_{ki}-\boldsymbol{X}_{lj}|}-(\boldsymbol{f}_k^{\mathrm{L}}-\boldsymbol{f}_{k\mathrm{e}}^{\mathrm{L}}). \quad (74)$$

The terms with spin only apply for the electrons, not for the nuclei. The total equation of motion in semi-relativistic approximation becomes

$$\partial_{t\mathrm{P}}(m_k\boldsymbol{v}_k+\boldsymbol{g}_k) = \boldsymbol{f}_k^{\mathrm{L}}+\boldsymbol{f}_k^{\mathrm{S}} \quad (75)$$

with both long range and short range forces at the right-hand side.

## b. *The energy equation*

The energy equation for a composite particle is obtained by multiplying equation (66) by $\partial_{t\mathrm{P}}\boldsymbol{X}_{ki}$ and summing over $i$. Then one finds, by introducing the relative positions $\boldsymbol{r}_{ki}$ and using (67), for the left-hand side an expression which is the analogue (in fact a Weyl transform) of (IV.A121). If one defines subsequently, just as in (IV.A122), a quantity $\dot{\boldsymbol{r}}'_{ki}$ by means of the definition

$$\dot{\boldsymbol{r}}'_{ki} \equiv \partial_{t\mathrm{P}}\boldsymbol{r}_{ki}+\tfrac{1}{2}c^{-2}\boldsymbol{v}_k^2\partial_{t\mathrm{P}}\boldsymbol{r}_{ki}+\tfrac{1}{2}c^{-2}\boldsymbol{v}_k\cdot(\partial_{t\mathrm{P}}\boldsymbol{r}_{ki})\boldsymbol{v}_k$$
$$+c^{-2}\boldsymbol{v}_k\cdot(\partial_{t\mathrm{P}}\boldsymbol{r}_{ki})\partial_{t\mathrm{P}}\boldsymbol{r}_{ki}+\frac{e_{ki}}{m_{ki}c^2}\boldsymbol{v}_k\cdot\boldsymbol{r}_{ki}\boldsymbol{e}_t(\boldsymbol{X}_{ki},t), \quad (76)$$

one finds for the left-hand side of the energy equation the Poisson bracket $\partial_{t\mathrm{P}}$ of a quantity that has the form (IV.A124). At the right-hand side of the energy equation appears the sum of an intra-atomic, an interatomic and an external field contribution. For the first of these one gets, by making use of the field expressions (46), a result that is the sum of an expression like (IV.A125) and extra terms depending on spins and on the delta function

$\delta(\boldsymbol{r}_{ki}-\boldsymbol{r}_{kj})$, namely (up to terms bilinear in the charges):

$$-\partial_{tP}\left[u_k+\boldsymbol{v}_k\cdot\boldsymbol{g}_k-\sum_{i,j(i\neq j)}\left\{\frac{e_{ki}\,e_{kj}}{8\pi|\boldsymbol{r}'_{ki}-\boldsymbol{r}'_{kj}|}+\frac{e_{ki}\,e_{kj}\,\hbar^2}{8m_{ki}^2c^2}\,\delta(\boldsymbol{r}'_{ki}-\boldsymbol{r}'_{kj})\right\}\right] \tag{77}$$

(in the last term the summation over $i$ is confined to the electrons), where we employed the abbreviation $u_k$ given by

$$u_k\equiv\sum_{i,j(i\neq j)}\left\{1+\frac{\hbar}{m_{ki}c^2}\,\dot{\boldsymbol{r}}'_{ki}\cdot(\boldsymbol{\sigma}_{ki}\wedge\boldsymbol{\nabla}_{ki})+\frac{\hbar^2}{4m_{ki}\,m_{kj}\,c^2}(\boldsymbol{\sigma}_{ki}\wedge\boldsymbol{\nabla}_{ki})\cdot(\boldsymbol{\sigma}_{kj}\wedge\boldsymbol{\nabla}_{ki})\right\}\frac{e_{ki}\,e_{kj}}{8\pi|\boldsymbol{r}'_{ki}-\boldsymbol{r}'_{kj}|} \tag{78}$$

(in the second term the summation over $i$ is confined to the electrons; the same applies to both $i$ and $j$ in the third term) and $\boldsymbol{g}_k$ given in (69). (In (78) we included the Coulomb energy although it drops out in (77); this will turn out to be convenient in the following.) As the energy law, up to order $c^{-2}$, we find now an equation which is the counterpart of (IV.A126), namely

$$\begin{aligned}
&\partial_{tP}\Bigg[\tfrac{1}{2}m_k\,\boldsymbol{v}_k^2+\tfrac{3}{8}c^{-2}m_k\,\boldsymbol{v}_k^4+\sum_i\left(\tfrac{1}{2}m_{ki}\,\dot{\boldsymbol{r}}'^2_{ki}+\tfrac{1}{4}c^{-2}m_{ki}\,\dot{\boldsymbol{r}}'^2_{ki}\boldsymbol{v}_k^2+\tfrac{3}{8}c^{-2}m_{ki}\,\dot{\boldsymbol{r}}'^4_{ki}\right)\\
&\quad+\tfrac{1}{2}c^{-2}\sum_{i,j(i\neq j)}\frac{e_{ki}\,e_{kj}}{8\pi|\boldsymbol{r}'_{ki}-\boldsymbol{r}'_{kj}|}\left\{\dot{\boldsymbol{r}}'_{ki}\cdot\mathbf{T}(\boldsymbol{r}'_{ki}-\boldsymbol{r}'_{kj})\cdot\dot{\boldsymbol{r}}'_{kj}+\boldsymbol{v}_k^2\right\}\\
&\quad-c^{-2}\sum_{i,j(i\neq j)}{}'\,\frac{e_{ki}\,e_{kj}\,\hbar^2}{8m_{ki}^2}\,\delta(\boldsymbol{r}'_{ki}-\boldsymbol{r}'_{kj})+u_k+\boldsymbol{v}_k\cdot\boldsymbol{g}_k\Bigg]\\
&=\sum_{k,i}e_{ki}(\partial_{tP}\boldsymbol{X}_{ki})\cdot\boldsymbol{e}(\boldsymbol{X}_{ki},t)+\sum_{k,i}{}'(\partial_{tP}\boldsymbol{X}_{ki})\cdot\left[\frac{e_{ki}\,\hbar}{2m_{ki}c}\{\boldsymbol{\nabla}_{ki}\,\boldsymbol{b}(\boldsymbol{X}_{ki},t)\}\cdot\boldsymbol{\sigma}_{ki}\right.\\
&\quad\left.+\frac{e_{ki}\,\hbar}{2m_{ki}c^2}\{\boldsymbol{\nabla}_{ki}\,\boldsymbol{e}(\boldsymbol{X}_{ki},t)\}\cdot\{(\partial_{tP}\boldsymbol{X}_{ki})\wedge\boldsymbol{\sigma}_{ki}\}+\frac{e_{ki}\,\hbar^2}{8m_{ki}^2c^2}\,\Delta_{ki}\,\boldsymbol{e}(\boldsymbol{X}_{ki},t)\right]\\
&\quad+c^{-2}\partial_{tP}\left[\sum_i e_{ki}\left\{\frac{\bar{\boldsymbol{s}}_k}{m_k}\wedge\boldsymbol{e}(\boldsymbol{X}_{ki},t)\right.\right.\\
&\qquad\left.\left.+2(\partial_{tP}\boldsymbol{r}_{ki})\cdot\boldsymbol{e}(\boldsymbol{X}_{ki},t)\boldsymbol{r}_{ki}+\boldsymbol{v}_k\cdot\boldsymbol{r}_{ki}\,\boldsymbol{e}(\boldsymbol{X}_{ki},t)\right\}\cdot\boldsymbol{v}_k\right],
\end{aligned} \tag{79}$$

where the tensor $\mathbf{T}$ has been defined in (5) and where $\boldsymbol{e}$ and $\boldsymbol{b}$ are the sums of the external fields and the interatomic fields due to the other atoms.

If the external fields change slowly a multipole expansion of the corresponding terms at the right-hand side of (79) may be performed. The dipole terms read in the semi-relativistic approximation

$$\psi_{ke}^{L} = e_k \boldsymbol{v}_k \cdot \boldsymbol{E}_e(\boldsymbol{X}_k, t) + \boldsymbol{v}_k \cdot \{\nabla_k \boldsymbol{E}_e(\boldsymbol{X}_k, t)\} \cdot (\boldsymbol{\mu}_k^{(1)} - c^{-1} \boldsymbol{v}_k^{(1)} \wedge \boldsymbol{v}_k)$$
$$+ \partial_{tP}(\boldsymbol{\mu}_k^{(1)} - c^{-1} \boldsymbol{v}_k^{(1)} \wedge \boldsymbol{v}_k) \cdot \boldsymbol{E}_e(\boldsymbol{X}_k, t) - (\boldsymbol{v}_k^{(1)} + c^{-1} \boldsymbol{\mu}_k^{(1)} \wedge \boldsymbol{v}_k) \cdot \frac{\partial \boldsymbol{B}_e(\boldsymbol{X}_k, t)}{\partial t}$$
$$+ \partial_{tP}\{\boldsymbol{v}_{k,\mathrm{spin}}^{(1)} \cdot \boldsymbol{B}_e(\boldsymbol{X}_k, t)\} + 2c^{-1} \partial_{tP}\{(\boldsymbol{v}_{k,\mathrm{orb}}^{(1)} \wedge \boldsymbol{v}_k) \cdot \boldsymbol{E}_e(\boldsymbol{X}_k, t)\}, \tag{80}$$

where we employed the Maxwell equation $\nabla \wedge \boldsymbol{E}_e = -\partial_0 \boldsymbol{B}_e$ for the external fields. Only the terms that are linear in the charges are to be retained here. (For convenience terms with $\partial_{tP}(\boldsymbol{v}_k^{(1)} \wedge \boldsymbol{v}_k)$ have been added, although strictly spoken they are negligible in the present approximation.) For the semi-relativistic (Weyl-transformed) energy equation of a single composite particle in a slowly varying external field we found thus

$$\partial_{tP}(\tfrac{1}{2} m_k \boldsymbol{v}_k^2 + \boldsymbol{v}_k \cdot \boldsymbol{g}_k + t_k + u_k) = \psi_{ke}^{L} \tag{81}$$

with the internal kinetic energy

$$t_k = \tfrac{1}{2} \sum_i m_{ki} \dot{\boldsymbol{r}}_{ki}'^2. \tag{82}$$

At the right-hand side of (81) an expression appears which is equal to (80). From the latter form one finds by comparison with the non-relativistic result (VI.105–106) which additional terms arise in the semi-relativistic theory: in the first place terms due to moving magnetic dipole moments and moreover terms with the spin parts of the electric and magnetic dipoles. Furthermore the fields are taken at the centre $\boldsymbol{X}_k$ instead of $\boldsymbol{R}_k$.

As compared to the semi-relativistic classical result (IV.A127) the present result contains as extra terms with the spin:

$$\boldsymbol{v}_k \cdot (\nabla_k \boldsymbol{E}_e) \cdot (\boldsymbol{\mu}_{k,\mathrm{spin}}^{(1)} - c^{-1} \boldsymbol{v}_{k,\mathrm{spin}}^{(1)} \wedge \boldsymbol{v}_k) + \boldsymbol{v}_k \cdot (\nabla_k \boldsymbol{B}_e) \cdot \boldsymbol{v}_{k,\mathrm{spin}}^{(1)}. \tag{83}$$

(In the semi-relativistic limit the term that couples the spin electric dipole moment in motion with the magnetic field does not contribute, and neither does the Poisson bracket derivative of the spin electric or magnetic dipole moment.) The form of these terms is consistent with the results of chapter VIII: v. problem 6 of that chapter.

If particles forming a set move in each other's fields the energy equation for these particles contains also an interatomic field contribution that may be developed in a double multipole series if the atoms are sufficiently far apart. We write, for brevity's sake, only the terms which couple the electric multipoles with each other and the magnetic ones with each other. They follow by using the expressions (45) and (46) for the Weyl transforms of the fields:

$$\psi_k^{\mathrm{L}} - \psi_{k\mathrm{e}}^{\mathrm{L}} = -\sum_{l(\neq k)} \sum_{n,m=0}^{\infty} \{\boldsymbol{v}_k \cdot \nabla_k \boldsymbol{\mu}_k^{(n)} \vdots \nabla_k^n \boldsymbol{\mu}_l^{(m)} \vdots \nabla_l^m$$
$$+ (\partial_{t\mathrm{P}} \boldsymbol{\mu}_k^{(n)}) \vdots \nabla_k^n \boldsymbol{\mu}_l^{(m)} \vdots \nabla_l^m\} \frac{1}{4\pi|\boldsymbol{X}_k - \boldsymbol{X}_l|}$$
$$-\sum_{l(\neq k)} \sum_{n,m=1}^{\infty} [(\nabla_k^{n-1} \vdots \boldsymbol{\nu}_k^{(n)} \wedge \nabla_k) \cdot \{\nabla_l^{m-1} \vdots (\partial_{t\mathrm{P}} \boldsymbol{\nu}_l^{(m)}) \wedge \nabla_l\}$$
$$+ \boldsymbol{v}_l \cdot \nabla_l (\nabla_k^{n-1} \vdots \boldsymbol{\nu}_k^{(n)} \wedge \nabla_k) \cdot (\nabla_l^{m-1} \vdots \boldsymbol{\nu}_l^{(m)} \wedge \nabla_l)$$
$$- (\boldsymbol{v}_k - \boldsymbol{v}_l) \cdot \nabla_k (\nabla_k^{n-1} \vdots \boldsymbol{\nu}_{k,\mathrm{spin}}^{(n)} \wedge \nabla_k) \cdot (\nabla_l^{m-1} \vdots \boldsymbol{\nu}_l^{(m)} \wedge \nabla_l)$$
$$- \{\nabla_k^{n-1} \vdots (\partial_{t\mathrm{P}} \boldsymbol{\nu}_{k,\mathrm{spin}}^{(n)} \wedge \nabla_k)\} \cdot (\nabla_l^{m-1} \vdots \boldsymbol{\nu}_l^{(m)} \wedge \nabla_l)$$
$$- (\nabla_k^{n-1} \vdots \boldsymbol{\nu}_{k,\mathrm{spin}}^{(n)} \wedge \nabla_k) \cdot \{\nabla_l^{m-1} \vdots (\partial_{t\mathrm{P}} \boldsymbol{\nu}_l^{(m)}) \wedge \nabla_l\}] \frac{1}{4\pi|\boldsymbol{X}_k - \boldsymbol{X}_l|}. \tag{84}$$

In the general case of arbitrary separations between the atoms, the interatomic contribution to the energy law may be written as a sum of a long range part which is given by (80) with (84) and a short range part $\psi_k^{\mathrm{S}}$. The latter is equal to the difference of the unexpanded and expanded interatomic field contributions. Again we write only those terms that give upon expansion the long range terms (84). One finds for these terms:

$$\psi_k^{\mathrm{S}} = \sum_{l(\neq k)} \sum_{i,j} \Big\{ -(\boldsymbol{v}_k + \partial_{t\mathrm{P}} \boldsymbol{r}_{ki}) \cdot \nabla_{ki} - c^{-2} (\partial_{t\mathrm{P}} \boldsymbol{r}_{ki}) \cdot (\partial_{t\mathrm{P}} \boldsymbol{r}_{lj}) (\boldsymbol{v}_l + \partial_{t\mathrm{P}} \boldsymbol{r}_{lj}) \cdot \nabla_{lj}$$
$$- \frac{\hbar}{2m_{lj}c^2} (\boldsymbol{v}_l + \partial_{t\mathrm{P}} \boldsymbol{r}_{lj}) \cdot \nabla_{ki} (\partial_{t\mathrm{P}} \boldsymbol{r}_{ki}) \cdot (\boldsymbol{\sigma}_{lj} \wedge \nabla_{ki})$$
$$+ \frac{\hbar}{2m_{lj}c^2} (\boldsymbol{v}_k + \partial_{t\mathrm{P}} \boldsymbol{r}_{ki}) \cdot \nabla_{ki} (\partial_{t\mathrm{P}} \boldsymbol{r}_{lj}) \cdot (\boldsymbol{\sigma}_{lj} \wedge \nabla_{ki})$$
$$- \frac{\hbar}{2m_{ki}c^2} (\boldsymbol{v}_k + \partial_{t\mathrm{P}} \boldsymbol{r}_{ki}) \cdot \nabla_{ki} \partial_{t\mathrm{P}} (\boldsymbol{r}_{ki} - \boldsymbol{r}_{lj}) \cdot (\boldsymbol{\sigma}_{ki} \wedge \nabla_{ki})$$
$$- \frac{\hbar^2}{4m_{ki}m_{lj}c^2} (\boldsymbol{v}_k + \partial_{t\mathrm{P}} \boldsymbol{r}_{ki}) \cdot \nabla_{ki} (\boldsymbol{\sigma}_{ki} \wedge \nabla_{ki}) \cdot (\boldsymbol{\sigma}_{lj} \wedge \nabla_{ki}) \Big\}$$
$$\frac{e_{ki} e_{lj}}{4\pi|\boldsymbol{X}_{ki} - \boldsymbol{X}_{lj}|} - (\psi_k^{\mathrm{L}} - \psi_{k\mathrm{e}}^{\mathrm{L}}). \tag{85}$$

(Again sums of terms in which spins occur are extended only over the electrons.)

The energy equation for an atom that is part of a set of atoms has been found now as:

$$\partial_{tP}(\tfrac{1}{2}m_k \boldsymbol{v}_k^2 + \boldsymbol{v}_k \cdot \boldsymbol{g}_k + t_k + u_k) = \psi_k^{\mathrm{L}} + \psi_k^{\mathrm{S}} \tag{86}$$

with long range and short range power terms at the right-hand side.

### c. *The angular momentum equation*

The angular momentum equation for a stable group of spin particles may be derived along similar lines. One finds then an expression for the sum of 1st: the Poisson bracket of the Weyl transform of the orbital angular momentum

$$\boldsymbol{s}_k^{(1)} = \sum_i m_{ki} \boldsymbol{r}'_{ki} \wedge \dot{\boldsymbol{r}}'_{ki} \tag{87}$$

(v. (61) and (76) for the expressions $\boldsymbol{r}'_{ki}$ and $\dot{\boldsymbol{r}}'_{ki}$) with the Weyl transform $H$ of the Hamiltonian, 2nd: a commutator of the spin angular momentum

$$\boldsymbol{s}_k^{(2)} = \sum_i{}' \tfrac{1}{2}\hbar \boldsymbol{\sigma}_{ki} \tag{88}$$

of the particles with $H$, 3rd: the Poisson bracket with $H$ of a term

$$\boldsymbol{s}_k^{(3)} = \sum_i{}' \tfrac{1}{2}\hbar(\boldsymbol{P}_{ki} \wedge \boldsymbol{\sigma}_{ki}) \wedge \boldsymbol{P}_{ki}/2m_{ki}^2 c^2, \tag{89}$$

due to the fact that the spins are in motion (v. (53)), and 4th: the Poisson bracket of an intra-atomic field contribution

$$\boldsymbol{s}_k^{(4)} = -\sum_{i,j(i \neq j)} \frac{\hbar}{2m_{kj}c^2} \boldsymbol{r}'_{ki} \wedge (\boldsymbol{\sigma}_{kj} \wedge \boldsymbol{\nabla}_{ki}) \frac{e_{ki}e_{kj}}{4\pi|\boldsymbol{r}'_{ki} - \boldsymbol{r}'_{kj}|}. \tag{90}$$

One finds for the special case of a single composite particle in a slowly changing external field $(\boldsymbol{E}_{\mathrm{e}}, \boldsymbol{B}_{\mathrm{e}})$:

$$\begin{aligned}\partial_{tP}(\boldsymbol{s}_k^{(1)} + \boldsymbol{s}_k^{(3)} + \boldsymbol{s}_k^{(4)}) + \frac{i}{\hbar}[H, \boldsymbol{s}_k^{(2)}] \\ = \boldsymbol{\mu}_k^{(1)} \wedge (\boldsymbol{E}_{\mathrm{e}} + c^{-1}\boldsymbol{v}_k \wedge \boldsymbol{B}_{\mathrm{e}}) + \boldsymbol{\nu}_k^{(1)} \wedge (\boldsymbol{B}_{\mathrm{e}} - c^{-1}\boldsymbol{v}_k \wedge \boldsymbol{E}_{\mathrm{e}}) + c^{-1}\boldsymbol{v}_k \wedge (\boldsymbol{\nu}_{k,\mathrm{spin}}^{(1)} \wedge \boldsymbol{E}_{\mathrm{e}}) \\ - \boldsymbol{v}_k \wedge \boldsymbol{g}_k\end{aligned} \tag{91}$$

(with $\boldsymbol{v}_k \equiv \partial_{tP}\boldsymbol{X}_k$ and $\boldsymbol{g}_k$ given by (69)) up to dipole terms. As compared to the non-relativistic quantum result (VI.111–112) one finds two extra terms that couple magnetic dipole moments with the external electric field and a term with the spin momentum $\boldsymbol{g}_k$. The fields are taken at the position of the central point $\boldsymbol{X}_k$ with respect to which also the semi-relativistic dipole moments are defined (v. (65)).

Comparison with the classical semi-relativistic equation (IV.A136) shows which spin terms are to be added in the present case. In the first place one has at the left-hand side the spin contributions (88–90) to the inner angular momentum of the composite particle. At the right-hand side four spin terms are added, namely a term $-\boldsymbol{v}_k \wedge \boldsymbol{g}_k$ with the spin momentum $\boldsymbol{g}_k$ (69), and three terms

$$\boldsymbol{\mu}^{(1)}_{k,\mathrm{spin}} \wedge \boldsymbol{E}_{\mathrm{e}} + \boldsymbol{v}^{(1)}_{k,\mathrm{spin}} \wedge \boldsymbol{B}_{\mathrm{e}} + c^{-1}(\boldsymbol{v}_k \wedge \boldsymbol{v}^{(1)}_{k,\mathrm{spin}}) \wedge \boldsymbol{E}_{\mathrm{e}}\,. \tag{92}$$

The form of the last two terms is the same as that of (54).

The extension to the case of a set of composite particles in each other's fields is straightforward and will not be given here.

CHAPTER X

# Semi-relativistic quantum statistics of spin media

## 1 Introduction

By means of a quantum-statistical averaging procedure with Wigner functions the semi-relativistic macroscopic laws for spin media will be obtained in this chapter, on the basis of the microscopic results found in the preceding. This will lead to laws from which one may infer – by comparison with the non-relativistic results of chapter VII – which new terms arise if the spin of the particles is taken into account. At the end of the chapter the magnetostriction phenomenon will be studied on the basis of a simple model of a magnetic medium.

## 2 The Wigner function in statistics; particles with spin

In quantum-statistical mechanics the average value of a dynamical quantity, represented by an operator[1] $A$, is usually written as

$$\bar{A}(t) = \mathrm{Tr}\,\{P(t)A\}, \tag{1}$$

where $P(t)$ is the density operator

$$P(t) = \sum_{\gamma} w_{\gamma} |\psi(t)\rangle\langle\psi(t)| \tag{2}$$

that describes the macroscopic mixed state. The states $|\psi_{\gamma}(t)\rangle$, which form a complete orthonormal set, are weighted by the numbers $w_{\gamma}$, which are normalized to unity ($\sum_{\gamma} w_{\gamma} = 1$).

The average (1) of the operator $A$ may be written in a different form, if one introduces Weyl transforms. As in section 3c of chapter VIII we denote the Weyl transform of the operator $A$ by the symbol $a_{\kappa_1 \ldots \kappa_N \lambda_1 \ldots \lambda_N}(1, \ldots, N)$ depending on indices $\kappa_i$ and $\lambda_i$ ($i = 1, \ldots, N$), which in the present semi-relativistic description of systems of spin particles take the values 1 and 2.

[1] In this section capitals denote operators, lower case symbols their Weyl transforms.

The arguments 1, ..., $N$ stand for the momentum and coordinate variables in a phase space of dimensionality $6^N$ for an $N$-particle system. The Wigner function is, apart from a factor $h^{-3N}$, equal to the Weyl transform of the density operator (2), so that it may be written as

$$\rho_{\kappa_1 \ldots \kappa_N \lambda_1 \ldots \lambda_N}(1, \ldots, N; t) = \sum_{\gamma} w_{\gamma} \rho_{\gamma, \kappa_1 \ldots \kappa_N \lambda_1 \ldots \lambda_N}(1, \ldots, N; t) \tag{3}$$

with partial Wigner functions $\rho_\gamma$ given by (cf. (VI.A166)):

$$\rho_{\gamma, \kappa_1 \ldots \kappa_N \lambda_1 \ldots \lambda_N}(1, \ldots, N; t) \equiv \rho_{\gamma, \kappa_1 \ldots \kappa_N \lambda_1 \ldots \lambda_N}(\boldsymbol{p}_1, \boldsymbol{q}_1, \ldots, \boldsymbol{p}_N, \boldsymbol{q}_N; t)$$
$$= h^{-3N} \int \mathrm{d}\boldsymbol{v}_1 \ldots \mathrm{d}\boldsymbol{v}_N \exp\left(\frac{i}{\hbar} \sum_{i=1}^{N} \boldsymbol{p}_i \cdot \boldsymbol{v}_i\right) \psi_{\gamma, \kappa_1 \ldots \kappa_N}(\boldsymbol{q}_1 - \tfrac{1}{2}\boldsymbol{v}_1, \ldots, \boldsymbol{q}_N - \tfrac{1}{2}\boldsymbol{v}_N; t)$$
$$\psi^*_{\gamma, \lambda_1 \ldots \lambda_N}(\boldsymbol{q}_1 + \tfrac{1}{2}\boldsymbol{v}_1, \ldots, \boldsymbol{q}_N + \tfrac{1}{2}\boldsymbol{v}_N; t). \tag{4}$$

In terms of this Wigner function one may write the average (1) as (cf. VI.A167)

$$\bar{A}(t) = \bar{a}(t) \equiv \sum_{\kappa_1 \ldots \kappa_N \lambda_1 \ldots \lambda_N} \int \mathrm{d}1 \ldots \mathrm{d}N \, \rho_{\kappa_1 \ldots \kappa_N \lambda_1 \ldots \lambda_N}(1, \ldots, N; t)$$
$$a_{\lambda_1 \ldots \lambda_N \kappa_1 \ldots \kappa_N}(1, \ldots, N). \tag{5}$$

The right-hand side contains a trace over the matrix indices, which will be denoted by the symbol Sp (to distinguish it from the trace Tr in Hilbert space which is meant in (1)). Thus (5) may be written in the form

$$\bar{A}(t) = \bar{a}(t) \equiv \mathrm{Sp} \int \mathrm{d}1 \ldots \mathrm{d}N \, \rho(1, \ldots, N; t) a(1, \ldots, N). \tag{6}$$

The Wigner function is normalized

$$\mathrm{Sp} \int \mathrm{d}1 \ldots \mathrm{d}N \, \rho(1, \ldots, N; t) = 1; \tag{7}$$

as a result of the normalization $\sum_\gamma w_\gamma = 1$ (or $\mathrm{Tr}\, P = 1$).

The time evolution of the Wigner function is governed by an equation of which (VI.A169) is the special case valid for a single particle in a pure state. It reads

$$\frac{\partial \rho_{\kappa_1 \ldots \kappa_N \lambda_1 \ldots \lambda_N}(1, \ldots, N; t)}{\partial t} = \frac{1}{\hbar} \sin \left\{ \frac{\hbar}{2} \sum_{i=1}^{N} \left( \frac{\partial^{(h)}}{\partial \boldsymbol{q}^i} \cdot \frac{\partial^{(\rho)}}{\partial \boldsymbol{p}^i} - \frac{\partial^{(h)}}{\partial \boldsymbol{p}^i} \cdot \frac{\partial^{(\rho)}}{\partial \boldsymbol{q}^i} \right) \right\}$$
$$\sum_{\mu_1 \ldots \mu_N} (h_{\kappa_1 \ldots \kappa_N \mu_1 \ldots \mu_N} \rho_{\mu_1 \ldots \mu_N \lambda_1 \ldots \lambda_N} + \rho_{\kappa_1 \ldots \kappa_N \mu_1 \ldots \mu_N} h_{\mu_1 \ldots \mu_N \lambda_1 \ldots \lambda_N})$$
$$- \frac{i}{\hbar} \cos \left\{ \frac{\hbar}{2} \sum_{i=1}^{N} \left( \frac{\partial^{(h)}}{\partial \boldsymbol{q}^i} \cdot \frac{\partial^{(\rho)}}{\partial \boldsymbol{p}^i} - \frac{\partial^{(h)}}{\partial \boldsymbol{p}^i} \cdot \frac{\partial^{(\rho)}}{\partial \boldsymbol{q}^i} \right) \right\}$$
$$\sum_{\mu_1 \ldots \mu_N} (h_{\kappa_1 \ldots \kappa_N \mu_1 \ldots \mu_N} \rho_{\mu_1 \ldots \mu_N \lambda_1 \ldots \lambda_N} - \rho_{\kappa_1 \ldots \kappa_N \mu_1 \ldots \mu_N} h_{\mu_1 \ldots \mu_N \lambda_1 \ldots \lambda_N}), \tag{8}$$

where the Weyl transform $h_{\kappa_1\ldots\kappa_N\lambda_1\ldots\lambda_N}$ of the Hamiltonian depends on the coordinates and momenta of all particles (i.e. on 1, ..., $N$), while the Wigner function $\rho_{\kappa_1\ldots\kappa_N\lambda_1\ldots\lambda_N}$ depends moreover on the time $t$.

As a consequence of the time behaviour (8) of the Wigner function the time derivative of the expectation value (6) of an operator is given by (cf. (VI.A170))

$$\frac{\mathrm{d}\bar{a}(t)}{\mathrm{d}t} = \frac{1}{\hbar}\,\mathrm{Sp}\int \mathrm{d}1 \ldots \mathrm{d}N\,\rho\left[\sin\left\{\frac{\hbar}{2}\sum_{i=1}^{N}\left(\frac{\partial^{(a)}}{\partial \boldsymbol{q}_i}\cdot\frac{\partial^{(h)}}{\partial \boldsymbol{p}_i}-\frac{\partial^{(a)}}{\partial \boldsymbol{p}_i}\cdot\frac{\partial^{(h)}}{\partial \boldsymbol{q}_i}\right)\right\}(ah+ha)\right.$$
$$\left.-i\cos\left\{\frac{\hbar}{2}\sum_{i=1}^{N}\left(\frac{\partial^{(a)}}{\partial \boldsymbol{q}_i}\cdot\frac{\partial^{(h)}}{\partial \boldsymbol{p}_i}-\frac{\partial^{(a)}}{\partial \boldsymbol{p}_i}\cdot\frac{\partial^{(h)}}{\partial \boldsymbol{q}_i}\right)\right\}(ah-ha)\right], \tag{9}$$

where $a_{\kappa_1\ldots\kappa_N\lambda_1\ldots\lambda_N}$ and $h_{\kappa_1\ldots\kappa_N\lambda_1\ldots\lambda_N}$ depend on all coordinates and momenta (1, ..., $N$) and $\rho_{\kappa_1\ldots\kappa_N\lambda_1\ldots\lambda_N}$ moreover on the time $t$.

In the following it will be convenient to employ reduced Wigner functions, which are generalizations of the reduced Wigner functions for spinless particles.

The average of a sum of one-point functions (i.e. of quantities that depend on the coordinates and momenta of a single particle)

$$A \rightleftarrows a_{\kappa_1\ldots\kappa_N\lambda_1\ldots\lambda_N}(1, \ldots, N) = \sum_{i=1}^{N}\prod_{j(\neq i)}\delta_{\kappa_j\lambda_j}\,a_{i,\kappa_i\lambda_i}(i) \tag{10}$$

may be written as

$$\bar{A}(t) = \mathrm{Sp}\int a_1(1) f_1(1; t)\mathrm{d}1 \tag{11}$$

with the one-point reduced Wigner function defined as

$$f_{1,\kappa\lambda}(1; t) = N\sum_{\kappa_2\ldots\kappa_N}\int \rho_{\kappa\kappa_2\ldots\kappa_N\lambda\kappa_2\ldots\kappa_N}(1, 2, \ldots, N)\mathrm{d}2 \ldots \mathrm{d}N, \tag{12}$$

normalized to $N$.

In the same fashion one may employ two-point reduced Wigner functions

$$f_{2,\kappa_1\lambda_1\kappa_2\lambda_2}(1, 2; t) = N(N-1)\sum_{\kappa_3\ldots\kappa_N}\int \rho_{\kappa_1\kappa_2\ldots\kappa_N\lambda_1\lambda_2\kappa_3\ldots\kappa_N}(1, 2, \ldots, N)\mathrm{d}3 \ldots \mathrm{d}N, \tag{13}$$

normalized to $N(N-1)$, to write the average of a two-point function in a compact way.

The two-point correlation function is defined as

$$c_{2,\kappa_1\kappa_2\lambda_1\lambda_2}(1, 2; t) = f_{2,\kappa_1\kappa_2\lambda_1\lambda_2}(1, 2; t) - f_{1,\kappa_1\lambda_1}(1; t) f_{1,\kappa_2\lambda_2}(2; t), \tag{14}$$

normalized to $-N$.

## 3 *The Maxwell equations*

The macroscopic field equations will follow from the atomic equations of chapter IX, section 5. Let us multiply these equations, valid for Weyl transforms, with a Wigner function, integrate over all coordinates and momenta, and take the spur. The resulting equations are still not in the form of the Maxwell equations, since they contain terms as $\overline{\partial_{0P}\boldsymbol{p}}$ instead of $\partial_0 \overline{\boldsymbol{p}}$, where $\boldsymbol{p}$ is the Weyl transform of the atomic polarization (and likewise two terms with the electric and magnetic fields). However they may be brought into the desired form, if one employs the identity (9) for the polarization density and the fields. Let us consider first the time derivative $\partial_0 \boldsymbol{p}$. With the help of the expressions (IX.63) with (IX.64–65) for the Weyl transform of the polarization and (IX.37) for the Weyl transform of the Hamiltonian of the system we find that in the right-hand side of (9) the cosine term does not contribute, while the sine term reduces to the average of the Poisson bracket $\overline{\partial_{0P}\boldsymbol{p}}$. This shows that indeed $\overline{\partial_{0P}\boldsymbol{p}}$ may be replaced by $\partial_0 \overline{\boldsymbol{p}}$.

The derivatives of the electromagnetic fields may be treated along similar lines by using the Weyl transforms of the atomic fields (v. (IX.46)). Then again one proves that $\overline{\partial_{0P}\boldsymbol{e}}$ and $\overline{\partial_{0P}\boldsymbol{b}}$ are equal to $\partial_0 \overline{\boldsymbol{e}}$ and $\partial_0 \overline{\boldsymbol{b}}$. If one writes $\boldsymbol{E}$, $\boldsymbol{B}$, $\varrho^e$, $\boldsymbol{J}$, $\boldsymbol{P}$, and $\boldsymbol{M}$ for the averages $\overline{\boldsymbol{e}}$, $\overline{\boldsymbol{b}}$, $\overline{\rho^e}$, $\overline{\boldsymbol{j}}$, $\overline{\boldsymbol{p}}$ and $\overline{\boldsymbol{m}}$ of the corresponding atomic quantities one recovers indeed the Maxwell equations:

$$\begin{aligned} \nabla\cdot\boldsymbol{E} &= \varrho^e - \nabla\cdot\boldsymbol{P}, \\ -\partial_0\boldsymbol{E} + \nabla\wedge\boldsymbol{B} &= c^{-1}\boldsymbol{J} + \partial_0\boldsymbol{P} + \nabla\wedge\boldsymbol{M}, \\ \nabla\cdot\boldsymbol{B} &= 0, \\ \partial_0\boldsymbol{B} + \nabla\wedge\boldsymbol{E} &= 0. \end{aligned} \tag{15}$$

The sources $\varrho^e$, $\boldsymbol{J}$, $\boldsymbol{P}$ and $\boldsymbol{M}$ are again found as statistical averages of (IX.58) and (IX.63) and thus ultimately expressed in terms of the atomic charges $e_k$ and electromagnetic multipole moments $\boldsymbol{\mu}_k^{(n)}$ and $\boldsymbol{\nu}_k^{(n)}$. They contain the complete semi-relativistic contributions due to the spins of the constituent electrons of the atoms.

## 4 *The momentum and energy equations*

### a. *Conservation of rest mass*

The atomic rest mass density is given by its Weyl transform

$$\rho \equiv \sum_k m_k \, \delta(\boldsymbol{X}_k - \boldsymbol{R}) \tag{16}$$

with $\boldsymbol{X}_k$ the Weyl transform (IX.68) of the position operator for the atom as a whole. The definition of the delta function is analogous to that of (IX.52). The quantity $m_k$ is the total rest mass $\sum_i m_{ki}$ of the atom. The Weyl transform $\rho$ satisfies the conservation law

$$\partial_{t\mathrm{P}}\, \rho = -\nabla \cdot \{ \sum_k m_k \, \boldsymbol{v}_k \, \delta(\boldsymbol{X}_k - \boldsymbol{R}) \} \tag{17}$$

with $\boldsymbol{v}_k \equiv \partial_{t\mathrm{P}} \boldsymbol{X}_k$ and $\partial_{t\mathrm{P}}\, \rho$ the Poisson bracket of the density $\rho$ and the Weyl transform of the Hamiltonian. We used the fact that $\boldsymbol{v}_k$ commutes with $\boldsymbol{X}_k$ in semi-relativistic approximation, since the non-diagonal matrix part of $\boldsymbol{X}_k$ is of order $c^{-2}$ while the part of order $c^0$ in $\boldsymbol{v}_k$ is diagonal.

A macroscopic conservation law is obtained from (17) by multiplying it by a Wigner function, integrating over phase space and taking the spur of the matrices. Then one obtains for the left-hand side of (17) the expression $\overline{\partial_{t\mathrm{P}}\, \rho}$, which may be shown to be equal to $\partial_t \bar{\rho}$. Indeed one may conclude this from (9) by inspection of the expressions (16) for $\rho$ and (IX.37) for the Weyl transform of the Hamiltonian, if one confines oneself to semi-relativistic terms. Thus the macroscopic conservation law of rest mass gets the usual form

$$\frac{\partial \varrho}{\partial t} = -\nabla \cdot (\varrho \boldsymbol{v}), \tag{18}$$

where the mass density and the mass flow density are given by

$$\varrho = \bar{\rho} = \mathrm{Sp} \int m_1 \, \delta(\boldsymbol{X}_1 - \boldsymbol{R}) f_1(1;t) \mathrm{d}1, \tag{19}$$

$$\varrho \boldsymbol{v} = \mathrm{Sp} \int m_1 \, \boldsymbol{v}_1 \, \delta(\boldsymbol{X}_1 - \boldsymbol{R}) f_1(1;t) \mathrm{d}1. \tag{20}$$

Here we introduced the one-point reduced Wigner functions (12). The expressions (19) with (20) serve to define the macroscopic velocity $\boldsymbol{v}(\boldsymbol{R}, t)$.

### b. *The momentum balance*

The momentum law will follow by multiplying the atomic law (IX.75) by

$\delta(\boldsymbol{X}_k - \boldsymbol{R})$, summing over $k$ and averaging with a Wigner function. The left-hand side may be written as

$$\mathrm{Sp}\int \sum_k \partial_{t\mathrm{P}}\{(m_k \boldsymbol{v}_k + \boldsymbol{g}_k)\delta(\boldsymbol{X}_k - \boldsymbol{R})\}\rho(1, \ldots, N; t)\mathrm{d}1 \ldots \mathrm{d}N$$
$$+\nabla\cdot \mathrm{Sp}\int \sum_k \boldsymbol{v}_k(m_k \boldsymbol{v}_k + \boldsymbol{g}_k)\delta(\boldsymbol{X}_k - \boldsymbol{R})\rho(1, \ldots, N; t)\mathrm{d}1 \ldots \mathrm{d}N, \quad (21)$$

where we used the fact that in the semi-relativistic approximation $\boldsymbol{v}_k$ commutes with the expression between the brackets, because both $\boldsymbol{v}_k$ and $\boldsymbol{X}_k$ are of the form of a diagonal matrix of order $c^0$ and $c^{-1}$ plus a non-diagonal matrix of order $c^{-2}$.

In the first term of (21) we may apply formula (9) to write it as the time derivative of an average. Indeed it follows from the forms of $\boldsymbol{X}_k$ (IX.68), $\boldsymbol{g}_k$ (IX.69), $\boldsymbol{v}_k$ and the Weyl transform of the Hamilton operator (IX.37) that the sine and cosine terms occurring in (9) reduce to a Poisson bracket. Then this first term becomes $\partial(\varrho\boldsymbol{v}+\boldsymbol{g})/\partial t$ with

$$\boldsymbol{g} \equiv \mathrm{Sp}\int \boldsymbol{g}_1\,\delta(\boldsymbol{X}_1 - \boldsymbol{R})f_1(1; t)\mathrm{d}1. \quad (22)$$

In the second term of (21) the velocity $\boldsymbol{v}_k$ will be split into the macroscopic velocity at the position $\boldsymbol{R}$, defined by (20) with (19), and a fluctuation term:

$$\boldsymbol{v}_k = \boldsymbol{v}(\boldsymbol{R}, t) + \hat{\boldsymbol{v}}_k(\boldsymbol{R}, t). \quad (23)$$

Then we may write this second term of (21) in the form

$$\nabla\cdot(\varrho\boldsymbol{v}\boldsymbol{v} + \boldsymbol{v}\boldsymbol{g} + \mathsf{P}^{\mathrm{K}}) \quad (24)$$

with the kinetic pressure

$$\mathsf{P}^{\mathrm{K}} \equiv \mathrm{Sp}\int \hat{\boldsymbol{v}}_1(m_1\hat{\boldsymbol{v}}_1 + \boldsymbol{g}_1)\delta(\boldsymbol{X}_1 - \boldsymbol{R})f_1(1; t)\mathrm{d}1 \quad (25)$$

and the momentum density $\boldsymbol{g}$ (22). Thus we have found that the momentum balance equation reads

$$\frac{\partial(\varrho\boldsymbol{v}+\boldsymbol{g})}{\partial t} = -\nabla\cdot(\varrho\boldsymbol{v}\boldsymbol{v} + \boldsymbol{v}\boldsymbol{g} + \mathsf{P}^{\mathrm{K}}) + \boldsymbol{F}^{\mathrm{L}} + \boldsymbol{F}^{\mathrm{S}} \quad (26)$$

with the long range and short range force densities:

$$\boldsymbol{F}^{\mathrm{L,S}} = \mathrm{Sp}\int \sum_k \boldsymbol{f}_k^{\mathrm{L,S}}\delta(\boldsymbol{X}_k - \boldsymbol{R})\rho(1, \ldots, N; t)\mathrm{d}1 \ldots \mathrm{d}N. \quad (27)$$

To specify the equation (26) completely, we shall now study the explicit expressions that result if (IX.71) with (IX.73) and (IX.74) are introduced into (27). The external field terms of the long range force density follow from (IX.71). We find, using the macroscopic charge–current and polarization densities $\varrho^{\mathrm{e}}$, $\boldsymbol{J}$, $\boldsymbol{P}$ and $\boldsymbol{M}$ of section 3, and limiting ourselves to dipole substances, for these terms

$$\boldsymbol{F}_{\mathrm{e}}^{\mathrm{L}} = \varrho^{\mathrm{e}}\boldsymbol{E}_{\mathrm{e}}+c^{-1}\boldsymbol{J}\wedge\boldsymbol{B}_{\mathrm{e}}+(\nabla\boldsymbol{E}_{\mathrm{e}})\cdot\boldsymbol{P}+(\nabla\boldsymbol{B}_{\mathrm{e}})\cdot\boldsymbol{M}$$
$$+c^{-1}\,\mathrm{Sp}\int\sum_k \partial_{t\mathrm{P}}\{\boldsymbol{\mu}_k^{(1)}\wedge\boldsymbol{B}_{\mathrm{e}}(\boldsymbol{X}_k,t)-\boldsymbol{v}_{k,\mathrm{orb}}^{(1)}\wedge\boldsymbol{E}_{\mathrm{e}}(\boldsymbol{X}_k,t)\}\delta(\boldsymbol{X}_k-\boldsymbol{R})$$
$$\rho(1,\ldots,N;t)\mathrm{d}1\ldots\mathrm{d}N. \quad (28)$$

The last term may be written as

$$c^{-1}\,\mathrm{Sp}\int\sum_k \partial_{t\mathrm{P}}[\{\boldsymbol{\mu}_k^{(1)}\wedge\boldsymbol{B}_{\mathrm{e}}(\boldsymbol{X}_k,t)-\boldsymbol{v}_{k,\mathrm{orb}}^{(1)}\wedge\boldsymbol{E}_{\mathrm{e}}(\boldsymbol{X}_k,t)\}\delta(\boldsymbol{X}_k-\boldsymbol{R})]$$
$$\rho(1,\ldots,N;t)\mathrm{d}1\ldots\mathrm{d}N$$
$$+c^{-1}\nabla\cdot\,\mathrm{Sp}\int\sum_k \boldsymbol{v}_k\{\boldsymbol{\mu}_k^{(1)}\wedge\boldsymbol{B}_{\mathrm{e}}(\boldsymbol{X}_k,t)-\boldsymbol{v}_{k,\mathrm{orb}}^{(1)}\wedge\boldsymbol{E}_{\mathrm{e}}(\boldsymbol{X}_k,t)\}\delta(\boldsymbol{X}_k-\boldsymbol{R})$$
$$\rho(1,\ldots,N;t)\mathrm{d}1\ldots\mathrm{d}N, \quad (29)$$

since $c^{-1}\boldsymbol{v}_k$ commutes with $\boldsymbol{X}_k$ and with $\boldsymbol{\mu}_k^{(1)}\wedge\boldsymbol{B}_{\mathrm{e}}-\boldsymbol{v}_{k,\mathrm{orb}}^{(1)}\wedge\boldsymbol{E}_{\mathrm{e}}$ in semi-relativistic approximation. In the first term of (29) one applies the identity (9). By inspection of the integrand of (29) and the Weyl transform of the Hamiltonian (IX.37), it follows that the cosine term does not contribute and of the sine term only the Poisson bracket, so that one may write for (29)

$$c^{-1}\frac{\partial}{\partial t}(\boldsymbol{P}\wedge\boldsymbol{B}_{\mathrm{e}}-\boldsymbol{M}_{\mathrm{orb}}\wedge\boldsymbol{E}_{\mathrm{e}})+c^{-1}\nabla\cdot\{\boldsymbol{v}(\boldsymbol{P}\wedge\boldsymbol{B}_{\mathrm{e}}-\boldsymbol{M}_{\mathrm{orb}}\wedge\boldsymbol{E}_{\mathrm{e}})\}$$
$$+c^{-1}\nabla\cdot\,\mathrm{Sp}\int\sum_k \hat{\boldsymbol{v}}_k(\boldsymbol{\mu}_k^{(1)}\wedge\boldsymbol{B}_{\mathrm{e}}-\boldsymbol{v}_{k,\mathrm{orb}}^{(1)}\wedge\boldsymbol{E}_{\mathrm{e}})\delta(\boldsymbol{X}_k-\boldsymbol{R})$$
$$\rho(1,\ldots,N;t)\mathrm{d}1\ldots\mathrm{d}N, \quad (30)$$

where the macroscopic orbital magnetic dipole density

$$\boldsymbol{M}_{\mathrm{orb}} \equiv \mathrm{Sp}\int \boldsymbol{v}_{1,\mathrm{orb}}^{(1)}\delta(\boldsymbol{X}_1-\boldsymbol{R})f_1(1;t)\mathrm{d}1 \quad (31)$$

and the velocity fluctuation $\hat{\boldsymbol{v}}_k$ of (23) have been introduced.

The interatomic field contribution to the long range force density follows from (IX.73). It may be split into an uncorrelated and a correlated part, if one employs the definition of the two-point Wigner correlation function (14). One finds in this fashion for the sum of the interatomic and external field contributions i.e. for the total long range force density

$$\boldsymbol{F}^{\mathrm{L}} = \varrho^{\mathrm{e}}\boldsymbol{E}+c^{-1}\boldsymbol{J}\wedge\boldsymbol{B}+(\nabla\boldsymbol{E})\cdot\boldsymbol{P}+(\nabla\boldsymbol{B})\cdot\boldsymbol{M}+c^{-1}\frac{\partial}{\partial t}(\boldsymbol{P}\wedge\boldsymbol{B}-\boldsymbol{M}_{\mathrm{orb}}\wedge\boldsymbol{E})$$
$$+c^{-1}\nabla\cdot\{\boldsymbol{v}(\boldsymbol{P}\wedge\boldsymbol{B}-\boldsymbol{M}_{\mathrm{orb}}\wedge\boldsymbol{E})\}-\nabla\cdot\mathbf{P}^{\mathrm{F}}+\boldsymbol{F}^{\mathrm{C}}, \quad (32)$$

where the Maxwell fields $\boldsymbol{E}$ and $\boldsymbol{B}$ appear. The field dependent part of the pressure is given by:

$$\mathbf{P}^{\mathrm{F}} \equiv -c^{-1}\,\mathrm{Sp}\int \hat{\boldsymbol{v}}_1(\boldsymbol{\mu}_1^{(1)}\wedge\boldsymbol{B}-\boldsymbol{v}_{1,\mathrm{orb}}^{(1)}\wedge\boldsymbol{E})\delta(\boldsymbol{X}_1-\boldsymbol{R})f_1(1;t)\mathrm{d}1 \quad (33)$$

and the correlation force density by

$$\boldsymbol{F}^{\mathrm{C}} = \mathrm{Sp}\int \boldsymbol{f}^{\mathrm{C}}(1,2)\delta(\boldsymbol{X}_1-\boldsymbol{R})c_2(1,2;t)\mathrm{d}1\,\mathrm{d}2, \quad (34)$$

where we introduced the abbreviation:

$$\boldsymbol{f}^{\mathrm{C}}(1,2) \equiv -\{\sum_{n,m=0}^{\infty} \boldsymbol{\mu}_1^{(n)} \vdots \nabla_1^n\,\boldsymbol{\mu}_2^{(m)} \vdots \nabla_2^m\nabla_1$$
$$+\sum_{n,m=1}^{\infty}(\nabla_1^{n-1}\vdots\boldsymbol{v}_1^{(n)}\wedge\nabla_1)\cdot(\nabla_2^{m-1}\vdots\boldsymbol{v}_2^{(m)}\wedge\nabla_1)\nabla_1\}\frac{1}{4\pi|\boldsymbol{X}_1-\boldsymbol{X}_2|}. \quad (35)$$

An expression for the short range force density $\boldsymbol{F}^{\mathrm{S}}$ (27) in the momentum balance (26) follows if one inserts (IX.74):

$$\boldsymbol{F}^{\mathrm{S}} = \mathrm{Sp}\int \boldsymbol{f}^{\mathrm{S}}(1,2)\delta(\boldsymbol{X}_1-\boldsymbol{R})f_2(1,2;t)\mathrm{d}1\,\mathrm{d}2 \quad (36)$$

with the abbreviation

$$\boldsymbol{f}^{\mathrm{S}}(1,2) \equiv -\sum_{i,j}\Big[1-c^{-2}(\partial_{t\mathrm{P}}\boldsymbol{r}_{1i})\cdot(\partial_{t\mathrm{P}}\boldsymbol{r}_{2j})$$
$$+c^{-2}\{\partial_{t\mathrm{P}}(\boldsymbol{r}_{1i}-\boldsymbol{r}_{2j})\}\cdot\left\{\left(\frac{\hbar}{2m_{1i}}\boldsymbol{\sigma}_{1i}+\frac{\hbar}{2m_{2j}}\boldsymbol{\sigma}_{2j}\right)\wedge\nabla_{1i}\right\}$$
$$+c^{-2}\frac{\hbar^2}{4m_{1i}m_{2j}}(\boldsymbol{\sigma}_{1i}\wedge\nabla_{1i})\cdot(\boldsymbol{\sigma}_{2j}\wedge\nabla_{1i})\Big]\nabla_{1i}\frac{e_{1i}e_{2j}}{4\pi|\boldsymbol{X}_{1i}-\boldsymbol{X}_{2j}|}-\boldsymbol{f}^{\mathrm{C}}(1,2). \quad (37)$$

Owing to the antisymmetric character of (37) with respect to the interchange of 1 and 2, one may write (36) as

$$\boldsymbol{F}^{\mathrm{S}} = \tfrac{1}{2}\mathrm{Sp}\int \boldsymbol{f}^{\mathrm{S}}(1,2)\{\delta(\boldsymbol{X}_1-\boldsymbol{R})-\delta(\boldsymbol{X}_2-\boldsymbol{R})\}f_2(1,2;t)\mathrm{d}1\,\mathrm{d}2. \quad (38)$$

Then, since $f^S(1, 2)$ diminishes rapidly with increasing interatomic distances we may apply an Irving–Kirkwood procedure to write (38) as a divergence[1]:

$$\boldsymbol{F}^S = -\nabla\cdot\mathbf{P}^S \tag{39}$$

with the short range pressure

$$\mathbf{P}^S \equiv \tfrac{1}{2}\,\mathrm{Sp}\int (\boldsymbol{X}_1-\boldsymbol{X}_2)\boldsymbol{f}^S(1, 2)\delta\{\tfrac{1}{2}(\boldsymbol{X}_1+\boldsymbol{X}_2)-\boldsymbol{R}\}f_2(1, 2; t)\mathrm{d}1\,\mathrm{d}2. \tag{40}$$

We are left with the correlation force density $\boldsymbol{F}^C$ (34). For systems of neutral atoms in which no long range Wigner correlations exist, so that a correlation length may be defined, one may apply an Irving–Kirkwood procedure to the correlation terms as well. One gets then

$$\boldsymbol{F}^C = -\nabla\cdot\mathbf{P}^C \tag{41}$$

with a correlation pressure given by

$$\mathbf{P}^C \equiv \tfrac{1}{2}\,\mathrm{Sp}\int (\boldsymbol{X}_1-\boldsymbol{X}_2)\boldsymbol{f}^C(1, 2)\delta\{\tfrac{1}{2}(\boldsymbol{X}_1+\boldsymbol{X}_2)-\boldsymbol{R}\}c_2(1, 2; t)\mathrm{d}1\,\mathrm{d}2. \tag{42}$$

Hence we have found now the macroscopic momentum balance equation

$$\begin{aligned}\frac{\partial(\varrho\boldsymbol{v}+\boldsymbol{g})}{\partial t} &= -\nabla\cdot(\varrho\boldsymbol{v}\boldsymbol{v}+\boldsymbol{v}\boldsymbol{g}+\mathbf{P})+(\nabla\boldsymbol{E})\cdot\boldsymbol{P}+(\nabla\boldsymbol{B})\cdot\boldsymbol{M}\\ &\quad +c^{-1}\frac{\partial}{\partial t}(\boldsymbol{P}\wedge\boldsymbol{B}-\boldsymbol{M}_{\mathrm{orb}}\wedge\boldsymbol{E})+c^{-1}\nabla\cdot\{\boldsymbol{v}(\boldsymbol{P}\wedge\boldsymbol{B}-\boldsymbol{M}_{\mathrm{orb}}\wedge\boldsymbol{E})\}\end{aligned} \tag{43}$$

with the total pressure tensor

$$\mathbf{P} = \mathbf{P}^K+\mathbf{P}^F+\mathbf{P}^C+\mathbf{P}^S, \tag{44}$$

valid for systems of neutral atoms without long range correlations. The Lorentz force density, which occurs in (32), is absent for systems with neutral atoms.

As compared with the non-relativistic theory, namely with (VII.64), the momentum density contains an additional term $\boldsymbol{g}$ specified in (22) with (IX.69). The pressure tensor includes, apart from a similar additional term in its kinetic part (25), additional terms due to the interaction of magnetic

[1] In contrast with the procedure used in earlier chapters, where the distribution functions were developed, we employ here the equivalent methods of developing the delta functions. This has a formal advantage because the delta functions have to be understood as (IX.52). Integration over $\boldsymbol{R}_1$ would lead then to slightly more complicated expressions, which are now avoided.

dipoles with each other, as (40) and (42) with (35) and (37) show. In the terms that depend solely on the Maxwell fields $(\boldsymbol{E}, \boldsymbol{B})$, the polarization densities $(\boldsymbol{P}, \boldsymbol{M})$ and the macroscopic velocity $\boldsymbol{v}$ two extra terms appear, which couple the orbital magnetization density $\boldsymbol{M}_{\text{orb}}$ to the electric field. These terms may be written in alternative form by using mass conservation, namely as

$$-c^{-1}\varrho\frac{\mathrm{d}}{\mathrm{d}t}\{v(\boldsymbol{M}_{\text{orb}}\wedge\boldsymbol{E})\} \tag{45}$$

(with $v$ the specific volume and $\mathrm{d}/\mathrm{d}t$ the material time derivative $\partial/\partial t+\boldsymbol{v}\cdot\nabla$): the magnetodynamic effect on the macroscopic level. It contains only the orbital part of the magnetization, since the spin magnetization is assumed to be completely normal (without an anomalous contribution). Furthermore the Maxwell fields and the polarization densities also contain terms due to the presence of spins.

By employing the Maxwell equations one may write the balance equation (43) in the form of a conservation law:

$$\begin{aligned}\frac{\partial}{\partial t}\{\varrho\boldsymbol{v}+\boldsymbol{g}+c^{-1}\boldsymbol{E}\wedge(\boldsymbol{B}-\boldsymbol{M}_{\text{orb}})\}\\ +\nabla\cdot\{\varrho\boldsymbol{v}\boldsymbol{v}+\boldsymbol{v}\boldsymbol{g}+\mathsf{P}-\boldsymbol{D}\boldsymbol{E}-\boldsymbol{B}\boldsymbol{H}-c^{-1}\boldsymbol{v}(\boldsymbol{P}\wedge\boldsymbol{B}-\boldsymbol{M}_{\text{orb}}\wedge\boldsymbol{E})\\ +(\tfrac{1}{2}E^2+\tfrac{1}{2}B^2-\boldsymbol{M}\cdot\boldsymbol{B})\mathsf{U}\}=0.\end{aligned} \tag{46}$$

The method outlined may be extended to plasmas and to systems with long range correlations. In the latter case one has to introduce a mean correlation function $\tilde{c}_2(1, 2; t)$, as discussed for instance in chapter II, section 5*h*.

### c. *The energy balance*

The energy balance equation on the macroscopic level will be derived from its atomic counterpart (IX.86). Multiplying the latter by $\delta(\boldsymbol{X}_k-\boldsymbol{R})$, summing over $k$ and averaging with a Wigner function yield an equation of which the left-hand side may be written as:

$$\begin{aligned}\mathrm{Sp}\int\sum_k \partial_{t\mathrm{P}}\{(\tfrac{1}{2}m_k\boldsymbol{v}_k^2+\boldsymbol{v}_k\cdot\boldsymbol{g}_k+t_k+u_k)\delta(\boldsymbol{X}_k-\boldsymbol{R})\}\rho(1,\ldots,N;t)\mathrm{d}1\ldots\mathrm{d}N\\ +\nabla\cdot\mathrm{Sp}\int\sum_k \boldsymbol{v}_k(\tfrac{1}{2}m_k\boldsymbol{v}_k^2+\boldsymbol{v}_k\cdot\boldsymbol{g}_k+t_k+u_k)\delta(\boldsymbol{X}_k-\boldsymbol{R})\rho(1,\ldots,N;t)\mathrm{d}1\ldots\mathrm{d}N.\end{aligned} \tag{47}$$

In the first term of this expression we apply the identity (9). In the present case it leads to the following form

$$\frac{\partial}{\partial t}\operatorname{Sp}\int\sum_k(\tfrac{1}{2}m_k\boldsymbol{v}_k^2+\boldsymbol{v}_k\cdot\boldsymbol{g}_k+t_k+u_k)\delta(\boldsymbol{X}_k-\boldsymbol{R})\rho(1,\ldots,N;t)\mathrm{d}1\ldots\mathrm{d}N$$
$$+\frac{\hbar^2}{24}\operatorname{Sp}\int\Bigg[\Bigg\{\sum_{l,j}3\frac{\partial}{\partial\boldsymbol{R}_{lj}}\cdot\frac{\partial^{(H)}}{\partial\boldsymbol{P}_{lj}}\Big(\sum_{m,p}\frac{\partial}{\partial\boldsymbol{P}_{mp}}\cdot\frac{\partial^{(H)}}{\partial\boldsymbol{R}_{mp}}\Big)^2-\Big(\sum_{l,j}\frac{\partial}{\partial\boldsymbol{P}_{lj}}\cdot\frac{\partial^{(H)}}{\partial\boldsymbol{R}_{lj}}\Big)^3\Bigg\}$$
$$H(1,\ldots,N;t)\sum_k(\tfrac{1}{2}m_k\boldsymbol{v}_k^2+\boldsymbol{v}_k\cdot\boldsymbol{g}_k+t_k+u_k)\delta(\boldsymbol{X}_k-\boldsymbol{R})\Bigg]$$
$$\rho(1,\ldots,N;t)\mathrm{d}1\ldots\mathrm{d}N,\quad(48)$$

where $H(1, \ldots, N; t)$ is the Weyl transform of the Hamiltonian. The superscripts $(H)$ at the differential operators indicate that they act only on the Hamiltonian, while the other operators act on all functions under the brackets save for the Hamiltonian. Using the explicit expression (IX.37) for $H(1, \ldots, N; t)$, one finds in the semi-relativistic approximation for the second term of (48) the divergence

$$\nabla\cdot\boldsymbol{J}_q^{\mathrm{K}'}\quad(49)$$

of the vector

$$\boldsymbol{J}_q^{\mathrm{K}'}\equiv\operatorname{Sp}\int\frac{\hbar^2}{8}\sum_{k,i}\frac{e_{ki}}{m_k m_{ki}c}\{\Delta_{ki}\boldsymbol{A}_{\mathrm{e}}(\boldsymbol{R}_{ki},t)\}\delta(\boldsymbol{X}_k-\boldsymbol{R})\rho(1,\ldots,N;t)\mathrm{d}1\ldots\mathrm{d}N$$
$$-\operatorname{Sp}\int\frac{\hbar^2}{16}\sum_k\sum_{i,j(i\neq j)}\frac{e_{ki}e_{kj}\hbar}{m_k m_{ki}c^2}\Big(\frac{1}{m_{ki}}+\frac{1}{m_{kj}}\Big)\boldsymbol{\sigma}_{ki}\wedge\nabla_{ki}\delta(\boldsymbol{X}_{ki}-\boldsymbol{X}_{kj})\delta(\boldsymbol{X}_k-\boldsymbol{R})$$
$$\rho(1,\ldots,N;t)\mathrm{d}1\ldots\mathrm{d}N$$
$$+\operatorname{Sp}\int\frac{\hbar^2}{16}\sum_{k,l(k\neq l)}\sum_{i,j}\frac{e_{ki}e_{kj}\hbar}{m_k m_{ki}m_{lj}c^2}\boldsymbol{\sigma}_{lj}\wedge\nabla_{ki}\delta(\boldsymbol{X}_{ki}-\boldsymbol{X}_{lj})\delta(\boldsymbol{X}_k-\boldsymbol{R})$$
$$\rho(1,\ldots,N;t)\mathrm{d}1\ldots\mathrm{d}N.\quad(50)$$

The first term with the vector potential of the external field has been found already in the non-relativistic treatment of chapter VII. It was shown there that such a term does not spoil the gauge invariance. The second term contains semi-relativistic contributions connected with the presence of spin. Collecting the results (47–50) we find the macroscopic energy balance in the form

$$\frac{\partial}{\partial t}(\tfrac{1}{2}\varrho\boldsymbol{v}^2+\boldsymbol{v}\cdot\boldsymbol{g}+\varrho u^{\mathrm{K}})=-\nabla\cdot\{\boldsymbol{v}(\tfrac{1}{2}\varrho\boldsymbol{v}^2+\boldsymbol{v}\cdot\boldsymbol{g}+\varrho u^{\mathrm{K}})+\mathsf{P}^{\mathrm{K}}\cdot\boldsymbol{v}+\boldsymbol{J}_q^{\mathrm{K}}\}+\Psi^{\mathrm{L}}+\Psi^{\mathrm{S}}.\quad(51)$$

It contains a kinetic contribution to the energy density

$$\varrho u^{\mathrm{K}}\equiv\operatorname{Sp}\int(\tfrac{1}{2}m_1\hat{\boldsymbol{v}}_1^2+\hat{\boldsymbol{v}}_1\cdot\boldsymbol{g}_1+t_1+u_1)\delta(\boldsymbol{X}_1-\boldsymbol{R})f_1(1;t)\mathrm{d}1\quad(52)$$

and a kinetic heat flow

$$\boldsymbol{J}_q^{\mathrm{K}} \equiv \mathrm{Sp} \int \hat{\boldsymbol{v}}_1(\tfrac{1}{2}m_1 \hat{\boldsymbol{v}}_1^2 + \hat{\boldsymbol{v}}_1 \cdot \boldsymbol{g}_1 + t_1 + u_1)\delta(\boldsymbol{X}_1 - \boldsymbol{R}) f_1(1; t) \mathrm{d}1 + \boldsymbol{J}_q^{\mathrm{K}'}. \qquad (53)$$

The long range and short range power densities $\Psi^{\mathrm{L}}$ and $\Psi^{\mathrm{S}}$ stand for the expressions

$$\Psi^{\mathrm{L,S}} \equiv \mathrm{Sp} \int \sum_k \psi_k^{\mathrm{L,S}} \delta(\boldsymbol{X}_k - \boldsymbol{R}) \rho(1, \ldots, N; t) \mathrm{d}1 \ldots \mathrm{d}N. \qquad (54)$$

The long range power density $\Psi^{\mathrm{L}}$ is found explicitly by considering first the external, then the interatomic field contributions (splitting the latter into an uncorrelated and a correlated part with the help of (14)) and using the identity (9). Then one finds in the present semi-relativistic case

$$\Psi^{\mathrm{L}} = \boldsymbol{J} \cdot \boldsymbol{E} + \frac{\partial \boldsymbol{P}}{\partial t} \cdot \boldsymbol{E} + \nabla \cdot (\boldsymbol{v} \boldsymbol{P} \cdot \boldsymbol{E}) - \boldsymbol{M} \cdot \frac{\partial \boldsymbol{B}}{\partial t} + \frac{\partial}{\partial t} (\boldsymbol{M}_{\mathrm{spin}} \cdot \boldsymbol{B}) + \nabla \cdot (\boldsymbol{v} \boldsymbol{M}_{\mathrm{spin}} \cdot \boldsymbol{B})$$
$$+ 2c^{-1} \frac{\partial}{\partial t} \{\boldsymbol{v} \cdot (\boldsymbol{E} \wedge \boldsymbol{M}_{\mathrm{orb}})\} + 2c^{-1} \nabla \cdot \{\boldsymbol{v}\boldsymbol{v} \cdot (\boldsymbol{E} \wedge \boldsymbol{M}_{\mathrm{orb}})\} - \frac{\partial \varrho u^{\mathrm{F}}}{\partial t}$$
$$- \nabla \cdot (\boldsymbol{v} \varrho u^{\mathrm{F}} + \mathbf{P}^{\mathrm{F}} \cdot \boldsymbol{v} + \boldsymbol{J}_q^{\mathrm{F}}) + \Psi^{\mathrm{C}}. \qquad (55)$$

Here the macroscopic fields $\boldsymbol{E}$, $\boldsymbol{B}$, the macroscopic current density $\boldsymbol{J}$, the polarization densities $\boldsymbol{P}$, $\boldsymbol{M} = \boldsymbol{M}_{\mathrm{orb}} + \boldsymbol{M}_{\mathrm{spin}}$ appear and moreover – also separately – the macroscopic orbital and spin magnetic dipole densities $\boldsymbol{M}_{\mathrm{orb}}$ (31) and

$$\boldsymbol{M}_{\mathrm{spin}} = \mathrm{Sp} \int \boldsymbol{\nu}_{1,\mathrm{spin}}^{(1)} \delta(\boldsymbol{X}_1 - \boldsymbol{R}) f_1(1; t) \mathrm{d}1. \qquad (56)$$

The symbol $\varrho u^{\mathrm{F}}$ stands for that part of the energy density $\varrho u$ which depends explicitly on the Maxwell fields. It is defined as

$$\varrho u^{\mathrm{F}} = -2c^{-1} \left\{ \int \boldsymbol{\nu}_{1,\mathrm{orb}}^{(1)} \wedge \hat{\boldsymbol{v}}_1 \, \delta(\boldsymbol{X}_1 - \boldsymbol{R}) f_1(1; t) \mathrm{d}1 \right\} \cdot \boldsymbol{E}. \qquad (57)$$

The field dependent part $\mathbf{P}^{\mathrm{F}}$ of the pressure tensor has been given in (33), while that of the heat flow is defined as

$$\boldsymbol{J}_q^{\mathrm{F}} = - \left\{ \mathrm{Sp} \int \hat{\boldsymbol{v}}_1 (\boldsymbol{\mu}_1^{(1)} - c^{-1} \boldsymbol{\nu}_1^{(1)} \wedge \hat{\boldsymbol{v}}_1) \delta(\boldsymbol{X}_1 - \boldsymbol{R}) f_1(1; t) \mathrm{d}1 \right\} \cdot (\boldsymbol{E} + c^{-1} \boldsymbol{v} \wedge \boldsymbol{B})$$
$$- \left\{ \mathrm{Sp} \int \hat{\boldsymbol{v}}_1 \, \boldsymbol{\nu}_{1,\mathrm{spin}}^{(1)} \delta(\boldsymbol{X}_1 - \boldsymbol{R}) f_1(1; t) \mathrm{d}1 \right\} \cdot (\boldsymbol{B} - c^{-1} \boldsymbol{v} \wedge \boldsymbol{E})$$
$$- 2c^{-1} \left\{ \mathrm{Sp} \int \hat{\boldsymbol{v}}_1 \, \boldsymbol{\nu}_{1,\mathrm{orb}}^{(1)} \wedge \hat{\boldsymbol{v}}_1 \, \delta(\boldsymbol{X}_1 - \boldsymbol{R}) f_1(1; t) \mathrm{d}1 \right\} \cdot \boldsymbol{E}. \qquad (58)$$

Finally the power density due to the correlations in the system is given by

$$\Psi^{\mathrm{C}} = \mathrm{Sp}\int \{\psi^{\mathrm{C}}(1,2)+\boldsymbol{v}\cdot\boldsymbol{f}^{\mathrm{C}}(1,2)\}\delta(\boldsymbol{X}_1-\boldsymbol{R})c_2(1,2;t)\mathrm{d}1\,\mathrm{d}2, \tag{59}$$

where we employed the abbreviations $\boldsymbol{f}^{\mathrm{C}}$ (35) and

$$\begin{aligned}
\psi^{\mathrm{C}}(1,2) \equiv & -\sum_{n,m=0}^{\infty}\{\hat{\boldsymbol{v}}_1\cdot\boldsymbol{\nabla}_1\,\boldsymbol{\mu}_1^{(n)}\vdots\boldsymbol{\nabla}_1^n\,\boldsymbol{\mu}_2^{(m)}\vdots\boldsymbol{\nabla}_2^m \\
& +(\partial_{t\mathrm{P}}\,\boldsymbol{\mu}_1^{(n)})\vdots\boldsymbol{\nabla}_1^n\,\boldsymbol{\mu}_2^{(m)}\vdots\boldsymbol{\nabla}_2^m\}\,\frac{1}{4\pi|\boldsymbol{X}_1-\boldsymbol{X}_2|} \\
& +\sum_{n,m=1}^{\infty}[(\boldsymbol{\nabla}_1^{n-1}\vdots\boldsymbol{\nu}_1^{(n)}\wedge\boldsymbol{\nabla}_1)\cdot\{\boldsymbol{\nabla}_2^{m-1}\vdots(\partial_{t\mathrm{P}}\,\boldsymbol{\nu}_2^{(m)})\wedge\boldsymbol{\nabla}_1\} \\
& -\hat{\boldsymbol{v}}_2\cdot\boldsymbol{\nabla}_1(\boldsymbol{\nabla}_1^{n-1}\vdots\boldsymbol{\nu}_1^{(n)}\wedge\boldsymbol{\nabla}_1)\cdot(\boldsymbol{\nabla}_2^{m-1}\vdots\boldsymbol{\nu}_2^{(m)}\wedge\boldsymbol{\nabla}_1) \\
& -(\boldsymbol{v}_1-\boldsymbol{v}_2)\cdot\boldsymbol{\nabla}_1(\boldsymbol{\nabla}_1^{n-1}\vdots\boldsymbol{\nu}_{1,\mathrm{spin}}^{(n)}\wedge\boldsymbol{\nabla}_1)\cdot(\boldsymbol{\nabla}_2^{m-1}\vdots\boldsymbol{\nu}_2^{(m)}\wedge\boldsymbol{\nabla}_1) \\
& -\{\boldsymbol{\nabla}_1^{n-1}\vdots(\partial_{t\mathrm{P}}\,\boldsymbol{\nu}_{1,\mathrm{spin}}^{(n)})\wedge\boldsymbol{\nabla}_1\}\cdot(\boldsymbol{\nabla}_2^{m-1}\vdots\boldsymbol{\nu}_2^{(m)}\wedge\boldsymbol{\nabla}_1) \\
& -(\boldsymbol{\nabla}_1^{n-1}\vdots\boldsymbol{\nu}_{1,\mathrm{spin}}^{(n)}\wedge\boldsymbol{\nabla}_1)\cdot\{\boldsymbol{\nabla}_2^{m-1}\vdots(\partial_{t\mathrm{P}}\,\boldsymbol{\nu}_2^{(m)})\wedge\boldsymbol{\nabla}_1\}]\,\frac{1}{4\pi|\boldsymbol{X}_1-\boldsymbol{X}_2|}\,.
\end{aligned} \tag{60}$$

The short range power density follows from (IX.85). One finds

$$\Psi^{\mathrm{S}} = \mathrm{Sp}\int \{\psi^{\mathrm{S}}(1,2)+\boldsymbol{v}\cdot\boldsymbol{f}^{\mathrm{S}}(1,2)\}\delta(\boldsymbol{X}_1-\boldsymbol{R})f_2(1,2;t)\mathrm{d}1\,\mathrm{d}2 \tag{61}$$

with the abbreviations $\boldsymbol{f}^{\mathrm{S}}$ (37) and

$$\begin{aligned}
\psi^{\mathrm{S}}(1,2) \equiv \sum_{i,j}\Big\{ & -(\hat{\boldsymbol{v}}_1+\partial_{t\mathrm{P}}\,\boldsymbol{r}_{1i})\cdot\boldsymbol{\nabla}_{1i}+c^{-2}(\partial_{t\mathrm{P}}\,\boldsymbol{r}_{1i})\cdot(\partial_{t\mathrm{P}}\,\boldsymbol{r}_{2j})(\hat{\boldsymbol{v}}_2+\partial_{t\mathrm{P}}\,\boldsymbol{r}_{2j})\cdot\boldsymbol{\nabla}_{1i} \\
& -\frac{\hbar^2}{2m_{2j}c^2}(\hat{\boldsymbol{v}}_2+\partial_{t\mathrm{P}}\,\boldsymbol{r}_{2j})\cdot\boldsymbol{\nabla}_{1i}(\partial_{t\mathrm{P}}\,\boldsymbol{r}_{1i})\cdot(\boldsymbol{\sigma}_{2j}\wedge\boldsymbol{\nabla}_{1i}) \\
& +\frac{\hbar}{2m_{2j}c^2}(\hat{\boldsymbol{v}}_1+\partial_{t\mathrm{P}}\,\boldsymbol{r}_{1i})\cdot\boldsymbol{\nabla}_{1i}(\partial_{t\mathrm{P}}\,\boldsymbol{r}_{2j})\cdot(\boldsymbol{\sigma}_{2j}\wedge\boldsymbol{\nabla}_{1i}) \\
& -\frac{\hbar}{2m_{1i}c^2}(\hat{\boldsymbol{v}}_1+\partial_{t\mathrm{P}}\,\boldsymbol{r}_{1i})\cdot\boldsymbol{\nabla}_{1i}\,\partial_{t\mathrm{P}}(\boldsymbol{r}_{1i}-\boldsymbol{r}_{2j})\cdot(\boldsymbol{\sigma}_{1i}\wedge\boldsymbol{\nabla}_{1i}) \\
& -\frac{\hbar^2}{4m_{1i}m_{2j}c^2}(\hat{\boldsymbol{v}}_1+\partial_{t\mathrm{P}}\,\boldsymbol{r}_{1i})\cdot\boldsymbol{\nabla}_{1i}(\boldsymbol{\sigma}_{1i}\wedge\boldsymbol{\nabla}_{1i})\cdot(\boldsymbol{\sigma}_{2j}\wedge\boldsymbol{\nabla}_{1i})\Big\}\,\frac{e_{1i}e_{2j}}{4\pi|\boldsymbol{X}_{1i}-\boldsymbol{X}_{2j}|} \\
& -\psi^{\mathrm{C}}(1,2).
\end{aligned} \tag{62}$$

If the system is sufficiently homogeneous and if no long range correlations are present, one may employ an Irving–Kirkwood procedure and the identity (9) to write both the correlation and short range power density as the sum of a divergence and a time derivative. In this way one finds for the correlation power density

$$\Psi^{\mathrm{C}} = -\nabla\cdot(\boldsymbol{v}\varrho u^{\mathrm{C}}+\mathbf{P}^{\mathrm{C}}\cdot\boldsymbol{v}+\boldsymbol{J}_q^{\mathrm{C}})-\frac{\partial\varrho u^{\mathrm{C}}}{\partial t}. \tag{63}$$

Here $u^{\mathrm{C}}$ is the correlation contribution to the internal energy density

$$\varrho u^{\mathrm{C}} \equiv \mathrm{Sp}\int \varphi^{\mathrm{C}}(1,2)\delta\{\tfrac{1}{2}(\boldsymbol{X}_1+\boldsymbol{X}_2)-\boldsymbol{R}\}c_2(1,2;t)\mathrm{d}1\,\mathrm{d}2 \tag{64}$$

with the abbreviation

$$\begin{aligned}\varphi^{\mathrm{C}}(1,2) \equiv \{&\sum_{n,m=0}^{\infty}\boldsymbol{\mu}_1^{(n)}:\nabla_1^n\,\boldsymbol{\mu}_2^{(m)}:\nabla_2^m\\ &-\sum_{n,m=1}^{\infty}(\nabla_1^{n-1}:\boldsymbol{\nu}_{1,\mathrm{orb}}^{(n)}\wedge\nabla_1)\cdot(\nabla_2^{m-1}:\boldsymbol{\nu}_{2,\mathrm{orb}}^{(m)}\wedge\nabla_1)\\ &+\sum_{n,m=1}^{\infty}(\nabla_1^{n-1}:\boldsymbol{\nu}_{1,\mathrm{spin}}^{(n)}\wedge\nabla_1)\cdot(\nabla_2^{m-1}:\boldsymbol{\nu}_{2,\mathrm{spin}}^{(m)}\wedge\nabla_1)\}\,\frac{1}{8\pi|\boldsymbol{X}_1-\boldsymbol{X}_2|}.\end{aligned} \tag{65}$$

The correlation internal energy density (64) consists hence of three contributions: due to electric multipole–electric multipole, orbital magnetic multipole–orbital magnetic multipole and spin magnetic multipole–spin magnetic multipole contributions. No cross-terms between orbital and spin magnetic multipoles occur, in contrast with the situation for the correlation pressure (42) with (35) and the correlation part of the heat flow which is given by:

$$\begin{aligned}\boldsymbol{J}_q^{\mathrm{C}} \equiv \mathrm{Sp}\int &\{\tfrac{1}{2}(\hat{\boldsymbol{v}}_1+\hat{\boldsymbol{v}}_2)\varphi^{\mathrm{C}}(1,2)+\tfrac{1}{2}(\boldsymbol{X}_1-\boldsymbol{X}_2)\psi^{\mathrm{C}}(1,2)\}\\ &\delta\{\tfrac{1}{2}(\boldsymbol{X}_1+\boldsymbol{X}_2)-\boldsymbol{R}\}c_2(1,2;t)\mathrm{d}1\,\mathrm{d}2.\end{aligned} \tag{66}$$

Likewise we find for the short range power density

$$\Psi^{\mathrm{S}} = -\nabla\cdot(\boldsymbol{v}\varrho u^{\mathrm{S}}+\mathbf{P}^{\mathrm{S}}\cdot\boldsymbol{v}+\boldsymbol{J}_q^{\mathrm{S}})-\frac{\partial\varrho u^{\mathrm{S}}}{\partial t}, \tag{67}$$

where the short range contribution to the energy density is

$$\varrho u^{\mathrm{S}} \equiv \mathrm{Sp}\int \varphi^{\mathrm{S}}(1,2)\delta\{\tfrac{1}{2}(\boldsymbol{X}_1+\boldsymbol{X}_2)-\boldsymbol{R}\}f_2(1,2;t)\mathrm{d}1\,\mathrm{d}2 \tag{68}$$

with the abbreviation

$$\varphi^{S}(1,2) \equiv \sum_{i,j}\left\{1+c^{-2}(\partial_{tP}\boldsymbol{r}_{1i})\cdot(\partial_{tP}\boldsymbol{r}_{2j})+\frac{\hbar}{m_{1i}c^{2}}(\partial_{tP}\boldsymbol{r}_{1i})\cdot(\boldsymbol{\sigma}_{1i}\wedge\boldsymbol{\nabla}_{1i})\right.$$
$$\left.+\frac{\hbar^{2}}{4m_{1i}m_{2j}c^{2}}(\boldsymbol{\sigma}_{1i}\wedge\boldsymbol{\nabla}_{1i})\cdot(\boldsymbol{\sigma}_{2j}\wedge\boldsymbol{\nabla}_{1i})\right\}\frac{e_{1i}e_{2j}}{8\pi|\boldsymbol{X}_{1i}-\boldsymbol{X}_{2j}|}-\varphi^{C}(1,2) \quad (69)$$

and where the short range part of the heat flow is

$$\boldsymbol{J}_{q}^{S} \equiv \mathrm{Sp}\int\{\tfrac{1}{2}(\hat{\boldsymbol{v}}_{1}+\hat{\boldsymbol{v}}_{2})\varphi^{S}(1,2)+\tfrac{1}{2}(\boldsymbol{X}_{1}-\boldsymbol{X}_{2})\psi^{S}(1,2)\}\delta\{\tfrac{1}{2}(\boldsymbol{X}_{1}+\boldsymbol{X}_{2})-\boldsymbol{R}\}$$
$$f_{2}(1,2;t)\mathrm{d}1\,\mathrm{d}2. \quad (70)$$

Collecting the results, we obtain as the energy balance equation for a system with short range Wigner correlations as for instance a fluid of neutral atoms

$$\frac{\partial}{\partial t}(\tfrac{1}{2}\varrho\boldsymbol{v}^{2}+\boldsymbol{v}\cdot\boldsymbol{g}+\varrho u) = -\boldsymbol{\nabla}\cdot\{\boldsymbol{v}(\tfrac{1}{2}\varrho\boldsymbol{v}^{2}+\boldsymbol{v}\cdot\boldsymbol{g}+\varrho u)+\mathsf{P}\cdot\boldsymbol{v}+\boldsymbol{J}_{q}\}$$
$$+\frac{\partial\boldsymbol{P}}{\partial t}\cdot\boldsymbol{E}-\boldsymbol{M}\cdot\frac{\partial\boldsymbol{B}}{\partial t}+\frac{\partial}{\partial t}\{\boldsymbol{M}_{\mathrm{spin}}\cdot\boldsymbol{B}+2c^{-1}\boldsymbol{v}\cdot(\boldsymbol{E}\wedge\boldsymbol{M}_{\mathrm{orb}})\}$$
$$+\boldsymbol{\nabla}\cdot[\boldsymbol{v}\{\boldsymbol{P}\cdot\boldsymbol{E}+\boldsymbol{M}_{\mathrm{spin}}\cdot\boldsymbol{B}+2c^{-1}\boldsymbol{v}\cdot(\boldsymbol{E}\wedge\boldsymbol{M}_{\mathrm{orb}})\}]. \quad (71)$$

(The Joule heat term $\boldsymbol{J}\cdot\boldsymbol{E}$ of (55) is absent here because the atoms were taken to be neutral.) This energy balance contains the total specific internal energy

$$u = u^{K}+u^{F}+u^{C}+u^{S} \quad (72)$$

and the total heat flow

$$\boldsymbol{J}_{q} = \boldsymbol{J}_{q}^{K}+\boldsymbol{J}_{q}^{F}+\boldsymbol{J}_{q}^{C}+\boldsymbol{J}_{q}^{S}. \quad (73)$$

As compared to the non-relativistic theory (v. VII.65)) the left-hand side of (71) contains a new term $\boldsymbol{v}\cdot\boldsymbol{g}$ with the macroscopic velocity and the momentum density $\boldsymbol{g}$ (22) due to the intra-atomic fields. It also appears at the right-hand side in the transport term. Furthermore at the right-hand side two extra terms are present with the macroscopic spin magnetization $\boldsymbol{M}_{\mathrm{spin}}$ and moreover two terms with the macroscopic orbital magnetization $\boldsymbol{M}_{\mathrm{orb}}$, which are coupled to the magnetic and electric fields respectively. Other extra semi-relativistic terms are contained in the expressions for the pressure, the specific energy and the heat flow.

With the use of the Maxwell equations we may cast the balance equation (71) into the form of the following energy conservation law

$$\frac{\partial}{\partial t}\{\tfrac{1}{2}\varrho \boldsymbol{v}^2+\boldsymbol{v}\cdot\boldsymbol{g}+\varrho u+\tfrac{1}{2}\boldsymbol{E}^2+\tfrac{1}{2}\boldsymbol{B}^2-\boldsymbol{M}_{\text{spin}}\cdot\boldsymbol{B}-2c^{-1}\boldsymbol{v}\cdot(\boldsymbol{E}\wedge\boldsymbol{M}_{\text{orb}})\}$$
$$+\nabla\cdot[\boldsymbol{v}(\tfrac{1}{2}\varrho \boldsymbol{v}^2+\boldsymbol{v}\cdot\boldsymbol{g}+\varrho u)+\mathbf{P}\cdot\boldsymbol{v}+\boldsymbol{J}_q+c\boldsymbol{E}\wedge\boldsymbol{H}$$
$$-\boldsymbol{v}\{\boldsymbol{P}\cdot\boldsymbol{E}+\boldsymbol{M}_{\text{spin}}\cdot\boldsymbol{B}+2c^{-1}\boldsymbol{v}\cdot(\boldsymbol{E}\wedge\boldsymbol{M}_{\text{orb}})\}]=0. \quad (74)$$

The energy laws for plasmas and for systems with long range correlations may be treated along similar lines, as was done in non-relativistic classical theory. In the case of long range correlations one has to introduce mean correlation functions (v. chapter II, section 5*h*).

### d. *The angular momentum balance*

The macroscopic inner angular momentum law will follow from its atomic counterpart (IX.91). In the first place we take half the anticommutator of that equation with $\delta(\boldsymbol{X}_k-\boldsymbol{R})$. This procedure is followed since the Weyl-transformed atomic inner angular momentum (more precisely its spin contribution) does not commute with the delta function. Then one takes the sum over the atoms $k$ and averages with a Wigner function. As a result, one obtains after calculations of the type discussed earlier for the momentum and energy laws, for the balance equation of inner angular momentum

$$\frac{\partial \boldsymbol{S}}{\partial t}=-\nabla\cdot(\boldsymbol{v}\boldsymbol{S}+\mathbf{J}_s)-\boldsymbol{P}_{\text{A}}-\boldsymbol{v}\wedge\boldsymbol{g}+\boldsymbol{P}\wedge\boldsymbol{E}+\boldsymbol{M}\wedge\boldsymbol{B}+c^{-1}\boldsymbol{v}\wedge(\boldsymbol{P}\wedge\boldsymbol{B}-\boldsymbol{M}_{\text{orb}}\wedge\boldsymbol{E}). \quad (75)$$

Here the inner angular momentum density $\boldsymbol{S}$ is defined as

$$\boldsymbol{S}(\boldsymbol{R},t)=\text{Sp}\int\tfrac{1}{2}\{\boldsymbol{s}_1,\delta(\boldsymbol{X}_1-\boldsymbol{R})\}f_1(1;t)\text{d}1 \quad (76)$$

with the atomic inner angular momentum $\boldsymbol{s}_1$, which is the sum of the four quantities defined in (IX.87–90). At the right-hand side appears the divergence of a convection term $\boldsymbol{v}\boldsymbol{S}$ and an inner angular momentum flow $\mathbf{J}_s$, which may be expressed as a statistical average of atomic quantities. Furthermore as source terms one has in the first place the antisymmetric part $\boldsymbol{P}_{\text{A}}\equiv\boldsymbol{\epsilon}:\mathbf{P}$ (with $\boldsymbol{\epsilon}$ the Levi-Civita tensor) of the pressure tensor $\mathbf{P}$ (44); secondly the vector product of the macroscopic velocity $\boldsymbol{v}$ and the momentum density contribution $\boldsymbol{g}$ (22), and thirdly four terms with the polarizations and fields.

As compared to the non-relativistic equation (VII.70) with (VII.73) two new terms appear here: a term due to the spin momentum density and a semi-relativistic term due to orbital magnetic dipoles in motion. Moreover the quantities $\boldsymbol{S}$, $\mathbf{J}_s$, $\boldsymbol{P}_{\text{A}}$, $\boldsymbol{P}$ and $\boldsymbol{M}$ all contain spin contributions.

The form (75) of the inner angular momentum equation allows us to deduce the conservation law of total angular momentum. In fact the balance equation of orbital angular momentum $\boldsymbol{L}$, which follows from (46) by vector multiplication with $\boldsymbol{R}$, is:

$$\frac{\partial \boldsymbol{L}}{\partial t} = -\nabla\cdot(\boldsymbol{v}\boldsymbol{L}+\mathbf{J}_l)+\boldsymbol{P}_\mathrm{A}+\boldsymbol{v}\wedge\boldsymbol{g}-\boldsymbol{D}\wedge\boldsymbol{E}-\boldsymbol{B}\wedge\boldsymbol{H}-c^{-1}\boldsymbol{v}\wedge(\boldsymbol{P}\wedge\boldsymbol{B}-\boldsymbol{M}_\mathrm{orb}\wedge\boldsymbol{E}) \tag{77}$$

with the orbital angular momentum defined as

$$\boldsymbol{L}(\boldsymbol{R}, t) = \boldsymbol{R}\wedge\{\varrho\boldsymbol{v}+\boldsymbol{g}+c^{-1}\boldsymbol{E}\wedge(\boldsymbol{B}-\boldsymbol{M}_\mathrm{orb})\} \tag{78}$$

and the conduction part of the orbital angular momentum flow

$$\mathbf{J}_l = -\mathbf{P}\wedge\boldsymbol{R}-\boldsymbol{D}(\boldsymbol{R}\wedge\boldsymbol{E})-\boldsymbol{B}(\boldsymbol{R}\wedge\boldsymbol{H})-c^{-1}\boldsymbol{v}\boldsymbol{R}\wedge(\boldsymbol{D}\wedge\boldsymbol{B}) + \boldsymbol{\epsilon}\cdot\boldsymbol{R}(\tfrac{1}{2}E^2+\tfrac{1}{2}B^2-\boldsymbol{M}\cdot\boldsymbol{B}). \tag{79}$$

The total angular momentum law follows by adding (75) and (77):

$$\frac{\partial(\boldsymbol{L}+\boldsymbol{S})}{\partial t} = -\nabla\cdot\{\boldsymbol{v}(\boldsymbol{L}+\boldsymbol{S})+\mathbf{J}_l+\mathbf{J}_s\}, \tag{80}$$

which has indeed the form of a conservation law.

Laws for plasmas and systems with long range correlations may be derived in a similar way.

## 5 *The laws of thermodynamics*

### a. *The first law*

The first law of thermodynamics follows if one subtracts the momentum equation (43) multiplied by the macroscopic velocity $\boldsymbol{v}$ from the energy law (71). If one employs the notation $\mathrm{d}/\mathrm{d}t$ for the material time derivative $\partial/\partial t+\boldsymbol{v}\cdot\nabla$ and the mass conservation law (18), one gets thus

$$\varrho\frac{\mathrm{d}u}{\mathrm{d}t} = -\nabla\cdot\boldsymbol{J}_q-\tilde{\mathrm{P}}:\nabla\boldsymbol{v}-\boldsymbol{g}\cdot\frac{\mathrm{d}\boldsymbol{v}}{\mathrm{d}t}+\varrho\frac{\mathrm{d}(v\boldsymbol{P}')}{\mathrm{d}t}\cdot\boldsymbol{E}'-\boldsymbol{M}'\cdot\frac{\mathrm{d}\boldsymbol{B}'}{\mathrm{d}t} + \varrho\frac{\mathrm{d}}{\mathrm{d}t}(v\boldsymbol{M}'_\mathrm{spin}\cdot\boldsymbol{B}')-c^{-1}\frac{\mathrm{d}\boldsymbol{v}}{\mathrm{d}t}\cdot(\boldsymbol{E}'\wedge\boldsymbol{M}'_\mathrm{spin}), \tag{81}$$

where $v$ is the specific volume and where the tilde indicates the transposed of a tensor. The electromagnetic quantities are all counted in the local rest frame:

$$\begin{aligned} \boldsymbol{E}' &= \boldsymbol{E}+c^{-1}\boldsymbol{v}\wedge\boldsymbol{B}, & \boldsymbol{B}' &= \boldsymbol{B}-c^{-1}\boldsymbol{v}\wedge\boldsymbol{E}, \\ \boldsymbol{P}' &= \boldsymbol{P}-c^{-1}\boldsymbol{v}\wedge\boldsymbol{M}, & \boldsymbol{M}' &= \boldsymbol{M}+c^{-1}\boldsymbol{v}\wedge\boldsymbol{P}. \end{aligned} \tag{82}$$

($\boldsymbol{M}_{\text{spin}}$ could be replaced by $\boldsymbol{M}'_{\text{spin}}$ since the difference is of order $c^{-3}$.)

The law (81) is valid for fluid systems of neutral atoms. Amorphous or polycrystalline solids, plasmas and systems with long range correlations may be treated in a similar way.

b. *The second law*

To derive the second law for a fluid system of neutral atoms in which only short range Wigner correlations are present we shall apply the canonical ensemble method to a nearly uniform sample of ellipsoidal shape which is at rest in a large polarized system. Then the external fields due to the surroundings of the sample are uniform and given by (II.220) with (II.219). The Hamilton operator $H_{\text{op}}$ which is to be used, is derived in appendix I and given by (A1) with (A9), with atomic charges $e_k = 0$. We denote its Weyl transform as $H(1, \ldots, N)$.

The partial derivatives of the free energy $F^*$ with respect to the external fields may be expressed as averages involving the partial derivatives of the Weyl transform of the Hamiltonian:

$$\begin{aligned} \frac{\partial F^*}{\partial \boldsymbol{E}_{\text{e}}} &= \text{Sp}\int \frac{\partial H}{\partial \boldsymbol{E}_{\text{e}}}\,\rho(1, \ldots, N)\text{d}1 \ldots \text{d}N, \\ \frac{\partial F^*}{\partial \boldsymbol{B}_{\text{e}}} &= \text{Sp}\int \frac{\partial H}{\partial \boldsymbol{B}_{\text{e}}}\,\rho(1, \ldots, N)\text{d}1 \ldots \text{d}N, \end{aligned} \tag{83}$$

where $\rho(1, \ldots, N)$ is the Wigner function of the canonical ensemble. From the explicit form of the Weyl transform of the Hamilton operator (A1) with (A9) it follows that

$$\begin{aligned} \frac{\partial F^*}{\partial \boldsymbol{E}_{\text{e}}} &= -\text{Sp}\int \left(\bar{\boldsymbol{\mu}}^{(1)}_{1,\text{orb}} - \tfrac{1}{2}c^{-1}\boldsymbol{v}^{(1)}_{1,\text{spin}} \wedge \frac{\boldsymbol{P}_1}{m_1} + \tfrac{1}{2}\boldsymbol{\mu}^{(1)}_{1,\text{spin}}\right) f_1(1)\text{d}1, \\ \frac{\partial F^*}{\partial \boldsymbol{B}_{\text{e}}} &= -\text{Sp}\int \left\{\boldsymbol{v}^{(1)}_1 + \tfrac{1}{2}c^{-1}\boldsymbol{\mu}^{(1)}_{1,\text{orb}} \wedge \frac{\boldsymbol{P}_1}{m_1} - \tfrac{1}{2}c^{-1}(\partial_{t\text{P}}\,\boldsymbol{\mu}^{(1)}_{1,\text{orb}}) \wedge \boldsymbol{R}_1\right\} f_1(1)\text{d}1. \end{aligned} \tag{84}$$

The bar in the first of these formulae denotes an orbital electric dipole moment that is defined in a non-relativistic fashion, i.e. in terms of the coordinates $\boldsymbol{R}_{ki}$ of the constituent particles (v. appendix I). This non-relativistic dipole moment may be expressed in terms of the semi-relativistic ones by means of the identity (A10). Then the first relation of (84) gets the form

$$\frac{\partial F^*}{\partial \boldsymbol{E}_e} = -\mathrm{Sp}\int \left(\boldsymbol{\mu}_1^{(1)} - c^{-1}\boldsymbol{\nu}_1^{(1)} \wedge \frac{\boldsymbol{P}_1}{m_1}\right) f_1(1)\mathrm{d}1 = -V\boldsymbol{P}, \tag{85}$$

where the uniformity of the system has been taken into account, so that the volume $V$ times the polarization $\boldsymbol{P}$ appears. The second relation of (84) may be transformed if one uses the stationarity of the canonical ensemble in the same way as has been done in the non-relativistic treatment (v. (VII.89)). Then it gets the form:

$$\frac{\partial F^*}{\partial \boldsymbol{B}_e} = -\mathrm{Sp}\int \left(\boldsymbol{\nu}_1^{(1)} + c^{-1}\boldsymbol{\mu}_1^{(1)} \wedge \frac{\boldsymbol{P}_1}{m_1}\right) f_1(1)\mathrm{d}1 = -V\boldsymbol{M} \tag{86}$$

with $\boldsymbol{M}$ the macroscopic magnetization. (A term with spin electric dipole moments $\boldsymbol{\mu}_{1,\mathrm{spin}}^{(1)}$ in motion is added for reasons of elegance, although it leads to a term of order $c^{-3}$ only.)

Since the change of free energy depends on changes in temperature, external fields and position of the boundary (specified by the deformation tensor $\delta\boldsymbol{\epsilon}$, see for instance (VII.92)), one may write

$$\delta F^* = -S\delta T - V\boldsymbol{P}\cdot\delta\boldsymbol{E}_e - V\boldsymbol{M}\cdot\delta\boldsymbol{B}_e + \mathbf{A} : \delta\boldsymbol{\epsilon}, \tag{87}$$

where $S = -\partial F^*/\partial T$ is the entropy. The second law will follow by establishing the statistical expressions for the free energy $F^*$ and the tensor $\mathbf{A}$.

The free energy may be obtained from the average of the Hamilton operator of the system. From the expression (A6) with (A11) for the Hamiltonian one finds in semi-relativistic approximation, by comparison with the expressions (52), (57), (64) and (68) for the various parts of the internal energy, for the average Hamiltonian $\langle H \rangle \equiv U^*$:

$$U^* = U + V\{\tfrac{1}{2}(\boldsymbol{PP} - \boldsymbol{MM}) : \mathbf{L} - \boldsymbol{P}\cdot\boldsymbol{E}_e + \tfrac{1}{2}\boldsymbol{M}^2 - \boldsymbol{M}_{\mathrm{spin}}\cdot(\boldsymbol{B}_e + \boldsymbol{M}) + \boldsymbol{M}_{\mathrm{spin}}\boldsymbol{M} : \mathbf{L}\}. \tag{88}$$

Here $U$ is the internal energy $V\varrho u$ with $\varrho$ the mass density and $\mathbf{L}$ the depolarizing tensor defined in (II.219).

Finally we consider the tensor $\mathbf{A}$, which is given in the appendix II (A23) as the statistical expression

$$\mathbf{A} = -\mathrm{Sp}\int \sum_k \partial_{t\mathrm{P}}(\boldsymbol{X}_k \boldsymbol{P}_k)\rho(1, \ldots, N)\mathrm{d}1 \ldots \mathrm{d}N, \tag{89}$$

where $\partial_{t\mathrm{P}}$ denotes the Poisson bracket with the Weyl transform $H$ of the Hamiltonian (A1) with (A9). With $\boldsymbol{v}_k = \partial_{t\mathrm{P}} \boldsymbol{X}_k$ we may write (89) as

$$\mathbf{A} = -\mathrm{Sp} \int \sum_k (\boldsymbol{v}_k \boldsymbol{P}_k + \boldsymbol{X}_k \partial_{t\mathrm{P}} \boldsymbol{P}_k) \rho(1, \ldots, N) \mathrm{d}1 \ldots \mathrm{d}N. \tag{90}$$

The velocity $\boldsymbol{v}_k$ may be calculated with the help of the explicit expressions for $\boldsymbol{X}_k$ and $H$ (v. (IX.40)). One finds then, in semi-relativistic approximation,

$$\begin{aligned} \boldsymbol{P}_k = {} & m_k \boldsymbol{v}_k + \boldsymbol{g}_k - \tfrac{1}{2} c^{-1} \boldsymbol{\mu}_k^{(1)} \wedge \boldsymbol{B}_\mathrm{e} + c^{-1} \boldsymbol{v}_{k,\mathrm{orb}}^{(1)} \wedge \boldsymbol{E}_\mathrm{e} \\ & + c^{-2} \sum_{l(\neq k)} \sum_{i,j} \{ (\partial_{t\mathrm{P}} \boldsymbol{r}_{lj}) \cdot \mathbf{T}(\boldsymbol{R}_{ki} - \boldsymbol{R}_{lj}) + 2 \boldsymbol{r}_{ki} (\partial_{t\mathrm{P}} \boldsymbol{r}_{ki}) \cdot \nabla_{ki} \} \frac{e_{ki} e_{lj}}{8\pi |\boldsymbol{R}_{ki} - \boldsymbol{R}_{lj}|} \\ & - c^{-2} \sum_{l(\neq k)} \sum_{i,j}{}' \frac{\hbar}{2m_{lj}} \boldsymbol{\sigma}_{lj} \wedge \nabla_{ki} \frac{e_{ki} e_{lj}}{4\pi |\boldsymbol{R}_{ki} - \boldsymbol{R}_{lj}|}, \end{aligned} \tag{91}$$

where $\boldsymbol{g}_k$ has been given in (IX.69). The quantity $\partial_{t\mathrm{P}} \boldsymbol{P}_k$ in the second term of (90) is equal to $-\partial H / \partial \boldsymbol{R}_k$ and may hence be evaluated explicitly. In this way one finds for the tensor $\mathbf{A}$ in semi-relativistic approximation, by comparing with the expressions (25), (33), (40) and (42) for the pressure tensor,

$$\mathbf{A} = -V \{ \mathbf{P} + \tfrac{1}{2} \mathbf{K} : (\boldsymbol{PP} + \boldsymbol{MM}) - \tfrac{1}{2} \boldsymbol{M}^2 \mathbf{U} \}, \tag{92}$$

where $\mathbf{K}$ is the tensor defined in (II.236). In the course of the derivation of (92) one needs to apply the stationarity of the canonical ensemble.

Substituting the expression (88) for $U^*$, which is equal to $F^* + TS$, and the expression (92) for the tensor $\mathbf{A}$ into the relation (87) one finds, upon dividing through the conserved total mass $\varrho V$ of the system:

$$\begin{aligned} & \delta [u + v \{ \tfrac{1}{2} (\boldsymbol{PP} - \boldsymbol{MM}) : \mathbf{L} + \tfrac{1}{2} \boldsymbol{M}^2 - \boldsymbol{M}_\mathrm{spin} \cdot (\boldsymbol{B}_\mathrm{e} + \boldsymbol{M}) + \boldsymbol{M}_\mathrm{spin} \boldsymbol{M} : \mathbf{L} \} ] \\ & \quad = T \delta s + \boldsymbol{E}_\mathrm{e} \cdot \delta (v \boldsymbol{P}) - v \boldsymbol{M} \cdot \delta \boldsymbol{B}_\mathrm{e} \\ & \qquad\qquad - v \{ \mathbf{P} + \tfrac{1}{2} \mathbf{K} : (\boldsymbol{PP} + \boldsymbol{MM}) - \tfrac{1}{2} \boldsymbol{M}^2 \mathbf{U} \} : \delta \boldsymbol{\epsilon}, \end{aligned} \tag{93}$$

where $v$ and $s$ are the specific volume and entropy.

If one makes use of the relation (II.220) one may eliminate the external fields in favour of the Maxwell fields. With the help of (II.246) one finds then for the change of the entropy the relation

$$T \delta s = \delta u + v \mathbf{P} : \delta \boldsymbol{\epsilon} - \boldsymbol{E} \cdot \delta (v \boldsymbol{P}) + v \boldsymbol{M} \cdot \delta \boldsymbol{B} - \delta (v \boldsymbol{M}_\mathrm{spin} \cdot \boldsymbol{B}). \tag{94}$$

One notices that the specific internal energy appears in combination with a magnetic energy term $-v \boldsymbol{M}_\mathrm{spin} \cdot \boldsymbol{B}$, just as in the first law (81).

For an isotropic fluid one may argue that under certain conditions (v. chapter II, section 7*b*) the pressure tensor $\mathbf{P}$ reduces in equilibrium to a multiple of the unit tensor $p\mathsf{U}$ with $p$ the (scalar) pressure. Then we may finally write (94) for such systems in the form of a Gibbs relation

$$T\mathrm{d}s = \mathrm{d}(u - v\boldsymbol{M}_{\mathrm{spin}}\cdot\boldsymbol{B}) + p\,\mathrm{d}v - \boldsymbol{E}\cdot\mathrm{d}(v\boldsymbol{P}) + v\boldsymbol{M}\cdot\mathrm{d}\boldsymbol{B}, \tag{95}$$

where we have written differentials instead of variations, since only state variables are involved now.

The derivation of the second law for amorphous or polycrystalline solids, neutral plasmas and systems with long range correlations such as crystalline solids may be performed in a way similar to that given above for fluid systems with short range correlations. In the following we shall indicate which new features arise for the case with long range correlations. Just as in the non-relativistic treatment of chapter II, we consider a sample at rest without macroscopic charge and current densities although we allow for the atoms carrying electric charges. Under those circumstances the system is characterized by a free energy with partial derivatives with respect to the external fields

$$\begin{aligned}\frac{\partial F^*}{\partial \boldsymbol{E}_{\mathrm{e}}} &= -\int \boldsymbol{P}(\boldsymbol{R})\mathrm{d}\boldsymbol{R} \equiv -V\overline{\boldsymbol{P}},\\ \frac{\partial F^*}{\partial \boldsymbol{B}_{\mathrm{e}}} &= -\int \boldsymbol{M}(\boldsymbol{R})\mathrm{d}\boldsymbol{R} \equiv -V\overline{\boldsymbol{M}}.\end{aligned} \tag{96}$$

The right-hand sides are written as integrals (or in terms of volume averages (II.280) indicated by bars over the symbols) since the polarizations of the system are not necessarily uniform in the present case.

For the average Hamiltonian $U^* \equiv \langle H \rangle$ we find, using again the fact that the charge density vanishes,

$$\begin{aligned}U^* = U - \int \{&\boldsymbol{P}(\boldsymbol{R}_1)\boldsymbol{P}(\boldsymbol{R}_2) - \boldsymbol{M}(\boldsymbol{R}_1)\boldsymbol{M}(\boldsymbol{R}_2) + 2\boldsymbol{M}(\boldsymbol{R}_1)\boldsymbol{M}_{\mathrm{spin}}(\boldsymbol{R}_2)\}\\ &: \nabla_1\nabla_1 \frac{1}{8\pi|\boldsymbol{R}_1 - \boldsymbol{R}_2|}\,\mathrm{d}\boldsymbol{R}_1\,\mathrm{d}\boldsymbol{R}_2 - V\overline{\boldsymbol{P}}\cdot\boldsymbol{E}_{\mathrm{e}} + \tfrac{1}{2}V\overline{\boldsymbol{M}^2}\\ &- V\overline{\boldsymbol{M}}_{\mathrm{spin}}\cdot\boldsymbol{B}_{\mathrm{e}} - V\overline{\boldsymbol{M}_{\mathrm{spin}}\cdot\boldsymbol{M}}.\end{aligned} \tag{97}$$

Here $U$ is the total internal energy $\int \varrho u\,\mathrm{d}V$, with $\varrho u$ the internal energy density, which occurs in the local energy balance equation for a system with long range correlations. (This balance equation has the same form as that for systems with short range correlations, v. (71), with the only difference that the correlation contributions to the heat flow and the pressure tensor contain mean correlation functions. A similar situation arose already in the classical non-relativistic theory.)

For the change $\delta_\varepsilon F^*$ of the free energy under deformations one finds by a reasoning analogous to that which led to formula (II.289) of the non-relativistic treatment of chapter II the result

$$\delta_\varepsilon F^* = -\int \mathbf{P}(\mathbf{R}) : \delta\mathbf{e}(\mathbf{R})\mathrm{d}\mathbf{R}$$

$$-\int \{\mathbf{R}_1 \cdot \delta\tilde{\boldsymbol{\epsilon}}(\mathbf{R}_1) - \mathbf{R}_2 \cdot \delta\tilde{\boldsymbol{\epsilon}}(\mathbf{R}_2)\} \cdot [\mathbf{P}(\mathbf{R}_1)\mathbf{P}(\mathbf{R}_2) : \nabla_1 \nabla_1$$

$$-\{\mathbf{M}(\mathbf{R}_1) \wedge \nabla_1\} \cdot \{\mathbf{M}(\mathbf{R}_2) \wedge \nabla_1\}]\nabla_1 \frac{1}{8\pi|\mathbf{R}_1 - \mathbf{R}_2|} \mathrm{d}\mathbf{R}_1 \,\mathrm{d}\mathbf{R}_2 \quad (98)$$

with $\delta\mathbf{e}$ the deformation gradient tensor (II.A51). As compared to the classical non-relativistic result (II.289) an additional term quadratic in the magnetization appears here. It may be written in the alternative form

$$-\int \{\mathbf{R}_1 \cdot \delta\tilde{\boldsymbol{\epsilon}}(\mathbf{R}_1) - \mathbf{R}_2 \cdot \delta\tilde{\boldsymbol{\epsilon}}(\mathbf{R}_2)\} M(\mathbf{R}_1)M(\mathbf{R}_2) \vdots \nabla_1 \nabla_1 \nabla_1 \frac{1}{8\pi|\mathbf{R}_1 - \mathbf{R}_2|} \mathrm{d}\mathbf{R}_1 \,\mathrm{d}\mathbf{R}_2$$

$$+\tfrac{1}{2}\int \{\mathrm{Tr}\,\delta\mathbf{e}(\mathbf{R}_1)\}\{M(\mathbf{R}_1)\}^2 \mathrm{d}\mathbf{R}_1. \quad (99)$$

The entropy law that follows from (96–99) reads:

$$T\delta S = \delta U + \int \mathbf{P}(\mathbf{R}) : \delta\mathbf{e}(\mathbf{R})\mathrm{d}\mathbf{R} - \mathbf{E}_\mathrm{e} \cdot \delta(V\overline{\mathbf{P}})$$

$$+ V\overline{\mathbf{M}} \cdot \delta\mathbf{B}_\mathrm{e} - \delta \left[ \int \{\mathbf{P}(\mathbf{R}_1)\mathbf{P}(\mathbf{R}_2) - \mathbf{M}(\mathbf{R}_1)\mathbf{M}(\mathbf{R}_2) \right.$$

$$+ 2M(\mathbf{R}_1)M_\mathrm{spin}(\mathbf{R}_2)\} : \nabla_1 \nabla_1 \frac{1}{8\pi|\mathbf{R}_1 - \mathbf{R}_2|} \mathrm{d}\mathbf{R}_1 \,\mathrm{d}\mathbf{R}_2$$

$$\left. -\tfrac{1}{2}V\overline{\mathbf{M}^2} + V\overline{\mathbf{M}_\mathrm{spin} \cdot (\mathbf{B}_\mathrm{e} + \mathbf{M})} \right]$$

$$+ \int \{\mathbf{R}_1 \cdot \delta\tilde{\boldsymbol{\epsilon}}(\mathbf{R}_1) - \mathbf{R}_2 \cdot \delta\tilde{\boldsymbol{\epsilon}}(\mathbf{R}_2)\}\{\mathbf{P}(\mathbf{R}_1)\mathbf{P}(\mathbf{R}_2) + \mathbf{M}(\mathbf{R}_1)\mathbf{M}(\mathbf{R}_2)\} \vdots \nabla_1 \nabla_1 \nabla_1$$

$$\frac{1}{8\pi|\mathbf{R}_1 - \mathbf{R}_2|} \mathrm{d}\mathbf{R}_1 \,\mathrm{d}\mathbf{R}_2 - \tfrac{1}{2}V\overline{(\mathrm{Tr}\,\delta\mathbf{e})\mathbf{M}^2}. \quad (100)$$

Along similar lines as followed in chapter II to obtain (II.302) from (II.296) one may introduce the Maxwell fields instead of the external fields, using (II.220). Then one gets for the entropy law

$$T\delta S = \delta U + V\overline{\mathbf{P} : \delta\mathbf{e}} - V\overline{\mathbf{E} \cdot \delta\mathbf{P}} + V\overline{\mathbf{M} \cdot \delta\mathbf{B}} - V\overline{\delta(\mathbf{M}_\mathrm{spin} \cdot \mathbf{B})}$$

$$- V\overline{(\mathrm{Tr}\,\delta\mathbf{e})(\mathbf{P} \cdot \mathbf{E} + \mathbf{M}_\mathrm{spin} \cdot \mathbf{B})}, \quad (101)$$

again with a bar notation for volume averages. By performing a partial integration in the integral represented by the second term at the right-hand side this equation may be written in the form (cf. (II.304)):

$$T\delta S = \delta U + \int^{S} \boldsymbol{n}\cdot\{\mathbf{P}\cdot\delta\boldsymbol{\epsilon}\cdot\boldsymbol{R} - \delta\boldsymbol{\epsilon}\cdot\boldsymbol{R}(\boldsymbol{P}\cdot\boldsymbol{E} + \boldsymbol{M}_{\text{spin}}\cdot\boldsymbol{B})\}\mathrm{d}S - \int \{\boldsymbol{E}\cdot\delta_0\boldsymbol{P} - \boldsymbol{M}\cdot\delta_0\boldsymbol{B} + \delta_0(\boldsymbol{M}_{\text{spin}}\cdot\boldsymbol{B})\}\mathrm{d}\boldsymbol{R}, \quad (102)$$

where we used the equation of motion (43) which reads in the present case

$$\nabla\cdot\mathbf{P} = (\nabla\boldsymbol{E})\cdot\boldsymbol{P} + (\nabla\boldsymbol{B})\cdot\boldsymbol{M}. \quad (103)$$

If the system is not rotated and if it is only slightly deformed, one may replace $\delta\boldsymbol{\epsilon}$ by the change $\delta\boldsymbol{\eta}$ of state variable $\boldsymbol{\eta}$ according to (II.311). Furthermore, if the system is chosen to have ellipsoidal shape, the uniform external fields cause polarizations and Maxwell fields that are nearly uniform. Then one finds from the entropy law (102) the Gibbs relation

$$T\mathrm{d}S = \mathrm{d}(U - V\boldsymbol{M}_{\text{spin}}\cdot\boldsymbol{B}) + \int^{S} \boldsymbol{n}\cdot\mathbf{P}(\boldsymbol{R})\cdot\mathrm{d}\boldsymbol{\eta}(\boldsymbol{R})\cdot\boldsymbol{R}\,\mathrm{d}S - \boldsymbol{E}\cdot\mathrm{d}(V\boldsymbol{P}) + V\boldsymbol{M}\cdot\mathrm{d}\boldsymbol{B}, \quad (104)$$

where we have written differentials since now the (uniform) Maxwell fields and the polarizations may be considered as state variables. Comparison with (II.321) shows that the only difference is that a magnetic spin term is present in the Gibbs relation (104) (v. also (95) for fluids).

## 6 *Applications*

On the basis of the energy–momentum equations and the laws of thermodynamics, derived in the preceding in the semi-relativistic approximation for systems with spin, one may derive now expressions for the pressure in the presence and in the absence of electromagnetic fields, i.e. the Kelvin and Helmholtz pressures (cf. chapter II, section 8). A special application of the general theory, namely the calculation of the magnetostriction in a simple model for a crystalline solid, will be treated in the following. In chapter II it was shown that the total electrostriction could be split into a so-called 'form effect' due to the outward and Liénard pressures and the proper electrostrictive phenomenon. A similar splitting may be made in the case of magnetostriction. Just as in (II.239–241) one may find the Liénard pressure, which gives the difference between the pressure just inside and just outside the boundary of a magnetized system. One finds

$$\boldsymbol{n}\cdot(\mathbf{P}-\mathbf{P}_{\text{out}}) = -\tfrac{1}{2}\boldsymbol{n}(\boldsymbol{M}\cdot\boldsymbol{n})^2+\tfrac{1}{2}\boldsymbol{n}M^2. \tag{105}$$

From this expression one obtains for the normal component of the pressure tensor $\mathbf{P}_0$ in the absence of fields, but with the same deformation at the surface and at the same temperature

$$\boldsymbol{n}\cdot\mathbf{P}_0(\boldsymbol{R}) = \boldsymbol{n}\cdot\mathbf{P}_{\text{out}}(\boldsymbol{R})-\tfrac{1}{2}\boldsymbol{n}(\boldsymbol{M}\cdot\boldsymbol{n})^2-\tfrac{1}{2}\boldsymbol{B}\boldsymbol{B}:\boldsymbol{\chi}_1(\boldsymbol{R},T)-\tfrac{1}{2}\boldsymbol{n}\boldsymbol{M}\cdot\boldsymbol{H} \tag{106}$$

(cf. (II.390)), for a uniformly magnetized ellipsoid of which the magnetization fulfils a linear relation (II.384) with (II.385). The magnetization of a paramagnetic substance is described by such a law, at least in a certain approximation. The first two terms of (106) are conventionally called the pressure corresponding to the form effect, while the (uniform) proper magnetostriction is determined by the pressure tensor

$$-\tfrac{1}{2}\boldsymbol{B}\boldsymbol{B}:\hat{\boldsymbol{\chi}}_1(T)-\tfrac{1}{2}\boldsymbol{M}\cdot\boldsymbol{H}\mathbf{U}, \tag{107}$$

where (II.391) has been used. The central quantity that is to be calculated is $\hat{\boldsymbol{\chi}}_1(T)$, which fulfils, according to its definition, the relation

$$\chi^{ij}(\boldsymbol{\eta},T) = \chi_0^{ij}(T)+\hat{\chi}_1^{ijlk}(T)\eta_{kl} \tag{108}$$

for uniform and small deformations, determined by $\eta_{kl}$.

Let us study the proper magnetostriction for the following model[1] of a magnetizable crystalline solid: spin magnetic dipoles $\boldsymbol{v}_k$ are situated on the lattice points $k$ of a simple cubic lattice. The Hamilton operator for such a system in a uniform time-independent, external field $\boldsymbol{B}_e$ follows from (A1) with (A9). Due to the combined action of the Pauli exclusion principle and the Coulomb interactions an effective 'exchange coupling' between spin magnetic moments of neighbouring atoms arises, so that the total effective spin Hamilton operator has the form

$$H_{\text{op}}(1,\ldots,N) \rightleftarrows -\sum_{k,l(k\neq l)}(\boldsymbol{v}_k\wedge\boldsymbol{\nabla}_k)\cdot(\boldsymbol{v}_l\wedge\boldsymbol{\nabla}_l)\frac{1}{8\pi|\boldsymbol{R}_k-\boldsymbol{R}_l|} +\tfrac{1}{2}\sum_{k,l(k\neq l)}v_{kl}\,\boldsymbol{v}_k\cdot\boldsymbol{v}_l-\sum_k\boldsymbol{v}_k\cdot\boldsymbol{B}_e \tag{109}$$

with $v_{kl}$ the exchange interaction which is different from zero for neighbouring atoms $k$ and $l$ only.

[1] L. Néel, J. Phys. Radium **15**(1954)225.

The magnetization of such a spin system will be calculated now in the high temperature limit[1], i.e. by writing first a series expansion of the energy in powers of $\beta = (kT)^{-1}$:

$$F = -\beta^{-1} \ln \{\mathrm{Tr}\, \mathrm{e}^{-\beta H_{\mathrm{op}}}\} = -\beta^{-1} \ln \{\mathrm{Tr}\, (I - \beta H_{\mathrm{op}} + \tfrac{1}{2}\beta^2 H_{\mathrm{op}}^2 + \ldots)\}. \quad (110)$$

Here Tr $I$ is the trace of the unit operator in Hilbert space; it is equal to the sum $(2s+1)N$ of all states ($s$ is the spin of the atoms). From the form (109) of the Hamilton operator it follows that its trace vanishes, so that one may write for the free energy (110) up to order $\beta^2$:

$$F = -\beta^{-1} \ln (\mathrm{Tr}\, I) - \tfrac{1}{2}\beta \frac{\mathrm{Tr}\, (H_{\mathrm{op}}^2)}{\mathrm{Tr}\, I} + \tfrac{1}{6}\beta^2 \frac{\mathrm{Tr}\, (H_{\mathrm{op}}^3)}{\mathrm{Tr}\, I}. \quad (111)$$

The field dependent part of the second term is equal to

$$-\tfrac{1}{6}\beta N \nu^2 \boldsymbol{B}_{\mathrm{e}}^2 \quad (112)$$

with $\nu \equiv (e\hbar/mc)\{s(s+1)\}^{\frac{1}{2}}$ the magnitude of the atomic magnetic dipole moment (with $g = 2$, since the orbital magnetic moments do not play a role here). Furthermore the field dependent part of the third term of (111) is

$$-\tfrac{1}{18}\beta^2 \nu^4 \sum_{k,l(k \neq l)} \left\{ (\boldsymbol{B}_{\mathrm{e}} \wedge \boldsymbol{\nabla}_k) \cdot (\boldsymbol{B}_{\mathrm{e}} \wedge \boldsymbol{\nabla}_l) \frac{1}{4\pi |\boldsymbol{R}_k - \boldsymbol{R}_l|} - v_{kl} \boldsymbol{B}_{\mathrm{e}}^2 \right\}. \quad (113)$$

For an ellipsoidal system the sum over $l$ for fixed $k$ may be split into a part that is a sum over the lattice points inside a sphere around $k$ (containing many atoms, but small compared to the system as a whole) and a remaining term which may be approximated by an integral. Then one finds for (113)

$$\tfrac{1}{18}\beta^2 \frac{N^2}{V} \nu^4 (\mathbf{L} - \tfrac{1}{3}\mathbf{U} + \mathbf{S}_1) : \boldsymbol{B}_{\mathrm{e}} \boldsymbol{B}_{\mathrm{e}}, \quad (114)$$

where $\mathbf{L}$ is the depolarizing tensor (II.219) and where we employed the (approximate) uniformity of the ellipsoidal system. Furthermore we introduced the lattice sum $\mathbf{S}_1$ defined as:

$$\mathbf{S}_1 \equiv -\frac{V}{N} \sum_{l(\neq k)}^{\mathrm{sph}} \left( \boldsymbol{\nabla}_k \boldsymbol{\nabla}_k \frac{1}{4\pi |\boldsymbol{R}_k - \boldsymbol{R}_l|} - v_{kl} \mathbf{U} \right). \quad (115)$$

(For undeformed cubic lattices the first term in this lattice sum does not contribute. However since we shall consider deformations in the following, it may not be suppressed here.)

[1] J. H. Van Vleck, J. Chem. Phys. **5**(1937)320.

The magnetization up to order $\beta^2$ follows from the field dependent part of the free energy, given by the sum of (112) and (114):

$$\boldsymbol{M} = -\frac{1}{V}\frac{\partial F}{\partial \boldsymbol{B}_e} = \tfrac{1}{3}\beta \frac{N}{V} v^2 \boldsymbol{B}_e - \tfrac{1}{9}\beta^2 \left(\frac{N}{V}\right)^2 v^4 (\mathbf{L} - \tfrac{1}{3}\mathbf{U} + \mathbf{S}_1)\cdot \boldsymbol{B}_e . \tag{116}$$

Introducing the Maxwell field $\boldsymbol{B}$ by means of (II.220) one finds up to order $\beta^2$:

$$\boldsymbol{M} = \boldsymbol{\chi}\cdot\boldsymbol{B} \tag{117}$$

with the susceptibility tensor

$$\boldsymbol{\chi}(T) = \tfrac{1}{3}\beta \frac{N}{V} v^2 \mathbf{U} - \tfrac{2}{27}\beta^2 \left(\frac{N}{V}\right)^2 v^4 \mathbf{U} - \tfrac{1}{9}\beta^2 \left(\frac{N}{V}\right)^2 v^4 \mathbf{S}_1 . \tag{118}$$

To find the magnetostriction we have to calculate the change of this susceptibility tensor under uniform deformations $\delta\boldsymbol{R} = \delta\boldsymbol{\eta}\cdot\boldsymbol{R}$ with a symmetric tensor $\delta\boldsymbol{\eta}$. The change of volume is then given by $\delta V = V \operatorname{Tr} \delta\boldsymbol{\eta}$. Furthermore the change of the lattice sum $\mathbf{S}_1$ (115) under such deformations may be written as

$$\delta\left(\frac{N}{V}\mathbf{S}_1\right) = -\frac{N}{V}(\delta\mathbf{L})^{\mathrm{sph}} - \sum_{l(\neq k)}^{\mathrm{sph}} \left\{(\boldsymbol{R}_k - \boldsymbol{R}_l)\cdot\delta\boldsymbol{\eta}\cdot\boldsymbol{\nabla}_k \boldsymbol{\nabla}_k \boldsymbol{\nabla}_k \frac{1}{4\pi|\boldsymbol{R}_k - \boldsymbol{R}_l|} - \delta v_{kl}\,\mathbf{U}\right\}. \tag{119}$$

The first term follows from (II.A25). Performing the differentiations in the second term and using the relations

$$\begin{aligned} \overline{\alpha_i \alpha_j} &= \tfrac{1}{3}\delta_{ij}, \\ \overline{\alpha_i \alpha_j \alpha_k \alpha_l} &= -\tfrac{1}{2}\delta_{ij}\delta_{ik}\delta_{il} + \tfrac{5}{2}\delta_{ij}\delta_{ik}\delta_{il}\overline{\alpha_1^4} + (\delta_{ij}\delta_{kl} + \delta_{ik}\delta_{jl} + \delta_{il}\delta_{jk})(\tfrac{1}{6} - \tfrac{1}{2}\overline{\alpha_1^4}) \end{aligned} \tag{120}$$

for averages over angles of the direction cosines $\alpha_i \equiv R_i/R$ $(i = 1, 2, 3)$ (with respect to the cubic axes) of the radius vectors $\boldsymbol{R}$ $(\equiv \boldsymbol{R}_l - \boldsymbol{R}_k)$ from a fixed lattice point $k$ to the other points $l$, one finds for the $ij$-component of (119):

$$\delta\left(\frac{N}{V} S_1^{ij}\right) = \frac{N}{V}\{\tfrac{2}{5}\delta\eta^{ij} - \tfrac{2}{15}\delta^{ij} \operatorname{Tr} \delta\boldsymbol{\eta} + S_2(\tfrac{3}{2}\delta^{ij} \operatorname{Tr} \delta\boldsymbol{\eta} + 3\delta\eta^{ij} - \tfrac{15}{2}\delta^{ij}\delta\eta^{ii}) + \mathbf{S}_3 : \delta\boldsymbol{\eta}\delta^{ij}\} \tag{121}$$

with the lattice sums

$$\begin{aligned} S_2 &\equiv \frac{V}{N}\sum_{\boldsymbol{R}}^{\mathrm{sph}} (1 - 5\alpha_1^4)\frac{1}{4\pi R^3}, \\ \mathbf{S}_3 &\equiv \frac{V}{N}\sum_{\boldsymbol{R}}^{\mathrm{sph}} \frac{\partial v(\boldsymbol{R})}{\partial \boldsymbol{\eta}}, \end{aligned} \tag{122}$$

where $v^{kl}$ has been written now as a function $v(\boldsymbol{R})$. Substituting the result (121) into the formula that follows from (118) by variation one finds that $\delta\boldsymbol{\chi}$ is given by an expression like the second term of (108) with a fourth rank tensor $\hat{\boldsymbol{\chi}}$:

$$\hat{\chi}^{ijkl}(T) = -\tfrac{1}{3}\beta \frac{N}{V} v^2\delta^{ij}\delta^{kl} + \beta^2 \left(\frac{N}{V}\right)^2 v^4\{\tfrac{22}{135}\delta^{ij}\delta^{kl} - \tfrac{2}{45}\delta^{ik}\delta^{jl} + \tfrac{1}{9}S_1^{ij}\delta^{kl}$$

$$-\tfrac{1}{6}S_2(\delta^{ij}\delta^{kl} + 2\delta^{ik}\delta^{jl} - 5\delta^{ij}\delta^{ik}\delta^{il}) - \tfrac{1}{9}S_3^{kl}\delta^{ij}\}. \quad (123)$$

For the pressure (107) that causes the proper magnetostriction one obtains now with (117–118), again up to order $\beta^2$,

$$\tfrac{1}{5}M^iM^j + \tfrac{1}{10}\boldsymbol{M}^2\delta^{ij} + \tfrac{3}{4}S_2(\boldsymbol{M}^2\delta^{ij} + 2M^iM^j - 5M^{i2}\delta^{ij}) + \tfrac{1}{2}S_3^{ij}\boldsymbol{M}^2. \quad (124)$$

The first three terms together form the dipole–dipole contribution, which is often called the classical magnetostriction[1], while the last term represents the exchange interaction contribution. The lattice sum $S_2$, which is a purely geometrical quantity, may easily be computed. The lattice sum $\mathbf{S}_3$ contains the partial derivatives of the exchange quantities $v_{kl}$ with respect to the components of the deformation tensor; only rough estimates can be given for its magnitude.

From the expression (124) for the pressure that causes the proper magnetostriction one may calculate the corresponding strains along the standard lines of elasticity theory (for the field free case, since (124) is a contribution to the Helmholtz pressure tensor). Then one obtains expressions for the conventionally defined[2] magnetostriction constants. The results, generalized for the proper lattices, are not in agreement with experiments. The reason for this is that effects due to spin–orbit coupling have not been taken into account[3]. The preceding treatment was meant only to give an introduction to the theory of magnetostriction.

[1] N. Akulov, Z. Physik **52**(1928)389; R. Becker, Z. Physik **62**(1930)253; F. C. Powell, Proc. Cambr. Phil. Soc. **27**(1931)561.

[2] E.g. R. Becker and W. Döring, Ferromagnetismus (Springer-Verlag, Berlin 1939) p. 270ff.

[3] V. for instance: J. H. Van Vleck, Phys. Rev. **52**(1937)1178; C. Kittel, Rev. Mod. Phys. **21**(1949)541.

# The Hamilton operator for a set of composite particles with spin

The Hamilton operator for a set of nuclei and electrons with spin in the $c^{-2}$ approximation has been given in (IX.37). We introduce new canonical variables with the help of the transformation formulae (VII.A45) and (VII.A46). (Here the non-relativistic central point $\boldsymbol{R}_k$ is employed, not the central point $\boldsymbol{X}_k$, since the latter is not a canonical variable.) The Weyl transform of the Hamilton operator then becomes

$$H_{\mathrm{op}}(1, \ldots, N; t) \rightleftarrows \sum_k \left\{ \frac{\boldsymbol{P}_k^2}{2m_k} + \sum_{i=1}^{f-1} \frac{\boldsymbol{p}_{ki}^2}{2m_{ki}} - \sum_{i,j=1}^{f-1} \frac{\boldsymbol{p}_{ki}\cdot\boldsymbol{p}_{kj}}{2m_k} - c^{-2} \sum_i \frac{\boldsymbol{P}_{ki}^4(\boldsymbol{p})}{8m_{ki}^3} \right\}$$

$$+ \sum_{ki,lj(ki\neq lj)} \frac{e_{ki}\, e_{lj}}{8\pi|\boldsymbol{R}_{ki}(\boldsymbol{q})-\boldsymbol{R}_{lj}(\boldsymbol{q})|} \left[ 1 - \frac{\boldsymbol{P}_{ki}(\boldsymbol{p})\cdot \mathrm{T}\{\boldsymbol{R}_{ki}(\boldsymbol{q})-\boldsymbol{R}_{lj}(\boldsymbol{q})\}\cdot \boldsymbol{P}_{lj}(\boldsymbol{p})}{2m_{ki}\, m_{lj}\, c^2} \right]$$

$$+ \sum_{ki,lj(ki\neq lj)}' \frac{e_{ki}\, e_{lj}\, \hbar}{4m_{ki}\, c^2} \left[ \left\{ \frac{\boldsymbol{P}_{ki}(\boldsymbol{p})}{m_{ki}} \wedge \boldsymbol{\sigma}_{ki} \right\} \cdot \boldsymbol{\nabla}_{ki} - 2 \left\{ \frac{\boldsymbol{P}_{lj}(\boldsymbol{p})}{m_{lj}} \wedge \boldsymbol{\sigma}_{ki} \right\} \cdot \boldsymbol{\nabla}_{ki} \right] \frac{1}{4\pi|\boldsymbol{R}_{ki}(\boldsymbol{q})-\boldsymbol{R}_{lj}(\boldsymbol{q})|}$$

$$+ \sum_{ki,lj(ki\neq lj)}'' \frac{e_{ki}\, e_{lj}\, \hbar^2}{8m_{ki}\, m_{lj}\, c^2} (\boldsymbol{\sigma}_{ki} \wedge \boldsymbol{\nabla}_{ki})\cdot(\boldsymbol{\sigma}_{lj} \wedge \boldsymbol{\nabla}_{ki}) \frac{1}{4\pi|\boldsymbol{R}_{ki}(\boldsymbol{q})-\boldsymbol{R}_{lj}(\boldsymbol{q})|}$$

$$- \sum_{ki,lj(ki\neq lj)}' \frac{e_{ki}\, e_{lj}\, \hbar^2}{8m_{ki}^2\, c^2} \delta(\boldsymbol{R}_{ki}-\boldsymbol{R}_{lj}) + H_{\mathrm{e}}(1, \ldots, N; t). \tag{A1}$$

The symbols $\boldsymbol{P}_{ki}(\boldsymbol{p})$ and $\boldsymbol{R}_{ki}(\boldsymbol{q})$ stand for the right-hand sides of (VII.A46). The quantity $H_{\mathrm{e}}(1, \ldots, N; t)$ represents the external field terms in (IX.37). If the external potentials are expanded around $\boldsymbol{R}_k$ and only first derivatives are retained, one gets

$$H_{\mathrm{e}}(1, \ldots, N; t) = \sum_k e_k \left\{ \varphi_{\mathrm{e}}(\boldsymbol{R}_k, t) - c^{-1} \frac{\boldsymbol{P}_k}{m_k} \cdot \boldsymbol{A}_{\mathrm{e}}(\boldsymbol{R}_k, t) \right\}$$

$$+ \sum_{k,i} e_{ki} \left[ \{\boldsymbol{R}_{ki}(\boldsymbol{q}) - \boldsymbol{R}_k\} \cdot \boldsymbol{\nabla}_k \left\{ \varphi_{\mathrm{e}}(\boldsymbol{R}_k, t) - c^{-1} \frac{\boldsymbol{P}_k}{m_k} \cdot \boldsymbol{A}_{\mathrm{e}}(\boldsymbol{R}_k, t) \right\} \right.$$

$$-c^{-1}\left\{\frac{\boldsymbol{P}_{ki}(\boldsymbol{p})}{m_{ki}}-\frac{\boldsymbol{P}_k}{m_k}\right\}\cdot\boldsymbol{A}_{\mathrm{e}}(\boldsymbol{R}_k,t)$$

$$-c^{-1}\{\boldsymbol{R}_{ki}(\boldsymbol{q})-\boldsymbol{R}_k\}\cdot\nabla_k\boldsymbol{A}_{\mathrm{e}}(\boldsymbol{R}_k,t)\cdot\left\{\frac{\boldsymbol{P}_{ki}(\boldsymbol{p})}{m_{ki}}-\frac{\boldsymbol{P}_k}{m_k}\right\}\Bigg]$$

$$-\sum_{k,i}{}'\frac{e_{ki}\hbar}{2m_{ki}c}\boldsymbol{\sigma}_{ki}\cdot\left\{\boldsymbol{B}_{\mathrm{e}}(\boldsymbol{R}_k,t)-\frac{\boldsymbol{P}_{ki}(\boldsymbol{p})}{2m_{ki}c}\wedge\boldsymbol{E}_{\mathrm{e}}(\boldsymbol{R}_k,t)\right\}. \qquad \text{(A2)}$$

A different form for the Weyl transform of the Hamilton operator is obtained, if one splits off the kinetic energy

$$\sum_{k,i}\{\tfrac{1}{2}m_{ki}(\partial_{t\mathrm{P}}\boldsymbol{X}_{ki})^2+\tfrac{3}{8}c^{-2}m_{ki}(\partial_{t\mathrm{P}}\boldsymbol{X}_{ki})^4\}, \qquad \text{(A3)}$$

which, according to the derivation of the energy law in chapter IX, section 6*b*, may be written in semi-relativistic approximation as:

$$\sum_k(\tfrac{1}{2}m_k\boldsymbol{v}_k^2+t_k)+2c^{-2}\sum_{k,l(k\neq l)}\sum_{i,j}e_{ki}e_{lj}\boldsymbol{r}_{ki}\cdot\boldsymbol{v}_k(\partial_{t\mathrm{P}}\boldsymbol{r}_{ki})\cdot\nabla_{ki}\frac{1}{4\pi|\boldsymbol{X}_{ki}-\boldsymbol{X}_{lj}|}$$

$$-2c^{-2}\sum_{k,i}e_{ki}\boldsymbol{r}_{ki}\cdot\boldsymbol{v}_k(\partial_{t\mathrm{P}}\boldsymbol{r}_{ki})\cdot\boldsymbol{E}_{\mathrm{e}}(\boldsymbol{X}_{ki},t), \qquad \text{(A4)}$$

where we introduced the abbreviations $\boldsymbol{v}_k \equiv \partial_{t\mathrm{P}}\boldsymbol{X}_k$ and $t_k$ (IX.82). The kinetic energy may be expressed in terms of the canonical variables introduced above. Then one finds for (A3) up to terms of order $c^{-2}$ and bilinear in the charges:

$$\sum_{k,i}\left\{\frac{\boldsymbol{P}_k^2}{2m_k}+\sum_{i=1}^{f-1}\frac{\boldsymbol{p}_{ki}^2}{2m_{ki}}-\sum_{i,j=1}^{f-1}\frac{\boldsymbol{p}_{ki}\cdot\boldsymbol{p}_{kj}}{2m_k}-c^{-2}\sum_i\frac{\boldsymbol{P}_{ki}^4(\boldsymbol{p})}{8m_{ki}^3}\right\}$$

$$-\sum_{ki,lj(ki\neq lj)}\frac{e_{ki}e_{lj}\boldsymbol{P}_{ki}(\boldsymbol{p})\cdot\mathrm{T}\{\boldsymbol{R}_{ki}(\boldsymbol{q})-\boldsymbol{R}_{lj}(\boldsymbol{q})\}\cdot\boldsymbol{P}_{lj}(\boldsymbol{p})}{8\pi m_{ki}m_{lj}c^2|\boldsymbol{R}_{ki}(\boldsymbol{q})-\boldsymbol{R}_{lj}(\boldsymbol{q})|}$$

$$-\sum_{ki,lj(ki\neq lj)}{}'\frac{e_{ki}e_{lj}\hbar}{2m_{ki}m_{lj}c^2}\{\boldsymbol{P}_{lj}(\boldsymbol{p})\wedge\boldsymbol{\sigma}_{ki}\}\cdot\nabla_{ki}\frac{1}{4\pi|\boldsymbol{R}_{ki}(\boldsymbol{q})-\boldsymbol{R}_{lj}(\boldsymbol{q})|}$$

$$-c^{-1}\sum_{k,i}e_{ki}\boldsymbol{P}_{ki}(\boldsymbol{p})\cdot\boldsymbol{A}_{\mathrm{e}}\{\boldsymbol{R}_{ki}(\boldsymbol{q}),t\}. \qquad \text{(A5)}$$

By employing the identity that follows by equating the expressions (A4) and (A5) one finds for the Hamiltonian (A1) with (A2) in semi-relativistic approximation:

$$H_{\text{op}}(1, \ldots, N; t) \rightleftarrows \sum_k (\tfrac{1}{2}m_k \boldsymbol{v}_k^2 + \boldsymbol{v}_k \cdot \boldsymbol{g}_k + t_k + u_k)$$

$$+ \sum_{k,l(k\neq l)} \sum_{i,j} \frac{e_{ki}\, e_{lj}}{8\pi|\boldsymbol{X}_{ki}-\boldsymbol{X}_{lj}|} \{1 + \tfrac{1}{2}c^{-2}(\partial_{tP}\boldsymbol{X}_{ki}) \cdot \mathbf{T}(\boldsymbol{X}_{ki}-\boldsymbol{X}_{lj}) \cdot (\partial_{tP}\boldsymbol{X}_{lj})\}$$

$$+ 2c^{-2} \sum_{k,l(k\neq l)} \sum_{i,j} e_{ki}\, e_{lj}\, \boldsymbol{r}_{ki} \cdot \boldsymbol{v}_k (\partial_{tP}\boldsymbol{r}_{ki}) \cdot \nabla_{ki} \frac{1}{4\pi|\boldsymbol{X}_{ki}-\boldsymbol{X}_{lj}|}$$

$$+ \sum_{k,l(k\neq l)} \sum_{i,j}{}' \frac{e_{ki}\, e_{lj}\, \hbar}{2m_{ki}\, c^2} (\partial_{tP}\boldsymbol{X}_{ki}) \cdot (\boldsymbol{\sigma}_{ki} \wedge \nabla_{ki}) \frac{1}{4\pi|\boldsymbol{X}_{ki}-\boldsymbol{X}_{lj}|}$$

$$+ \sum_{k,l(k\neq l)} \sum_{i,j}{}'' \frac{e_{ki}\, e_{lj}\, \hbar^2}{8m_{ki}\, m_{lj}\, c^2} (\boldsymbol{\sigma}_{ki} \wedge \nabla_{ki}) \cdot (\boldsymbol{\sigma}_{lj} \wedge \nabla_{ki}) \frac{1}{4\pi|\boldsymbol{X}_{ki}-\boldsymbol{X}_{lj}|}$$

$$+ H'_{\text{e}}(1, \ldots, N; t), \quad \text{(A6)}$$

where we introduced $\boldsymbol{X}_{ki}$ and $\partial_{tP}\boldsymbol{X}_{ki}$ instead of $\boldsymbol{R}_{ki}$ and $\boldsymbol{P}_{ki}$ respectively and moreover the abbreviations $\boldsymbol{g}_k$ (IX.69) and $u_k$ (IX.78). The external field terms represented by $H'_{\text{e}}(1, \ldots, N; t)$, are found to be, upon Taylor expansion,

$$H'_{\text{e}}(1, \ldots, N; t) = \sum_k e_k\, \varphi_{\text{e}}(\boldsymbol{X}_k, t)$$

$$+ \sum_{k,i} e_{ki}\, \boldsymbol{r}_{ki} \cdot \nabla_k\, \varphi_{\text{e}}(\boldsymbol{X}_k, t) - \sum_{k,i}{}' \frac{e_{ki}\, \hbar}{2m_{ki}\, c} \boldsymbol{\sigma}_{ki} \cdot \{\boldsymbol{B}_{\text{e}}(\boldsymbol{X}_k, t)$$

$$- c^{-1}(\partial_{tP}\boldsymbol{X}_{ki}) \wedge \boldsymbol{E}_{\text{e}}(\boldsymbol{X}_k, t)\} - 2c^{-2} \sum_{k,i} e_{ki}\, \boldsymbol{r}_{ki} \cdot \boldsymbol{v}_k (\partial_{tP}\boldsymbol{r}_{ki}) \cdot \boldsymbol{E}_{\text{e}}(\boldsymbol{X}_k, t), \quad \text{(A7)}$$

where only terms up to first derivatives of the potentials have been retained.

For uniform and time-independent external fields the potentials may be chosen as

$$\varphi_{\text{e}}(\boldsymbol{R}) = -\boldsymbol{R} \cdot \boldsymbol{E}_{\text{e}}, \qquad \boldsymbol{A}_{\text{e}}(\boldsymbol{R}) = \tfrac{1}{2}\boldsymbol{B}_{\text{e}} \wedge \boldsymbol{R}. \quad \text{(A8)}$$

Then the external field term $H_{\text{e}}(1, \ldots, N; t)$ given by (A2) becomes

$$H_{\text{e}}(1, \ldots, N; t) = -\sum_k \left\{ e_k\, \boldsymbol{R}_k \cdot \left( \boldsymbol{E}_{\text{e}} + \tfrac{1}{2}c^{-1} \frac{\boldsymbol{P}_k}{m_k} \wedge \boldsymbol{B}_{\text{e}} \right) \right.$$

$$+ \bar{\boldsymbol{\mu}}_{k,\text{orb}}^{(1)} \cdot \left( \boldsymbol{E}_{\text{e}} + \tfrac{1}{2}c^{-1} \frac{\boldsymbol{P}_k}{m_k} \wedge \boldsymbol{B}_{\text{e}} \right) + \tfrac{1}{2}c^{-1}(\partial_{tP}\boldsymbol{\mu}_{k,\text{orb}}^{(1)}) \cdot (\boldsymbol{B}_{\text{e}} \wedge \boldsymbol{R}_k)$$

$$\left. + \boldsymbol{v}_{k,\text{orb}}^{(1)} \cdot \boldsymbol{B}_{\text{e}} + \boldsymbol{v}_{k,\text{spin}}^{(1)} \cdot \left( \boldsymbol{B}_{\text{e}} - \tfrac{1}{2}c^{-1} \frac{\boldsymbol{P}_k}{m_k} \wedge \boldsymbol{E}_{\text{e}} \right) + \tfrac{1}{2}\boldsymbol{\mu}_{k,\text{spin}}^{(1)} \cdot \boldsymbol{E}_{\text{e}} \right\}. \quad \text{(A9)}$$

Here we introduced the electric and magnetic orbital and spin dipole moments, defined in (IX.65) with the choice $\boldsymbol{X}_k$ for the privileged point. In the first instance we encounter here dipole moments containing $\boldsymbol{R}_{ki}$ and $\boldsymbol{R}_k$, i.e.

dipole moments such as defined in non-relativistic theory and denoted by a bar. Since however the difference between $\boldsymbol{X}_{ki}$ and $\boldsymbol{R}_{ki}$ (and hence between $\boldsymbol{X}_k$ and $\boldsymbol{R}_k$) is of order $c^{-2}$, it follows by inspection of the various terms in (A9) and of the definitions (IX.65) that the difference between the two kinds of multipole moments is significant only in the second term of $H_e(1, \ldots, N; t)$. The orbital electric dipole moment $\bar{\boldsymbol{\mu}}^{(1)}_{k,\text{orb}}$ which is defined as $\sum_i e_{ki}(\boldsymbol{R}_{ki}-\boldsymbol{R}_k)$ may be related to the semi-relativistic dipole moments. For the case of neutral atoms ($e_k = 0$) one finds from (IX.65) with (IX.38, 61) in semi-relativistic approximation:

$$\bar{\boldsymbol{\mu}}^{(1)}_{k,\text{orb}} = \boldsymbol{\mu}^{(1)}_{k,\text{orb}}+\tfrac{1}{2}\boldsymbol{\mu}^{(1)}_{k,\text{spin}}-c^{-1}(\boldsymbol{\nu}^{(1)}_{k,\text{orb}}+\tfrac{1}{2}\boldsymbol{\nu}^{(1)}_{k,\text{spin}})\wedge \frac{\boldsymbol{P}_k}{m_k}\,. \tag{A10}$$

The term $H'_e(1, \ldots, N; t)$ that is given in (A7) may likewise be written in a different way, if one introduces the potentials (A8) and the multipoles (IX.65). One finds then

$$\begin{aligned} H'_e(1, \ldots, N; t) = -\sum_k \{e_k \boldsymbol{X}_k\cdot\boldsymbol{E}_e+(\boldsymbol{\mu}^{(1)}_k-c^{-1}\boldsymbol{\nu}^{(1)}_k\wedge \boldsymbol{v}_k)\cdot\boldsymbol{E}_e \\ +\boldsymbol{\nu}^{(1)}_{k,\text{spin}}\cdot\boldsymbol{B}_e+2c^{-1}(\boldsymbol{\nu}^{(1)}_{k,\text{orb}}\wedge \boldsymbol{v}_k)\cdot\boldsymbol{E}_e\}. \end{aligned} \tag{A11}$$

# Change of free energy under deformations for a spin particle system

The free energy $F^*$ follows from the partition sum (VII.A58). The Hamilton operator $H_{op}^{W}$ contains a wall potential $U_{op}^{W}$ that depends on the position of the boundary, so that it changes under deformations. The change of free energy is therefore given by (VII.A63) or, in terms of the Wigner function of the canonical ensemble and the Weyl transform $U^{W}$ of the wall potential, by

$$\delta_{\varepsilon} F^* = \mathrm{Sp} \int \delta U^{W} \rho(1, \dots, N) \mathrm{d}1 \dots \mathrm{d}N. \tag{A12}$$

The wall potential is a sum of functions $U_k^{W}$ which depend on $\boldsymbol{X}_k$ given in (IX.68) with (IX.38, 39). It is to be understood in the same fashion as for instance (IX.47):

$$U^{W} = \sum_k U_k^{W}(\boldsymbol{X}_k) \equiv \sum_k \{U_k^{W}(\boldsymbol{R}_k) + (\boldsymbol{X}_k - \boldsymbol{R}_k)\cdot\nabla_k U_k^{W}(\boldsymbol{R}_k)\}, \tag{A13}$$

where $\boldsymbol{X}_k - \boldsymbol{R}_k$ is of the order $c^{-2}$. The wall potential $U_k^{W}(\boldsymbol{R}_k)$ is a function which is zero if $\boldsymbol{R}_k$ is situated in the interior of the system, increases rapidly at the position of the wall and becomes infinite outside.

If the position of the wall changes according to (VII.A60), one may use as a new partial wall potential $U_k^{W}(\boldsymbol{R}_k)$ the old wall potential $U_k^{W}$ with argument $\{\mathbf{U} - \delta\boldsymbol{\epsilon}(\boldsymbol{R}_k)\}\cdot\boldsymbol{R}_k$, as in (VII.A61). Then the new wall potential $U^{W'}$ is

$$U^{W'} = \sum_k U_k^{W}[\{\mathbf{U} - \delta\boldsymbol{\epsilon}(\boldsymbol{R}_k)\}\cdot\boldsymbol{R}_k] + \sum_k (\boldsymbol{X}_k - \boldsymbol{R}_k)\cdot\nabla_k U_k^{W}[\{\mathbf{U} - \delta\boldsymbol{\epsilon}(\boldsymbol{R}_k)\}\cdot\boldsymbol{R}_k]. \tag{A14}$$

Hence the change of the wall potential is

$$\delta U^{W} = -\sum_k [\boldsymbol{R}_k\cdot\delta\tilde{\boldsymbol{\epsilon}}(\boldsymbol{R}_k)\cdot\nabla_k U_k^{W}(\boldsymbol{R}_k) + (\boldsymbol{X}_k - \boldsymbol{R}_k)\cdot\nabla_k\{\boldsymbol{R}_k\cdot\delta\tilde{\boldsymbol{\epsilon}}(\boldsymbol{R}_k)\cdot\nabla_k U_k^{W}(\boldsymbol{R}_k)\}] \tag{A15}$$

or, performing the differentiation in the second term,

$$\delta U^{W} = -\sum_k \boldsymbol{X}_k\cdot\delta\tilde{\boldsymbol{\epsilon}}(\boldsymbol{X}_k)\cdot\nabla_k U_k^{W}(\boldsymbol{X}_k) \tag{A16}$$

up to order $c^{-2}$. Substituting this result into (A12) one gets for the change of the free energy:

$$\delta_\varepsilon F^* = -\mathrm{Sp}\int \{\sum_k \boldsymbol{X}_k\cdot\delta\tilde{\boldsymbol{\epsilon}}(\boldsymbol{X}_k)\cdot\boldsymbol{\nabla}_k U_k^{\mathrm{W}}(\boldsymbol{X}_k)\}\rho(1, \ldots, N)\mathrm{d}1 \ldots \mathrm{d}N. \qquad \text{(A17)}$$

This expression bears a strong similarity to the non-relativistic form (VII. A65), the sole difference being that $\boldsymbol{R}_k$ is replaced by $\boldsymbol{X}_k$ (and a spur added since the Wigner function and the quantity in front of it are matrices).

To bring the right-hand side of (A17) into a more convenient form, we proceed in a way analogous to that followed in non-relativistic theory. Since the canonical ensemble is stationary, one has the identity

$$0 = \frac{\partial}{\partial t}\,\mathrm{Sp}\int \sum_k \boldsymbol{P}_k\cdot\delta\boldsymbol{\epsilon}(\boldsymbol{X}_k)\cdot\boldsymbol{X}_k\,\rho(1, \ldots, N)\mathrm{d}1 \ldots \mathrm{d}N, \qquad \text{(A18)}$$

because the Wigner function is time-independent. With the use of the relation (9) one may write this identity in semi-relativistic approximation as:

$$\begin{aligned}0 = {} & \mathrm{Sp}\int \sum_k \partial_{t\mathrm{P}}\{\boldsymbol{P}_k\cdot\delta\boldsymbol{\epsilon}(\boldsymbol{X}_k)\cdot\boldsymbol{X}_k\}\rho(1, \ldots, N)\mathrm{d}1 \ldots \mathrm{d}N \\ & -\mathrm{Sp}\int \sum_k \Big[(\boldsymbol{\nabla}_k U^{\mathrm{W}})\cdot\delta\boldsymbol{\epsilon}(\boldsymbol{X}_k)\cdot\boldsymbol{X}_k + \frac{1}{m_k^2c^2}\{(\sum_i \tfrac{1}{2}\hbar\boldsymbol{\sigma}_{ki})\wedge\boldsymbol{\nabla}_k U^{\mathrm{W}}\}\cdot\boldsymbol{\nabla}_k \\ & \{\boldsymbol{P}_k\cdot\delta\boldsymbol{\epsilon}(\boldsymbol{X}_k)\cdot\boldsymbol{X}_k\}\Big]\,\rho(1, \ldots, N)\mathrm{d}1 \ldots \mathrm{d}N. \qquad \text{(A19)}\end{aligned}$$

The first term contains the Poisson bracket of $\sum_k \boldsymbol{P}_k\cdot\delta\boldsymbol{\epsilon}(\boldsymbol{X}_k)\cdot\boldsymbol{X}_k$ with the Weyl transform of the Hamilton operator without wall potential, while the second comes about as a result of the wall potential part of the total Hamiltonian. It contains the force $-\boldsymbol{\nabla}_k U^{\mathrm{W}}$, which the wall exerts on atom $k$. Combining (A17) and (A19) we find for the change of the free energy under deformation

$$\begin{aligned}\delta_\varepsilon F^* = {} & -\mathrm{Sp}\int \sum_k \partial_{t\mathrm{P}}\{\boldsymbol{P}_k\cdot\delta\boldsymbol{\epsilon}(\boldsymbol{X}_k)\cdot\boldsymbol{X}_k\}\rho(1, \ldots, N)\mathrm{d}1 \ldots \mathrm{d}N \\ & +\mathrm{Sp}\int \sum_k \frac{1}{m_k^2c^2}\{(\sum_i \tfrac{1}{2}\hbar\boldsymbol{\sigma}_{ki})\wedge(\boldsymbol{\nabla}_k U^{\mathrm{W}})\}\cdot\boldsymbol{\nabla}_k\{\boldsymbol{P}_k\cdot\delta\boldsymbol{\epsilon}(\boldsymbol{X}_k)\cdot\boldsymbol{X}_k\} \\ & \rho(1, \ldots, N)\mathrm{d}1 \ldots \mathrm{d}N. \qquad \text{(A20)}\end{aligned}$$

This result may be compared to (VII.A68) with (II.A51) of the non-relativistic theory. In the first place the central point $\boldsymbol{X}_k$ instead of $\boldsymbol{R}_k$ appears. Furthermore a second term occurs here in the expression for $\delta_\varepsilon F^*$. However this term is proportional to the surface of the system, since the order of magnitude of the factor of $\delta\boldsymbol{\epsilon}$ can be estimated by writing the product of the Compton wave length $\hbar/m_k c$ of the composite particle, the pressure times the surface of the walls and $c^{-1}$ times the average velocity ($\sim \boldsymbol{P}_k/m_k$) of the composite particles. The first term of (A20) is proportional to the volume of the system as is shown in section 5*b* of this chapter. Since in the thermodynamical treatment given here surface effects have been neglected throughout, one may write for (A20):

$$\delta_\varepsilon F^* = -\mathrm{Sp} \int \sum_k \partial_{t\mathrm{P}} \{ \boldsymbol{P}_k \cdot \delta\boldsymbol{\epsilon}(\boldsymbol{X}_k) \cdot \boldsymbol{X}_k \} \rho(1, \ldots, N) \mathrm{d}1 \ldots \mathrm{d}N. \qquad \text{(A21)}$$

In the case of uniform deformations $\delta\boldsymbol{\epsilon}$ one finds from this expression

$$\delta_\varepsilon F^* = \mathbf{A} : \delta\boldsymbol{\epsilon} \qquad \text{(A22)}$$

with the tensor

$$\mathbf{A} \equiv -\mathrm{Sp} \int \sum_k \partial_{t\mathrm{P}} (\boldsymbol{X}_k \boldsymbol{P}_k) \rho(1, \ldots, N) \mathrm{d}1 \ldots \mathrm{d}N, \qquad \text{(A23)}$$

which is the expression (89) of the main text.

# Author index

# Subject index